Geomorphology and Natural Hazards

Advanced Textbook Series

Geomorphology and Natural Hazards

Understanding Landscape Change for Disaster Mitigation

Tim R. Davies
University of Cantebury
New Zealand

Oliver Korup
Universität Potsdam
Germany

John J. Clague
Simon Fraser University
Canada

This Work is a co-publication of the American Geophysical Union and John Wiley and Sons Ltd.

ADVANCING EARTH
AND SPACE SCIENCE

Published under the aegis of the AGU Publications Committee

Brooks Hanson, Executive Vice President, Science
Carol Frost, Chair, Publications Committee
For details about the American Geophysical Union visit us at www.agu.org.

Wiley Global Headquarters
111 River Street, Hoboken, NJ 07030, USA

For details of our global editorial offices, customer services, and more information about Wiley products visit us at www.wiley.com.

Limit of Liability/Disclaimer of Warranty

Library of Congress Cataloging-in-Publication Data

Names: Davies, Tim R., 1970- author. | Korup, Oliver, author. | Clague,
    John, J., author.
Title: Geomorphology and Natural Hazards : Understanding Landscape Change for Disaster Mitigation / Tim R.
    Davies, Oliver Korup, John J. Clague
Description: First edition. | Hoboken, NJ : Wiley-American Geophysical Union,
    2021. | Series: Advanced Textbook Series | Includes bibliographical references and
    index. | Description based on print version record and CIP data provided
    by publisher; resource not viewed.
Identifiers: LCCN 2020024258 (print) | LCCN 2020024259 (ebook) | ISBN
    9781118648605 (epub) | ISBN 9781118648612 (adobe pdf) | ISBN 9781119990314
    (paperback) |
Subjects: LCSH: Natural disasters–Research. | Geomorphology.
Classification: LCC GB5005 (ebook) | LCC GB5005 D38 2021 (print) | DDC
    363.34–dc23
LC record available at https://lccn.loc.gov/2020024258 LC record available at https://lccn.loc.gov/2020024259

Cover Design: Wiley
Cover Image: Landslide in Cusco, Peru, 2018
Ministerio de Defensa del Perú (CC BY 2.0)

Set in 9.5/12.5pt STIXTwoText by SPi Global, Chennai, India
Printed and bound by CPI Group (UK) Ltd, Croydon, CR0 4YY

C9781119990314_170321

Contents

Preface

In spite of ever-increasing research into natural hazards, the reported damage from naturally-triggered continues to rise, increasingly disrupting human activities. We, as scientists who study the way in which the part of Earth most relevant to society—the surface—behaves, are disturbed and frustrated by this trend. It appears that the large amounts of funding devoted each year to research into reducing the impacts of natural disasters could be much more effective in producing useful results. At the same time we are aware that society, as represented by its decision makers, while increasingly concerned at the impacts of natural disasters on lives and economies, is reluctant to acknowledge the intrinsic activity of Earth's surface and to take steps to adapt societal behaviour to minimise the impacts of natural disasters. Understanding and managing natural hazards and disasters are beyond matters of applied earth science, and also involve considering human societal, economic and political decisions.

In this book we attempt to address this multidisciplinary problem directly, based on our experiences in earth science, and also in attempting to apply earth science to hazard and risk management in real-life situations. We acknowledge that other books offer exhaustive material on natural hazards and disasters, or manuals on integrated risk management. We recommend these alternatives for learning the basics about the many natural processes that may cause harm to human activity. Also, the breadth of textbooks devoted to specific natural

hazards such as earthquakes, volcanoes, landslides, or floods motivates us to recapitulate only briefly key points from these works, while allowing us to focus more on their geomorphic consequences and implications. The same applies for the theoretical basics of geomorphological processes that are the focus of this book. Instead, we examine many practical issues that arise when dealing with potentially damaging geomorphic processes as a direct or indirect consequence of natural disasters. We choose this avenue because we feel that current textbooks on natural hazards and disasters fail to adopt a holistic and general focus. We find that little synthesised material comprehensively addresses geomorphic hazards and risks, and their mitigation.

Traditionally, and still to a large extent today, hazard management consists of constructing physical works or structural countermeasures to modify the troublesome and potentially destructive processes that operate at Earth's surface. The engineering profession is tasked with the design and construction of these works. Engineering—and in particular hazards engineering—is essentially a societal profession, in that engineers carry out their work in the service of society. When society is threatened or damaged by a natural event, engineers are paid to solve the problem so that societal activity can, as much as possible, continue uninterrupted and unchanged. For millennia, during which low human population levels meant overall lower levels of risk, the vulnerability and adaptive capacity of society to natural hazards may have been different. Still, engineering was dramatically successful in mitigating hazards: floodplains were drained, channelised, and settled; sea-walls kept extreme tides from inundating coastal flats; and river control works channelised sediment across inhabited fans.

Today this situation is changing markedly. Human numbers are continuously increasing and our species is increasingly modifying the planet's surface. Society is becoming increasingly complex and sophisticated and thus less able to adjust its behaviour; economic pressures reduce wasteful system redundancy; and society increasingly—and justifiably—expects the money it spends on risk reduction to protect it from disasters. Whether contemporary climate change is the dominant driver of the observed increase in disaster costs is unclear, but it is certainly a potentially important factor that is some extent also the result of human activity. It is clear that traditional hazard management strategies have become inadequate, and their adequacy will decrease further into the future. A key element of this situation is that society now is expanding into areas for which we have little or unreliable knowledge about the rates of geomorphic processes. These areas may be prone to large and commensurately rare events that, owing to their rarity, are less well described and understood than their more moderate and familiar counterparts. Such events are more powerful and harder to design against, so the reliability of engineering countermeasures is reduced, which must eventually lead to an increase in disasters.

In this book we go beyond the view that natural hazards and disasters have adverse implications for human assets by definition. We argue that understanding the forms and processes of Earth's surface—encapsulated in the science of geomorphology—is essential to assess natural hazards and gauge the consequences of natural disasters on Earth's surface. These consequences involve the often rapid erosion, transport, and deposition of rock debris, soil, biomass, human waste, nutrients, and pathogens, thereby changing or setting the boundary conditions for subsequent hazardous processes. We call for a more detailed view on natural disasters by identifying those processes in a chain of harmful events that produce most damage. Often we find that most damage by earthquakes or storms, for example, is due to landslides instead of seismic shaking or intensive rainfall. By doing so we acknowledge that Earth is an intrinsically active—and therefore hazardous—planet. Occasional intense events that disturb Earth's surface are inevitable, and if society ignores such events, natural disasters and catastrophes will inevitably and repeatedly happen.

We acknowledge that there must be a physical limit to the intensity of a given surface event that can be controlled reliably by engineering works, and therefore suggest that structural works stay within those limits. We particularly underline several lines of empirical evidence and reasons that show that structural interventions may make a disaster-prone situation worse. We also argue that in many situations an extraordinarily large or severe event, although unlikely, can happen, thus both procedures and structures must be put in place to reduce the death and damage that this event can cause. This last point is crucial and fundamental: the extreme events of nature cannot be controlled, but they can be avoided in some cases, and their negative consequences reduced in many cases. Therefore, to reduce the impacts of such events, society must adapt so that their damage is reduced to acceptable levels. This is our key message.

In pointing out some limitations of traditional engineering approaches to control hazards, we refrain from denigrating the engineering profession. One of us was trained and has practised as an engineer, and we understand and sympathise with the aspirations of engineering to improve the lot of society. Nevertheless, we encourage the engineering profession to seek to know and understand its limitations, and we encourage engineers and geomorphologists to understand how they can interact with each other, and with society, to provide better information on threatening events and the options available to manage the threats.

Acknowledging that natural hazards are by definition estimates that involve uncertainty requires that society wilfully adjust its behaviour to nature's. This, in turn, requires that natural systems be adequately known. We must be able to foresee what sizes and types of surface changes can potentially harm human assets (including our natural environment). And we need to know how to make that information available and useful to society. Whether, or to what extent, society acts on that knowledge depends on its nature and aspirations. We are uninformed, except through experience, about the nature and aspirations of society, but recognising that society does have a nature and aspirations is crucial to the way that information is acquired and presented.

In attempting to reduce the impact of hazardous surface processes, we must recognise that two systems interact to create a disaster: the powerful and complex surface geological processes of Earth; and the less powerful but also complex human system, which operates through society and occupies Earth's surface. We have only limited control over nature, and especially over its rare and highly energetic processes. However, we increasingly understand the rules by which the natural system operates, even though that understanding could lead more often to better predictions. In contrast, we have in principle a measure of control over the human system, although we have little understanding of its operation in social, cultural, political and economic terms. However, we believe that by approaching the problem from an applied geomorphological perspective, we can shed some light on what can and cannot be achieved in the way of hazard mitigation and disaster reduction in a range of situations in the future. Whether society has the will to respond to this illumination is beyond our influence, but we sincerely hope that, if future disasters are considered in terms of the concepts we set out herein, illumination might give rise to realisation, acceptance and ultimately action.

Acknowledgements

Reading through several thousand scientific publications to collect material for a book seems like a futile task in a time of rapidly increasing publication numbers. Deciding which publications to include here was tough, as was keeping track with the many new natural disasters that occurred when we were writing this book. By the time you are reading this book, many of the numbers, especially those concerning projections and predictions, will most likely have changed with new research results arriving, refining, or perhaps even refuting previous work. While you may find parts of this book outdated, perhaps consider it instead as a document of how swiftly our scientific understanding of the vibrant field of geomorphic footprints of natural hazards and disasters changes. At the very least, we hope that the contents of this book distill some of the more persistent findings that a solid understanding of the geomorphic footprints of natural hazards and disasters rests on.

We acknowledge all the hard work that researchers have carried out to better understand natural hazards and to reduce the risks from natural disasters. We have also been involved with many communities, government officials, scientists, technologists, planners, and people affected over several decades in hazard assessment and planning to mitigate disasters. We have learned much from these interactions, and express our gratitude to all involved.

1

Natural Disasters and Sustainable Development in Dynamic Landscapes

1.1 Breaking News

Natural disasters are making the headlines in the news more and more frequently. Scarcely a month goes by without a major earthquake, a volcanic eruption or a huge flood, with dramatic footage of fallen buildings, billowing ash clouds and devastated victims on the evening news. Thousands of videos and blogs posted to online portals illustrate in unprecedented and disturbing detail the destructive forces of earthquakes, storms, floods or landslides, together with their impacts on persons or entire communities. Interactive learning platforms and serious games offer various immersive perspectives on what it means to manage natural hazards, risks, and disasters. Many universities offer full-fledged graduate courses specialising in natural hazards and risk management. The entertainment industry regularly produces natural disaster movies that conjure the end of the world by gargantuan tsunamis or at least the demise of someone's favourite city by an unexpected volcanic eruption. In the real world, every few years something truly catastrophic captivates both public attention and political opinion for weeks – the Indian Ocean tsunami, Hurricanes Katrina, Sandy, and Harvey, the Pakistan floods, the Wenchuan, Christchurch, and Tohoku earthquakes – and we contribute willingly to relieving the suffering of the victims.

The increase in reported disasters seems alarming and rapidly growing (Figure 1.1).

Most news reports deliver the numbers of people killed or injured or assets destroyed, but rarely illuminate in detail the causes, consequences, or whether these losses could have been predicted, let alone avoided. The statistics of disasters can be sobering. Natural disasters claimed more than 31 million lives in the twentieth century, and more than 4.1 billion people were affected, which was the world's population count in the early 1970s. Estimates of the overall insured economic losses exceed US$ 1019 billion (Figure 1.2) (www.emdat.be, last accessed December 2014). The number and costs of natural disasters appear to be rising exponentially, although disaster deaths have been decreasing in recent decades. The years from 2000 to 2010 saw more than 1.1 million people killed in natural disasters, and more than 2.5 billion people affected. Hence, more than one out of three persons on Earth on average has had to deal with natural disasters in some way recently. The financial damage in the wake of twenty-first century natural disasters has been estimated at US$ 1022 billion, which is already more than the total damage of the past century.

Moreover, past estimates of fatalities by natural hazards such as landslides have probably been too low (Froude and Petley 2018). If we want to learn from these losses, we need adjust them first for growing population, increasing welfare, economic inflation, and improvements in engineered infrastructure and planning for natural disasters (Vranes

Geomorphology and Natural Hazards: Understanding Landscape Change for Disaster Mitigation, Advanced Textbook Series,
First Edition. Tim R. Davies, Oliver Korup, and John J. Clague.

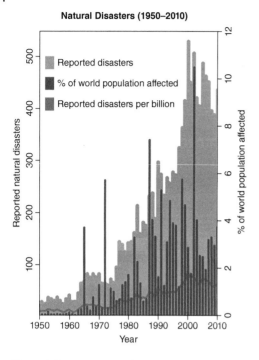

Natural Disasters (1950–2010)

Figure 1.1 The number of reported natural disasters is on the rise worldwide and seems to follow a strongly nonlinear trend between 1950 and 2010 (orange bars). This trend mimics the similar nonlinear growth in the world's population, and normalizing for this effect shows that natural disasters increase much less rapidly (red line). The percentage of the world's total population affected by natural disasters (pink bars) has also been growing, although with much more variability. Natural disaster data are from the EM-DAT database, and population data are from the United Nations World Population Prospects, The 2012 Revision. https://www.un.org/en/development/desa/publications/world-population-prospects-the-2012-revision.html. Data accessed 24 April 2015.

and Pielke 2009). Bangladesh, for example, has a population of more than 150 million people who are vulnerable to tropical cyclones, flooding, and earthquakes. Between the 1960s and 1980s, the country had the world's highest mortality from storm-induced disasters, even though it was struck by fewer cyclones than India or Indonesia. However, mortality rates have dropped since the 1980s thanks to construction of cyclone shelters and improvements

in storm forecasting (Figure 1.3) (Cash et al. 2013).

This and many other observations remind us that Earth is a dangerous planet to live on. However, because alternative planets are currently unavailable, abandoning ship is hardly an option. Is the continuous increase in deaths, destruction and misery, and all the financial costs due to disasters inevitable and something we must simply suffer from? Or is there something we can do about it?

Scientific interest in natural hazards and disasters is similarly growing at exponential rates. However, the publication count on this topic is dwarfed by the huge number of articles on climate change or global warming (Figure 1.4). This trend is surprising, given that many scientists accept and stress the many connections between contemporary global warming and increasing numbers of extreme weather events. In 2014, international publishers released an average of 44 scientific publications per day(!) with the term 'climate change' in the title or abstract; this is more than ten times the number of publications with the term 'natural disaster' similarly in the title or the abstract, and nearly 30 times the number of publications that mention 'natural hazard' (www.scopus.com). PLoS ONE, currently ranked as the world's largest journal, has published more than 5000 articles on climate change, but fewer than 300 on natural disasters since the journal was founded in 2006 (data accessed 25 April 2015). The term 'climate risk' rarely refers to risks, but rather hazards that respond to changes in Earth's weather and its climate system (Moss et al. 2013). This focus on a seemingly single issue has been criticised for three reasons: (i) climate change seems a distant threat to many people in spite of current publicity and interest in the topic; (ii) a single focus may hinder an integrative view of mitigation and adaptation strategies; and (iii) the culturally and socially diverse views and perceptions of risk may be insufficiently captured (Luers and Sklar 2013). More integrative considerations of climate hazards and risks

Topics Geo – World map of natural catastrophes 2017   Munich RE

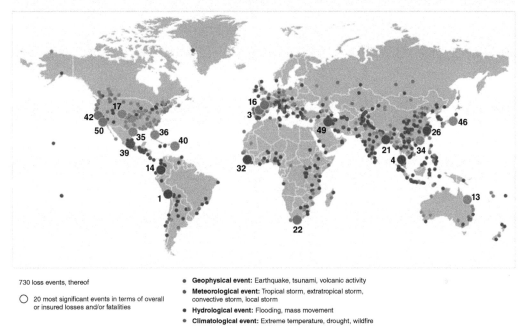

730 loss events, thereof

○ 20 most significant events in terms of overall or insured losses and/or fatalities

● **Geophysical event:** Earthquake, tsunami, volcanic activity
● **Meteorological event:** Tropical storm, extratropical storm, convective storm, local storm
● **Hydrological event:** Flooding, mass movement
● **Climatological event:** Extreme temperature, drought, wildfire

Figure 1.2 Global overview of (insured) natural disasters by Munich Re. From MunichRe (2018).

might couple biophysical controls and social values.

Many national and international research programmes have, for many years, been funded to investigate and reduce the impacts of natural disasters. For example, 1990–1999 was declared by UNESCO as the International Decade for Natural Disaster Reduction (IDNDR), and a concerted, large-scale international research effort was made to lessen loss of life, injury, and economic damage from natural disasters. However, the programme had little if any effect. Every year, major aid programmes provide developing countries with flood protection and soil erosion control measures. Sadly, the all-too-common result is subsequent neglect and rapid deterioration, with little positive effect. The large sums spent researching and reducing disasters appear to be having little effect.

This bleak outcome is unsurprising. The number of people and their assets affected by disasters is increasing in part because the total population and the total value of human assets are rising. As time goes by we have more people and more to lose, so even if the number of extreme natural events remains unchanged, we can expect that life loss and costs will also increase with time. The rapidly increasing impacts of disasters only worsen this effect. Disasters disrupt commerce and this is an additional cost that also increases with time as commercial activity increases.

Even without natural disasters increasing in intensity or frequency, the number of people in harm's way and the value of vulnerable assets and activities are increasing (Figure 1.5). Of course, it is possible that the number or intensity of disastrous natural events may indeed be on the rise, either because the Earth's surface is rarely in a steady state over periods that are of interest to humans, or because humans themselves are generating more weather extremes by dumping their waste products, specifically greenhouse gases, into the atmosphere. Among our biggest problems in the

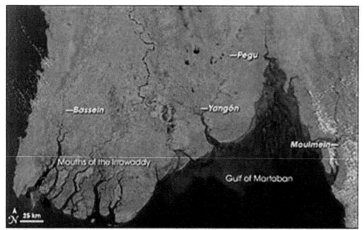

April 15, 2008

May 5, 2008

Figure 1.3 ASTER satellite images before and after Tropical Cyclone Nargis hit the coast of Myanmar (Burma) near the Irrawaddy delta in 2008, killing at least 85 000 people according to official records. Moreover, the storm destroyed 783 000 ha of agricultural land that most of the local farmers depend on heavily (NASA Images, www.nasa.gov).

twenty-first century is air pollution. High concentrations of fine particulate matter with a diameter smaller than 2.5 μm may be responsible for some 3.3 million of premature deaths worldwide in 2010 (Lelieveld et al. 2015).

When we compare the documented increases of population and global gross domestic product, the effect of changing natural hazards is either minor so far or has been largely underestimated. From this perspective, the increase in natural disasters is largely tied to rapid population growth. As we occupy more and more of our limited planetary surface, and occupy these areas for longer times, we increase the risk of being affected by extreme natural events that are inevitable. What we call

natural disasters or catastrophes are part of the dynamics of Planet Earth. Its physical systems have been behaving in much the same fashion for millions of years, even after *Homo sapiens* evolved. We cannot prevent earthquakes, volcanic eruptions, catastrophic landslides, hurricanes or blizzards; so it looks like we are destined to live with our unruly planet for the foreseeable future.

In 1989, the American geologist and author John McPhee wrote a fascinating book called *The Control of Nature*, in which he recounted efforts to control Los Angeles debris flows, the Mississippi River, and an Icelandic lava eruption (McPhee 1989). The book also highlighted some of the aspects to consider

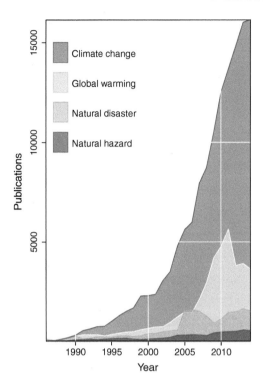

Figure 1.4 The number of scientific publications recorded in Elsevier's SCOPUS database (www.scopus.com) has grown exponentially across all disciplines over the past three decades. Publications with 'climate change' or 'global warming' in their titles or abstracts far outnumber publications with 'natural disaster' or 'natural hazard' similarly listed. Source: Data from Elsevier's SCOPUS database (www.scopus.com). Data accessed 24 April 2015.

when manipulating all but the minor and short-lived processes of nature, in spite of the power and ingenuity increasingly available to humankind. Readers of that excellent book gain the impression that, in order to live in some very desirable places on Earth, society has to spend large sums of money on an everlasting basis maintaining some sort of protection against disasters. The protection, moreover, is statistical and thus uncertain, and so may fail at any time.

This train of logic leads to the rather depressing conclusion that catastrophes cannot be prevented and will be inevitably visited on humankind. If, as appears to be likely,

human numbers continue to grow and we generate more and more commercial activity, this outcome will be realized. Must we therefore accept and resign ourselves to the continuation of these trends, and their consequences – shattered dreams, misery and desperation? We believe otherwise, hence this book.

1.2 Dealing with Future Disasters: Potentials and Problems

The extremes of nature are too powerful to control reliably, and research to date seems to have had negligible effect on natural disaster reduction (Table 1.1). Also, human exposure to extreme events must increase with increases in population and economic activity, as more people need access to natural resources to sustain their livelihoods. We contend, however, that by better using our understanding of the dynamics of the Earth we can design ways in which society can continue to develop, while becoming less vulnerable to natural disasters (Figure 1.6). Here we accept that we can neither predict nor control fully the high-energy natural processes that give rise to disasters, and instead focus on ways in which society can alter its own behaviour so as to become less vulnerable, and more resilient, to future disasters. This requires knowing the types of natural events that can cause disastrous impacts in specific locations, and it is this knowledge that we deal with herein.

In recent years society has, to an extent, accepted this point of view. The days when civil engineering was defined as some art of governing the sources and forces of Nature for sole convenience of man have all but gone. Nevertheless, the tradition of using engineered countermeasures to mitigate physical disasters continues to be the *modus operandi* of disaster management in many organizations. Building structural countermeasures, instead of reducing disaster costs, can thus leads to increases

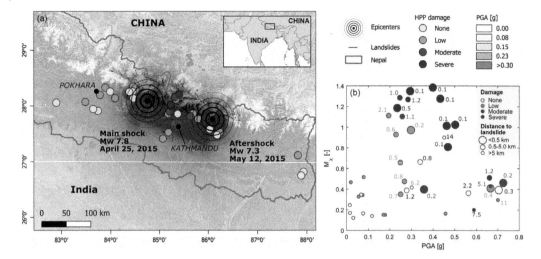

Figure 1.5 Map of Nepal including peak ground acceleration derived from U.S. Geological Survey ShakeMap, landslides mapped by a team from Durham University and the British Geological Survey, and damage scales of hydropower projects (HPPs). (b) HPP damage and distance from locations where landslide runout paths intersect the river network. The marker size and numbers refer to HPP distances (in km) from these landslides. The markers without numbers refer to HPPs without any landslides nearby (>15 km). From Schwanghart et al. (2018).

Table 1.1 Summary of major volcanic disasters in the twentieth century together with estimates of the mortality, financial loss, and total number of people affected involved. Note the variety of processes associated with volcanoes. After Witham (2005). Numbers in brackets give the percentage of events caused by each phenomenon for each impact.

Phenomenon	Killed (% of events)	Injured (% of events)	Homeless (% of events)	Evacuated/affected (% of events)
Debris flows/avalanches	741 (2.4)	267 (3.7)	4600 (2.5)	28950 (1.6)
Epidemic	5180 (0.7)			
Famine				
Gas/acid rain	2016 (14.5)	2860 (6.6)		58138 (3.6)
Volcanic unrest				33000 (2.8)
Other indirect	167 (4.8)	161 (3.7)		1000 (0.4)
Jökulhlaups				300 (0.4)
Lava	664 (4.5)	56 (6.6)	21490 (33.3)	113052 (13.3)
Primary lahars	29937 (12.5)	5022 (5.9)	91400 (12.3)	1078331 (10.5)
Secondary lahars/flooding	797 (7.3)	178 (5.1)	1925 (6.2)	84415 (4.4)
Pyroclastic currents	44928 (13.5)	2762 (15.4)	72481 (23.5)	521859 (11.7)
Seismicity	391 (2.4)	66 (2.9)	1448 (2.5)	165700 (10.1)
Tephra	6047 (29.1)	4321 (43.4)	97513 (22.2)	3 103580 (36.7)
Tsunami (waves)	661 (2.4)	300 (1.5)		
Unknown	195 (5.9)	20 (5.1)	600 (1.2)	93581 (5.6)

Figure 1.6 Structural vulnerability refers to the fraction of damage expected from a given impact; this building collapsed during strong seismic shaking. (Oliver Korup)

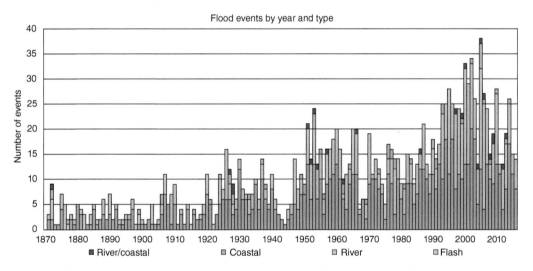

Figure 1.7 Time series of reported damaging floods colour-coded by flood type in 37 countries throughout Europe since 1870 in the HANZE database. From Paprotny et al. (2018).

in average annual damage costs. Constructing impressive and expensive structural countermeasures to deal, for example, with flood hazards, encourages people to invest heavily in thus protected areas, in the belief that they are completely safe (Figure 1.7). When, inevitably, an extraordinarily large flood occurs, it will cause more damage than would have been the case without any countermeasures, because in that case the investments would have been much smaller. Structures are mostly built to control frequent instead of rare events, because it is the most, and often only, economic way to do so.

Structural countermeasures also interfere with natural processes, generating a response that tries to restore the system to its original natural state. Some rivers, for example, are dammed to generate electricity or provide water for irrigation or domestic use.

The impounded water, however, reduces the gradient of the river channel upstream, while increasing it below the dam. As a result, local erosion commonly occurs immediately downstream of the structure. The effects of the dam on the river profile thus extend both up- and downstream, and river processes work towards establishing the former longitudinal profile. Thus nature 'fights back', leading to different and possibly unanticipated system behaviour that exceeds what countermeasures were designed for.

The approach we use in this book begins with accepting that, irrespective of future technological developments, it is unwise to try to change the extreme behaviour of natural systems. For example, even if we succeed for a time in dampening high flood levels on a river by repeatedly raising levees, the thus confined river as a system might react by increasing local bed aggradation, so that flooding levels increase commensurately. The normal and understandable response of a flooded community is to demand that the authorities stop the river from flooding. Often, decision makers involved are all too willing to try to do so, because constructing dykes generates both work and votes. Also, it is statistically very unlikely that a flood event so large as to defeat the new engineered works will occur within the political memory of the community. Thus, however logical it may be, the approach we propose is far from a simple process. In a sense, we know where we are, but where we want to be is a potentially contentious issue. Even if we agree as to where we want to be, how we get there from here in the real world is a problem.

Where do we want to be? The answer to this question depends on the ultimate goals of protection and safety from natural hazards that we collectively desire and are willing to pay for. How much risk are we willing to tolerate, both at the personal and societal levels? Do we wish to live in a society in which the siting of assets, and commercial and other activities, are regulated with the intent of restricting development and occupation of areas known to be vulnerable to extreme natural events? An important caveat is that society will put up with some risk, commonly referred to as 'acceptable' or 'tolerable' risk. We also want society to be able to anticipate the effects of a given disaster and to deliberately adapt its behaviour so that it can quickly and efficiently recover from a disaster should one occur. In many ways these two aspirations are one and the same, but it is useful to consider them separately. Importantly, both explicitly accept that disasters will continue to occur.

Why is it so difficult to get there from here? Most economic activity, and the societal network that supports it, is designed for maximum short-term profit under ideal conditions (that is, assuming without any disasters); it is sophisticated and intricately interlinked to that end. The result is a highly sophisticated social – commercial system with a minimum of 'wasteful' redundancy. By its nature, this system is vulnerable to failure; a single component can cause a widespread failure cascade (Figure 1.8). Examples are the 2008 financial crisis, the Fukushima nuclear power plant meltdown following the 2011 Tohoku earthquake, and the electricity blackouts during the 1998 ice storm in Ontario, Quebec and the northeastern USA. One complication is that the timescale of strategic thinking in politics and commerce is rarely longer than about five years, thus planning for things that are unlikely to happen in the time frame relevant to a politician is seen as a waste of money or votes, even though economic cost – benefit analyses show that disaster planning and investments have longer-term financial benefits. Some of these issues are now well recognized and spelled out in international efforts to reduce natural disaster risk, such as the current Sendai Framework for Disaster Risk Reduction (https://www.unisdr.org/we/coordinate/sendai-framework). Persuading captains of industry, politicians and the public that a slight reduction in profit in the short term will lead to large savings in future disaster costs is a difficult task, in spite of the simple arithmetic involved. A common response to such attempts is that 'technology will find a way to solve the problem' (Figure 1.9). A layperson's faith in the ability of science to

Figure 1.8 The earthquake hazard cascade in Beichuan, Sichuan province, China, after the Ms 8 Wenchuan earthquake in 2008. Buildings collapsed or were severely damaged due to the strong ground shaking. The shaking also triggered several landslides that invaded the town. A large landslide dam upstream of the town had to be artificially breached, sending floodwaters and sediment through parts of the city. Monsoon rains mobilized more landslide debris from hillslopes months after the earthquake, triggering a series of debris flows that caused massive aggradation of up to several meters. (Tim Davies)

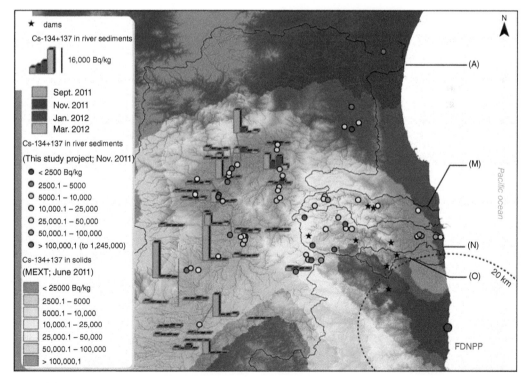

Figure 1.9 The interface between geomorphology and a na-tech disaster – fallout recorded by soil and river sediments following destruction of the Fukushima nuclear power plant by the tsunami of the 2011 Great Tohoku earthquake. $^{134+137}$Cs activity measured in river sediments and in soils. A: Abukuma catchment; M: Mano catchment; N: Nitta catchment; O: Ota catchment). From Chartin et al. (2013).

come up with miracle solutions should also be considered.

We believe that the key to progress in disaster reduction is that we know and accept that future disasters will occur and that their costs can be reduced by strategic direction of investments now. People are aware to varying degrees that natural disasters happen, although rarely in any given place. The potential for a disaster to affect them personally is almost always so small that inertia overcomes any desire to take action. People might believe that, after having experienced a 100-year event, they (and their community) might be OK for another 99 years. One opinion about the 2010/2011 earthquakes at Christchurch, New Zealand, was that they were 'maximum credible events', the implication being that strong ground shaking has a known upper limit. The problem goes away and the teachable moment for society has been lost.

Among the glimmers of hope is the traction that the environment and sustainability movements have gained among both the public and politicians in recent decades. People in some cases have been willing to pay more for sustainably and ethically produced goods, to sort rubbish before putting it out for collection, and to quit smoking in large numbers when the risks are clear to them. Disaster management is a key component of sustainable development, and by demonstrating this connection we can foster disaster consciousness and disaster preparedness.

1.3 The Sustainable Society

Many definitions have been proposed for sustainability over the years, but our definition is straightforward and we think acceptable to all: *an activity is sustainable if it can continue for a specified time period at a specified intensity without unacceptable consequences.* Applying this definition to society, the activity of concern is how humans use the Earth's resources, including its surface and atmosphere for waste disposal. The maximum allowable intensity is the rate of use of resources and waste disposal that meets the sustainability criterion rather than simply the needs of future generations, which may be variable and potentially different from current needs. Unacceptable consequences could be, for example, lack of oxygen caused by completely deforesting of the planet, or the death of grass due to failed genetic manipulation, or even extinction of the Sumatran tiger because that eliminates the need for Sumatran Tiger Safaris Inc., which is unacceptable to the shareholders and potential customers. These are the conventional environmental aspects of sustainability. The political dimension at the national and global scale is encapsulated by a set of 17 Sustainable Development Goals that the United Nations (www .un.org/sustainabledevelopment) adopted in 2015 as part of the 2030 Agenda for Sustainable Development:

Goal 1 End poverty in all its forms everywhere

Goal 2 End hunger, achieve food security and improved nutrition and promote sustainable agriculture

Goal 3 Ensure healthy lives and promote well-being for all at all ages

Goal 4 Ensure inclusive and quality education for all and promote lifelong learning

Goal 5 Achieve gender equality and empower all women and girls

Goal 6 Ensure access to water and sanitation for all

Goal 7 Ensure access to affordable, reliable, sustainable and modern energy for all

Goal 8 Promote inclusive and sustainable economic growth, employment and decent work for all

Goal 9 Build resilient infrastructure, promote sustainable industrialization and foster innovation

Goal 10 Reduce inequality within and among countries

Goal 11 Make cities inclusive, safe, resilient and sustainable

Goal 12 Take urgent action to combat climate change and its impacts

Goal 13 Conserve and sustainably use the oceans, seas and marine resources

Goal 14 Sustainably manage forests, combat desertification, halt and reverse land degradation, halt biodiversity loss

Goal 15 Sustainably manage forests, combat desertification, halt and reverse land degradation, halt biodiversity loss

Goal 16 Promote just, peaceful and inclusive societies

Goal 17 Revitalize the global partnership for sustainable development

Most of these ambitious goals have direct ties to how people are exposed or vulnerable to natural disasters. An often overlooked, unacceptable consequence is that a disaster reduces societal actions to an unacceptable level. Irrespective of its rate of use of resources or how it cares for waste management, society cannot be sustainable, by our definition, if a natural disaster causes an unacceptable reduction of activity. Thus, resilience to natural and other types of disasters is both a desirable and necessary attribute of a sustainable society (Klein et al. 2003). 'Resilience' to natural disasters is a widely-used term that the UNDRR defines as:

> The ability of a system, community or society exposed to hazards to resist, absorb, accommodate, adapt to, transform and recover from the effects of a hazard in a timely and efficient manner, including through the preservation and restoration of its essential basic structures and functions through risk management.

Take note that this definition is one of many views: Zhou et al. (2009) compiled some thirty different definitions of resilience, and Alexander (2013) cautioned against overusing and overinterpreting this term. Ayyub (2014) listed seven different views of resilience, and emphasized the need for objective and reproducible metrics. His proposed approach to measure resilience assumes that 'incidents'

(or disasters) occur at a given rate and independently of each other, and takes into account the duration of both the damaging incidence and the subsequent recovery. Another interesting feature of this approach is an ageing effect that specifies that the ability to handle disasters may decrease with time.

One view is that resilience can be achieved by disaster risk reduction, that is, reducing probabilistic risk. For hazards that are likely to occur frequently in the period targeted for disaster mitigation measures, reducing risk may indeed be the appropriate way to achieve resilience. Yet several studies have pointed out that this approach may become inaccurate and, at worst, misleading or ineffective when applied to rare events (Park et al. 2013; Davies and Davies 2018). Reducing the disaster risk from such hazards by trying to reduce further their probability of occurrence may be neither noticeable nor pragmatic in terms of measurable benefits. The main motivation to increase resilience is to reduce disaster impacts. While in some cases this can be done by way of risk reduction, in other cases probabilistic risk may be inappropriate.

In the same way, development can only be sustainable if it is constrained by the requirement to avoid disasters and to develop and follow plans for recovery from foreseen disasters in a timely manner. The key word here is 'foreseen'. Preparing for unforeseen or unexpected disasters may be impractical given the many uncertainties involved. Nor will society have had the option to limit its exposure to the disaster. A crucial factor in sustainability, then, is the *ability to foresee natural disasters*.

This foresight relies on the geosciences, because extreme natural events are geoscientific phenomena and geoscientific research is required to find out what they are, where they can occur and how big they might be. A disaster requires a community at risk, thus foreseeing a disaster also requires an understanding of the characteristics and mechanisms that make this community disaster prone. Scientists who are identifying a disaster-prone community only make the first step. What is also required is that the disaster be foreseen, that is recognized and accepted as a pending reality, so that the community can choose whether and how to adjust its organization and behaviour to reduce the risks to acceptable levels. Reducing disaster impacts thus requires that communities become aware of potential disasters, and that requires a combination of geoscience and social science knowledge that is understood and accepted by communities. In this book we emphasize the role of geoscience and of geoscientists in this endeavour.

1.4 Benefits from Natural Disasters

Documenting past and likely future consequences of natural disasters is but the first step in developing solutions to many of the problems we are facing in the twenty-first century. The wish to strengthen adaptive capacity is a key strategy in the multi-faceted discussion about the connections between climate change, climate risks, and natural disasters (Moss et al. 2013). Yet communication among the many research communities concerned with climate change and natural hazards must be improved to better coordinate findings and develop joint strategies. Strengthening resilience against natural disasters is one possible avenue for improving this cooperation (Klein et al. 2003). Climate change is likely to undermine or destroy the livelihoods of millions of people. Resettlement of 'climate refugees' is far from a future scenario, as it has already begun in many places. In Vietnam, for example, more than 200 000 people have been resettled away from the nation's major river delta as the sea level has been rising, and a similar fate awaits the 380 000 inhabitants of the Maldives, as these islands will probably vanish with rising sea level by the end of the twenty-first century (López-Carr and Marter-Kenyon 2015). A resilience-based approach to engineering systems and solutions

of difficult natural problems (Park et al. 2013) offers a complement to the current risk-based paradigm (see Chapter 18).

The saying that adversity creates opportunity holds for natural disasters. Despite the long list of adverse and harmful consequences of natural disasters, some positive aspects are easily neglected when speaking of death tolls, financial damages, and long-term losses in disaster-struck regions. From the geological perspective, earthquake-induced uplift creates new land, including areas where flat terrain is precious. For example, most of the downtown area of New Zealand's capital of Wellington is situated on a shore platform that was raised out of the sea during the 1855 Wairarapa earthquake.

Volcanic ash can enrich soil layers with nutrients and form andosols. Enhanced plant growth is a direct benefit of this natural fertilization. However, thick ash cover completely seals the underlying soil, effectively sterilizing the ground surface such that agricultural use is impossible for several years to decades. Some volcanic eruptions may be beneficial for tree growth if elevated atmospheric aerosol inputs scatter sunlight; detailed studies of tree rings added after 23 major pyroclastic eruptions in the past 1,000 years, however, show that negative short-term cooling effects likely outweigh the positive effects of sunlight scattering, at least in in Northern Hemisphere forests (Krakauer and Randerson 2003). Volcanism has many other benefits, such as the provision of hydrothermal energy, which is the reason Iceland's capital of Reykjavik has a natural floor-heated pavement.

From an ecological perspective, for example, many ecosystems are prone to episodic disturbances. Species can adapt to, or even depend on, these disturbances. Wildfires can destroy living vegetation, but also clear the ground for new plants and promote germination. Case studies that balance in detail the negative and positive consequences of wildfires sometimes offer surprising insights, for example that wildfires may also improve the habitat quality of certain species of salmon by changing the delivery of fine sediment and wood to streams (Flitcroft et al. 2016). Dust storms are a major source of terrestrial sediment and nutrients, and partly fertilize oceans and remote islands. Saharan dust provides nutrients to the Amazon rainforest. The airborne transport of biogeochemical materials may have helped to boost the biodiversity of remote island chains, such as the Hawaiian archipelago, and highlights how wind-driven dust transport is prominent in global biogeochemical cycles (Okin et al. 2004). Much of the iron-rich mineral dust entering the oceans, however, is unavailable for marine biota, and fertilizing effects are at best local, at least as far as marine biological productivity is concerned (Doney 2010).

Natural climatic oscillations such as the El Niño–Southern Oscillation (ENSO) also provide some benefits. El Niño phases tend to suppress the development of Atlantic tropical storms. The strong 1997–1998 El Niño resulted in a net benefit of $20 billion to the United States' economy because of the reduced number of land-falling hurricanes and the unusually warm winter in the Midwest. However, this decrease in Atlantic tropical cyclone activity coincides with an increase in typhoons in the eastern and central Pacific.

Contemporary warming has also increased net primary production in many areas of the world because higher temperatures lower some of the constraints on plant growth. Nemani et al. (2003) concluded that the observed increase of 6% in the global net primary production between 1982 and 1999 is the result of warmer temperatures. Rainforests in the Amazon seem to have benefitted in particular from this warming, which was accompanied by lesser cloud cover and higher solar irradiation. Overall, however, the Intergovernmental Panel on Climate Change (IPCC) has concluded that primary agricultural production on a global scale will be negatively impacted by warming, so these benefits to productivity are likely to be offset towards an ultimately negative outcome.

Floods on the Nile River are a classic example of how entire civilisations depend on regular water and nutrient supply by rivers. Some of these floods have been disastrous, and so has been their absence:

> The first Old World civilizations, along the Huang He, Indus, Nile, Tigris and Euphrates rivers were almost entirely on alluvium. They were 'hydraulic' [...] or 'potamic' in the sense that they were in relatively dry environments and farming depended on natural inundation or controlled irrigation from river water. [...] Floods also brought nutrient-rich sediments. This provided the potential for a prosperous agriculture and for organised societies to develop urban cultures in which deified rulers, writing, and artistic creativity flourished. (Macklin and Lewin 2015)

Several large floods on the Yangtze River and other major rivers of the world have increased primary productivity in near-coastal oceans by enhancing the growth of phytoplankton. However, river-borne sediment plumes can flush excess agricultural fertilizers and trigger algal blooms that lead to hypoxia (Gong et al. 2011). The boosts to microbial and algal growth are short lived and localized, and represent peaks in productivity that decimated marine food webs fail to take care of (McCauley et al. 2015). Tropical cyclones may episodically flush coastal lagoons, causing short-lived spikes in particulate suspended matter, water opacity, and nutrient loads leading to eutrophication (Jennerjahn 2012).

Landslide, moraine, and volcanic dams, if stable for millennia or longer, may impound valuable freshwater resources, particularly in semiarid or arid mountain belts (Strom 2010). These lakes also attenuate floods, thus providing some level of flood protection downstream until they become infilled with sediment. Landslides into naturally dammed lakes, however, may set off destructive displacement waves that overtop the dams and initiate catastrophic incision. Some natural dams serve as foundations for hydropower stations. The lakes behind natural dams may eventually become infilled, providing flat fertile land for cultivation in otherwise steep terrain.

People also adapt to disasters and try to make the best out of the situation they face, especially if few other options, like leaving the area, are available. Farmers in the steeplands of Papua New Guinea, for example, have long taken advantage of landslides, which modify the properties of soil and the topography of hillslopes, for agricultural use. For example, they plant carefully selected crops and mixed gardens on deposits of shallow rotational landslides (Humphreys and Brookfield 1991). Agriculture may have developed as early as 9000 years ago in the highlands of Papua New Guinea, and evidence from sediments in swamps and caves points at rates of soil erosion that were lower for most of this period than that following contact with Europeans:

> It is a remarkable fact that traditional swidden and wetland agriculture operated in the ecologically fragile highlands of Papua New Guinea for over 8000 years, eventually supporting almost a million people, without serious environmental degradation. This situation only changed when indigenous environmental relations were disrupted, firstly with the introduction of a new exotic domesticate – the sweet potato – and secondly with the advent of the twentieth-century cash economy. (Roberts 2014)

This observation is at odds with the documented effects of agricultural practice on soils and sediment flux in other regions, so that we caution against making generalized statements regarding these intimate links between land use, vegetation, soils, and geomorphology. Nevertheless, the lesson from the highlands of Papua New Guinea also demonstrates how important and useful it is to obtain detailed local records of past geomorphic activity in response to human disturbances.

Aims of this book This is an advanced textbook. We assume that readers are already familiar with the basics of geomorphology or Earth sciences in general. Our objective is to raise your awareness that natural disasters are inevitable and result in far more than deaths and economic loss. We have assembled a variety of lines of independent evidence that show that natural disasters also cause substantial geomorphic changes that range from catastrophic reshaping of landscapes to very high fluxes of water, sediment, and biogeochemical constituents that continue to impact people long after a given disaster has passed (Figure 1.10). We therefore also emphasize the indirect and intangible losses caused by natural disasters. Measuring such losses requires detailed knowledge of underlying geomorphic processes and the response times of processes that impact landscapes. We are convinced that geomorphology – a rapidly evolving and increasingly interdisciplinary field – is essential for sustainably managing future natural hazards, risks, and disasters. We share the view that fostering a quantitative understanding of landscape and landscape-scale processes is an important unfilled niche in the global environmental change debate.

Figure 1.10 Pulsed sediment transport can damage infrastructure. Dealing with the problem requires understanding of sediment delivery mechanisms, sediment volumes, duration, and spatial reach. Top left: Bridge destroyed by a lahar, Chile. Top right: Cascade of check-dams and sediment retention basins in a mountain stream near Nikko, Japan. Lower left: Collapsed sediment retention basin in a steep mountain watershed, Taiwan. Bottom right: A series of check-dams filled with sediment, Taiwan (all photos by Oliver Korup).

1.5 Summary

(i) We can never be free of disasters because of the dynamic nature of the Earth's surface and the continuing growth of human numbers.

(ii) The extreme dynamics of natural systems, which are responsible for disasters, cannot be controlled, thus disaster impacts can only be reduced if society adapts to nature.

(iii) Reduction of disaster impacts is a crucial component of sustainable development.

(iv) Developing community resilience to disasters requires accepting that disasters are inevitable; otherwise resilience is seen to be unnecessary.

(v) Political time frames are so short that politicians often ignore the inevitability of disasters. Therefore, the community, which has the power to select its political representatives, must accept the inevitability of natural disasters and insist on appropriate action.

(vi) Developing disaster resilience also means effectively communicating geomorphic information to the community at risk.

References

Alexander DE 2013 Resilience and disaster risk reduction: an etymological journey. *Natural Hazards and Earth System Sciences* **13**(11), 2707–2716.

Ayyub BM 2014 Systems resilience for multihazard environments: definition, metrics, and valuation for decision making. *Risk Analysis* **34**(2), 340–355.

Cash RA, Halder SR, Husain M, et al. 2013 Reducing the health effect of natural hazards in Bangladesh. *The Lancet* **382**(9910), 2094–2103.

Chartin C, Evrard O, Onda Y, et al. 2013. Tracking the early dispersion of contaminated sediment along rivers draining the Fukushima radioactive pollution plume. *Anthropocene* **1**, 23–34.

Davies TRH, McSaveney MJ and Clarkson PJ 2003 Anthropic aggradation of the Waiho River, Westland, New Zealand: microscale modelling. *Earth Surface Processes and Landforms* **28**(2), 209–218.

Davies TRH and Davies AJ 2018 Increasing communities' resilience to disasters: An impact-based approach. *International Journal of Disaster Risk Reduction* **31**, 742–749.

Doney SC 2010 The growing human footprint on coastal and open-ocean biogeochemistry. *Science* **328**(5985), 1512–1516.

Flitcroft RL, Falke JA, Reeves GH, et al. 2016 Wildfire may increase habitat quality for spring Chinook salmon in the Wenatchee River subbasin, WA, USA. *Forest Ecology and Management* **359**, 126–140.

Froude MJ and Petley D 2018 Global fatal landslide occurrence from 2004 to 2016. *Natural Hazards and Earth System Sciences* **18**, 2161–2181.

Gong GC, Liu KK, Chiang KP, et al. 2011 Yangtze River floods enhance coastal ocean phytoplankton biomass and potential fish production. *Geophysical Research Letters* **38**(13), 1–6.

Humphreys GS and Brookfield H 1991 The use of unstable steeplands in the mountains of Papua New Guinea. *Mountain Research and Development* **11**(4), 295–318.

Jennerjahn TC 2012 Biogeochemical response of tropical coastal systems to present and past environmental change. *Earth-Science Reviews* **114**(1-2), 19–41.

Klein RJT, Nicholls RJ and Thomalla F 2003 Resilience to natural hazards: How useful is this concept?. *Environmental Hazards* **5**(1), 35–45.

Krakauer NY and Randerson JT 2003 Do volcanic eruptions enhance or diminish net primary production? Evidence from tree rings. *Global Biogeochemical Cycles* **17**(4), 1–11.

Lelieveld J, Evans JS, Fnais M, et al. 2015 The contribution of outdoor air pollution sources to premature mortality on a global scale. *Nature* **525**(7569), 367–371.

López-Carr D and Marter-Kenyon J 2015 Human adaptation: Manage climate-induced resettlement. *Nature* **517**(7543), 265–267.

Luers AL and Sklar LS 2013 The difficult, the dangerous, and the catastrophic: Managing the spectrum of climate risks. *Earth's Future* **2**, 114–118.

Macklin MG and Lewin J 2015 The rivers of civilization. *Quaternary Science Reviews* **114**(C), 228–244.

McCauley DJ, Pinsky ML, Palumbi SR, et al. 2015 Marine defaunation: Animal loss in the global ocean. *Science* **347**(6219), 1255641–1–1255641–7.

McPhee J 1989 *The Control of Nature*. Farrar, Stratus, and Giroux.

Moss RH, Meehl GA, Lemos MC, et al. 2013 Hell and high water: practice-relevant adaptation science. *Science* **342**(6159), 696–698.

MunichRe 2018 A stormy year – Natural disasters in 2017. https://www.munichre .com/topics-online/en/climate-change-and-natural-disasters/natural-disasters/topics-geo-2017.html (Accessed 16 June 2020).

Nemani RR, Keeling CD, Hashimoto H, et al. 2003 Climate-driven increases in global terrestrial net primary production from 1982 to 1999. *Science* **300**(5625), 1560–1563.

Okin GS, Mahowald N, Chadwick OA and Artaxo P 2004 Impact of desert dust on the biogeochemistry of phosphorus in terrestrial ecosystems. *Global Biogeochemical Cycles* **18**(2), 1–9.

Park J, Seager TP, Rao PSC, et al. 2013 Integrating risk and resilience approaches to catastrophe management in engineering systems. *Risk Analysis* **33**(3), 356–367.

Parotny D, Morales-Nápoles O, and Jonkman SN 2018 HANZE: A pan-European database of exposure to natural hazards and damaging historical floods since 1870. *Earth System Science Data* **10**, 565–581.

Roberts N 2014 *The Holocene*. Wiley Blackwell.

Schwanghart W 2016 Topographic and seismic constraints on the vulnerability of Himalayan hydropower. *Geophysical Research Letters* **45**(17), 8985–8992.

Strom A 2010 Landslide dams in Central Asia region. *Journal of the Japan Landslide Society* **47**(6), 309–324.

Vranes K and Pielke, Jr, R 2009 Normalized earthquake damage and fatalities in the United States: 1900–2005. *Natural Hazards Review* **10**(3), 84–101.

Witham CS 2005 Volcanic disasters and incidents; a new database. *Journal of Volcanology and Geothermal Research* **148**, 191–233.

Zhou H, Wang J, Wan J and Jia H 2009 Resilience to natural hazards: a geographic perspective. *Natural Hazards* **53**(1), 21–41.

2

Defining Natural Hazards, Risks, and Disasters

2.1 Hazard Is Tied To Assets

A natural hazard occurs at the interface between human activities or assets and natural processes operating at, above, or below the Earth's surface. Some of the most energetic of these processes are caused by forces within the Earth, mainly in the crust and upper mantle. Such processes include earthquakes and volcanic eruptions resulting from tectonic plate motions. The other family of processes is driven by forces acting on and above the Earth's surface. These processes result from gravitational forces and the resulting fluxes of wind, water, and ice that erode, transport, and deposit sediment and its biogeochemical constituents. The main agents that drive these transfers are landslides, windstorms, rivers, waves, and glaciers. The spectrum of natural hazards also includes biological processes. We might regard locust swarms, famine, outbreaks of bird or avian flus, or the spread of other diseases as natural hazards. In this book, however, we focus mostly on abiotic natural hazards that have the potential to harm humans and the environment, including health, life, infrastructure or natural resources. Importantly, the notion of hazard is irrelevant without any lives or other human assets at stake.

This use of the term 'hazard' is independent of how slow or rapid the possible negative process is. Droughts are the single natural hazard that affects the most people worldwide, and may take weeks to months or years to develop and then ease. Yet the onset and termination of a given drought takes some effort to identify simply because it is a gradual phenomenon linked to potentially slow decreases in regional water availability or increases in water demand. Likewise, the magnitude of a drought can be problematic to measure. Temperature alone is meaningless without considering the regional water balance or the preceding trends in temperature. Water deficits may be defined in different ways, depending on water demand, and also on what plant species are being grown. Similarly, the movement of sediment throughout river systems is much slower than that of water. Rivers are natural conveyor belts for disturbances, and transmit these disturbances both upstream and downstream. Sedimentary hazards such as rapid or catastrophic channel sedimentation or rapid incision can have detrimental effects on river traffic, bridges, hydropower schemes, or watergates. In the same slowly progressing manner as drought, the motion of such sedimentary hazards can be slow and may require months or even years before impacting river reaches upstream or downstream reaches of the initial disturbance. The same principle applies to other natural hazards such as sea-level rise, gradual coastal erosion, and soil degradation. These are examples of slow-onset, chronic or elusive natural hazards. Other examples include the long-term exposure to toxic Earth

Geomorphology and Natural Hazards: Understanding Landscape Change for Disaster Mitigation, Advanced Textbook Series,
First Edition. Tim R. Davies, Oliver Korup, and John J. Clague.

materials that may occur in rock outcrops or be dispersed by rivers, wind or other natural processes (Skinner 2007). Even if causes and effects of slow-onset hazards can be catastrophic, their origin and onset may remain vaguely defined.

Taking yet another angle, Finkl and Makowski (2013) suggested classifying natural hazards based on how the public perceives them:

> […] hazards can be categorized as *apparent or obvious* (undeniably in the public's eye), *incipient or cryptic* (unseen to the public eye and intermittent in frequency), and *misunderstood or uncomprehended* (public is unaware through a low level of consciousness)

The essential anthropocentric aspect of natural hazards distinguishes naturally occurring processes without potential for harm from those that can inflict damage. We adopt this as a straightforward working definition before a more comprehensive discussion about whether natural hazards still link to fully natural processes. Along similar lines, a debate revolves around whether humans have left a globally detectable imprint in the geological record. Yet most of this dispute focuses on suitable geological markers that would formalize and justify a new geological epoch that some propose to name the 'Anthropocene'. Natural disasters have featured surprisingly little in this debate, although some scientists argue that natural disasters are far from natural, given that many allegedly natural processes have a large human footprint. River flooding, for example, has remained a major natural hazard in central Europe despite (or because of) widespread river training and regulation works. These protective works, together with widespread building and infrastructure development, have altered runoff and discharge regimes such that many floods are partly human made (Criss and Shock 2001). Deforestation in many parts of the world has so reduced the stability of soil

such that landslides or wave erosion by tropical cyclones have been exacerbated. So how 'natural' are the natural hazards and disasters that we have to deal with? In this book, we use a broad definition of natural hazards and disasters, while acknowledging the growing effect of humans on the severity of disasters. We nevertheless distinguish natural hazards and disasters from technological or purely human made ones such as oil spills, dam failures, ground subsidence following mining, and so on.

2.1.1 Frequency and Magnitude

Car-sized boulders frequently tumbling down a steep slope in a remote mountain valley can be a risk only if someone is around at least some of the time, or if some assets are located in that valley. If a handful of mountaineers enters the valley every year, the risk becomes nonzero: someone might be in the wrong place at the wrong time, but in mountaineering such objective risks are considered and accepted as a matter of course. The risk may be minute, but it is real and nonzero by definition, as humans put themselves at risk by venturing into areas where the level of geological activity is nonzero. Building a busy highway through the same mountain valley will considerably increase the risk to people, and constructing a big hotel in the runout path of the boulders can be a bad idea, given that people should know that boulders could tumble down the hillslope in the future because such past events are documented by boulders lying around.

Such a qualitative perspective may be intuitive and easy to understand, but is often of limited value. A more quantitative approach involves expressing natural hazards as *probabilities of potentially harmful processes*. The probabilities are generally specified for an area and interval of interest; for example, we estimate a 10% chance that a given length of mountain road will be damaged by falling boulders in any given year. This probabilistic approach has several advantages over a purely qualitative one. For example, probabilities allow us to put numbers to uncertainties about future events, such as how likely it is that New Orleans will be hit by a tropical cyclone of category V – the strongest category that Hurricane Katrina attained in 2005 – once again in the twenty-first century. Similarly, probabilities also allow us to express in numbers how likely it is that an asteroid of 100 metres or more in diameter will hit central Europe by the time you reach the end of this chapter (Figure 2.1). Weather forecasts on the TV or radio might predict a 60% chance that it will rain on the following day, sometimes referring to this prediction as the 'probability of precipitation' or even 'risk of rain'. Yet the method

and idea behind forecasting rain is slightly different from the probability predictions provided above. Getting a bit wet in the rain is likely to have less impact on you than having your house flooded by a storm surge or being struck by a falling asteroid. Nevertheless, the objective of assessing risk is to determine the expected damage. We use the term 'expected' here in a statistical sense, so that we treat risk as the sum of all possible damage outcomes, each weighted by its probability of happening within a given study area and period.

Sticking with the example of forecasting rain, another key question is: 'How hard will it rain?' We are interested in the *magnitude*

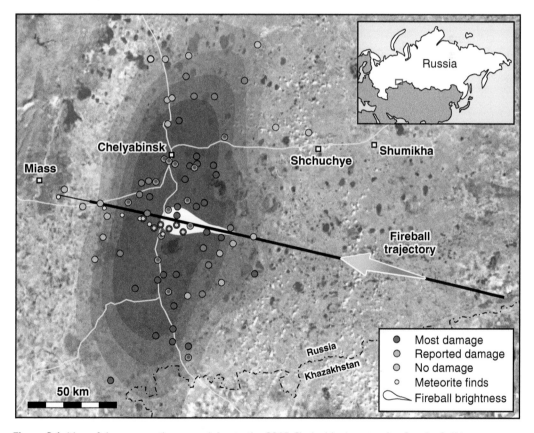

Figure 2.1 Map of damage on the ground due to the 2013 Chelyabinsk meteorite, Russia. Solid orange circles indicate locations of reported damage; grey circles indicate no damage. Solid red circles show the most damaged villages in each district, as reported by the government. Contoured greyscale shows modelled kinetic energies and overpressures due to the fireball, innermost to outermost: 300 kt (equivalent to kilotons of TNT) and >1000 Pa; 520 kt and >1000 Pa; 300 kt and p >500 Pa; and 520 kt and >500 Pa. Also shown are locations of meteorite finds (yellow points) and the fireball trajectory (black line), moving from 97 km altitude on the right to 14 km altitude on the left. Modified from Popova et al. (2013).

and *intensity* of a potentially negative outcome. Natural hazards researchers reserve the term 'magnitude' mostly for physical measures of the size, strength or energy of a natural phenomenon. Examples include the maximum wind speed of a storm, the maximum height of a tsunami wave, or the seismic energy released during an earthquake (Figure 2.2). Many of these magnitudes can be either measured directly or estimated from the geological record based on the assumption that bigger events leave larger and longer lasting signatures. But even some of the bigger events in recorded history have only indirectly or inaccurately measured magnitudes such as wind speed or earthquake magnitude. In 1960, for example, seismic stations in Chile were damaged, and failed to record the maximum magnitude of what has been the largest (M ~9.5) so far documented earthquake in human history (Kerr 2011). A useful approach to reconstructing the approximate magnitude of previously poorly-documented events is to use the spatial pattern of observed impacts or

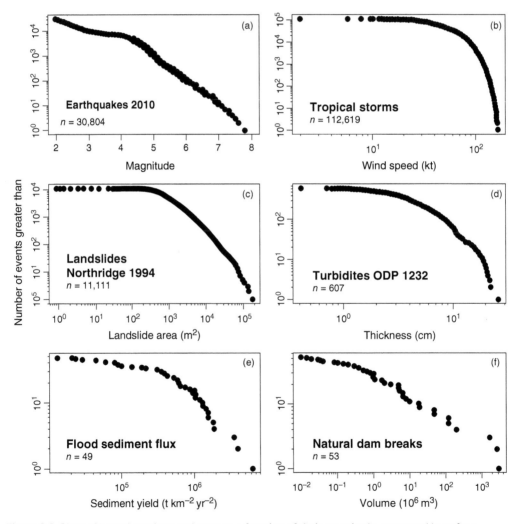

Figure 2.2 Plots of hazard numbers and rates as a function of their magnitude, expressed here for a specified region and period. Note that, for each hazard, there are many more data for smaller magnitude events than for larger ones, which is typical for geomorphological and meteorological phenomena.

resulting damage as a proxy. The underlying idea is that overall damage generally decreases away from the source of the disturbance. Here the concept of Intensity offers empirical or experience-based measures of the effects of a given event. An example is the Modified Mercalli Intensity Scale for earthquakes, which is based on a series of hierarchically structured phenomena that may be observed during earthquakes, ranging from subtly swinging lamps to falling chimneys through to widespread destruction. Other proxies for intensity include the number of houses swept away during a tornado or the length of road buried by a landslide.

Looking more closely at how the magnitude of many geological and meteorological processes on Earth varies, we find an interesting tendency. Events with lower magnitudes are much more frequent than those with larger magnitudes. Regardless of whether we are studying earthquakes, tropical storms, landslides or floods, we find a strikingly systematic relationship between the abundance and the magnitude of events. This relationship is inverse and distinctly nonlinear, and often extends over several orders of magnitude in both frequency and magnitude. Rare, but large events are in the right-hand tails of these distributions, and include, for example, 1000-year events that people seldom tend to think about, or forget, during their everyday lives simply because they are so rare that few people have experienced them. However, these rare events do occur, and are among the most destructive events in human history (Figure 2.3). Some extreme events are more frequent than the trends predict. These 'dragon-king' events appear to result from dynamic systems when an event occupies the entire space in the

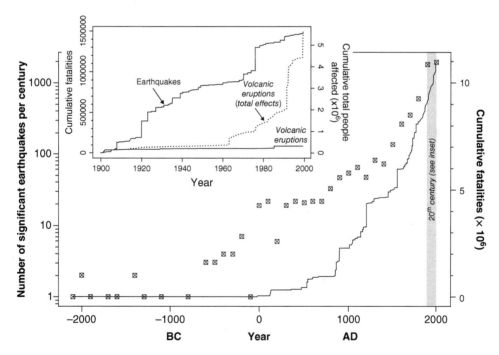

Figure 2.3 Number of significant earthquakes per century for the past 4000 years (crossed squares), and the cumulative minimum number of associated fatalities (black line) (US National Geophysics Data Center). All events recorded here caused at least US$ 1 million damage, claimed more than 10 lives, had a magnitude >7.5 or Mercalli Intensity >X, or triggered a tsunami. Inset shows cumulative fatalities caused by 2522 earthquakes during the twentieth century compared to those caused by 491 eruptions (Whitham, 2005); the total number of people affected by the eruptions is also shown. From Korup and Clague (2009).

system that generates it. They are the very rare and large events that lurk beyond systematic frequency-magnitude relationships that we reconstruct from many previous observations (see Chapter 5).

The inverse relationship between frequency and magnitude has an important advantage for use in hazard appraisals. If cast in mathematical form – as a probability distribution that assigns weights to each possible magnitude – we can conveniently estimate the long-term frequency of future events of a given magnitude or larger. Empirical frequency-magnitude curves form the basis for probabilistic hazard estimates and risk analyses. The underlying and often rather simplistic assumption, however, is that the patterns of frequency and magnitude of previous events are valid for the future and for the area of interest: we assume that the data are stationary. We can easily question this assumption. With regard to hydrometeorological hazards, and using a phrase from the investment community, 'past performance is no guarantee of future return'; changing climate is by definition imposing non-stationarity on time series of floods, storms, and sea-level rise. Geologists and geophysicists emphasize that how often volcanoes erupt can vary over decades and centuries. The presumption that mantle convection and tectonic plate movements are constant over periods that are of interest for decision making is convenient, but needs testing. The key thing to remember is that these probabilities can be seen as statistical measures that tell you how likely something is to happen. Ultimately, it is a question of when – instead of if – a natural disaster will occur. It is important to understand that statistics mostly predict what we expect to happen on average and with the least uncertainty, and over a long enough period that the probabilities will approximately match the frequency of observed events; yet these predictions can never inform us exactly what events will occur in a short time period, nor when the next major event will happen.

So strictly speaking, probability-based estimates of occurrences of natural hazards based on historical or geological data are only valid for a specified area and period of time, that is the study area and time from which data have been derived. This is mainly because the necessary completeness and detail of the geological archives required for reconstructing former events and computing their probabilities are limited. Also, the frequency-magnitude characteristics of specific processes differ between regions because of differences in climate, tectonic activity, anthropogenic interference (think of dams or other structural hazard countermeasures), and land-use practices. Extrapolating natural hazard estimates beyond a given study area and time horizon is possible, but the resulting predictions become increasingly unreliable the farther we extrapolate empirical data. Such extrapolations contain uncertainties that may lead to misestimates of risk.

2.1.2 Hazard Cascades

Natural hazards may occur singly and also in cascading fashion. Many natural disasters have resulted from a chain of coupled hazardous processes, when things have gone from bad to worse, also involving human made disasters in the chain of events. The 2011 Tohoku-Oki M9.0 earthquake in Japan triggered a huge tsunami that devastated hundreds of kilometres of coastline along the eastern seaboard of Honshu, and disabled the generators of the nuclear power plant at Fukushima-daichi, triggering a partial melt down. Risk researchers have proposed the term 'na-tech' disaster to describe this functional link between disasters that are partly natural and partly technological. To account for this mixture, the probabilities of future occurrences of such hazard chains can be combined in what are known as 'event trees'. These systematically link the probabilities of harmful events occurring conditioned on the likelihood of preceding events.

The strategy of assigning probabilities to natural hazards becomes problematic where we need to account for far-reaching and long-lasting impacts that often elude local hazard assessments. Volcanic eruptions, tsunami or dust storms originate from point sources, but their impacts may be hemispheric or even global. The 2010 eruption of the Icelandic volcano Eyjafjöll started as a local event. However, the wind-driven dispersal of its ash plume resulted in widespread disruption to air traffic in northern and central Europe, because of the apparent risk that volcanic glass poses to the operation of aircraft jet engines. The ash plume also reached much higher into the atmosphere than for any other known eruption of comparable volume (Gudmundsson et al. 2011). Airlines lost hundreds of millions of dollars due to flight groundings and cancellations, and the inconvenience to passengers was unprecedented. The volcanic eruptions themselves were considered minor compared to those documented in Icelandic history, yet they caused a major breakdown of the sophisticated international airline traffic network, which turned out to be highly vulnerable to such disruption.

The 2004 Indian Ocean tsunami affected the coasts of 15 countries, and killed more than 283 000 people. While most persons died in Banda Aceh, Indonesia, close to the earthquake epicentre, tens of thousands of additional deaths could have been avoided, given that the tsunami needed up to eleven hours to cross the Indian Ocean. Tsunamis are a good example of hazard chains as they always require an external trigger, such as an earthquake, volcanic eruption, landslide or asteroid impact, that rapidly displaces the water column, thus producing the waves. In rivers, several types of floods may occur 'out-of-the-blue' without any preceding rain due to the sudden failure of natural dams along rivers. These natural impoundments may result from landslides, glaciers or lava flows. Incidentally, many of the world's largest known floods have resulted from natural dam breaches instead of intensive rainstorms or snowmelt.

2.2 Defining and Measuring Disaster

A natural disaster is a particularly destructive outcome of one or several processes that disturb the Earth's surface. Natural disasters occur when large numbers of people are killed and injured or when economic assets are damaged or destroyed during an event. For the most grave of these disasters, some prefer using the term 'natural catastrophe'. The amount of damage and loss of life involved in natural disasters affects anything from many communities to whole nations, rather than a group of persons. According to the United Nations Office for Disaster Risk Reduction (UNDRR) a disaster is:

> A serious disruption of the functioning of a community or a society causing widespread human, material, economic or environmental losses which exceed the ability of the affected community or society to cope using its own resources (UNDRR, 2016).

From this definition it becomes clear that a natural disaster causes more than just immediate material damage; it further affects how people deal with its aftermath. The current (2017) definition that the UNDRR offers is:

> A serious disruption of the functioning of a community or a society at any scale due to hazardous events interacting with conditions of exposure, vulnerability and capacity, leading to one or more of the following: human, material, economic and environmental losses and impacts. Annotations: The effect of the disaster can be immediate and localized, but is often widespread and could last for a long period of time. The effect may test or exceed the capacity of a community or society to cope using its own resources, and therefore may require assistance from external sources, which could include neighbouring jurisdictions,

or those at the national or international levels. (www.preventionweb.net)

Scientists, planners, and decision makers may have differing views of what constitutes a natural disaster. For example, the European Union Solidarity Fund (EUSF, https://ec.europa.eu/regional_policy/EN/funding/solidarity-fund/), a financing instrument 'to respond to major natural disasters and express European solidarity to disaster-stricken regions within Europe' uses a strict definition for providing relief funds to 'Member States' following a disaster. EUSF provides support in case the '…total direct damage caused by the disaster exceeds €3 billion (at 2002 prices) or 0.6% of the country's gross national income, whichever is lower.' Neighbouring member states affected by the same disaster can also receive aid, even if the amount of damage is below the specified threshold.

We may argue about the usefulness and shortcomings of this and similar, purely monetary, definitions of natural disasters. For instance, imagine the case of a natural disaster such as a drought that would affect millions of people but cause little material damage. Nevertheless, defining quantitative criteria or thresholds for natural disasters is essential for supporting agencies and governments in deciding whether to provide support. Furthermore, quantitative criteria are indispensable for creating databases or inventories of natural disasters. Being able to reliably record damage from natural disasters is a prerequisite for any meaningful subsequent data analysis. Consider, for example, the NOAA National Geophysical Data Center, which maintains a global catalogue of 'significant earthquakes' (http://www.ngdc.noaa.gov). To be included in this database, an earthquake must meet at least one of the following criteria:

- >US$1 million damage
- >10 fatalities
- Magnitude >7.5
- Mercalli Intensity >X, or
- Triggered a tsunami.

These criteria mainly address two major disaster-related outcomes (mortality and financial loss), which appears reasonable. Only the last criterion regarding the tsunami needs a lower limit, given that small tsunami might have maximum wave heights of only a few centimetres! Multiple and well-balanced criteria for natural disasters are useful, but require commensurately more effort should we wish to include information of past events in large disaster databases.

2.3 Trends in Natural Disasters

Scientists have compiled many natural disaster databases to study trends of recurrence and damage. The most frequently used database seems to be EM-DAT (https://www.emdat.be/), an online catalogue created and maintained by the University of Leuven, Belgium. Many databases have a national or regional scope, but it is encouraging to see that most of these data are becoming publicly available (Paprotny et al. 2018). The records in these and many other databases show a distinct increase in the number of reported natural disasters in recent years. This increase appears to have been linear during the past three or four decades (Figure 2.1), but it is nearly exponential when the time series is extended back to the beginning of the twentieth century. This observation might make us think that the world has become a place less safe with respect to hazardous natural processes. However, the increase in reported natural disasters reflects several tightly linked developments.

For one, our ability to report and communicate natural disasters grew rapidly throughout the twentieth century. Consider the time it took to report a tropical flood disaster in the early 1900s compared to today, when a large fraction of the population has immediate access to radio, television, mobile phones, cameras, and the Internet. Hence some of the observed increase in natural disasters is tied to improved communication, and thus partly reporting bias. Moreover, the world population and the infrastructure to sustain it have grown widely during the past century, meaning that more people than ever before are exposed to natural hazards. At the time of writing, more than 7.4 billion people live on Earth, having more than doubled their number during the past five decades. This increase tracks the rise of reported natural disasters during the twentieth and early twenty-first centuries.

Accordingly people have had to move into many coastal, hilly, or mountainous areas that were formerly regarded as too inaccessible or too dangerous to populate. In such circumstances it is easy to imagine that the potential for damage and injury, and thus disasters, will rise even if the frequency and magnitude of hazardous natural processes do not. The financial costs of disasters inevitably increase as economic growth increases. Since economic growth is, for some reason, the *sine qua non* of modern civilization, increasing disaster costs appear to be built in to our economic system. What remains unknown, however, is the number of uninsured losses that rarely make their way into disaster statistics.

The global trend of increasing natural disasters on record poorly reflects regional patterns however. For example, natural disasters between 1970 and 2010 caused a disproportionately high number of deaths in Africa, whereas numbers were lower in Asia and America (Bank 2010). In terms of insured financial losses, the eastern United States and many mid-income countries have been impacted the most, Africa the least. This trend is emblematic of the observation that fatalities from natural disasters are usually highest in poorer countries, while damage costs from natural disasters are rising most rapidly in the more affluent nations, whereas fatalities are mostly higher in poorer countries. Earthquakes claimed 3.3 million lives, and were the most deadly disasters during the 1970–2010 period, on all continents except for Africa. Regional studies allow more detailed insights regarding earlier decades. In Turkey,

for example, at least 90 000 people have lost their lives in 76 earthquakes since the beginning of the twentieth century, and about seven million people were affected in total; the associated direct losses amounted to at least US$ 25 billion (Erdik 2013). Still, losses may have been comparable in earlier times. Two large earthquakes in Antioch (today's city of Antakya in southern Turkey) in 115 AD and 526 AD may have claimed more than 500 000 lives alone, mainly in large cities located close to major active faults. The 1923 M7.9 Kanto earthquake destroyed much of Tokyo, claiming 105 000 lives. That greater Tokyo metropolitan area is now the largest on Earth with an estimated population of 36 million who are at risk from strong earthquakes (Sato et al. 2005).

A similar trend of increasing seismic events and their impacts is documented in the NOAA National Geophysical Data Center catalogue of 'significant earthquakes', and by a database on twentieth century volcanic disasters. According to the NOAA database, the number of these earthquakes (defined as we presented above) per century in the past 4000 years has risen by three orders of magnitude. Without any supporting evidence whatsoever that the Earth has seen a commensurate increase in the rate of tectonic activity over this period, the observed increase arises more likely from better reporting and more detailed knowledge about younger events instead of higher earthquake activity. The farther we look back in time, the more incomplete and biased the record of past natural disasters becomes. To address this problem, archaeoseismologists use the tools of both seismology and archaeology for teasing information on prehistoric earthquakes from the way that buildings were damaged by seismic shaking (Sintubin 2011).

2.4 Hazard is Part of Risk

We can express a hazard by the probability that a harmful event will occur. Impact describes the damage from that event, while risk describes the product of the impact and its probability. The notion of 'disaster risk reduction' summarizes efforts to lessen the loss of life and damage caused by natural disasters. The UNISDR (www.unisdr.org/we/inform/terminology) holds that

> The word *risk* has two distinctive connotations: in popular usage the emphasis is usually placed on the concept of chance or possibility, such as in *the risk of an accident*; whereas in technical settings the emphasis is usually placed on the consequences, in terms of *potential losses* for some particular cause, place and period.

The UNDRR revised this definition after the Third United Nations World Conference on Disaster Risk Reduction in Sendai, Japan in 2015. This revised version defines risk only in the context of disasters. Accordingly, disaster risk is

> The potential loss of life, injury, or destroyed or damaged assets which could occur to a system, society or a community in a specific period of time, determined probabilistically as a function of hazard, exposure, vulnerability and capacity.

This definition interprets risk unequivocally as the product of probability of occurrence and consequence (Figure 2.4).

Putting numbers to risk involves several factors, which together constitute the risk equation. Among the several variants of this equation (Jonkman et al. 2003) we prefer the following because its components are easily understood:

$$R = H \times V \times E \times A \qquad (2.1)$$

where R is risk, H is hazard, V is vulnerability, E are the elements at risk, and A is risk aversion (Figure 2.5). Hazard is a dimensionless probability of occurrence, and refers to a fixed period such as any given year. Vulnerability denotes the percentage of the maximum possible loss

given a specified impact. A vulnerability of 0 means completely exempt from damage, whereas 1 means total destruction or loss. The elements of risk enter the risk equation as values that we can measure in either monetary terms or human lives. The factor of risk aversion is concerned with how persons or groups perceive different risks, and is similar to a volume knob in the risk to emphasize or tone down the overall expected losses in Eq (2.1). Thus, if including risk aversion in this equation, we refer to R as *perceived risk*. The risk from natural natural hazards is therefore a measure of the expected loss from an event of a given size (Figure 2.6). This anticipated annual loss is often expressed in monetary value, using for example units [US\$ yr^{-1}] or [€yr^{-1}]; some risk applications, however, explicitly concern mortality rates or the expected number of lives lost per unit time.

2.4.1 Vulnerability

Modern risk research recognizes many different types of vulnerability (Adger 2006). Among some of the most investigated is structural vulnerability, which refers to potential damage to buildings, bridges, roads, and other engineered infrastructure by direct or indirect impacts. Socioeconomic or demographic vulnerability concerns the loss potential of nations, groups of people, or gender. Economic vulnerability involves monetary losses from natural disasters including losses from reduction of commerce until normality returns. Social vulnerability entails potential loss of societal functionality of all types. One example was the frequent looting in the aftermath of Hurricane Katrina, but the spectrum of impacts of societal damage is vast. Gender and status in social groups can play a crucial role. Drawing on three decades worth of data from 141 countries, Neumayer and Plümper (2007) argued that girls and women were on average more vulnerable to natural disasters and had a higher disaster mortality than boys and men, largely because of their everyday socioeconomic status, and

Hazard scenarios

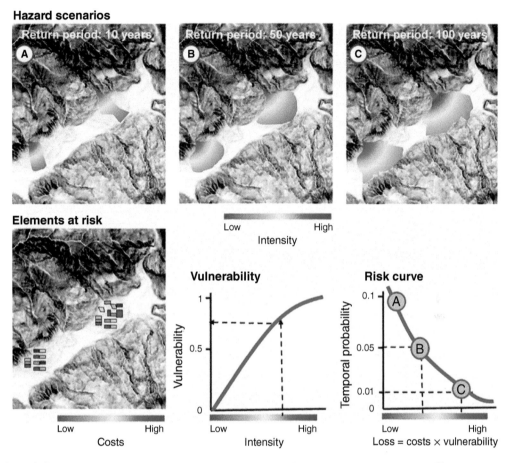

Figure 2.4 Three hypothetical landslide scenarios (A, B, and C) with different return periods. Each scenario includes an intensity map (e.g. impact pressure). Each element at risk (lower left) is characterized by type, location, and replacement cost. The vulnerability of each element at risk is determined using a vulnerability curve for that particular type of structure and the scenario hazard intensity. A risk curve (lower right) shows the temporal probabilities of the three scenarios plotted against loss. Losses are determined by multiplying the vulnerabilities by the replacement costs for all exposed elements at risk. After defining a number of points, a risk curve can be drawn. The area under the risk curve represents the annualized losses. From Corominas et al. (2014).

despite their generally higher life expectancy. Socioeconomic and demographic vulnerability may have chiefly contributed to reported losses from natural hazards in developing countries in particular (Alcántara-Ayala 2002). However, social vulnerability to natural disasters has also been changing in countries such as the United States by becoming more variable between regions in the past few decades, and mostly reflecting changes in urban density, race and ethnicity, and socioeconomic status (Cutter and Finch 2008).

Social media have emerged rapidly as a form of rapid communication, and may have measurable impacts on vulnerability to natural disasters. Alexander (2014) summarized some of the basic functions of social media in disasters and crises. Social media:

- Have a listening function and allow single persons and groups to democratically express their views and opinions.
- Allow monitoring a situation with a multitude of inputs that should also ideally be capable of correcting false information.

Natural Hazard	× Vulnerability	× Elements at risk	(× Aversion)	= Risk
		Consequence	Exposure	
(Annual) Probability of a damaging natural process [yr⁻¹]	*% Loss of total value* [1] *(socioeconomic, political, structural, ecological, ...)*	*Lives, buildings, infrastructure, intangible losses* [€]	*Perception* [1]	*Annual expected loss.* [€ yr⁻¹]
Geosciences Mathematics Physics	Engineering, Social, and Economic Sciences, Ecology	Economic Sciences	Psychology Planning Political Sciences	Multi-disciplinary

Figure 2.5 'Risk' is commonly defined as the product of the four factors: hazard, vulnerability, elements at risk, and aversion. Understanding and characterization of each of the four factors requires expertise from different scientific fields, consequently studies of 'risk' are inherently multidisciplinary.

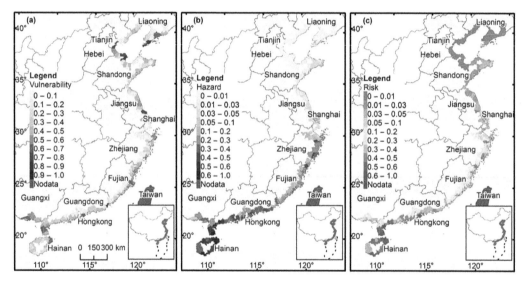

Figure 2.6 Regional model of risk from tropical cyclones in China. (a) Vulnerability of people and property at risk. (b) Storm surge hazard. (c) Storm surge risk for coastal countries and districts. Levels of vulnerability hazard and risk are normalised and range from 0 to 1. From Gao et al. (2014).

- Can be integrated into emergency planning and crisis management, for example if people are being warned of states of emergency or pending disasters.
- Can promote crowd-sourcing projects and collaborative development such as real-time mapping of damage in disaster areas.
- Create social cohesion, a sense of belonging to a specific community, promote therapeutic initiatives, and support voluntary organization.
- Further causes by launching appeals for donations or other kinds of support.
- Aid research by providing large amounts of social network data, e.g. for regional, national or worldwide 'sentiment analyses'.

We might add that social media can also influence vulnerability by spreading untested, inaccurate or false information. Such 'fake news' may give a false impression of a natural hazard, risk, or disaster. Environmental or ecological vulnerability refers to irreversible losses in natural resources. The concept of ecosystem services, which comprise, among others, things such as access to clean water, the protection and hydrological functions of natural forests, and the functioning of the food web, also aims to measure losses from natural disasters. The many other types of vulnerability require close attention, depending on the type of risk in question.

Resilience measures the extent to which people, communities, assets or economies are capable of recovering from external disturbance and concomitant losses, while maintaining their functionality, and avoiding catastrophic damage (Figure 2.7). The term is borrowed from mathematics (Adger 2006), and has a similar connotation in ecology, where it deals with how ecosystems respond to disturbances. In the case of engineering systems, Park et al. (2013) see resilience as a set of recursive processes that involve sensing, anticipation, learning, and adaptation. Although their analysis referred specifically to engineering systems the concepts are readily applicable to societal systems.

How do we measure vulnerability? The empirical approach involves careful analyses of the records of damage following a natural disaster of a specific magnitude or intensity. The number and magnitude of insurance claims afford a good, if relative, overview of the damages incurred, as do detailed accounts in historic archives, newspapers, or field observations of structural damage. Field-based mapping of damage to buildings and infrastructure is a key method of estimating empirically the damage from a natural disaster, and expressing this damage as a fraction of the total value of the object considered. The ability of a natural event of specific type and magnitude to damage specific structure types is often described by stage-damage curves or 'fragility curves' (Figure 2.8) (Gokon et al. 2014). For example, a tsunami two metres high will cause damage amounting to half of the value of a highway bridge made of

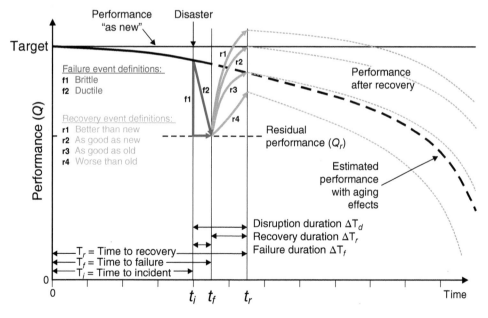

Figure 2.7 Definitions of resilience metrics in the context of the performance of a built structure (for example, a dyke, road, or bridge) before, during, and after a natural disaster. Modified from Ayyub (2014).

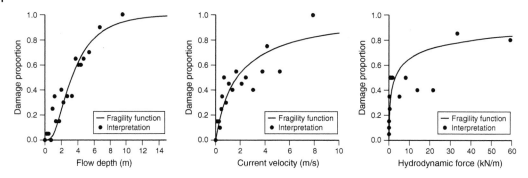

Figure 2.8 Fragility functions for buildings in the area impacted by the 2009 Samoa earthquake and tsunami. Left: Maximum flow depth. Middle: Maximum current velocity. Right: Maximum hydrodynamic force. Modified from Gokon et al. (2014).

concrete with its deck 2.5 m above mean water level. Most physical impacts attenuate with increasing distance from the source, so it is valuable to determine both the magnitude and the local intensity of a given event. Numerical modelling offers new and nondestructive ways to simulate physical damage to structures as a function of prescribed impact forces and stresses, for example how a bridge of known geometry and material properties will respond to the estimated impact forces of flood debris. The underlying concept of balancing resisting and impacting forces can also be used to approximate, for example, the flood-water velocities that a person can withstand without being swept away (Milanesi et al. 2015).

Measuring vulnerabilities that concern nonstructural or immaterial values calls for a different set of methods. Interviews, questionnaires, online surveys, or bulk socioeconomic indicators can be used to estimate vulnerabilities related to income, financial coping capacity, awareness, preparedness, and many other aspects. Such data can be collected for individuals or groups, and require techniques from empirical political and social research. Peduzzi et al. (2009) used as many as 32 socioeconomic indicators for estimating human vulnerability, including data on gross domestic product, inflation of food prices, percentage of urban population, mortality rate among children, and illiteracy rate. Besides these long-established metrics, newer ones such as

the average number of cell phones per capita might add information about how well people can connect with each other and share news during and after a natural disaster. Whether people have access to this information might influence communication and decisions during such crises, and hence influence the vulnerability of people (Table 2.1).

2.4.2 Elements at Risk

Vulnerability is intimately linked to the number and values of the elements at risk, which include human health and safety, property, the environment and financial interests. We first consider human health and safety and then turn to the built environment. In western society, human health and safety generally take precedence over all other elements. As an example, the Swiss Federal Office of the Environment (www.bafu.admin.ch) proposes the following hierarchy of elements at risk relating to industrial activities:

1) human life
2) personal injury
3) surface water pollution
4) groundwater pollution
5) agricultural land usability
6) material losses.

An important issue for risk assessment is whether the risk to human life is voluntary, that is within one's control (e.g. smoking or

Table 2.1 Summary of vulnerability ranges and recommended values for death from landslide debris in similar situations, Hongkong. From Dai et al. (2004)

Case	Vulnerability of person Range in data	Recommended value	Comments
Person in open space			
1. If struck by a rockfall	0.1–0.7	0.5[a)]	May be injured but unlikely to cause death
2. If buried by debris	0.8–1.0	1.0	Death by asphyxia
3. If not buried	0.1–0.5	0.1	High chance of survival
Person in a vehicle			
1. If the vehicle is buried/crushed	0.9–1.0	1.0	Death is almost certain
2. If the vehicle is damaged only	0–0.3	0.3	Death is highly likely
Person in a building			
1. If the building collapses	0.9–1.0	1.0	Death is almost certain
2. If the building is inundated with debris and the person buried	0.8–1.0	1.0	Death is highly likely
3. If the building is inundated with debris and the person not buried	0-0.5	0.2	High chance of survival
4. If the debris strikes the building only	0-0.1	0.05	Virtually no danger[a]

a) Better considered in more detail, i.e. the proximity of person to the part of the building affected by sliding.

sky diving), or involuntary and thus outside of one's control (e.g. being struck by lightning or being killed in a train crash). Risks from natural disasters are considered involuntary risks, and society has a lower tolerance for involuntary risks than voluntary ones. Any risk assessment must also consider whether it concerns a single person or a group of persons. The terms 'individual risk' or 'group risk', 'political risk', or 'societal risk' refer to this distinction.

Estimating the values of and direct damages to the built environment is more straightforward. Experts can readily assess, with some uncertainty, the monetary value of a building, road or pipeline. However, losses may also be defined in a broader sense. For example, the United Nations Framework Convention on Climate Change (UNFCCC) treats losses and damage as 'the actual and/or potential

manifestation of impacts associated with climate change in developing countries that negatively affect human and natural systems' (James et al. 2014). Additional costs arise in the wake of a natural disaster, including:

- Societal impacts, including personal stress, injury or loss of life, disruption to lifestyle, and demands on social and medical services.
- Economic impacts, including business interruptions, damage to property and other infrastructure, loss of income, loss of income generators such as productive land, interruption of commerce, and clean-up costs,
- Environmental impacts, including pollution (sewage, chemicals, debris), loss of cultural values, loss of habitats, modification of environments.

These costs are indirect, because they are secondary and arise from loss of industrial or agricultural activity or trauma and cannot easily be compared to material costs. Indirect losses will need more time to estimate, as many of them arise from consequences that unfold in the weeks to months or years after a disaster. Yet these indirect losses may be high if key structures are damaged or lost. The disruption of lifelines, notably traffic arteries, communication lines, power and pipelines or critical infrastructure (hospitals, power plants, water reservoirs, and bridges) often entails lower direct material damage than the indirect damage resulting from services that are in desperate need, but unavailable, during a state of emergency. Similarly, the destruction of a factory causes material loss, but also loss of production, delivery, work places, and jobs. Loss of industrial productivity may reverberate through communities long after the disaster. Reinsurance groups such as the SwissRe use the value of working days lost in the case of disasters, estimated from a global index and normalized by the national economy of the home country, to rank the exposure of the world's largest cities to earthquakes, storms, tsunamis, and floods (www.swissre.com). The spread of diseases following natural disasters because of contaminated water (Chen et al. 2012) likewise creates indirect costs, as do immediate disaster assistance and clean-up. Studying the impacts of weather-related disasters nation by nation, Lesk et al. (2016) found that droughts and extreme heat waves reduced cereal production by nearly 10% between 1964 and 2007. On average, developed countries suffered 8–11% more losses than developing ones. The authors argued that this trend partly reflected the reliance of wealthier nations on larger monocultures, fewer strategies to diversify crops and minimize risk, and the seasonal timing of droughts.

Disaster costs are trickier to estimate for values of cultural or heritage structures such as museums, theatres, and parks. Destruction of such assets results in both material and immaterial losses such as cultural identity. Such losses are referred to as intangible losses. Psychological trauma and loss of personal memorabilia are among the most important intangible losses of natural disasters (Bartels and VanRooyen 2011). Other intangible losses include those to ecosystem services and iconic landmarks. Some intangible losses can be approximated by assigning monetary values to their role in intact ecosystems. For example, Chambers et al. (2007) estimated that Hurricane Katrina's total damage to vegetation was equal to a total biomass loss of 92–112 Mt C, or about 50–140% of the net annual US carbon sink. This ecological damage can be seen in terms of national and international trading in carbon certificates. When fully considering the consequences on ecosystems, many direct and indirect benefits also arise from of extreme natural events, for example, creation of new nutrients and soil, flushing of rivers of pollutants, replenishing wetlands, creation of new land created by uplift during earthquakes, and creation of fertile volcanic soils and many others.

Coming back to the idea of risk as an economic value inevitably leads to the question of how to treat people as elements at risk in the risk equation. Can the value of a human life be measured? Direct losses such as injury, impairment or loss of life can be expressed in numbers, but is it possible and ethically justifiable to assign a price tag to a person's existence? Indeed every health or life insurer does just this. When applying for such insurance, you are likely to be asked to fill in a questionnaire about certain aspects of your health, for example whether you are a smoker, whether you drive a motorcycle or whether you like surfing, paragliding or extreme mountain climbing. In essence, these questions are used to gauge your personal vulnerability. Essentially all insurers, transport authorities, and other agencies concerned with risk management regularly assign a value to health or life. As an example of life value, highway agencies in Switzerland and New Zealand cost traffic-related deaths at about 2.5 million Swiss francs and 4.2 million NZ dollars, respectively, when carrying out cost–benefit analyses of roading projects.

Several methods tackle the problem of determining the monetary value of health and life (Jonkman et al. 2003). One approach is to estimate the 'value of a statistical life', an economic measure that takes into account aspects of wage and risk during a given occupation (Viscusi and Aldy 2003); we refer interested readers to the economics literature for a more in-depth coverage of this topic. This economic metric is commonly based on a person's willingness to pay for added safety and security, and their willingness to accept risk. The willingness to pay expresses the person's readiness to invest a specific amount of money to acquire a specified amount of added safety or security from a threat, in this case a natural hazard. In contrast, we can measure the willingness of a person to accept a higher level of risk, for example in moving closer to the flank of an active volcano.

2.4.3 Risk Aversion

Quantitative risk assessment is reliant on putting numbers to the probability of occurrence, the vulnerability, and the elements at risk (Eq 2.1). Some assessments focussing on perceived risk include another factor in the risk equation – risk aversion. This is a measure of the psychological perception of risk from natural disasters, and incorporates aversion to rare but potentially very destructive events. Risk aversion can be understood by considering the notion that 300 people killed in road traffic over a year is generally less disturbing to the public than 300 people killed in a single airplane crash. Note that if 300 people died in a single traffic accident the public reaction would also be extreme. Although great efforts are made to reduce that number in either case,

the loss is exactly the same, but most people would perceive the latter as more grave and less acceptable (especially if the event had been anticipated). Another example of how risk aversion may change rapidly is Germany's policy change with respect to nuclear power in the aftermath of the 2011 Tohoku earthquake and the partial meltdown of the Fukushima nuclear power plant in Japan. Seen objectively, the hazard of an accident at any of Germany's nuclear power plants remained unchanged after the earthquake in Japan. The only factor in the risk equation that changed was that of risk aversion. Suddenly nuclear power seemed more risky than before. Human perception of natural hazards, risks, and disaster is prone to biases, so that much research focuses on objectively measuring how people perceive risk. Aversion can also result from knowing too much or too little about a given risk. Consider the blank spots on maps of natural hazards and risks: Are these spots safe places or simply places we do not know about (Osuteye et al. 2017)?

How people perceive hazards and risks in their immediate surroundings can also determine how readily they adapt (Uprety et al. 2017). In a study of perceptions of flood risk among some 1000 homeowners in the Netherlands, Botzen et al. (2009) found that differences in responses were consistently in line with independently derived risk levels, such that people living closer to rivers were more aware of the possibility of harmful floods than those living farther away. Yet homeowners living in areas that were unprotected by dykes or other structural measures tended to underestimate the flood risk, similarly to residents who were poorly informed about the underlying causes of these risks. Surprisingly, older and more highly educated people also appeared to be less aware about floods. In particular, people who lived on floodplains behind dykes felt that they were safe from flooding. Such perceptions arise partly from how specialists and decision makers communicate technical terms about risk to the public. Scientists routinely use technical terms such as the 100-year flood, but many people fail to understand what this term means (Michel-Kerjan and Kunreuther 2011).

2.4.4 Risk is a Multidisciplinary Expectation of Loss

The preceding definitions of natural hazard, vulnerability, the elements at risk, and risk aversion show that risk is a multidisciplinary metric. Risk expresses how much loss we expect statistically for a specified region and period. Given the different factors in the risk equation, it should come as little surprise that evaluating risk from natural disasters is an exercise that is thoroughly inter- and trans-disciplinary (Figure 2.9). While estimating probabilities of specific hazardous events lies within the realm of the natural sciences, including the geosciences, mathematics, and physics, determining vulnerability and elements at risk is among the key tasks in engineering, economic, social, and ecological sciences. When risk aversion is included, psychological, planning, and political sciences must be involved. Risk has become a

multidisciplinary metric and thus calls for effective collaboration across the board. That is why most modern disaster risk estimates and country risk profiles are based on many compound metrics trying to characterize the natural, socioeconomic, demographic, and environmental setting (Peduzzi et al. 2009).

Several risk indices circumvent the need to compute directly the expected number of fatalities or monetary values lost per year via the risk equation. The World Risk Report 2012 (www.WorldRiskReport.org), for example, proposed a global risk index that combines social, economic, ecological, and physical parameters that address four compound metrics of exposure, susceptibility, coping capacity, and adaptive capacity. These result in an index for ranking the world's nations according to the risk they face from natural disasters. This ranking placed the island nations of Vanuatu, Tonga, and the Philippines among the three most 'risky', owing to their proximity to the ocean, and thus exposure to tropical cyclones, sea-level rise, and flooding. The world risk index thus goes beyond the conventional risk equation by considering and synthesizing

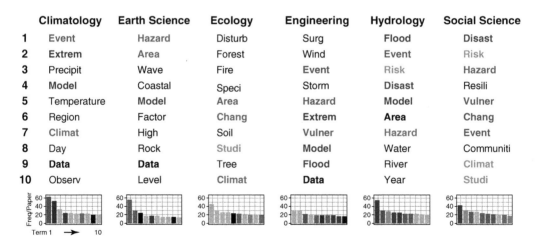

	Climatology	Earth Science	Ecology	Engineering	Hydrology	Social Science
1	Event	Hazard	Disturb	Surg	Flood	Disast
2	Extrem	Area	Forest	Wind	Event	Risk
3	Precipit	Wave	Fire	Event	Risk	Hazard
4	Model	Coastal	Speci	Storm	Disast	Resili
5	Temperature	Model	Area	Hazard	Model	Vulner
6	Region	Factor	Chang	Extrem	Area	Chang
7	Climat	High	Soil	Vulner	Hazard	Event
8	Day	Rock	Studi	Model	Water	Communiti
9	Data	Data	Tree	Flood	River	Climat
10	Observ	Level	Climat	Data	Year	Studi

Figure 2.9 The top ten most frequently occurring word roots in papers reviewed from six disciplines (climatology, earth science, ecology, engineering, hydrology, social science). Words in bold occur in more than one column; each is represented by a different color to aid in visualizing similarities across the disciplines. The histograms show the frequencies of occurrence of these top ten word roots, normalized by the total number of papers examined in that discipline. From McPhillips et al. (2018).

composite indicators of exposure to natural hazards, projected impacts of climate change, and vulnerable societies.

2.5 Risk Management and the Risk Cycle

Generally speaking, the purpose of risk analysis is to determine the probability that a specific hazard will cause specific harm (Salvati et al., 2010). By analysing risk we pave the way for designing measures to manage risk. These measures are a toolkit for reducing impacts from natural disasters to an acceptable level, which is the ultimate goal of risk management. The aim of any risk management is to reduce risk, i.e. the *expected* damage from an event (Figure 2.10). This approach applies equally to natural disasters, financial crises, the structural integrity of nuclear power plants, or

medicinal applications. In terms of the risk equation, we could consider reducing any to all of its factors, so that we reduce risk to an acceptable or tolerable level. Now, should we concentrate on reducing the probability that a damaging event will happen, should we instead work on reducing vulnerability, or should we try to modify the burden of loss? Smith and Petley (2009) refer to these three options as protection, mitigation, and adaptation, respectively:

- *Protection* involves actively interfering with the physical and chemical processes that may cause harm. The aim is to reduce the frequency, magnitude, or impact of these processes. Physical protection measures often rely on engineering solutions such as river dykes, sediment retention basins, tsunami sea walls, and rockfall nets, which all directly interfere with the hazardous process.

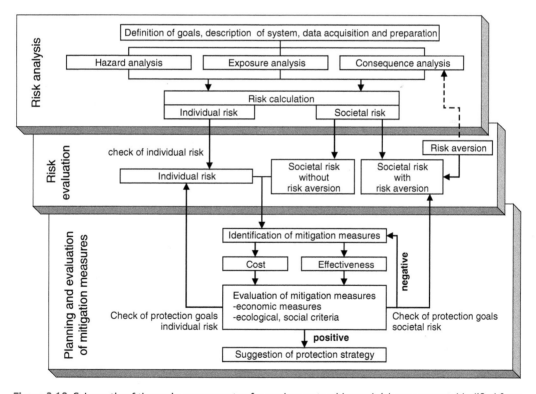

Figure 2.10 Schematic of the main components of a modern natural hazard risk assessment. Modified from Bründl et al. (2009).

- *Adaptation* offers a counterpart to physical protection, and aims at making people less vulnerable to natural disasters. Ways to achieve this include early warning systems, evacuation drills and routes, retrofitting for earthquake-safe buildings, or delineating hazard zones in land-use planning.
- *Mitigation* builds on the idea to distribute or share the expected or actual losses from natural disasters; this generally works either via insurance measures or direct financial aid, respectively.

Risk management is a process involving two major stages (Bründl et al. 2009). Firstly, risk analysis is concerned with the mathematical derivation of a given set of risk values according to a specified risk equation. Secondly, risk evaluation involves assessing computed risk values against the background of personal and societal risk perception and of risk acceptability in terms of deaths and costs. The latter

is often approached by way of cost–benefit analysis (CBA). We tend to systematically under- or overestimate certain risks, and this cognitive distortion enters risk assessments. Yet risk analysis and evaluation are only parts of the management process; risk treatment is also needed.

On average, only a few percent of the annual humanitarian assistance is allocated to prevention, yet each dollar invested in risk reduction saves multiples in economic losses from disasters. One way to improve how we use disaster funds is to coordinate better the investments in prevention, intervention, and recovery. This approach is referred to as the risk cycle, and prevention, intervention, and recovery refer to the phases before, during, and after a given natural disaster (Figure 2.11). We would think that the best risk management strategy would pay attention to and support these three phases of the risk cycle in proportion to their return in

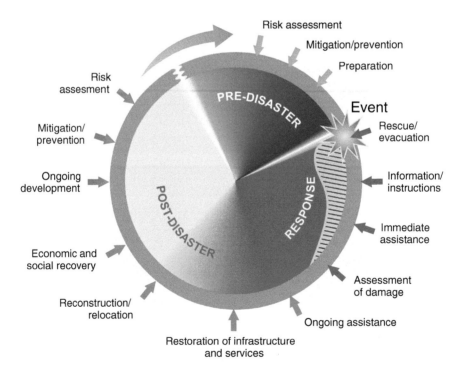

Figure 2.11 The risk cycle. Sustainable management of natural hazards and risk is based on strategies that address prevention, intervention, and recovery. These three stages coincide with the time before, during, and after a natural disaster. All of these strategies require adequate attention and funding. Modified from Health Systems Info (http://healthsysteminfo.blogspot.com/ 2010/12/disaster-management-in- general.html).

reduced injury and costs. However, more and more risk managers have begun to realize that the prevention strategy in particular deserves more attention, given that the return on this sort of investment might be greatest. Improved disaster preparedness can be achieved through inexpensive and simple means such as informing, training or educating people. Knowing what to do during an earthquake saves lives, and training people to react accordingly can be made easy via warning signs, films or safety drills.

Dealing with natural hazards and disasters almost always has some political dimension. The following two end-member examples highlight how some of the consequences have played out at the community and international levels. In Switzerland every community is required by law to produce maps showing natural hazards. The maps are based on a unified system that identifies the intensity (high, medium or low) and probability (high, medium, low or very low) of each potentially harmful process. Many of the data for this classification come from detailed field mapping of 'silent witnesses'. These are landforms and landform elements that contain information about the type, intensity, and likely timing of former events that might be damaging if they happened again. Probabilities are based on return periods, which are average times it takes for a process of a given magnitude to recur. Quantitative threshold values, such as impact force, flow depth or velocity, are useful for defining the appropriate categories in the classification. An intensity–probability matrix helps delineate three different zones that are legally binding for the construction of buildings.

In contrast, at the international level the 2004 tsunami in the Indian Ocean spurred important legislated initiatives, although the action mandated by the legislation at the international level is considerably less in evidence. Nevertheless, action plans such as the Hyogo Framework for Action designed for the decade from 2005 to 2015, followed by the 15-year

Sendai Framework for Action in 2015, helped raise awareness and prompted basic mitigation efforts.

2.6 Uncertainties and Reality Check

It is essential to measure and communicate the inevitable uncertainties that arise in any risk assessment. Recall that quantitative risk analysis uses probability to express uncertainties. These uncertainties are either epistemic or aleatoric. Epistemic uncertainties arise from the reductionist nature of models that attempt to portray, reduce, and thus insufficiently represent, reality. They involve things unknown to us but things that we believe we can learn by doing more research. For example, the most sophisticated numerical rock-fall runout model may produce very realistic results, but the model will be unable to replicate every minute detail of a block's path across the ground.

Aleatoric or statistical uncertainty occurs in most physical experiments that we carry out. It refers to random processes that we can cast in quantitative terms, although their eventual outcome will remain unknown to us. For example, consider an experiment in which we try to simulate rock fall by letting a tennis ball fall along an inclined plane of specified angle and roughness from a fixed initial height. If we repeated this experiment thirty times, we would find that in each of these runs the tennis ball comes to rest at a slightly different location because of minor irregularities on the inclined surface or the tennis ball itself. Similarly, radioactive decay of elements can be measured by quantities such as the half-life, which is the time needed for the number of radioactive atoms to reduce to half of the original value. We can predict how many atoms will decay during this interval on average. Yet which atoms will decay next remains elusive. This process may only appear random to us because we lack the knowledge or tools to accurately predict the timing of each isotopic decay. Hence the distinction between epistemic and aleatoric uncertainties depends on how well we believe that we can learn more about the underlying unknowns in the future.

Putting numbers to these uncertainties is an essential, but also the most engaging, part of any hazard or risk analysis. In the geosciences we derive most knowledge about prehistoric disasters from geological archives instead of direct instrumental readings, so that we have to account for uncertainties when we reconstruct past disasters (Figure 2.12). Hence, the inferred timing of past events is as accurate as the dating method we apply, and their magnitude and intensity are as accurate as we are able to infer from the resulting sediments and landforms. But even when detailed instrumental data are available, hazard and risk products such as maps remain prone to change; they become obsolete and need refinement because of climate and land-use change, legacy effects of precursory natural disasters, changes in administrative regulations for dealing with

natural resources, increases in knowledge, and so on. Therefore, it is essential to decide on a meaningful lifespan of hazard and risk products, or to define a timespan over which the hazard and risk predictions remain accurate and reliable.

The 2011 Tohoku earthquake has stimulated a discussion of how realistic the forecasts of Japanese seismologists have been during the past several decades. Most of the damaging earthquakes in Japan over the past 30 years have happened in or near areas that were thought to have rather low or moderate seismic risk (Geller 2011). Furthermore, the 12 deadliest earthquakes between 2000 and 2011 had intensities that exceeded predictions by the Global Seismic Hazard Assessment Program (GSHAP) (Bela 2014). The Japanese Meteorological Agency is in charge of regularly issuing probabilistic ground-motion predictions of this type, and each revision draws on new data collected by a dense seismograph network. It is important to review the practice of natural hazard and risk mitigation regularly, because one or several factors in the risk equation may have changed since the last

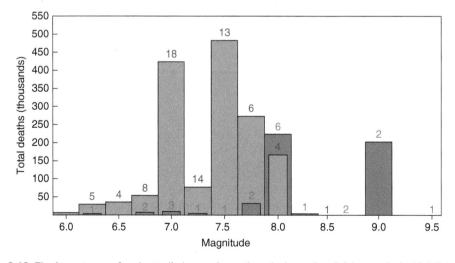

Figure 2.12 The importance of understudied areas in earthquake hazard and risk appraisals, highlighting the high loss of life from earthquakes in continental interiors over the past century. Earthquakes in continental interiors (orange/red) killed significantly more people than earthquakes at plate boundaries (blue). The earthquakes are grouped into bins of width 0.25 in magnitude, and the number of earthquakes in each bin is shown above each bar. From England and Jackson (2011).

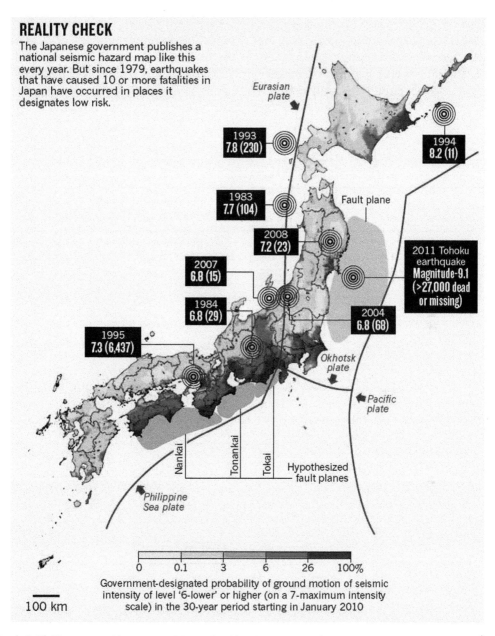

REALITY CHECK

The Japanese government publishes a national seismic hazard map like this every year. But since 1979, earthquakes that have caused 10 or more fatalities in Japan have occurred in places it designates low risk.

Eurasian plate

1993
7.8 (230)

1994
8.2 (11)

1983
7.7 (104)

Fault plane

2008
7.2 (23)

2011 Tohoku earthquake
Magnitude-9.1
(>27,000 dead or missing)

2007
6.8 (15)

1984
6.8 (29)

2004
6.8 (68)

1995
7.3 (6,437)

Okhotsk plate

Pacific plate

Nankai

Tonankai

Tokai

Hypothesized fault planes

Philippine Sea plate

| 0 | 0.1 | 3 | 6 | 26 | 100% |

Government-designated probability of ground motion of seismic intensity of level '6-lower' or higher (on a 7-maximum intensity scale) in the 30-year period starting in January 2010

100 km

Figure 2.13 The geographic pattern of damaging historic earthquakes in Japan (black circles) is inconsistent with the predicted seismic hazard (colour scale), calling for a reassessment of hazard and risk appraisals. From Geller (2011).

assessment (Figure 2.13). The importance of this practice is that geophysical analyses of the 2011 Tohoku earthquake cannot exclude a future large earthquake happening farther south, closer to the megacity of Tokyo (Simons et al. 2011).

2.7 A Future of More Extreme Events?

Another source of major uncertainty concerns climate change, which is a strong incentive to check and refine current estimates of natural hazards. With financial losses from extreme weather events having risen rapidly over the past few decades, countless mitigation efforts have been proposed to address this problem. Many projections for the twenty-first century and beyond emphasize an increase in meteorological extremes, for example concerning a higher likelihood of winter rainfall totals in the Northern hemisphere or higher summer monsoon rainfall across Asia (Palmer and Räisänen 2002). Yet the technical definitions of 'extreme events' vary widely depending on which scientific discipline uses them, though this is also often the case within a discipline. A survey of ten years' worth of hazard- and disaster-related publications in climatology, earth sciences, ecology, engineering, hydrology, and social sciences referring to 'extreme events' showed that only half of the studies offering a definition did so explicitly and without the need to consult independent sources (McPhillips et al. 2018). A similar phenomenon seems to hold for 'risk' studies in the geosciences: the term is highly popular, but less well explained, let alone measured. Some more rigorous and consistent handling would help to avoid this confusion, especially as changes in risk from natural hazards have begun to receive a lot of attention from scientists, practitioners, and decision makers. McPhillips et al. (2018) recommended that:

- We must recognize where our own efforts related to the study and management of extreme events fall relative to other disciplines and work to communicate across these boundaries.
- Definitions of extreme events should not be conflated with their impacts or effects.
- Thresholds used to define an extreme event can be based on probabilities of occurrence or on the point where they have potential consequences or impacts.
- We all should be more explicit in defining what we mean by an extreme event or an extreme impact.

Keeping these caveats in mind, we can identify at least three major questions bearing on the feedbacks between climate change and natural hazards:

- To what degree does climate change affect the magnitude and frequency of potentially adverse natural processes?
- To what degree does climate change affect each factor in the risk equation?
- Can any of the observed changes in natural hazards and risk be measured and attributed to contemporary climate change?

By definition, natural hazards change if either their magnitude or frequency changes. Larger events are rarer than smaller events over a sufficiently long observation period. It is convenient to use the exceedance probability of an event of a given magnitude as a probabilistic measure of the hazard. Thus any change in the frequency–magnitude relationship results in a change in hazard. Most of the scenarios of climate change point to an increase in extreme weather events, according to predictions of the Intergovernmental Panel on Climate Change (IPCC, https://ipcc.ch), the world's largest collective of scientists concerned with causes and effects of contemporary global warming. Projected increases in droughts, heat waves, and floods may trigger increased water stress, the frequency and size of wildfires, decreases in food production, adverse health effects, increased flood risk, higher sea levels, and damage to infrastructure. But how can climate-induced changes in the frequency or magnitude of these events be distinguished from noise or natural variability in the data? Attribution of disasters and catastrophes to climate change requires rigorous testing. This again necessarily involves either large quantities of data therefore extending over long time periods, or robust enough models for running climate scenarios both with and without atmospheric warming. In either case an important goal is to detect the fraction of observed changes in natural systems that are attributable to, for example, anthropogenic greenhouse gas emissions, and that cannot be explained by a naturally changing climate alone.

Nonetheless, natural climate oscillations such as the El Niño Southern Oscillation (ENSO) are responsible for some extreme weather in many parts of the world. The extensive flooding in northeastern Australia in 2010, for example, was linked to a pronounced La Niña phase of ENSO. Climate change can also have some positive consequences. McPhaden et al. (2006) estimated, for example, that the 1997/1998 El Niño resulted in a net benefit of $20 billion to the US economy because of the reduced number of land-falling hurricanes and the unusually warm winter in the midwest. Many of the observed changes in natural systems since the 1970s are especially pronounced in areas of rising temperatures, thus natural climate variability is highly unlikely to be solely responsible for these changes (Rosenzweig et al. 2008). Climate-change attribution also involves examining extreme events that many climate scientists believe are symptomatically increasing with global warming. Three points of critique have emerged from this undertaking (James et al. 2014). Firstly, extreme events occur under a natural climate regime as part of its intrinsic variability. Secondly, detecting clear trends is hampered by the rarity of extreme events and the commensurately limited database. Thirdly, contemporary numerical climate models are incapable of adequately reproducing the dynamics associated with extreme weather events.

These uncertainties propagate as we look into changes in risk from a given type of hazardous process. The number of factors in the risk equation determines the number of ways we can alter risk. Risk might change because of changes in hazard, vulnerability, the number of elements at risk, risk aversion, or any combination of these four factors. If reducing disaster impacts is the ultimate objective, however, then surely vulnerability and elements at risk might be among the more useful ones to reduce. Only one of these factors need be changed to change risk, though all can respond

to climate. Vulnerability may decrease through changes in countermeasures, awareness and capacity building, socioeconomic development or political legislation. The elements at risk may change due to changes in demographics, migration, the spread of development into hazardous areas or socioeconomic development. Hence, to attribute changes in risk to global warming or global environmental change, we have to analyse in depth how these changes would affect each factor in the risk equation. One aspect of global change is well documented and hardly debated anymore: the growing number of people on the planet, and the concomitant increase in the value of assets and commerce at risk from natural disasters. In relative magnitude, population growth greatly exceeds climate change – whether natural or anthropogenic – as a likely cause of increased disasters in the future.

2.8 Read More About Natural Hazards and Disasters

Several introductory textbooks deliver extensive and richly illustrated accounts of natural hazards and disasters, including Hyndman and Hyndman (2016) and Keller and Devecchio (2014). Both books feature many case studies and real-world applications of geomorphology. The books by Bryant (2004) and Smith and Petley (2009) do without much spectacular photographic material, but offer compact and highly informative summaries on the topic. These books cover all you need to get started with the basics. The *Encyclopedia of Natural Hazards* (Bobrowsky 2013) offers quick access to more specialised topics ranging from 'aa-lava' to 'zoning', including well-illustrated case studies, as well as some hazards rarely covered in detail in standard textbooks.

The rapidly growing field of more specialized books and edited volumes on Earth surface processes, hazards and risks offers many different technical levels, and is beyond a short summary. For example, Stein and Stein (2014) succinctly capture the key concepts of natural hazards and risks from the viewpoint of seismologists. Lockwood and Hazlett (2010) provide a well-illustrated and field-oriented treatise on volcanoes, and Oppenheimer (2011) offers a captivating narrative of the major known volcanic eruptions that have had major global impacts in the past. Storm surges, sea-level rise, flooding, saltwater intrusions, and erosion or sedimentation are among the key topics of a volume that Finkl (2013) edited on coastal hazards. The engaging narrative by Atwater et al. (2015) describes how they pulled together vast amounts of evidence to reconstruct the origin and impacts of a large tsunami that struck both Japan and the Pacific Northwest at the beginning of the eighteenth century. Cold regions have their own characteristic set of hazards that are largely linked to thermal regimes, including permafrost, drifting sea ice, and glacial lake outbursts; these and other processes are the focus of the introductory book by Whiteman (2011). The book on periglacial environments and cold region landforms by French (2013) also features many practical issues and hazards with permafrost. Scott et al. (2014) give a well illustrated and multidisciplinary perspective of both natural and human-induced fires on Earth, and Sheffield and Wood (2011) offer a systematic overview on droughts and their impacts. A very accessible account on landslides is the book by Sidle and Ochiai (2006), which provides many interesting case studies and implications for land use. Clague and Stead (2012) and Davies (2015) provide state-of-the-art overviews on landslides and related mass movements, where and why they occur, what effects they have, and how they can be modelled.

Resources on the Web The number of websites and other Internet resources on natural hazards is growing rapidly. The following is only a small fraction of sites that provide key information about natural hazards and how to quantify them. The sites listed below are exclusively nonprofit initiatives, and many depend on the active contributions from scientists around the world. Many commercial websites also provide free trial software in the fields of risk modelling, slope-stability analyses, hydrodynamic computing, and statistical analyses.

EM-DAT: The International Disaster Database is hosted by the Centre of Research on the Epidemiology of Disasters (CRED) at the Université Catholique de Louvain, Brussels, Belgium (www.emdat .be/database). This is one of the most cited databases by those trying to detect historic trends of natural disasters. You can download disaster information by geographical region, historic period, and disaster type. The website also hosts prefabricated and ready-to-use figures that show some interesting historic trends.

The US Geological Survey has a large portal with detailed information on different natural hazards (www.usgs.gov). It is a great platform for exploring earthquakes, volcanic hazards, landslides, tropical cyclones, and many other hazards. For example, a real-time earthquake map shows the most recent seismic activity around the world. You can click on a specific earthquake, access details about its tectonic and seismological setting, and download its key parameters, including the local shaking intensities based on public feedback. You can also obtain a real-time assessment of its potential damage impact via PAGER (Prompt Assessment of Global Earthquakes for Response), which is an automated system that provides information on the impact of possibly damaging earthquakes around the world to emergency responders, government and aid agencies, the media, and the general public. PAGER assesses earthquake impacts by comparing the population exposed to each level of shaking intensity with models of economic losses and fatalities based on past earthquakes in different countries or regions of the world (earthquake.usgs.gov/earthquakes/map).

NOAA's National Geophysical Data Center is an excellent site for natural hazard data, images, and education (www.ngdc.noaa .gov/hazard/hazards.shtml). There you will find, among a wealth of information and resources, a natural hazards interactive online map, *.kmz files for importing into GoogleEarth™ or Geographic Information Systems, *.pdf versions of full-size colour posters for printing, and quizzes.

The World Bank hosts a database of dozens of socioeconomic, administrative, political, and environmental indicators on a country-by-country basis (data.worldbank.org/indicator). This site provides additional background information on climate change and natural hazards with regional to global perspectives.

The Global Volcanism Program at the Smithsonian National Museum of Natural History is the authoritative source on volcanoes worldwide (www.volcano.si.edu/ index.cfm). The site hosts a comprehensive archive of Earth's volcanoes that have been active in the Holocene. The database consists of coordinates, photos, eruption dates and characteristics that may be downloaded in various formats.

Stromboli online is an educational website that has spectacular photographs and footage of many active volcanoes (www .swisseduc.ch/stromboli/index-en.html). The site also offers virtual field trips and small applets that allow visitors to simulate flight paths of volcanic bombs at Stromboli volcano.

Dave's Landslide Blog is administered by Professor Dave N. Petley at the University of Sheffield, UK (blogs.agu.org/landslideblog). This blog is part of the American Geophysical Union's Blogosphere. Dave provides updated information on landslides and related mass-wasting phenomena throughout the world.

The Pacific Tsunami Warning Center is part of NOAA's National Weather Service and the site to visit for real-time advisories regarding tsunamis worldwide (https:// www.tsunami.gov/). The site features a global map and a georeferenced list of the latest tsunami warnings.

The US Drought Portal hosts information on droughts in the United States (www .drought.gov). The site gives national forecasts and updated drought warnings based on recent and current drought conditions, also offering downloads of these data.

Natural Hazards and Earth System Sciences is an open-access journal devoted to research on natural hazards and risk (www.nat-hazards-earth-syst-sci.net). The journal has many special issues that offer quick and useful overviews of different hazard topics.

The Intergovernmental Panel on Climate Change (IPCC) website is the definitive source of information on climate change and contemporary global warming (www .ipcc.ch). The website offers free access to the IPCC's Assessment Reports on the current and projected future state of climate change, along with a lot of supplementary materials.

The World Meteorological Organization is a good site to access the latest news about weather warnings on a global scale (www.wmo.int). The site hosts information and apps on weather, climate and related hazards and risks.

References

Adger WN 2006 Vulnerability. *Global Environmental Change* **16**(3), 268–281.

Alcántara-Ayala I 2002 Geomorphology, natural hazards, vulnerability and prevention of natural disasters in developing countries. *Geomorphology* **47**, 107–124.

Alexander DE 2014 Social media in disaster risk reduction and crisis management. *Science and Engineering Ethics* **20**(3), 717–733.

Atwater BF, Musumi-Rokkaku S, Satake K, et al. 2015 *The Orphan Tsunami of 1700: Japanese Clues to a Parent Earthquake in North America*. University of Washington Press.

Ayyub BM 2014 Systems resilience for multihazard environments: definition, metrics, and valuation for decision making. *Risk Analysis* **34**(2), 340–355.

Bank TW (ed) 2010 *Natural Hazards, UnNatural Disasters*. The World Bank and The United Nations.

Bartels SA and VanRooyen MJ 2011 Medical complications associated with earthquakes. *The Lancet* **379**(9817), 748–757.

Bela J 2014 Too generous to a fault? Is reliable earthquake safety a lost art? Errors in expected human losses due to incorrect seismic hazard estimates. *Earth's Future* **2**, 569–578.

Bobrowsky PT (ed) 2013 *Encyclopedia of Natural Hazards (Encyclopedia of Earth Sciences Series)*. Springer.

Botzen WJW, Aerts JCJH and van den Bergh JCJM 2009 Dependence of flood risk perceptions on socioeconomic and objective risk factors. *Water Resources Research* **45**(10), 1–15.

Bründl M, Romang H, Bischof N and Rheinberger CM 2009 The risk concept and its application in natural hazard risk management in Switzerland. *Natural Hazards and Earth System Sciences* **9**, 801–813.

Bryant E 2004 *Natural Hazards* 2nd edn. Cambridge University Press.

Chambers JQ, Fisher JI, Zeng H, et al. 2007 Hurricane Katrina's Carbon Footprint on U.S. Gulf Coast Forests - SI. *Science* **318**(5853), 1107–1107.

Chen MJ, Lin CY, Wu YT, et al. 2012 Effects of extreme precipitation to the distribution of infectious diseases in Taiwan. *PLoS ONE* **7**(6), 1–8.

Clague JJ and Stead D (eds) 2012 *Landslides. Types, Mechanisms and Modelling*. Cambridge University Press.

Corominas J, van Westen CP, Frattini P, et al. 2014. Recommendations for the quantitative analysis of landslide risk. *Bulletin of Engineering Geology and the Environment* **73**, 209–263.

Criss RE and Shock EL 2001 Flood enhancement through flood control. *Geology* **10**, 875–878.

Cutter SL and Finch C 2008 Temporal and spatial changes in social vulnerability to natural hazards. *Proceedings of the National Academy of Sciences* **105**(7), 2301–2306.

Dai F, Lee C, and Ngai YY 2002 Landslide risk assessment and management: an overview. *Engineering Geology* **64**, 65–87.

Davies TRH (ed) 2015 *Landslide Hazards, Risks, and Disasters*. Elsevier Inc.

England P and Jackson S 2011 Uncharted seismic risk. *Nature Geoscience* **4**, 348–349.

Erdik M 2013 Earthquake risk in Turkey. *Science* **341**(6147), 724–725.

Finkl CW (ed) 2013 *Coastal Hazards*. Springer Science & Business Media.

Finkl CW and Makowski C 2013 The Southeast Florida Coastal Zone (SFCZ): A cascade of natural, biological, and human- induced hazards In *Coastal Hazards* (ed. Finkl CW) Springer Science & Business Media, pp. 3–56.

French HM 2013 *The Periglacial Environment*. Wiley Blackwell.

Gao Y, Wang H, Liu GM, et al. 2014 Risk assessment of tropical storm surges for coastal regions of China. *Journal of Geophysical Research: Atmospheres* **119**, 5364–5374.

Geller RJ 2011 Shake-up time for Japanese seismology. *Nature* **472**(7344), 407–409.

Gokon H, Koshimura S, Imai K, et al. 2014 Developing fragility functions for the areas affected by the 2009 Samoa earthquake and tsunami. *Natural Hazards and Earth System Sciences* **14**(12), 3231–3241.

Gudmundsson MT, Thordarson T, Höskuldsson Á, et al. 2011 Ash generation and distribution from the April-May 2010 eruption of Eyjafjallajökull, Iceland. *Scientific Reports* **2**, 572–572.

Hyndman D and Hyndman D 2016 *Natural Hazards and Disasters*. Brooks Cole.

James R, Otto F, Parker H, et al. 2014 Characterizing loss and damage from climate change. *Nature Climate Change* **4**(11), 938–939.

Jonkman SN, van Gelder PHAJM and Vrijling JK 2003 An overview of quantitative risk measures for loss of life and economic damage. *Journal of Hazardous Materials* **A99**(1), 1–30.

Keller EA and Devecchio DE 2014 *Natural Hazards: Earth's Processes as Hazards, Disasters, and Catastrophes*. Taylor & Francis.

Kerr R 2011 More megaquakes on the way? That depends on your statistics. *Science* **332**(6028), 411.

Korup O and Clague JJ 2009 Natural hazards, extreme events, and mountain topography. *Quaternary Science Reviews* **28**, 977–990.

Lesk C, Rowhani P and Ramankutty N 2016 Influence of extreme weather disasters on global crop production. *Nature* **529**(7584), 84–87.

Lockwood JP and Hazlett RW 2010 *Volcanoes. Global Perspectives*. Wiley Blackwell.

McPhaden MJ, Zebiak SE and Glantz MH 2006 ENSO as an integrating concept in earth science. *Science* **314**(5806), 1740–1745.

McPhillips LE, Chang H, Chester MV, et al. 2018 Defining extreme events: A cross-disciplinary review. *Earth's Future* **6**(3), 441–455.

Michel-Kerjan E and Kunreuther H 2011 Redesigning flood insurance. *Science* **333**, 408–409.

Milanesi L, Pilotti M and Ranzi R 2015 A conceptual model of people's vulnerability to floods. *Water Resources Research* **51**(1), 182–197.

Neumayer E and Plümper T 2007 The gendered nature of natural disasters: The impact of catastrophic events on the gender gap in life expectancy, 1981–2002. *Annals of the Association of American Geographers* **97**(3), 551–566.

Oppenheimer C 2011 *Eruptions that Shook the World*. Cambridge University Press.

Osuteye E, Johnson C and Brown D 2017 The data gap: An analysis of data availability on disaster losses in sub-Saharan African cities. *International Journal of Disaster Risk Reduction* **26**, 24–33.

Palmer TN and Räisänen J 2002 Quantifying the risk of extreme seasonal precipitation events in a changing climate. *Nature* **415**, 512–514.

Paprotny D, Morales-Nápoles O and Jonkman SN 2018 HANZE: a pan-European database of exposure to natural hazards and damaging historical floods since 1870. *Earth System Science Data* **10**(1), 565–581.

Park J, Seager TP, Rao PSC, et al. 2013 Integrating risk and resilience approaches to catastrophe management in engineering systems. *Risk Analysis* **33**(3), 356–367.

Peduzzi P, Dao H, Herold C and Mouton F 2009 Assessing global exposure and vulnerability towards natural hazards: the Disaster Risk Index. *Natural Hazards and Earth System Sciences* **9**, 1149–1159.

Popova IP, Jenniskens P, Emel'yanenko V, et al. 2013 Chelyabinsk airburst, damage assessment, meteorite recovery, and characterization. *Science* **342**(6162), 1069–1073.

Rosenzweig C, Karoly D, Vicarelli M, et al. 2008 Attributing physical and biological impacts to anthropogenic climate change. *Nature* **453**(7193), 353–357.

Salvati P, Bianchi C, Rossi M, and Guzzetti F 2010 Societal landslide and flood risk in Italy. *Natural Hazards & Earth System Sciences*, **10**(3), 465–483.

Sato H, Hirata N, Koketsu K, et al. 2005 Earthquake source fault beneath Tokyo. *Science* **309**(5733), 462–464.

Schwartz 2006 UNDRR (United Nations Office for Disaster Risk Reduction) 2016 Report of the open-ended intergovernmental expert working group on indicators and terminology relating to disaster risk reduction.

Scott AC, Bowman DMJS, Bond WJ, et al. 2014 *Fire on Earth. An Introduction*. Wiley Blackwell.

Sheffield J and Wood EF 2011 *Drought. Past problems and future scenarios*. Earthscan.

Sidle RC and Ochiai H 2006 *Landslides. Processes, Prediction, and Land Use* Water Resources Monograph 18. American Geophysical Union.

Simons M, Minson SE, Sladen A, et al. 2011 The 2011 magnitude 9.0 Tohoku-Oki earthquake: Mosaicking the megathrust from seconds to centuries. *Science* **332**(6036), 1421–1425.

Sintubin M 2011 Archaeoseismology: Past, present and future. *Quaternary International* **242**(1), 4–10.

Skinner HCW 2007 The Earth, source of health and hazards: An introduction to medical geology. *Annual Review of Earth and Planetary Sciences* **35**(1), 177–213.

Smith K and Petley DN 2009 *Environmental Hazards: Assessing Risk and Reducing Disaster*. Routledge.

Stein S and Stein J 2014 *Playing Against Nature: Integrating Science and Economics to Mitigate Natural Hazards in an Uncertain World*. AGU Wiley.

Uprety Y, Shrestha UB, Rokaya MB, et al. 2017 Perceptions of climate change by highland communities in the Nepal Himalaya. *Climate and Development* **9**(7), 649–661.

Viscusi WK and Aldy JE 2003 The value of a statistical life: A critical review of market estimates throughout the world. *The Journal of Risk and Uncertainty* **27**(1), 5–76.

Whiteman CA 2011 *Cold Region Hazards and Risks*. Wiley Blackwell.

Witham CS 2005 Volcanic disasters and incidents; a new database. *Journal of Volcanology and Geothermal Research* **148**, 191–233.

3

Natural Hazards and Disasters Through the Geomorphic Lens

In this chapter we outline how geomorphology helps to better understand the consequences of natural hazards and disasters, particularly in how they change the Earth's surface and impact the livelihoods it sustains. For a broader and more introductory treatment of geomorphology, we refer readers to several instructive textbooks. Among our favourites are those by Anderson and Anderson (2010) and Willgoose (2018), as they explain in a clear manner many of the physical and chemical foundations of geomorphology. They also show nicely how to encapsulate those principles in a quantitative manner, so we refrain from providing exhaustive references in this chapter. The studies that we highlight here are merely pointers to a broad and growing body of work.

More specialized books include that by Leeder and Perez-Arlucea (2005), which offers an accessible overview of many of the processes responsible for shaping the Earth's surface. The books by Julien (2010) and Lu and Godt (2013) are good examples that emphasize some of the perspectives of engineering in rivers and on hillslopes, and host a wealth of quantitative relationships widely used in practice today. Despite these and many other inspiring references, we found little in the way of textbooks that deal with geomorphology and natural hazards; the few exceptions concern mostly edited volumes. Hence we

focus on appraising natural hazards, risks, and disasters through the broadest geomorphic lens. We maintain that geomorphology is more than the sum of its specialized subdisciplines. Focusing on specified fields of geomorphology dealing with, for example, rivers, hillslopes or glaciers is pedagogically convenient, but limits our understanding of landscapes and their dynamics in their entirety.

Geomorphic processes and human activities interact in changing landscapes. We have seen that natural disasters occur when the processes operating at or near the Earth's surface negatively impact a large number of human assets. Hence we need to understand and be able to anticipate, or better predict, these natural processes in order to reduce future adverse impacts. Geomorphology has a venerable history at the interface of geography and geology that allows us to study the dynamics of our planet's surface. Geomorphology attempts to recognize, classify, understand, and predict the processes and landforms responsible for shaping a given landscape. These processes are governed by a set of well-known principles of fluid and solid mechanics. But geomorphology goes beyond this mechanistic aspect by informing us about the frequency, magnitude, and consequences of these processes. Hence geomorphology has been contributing increasingly to a science-based management of natural hazards and disasters.

Geomorphology and Natural Hazards: Understanding Landscape Change for Disaster Mitigation, Advanced Textbook Series, First Edition. Tim R. Davies, Oliver Korup, and John J. Clague.
© 2021 John Wiley & Sons Ltd. Co-published 2021 by the American Geophysical Union and John Wiley & Sons Ltd.

Why care about geomorphology? In this book, we show that geomorphology entails observing, measuring, and predicting the dynamics of the Earth's surface. Houses, factories, roads, harbours, bridges, rice paddies, Internet cables, or clean drinking water all depend on stable ground. Here, 'stable' means a low probability of moving for a specified time period, as every portion of the Earth's surface will change at some time. Geomorphology allows us to understand and anticipate how gravity, wind, water, and ice compromise ground stability by transporting materials across the Earth's surface, thus changing its form. We argue that much of the damage from natural disasters results from such transport events, especially if the volumes of material involved are large, infrequent, and unexpected. We also argue that geomorphic consequences of some natural disasters can evolve into new, unexpected hazards as landscapes adjust to past disturbances.

3.1 Drivers of Earth Surface Processes

3.1.1 Gravity, Solids, and Fluids

The key processes of geomorphic systems are erosion, transport, and deposition of mass. Geomorphic change means that parts of a landscape respond to the motion of solid Earth materials driven by gravity following tectonic uplift, volcanic eruptions, or mantle dynamics. Gravity and moving fluids (water, ice, and air) redistribute these uplifted materials along the sloping land surface. Gravity is the main driver causing solid particles and fluids to move across the Earth's surface. Moving fluids also carry solids below the Earth's surface. Shifts of solid material from one place to another alter the elevation of the surface and thus the local surface gradient, which in turn alters the rate of mass transfer. Thus the shape of the land both responds to and

constrains rates of fluid and solid material transfer.

We distinguish between supply- and transport-limited conditions. In supply-limited conditions the flux of material is set by how much is available per unit time, so that eventually the shape of the land is determined by the transfer rate. In transport-limited conditions the shape of landforms or landscapes depends on how well surface processes are capable of moving soil, debris, and rock.

Other than transport by lithospheric plate motions, uplifted rock material must be broken into sufficiently small pieces before it can be transported by surface processes. Some of this rock breakage occurs during tectonic uplift, volcanic activity, and fracturing in the brittle lithosphere (Molnar et al. 2007). Once at the Earth surface, in situ mechanical, biological, and chemical weathering adds to the break up, as does fracturing and abrasion during transport. The size of rock fragments can range from blocks many cubic kilometres in volume, such as in the case of the Waikaremoana rock slide in New Zealand, to grains less than a few tens of nanometres (10^{-9} m) across (Davies et al. 2019). The finest fragments can agglomerate to form larger ones and may form cohesive clays after weathering; otherwise, the rock debris is noncohesive. The size distribution of fractured material affects the rate of transport resulting from a given intensity of 'input power' (Bagnold 1977).

Geomorphology provides insights into land surfaces, as well as forms and processes beneath oceans, lakes, and inland seas. The geomorphology of planets and moons in the solar system offers rigorous tests of concepts developed on Earth, although practical validation in extraterrestrial locations is currently limited, but perhaps more likely in the future. But even some terrestrial environments remain demanding to study. For example, landslide-generated tsunamis involve both terrestrial and marine environments in sequence. The differing densities and viscosities of air and water mean that geomorphic processes in air and under water also differ. For example, sand grains settle much more rapidly in still air than in still water because the density and viscous drag of water are so much greater than those of air. Nevertheless all the gravity-driven motions that we outline in more detail below can occur in both environments.

The rules that govern the dynamics of geomorphic systems are fairly simple. The first rule is that matter is conserved; that is it is neither created nor destroyed (at least in classical mechanics). This principle of mass conservation is the basis of water and sediment budgets and glacier mass balances, and is widely used in geomorphology. We can express mass conservation in its most basic form as a one-dimensional equation of sediment continuity, which relates the rate of change of a geomorphic surface elevation $\partial z/\partial t$ in units length per time, or $[L\,T^{-1}]$, to changes in the lateral sediment flux q_s per unit width of channel or hillslope $[L^2\,T^{-1}]$:

$$\frac{\partial z}{\partial t} = -\frac{1}{1-\lambda}\frac{\partial q_s}{\partial x} \qquad (3.1)$$

where z is the vertical coordinate of the geomorphic surface $[L]$, t is time $[T]$, λ is sediment porosity $[1]$, and sediment transport is along the coordinate x $[L]$. The local topographic gradient largely determines the sediment flux q_s; the steeper the slope, the higher the flux.

Conservation of momentum can be a useful principle for situations when velocity changes suddenly, for example a hydraulic jump in a river or the impact of an asteroid. In essence, Newton's Laws of Motion encapsulate how geomorphic systems work:

- Force = mass × acceleration
- Flow resistance in laminar flows ∝ empirical coefficient of viscosity × strain rate
- Flow resistance in turbulent flows ∝ empirical drag coefficient × (velocity)^2
- Frictional resistance of solids = empirical friction coefficient × effective normal stress.

Conservation of energy is a less useful principle than conservation of mass in geomorphology because all geomorphic systems dissipate energy, mainly through frictional heat. In some cases, such as laminar flows, we can calculate this dissipation, but in turbulent and granular flows, which are far more common, we can only estimate energy dissipation empirically. Nonetheless, several important equations or concepts in geomorphology, such as that of stream power, derive partly from energy considerations.

These physical principles allow us to predict some aspects of the future geometry of a river or a hillslope if we know its initial geometry and the forces acting on it. This logic is the basis for solving differential equations expressing rates of change. In spite of the simplicity of this set of rules, predicting landscape dynamics is often limited to simple and regional-scale cases, as interacting processes cause confounding nonlinear effects. Most observations of geomorphic processes also involve a lot of noise that limits the use of the mechanistic rules outlined above. One way out of this dilemma is to assume that some components of a particular process are constant or negligible with respect to the study area or period of interest. For example, in studying how a river reacts to a large-scale, sudden sediment input, the valley rock walls that the river has formed over a period of millions of years might remain stable during the years or decades that it takes to flush the excess sediment from the system. For even shorter intervals, the slope of a river reach can be regarded as constant, or at least an independent variable; accordingly, the short-term sediment transport capacity of a river is a function of slope and instantaneous discharge. Over longer intervals, however, the mean flow rate and sediment supply determine slope, which is then a dependent variable.

For example, the turbulent nature of river flows has long been thought to be fundamental to river bed or channel forms such as dunes or meanders. Recently, however, Lajeunesse et al. (2010) demonstrated that many turbulent-flow channel forms have exact laminar-flow equivalents. Along the same lines Davies et al. (2003)

found that aggregation rates along the Waiho River in New Zealand could be matched using a laminar-flow, microscale model in which the only factors common to the model and the field situation are the river's planform geometry and Froude number. The variability of flow rate, channel slope, sediment grain-size distribution, and its supply rate were all chosen randomly in the model.

A grain of any size will remain at rest until the forces acting on it cause it to move. These fluid forces and gravitational forces must be sufficient to overcome the forces tending to cause the grain to stay in place, including inertial forces, frictional forces, and cohesive forces. Commonly these detailed force balances are generalized to the relative magnitudes of shear stresses and material strength. Shear stress τ_0 is defined as

$$\tau_0 = \rho g h S_0 \qquad (3.2)$$

where g is gravitational acceleration [L T^{-2}], h is local flow depth [L], and S_0 is the energy slope [1]. Geomorphologists commonly use the topographic slope (e.g. the local inclination of the channel bed) to approximate S_0, although this is strictly valid only for uniform flows.

If the shear stress on a grain exceeds its ability to resist shear, it will begin to move. The shear strength acting to resist fluid and gravity shear parallel to a granular surface is called the 'critical shear stress'; it is assumed that motion occurs once this critical shear stress is exceeded. In fluids, we can also define critical shear stress via a critical shear velocity u_*^2 [L^{2}T^{-2}]:

$$\tau_0 = \rho u_*^2 \qquad (3.3)$$

Critical shear stress can be tricky to determine under natural conditions, especially in the case of granular beds of mixed particle sizes during turbulent flow. This motivates the widespread use of Eq (3.2), although other approximations use empirically derived constants such as the Darcy–Weisbach friction factor combined with measurements of average flow velocity.

3.1.2 Motion Mainly Driven by Gravity

Erosion and deposition entail changes of the land surface elevation caused mostly by moving solid particles (Eq 3.1). Particle motion is driven by gravity: firstly, directly, in causing solid material to move downslope because the downslope component of the solid's weight exceeds the frictional resistance to its motion across the substrate; and secondly, indirectly, because gravity causes water, ice, or air to flow across the land surface, exerting forces that move solid grains in the direction of flow.

For material to move across the land surface, there must be either a gravitational force, which requires a surface gradient sufficient to overcome friction, or a fluid pressure force, which is also gravitational in the case of water and ice, but is driven by thermal density – and thus pressure – differences in the case of air. Coarse rock debris moves down steep mountain slopes under gravity and can be transported on much lower gradients by flowing water. Fine rock debris, however, can also be carried over horizontal surfaces and even up steep slopes by strong winds.

Solid materials can move as single fragments unaffected by other moving fragments; as multiple fragments that interact intermittently without affecting their movement much; or as assemblages of many grains that interact with each other to create a 'granular flow'. Granular flows move very differently from single or multiple fragments because of these granular interactions. Gravity-driven falls, slides, bounces, rolls are some of the main forms of sediment transport on land and include rockfalls, rockslides, and snow avalanches.

In the simplest case, a rock detaches from a slope and accelerates downhill until friction reduces its energy and it slows to a halt on a gentler slope. The initial motion is likely to be by sliding, then rolling if the particle geometry is suitable, followed by bouncing if it accelerates enough and the slope is sufficiently steep. As the rock rolls, it gains both angular and linear momentum, both of which contribute to its final runout distance. During deceleration, the motion sequence may be reversed from bouncing to rolling to sliding to rest.

When a single mass of dry sediment moves down a planar slope under the influence of gravity, the rate of erosion q_d or transport q_t is a function of the balance between the downslope component of gravity and the upslope friction force resisting motion. The slope gradient S is the controlling variable:

$$q_d, q_t = f(S) \qquad (3.4)$$

The next level of complexity in gravity-driven transport involves multiple rock particles falling, sliding, bouncing, or rolling under gravity and intermittently interacting with each other. This happens in many rockslides, rockfalls, and ice falls. Particles may all have different paths without interacting or colliding with one another, in which case we simply have multiple instances of the single event. If the number of particles also is small, the number of interactions will be small, with little mechanical difference from the single particle case. However, if the particles repeatedly collide, their trajectories will change; although their runout distances might only be slightly altered, the lateral divergence of the debris from the fall line will increase. In addition, if one or more rock particles break during transport, the potential for divergence increases.

The next type of gravity-driven motion involves dry subaerial granular slides and flows with continuous intergranular interactions; examples include some landslides, rock avalanches, dry snow avalanches, block-and-ash flows, and pyroclastic flows. Many particles move as a granular flow, and the motion of grains is dominated by their interactions with one another and with the substrate, although gravity acts on each grain to drive the motion. Dry granular flow is mainly controlled by intergranular and basal friction, which depend on grain shape and size distribution. A further complication is that grains may break under the stresses generated by intergranular contact, which increases the number and decreases the size of the grains. Breakage also releases energy that, during shearing of the grain mass, is converted from kinetic energy to grain elastic strain energy. If this fragmentation is extensive, the energy thus transformed may affect internal friction. Such fragmentation in a pyroclastic flow may release large volumes of hot high-pressure gases that affect the flow. In all cases, very fine fragments can form a dust cloud accompanying the motion of the granular mass.

Dominantly gravity-driven motion of water-rich subaerial granular slides and flows with continuous intergranular interaction is common; examples include many landslides, rock–ice–snow avalanches, and wet snow avalanches. The presence of water in intergranular voids affects grain interactions considerably; porewater pressures can be generated by grain motion, which in turn affect grain motion. Hydrostatic uplift reduces intergranular contact stresses, and the downslope gravitational component due to the water itself comes into play. Kinetic sieving tends to move larger grains to flow boundaries, likely creating levees. Granular saturated subaerial masses containing fine sediment include debris flows, mudflows, sieve flows, and earthflows. A dense and sufficiently viscous slurry affects contacts between grains by reducing their excess density and changing how quickly pore fluids migrate. This can lead to a dramatic increase in mobility, accompanied by severe erosion of the substrate. Movement becomes viscous. which allows coarse material to be distributed throughout the flow depth instead of being concentrated at the base. Flows are unstable longitudinally and can break up to form a series of separate roll waves. Kinetic sieving can be intense with boulders moving to the front of a flow wave. The bouldery fronts of flow waves may be unsaturated, and thus travel more slowly than the saturated material that follows, amplifying the formation of separate, large roll waves. This phenomenon is called a debris flow, and may develop from a sediment-rich water-driven flow in a stream, but a fully developed debris flow is driven by the gravity component of the entire mass. When debris flows slow down, they do so *en masse*, rather than progressively according to particle size, as in water-driven flows. This is responsible for the characteristic lobate shape of the deposit.

Another important and mostly gravity-driven transport mode is the suspension of fine particles in turbulent air, as for example in sandstorms, powder snow avalanches, pyroclastic clouds, and dust clouds of rock avalanches. Here a large mass of fine grains becomes suspended in air and forms a fluid that is denser than the surrounding clear air, unless elevated temperature such as in hot eruptive columns adds buoyancy. This mixture thus moves downward as a density current. This type of flow may be associated with, or develop from, an underlying granular flow.

3.1.3 Motion Mainly Driven by Water

Water, the most widely available fluid for transporting solids on Earth, has zero shear strength and a very low viscosity, and can thus flow rapidly across a land surface on even slight gradients. Water has a density that is about 40% that of intact rock and, when flowing, applies drag and lift forces to rock particles. It can flow very long distances across the land and thus is capable of transporting solid debris from mountain ranges to distant oceans where it accumulates in sedimentary basins. Flowing water almost always is dominated by strong turbulent eddies, and this turbulence affects its ability to transport solid grains. Water also has the ability to carry fine grains in suspension, and if the concentration of this suspended sediment becomes sufficiently high, it can affect the density and viscosity of the fluid and thus its ability to transport coarser rock debris; this is particularly the case with debris flows.

When describing sediment transport in water flows, the term 'bedload' is used to refer to single or multiple particles that slide, bounce, or roll in contact with the substrate. The motion of the solid particles is driven by

gravity and water flow, and resisted by friction exerted by the bed. The force of moving water on a particle varies with the velocity of the flow past the particle, which in turn varies with the flow rate of water per unit channel width q_w $[L^2 T^{-1}]$ in the river and the local bed slope. The translational component of the force of gravity is also a function of the local slope. We can express the rates at which flowing water can erode and transport sediment as:

$$q_d, q_t = f(S, q_w) \tag{3.5}$$

If water flowing past a grain exerts a hydrodynamic lift or drag force large enough to overcome the grain's resistance to rolling or sliding, the grain begins to move along the bed. We can specify these forces as a function of grain size:

$$\mathbf{F_D} = \frac{1}{2} C_D \rho \mathbf{u}^2 A \tag{3.6}$$

where $\mathbf{F_D}$ is the drag force $[M\ L\ T^{-2}]$, C_D is an empirically determined drag coefficient [1], ρ is the fluid density $[M\ L^{-3}]$, $\mathbf{u}$ is the flow velocity $[L\ T^{-1}]$, and A $[L^2]$ is the projected area of the grain exposed normal to the direction of flow. For a spherical particle with diameter d, $A = \frac{\pi}{4} d^2$. We can express the lift force $\mathbf{F_L}$ $[M\ L\ T^{-2}]$ in a very similar way:

$$\mathbf{F_L} = \frac{1}{2} C_L \rho \mathbf{u}^2 A \tag{3.7}$$

where C_L is an empirically determined lift coefficient [1]. How can the forces acting on a single grain of sediment be relevant to studying natural hazards? Drag forces are important to consider when computing the possible trajectories of volcanic bombs – particles >64 mm in diameter – ejected from a vent. The aerodynamics of the projectile, together with its initial velocity and the gravitational force, control the parabolic shape of its flight path. We will also see that the dimensions of wave- or flood-transported large boulders, also termed megaclasts – particles >4.1 m in diameter (Terry and Goff 2014) – , scale with the drag and lift forces at work (Chapter 9.3). With some simplifying assumptions, we are thus able to estimate past flow velocities, discharge, or wave height needed to entrain or move those boulders. In many cases, solving adequate force, or momentum, balances adds to our knowledge of flows or waves by offering data that even modern instruments may fail to record because of the destructive nature of such events.

In rivers we refer to particles that move in the vicinity of the bed, although they may bounce, as bedload. Large numbers of grains can move simultaneously, but interactions between moving grains only negligibly affect the process. Bedload motion can create a series of repeating bedforms such as ripples, dunes, or antidunes. These bedforms, together with larger particles, generate resistance to flow and hence alter flow characteristics. Particularly shallow and steep gravel-bed channels that are characteristic of mountain ranges have highly variable flow resistance, so that its prediction remains error-prone, yielding mismatches of up to a factor of two with respect to observed values (Powell 2014). In all streams with abundant coarse sediment, bedload motion is the primary contributor to changing river-bed geometry, and hence also flow resistance.

Debris flows are among the most important, but also most puzzling, types of sediment transport. The theory of dense multiphase flows is fairly well developed (Iverson and George 2014), but predictive understanding of debris-flow dynamics is not. Debris flows are mixtures of coarse solids (boulders to sand), fine solids (sand to clay), plant material, and water that move downslope under the influence of gravity, typically in a channel. They commonly have bimodal particle-size distributions, with peaks in the sand–silt and gravel–boulder size ranges, and bulk densities of 1.4–2.2 t m^{-3}. Debris flows are mostly laminar, nonuniform, and unsteady, with episodic bouldery surges interspersed with slurry flows. Boulders tend to accumulate at surge fronts due to dispersive stresses, kinetic sieving, and the vertical velocity profile in the flow. The front of a surge moves more slowly than the following less bouldery material, so the surge remains coherent as it moves down the channel. Surges are deeper and faster than a steady flow. Debris flows erode channel fills and underlying bedrock, and can increase in bulk as bed material is entrained during motion. Fine sediment in debris flows results from such entrainment of bed material or from debris slides or slumps into a swollen torrent. Debris flows have higher solids concentration than clear-water floods and thus greater flow depths and velocity. They have higher physical impact forces than water floods for a given discharge. A detailed measuring campaign of debris flows in Jiangjia Ravine, Yunnan, China, for example, recorded 139 debris flows between 1961 and 2000: the highest impact pressure estimated from hydrodynamic principles were mostly of the order of several hundred kPa (Hong et al. 2015). Failure of landslide dams can trigger large debris flows, as can rain-induced rapid melt of snow in highly erodible terrain. Debris flows are mostly restricted to steep catchments: Welsh and Davies (2010) proposed that a Melton ratio (catchment relief divided by the square root of catchment area) of >0.5 might be useful for identifying catchments prone to debris flows. Debris flows can be fast (>5 m s^{-1}) and several metres to tens of metres thick. Debris flows can destroy buildings anywhere on a developed fan, as even small surges may halt in a channel and cause subsequent larger surges to flow onto the open fan surface. Control structures include retention basins that trap and store sediment. Such basins must be able to contain the largest debris flows; hence sediment also needs to removed from the basin to maintain its capacity. Some basins are designed so that trapped sediment is reworked through a slit or grid by subsequent rainstorms. Flexible ring-mesh barriers seem to be effective for small debris flows. In Japan check-dams or *sabo* are common, hard structures built into gulley beds to trap sediment and to prevent bed erosion and undercutting of hillslopes. Although effective for small and medium-size events, they may be vulnerable to domino-type collapse during large flows.

Sediment carried in suspension in flowing water is referred to as suspended load. Flowing water generates turbulence through interaction with the bed and if the upward components of turbulent flow are stronger than the downward components, then fine sediment, although always falling relative to the water around it, can be maintained in suspension indefinitely. Although suspended grains may fall to the bed, they are replaced by others lifted from the bed, and a statistically distinct concentration of grains is maintained in suspension. The suspended load may have effects on river-bed geometry if sediment loads are dominantly clay and silt, but is rarely of sufficient concentration to affect flow processes.

High concentrations of sediment carried in flowing water are referred to as slurry flows, hyperconcentrated flows, lahars, and debris floods. Here the concentration of suspended sediment is sufficiently high that the flow transports bedload, though without necessarily developing into a debris flow. Coarse sediment is still carried at the base of the flow as bedload, rather than being dispersed throughout it.

3.1.4 Motion Mainly Driven by Ice

Ice behaves as a highly viscous fluid under gravitational forces, shearing at less than $\sim 10^{-3} s^{-1}$. Consequently, ice flow is very slow and never turbulent. Ice has a nonzero shear strength, so can transport rock debris on its surface. On Earth, most flowing ice forms glaciers in mountain valleys, ice caps, and the large ice sheets in Greenland and Antarctica. Valley glaciers can flow at rates up to 10^{-5} m s^{-1} and can be several kilometres wide and many kilometres long. They can transport large quantities of rock, either on the ice surface, within the ice or at the base of the ice. The only other geomorphic fluid that flows without turbulence is lava. It can and does transport solid fragments but, like water carrying ice, flowing lava has a high likelihood of mass exchange between the fluid and the solid components.

When ice shears at rates greater than $\sim 10^{-3} s^{-1}$ it breaks into blocks and, despite its much lower density, can form slides and falls similar to rockslides and rockfalls. However, the landforms thus generated become unstable because the ice soon melts, so henceforth we consider ice to be a fluid agent, rather than a component, of solid transport.

Single, multiple, or granular masses of rock material are transported passively by moving ice. In glaciers, the relative location of the solids with respect to the ice defines whether we refer to supraglacial, englacial, or subglacial debris. Debris derived from rockslides or rockfalls and carried on a glacier is termed supraglacial load. Supraglacial debris in the accumulation zone of a glacier becomes buried by subsequent snowfall and moves below the ice surface to travel as shallow englacial load; as it moves into the ablation zone of the glacier, where surface ice is melting, it is mostly re-exposed as supraglacial debris. A glacier is also able to mobilize and drag debris at its bed; this type of transport is termed subglacial, although (locally pressurized) meltwater can also transport material through or beneath the ice. Subglacial debris can be elevated into the glacier along shear planes to become part of the englacial load.

3.1.5 Motion Driven Mainly by Air

Air has about 0.002 times the density of rock. Air thus must move very rapidly to transport mostly small fragments of rock. Nevertheless, masses of flowing air can be many kilometres deep, hundreds of kilometres wide, and can achieve velocities of up to $50\ m\ s^{-1}$, even higher in the strongest cyclones, hurricanes, and tornadoes. These storms can move large volumes of fine material long distances. The vast Asian loess deposits downwind of Pleistocene glacier systems, for example, attest to the geomorphic work of wind. Particles mobilized by flowing air bounce in intermittent contact with the substrate, forming the saltation load of transport by wind. Saltation results from the very low density of air compared to that of sediment. Airborne grains that return to the substrate at high velocities may deliver enough momentum to grains on the land surface to mobilize them and bring them up into the air column. Saltation is very common in airflows, much more so than in water flows. Saltation in air shows noticeable hysteresis: once initiated by a sufficient flow velocity, saltation can be maintained even if the flow velocity decreases.

Impact load is the carpet of surface grains mobilized by impact of grains in the saltation load. If the bed contains grains too large to be lifted into the air, these grains may nevertheless roll or slide along in continuous contact with the bed due to impacts from descending saltation grains. This motion is often called saltation creep. Finally, suspended load consists of multiple particles carried in suspension by turbulence in flowing air. Fine grains in saltation can be lifted high into the air to form dust storms or sandstorms. These flows have a density (gravity) component because they are behaving as a denser fluid below the clear air above.

3.2 Natural Hazards and Geomorphic Concepts

We now discuss several concepts that summarize some of the key insights that geomorphic research has gleaned during the past decades. We address mainly those insights relevant to natural hazards, risks, and disasters.

3.2.1 Landscapes are Open, Nonlinear Systems

Given the many diverse types of natural hazards and disasters, we might think that all we need to do is work out how specific parts of a given landscape operate. In particular, we would focus on the one facet of a landscape that appears to be most relevant, for example, we would study rivers to deal with floods, consider hillslope evolution to anticipate landslides, or do more research on glaciers to refine estimates of their contribution to sea-level rise. Each of these facets alone has attracted a body of research papers that is growing exponentially. Understanding how sediment moves along a river bed is far from trivial even if all outside influences are assumed to be negligible. Dealing with sediment storage, be it transient or persistent, is a major research challenge in itself (Hoffmann 2015). Yet, if a community has a flood problem, engaging a hydraulic engineer or a fluvial geomorphologist might solve only part of the problem, because the problem involves so much more than just understanding sediment transport (Figure 3.1).

Scientific fields such as system theory and the more applied landscape ecology offer a means to break down a landscape conceptually into components that exchange matter, energy, and information in the form of disturbance signals. Thus, geomorphic systems interact with each other by exchanging mass (rock, soil, solutes, water, ice, organic matter), energy (kinetic, potential, elastic, heat), and information (e.g. often surface level change, beach erosion following a tsunami). Atmospheric, tectonic, and biotic processes exchange matter and energy with geomorphic systems, and because of this exchange geomorphic systems are open systems. The laws of thermodynamics apply only to closed systems, thus any attempts to rigorously apply thermodynamics to geomorphic processes or systems have limited value. Seen that way, a system-based approach to geomorphology

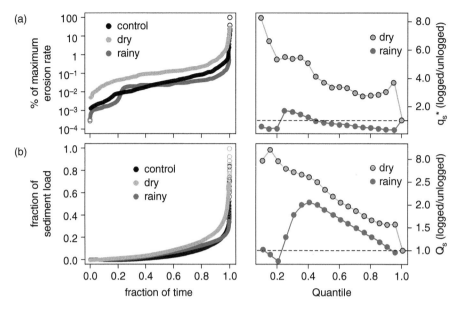

Figure 3.1 Seasonal logging, process response, and geomorphic work in small forested watersheds in Chile. (a) Fraction of instantaneous sediment transport rates normalized to catchment maximum as a function of monitoring time during which these rates were lower than shown. (b) Fraction of total sediment load normalized per catchment as a function of the fraction of total monitoring period for unlogged conditions as well as rainy- and dry-season logging. Right-hand panels show ratios of instantaneous transport rates q_s^* and total sediment loads Q_s per quantile for logged versus unlogged conditions. Black horizontal dashed lines are 1:1 ratios. From Mohr et al. (2014).

may seem pointless. However, an open-system mindset is invaluable because it promotes consideration of the full range of influences, from those that are close and prominent to those that are distant and subtle. Knowledge of the large-scale, long-term dynamics of a system may be adequate to identify likely first-order influences, paving the way for a more formal small-scale analysis (Figure 3.2).

Interacting processes comprise simpler sub-processes, so one way to understanding is to isolate and study the simplest constituent processes. However, this approach neglects how processes influence one another. Even linear processes can interact nonlinearly and thus complicate explanations or predictions of the entire system. In recent decades scientists have learned much about 'complex dynamic systems' or 'complex adaptive systems'. The underlying idea is that the more processes interact to generate an outcome, the more resources it will require to predict

that outcome at the cost of a potentially lower accuracy. The processes may be perfectly deterministic, but the overall system dynamics are not. We call this phenomenon 'emergence' – at any given level of a complex system, unexpected phenomena emerge that we cannot anticipate from knowledge of the lower-level dynamics of the system, thus precluding any accurate mechanistic prediction. We cannot simply study turbulence in a river, determine rock-mass strength in hillslopes, estimate the magnitude–frequency distribution of local earthquakes, or measure the root depths of vegetation to understand the shape of a valley. Studying these aspects can offer insights into geomorphic subsystems, but offers only limited understanding of how these subsystems interact: how does the mean root diameter of channel-bank vegetation affect turbulence in the river? Does seismicity change the strength of rock cliffs? Do earthquakes influence vegetation? Does vegetation affect slope stability?

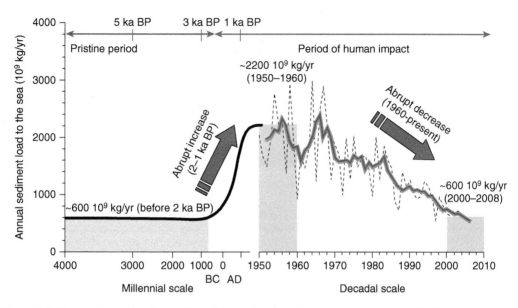

Figure 3.2 Changes in total sediment flux of five major rivers flowing into the western Pacific Ocean at two timescales: the millennial timescale (6000 yr BP to AD 1950) and decadal timescale (1950–2008). The historical decrease in sediment flux is due to human interventions in the river basins. From Wang et al. (2011).

Scientists use differential equations to describe such nonlinear systems, and some of these equations may attempt to capture chaotic dynamics. Hence, solutions to the system state are highly sensitive to details of initial conditions. Even for nearly identical starting points, system states can differ dramatically after the same length of time, thus compromising predictions, especially for longer periods. Application of physical relationships such as energy dissipation, numerical models of flow resistance and rheology can be vulnerable to emergence, and something unexpected may happen in landscapes, which models fail to predict. An approach that includes multiple model runs with inputs changed to cover a plausible range is useful, but again cannot fully predict emergent phenomena.

Yet some degree of anticipation of geomorphic events is possible, based largely on empirical data of how and at which rate landscapes have changed in the past. To be useful for hazard assessment, such data must relate to the type of event expected, its magnitude, and its location. These are the factors that geomorphic study can determine, by interpreting landscape features in terms of the processes that formed them. Dating of landscape features can reveal how many times a specific event magnitude is exceeded for an area and time.

3.2.2 Landscapes Adjust to Maximize Sediment Transport

River networks abound widely on Earth, so that running water is among the most prolific agents transporting sediment across the Earth's surface, and scientists have studied rivers and their ability to transport sediment much more intensely than most other systems. Some recent findings about the geometry of rivers also relate to the rate at which rivers can move sediment across landscapes (Figure 3.3). These results link up with recent advances in understanding the non-equilibrium thermodynamics of complex systems, and showing that some dynamics of rivers may be extrapolated to entire landscapes.

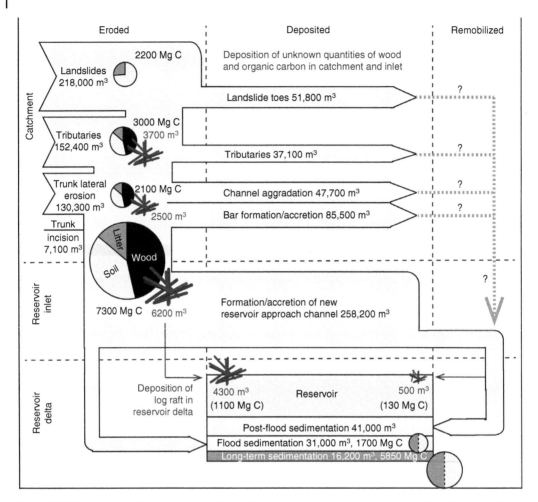

Figure 3.3 The fate of sediment, wood, and organic carbon eroded during an extreme flood, Colorado Front Range, USA. This diagram summarizes the flood and post-flood sediment, carbon, and wood flux for the North St Vrain Creek catchment. Pie diagrams represent organic carbon as soil, litter, and wood, and cartoonized wood jams represent volumes of large woody debris. Colours of boxes, arrows, and text are consistent for sediment (tan) and wood (brown), with pie diagram colors for soil (grey), litter (dark grey), and wood (black) as carbon. From Rathburn et al. (2017).

Engineers and fluvial geomorphologists have long recognized that the bed shapes and planforms of natural rivers are far from random. Considering the many geometric shapes that rivers could display in theory, we find only a few characteristic varieties in the field; examples are ripples, dunes, bars, and meanders (Kleinhans and van den Berg 2010). Early studies devoted to channel patterns proposed key principles, including those of least action, minimum unit stream power, minimum rate of energy dissipation, maximum shear stress, maximum sediment transport rate, and maximum transport efficiency. Some of these proposals were founded on linear thermodynamical analogies whose validity was the subject of intense debate. Many of these different principles are equivalent, in that they are based on maximized mean bed shear stress τ_b:

$$\tau_b = \rho g d S_e \qquad (3.8)$$

where ρ is the water density [kg m^{-3}], g is the acceleration due to gravity [m s^{-2}], d is the mean flow depth [m], and S_e the energy slope, which is commonly approximated by taking instead the local channel-bed slope S [m m^{-1}].

The level of shear stress that, over time, is most effective at altering the channel boundary shape is always greater than the mean shear stress. The tendency for the erodible bed of a turbulent water flow is to change shape over time so as to increase the mean bed shear stress acting on it. This tendency must, however, cease to operate if the bed configuration reaches such a state that any available alternative configuration would result in a *decrease* of mean bed shear stress. A river that is always changing shape to increase its bed shear stress must cease to change shape when a local maximum of bed shear stress is attained, or when some other constraint is activated.

This use of shear stress as a descriptor of river-channel dynamics is independent of any assumption about the nature of the processes by which channels change shape, and should apply to any turbulent flow; or, indeed, any stream whose boundary shear stress distribution is widely distributed. The geometry of meandering streams can be explained by this process (Davies and Sutherland, 1980, 1983) and applies equally well to channels that carry negligible sediment, for example surface tension streams, supraglacial meltwater channels, and density currents, as to those that alter their shape by eroding and depositing sediment.

Equation (3.8) shows that maximizing bed shear stress is equivalent to maximizing the product of flow depth (more correctly, cross-sectional area, but the end result is the same) and water surface slope. For a given flow rate, increasing the flow depth usually results in a decrease of the energy slope. However, in a river reach of sufficient length, in which uniform flow is inevitable because depth cannot change much, the required change of bed slope can occur only very slowly because the river

has to move a lot of bed sediment to achieve it. Thus, we can assume a constant reach-scale slope, whereby an increase in resistance or bed shear stress at any given flow rate means an increase in flow depth and thus a decrease in mean flow velocity.

Specific stream power ω [W m^{-2}] is closely related to bed shear stress, and can be seen as a measure of energy expenditure in a river channel:

$$\omega = \frac{\rho g Q S}{w} \tag{3.9}$$

where Q is discharge [m^{3}s^{-1}] and w is flow width [m]. Stream power is a widely used metric, and features prominently in numerical models of channel and landscape evolution. Yet detailed field measurements of stream power are few, and most quantitative studies resorted to some approximation. Systematic observations suggest a continuum of channel patterns linked to stream power: rivers with high stream power tend to have high channel width:depth ratios and actively forming bars, whereas rivers with very low stream power and low width:depth ratios tend to be straight or sinuous with little meandering activity (Kleinhans and van den Berg 2010). Davies and McSaveney (2006) show that the tendency for a self-adjusting river to attain a maximum bed shear stress is equivalent to a tendency to attain a minimum critical bed shear stress (the shear stress at which sediment transport begins) and thus to a maximum sediment transport rate, assuming that bedload transport rate is a function of excess stream power (Bagnold 1977). A self-formed river that has achieved a state of equilibrium will be carrying the most sediment it can with the available water, and any moderate artificial alteration of the form of the river will have the effect of reducing its ability to transport sediment – an important implication for river management.

Keep in mind several limitations to this assumption, including the use of empirical relationships that are mostly derived from controlled laboratory- or flume-based experiments, and uncorrected for correlations between some of the parameters that describe the channel geometry (Lammers and Bledsoe 2017). Systematic reviews and comparisons of the many bedload transport equations for rivers reveal many limitations. For example, Recking et al. (2012) summarized 16 common bedload transport relationships and reported that most equations based on threshold-of-motion were particularly inaccurate for flows that reached less than half of the required threshold conditions, yielding either zero values or highly overestimated rates of transport.

To sum up, river channel geometries are neither arbitrary nor random; the concept that river morphology is that which maximizes sediment transport (within the constraints of sediment and water supply) is both supported by available data and consistent with accepted sediment transport relationships. Can we use this principle to explain also the landscape forms generated by sediment motion other than in rivers, e.g. hillslope profiles formed by soil and rock movement under gravity, and windblown sand and soil morphologies?

To plausibly suggest that the result obtained for rivers from river-mechanical analysis also applies to other landform types, we need to relate the river analysis to a more general analysis. Here we revert to thermodynamics in the sense of recent understanding of nonlinear (or 'far-from equilibrium') thermodynamics. For example, Kleidon (2010) argued that complex natural system behaviour, when interpreted in thermodynamic terms, consistently corresponds to a state of maximum local entropy production. This is exactly equivalent to maximum resistance to flow in river systems, suggesting that landscapes will also tend to evolve towards a state of maximum sediment transport ability. Windblown ripples and dunes, including those formed by saltation-driven creep, have geometries similar to those formed under water, except that we miss the free surface to flatten the crests of large subaerial dunes, which are sharp like those of water-formed ripples.

3.2.3 Tectonically Active Landscapes Approach a Dynamic Equilibrium

Tectonic uplift, earthquakes, base-level change, weather, climate variation, and land-use changes are all drivers of geomorphic processes in active landscapes. The mechanistic interplay between rates of uplift, weathering, climate, and erosion in active landscapes has been an active research focus for several decades, though with an emphasis on longer-term evolution spanning thousands to millions of years (Bishop 2007; Burbank and Anderson 2011). Many of the underlying concepts, however, remain valid for shorter intervals. The landscape responds to these drivers through a range of processes, including increased or decreased sediment delivery from hillslopes to rivers, river aggradation or incision, and glacier advance or retreat. These processes, in turn, can trigger local responses such as undercutting of slopes by rivers, blocking of rivers by landslides, meltwater lakes forming in front of stagnating glaciers, or sudden slope failure during strong earthquakes (Figure 3.4)(Larsen and Montgomery 2012). At the scale of mountain ranges we now know that erosion can also promote tectonic uplift, whereas isostatic rebound can cause earthquakes and volcanic eruptions. Tectonically active landscapes grow in elevation and steepness, drawing higher precipitation, denudation and sediment fluxes (Finnegan et al. 2008). However, conservation of mass requires that, over the long term, uplift equals the downward transfer of solid and dissolved material. The consequence is that the relief of mountain ranges is ultimately controlled by gravity. Mountains cannot grow indefinitely, because erosion trims their height (Whipple et al. 1999; Thomson et al. 2010).

In the Southern Alps of New Zealand, uplift has been going on for ~10 Myr, and is presently increasing the mean elevation by up to 10 mm yr^{-1} (Figure 3.5). In the case of the 190 km^2 Waiho River catchment on the western side of the Southern Alps, this uplift

Figure 3.4 Top left: Threshold landscape of the western flank of the Southern Alps, New Zealand. Hillslopes prone to frequent landslides and stream incision can only steepen to inclinations set by their mechanical limits. Top right: Less steep hillslopes (foreground with agricultural terraces) in threshold landscapes may highlight slow-moving landslides that also contribute to landscape lowering. Bottom left: A valley fill (light brown and reddish colors) in Ladakh, India, decouples hillslopes from channel erosion. Bottom right: Lake sediments record natural sediment traps behind landslide or glacier dams (all photos by Oliver Korup).

generates ~1 km³ of additional solid material above base level, or roughly 2 Gt of mass, every 500 years. Thus, ~12 km³ of rock would have been added to this catchment since the sea level approached its present level about 6000 years ago. If tectonic uplift exceeded erosion by as little as 1 mm yr⁻¹ since the Southern Alps began to form, they would now be ~10 km high instead of ~2.5 km, their present mean elevation. This calculation may seem crude, but low-temperature chronometers of rock uplift and exhumation indicate that erosion rates and topographic relief have changed little in the western Southern Alps even during glacial-interglacial cycles (Herman et al. 2010).

Doing the same budget approach over shorter periods might ignore the fact that sediment is likely to be stored temporarily on hillslopes or valley floors. Such storage is frequently overlooked when estimating the overall mass flux of solid materials, but may match or exceed changes in sediment production or transfer. Viewed differently, storage alters the geometry of the land surface by matching changes in the production and transfer of sediment. Equation (3.4) states that storage occurs preferentially where the slope gradient is low. The form of the slope adjusts itself so that sediment can move down it at the rate it is supplied by denudation and transport. This principle also works for rivers and fundamentally constrains how water modifies landforms:

$$S = f(q_d, q_t, q_w) \qquad (3.10)$$

As a consequence, if a mountain range is experiencing uplift, the rate of erosion increases as elevation, precipitation, and steepness increase. For sufficiently high and steep mountain ranges, the rate of denudation can

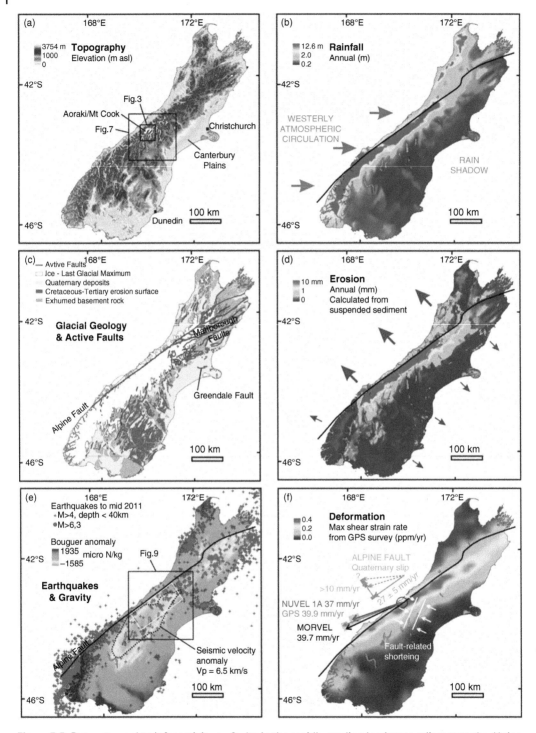

Figure 3.5 Data sets used to infer activity on faults in the rapidly eroding landscape adjacent to the Alpine Fault in the central Southern Alps, New Zealand. (a) Digital terrain model of topography. (b) Mean annual rainfall, modeled for the period 1971–2000. (c) Glacial geology and preserved Cretaceous–Tertiary erosion surface with extent of ice at the Last Glacial Maximum (white) and active faults with known evidence for rupture in the past 120 000 years. (d) Erosion calculated as mean ground lowering from a suspended sediment yield model. (e) Shallow (<40 km) seismicity for the period 1964–2011, overlain on Bouguer gravity and showing area of anomalously high seismic velocity. (f) Maximum rates of contemporary shear strain derived from GPS surveys between 1996 and 2008, together with Pacific–Australian plate vectors calculated at Fox Glacier. From Cox et al. (2012).

equal the rate of uplift. At that time increases in mass above sea level, mean elevation, and steepness all approach zero toward a landscape in dynamic equilibrium. The western Southern Alps appear to be in this state because the rates of denudation and uplift roughly match at up to ~10 mm yr^{-1}. Most hillslopes are straight from river to ridge, which must mean that eroding events are frequent (unless the straightness results from dipping layers of strong rocks, which is not the case). Albeit crudely, we can thus estimate the long-term average sediment output from a catchment in the absence of river gauging data. In the case of the Waiho River, the average annual sediment output estimated from the uplift rate and catchment area is roughly 2 Mm^3. This estimate includes all grain sizes, both bedload and suspended load. Between 1992 and 1998, aggradation of 3 Mm^3 (or 0.5 Mm^3 per year on average) was measured at the river's fanhead. If half of the sediment carried by the river is bedload, then half of the bedload entering that reach from upstream failed to be transported farther downstream. Bear in mind, though, that the concept of a dynamic equilibrium landscape is an idealization that rests on amplitudes and periods by definition (Figure 3.6). Hence, our success of validating this dynamic equilibrium with measurements hinges on when and for how long we measure with respect to these amplitude and periods. In mountain areas with negligible rock uplift, the rate of sediment supply will be the rate at which the landscape is lowering by denudation, which can be controlled largely by weathering. In such a landscape the topographic relief is gradually decreasing.

3.2.4 Landforms Develop Toward Asymptotes

We have seen that the rate at which a river transports sediment over the long term is determined by the rate of sediment production and transport within the catchment. Yet water and sediment inputs to a river are highly variable. From days to years and decades, it is more useful to say that the river is continuously adjusting its morphology towards equilibrium with the water and sediment flows it accommodates. But before it achieves equilibrium, the inputs have changed, so equilibrium is rarely achieved. A mismatch of supply and transport along a very steep river might induce a change in flow dynamics, bed elevation, or slope within a year to a decade, whereas the same perturbation along a low-gradient river might take several decades to more than a century to become apparent.

Geomorphic equations describe only processes and so cannot reveal the stages at which the processes cease to be effective in producing landforms. River meanders, for example, have been enjoying many decades of research attention (Vermeulen et al. 2014). Automated analyses of entire time series of satellite images of a given reach now allow highly instructive insights on the rates of meander and channel migration through time (Schwenk et al. 2017). Equations describe the processes by which a meander bend grows with time, but fail to address why and when a meander stops growing, which is what determines its 'equilibrium' form. An additional concept is needed to complete the picture: that concept is the constraint limiting how long the process operates. The constraint may be intrinsic, such as the depth of scour in thick alluvium accomplished by a self-limiting river, or extrinsic, such as when scour reaches a substrate resistant to erosion. An initially small perturbation such as a bend in a straight river might cause the outer banks of the bend to erode and the inner banks to accrete, causing the bend to increase in amplitude. The equation that describes this tendency sidesteps any information about what eventually causes the bend to stop growing. Meanders are of finite size, so the description by the equations is incomplete. If we are interested in the fully developed meander bend, an important part of the problem remains unsolved. Which processes constrain meander growth? As the bend grows in amplitude the shear stresses on the banks diminish because the

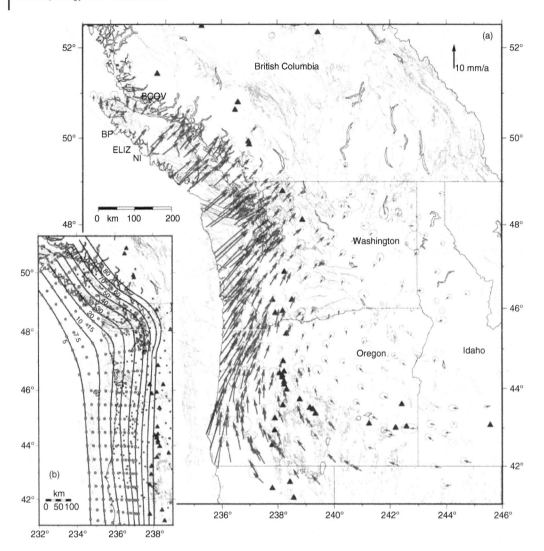

Figure 3.6 Active landscapes can change without notice, as evidenced by contemporary deformation of the lithosphere in the US Pacific Northwest. (a) Velocities of GPS sites in the North American reference frame. Red vectors are derived from continuous GPS sites, blue vectors from survey mode sites. Error ellipses are at the 70% confidence level. Triangles show locations of volcanoes. (b) Black contours parallel to the coast show the depth to the top of the subducting Juan de Fuca plate in kilometres. Gray dots show locations of fault nodes used in the inversions. Red (continuous) and blue (survey mode) dots show locations of GPS sites. From McCaffrey et al. (2007).

radius of curvature increases, so eventually the banks are able to resist the eroding stresses and growth ceases. Alternatively, the sinuosity of the growing meander set increases to the point that bends contact each other, at which point a cut-off occurs and that bend is abandoned while another starts to grow. Both of these are examples of intrinsic limits to the fundamental

tendency of the bend to grow. An extrinsic constraint might be the presence of high terraces some distance from the river; the growth of bends will be stopped or at least slowed by the terrace risers, leading to an equilibrium planform.

Continuing with the example of river meanders, the observed ratio of meander wavelength

L to channel width W falls within a narrow range. Over many scales and environments, this ratio is $4 < L/W < 40$. Why is this range so well-defined? The answer is clear when we try to draw such a river in plan: it turns out that for $L < 4W$ the planform is impossible to draw. Try it! The upper limit of L/W may be our ability to recognize what looks like a straight reach of river as a bend. Again, try drawing a river bend with $L/W = 100$; its curvature is so small that it looks straight. Despite mechanical explanations for $L/W \approx 10$, the geometric constraint limiting the extent to which L/W can depart from 10 may be free of any link whatsoever to the mechanics of rivers.

The constraint concept, together with the ideas of complexity and emergent phenomena, suggest that a clear one-to-one correspondence between process and form in geomorphology is rare. Similar forms can arise from different processes if constraints act in similar ways, a concept known as 'equifinality'. For example, avalanches of rock and wet snow can form deposits with very similar morphologies. Any mechanical analysis would suggest that the differing material properties, such as strength, density, elasticity, grain size, and water content,

should be important in determining the dynamics of motion and hence the deposit morphology. These properties are very different for rock and wet snow, so the suggestion appears incorrect. Because different forms can result from different processes, the traditional geomorphic search for relationships between process and form may yield ambiguous results.

A similar case applies to the geometry of hillslopes (Figure 3.7). Stability models predict that hillslopes are limited in both height and inclination by their strength, and many empirical studies drawing on engineering geological data and digital elevation models have supported this notion (Montgomery 2001). The idea of asymptotic or threshold hillslopes that cannot become much steeper regardless of the rates or erosion that attack them has become a key concept in geomorphology. One implication is that hillslope steepness may poorly capture erosion rates in tectonically active landscapes (Ouimet et al. 2009). Even soil-mantled hillslopes subjected to high erosion rates may trend towards a rectilinear threshold profile, assuming that material arriving at the foot of the hillslope is removed at the rate that is supplied. Roering et al. (2007)

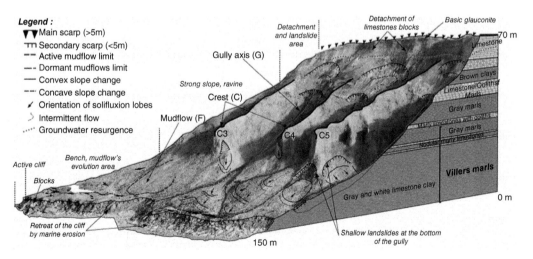

Figure 3.7 Terrestrial laser scanning and drone-based field surveys using structure-from-motion techniques allow detailed three-dimensional analyses of landforms such as this unstable coastal cliff in Normandy, France. From Medjkane et al. (2018).

proposed for this state that dimensionless hillslope erosion rate E^* depends nonlinearly on local surface inclination as:

$$E^* = \frac{2E\frac{\rho_r}{\rho_s}L_H}{KS_c} = \frac{2C_{HT}L_H}{S_c} \qquad (3.11)$$

where E is erosion rate $[LT^{-1}]$, ρ_r/ρ_s is the ratio of rock density to soil density [1], L_H is a characteristic hillslope length [L], K is the sediment transport rate $[L^2T^{-1}]$, S_c is the asymptotic (or critical) hillslope gradient [1], and C_{HT} is hilltop curvature $[L^{-1}]$. This model combines the aspect of an asymptotic relief with erosion rates, and outlines many uses of hilltop curvature for inferring, from topography alone, an estimate of how rapidly soil-mantled landscapes might wear down.

3.2.5 Landforms Record Recent Most Effective Events

Landscapes are archives of past disturbances. The legacy of regional impacts such as those from asteroid impacts or late Quaternary megafloods can reside longer in landscapes to the extent that they control to first order many contemporary geomorphic processes (Figure 3.8). Detecting these long-lasting imprints in the landscape can take decades of research, and new outburst pathways are still being found (Murton et al. 2010). In any case, the legacy of past disturbances becomes more diffuse as time goes by (Beven 2015). For example, the morphology of a recent landslide deposit records the small-scale processes that occurred as it halted, while its larger-scale

Figure 3.8 (a) View of the 8,000-m Annapurna massif that catastrophically shed several cubic kilometres of debris for several tens of kilometres downstream during Medieval times. (b) Today, terraces are cut into this >70-m thick valley fill consisting mainly of conglomerates and coarse sand layers (lithofacies F1) that sustain Nepal's second largest city of Pokhara. (c) Massive debris-flow deposits (F2) indicate rapid deposition and reworking. (d) Matrix-supported gravels (2.2 m) capped by sand with finer, light brown layers indicate rapid settling from suspension (F2). (e) and (f) Thick and massive layers of slackwater muds (F3) formed in lower tributaries that were swamped by the rapidly depositing sediment. (g) Panorama of the Seti Khola with sand and gravel layers capped by debris-flow deposits (F4). From Stolle et al. (2017).

form reflects processes that were active during its motion, and perhaps even aspects of its prefailure state. The morphology of an older landslide deposit will also reflect the reworking processes that have operated since its emplacement.

Incised alluvial fans are another example of how large and most recent disturbances may compromise how we intepret landforms. For many years, geomorphologists attributed trenched fan heads to changes in sediment supply caused by variations in climate or tectonics. However, steep mountain rivers that episodically receive sediment inputs from large landslides, but that are in dynamic equilibrium with smaller more frequent events, can also develop steep incised fan heads. The shape of a trenched fan head reflects the episodic input of large sediment pulses that aggrade the fan, while subsequent smaller floods begin to re-incise the fan (Davies and Korup 2007).

Whether rare and large disturbances leave the only imprints in landscapes is debated. The long-standing concept in fluvial geomorphology known as dominant or channel-forming discharge, which is defined as the flow most responsible for shaping the channel geometry, has been useful in providing engineers with a single discharge when designing culverts, bridges, bank protection, and other structures. However, associating the river form we see today with a dominant discharge may only be a snapshot of the morphology developed over centuries, perhaps preserved in infilled meander traces on floodplains. Analysing the magnitude and frequency of discharges can identify better the range of flow rates that have the greatest effect on river morphology and so are, in a sense, dominant (Figure 3.9). Yet we cannot be certain, nor can we find evidence, that these prominent discharges always give rise to equivalent channel patterns, particularly in braided rivers in mountain valleys, whose complex patterns have to date defied analysis or even characterisation.

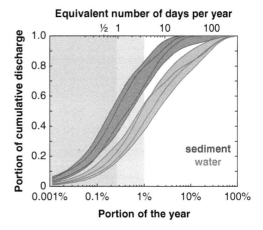

Figure 3.9 Cumulative water and suspended sediment discharge as a function of time over the course of the 2005 water year for small steep coastal watersheds in California. Roughly half of annual suspended sediment discharged from the watersheds happened in ~1 day. From Warrick et al. (2015).

3.2.6 Disturbances Travel Through Landscapes

The processes of erosion, sediment transport, and deposition transfer information about disturbances across landscapes. This information is largely in the form of topographic changes. Disturbances in one part of a landscape can affect processes operating in other parts, sometimes tens to thousands of kilometres away, and also sometimes for years to millennia afterwards. The switch of a river channel from its present outlet to the sea to another location can shorten the river's length by several kilometres, creating a steeper channel bed that will increase bed erosion. The steepened reach will propagate upstream, and years or even centuries later a riverside wetland tens of kilometres upstream might dry up when the river deepens its channel past the site. Basic relationships describing the connection between topography and erosion rates predict these dynamics. For example, the stream-power model for bedrock rivers states that local fluvial incision rate $\dot{E}$ [L T^{-1}] scales with local channel-bed slope S [1] and discharge Q [L^3 T^{-1}]. Assuming that Q also scales with upstream contributing catchment area A [L^2], we have:

$$\dot{E} = KA^m S^n \tag{3.12}$$

where K is a dimensional coefficient [$L^{1-2m}\ T^{-1}$] describing how sensitive the channel bed is to detachment-limited erosion, and m and n are positive scaling exponents that are assumed to be constant. Adding a source term that represents the uplift rate U [$L\ T^{-1}$] as a function of along-channel coordinate x and time t, we can rewrite this formulation as an advective differential equation:

$$\frac{\partial z}{\partial t} = U(x,t) - K(x,t)A(x,t)^m \left|\frac{\partial z}{\partial x}\right|^n \tag{3.13}$$

where z is the coordinate of channel-bed elevation (Kirby and Whipple 2001). For a given disturbance, Eq (3.13) describes the advection of a knickpoint that moves headward along a channel longitudinal profile, thus gradually adjusting to changes in $U(x,t)$ or $K(x,t)$ until the profile reaches a new equilibrium. We obtain the speed of knickpoint retreat $C = \partial x/\partial t$ by multiplying both sides of Eq (3.13) with $\partial x/\partial z$:

$$\frac{\partial x}{\partial t} = U(x,t)\frac{\partial x}{\partial z} - K(x,t)A(x,t)^m \left|\frac{\partial z}{\partial x}\right|^{n-1} \tag{3.14}$$

Thus disturbances in bedrock channels subjected to rapid uplift should migrate faster upstream than in channels with slow uplift, assuming that all other controls remain constant. In catchments with negligible uplift rate, $U(x,t) \to 0$, constant erodibility $K(x,t) = K$, and $n = 1$, the rate of knickpoint retreat depends only on $A(x,t)^m$ and thus diminishes in an upstream direction. Knickpoints or knickzones can also migrate up hillslopes, though more episodically driven by processes of mass wasting. Hence, to better appreciate how a landscape functions, we should consider its connectivity, including the invisible bits such as once active faults or glaciers that have melted but whose effects might still be present, and processes that are active beneath glaciers or far downstream. Various metrics of normalized river-channel steepness can help to map contrasts in uplift rates and substrate erodibility over entire mountain ranges (Chen and Willett 2016). We also need to consider the geologic history of the catchment: has postglacial isostatic rebound changed the pattern of topographic uplift and seismicity in the past ten thousand years, and is it possibly still affecting present-day processes? Look beyond the catchment boundaries to appreciate its tectonic, geologic, and climatic setting: are there active faults nearby that might rupture and trigger large coseismic landslides? Are there nearby mountain ranges that affect precipitation in the catchment?

How a landscape responds to a disturbance depends on how information is transmitted (Voller et al. 2012). It also depends on the speed of this transfer and the potential ways in which the information content can be changed (Figure 3.10). When a pebble is thrown into a pond, the water at the edge of the pond remains undisturbed until the ripples reach it. Information travels through a pond in the form of water waves at calculable rates, but waves of information travel through the landscape in the form of knickpoints, knicklines, or waves of fluids, sediment, and accompanying organic detritus (Cui et al. 2003). The upper reaches of a river remain unresponsive to a fall in sea level or uplift on a range-bounding fault until the resulting knickpoint propagates far enough upstream.

Disturbances as waves Flood waves in rivers emphasize the concept of travelling disturbances. A flood wave can be tracked as a temporary, local increase of the water surface. As the flood wave travels downstream, its shape measured along the channel can change by either increasing or decreasing the wave height. Without any other sources or sinks of water, mass conservation dictates that higher flood waves

(a)

(b)

(c)

Figure 3.10 Landscapes record the most recent and geomorphically most effective events. (a) Dry Falls, Washington, was created by catastrophic discharges from Glacial Lake Missoula in the Late Pleistocene (John Clague). (b) Grand Coulee, Washington; a major pathway of floodwaters from outburst floods from Glacial Lake Missoula (John Clague). (c) Bottom: Couplets of slackwater sediments deposited during catastrophic, Glacial Lake Missoula outburst floods (John Clague).

must be narrower, whereas lower flood waves must be wider in the direction of flow. This gradual widening is called attenuation and is characteristic of 'diffusive waves'. Waves that are instead increasing in height during motion are called 'dynamic waves'. Finally, waves that maintain their height (and shape) while travelling are called 'kinematic waves', and are an idealized model for propagating disturbances in many media, including oceans, glaciers, avalanches, or flowing mixtures of fluids and particles.

A local disturbance that alters the shape of the land surface causes a lateral movement of material by virtue of mass conservation, or sediment continuity, so that adjacent areas incrementally respond to this change. Like water waves, erosional waves or sediment waves can dissipate or amplify with distance, taking in stored sediments along the way (Craddock et al. 2010) or causing stream capture and shifting of drainage divides (O'Hara et al. 2019). Some processes of solid material motion can transmit information very long distances, in some cases quite rapidly (think of earthquakes), whereas others act at much shorter distances (think of riverbank collapse). It follows that knowing about adequate natural buffers that slow down or even halt the propagation of unwanted or potential harmful disturbances can be valuable when anticipating natural hazards or disasters (Figure 3.11) (see Chapter 17.2).

3.2.7 Scaling Relationships Inform Natural Hazards

Since the 1980s there has been growing recognition that many natural and human systems show self-organizing characteristics and that such characteristics can be simulated numerically even though they also have some theoretical foundations. Chaos theory is one of the earlier of these ideas, and there have several others dealing with complex dynamic systems. Complex nested geomorphic systems (Section 3.2.1) share distinct attributes with quite different systems, such as disease spread, financial market booms and crashes, weather, turbulence, heartbeats, and societal dynamics. The similarities of these nominally very

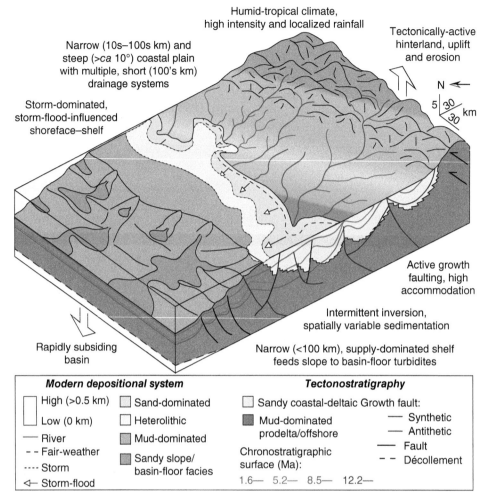

Humid-tropical climate,
high intensity and localized rainfall

Tectonically-active
hinterland, uplift
and erosion

Narrow (10s–100s km) and
steep (>*ca* 10°) coastal plain
with multiple, short (100's km)
drainage systems

N ←

5 | 30 / 30 km

Storm-dominated,
storm-flood-influenced
shoreface–shelf

Active growth
faulting, high
accommodation

Intermittent inversion,
spatially variable sedimentation

Rapidly subsiding
basin

Narrow (<100 km), supply-dominated shelf
feeds slope to basin-floor turbidites

Modern depositional system		*Tectonostratigraphy*	
☐ High (>0.5 km)	☐ Sand-dominated	☐ Sandy coastal-deltaic	Growth fault:
Low (0 km)	☐ Heterolithic	☐ Mud-dominated	— Synthetic
— River	☐ Mud-dominated	prodelta/offshore	— Antithetic
- - Fair-weather	☐ Sandy slope/	Chronostratigraphic	— Fault
···· Storm	basin-floor facies	surface (Ma):	- - Décollement
◁— Storm-flood		1.6— 5.2— 8.5— 12.2—	

Figure 3.11 An example of a conceptual model that connects concepts of tectonostratigraphy and natural hazards. The model synthesizes key factors responsible for storm-driven sediment pulses in a source-to-sink system in tectonically active northwestern Borneo. From Collins et al. (2017).

different systems might promise a common mathematical tool for characterizing or even predicting how these systems operate.

The statistics of natural and human complex systems relevant to natural disasters contain abundant mention of power-law relationships (Turcotte et al., 2002), which could mean that these systems operate close to a critical point at which dynamics can change dramatically. Many natural phenomena, including landslides, wildfires, avalanches, and river turbulence (Sachs et al. 2012) appear to follow power-law frequency–magnitude relationships, for reasons that appear to be connected with fractal distributions of geometric characteristics (see Chapter 5.4.2). Log-normal and power-law distributions are fundamentally related (Mitzenmacher 2004). Thus the consistency of dynamics that underlies the scaling of natural processes may have a fundamental importance in the way complex natural systems organize themselves in response to driving forces and constraints.

These dynamics serve as a warning about how sensitive interactions between human and natural systems can be. The dynamics also show the value of caution in initiating such interactions where they may occur for the first time. Many natural distributions appear to change when the scale of the phenomenon under consideration increases, so that it occupies a large proportion of the system space. System boundaries then affect the phenomenon, resulting in a higher-than-expected frequency of exceptionally large, so-called 'dragon-king', events (Sachs et al. 2012). These are the very natural events that trigger disasters, and warn us about extrapolating small-scale dynamics to larger scales. The prospect of increased understanding of the occurrence of extreme events is attractive, especially as it is very general in nature. Complex system dynamics may turn out to be as important to understanding extreme events as the introduction of statistical mechanics to understanding thermodynamics in the late eighteenth century.

References

Anderson RS and Anderson SP 2010 *Geomorphology: The Mechanics and Chemistry of Landscapes.* Cambridge University Press.

Bagnold RA 1977 Bedload transport by natural rivers. *Water Resources Research* **13**(2), 303–312.

Beven K 2015 What we see now: Event-persistence and the predictability of hydro-eco-geomorphological systems. *Ecological Modelling* **298**, 4–15.

Bishop P 2007 Long-term landscape evolution: linking tectonics and surface processes. *Earth Surface Processes and Landforms* **32**(3), 329–365.

Burbank DW and Anderson RS 2011 *Tectonic Geomorphology.* Wiley Blackwell.

Chen CY and Willett SD 2016 Graphical methods of river profile analysis to unravel drainage area change, uplift and erodibility contrasts in the Central Range of Taiwan. *Earth Surface Processes and Landforms* **41**(15), 2223–2238.

Collins DS, Johnson HD, Allison PA, et al. 2017 Coupled 'storm-flood' depositional model; application to the Miocene-modern Baram Delta Province, north-west Borneo. *Sedimentology* **64**, 1203–1235.

Cox SC, Stirling MW, Herman, F, et al. 2012 Potentially active faults in the rapidly eroding landscape adjacent to the Alpine Fault, central Southern Alps, New Zealand. *Tectonics* **31**(2), TC2011.

Craddock WH, Kirby E, Harkins NW, et al. 2010 Rapid fluvial incision along the Yellow River during headward basin integration. *Nature Geoscience* **3**(3), 209–213.

Cui Y, Parker G, Lisle TE, et al. 2003 Sediment pulses in mountain rivers: 1. Experiments. *Water Resources Research* **39**(9), 1–12.

Davies TRH and Korup O 2007 Persistent alluvial fanhead trenching resulting from large, infrequent sediment inputs. *Earth Surface Processes and Landforms* **32**(5), 725–742.

Davies TRH, McSaveney MJ and Clarkson PJ 2003 Anthropic aggradation of the Waiho River, Westland, New Zealand: microscale modelling. *Earth Surface Processes and Landforms* **28**(2), 209–218.

Davies TRH and McSaveney MJ 2006 Geomorphic constraints on the management of bedload-dominated rivers. *Journal of Hydrology (NZ)* **45**, 63–82.

Davies TRH and Sutherland AJ 1980 Resistance to flow past deformable boundaries. *Earth Surface Processes* **5**, 175–179.

Davies TRH and Sutherland AJ 1983 Extremal hypotheses for river behaviour. *Water Resources Research* **19**(1), 141–148.

Davies TRH, Reznichenko NV, and McSaveney MJ 2019 Energy budget for a rock avalanche: fate of fracture-surface energy. *Landslides* **17**, 3–13.

Finnegan NJ, Hallet B, Montgomery DR, et al. 2008 Coupling of rock uplift and river incision in the Namche Barwa-Gyala Peri massif, Tibet. *Geological Society of America Bulletin* **120**(1–2), 142–155.

Herman F, Rhodes EJ, Braun J, and Heiniger L 2010 Uniform erosion rates and relief amplitude during glacial cycles in the Southern Alps of New Zealand, as revealed from OSL-thermochronology. *Earth and Planetary Science Letters* **297**(1–2), 183–189.

Hoffmann T 2015 Sediment residence time and connectivity in non-equilibrium and transient geomorphic systems. *Earth-Science Reviews* **150**(C), 609–627.

Hong Y, Wang JP, Li DQ, et al. 2015 Statistical and probabilistic analyses of impact pressure and discharge of debris flow from 139 events during 1961 and 2000 at Jiangjia Ravine, China. *Engineering Geology* **187**(C), 122–134.

Iverson RM and George DL 2014 A depth-averaged debris-flow model that includes the effects of evolving dilatancy. I. Physical basis. *Proceedings of the Royal Society A: Mathematical, Physical and Engineering Sciences* **470**(2170), 20130819.

Julien PY 2010 *Erosion and Sedimentation*. Cambridge University Press.

Kirby E and Whipple K 2001 Quantifying differential rock-uplift rates via stream profile analysis. *Geology* **29**(5), 415–418.

Kleidon A 2010 A basic introduction to the thermodynamics of the Earth system far from equilibrium and maximum entropy production. *Philosophical Transactions of the Royal Society B: Biological Sciences* **365**(1545), 1303–1315.

Kleinhans MG and van den Berg JH 2010 River channel and bar patterns explained and predicted by an empirical and a physics-based method. *Earth Surface Processes and Landforms* **36**(6), 721–738.

Lajeunesse E, Malverti L, Lancien P, et al. 2010 Fluvial and submarine morphodynamics of laminar and near-laminar flows: a synthesis. *Sedimentology* **57**(1), 1–26.

Lammers RW and Bledsoe BP 2017 Parsimonious sediment transport equations based on Bagnold's stream power approach.

Earth Surface Processes and Landforms **43**(1), 242–258.

Larsen IJ and Montgomery DR 2012 Landslide erosion coupled to tectonics and river incision. *Nature Geoscience* **5**(7), 468–473.

Leeder MR and Perez-Arlucea M 2005 *Physical Processes in Earth and Environmental Sciences*. Blackwell.

Lu N and Godt JW 2013 *Hillslope Hydrology and Stability*. Cambridge University Press.

McCaffrey R, Qamar AI, King RW, et al. 2007. Fault locking, block rotation and crustal deformation in the Pacific Northwest. *Geophysics Journal International* **169**, 1315–1340.

Mitzenmacher M 2004 A brief history of generative models for power law and lognormal distributions. *Internet Mathematics* **1**(2), 226–251.

Mohr CH, Zimmermann A, Korup O, et al. 2014 Seasonal logging, process response and geomorphic work. *Earth Surface Dynamics* **2**, 117–125.

Molnar P, Anderson RS, and Anderson, SP 2007 Tectonics, fracturing of rock, and erosion. *Journal of Geophysical Research: Earth Surface* **112**(F03014).

Montgomery DR 2001 Slope distributions, threshold hillslopes, and steady-state topography. *American Journal of Science* **301**, 432–454.

Murton JB, Bateman MD, Dallimore SR, et al. 2010 Identification of Younger Dryas outburst flood path from Lake Agassiz to the Arctic Ocean. *Nature* **464**(7289), 740–743.

O'Hara D, Karlstrom L, and Roering JJ 2019 Distributed landscape response to localized uplift and the fragility of steady states. *Earth and Planetary Science Letters* **506**, 243–254.

Ouimet WB, Whipple KX, and Granger DE 2009 Beyond threshold hillslopes: Channel adjustment to base-level fall in tectonically active mountain ranges. *Geology* **37**(7), 579–582.

Powell DM 2014 Flow resistance in gravel-bed rivers: Progress in research. *Earth-Science Reviews* **136**(C), 301–338.

Rathburn SL, Bennett GL, Wohl EE, et al. 2017. The fate of sediment, wood, and organic carbon eroded during an extreme flood, Colorado Front Range, USA. *Geology* **45**, 499–502.

Recking A, Liébault F, Peteuil C, and Jolimet T 2012 Testing bedload transport equations with consideration of time scales. *Earth Surface Processes and Landforms* **37**(7), 774–789.

Roering JJ, Perron JT and Kirchner JW 2007 Functional relationships between denudation and hillslope form and relief. *Earth and Planetary Science Letters* **264**(1–2), 245–258.

Sachs MK, Yoder MR, Turcotte DL, et al. 2012 Black swans, power laws, and dragon-kings: Earthquakes, volcanic eruptions, landslides, wildfires, floods, and SOC models. *The European Physical Journal Special Topics* **205**(1), 167–182.

Schwenk J, Khandelwal A, Fratkin M, et al. 2017 High spatiotemporal resolution of river planform dynamics from Landsat: The RivMAP toolbox and results from the Ucayali River. *Earth and Space Science* **4**(2), 46–75.

Stolle A, Bernhardt A, Schwanghart W, et al. 2017 Catastrophic valley fills record large Himalayan earthquakes, Pokhara, Nepal. *Quaternary Science Reviews* **177**, 88–103.

Terry JP and Goff J 2014 Megaclasts: Proposed revised nomenclature at the coarse end of the Udden-Wentworth grain-size scale for sedimentary particles. *Journal of Sedimentary Research* **84**(3), 192–197.

Thomson SN, Brandon MT, Tomkin JH, et al. 2010 Glaciation as a destructive and constructive control on mountain building. *Nature* **467**(7313), 313–317.

Turcotte DL, Malamud BD, Guzzetti F and Reichenbach P 2002 Self-organization, the cascade model, and natural hazards. *Proceedings of the National Academy of Sciences* **99** (Suppl 1), 2530–2537.

Vermeulen B, Hoitink AJF, van Berkum SW and Hidayat H 2014 Sharp bends associated with deep scours in a tropical river: The river Mahakam (East Kalimantan, Indonesia). *Journal of Geophysical Research: Earth Surface* **119**, 1441–1454.

Voller VR, Ganti V, Paola C, and Foufoula-Georgiou E 2012 Does the flow of information in a landscape have direction?. *Geophysical Research Letters* **39**(1), 1–5.

Wang H, Saito Y, Zhang Y, et al. 201. Recent changes of sediment flux to the western Pacific Ocean from major rivers in East and Southeast Asia. *Earth-Science Reviews* **108**, 80–100.

Warrick JA, Melack JM, and Goodridge BM 2015 Sediment yields from small, steep coastal watersheds of California. *Journal of Hydrology: Regional Studies* **4**, 516–534.

Welsh A and Davies T 2010 Identification of alluvial fans susceptible to debris-flow hazards. *Landslides* **8**(2), 183–194.

Whipple KX, Kirby E and Brocklehurst SH 1999 Geomorphic limits to climate-induced increases in topographic relief. *Nature* **401**, 39–43.

Willgoose G 2018 *Principles of Soilscape and Landscape Evolution*. Cambridge University Press.

4

Geomorphology Informs Natural Hazard Assessment

Having introduced several basic concepts of modern geomorphology, we now consider how it can be applied to quantitative assessments of natural hazards. We show that this broad area of applied geomorphology contributes to mitigating the risks from natural hazards and to reducing the impacts of natural disasters. We also outline some of the disciplinary knowledge and understanding that are required for effectively using geomorphic concepts in this way.

4.1 Geomorphology Can Reduce Impacts from Natural Disasters

A natural disaster results from the interaction of two systems – the natural Earth system and the human societal system. Our focus in this book is to describe the extent to which we can anticipate how extreme events operate in landscapes and how we can incorporate that knowledge into societal adaptation and mitigation strategies. Initially this task requires sufficient understanding of the underlying Earth surface processes. At later stages it requires that this understanding and its implications be explained clearly to society, and that society commits to adopting measures that reduce the impacts of extreme natural events. We recognize that this task concerns more fields than just geomorphology alone.

Communicating science and getting science taken seriously by groups with conflicting short-term agendas has also become a must in times of social media. A useful decision support process also requires developing a range of likely hazard scenarios that include societal impacts, a triple-bottom-line analysis, and sufficient political acceptability of any measures proposed. Scientists also need to maintain credibility when the predicted extreme event fails to occur in the short term. Nonetheless, we recommend using our knowledge of Earth system processes to provide meaningful information to laypersons and decision makers. This is particularly true of the more extreme events that trigger disasters, in part because reliable data describing these rare events are sparse.

Geomorphology is also concerned with deriving frequency–magnitude relationships for processes such as floods or landslides. Such relationships are inevitably empirical and based on available data covering mainly minor or moderate events that may be problematic for reliably extrapolating to larger events. Even if such relationships can be defined for large events, the obverse problem is that so few of these large events will occur at a given location in a realistic planning time frame that their occurrence might mismatch the longer-term relationship (Davies 2015). We maintain that reconstructing and measuring extreme geomorphic events extends our knowledge about

Geomorphology and Natural Hazards: Understanding Landscape Change for Disaster Mitigation, Advanced Textbook Series,
First Edition. Tim R. Davies, Oliver Korup, and John J. Clague.

the range of impacts that we can expect. Sound knowledge of physics and chemistry – the sciences that underpin earth dynamics – has the advantage of encouraging geomorphologists to conceptualize processes in terms of conventional mechanical concepts.

The surface of the Earth is more dynamic than it usually seems to the casual observer. Firstly, the changes are often small and slow. Secondly, society and its infrastructure change much more rapidly, and the public focuses on these changes. The landscape changes at all scales and rates; small changes generally are more frequent and may have a larger net effect than large ones. The time it takes for surface activity to become noticeable varies greatly over different geological and climatic environments. In general, high rates of surface change are associated with geologically active terrains such as plate boundaries and active volcanic zones. In contrast, in intraplate areas such as much of Europe and Australia, much longer times are required for the surface to change noticeably. It follows that humans are certain to be affected from time to time by surface changes, but will be affected more and more often in some places than in others. This also explains why until recently much infrastructure was constructed on the implicit assumption that the landscape will never change.

In any area the dynamic processes that alter the Earth's surface vary greatly in intensity and frequency. Low-intensity events, which correspondingly produce slow or slight surface changes, are quite common, whereas high-intensity ones are infrequent. Most people are aware of the common low-intensity events because they have experienced them. Examples are small river floods that cause sedimentation or erosion of riverside land areas or small slippages of the land. Accordingly people naturally arrange their affairs to avoid these events.

Societal experience and therefore awareness of rare high-intensity events and their consequences are much poorer because few,

if any, people have experienced them. These events are natural and inevitable, and when they do occur they can disrupt society very seriously, causing large losses of life and assets. Such disruptions are called natural disasters. Insofar as the causative events are inevitable and people occupy much of Earth's surface, future disasters are inevitable.

Although people living in a given area may never have experienced an extreme event, evidence of their prehistorical occurrence is likely to remain in the landscape and can be recognized and interpreted by geomorphologists. This knowledge can inform society of the natural disasters it might experience in the future. It also provides opportunities for society to make informed decisions about the injury and damage it is prepared to accept in exchange for occupying land affected by these events, and about the measures it is prepared to undertake to reduce these. In particular, geomorphology can inform society about the likely effectiveness and costs of those measures.

We call this particular use of geomorphological science 'applied geomorphology'. In essence it uses predictions of future natural events derived from landscape evidence of events of the past to inform societal decisions on measures that might be taken to avoid or reduce the adverse effects of future event-triggered disasters. Scientists suggested the term 'engineering geomorphology' to address this application two decades ago. Bravard et al. (1999) proposed engineering geomorphology as an approach to manage channel erosion and bed-load transport in French gravel-bed rivers. The rationale for this approach is that river systems, especially those in populated areas, must satisfy social demands such as recreation, navigation, hydro-electric power production, and flood and erosion control. Meeting these demands requires a sound quantitative understanding of how rivers work, which is a core issue in geomorphology. Similarly, Burger et al. (1999) advocated engineering geomorphology as a merger of geomorphic and engineering techniques for identifying options and best practices for managing processes that might compromise the stability of roads and other infrastructure. Other applications concern questions of coastal cliff stability. The field of landslide research has traditionally focused on deterministic principles of slope stability, such as force and moment equilibria on potentially unstable hillslopes (Brunsden and Moore 1999; Guzzetti et al. 1999).

Commenting on other applications, Harbor (1999) noted that management of construction sites would benefit from collaboration between professional engineers and geomorphologists to minimize the environmental impacts of land disturbance. He proposed ten commandments of erosion control as a practical check list of things to consider when merging engineering and geomorphological expertise at construction sites. Coastal and ocean engineering dedicated to oil and gas production, and building and maintaining pipelines and cable routes on the seafloor also increasingly depend on detailed bathymetric data, which have shed new light on our understanding of the world's oceans (Prior and Hooper 1999). At the beginning of the twenty-first century, John Hutchinson delivered a lecture to the UK Geological Society, providing perhaps the first complete exposition of the fundamental importance of geomorphology to engineering (Hutchinson 1990).

The ten commandments of erosion control at construction sites Engineers and geomorphologists can work together to minimize unwanted environmental impacts at construction sites. Harbor (1999) emphasized this opportunity by proposing ten commandments for reducing erosion at these sites:

- Adapt the development to site conditions
- Retain as much existing vegetation as possible
- Minimize bare soil exposure by managing grading and construction timing
- Use natural and artificial cover on disturbed areas
- Divert runoff away from disturbed areas
- Minimize slope length and gradient
- Minimize runoff discharge and velocity
- Design drainageways and outlets to be non-erosive
- Trap sediment on site
- Inspect and maintain

We prefer using the term 'applied geomorphology' to emphasize that we go beyond technical or engineering solutions to problems arising from natural hazards or disasters. Particularly rare and destructive events, such as tropical cyclones, large rock avalanches, or tsunamis, have physical impact forces that may render most structural countermeasures useless to the degree that other strategies of risk reduction – such as avoidance – are preferable.

Case studies in applied geomorphology chiefly address natural hazards that produce distinct geomorphic changes. These hazards include river floods, debris flows, landslides, storms, and tsunamis that, if sufficiently large, could alter landforms and sediment fluxes for several years to several millennia. Such long-lived impacts lessen the design lives of most engineered structures such as bridges, dykes, road embankments, dams, or nuclear power plants. Above some critical threshold, however, the magnitude of such impacts will make engineering countermeasures too unreliable, too expensive, or both. Applied geomorphology provides the toolkit to estimate the frequency and potential maximum magnitude of such impacts, and their area of impact. This aim is analogous to that of the scientific discipline of palaeoseismology, which attempts to reconstruct the chronology and magnitude of earthquakes in pre-instrumental and pre-historic times. Yet applied geomorphology is equally or more concerned with modelling and predicting the consequences of an event than the probability or size of the event itself.

4.2 Aims of Applied Geomorphology

Geomorphology lies at the heart of many attempts to reduce the impacts of natural disasters. Most natural disasters have a distinct geomorphic impact, and exceptions confirm the rule. For example, large solar flares may temporarily disturb electromagnetic fields and compromise satellite systems, but leave the Earth's surface or the processes that shape it undisturbed. We also acknowledge that many other sciences contribute to reducing the impacts of natural disasters. We mention here, for example, the field of social sciences that is concerned with an effective mitigation of natural disasters by using detailed knowledge of how societal, economic, and political systems work. Here, however, we play out the strengths of geomorphology as a field in natural science, devoted to derive objectively what type, how frequent and large or intense natural processes can be in a specific location. We use geomorphology to estimate probabilities of exceedance (or nonexceedance), and to establish the size of event corresponding to these probabilities. Determining the 100-year flood is a classical example from the hydrological sciences, but has also some roots and many connections to geomorphology. The concept of a 100-year event is applicable to many natural hazards that have the potential to remobilize sediments and thus rework the Earth's surface, although some have criticised the use of this concept (Ludy and Kondolf 2012). We later describe our approach, founded in pragmatism, for integrating human factors into disaster reduction programs. We also are aware that while geomorphologists are scientists in the first place they have an obligation to act as communicators or interpreters of scientific findings.

Applied geomorphology for natural hazard assessment One of the roles of the geomorphologist is to develop and communicate information about the likely future behaviour of a landscape to engineers, decision makers, and the general public. The purpose is to independently and objectively inform societal decisions about whether, where, and how to develop infrastructure so that the risks from future natural disasters are reduced to acceptable levels. Geomorphology has at its core the toolkit to determine objectively the frequency and magnitude of sediment transport across the Earth's surfaces, and thus to quantify the hazard factor in the risk equation.

Applied geomorphologists should have a good understanding about what information the public needs, because few members of the public know enough about geomorphology to identify these needs. They must also be able to provide relevant geomorphic information in a form and manner suited for societal decision making, because it is very likely that neither the public nor the decision makers will be adequately familiar with geomorphic processes.

Further, because decision makers are politicians at some level, and commonly are elected democratically every few years by the populace, it is also necessary that the populace be provided with the same information as the politicians, again in a suitable form. Only in this way can political expediency in risk-related decision making be minimized. People need to be informed about the risks they are running, so that they can influence their political representatives with respect to their wishes.

We summarize this high-level description of applied geomorphology by outlining a procedure for acquiring geomorphic data for developing strategies to reduce the impacts of natural disasters:

(i) Develop an understanding of the medium- to long-term (century- to millennium-scale) geomorphic history of the area studied and consider the processes shaping its surrounding landscape.

(ii) Obtain information on the types, frequencies, and magnitudes of events or phases that punctuate the long-term evolution of the study area, for example tectonic uplift, earthquake deformation, river flooding and avulsion, landslides, glacier advances, or volcanic eruptions.

(iii) Measure the contribution and the possible consequences of these events or phases; for example, earthquakes might trigger landslides that could overload rivers with sediment and cause flooding.

(iv) Develop a set of realistic scenarios for the specified planning period that encompass a range of event types, magnitudes and associated probabilities.

(v) Consider also a worst-case scenario.

(vi) Work with the affected community, and its officials, to develop scenarios describing the impacts of the geomorphic scenarios on the community.

(vii) Investigate possibilities for ameliorating the consequences of these events and consider their geomorphic consequences.

4.3 The Geomorphic Footprints of Natural Disasters

In the following chapters, we review the key characteristics and geomorphic consequences of natural hazards and disasters. We intentionally diverge from the traditional approach that textbooks on natural hazards take, and refrain from reiterating, where possible, the basic concepts such as plate tectonics, meteorology or the hydrological cycle that are necessary to understand natural hazards. We refer the reader to some richly detailed textbooks on natural hazards to learn more about their underlying physical principles or societal consequences. Our motivation herein is to showcase a dedicated geomorphic viewpoint on natural hazards and disasters. Specifically, we review for each hazard:

- the key geomorphic impacts, especially those resulting from highly damaging past disasters,
- current knowledge about the use of frequency and magnitude for prediction,
- links to other hazards that may occur within a chain of events (what we call the *hazard cascade*),
- the geomorphic toolkit for reconstructing the times, size and consequences of past events,
- future projections about each hazard in a warming world.

We emphasize the dynamic linkages among different types of hazards (Figure 4.1). Such linkages are almost always hidden in natural

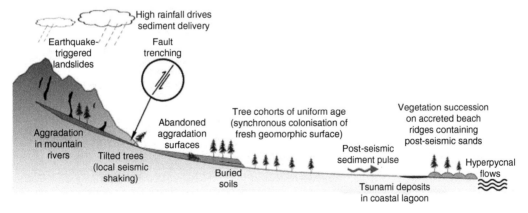

Figure 4.1 Example of a cascade of natural hazards. Earthquakes or rainstorms trigger landslides in mountains; mobilized sediment enters river channels that move the debris downstream, forming terraces, fans, and inundating floodplains, and generating high-density currents in water bodies. From Korup and Clague (2009).

disaster databases that tend to aggregate loss of life and damage, and attribute them to a single hazard type. Some scientists have used the term 'secondary' when referring to subsequent impacts, although this usage may falsely convey such impacts as minor. The more neutral term of 'consequent' hazards or impacts reflects better the sequential occurrence of these effects without any weighting or preference.

Multihazard, compound hazards, and hazard cascades When facing more than one natural hazard, we need a structured approach to account for possible interactions between each potentially damaging process. Multihazard (or multi-risk, if vulnerability and exposed elements are included) appraisals consider that various of these processes can occur in an area and interval of interest. Compound hazards (or risks) refer to those situations where two or more damaging processes occur at the same time, though without any physical connection. Finally, hazard cascades involve two or more damaging processes that are causally (and often physically or chemically) linked to each other, and thus often happen in sequence (Figure 4.2). In this book, we focus primarily on hazard cascades that involve geomorphic processes. You may find that the contents of

Figure 4.2 Hazard cascade caused by a large landslide at Mount Meager, British Columbia, in 2010. Top left: Track of the landslide; Lillooet River valley in the foreground and Capricorn Creek in the distance. Top right: Lake in Lillooet Valley impounded by the landslide. The lake overtopped the barrier within hours. Lower right: Confluence of Capricorn Creek (middle right) and Meager Creek (centre). The landslide entered the valley of Meager Creek at a high speed, ran up the opposing valley wall (note trimline, lower left) and split into two lobes that travelled up and down Meager Creek. A debris lobe deposited at the mouth of Capricorn Creek impounded a lake (middle) that had largely drained when this photo was taken. The outburst from this lake caused downstream flooding in Lillooet Valley (all photos by John Clague).

Figure 4.3 Ecological view of how a fire disturbance affects rivers, in this case based on links among drivers and responses identified in the Australian literature. Responses may be affected by the intensity or extent of fires, for example by increasing the extent and quantity of vegetation destroyed or by reducing the survival or abundance of species. Solid arrows show direct links; broken arrows indicate where one event coincides with, or increases the likelihood or intensity, of a linked event. Ovals show where effects are hypothesized. Δ means change in variable shown. Modified from Leigh et al. (2014).

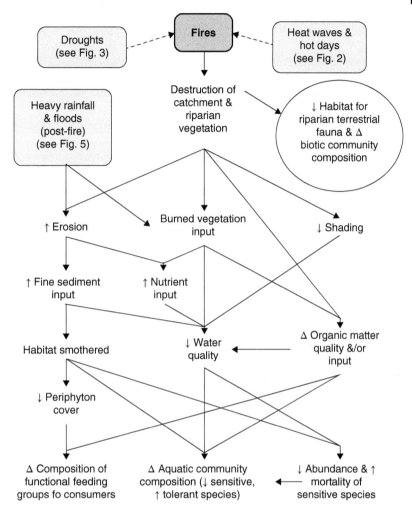

individual chapters of this book do show some overlap, and part of this overlap is intentional to show how the tectonic, climatic, geomorphic, and biological processes linked to natural hazards interact with each other (Figure 4.3).

A more diversified and integrative way of dealing with the many impacts arising from natural hazards and risks has gained increasing traction in research and application, and has led to various – and often inconsistently used – terms such as 'multihazard', 'multirisk', 'compound hazards', and 'hazard cascades'. We endorse this integrative approach because many natural disasters involve several types of adverse impacts wrought by fundamentally different, although physically and chemically coupled processes (Figure 4.4) (Gill and Malamud 2014). It is rare that a single process is responsible for a natural disaster. For example, earthquakes, volcanic eruptions, and storms may cause landslides, floods or tsunamis. Another characteristic of hazard cascades is that the consequences of a local or regional disaster can carry onward and affect regions far away and later in time. The selection of hazard cascades that we highlight in subsequent chapters only emphasizes that many more combinations of processes are possible, and some linkages or feedbacks probably await

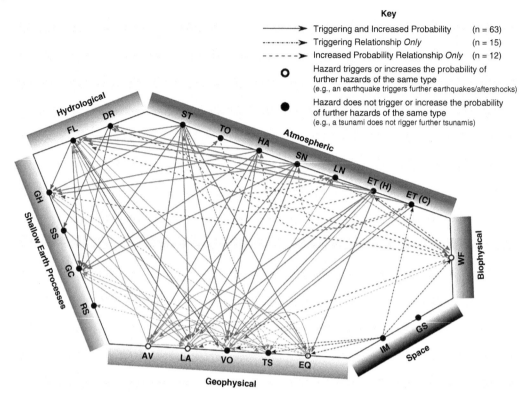

Figure 4.4 Network diagram showing potential hazard type linkages among natural hazards reported in the scientific literature: EQ = earthquake, TS = tsunami, VO = volcanic eruption, LA = landslide, AV = snow avalanche, RS = regional subsidence, GC = ground collapse, SS = soil (local) subsidence, GH = ground heave, FL = flood, DR = drought, ST = storm, TO =tornado, HA = hailstorm, SN = snowstorm, LN = lightning, ET (H) = extreme high temperatures, ET (C) = extreme cold temperatures, WF = wildfires, GS = geomagnetic storms, and IM = impact events. Line patterns show cases where both triggering and increased probability are possible (solid), cases where only a triggering relationship is possible (dash-dotted), and cases where only an increased probability relationship is possible (dashed). The node is hollow where a hazard might trigger or increase the probability of further hazards of the same type (e.g. earthquakes). From Gill and Malamud (2014).

detailed study or even discovery. Some of the hazard cascades that we examine in detail in this book are:

Earthquakes can set off a chain of events including surface deformation, tsunamis, landslides, or liquefaction that might impact landscapes for years to millennia, culminating in a seismic hazard cascade.

Volcanic eruptions involve processes that affect the Earth's atmosphere, its surface, and its water bodies by introducing large amounts of volcaniclastic materials that affect the ambient sediment and biogeochemical cycles.

Storms cause floods, storm surges, coastal erosion, and landslides, which themselves can initiate unwanted geomorphic and ecological changes that render landscapes unstable for years to decades.

Climate change is a key driver of dynamic processes in the atmosphere, hydrosphere, and cryosphere, influencing dust storms, droughts, wildfires, mass balances of ice sheets and glaciers, sea-level change, permafrost degradation, slope stability, and a host of geomorphic consequences.

Na-tech disasters are those involving both natural and human processes, in which

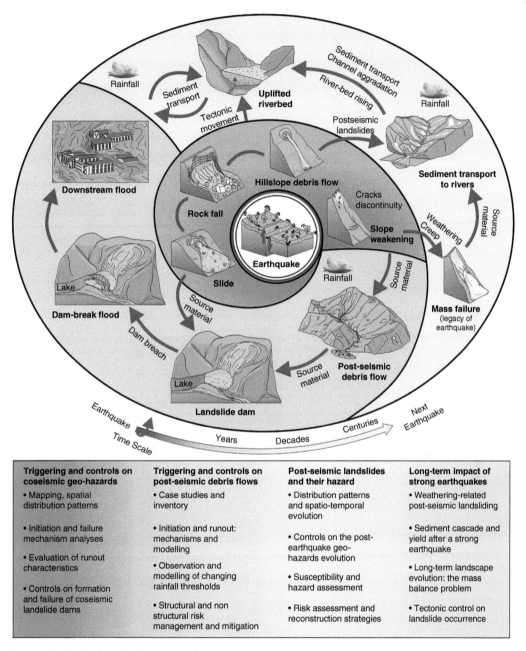

Figure 4.5 Chain of geologic hazards triggered by a strong continental earthquake showing causal relations among hazards. Red background shows different types of coseismic landslides; blue background indicates the postseismic cascade of hazards in days to years; and yellow background represents the long-term impact of an earthquake, years to decades later, and perhaps even longer. From Fan et al. (2019).

natural event rates or intensities trigger a chain of events that is amplified by land-use practices or built structures. Examples include the transport of human waste by tsunamis, the rainstorm-driven dispersal of artificial radionuclides in rivers, or the failure of hydropower dams following earthquakes.

Most of these hazard cascades share similar characteristics or sequences of events with others, and clear distinctions may be impossible (Figures 4.4 and 4.5). We may wish to see natural disasters as the consequences of coupled interacting processes in the Earth system, and a more holistic treatment of these phenomena may be the way forward. For each of these hazard cascades, we summarize how geomorphologists and related scientific disciplines measure the impacts of such extreme natural events on landscapes. The quantitative metrics are derived from what we refer to as the geomorphic footprints of natural disasters. We showcase studies that provide an overview of different geomorphic footprints, while highlighting ways of using landforms and sedimentary records along with other proxies to reconstruct the times, frequency and magnitude of natural disturbances and disasters (Table 4.1).

We also examine the rates and longevity of the geomorphic responses to natural disasters. Determining these important characteristics is essential where different processes overlap, and hazard impact timing and duration are important for societal planning to reduce impacts. The good news is that bookshelves (or servers) full of case studies demonstrate the ways in which many of these geomorphic disturbances adversely affect human land use, causing problems that may last for years or even centuries. Such lag effects commonly are ignored when calculating disaster losses, yet occasionally they are responsible for the lion's share of losses. We posit that applied geomorphology can be effectively used in reducing natural hazard impacts if researchers and practitioners are aware of the many possible geomorphic footprints of natural disasters.

Geomorphologists prefer to express and measure geomorphic footprints as changes to the land surface or as commensurate disturbances to the sediment budget, which balance the volumes and fluxes of erosion, sediment transport, and deposition in an area of interest (Hinderer 2012). A common approach is to compare rates of geomorphic activity before and after natural disturbances. This means that we need a standard set of metrics of magnitude and frequency that we can compare against each other (Table 4.2).

Comparing average rates of erosion, transport or sedimentation over different periods means that we have to include times of low or zero geomorphic activity, which gives rise to mean rates that decrease with increasing observation periods (Gardner et al. 1987). Short-term geomorphic response to disturbances such as earthquakes or volcanic eruptions may involve very high rates, but rates decrease to much lower background values over the long term.

4.4 Examples of Hazard Cascades

We offer some examples of natural hazard cascades that emphasize that it is important to identify both the underlying cause and resulting effects of landscape dynamics through geomorphic analysis (Figure 4.6). Yet simply carrying out the science fulfils only part of natural hazard assessments – the results must find their way effectively into the societal processes of politics and decision making. We contend that it is one of the roles of geomorphologists to make their insights easily available and understandable to the public, officials, and decision makers.

4.4.1 Megathrust Earthquakes, Cascadia Subduction Zone

The west coast of North America over a distance of 1100 km between northern California and central Vancouver Island is located along the Cascadia subduction zone, one of the many subduction zones that ring the Pacific Ocean. Here the oceanic Juan de Fuca plate is moving at a rate of about 45 mm yr^{-1} beneath the continental North American plate. Subduction earthquakes occur along the megathrust fault that separates the two plates (Clague 1997). During each of these earthquakes, slip of up

Table 4.1 Selected geomorphic and biogeochemical impacts associated with natural hazards and disasters.

Hazard	Geomorphic impacts	Biogeochemical impacts
Earthquakes	Fault-scarp diffusion, surface uplift and subsidence,	Changes in groundwater composition,
	soil liquefaction, landslides, river diversions,	streamflow changes
	tsunami, turbidity currents, post- and interseismic	
	surface deformation	
Volcanic eruptions		
effusive	Crater formation, lava flows	
explosive	Crater formation, pyroclastic flows	Emission of SO_2, CO_2, HF, volcanic smog
post-/intereruption	Volcanic flank collapse, debris avalanches, lahars,	
	jökulhlaups (subglacial volcanoes only)	
Landslides	Lowering of steep slopes and interfluves,	Biomass erosion and burial,
	blocking of rivers (landslide dams),	alteration of habitats
	catastrophic dam-break floods, sediment pulses,	
	coarsening or fining of ambient river sediment	
Land subsidence	Karst and other solution phenomena,	
	sinkhole collapse, flooding	
Windstorms	Sand and dust storms, abrasion and deposition,	Uprooting of trees,
	storm surge, wave erosion, coastal inundation,	nutrient and pathogen transport,
	meteotsunamis	groundwater salinization
Rainstorm floods	Channel and floodplain reworking,	Nutrient and pathogen transport and dispersal
	altered channel geometry and flood frequency,	
	sediment waves, floodplain aggradation,	
	channel avulsion, bank erosion	
Tsunamis	Coastal inundation, shoreline erosion, onshore	Biomass erosion and burial
	deposition, turbidity currents	
Sea-level changes	Coastal erosion, inundation	Groundwater salinization
Permafrost degradation	Increased slope instability, thermokarst,	Release of nutrients and CH_4
	gully and coastal erosion	
Wildfires	Changes to soil infiltration capacity,	Emission of ash,
	enhanced erosion, debris flows, sediment pulses,	high charcoal loads
	channel aggradation	

Note that many of these impacts may occur as coupled processes in hazard cascades.

Table 4.2 Selected ways to measure the physical magnitude or activity of natural hazards.

Hazard	Measured quantity	Unit	Interpretation
Earthquake	Magnitude	[1]	Energy released
	Seismic moment	$[M\,L^2T^{-2}]$	Deformation times force
Volcanic	Peak intensity	$[M\,T^{-1}]$	Maximum magma discharge rate
	Volcanic explosivity index (VEI)	[1]	Erupted mass
	Volcanic vigour	$[L^{-1}]$	Number of active Holocene volcanoes
	Volcanic vigour	$[T\,L^{-1}]$	Total duration of Holocene eruptions
	Volcanic vigour	$[L^{-1}]$	Total of eruptions above a threshold
	Maximum column height	[L]	buoyancy during explosive eruptions
	Area covered by volcanic deposits	$[L^2]$	
	Thickness of volcanic deposits	[L]	
Landslides	Area	$[L^2]$	Affected area
	Volume	$[L^3]$	Potential mobility
	Runout	[L]	Maximum reach
	Mobility ($H\,L^{-1}$)	[1]	Mobility
	Velocity	$[L\,T^{-1}]$	Physical impact
Tsunamis	Magnitude		Empirical estimate (seismic trigger)
	Velocity	$[L\,T^{-1}]$	Rate of spread
	Height	[L]	
	Runup	[L]	Momentum, coastal topography
	Horizontal inundation	[L]	
Storms	Wind speed	$[L\,T^{-1}]$	Beaufort, Saffir–Simpson, Fujita scales
	Accumlated cyclone energy (ACE)	$[L^2\,T^{-2}]$	Seasonal activity
	Rainfall amount	[L]	
	Rainfall duration	[T]	
	Rainfall intensity	$[L\,T^{-1}]$	
	Lightning frequency	$[L^{-2}\,T^{-1}]$	Thunderstorm activity
Floods	Discharge	$[L^3\,T^{-1}]$	
	Volume	$[L^3]$	
	Stage	[L]	Local flood level
	Duration	[T]	
	Stream power	$[M\,L\,T^{-1}]$	Energy expenditure
	Avulsion frequency	$[T^{-1}]$	Lateral channel mobility
Sedimentary	Sediment concentration	$[M\,L^{-3}]$	
	Sediment flux	$[M\,T^{-1}]$	
	Sediment yield	$[M\,L^{-2}\,T^{-1}]$	
	Volume eroded/deposited	$[L^3]$	
	Erosion/sedimentation rate	$[L\,T^{-1}]$	Average efficacy
Drought	Palmer drought severity index (PDSI)	[1]	
Wildfires	Fireline intensity	$[M\,L\,T^{-1}]$	Energy released
	Flame height	[L]	
	Area burned	$[L^2]$	Fire size
Snow and Ice	Volume	$[L^3]$	
	Rate of volume gain or loss	$[L\,T^{-1}]$	Glacier or permafrost mass balance, etc.
Sea-level change	Rate of sea-level rise/fall	$[L\,T^{-1}]$	Rapidity of change

Note that the term *intensity* mostly refers to the observed damage from a natural process.

Figure 4.6 Coupled geomorphic processes in bedrock landscapes. Left: Glacier lake within a bedrock hollow may attenuate glacial runoff and sediment, but can empty rapidly upon being hit an avalanche. Middle: The cutting of bedrock gorge involves river incision, side-wall undercutting and rockfall; all these processes feed back with one another. Right: Shallow soil landslide scar, an example of an erosion process that removes vegetation and root mats rather than bedrock, thus altering the local weathering conditions of the bedrock (all photos by Oliver Korup).

to a few tens of metres takes place over a large part of the fault plane. No such earthquake has been documented since the first European exploration of the region, though geological evidence points to a large earthquake in AD 1700 and for 19 others in the past 10 000 years.

The geological evidence for these very large earthquakes is of three types. Firstly, stratigraphic evidence in numerous coastal wetlands along the entire length of the subduction zone indicates repeated sudden coseismic subsidence. Specifically, during each earthquake, the vegetated surface of each coastal marsh is instantly lowered up to 2 m into the intertidal zone, where it then blanketed by tidal muds in the days, weeks, and months that follow. Secondly, layers of sand, which directly overlie the subsided marsh surfaces, record deposition by landward-surging tsunami waves. These tsunamis are an expected effect of subduction zone earthquakes because the sea floor some distance away from the coast is rapidly elevated by slip along the megathrust fault. Thirdly, strong ground motions during these earthquakes induce liquefaction of loose, water-saturated coastal sediments, producing sand dykes and sand blows or sand volcanoes that may be preserved in the geologic record.

The most recent of the great Cascadia earthquakes is precisely dated by Japanese accounts of an 'orphan tsunami' in AD 1700 (Atwater et al. 2015). This tsunami was unaccompanied by a local earthquake and struck a 1000-km length of the east coast of Honshu, Japan, with waves up to several metres high, damaging some towns. Japanese researchers concluded that the tsunami was triggered by the same earthquake that had caused the land to drop on the west coast of North America. They back-calculated the travel time of the tsunami to Japan (some nine hours), factored in the change in time at the International Date Line, and suggested that the earthquake occurred at about 9 pm Pacific Standard Time on 26 January 1700. Support for this age assignment is provided by tree-ring ages on the death dates of 'ghost cedars', standing trees that died when their roots were lowered by the last great earthquake from the forest edge into intertidal waters in a wetland in Washington (Figure 4.7). The trees remained standing, but their bases were soon buried by 1–2 m of tidal mud. By matching the ring patterns of the standing dead cedars with those of old-growth living trees in the area, scientists were able to date tree death to within about a two-year period centred on AD 1700.

Unfortunately, older great Cascadia earthquakes are much less precisely dated because they happened before any historical records were kept and because they are older than any living trees suitable for dendrochronological dating. The ages of these older earthquakes have been determined by radiocarbon dating, but calibrated radiocarbon ages have uncertainties of up to 100 years or more. Both

Figure 4.7 Left: Ghostly stems of conifers in a tidal marsh at the mouth of Copalis River, Washington. The trees were killed soon after the last great earthquake at the Cascadia subduction zone in AD 1700. They subsided 1–2 m, from a position at the forest edge into the intertidal zone, where they were killed by brackish tidal waters. The roots and lower part of the tree stems are buried beneath tidal mud that was deposited after the earthquake (Brian Atwater). Top: Peat of an AD 1700 tidal marsh abruptly overlain by tsunami sand (thin-bedded grey layer) and tidal mud, in a tidal marsh in southwest Washington. This stratigraphy is the characteristic signature of a great earthquake at the Cascadia subduction zone (Brian Atwater). Right: A stacked sequence of buried tidal marsh peats at Willapa Bay, Washington, recording a sequence of late Holocene subduction zone earthquakes (Brian Atwater).

turbidite stacks offshore and sediments from the floors of landslide-dammed lakes in the Olympic Mountains of Washington record some of the latest megathrust earthquakes (Figure 4.8) (Leithold et al. 2017). Yet further evidence of earthquake-triggered hillslope instability is mostly elusive in the coastal mountains along the subduction zone, except for a few active landslides that may have started moving in the last earthquake (Schulz et al. 2012).

4.4.2 Postseismic River Aggradation, Southwest New Zealand

The Alpine Fault in southwest New Zealand is another major plate boundary that separates the Australian from the Pacific continental plates. About three to four $M \sim 8$ earthquakes have occurred along the Alpine Fault per millennium. The fault also marks the abrupt western flank of the Southern Alps. Narrow coastal lowlands are fed by rivers draining these rapidly uplifting and eroding mountains and depositing outwash fans. Earthquake-triggered landslides are one prime candidate for delivering episodic pulses of sediment to these fans. Many fans have buried soils and woody debris that are a few hundred years old, indicating that sudden aggradation gave way to more stable periods during which soils could develop (Davies and Korup 2007). At one site along the Whataroa River, about 12 km from the sea, river bank erosion has exhumed a buried forest at a depth of ~5 m below the present ground surface. The site

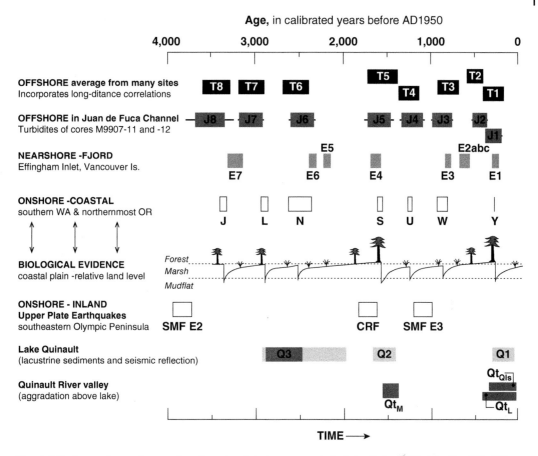

Figure 4.8 Comparisons of ages of earthquake disturbance events in Lake Quinalt, Washington (Q1–Q3) and regional paleoseismic history events derived from offshore and onshore datasets in the US Pacific Northwest. Offshore data include deep-sea turbidite records, both as an average of many sites along the Cascadia margin (T events) and from just the Juan de Fuca submarine channels (J events), as well as from Effingham Inlet on the west coast of Vancouver Island (E events). Onshore records include buried soils in coastal marsh deposits that record coseismic subsidence (events L–Y) and trenching of upper-plate fault scarps along the Saddle Mountain (SMF) and Canyon River (CRF) faults. Episodes of fluvial valley-bottom aggradation in the upper Quinalt River valley are denoted as either middle (Qt_M) or lower (Qt_L) alluvial terraces or landslide terrace deposits (Qt_{Qls}). From Leithold et al. (2018).

is some 13 km from the range front, and the trees have been radiocarbon-dated to 1650 AD, indicating that this part of the Whataroa fan (which has a total area of ~100 km²) has aggraded by several metres since then.

A rock avalanche that fell from Mount Adams into the nearby Poerua River in 1999 was the closest historical event to a natural simulation of massive aggradation following future earthquakes on the Alpine Fault. The landslide formed a temporary dam in the Poerua gorge upstream of the fault. The dam failed about one week later, supplying several

million cubic metres of sediment to the river, which carried it down the valley. The material input caused the river to aggrade and avulse across the valley floor. Aggradation at the range front peaked at about 15 m a decade after the landslide; since then the river has incised again into the outburst deposits. If widespread in the aftermath of a major earthquake, such rapid channel changes would put many farms, businesses, roads, and settlements at risk.

The volume of the Mount Adams rock avalanche was about 10 Mm³, but that involved in the aggradation of the lower Whataroa River

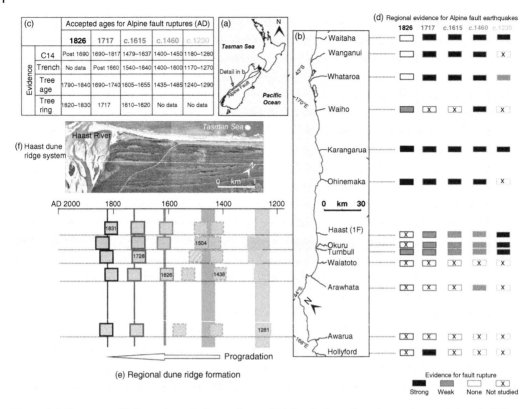

Figure 4.9 Summary of information relating to times of Alpine fault earthquakes and coastal dune building, South Westland, New Zealand. (a) Map of New Zealand, outlining the Alpine fault and South Westland. (b) Map of South Westland, showing major river catchments. (c) Summary of published data on times of Alpine fault earthquakes. Five earthquakes are recognized in the past 900 years, based on four sources: ruptures in fault trenches dated by radiocarbon (Trench), radiocarbon-dated landslides, and aggradation terraces (^{14}C), forest cohorts that regenerated after disturbance events (Tree age), and tree-ring growth anomalies in trees growing near the fault (Tree ring). Events are color-coded. (d) Summary of published evidence for Alpine fault earthquake impacts within South Westland river catchments for each of the five events. (e) Summary of times of stabilization of newly formed dune units at five river mouths. Boxes show the time period over which dune stabilization most likely took place. Vertical shaded columns show date ranges for Alpine fault earthquakes based on the most precise data from (c). (f) Aerial photo of the Haast River dune sequence. Source: Wells and Goff (2007).

in the seventeenth century was about 300 Mm^3 (Berryman et al. 2009). This volume is equivalent to an average of some 0.7 m of erosion over the entire 450-km^2 mountain catchment. A likely source for this material is that, during an earthquake in the Southern Alps in 1620 (Briggs et al. 2018), a large landslide or many small landslides happened in the Whataroa catchment, providing the sediment input that caused the extensive aggradation of the lowland. Aggradation on this scale is likely to disrupt infrastructure in the area, and thus

mobilized sediments may continue to shift for years to decades (Davies and Korup 2007). Trees on parallel beach ridges near the mouths of some of the major South Westland rivers have established as a result of past Alpine Fault earthquakes (Figure 4.9) (Wells and Goff 2007). These beach ridges appear to have formed several years to decades after major earthquakes, indicating that it could take that much time for the earthquake-generated sediment to move from the mountains to the coastline. Turbidite layers in several lowland lakes support and extend the history of past Alpine Fault earthquakes and testify how they mobilise large amounts of sediment (Howarth et al. 2016). The current estimate from these combined proxies is that fault ruptures on the southern onshore portion of the Alpine Fault happened about every 290 years during the past 7000 years on average, with the last event dated to AD 1717 (Cochran et al. 2017).

4.4.3 Explosive Eruptions and their Geomorphic Aftermath, Southern Volcanic Zone, Chile

Volcanoes along subduction zones such as that off the west coast of Chile are sources of another type of hazard cascade. There, explosive volcanic eruptions episodically deliver cubic kilometres of airborne tephra material over large areas of the Andes, their eastern foreland, and out to the Atlantic Ocean over a period of days to months. More than 50 eruptions, each involving at least 10 Mt of pyroclastic debris, have occurred in Chile's Southern Volcanic Zone in postglacial times (Watt et al. 2011). Pyroclastic flows from collapsing eruption columns blocked rivers and added more lakes to the myriad postglacial ones that formed after the Pleistocene Patagonian ice sheet had melted. High tephra loads have repeatedly impacted both the fjord landscape and its dense temperate rainforests by mantling hillslopes and river channels with tephra. Depending on their thickness and chemical composition, the fallout products can smother or poison vegetation, bury or even sterilize soils, and rapidly change the rates at which water infiltrates and runs off the land surface. Rain falling onto these bare and unconsolidated surfaces produces lahars that rework the highly mobile tephra cover and flush it into the drainage network. Some of the highest reported sediment yields in rivers were fed by reworked volcaniclastic materials, whereas stream and floodplain habitats downwind of the eruptions can suffer from the ash fall and dust storms

(Wilson et al. 2010; Pierson and Major 2014). Overloaded rivers catastrophically raised and shifted their channel beds, covering floodplains and terraces, and damming tributaries to form lakes (Figure 4.10). Sediment yields may remain elevated for years to decades. An important element in this volcanic hazard cascade involves feedbacks of erosion and sedimentation with the widespread stands of rainforest. Studies in the field of disturbance ecology have demonstrated that tephra can have a major impact on forests (Swanson et al. 2013). Damaged or removed tree cover can enhance stream flow and sediment transport (Birkinshaw et al. 2010), and also promote landslides in steep terrain, thus paving the way for more erosion once the anchoring and cohesive effects of the root network are lost. This decay usually takes several years to more than a decade, so that enhanced sediment and biomass inputs to rivers by posteruptive landslides may keep fluvial transport rates high long after the eruption (Korup et al. 2019). The eroded rainforests deliver large amounts of dead logs to the rivers, where they form nests and hydraulic flow obstacles that slowly migrate downstream during floods, and promote repeated channel instabilities.

4.4.4 Hotter Droughts Promote Less Stable Landscapes, Western United States

Earthquakes or volcanic eruptions aside, weather and climate can also trigger a number of hazard cascades. Contemporary atmospheric warming, for example, is projected to increase weather extremes in many regions, including the western United States. More pronounced and hotter droughts have become more than a scenario, as time series show an increasing 'browning' of vegetation together with a decrease in the greenness of forest cover (Millar and Stephenson 2015). These changes are monitored by satellite and reflect increasing mortality of trees in the region, most likely in response to drier conditions. At the same

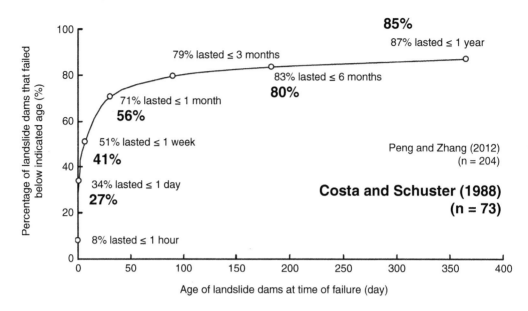

Figure 4.10 Cumulative distribution of the time of failure of landslide dams based on a global data set of 204 cases. A comparison of the two data sets indicates that the reported percentages are consistent for longer time intervals, but not for short ones where time for mitigation options may be limited. From Korup and Wang (2014).

time, insect infestations by species such as bark beetles have been increasing, taking advantage of the weakened trees and warmer climate. Another consequence in this hazard cascade is the increase in wildfires, fuelled by the higher availability of drier wood and hotter weather. The fires remove forest stands and their undergrowth, leaving bare soils vulnerable to erosion processes such as dry ravel, rilling, and gullying. Under sustained drought, wind erosion can increase and carry away topsoil and mobilize dunes (Munson et al. 2011). Rare rainstorms can deliver high-intensity runoff, as some burned soils have reduced water infiltration capacities, so that streamflows can increase together with the hazard of debris flows, channel aggradation, and poor water quality. Peak sediment yields frequently occur in the first years following fires and may add to the problem of flashy discharge in the semiarid parts of the western United States (Goode et al. 2012). This hazard cascade emphasizes again some of the many connections between natural hazards, geomorphology, and the biosphere that we need to understand and quantify if we wish to predict future changes to landscapes and ecosystems. Quantitative risk appraisals and ecological restoration depend on knowing in detail about feedbacks between biotic and abiotic processes.

References

Atwater BF, Musumi-Rokkaku S, Satake K, et al. 2015 *The Orphan Tsunami of 1700: Japanese Clues to a Parent Earthquake in North America*. University of Washington Press.

Berryman K, Almond P, Villamor P, et al. 2009 Alpine Fault ruptures shape the geomorphology of Westland, New Zealand: Data from the Whataroa catchment. Geomorphology

2009, ANZIAG 7th International Conference on Geomorphology, Melbourne.

Birkinshaw SJ, Bathurst JC, Iroumé A, and Palacios H 2010 The effect of forest cover on peak flow and sediment discharge-an integrated field and modelling study in central-southern Chile. *Hydrological Processes* **25**(8), 1284–1297.

Bravard JP, Landon N, Peiry JL, and Piégay H 1999 Principles of engineering geomorphology for managing channel erosion and bedload transport, examples from French rivers. *Geomorphology* **31**(1), 291–311.

Briggs J, Robinson T, and Davies T 2018 Investigating the source of the c. AD 1620 West Coast earthquake: implications for seismic hazards. *New Zealand Journal of Geology and Geophysics* **61**(3), 376–388.

Brunsden D and Moore R 1999 Engineering geomorphology on the coast: lessons from west Dorset. *Geomorphology* **31**(1), 391–409.

Burger KC, Degenhardt JJ, and Giardino JR 1999 Engineering geomorphology of rock glaciers. *Geomorphology* **31**(1), 93–132.

Clague JJ 1997 Evidence for large earthquakes at the Cascadia subduction zone. *Reviews of Geophysics* **35**(4), 439–460.

Cochran UA, Clark KJ, Howarth JD, et al. 2017 A plate boundary earthquake record from a wetland adjacent to the Alpine fault in New Zealand refines hazard estimates. *Earth and Planetary Science Letters* **464**, 175–188.

Davies T 2015 Developing resilience to naturally triggered disasters. *Environment Systems and Decisions* **35**, 237–251.

Davies TRH and Korup O 2007 Persistent alluvial fanhead trenching resulting from large, infrequent sediment inputs. *Earth Surface Processes and Landforms* **32**(5), 725–742.

Fan X, Scaringi G, Korup O, et al. 2019 Earthquake-induced chains of geologic hazards: Patterns, mechanisms, and impacts. *Reviews of Geophysics* **57**, 421–503.

Gardner TW, Jorgensen DW, Shuman C, and Lemieux CR 1987 Geomorphic and tectonic process rates: Effects of measured time interval. *Geology* **15**, 259–261.

Gill JC and Malamud BD 2014 Reviewing and visualizing the interactions of natural hazards. *Reviews of Geophysics* **52**, 680–722.

Goode JR, Luce CH, and Buffington JM 2012 Enhanced sediment delivery in a changing climate in semi-arid mountain basins: Implications for water resource management and aquatic habitat in the northern Rocky Mountains. *Geomorphology* **139–140**(C), 1–15.

Guzzetti F, Carrara A, Cardinali M, and Reichenbach P 1999 Landslide hazard evaluation: a review of current techniques and their application in a multi-scale study, Central Italy. *Geomorphology* **31**(1), 181–216.

Harbor J 1999 Engineering geomorphology at the cutting edge of land disturbance: erosion and sediment control on construction sites. *Geomorphology* **31**(1), 247–263.

Hinderer M 2012 From gullies to mountain belts: A review of sediment budgets at various scales. *Sedimentary Geology* **280**(C), 21–59.

Howarth JD, Fitzsimons SJ, Norris RJ, et al. 2016 A 2000 yr rupture history for the Alpine fault derived from Lake Ellery, South Island, New Zealand. *Geological Society of America Bulletin* **128**(3-4), 627–643.

Hutchinson JN 1990 The Fourth Glossop Lecture: Reading the ground: Morphology and geology in site appraisal. *Quarterly Journal of Engineering Geology and Hydrogeology* **34**(1), 7–50.

Korup O and Clague JJ 2009 Natural hazards, extreme events, and mountain topography. *Quaternary Science Reviews* **28**, 977–990.

Korup O and Wang G 2014 Mutliple landslide-damming episodes. In Davies TR [ed] *Landslide Hazards, Risks, and Disasters*, Elsevier, pp. 241–261.

Korup O, Seidemann J, and Mohr CH 2019 Increased landslide activity on forested hillslopes following two recent volcanic eruptions in Chile. *Nature Geoscience* **12**, 284–289.

Leigh C, Bush A, Harrison ET, et al. 2014 Ecological effects of extreme climatic events on riverine ecosystems: Insights from Australia. *Freshwater Biology* **60**, 2620–2638.

Leithold EL, Wegmann KW, Bohnenstiehl DR, et al., 2017 Slope failures within and upstream of Lake Quinault, Washington, as uneven responses to Holocene earthquakes along the Cascadia subduction zone. *Quaternary Research* **89**, 178–200.

Leithold EL, Wegmann KW, Bohnenstiehl DR, et al. 2017 Slope failures within and upstream of Lake Quinault, Washington, as uneven responses to Holocene earthquakes along the Cascadia subduction zone. *Quaternary Research* **89**, 178–200.

Ludy J and Kondolf GM 2012 Flood risk perception in lands "protected" by 100-year levees. *Natural Hazards* **61**(2), 829–842.

Millar CI and Stephenson NL 2015 Temperate forest health in an era of emerging megadisturbance. *Science* **349**, 823–826.

Munson SM, Belnap J, and Okin GS 2011 Responses of wind erosion to climate-induced vegetation changes on the Colorado Plateau. *Proceedings of the National Academy of Sciences* **8**(10), 3854–3859.

Pierson TC and Major JJ 2014 Hydrogeomorphic effects of explosive volcanic eruptions on drainage basins. *Annual Review of Earth and Planetary Sciences* **42**(1), 469–507.

Prior DB and Hooper JR 1999 Sea floor engineering geomorphology: recent achievements and future directions. *Geomorphology* **31**(1), 411–439.

Schulz WH, Galloway SL, and Higgins JD 2012 Evidence for earthquake triggering of large landslides in coastal Oregon, USA. *Geomorphology* **141–142**(C), 88–98.

Swanson FJ, Jones JA, Crisafulli CM, and Lara A 2013 Effects of volcanic and hydrologic processes on forest vegetation: Chaitén Volcano, Chile. *Andean Geology* **40**(2), 359–391.

Watt SFL, Pyle DM, Naranjo JA, et al. 2011 Holocene tephrochronology of the Hualaihue region (Andean southern volcanic zone, 42 ° S), southern Chile. *Quaternary International* **246**(1-2), 324–343.

Wells A and Goff J 2007 Coastal dunes in Westland, New Zealand, provide a record of paleoseismic activity on the Alpine fault. *Geology* **35**, 701–704.

Wilson TM, Cole JW, Stewart C, et al. 2010 Ash storms: impacts of wind-remobilised volcanic ash on rural communities and agriculture following the 1991 Hudson eruption, southern Patagonia, Chile. *Bulletin of Volcanology* **73**(3), 223–239.

5

Tools for Predicting Natural Hazards

Examining and interpreting the geomorphic and sedimentary legacy of natural processes and disasters forms the core toolkit of applied geomorphology. Part of this toolkit borrows methods from other disciplines in the geosciences and beyond. In this chapter we discuss ways of making use of this past evidence for anticipating and, ideally, predicting the nature and occurrence of natural hazard events, their impacts and associated risks. We discuss in some detail the nature of predictions and their underlying models, which are essential for making informed statements about natural phenomena. In particular, we show how the concept of probability is relevant to modern natural hazard and risk assessments. Technical risk assessments increasingly require, if not rely on, such probabilistic input. Yet this information needs to reach individual communities in an understandable form that allows them to make decisions about how react to these risks accordingly.

5.1 The Art of Prediction

The long list of losses due to natural disasters underpins at least three reasons that motivate scientists and practitioners to predict natural hazards:

Firstly, it is axiomatic that the goal of disaster risk management is to reduce damage and injury from natural processes. Revisiting Eq (2.1), we see that to reduce risk, we have the choice of reducing at least one of the factors in the risk equation – hazard H, vulnerability V, the elements at risk E, or risk aversion A. Any changes in all other factors must remain negligible to ensure that the net result is a risk that is lower than before the risk reduction effort. How can we lower risk most effectively? Risk reduction may involve addressing vulnerability V by better informing us about how to act before or during a natural disaster, so that we are less prone to be badly affected. Knowing how to react and what to do during a strong earthquake or during an approaching tsunami may save your life. Appropriately placed warning signals or signs with evacuation routes can be a cheap but effective way of reducing your vulnerability in this situation. One outcome of prediction is that it may lower our risk aversion A by raising our awareness of a given problem. The assumption here is that more knowledge leads to less aversion. This assumption, however, depends on the original risk aversion. Someone who was risk-affine (which is the opposite of averse) might become more risk-averse when the gravity of the situation becomes clear. On the other hand, with sufficient advance notice of a pending flood, for example, people will be able to evacuate. Of course, being confronted with predictions about natural hazards can also achieve the unwanted opposite effect and make people feel overconfident. This can be the case, for example, if a hazard model predicts very small probabilities of future disasters that

Geomorphology and Natural Hazards: Understanding Landscape Change for Disaster Mitigation, Advanced Textbook Series,
First Edition. Tim R. Davies, Oliver Korup, and John J. Clague.

nevertheless occur, or if seemingly infallible warning systems make people oblivious about a particular hazard. It is difficult to imagine what an annual exceedance probability of 10^{-6} really means. In such cases a higher risk aversion will, by definition, increase the overall perceived risk. We revisit the effects of these subjectively distorted views of real hazards and risk, also known as cognitive biases, in Section 5.4.1.

Secondly, we believe that making useful predictions about natural processes is an ultimate goal of scientific research. Countless observations and controlled experiments have amassed a broad body of knowledge from which we hope to first distill, and then integrate, the most fundamental principles of the physical and chemical processes that gradually shape – or suddenly disturb – landscapes. Even without a full understanding of every single process and feedback, we can offer simplified but realistic abstractions of nature to anticipate the rate and effects of many processes. This knowledge reduces our degree of surprise and makes us better prepared when we encounter such processes, and motivates us to do something to reduce their possibly negative consequences. Being able to predict a storm or a flood can be independent of a thorough physical understanding of the atmospheric and hydrological processes. Instead we can rely on knowledge (but a minimal understanding) of appropriate warning signs. Many people will likely pack an umbrella or raincoat when they go outside on a gloomy day. Most of us will seek shelter if caught out in the rain, simply because we have learned that we will become uncomfortable or might catch a cold after exposure to the wet and cold. We all make these and other risk-related decisions in our everyday lives without explicitly knowing or even worrying about the physical, mathematical. or statistical background. Sometimes we are right about our decision to bring a raincoat, and sometimes we are not; however, the consequences of a conservative decision are usually less serious than those of leaving the umbrella at home when it does rain. In order to anticipate and quantify risks we need to couple modelled predictions with functions of resulting loss (or utility, which is negative loss). With the toolkit of physics, mathematics, and statistics are we able to predict in a reproducible manner the timing and location of rain, and its intensity, duration, and likely impact. The flip-side is that such predictions need to be reliable, which is what really matters to those liable to be affected.

Thirdly, being able to predict natural processes satisfies our curiosity, and affects how we grow and how we experience and interact with our environment. Learning about how nature works is intrinsic to human evolution, so that we can hope to improve our predictions with an increasing body of evidence and knowledge.

In predicting natural hazard events, the simple questions are often the most demanding to answer: specific targeted questions are easier to answer. For example, asking when the next earthquake will happen may sound reasonable, but may also seem impossible to answer. Modern digital seismometers record several thousands of earthquakes per day, so a straightforward and correct reply is 'very soon' or 'just now'. If the question is when will the next earthquake occur in a given location then the only possible (but not particularly useful) answer is 'in the future'. Most of the thousands of daily earthquakes, however, are imperceptibly weak, thus we need to reformulate the question – What is the likelihood of the next earthquake of a particular magnitude? We need to ask this as the observed frequency of earthquakes depends on their magnitude. We may further refine our question by asking when a particular geographic region, say the San Francisco area, will experience a M >7 earthquake? We could also ask about the probability of at least one earthquake of this size occurring within the next ten years and so on. Similar questions that scientists attempt to answer with predictive models include:

- How many hours does a tsunami triggered by an M~8 earthquake offshore of Chile travel before reaching the coastline of Japan?

- How long will mountain roads in Austria have to be closed because of avalanches in a winter with average snowfall?
- What is the maximum expected storm-surge height of a Category V hurricane in the Gulf of Mexico?
- What is the discharge of the 100-year flood in an ungauged headwater catchment in Papua New Guinea?
- How likely is an elderly person to die from heatstroke in Berlin during an exceptionally hot summer?

Answering these more specific questions is often easier than it is for more general ones. Predictive models break down these questions by estimating the occurrence of potentially harmful processes. More detailed models attempt to capture changes or rates of change in, for example, seismicity or climate that control the frequency and magnitude of natural hazards. In the following, we take a look at the basic types of models that scientists use in research and practice today.

About prediction Much of science involves prediction. Scientists conduct experiments under controlled and reproducible conditions to learn more about natural phenomena. The ultimate aim of many of these experiments is to improve our ability to predict their outcomes. Prediction goes beyond making statements about future outcomes. Prediction can also mean making informed statements about unseen outcomes that have occurred in the past, for example in an unmonitored part of a study area. Defining prediction this way is an important point in many statistical models and is also central to geoscience, where space is often substituted for time. We may argue that natural hazards by definition refer only to possible future harmful impacts. Nonetheless, reconstructing the frequency of past events generates essential input for estimating the occurrence of future events.

The terms "forecasting" and "hindcasting" are related to prediction. Depending on their discipline, scientists have debated the difference between predicting and forecasting. In seismology, for example, being able to predict an earthquake successfully means knowing its exact magnitude, location, and time in advance, something that is currently intractable. In contrast, forecasting an earthquake means that we issue a probability of its future occurrence over a specified time and region. Every forecast is a prediction, whereas some predictions cannot be considered forecasts. This is because some predictions can also work without probabilities.

Finally, the term 'projection' refers to a forecast that depends on a predefined scenario and thus is based on conditional probabilities. The IPCC's scenarios of climate change fall into this category and offer projections of future change, assuming, for example, a prior change in the rates of carbon dioxide emissions together with a certain political strategy of dealing with those emissions.

We might think of prediction as one of the ultimate goals of science. We wish to understand the processes responsible for generating natural hazards at a level of confidence that allows us to make useful and informed quantitative statements about unobserved events. These unobserved events can reside in the future or in the past, or lie simply outside the scope of our study: for example, they can happen right now though beyond the range of our instruments.

Predicting that a natural disaster might happen differs from predicting its potential impact. Seasonal forecasts of droughts or floods linked to the ENSO are now routinely issued with growing accuracy, but remain inconclusive about consequential disasters. Even if the exact time and location of landfall of a newly formed tropical cyclone cannot be known, its trajectory becomes easier to forecast once the storm has fully developed. The tracks and intensities of tropical cyclones can be estimated with some reliability up to several days in advance. The same applies to floods in large (>1000 km^2) drainage basins, as topography and retention space on valley floors modify and attenuate the downstream propagation of peak discharge in foreseeable and physically predictable ways. More rapid runoff in smaller basins means that flash floods and debris flows have much shorter warning and prediction times.

In contrast, times of occurrence of large earthquakes are currently impossible to predict except for a few seconds to tens of seconds in advance thanks to the differing speeds at which seismic waves travel, which translates into differing arrival times that seismic stations record. Yet these few precious seconds can be enough to automatically shut down critical infrastructure and warn people to take cover. Japan's high-speed bullet trains will stop when ground acceleration recorded by a dense network of seismometers exceeds a critical threshold. Computer chips installed in gas pipelines shut down the flow of gas upon registering seismic ground motion above a defined threshold. This technology helps to reduce the incidence and spread of fires, especially in Japan's urban areas where traditional cooking on open stoves is still widespread. Nuclear power plants have safety shutdown procedures in the event of an earthquake or tsunami, although the 2011 Tohoku tsunami illustrates the consequences when the sizes of rare events are underestimated. Nonetheless, in only a few minutes, the likely propagation of a tsunami across an ocean basin can be selected from a range of premodelled scenarios. These scenarios incorporate potential tsunami sources along major earthquake zones, ocean-floor bathymetry, and the potential volumes of displaced water. This procedure enables early warning that, depending on the location, may be several minutes to almost a day before the tsunami makes landfall. Tsunami warnings following major earthquakes in Japan can reach the public in as little as three minutes following the earthquake shock.

Local impacts of spatially more dispersed natural hazards can be challenging to predict. Accurate forecasts of potentially damaging landslides are mostly limited to sites that, having been identified as potentially hazardous, are continuously monitored; actively deforming hillslopes and volcanic edifices fall into this category. For example, measurements have shown that the gradual acceleration of surface deformation associated with large deep-seated landslides can be used to predict with increasing accuracy the times of catastrophic slope failures. For some volcanic hazards, the warning times may span several days to weeks. Volcanic unrest can cause elevated surface deformation as the magma chamber inflates. Movement of magma to shallow depths in a volcano can be detected and measured by instruments such as tiltmeters, GPS, LiDAR, Interferometric Synthetic Aperture Radar (InSAR), and seismometers, while increasing gas emissions may signal pending eruption. A system of dedicated and regularly updated alert levels serves to warn nearby population centres about the state of volcanic unrest.

5.2 Types of Models for Prediction

A predictive model is a simplified abstraction of one or many natural phenomena. We formulate, test, and use models to learn, understand, and eventually anticipate how nature generates processes that can be harmful to humans and their assets. A model cannot encompass the full diversity and complexity of natural processes, thus the critique that models are too simple, insufficient, or unable to represent reality is beside the point. Statistician George E.P. Box (1919–2013) captured this reality in his famous statement 'essentially, all models are wrong, but some are useful'. What is important, therefore, is to know how wrong a given model can be in a given situation. Breiman (2001) recalled that, if a model is used on data to obtain quantitative insight, the resulting conclusions concern the model alone. We may take away from this that 'If the model is a poor emulation of nature, the conclusions may be wrong'.

Scientific models can be classified in many ways, depending on their eventual purpose, their mathematical foundations, or the disciplines in which they are used. To keep things simple, we distinguish in this chapter *statistical* and *deterministic* models, while acknowledging that the boundaries between them are diffuse. For example, scientists who predict rates of sediment transport by rivers that are in flood might use a mix of statistical and deterministic models. The empirical component of these models takes into account direct measurements of grain size, channel-bed and bank roughness, or flow velocity. These parameters, in turn, feed into deterministic (e.g., hydraulic) models of sediment transport. More complex processes such as turbulence in flows can be captured by numerical models or represented by statistical approaches.

The flood of digital data that is increasingly becoming available from all sectors of life (and well beyond science) has motivated a rapidly growing range of *data-driven* models. These models are somewhat in the statistical domain, but often bypass or extend the ways of traditional statistical methods and use computer algorithms to detect novel or unexpected patterns in data. Still, some of these data-driven models can also be coupled to physical-based expectations to obtain more interpretable results.

Models for predicting natural hazards Models are abstractions of nature, designed to simplify and capture the essence of natural phenomena with mathematical formulations. The 'laws of nature' are scientific models that have stood the test of time, having been repeatedly and consistently supported by data, and yielding robust and reproducible results. We can distinguish several types of model for quantitative hazard assessment. For example, most sophisticated hazard models require computer solutions, so numerical models have lost their distinctive character. To keep things simple, we discuss here statistical and deterministic models, and identify several subtypes:

- *Empirical models* offer predictions that rely on measured observations, for example storm wind speeds, tsunami run-up heights, or the runout of catastrophic landslides. Most models of natural hazards have an empirical component, but also belong to the group of statistical models. Scientists use statistical techniques to learn more about potential relations among observed data.
- *Heuristic models* are a special type of empirical model that rely on expert opinion. They involve subjectively assessing unobserved events based on previous observations or experience and thus have a strong empirical component. Expert opinion forms the ultimate step in any scientific analysis, as scientists need to judge the reliability of their predictions.
- *Probabilistic models* make use of probability theory to arrive at predictions of natural phenomena. Frequentist models use probabilities to infer information about fixed, although unknown, parameters of one or several probability distributions that generate random realizations of data that we measure. In contrast, Bayesian models treat the data that we measure as fixed and apply probability theory to infer the parameters of the generating distributions, which are assumed to be unknown. Probability theory allows scientists to express the degree of uncertainty about these distributions.
- *Deterministic models* are based on physical principles and natural laws, including conservation of mass, energy, and momentum. These models use

constitutive equations that contain physical quantities linked to observable phenomena. Deterministic models also include scaled analogue simulations of mass transport such as flume or sandbox experiments, or numerical studies intended to solve differential equations that encapsulate fundamental physical processes that otherwise elude any analytical solutions.

5.3 Empirical Models

One of the most straightforward and intuitive modelling approaches is to predict natural hazards from past observations. This approach is also a first and vital step in most hazard and risk assessments. Even if using more complex models in the later workflow, it is essential to have some data that we can use to test our model predictions. Empirical models hinge on the assumption that we can learn sufficiently from the past to be prepared for the future. Hence, the collective experience of past geomorphic impacts over a broad range of environmental conditions could be an indicator of possible future impacts. This idea reverses the statement of geologist Charles Lyell (1797–1850) that 'the present is the key to the past', a credo of the uniformitarianism view in geological sciences. Yet prediction concerns more than future events; instead it expresses the ability to make informed statements about unobserved events at any time.

Empirical modelling in natural hazards research often involves searching for parameters that can be easily measured and cast readily in mathematical form. The guiding principle is to make the most of the few measured data of a given natural hazard, risk, or disaster and to capture these data as reliable predictors for the future. From empirical models we can, for example, estimate flood peak discharge from catchment size and rainfall amount, predict fireline intensity from flame

height, or approximate avalanche deposit thickness from the detached snowpack volume. Scientists mostly choose regression and classification analyses as statistical tools for empirical modelling.

5.3.1 Linking Landforms and Processes

Many empirical models in geomorphology strive to identify causal links between processes and landforms. The aim is to identify certain characteristics of landforms as indicators of rates of erosion, transport, and deposition. We have argued for dynamic feedbacks between landscapes and the processes of erosion, sediment transport, and deposition. Hence we rarely, if ever, expect any landscape to be understandable in terms of a single set of processes. Likewise, it may be difficult to identify dependent and independent variables in a dynamic feedback (by definition). Even if the process of meandering is dominant in a river, the resulting meanders may differ greatly from textbook examples because of the influences of, for example, falling trees, previous cutoffs, river training, or simply nonlinear dynamics (Hooke 2007). Early researchers had to search long and hard before finding rivers with meanders sufficiently ideal for analysis. Beware of the pitfall of attributing textbook landforms to textbook processes. In many cases, landforms in the field are much more complicated than landforms in textbooks.

For example, scientists reconstructing the discharge of prehistoric or unobserved floods often use tractive stress relations for deposited boulders. However, many floods that transport boulders have high sediment loads, so clear-water tractive stress analyses are likely to be inaccurate. In 1995, a jökulhlaup from the Franz Josef Glacier in New Zealand deposited boulders several metres in diameter as a layer on top of stratified finer gravels (Davies et al. 2003b). Theory would require clear-water flows some tens of metres deep to move these large boulders. Instead, the boulders were ejected from beneath the glacier, along with

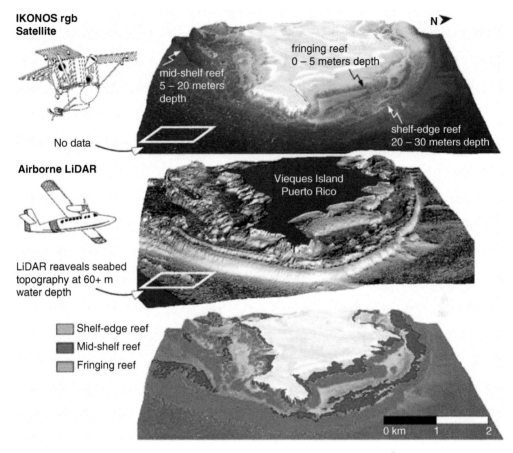

Figure 5.1 Reef map off the east point of the Vieques Island in Puerto Rico (bottom) made using IKONOS multispectral satellite data (top) and airborne bathymetric LiDAR soundings (middle). Although LiDAR lacks multispectral capability, it is able to resolve the seabed to depths exceeding 60 m (white rectangle) over exceptionally clear waters. The resulting thematic map is highly accurate and 3D. From Goodman et al. (2013).

huge quantities of finer sediment, when the pressurized subglacial drainage system blew a large plug of ice from the front of the glacier. Surveys showed that the boulders were nearly level with vegetation trimlines at the valley sides, and so were effectively transported at the water surface.

Similarly, several studies have use the bend geometry of debris-flow levees to infer the flow velocity from the superelevation, i.e. the height difference, between levee tops. Again, the equations used to estimate the velocities are based on clear-water hydrodynamics. By definition, debris flows involve large amounts of sediment and are distinctly non-Newtonian

fluids, so the calculated velocities are likely to be inaccurate. Calibrating flood models with field observations may be problematic, if the these observations hinge on crude estimates in the first place. This large uncertainty is characteristic of rare and large floods in particular, and makes it difficult to evaluate the performance of even sophisticated models.

Nonetheless, geomorphologists study specific landforms to learn more about the processes that shaped them. The discipline of geomorphometry offers one principled approach by being concerned with the quantitative analysis and interpretation of the Earth's surface (Hengl and Reuter 2009). Geomorphometry

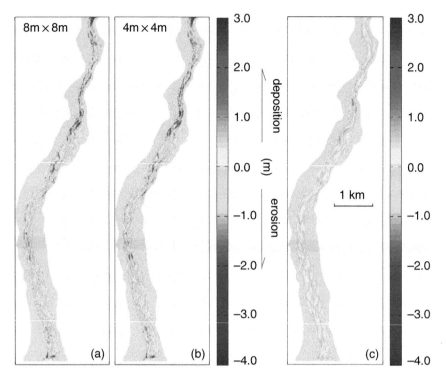

Figure 5.2 (a) Vertical channel changes along the Jökulsa River, Iceland, simulated with a 2D hydro-morphodynamic model. (b) DEM-of-difference (DoD) of field-measured changes. (c) Difference of modelled changes in (a) and DoD in (b). From Guan et al. (2015).

has been around for decades, but has flourished with the widespread and increasing availability of digital elevation models (DEMs) and computing power. Many tools in geomorphometry are useful to detect or characterize landforms diagnostic of natural hazards and disasters. Modern high-resolution DEMs obtained from terrestrial or airborne laser (LiDAR) scanning now have ground resolutions of a few centimetres or less. 'Bare-earth' images are routinely derived from LiDAR point clouds, greatly enhancing our ability to generate high-resolution elevation or terrain models without the obscuring effects of vegetation (Figure 5.1). A welcome side effect is being able to estimate the structure and volume of this vegetation cover. Unmanned aerial vehicles (UAVs) or 'drones' deliver thousands of images per survey and structure-from-motion software enables the creation of DEMs from these overlapping images. Although drones

can cover only small study areas, they are less costly than most airborne LiDAR platforms and thus offer the opportunity for a higher frequency of repeat measurements (Figure 5.2).

Dozens of topographic metrics are now routinely used to derive information about geomorphic activity from DEMs that discretize landscapes, mostly in the form of grids or point clouds. Many topographic data are also hidden in other environmental data sets used in natural hazards research. For example, regional temperature or precipitation data can be corrected for topographic effects, especially in mountainous areas that have few, if any, climate stations. Hence climatic data are often interpolated based on DEMs. This approach, however, compromises efforts to disentangle climatic from topographic influences, for example for many ice-related natural hazards.

Geomorphometry works mostly with derivatives of elevation data. The first partial derivative of a scalar elevation field is known as the gradient [m m^{-1}], which is a vector field pointing towards the greatest difference in elevation (Figure 5.3). The maximum slope gradient directly controls the transport of sediment on the Earth's surface, be it hillslopes, river channels, or submarine canyons. Countless studies have suggested strong links between erosion rates and (absolute) slope gradient – landscapes that are steeper on average favour commensurately higher rates of erosion and surface lowering. The nature of this relationship is likely nonlinear for soil-mantled hillslopes (Roering et al. 2007), although its exact generally applicable form is contentious. In practice, predictions of erosion rates or sediment yield are confounded by the strong correlation between local hillslope gradient and similar topographic metrics such as mean local relief, which expresses elevation ranges over large distances. For example, Vanmaercke et al. (2014) proposed that sediment yield across the African continent was largely a function of mean local relief, modelled peak ground acceleration, the percentage of tree cover, and surface runoff, but noted that these and several other possible predictors of sediment yields were linearly correlated.

The second partial derivative of elevation is curvature [m^{-1}] and describes the local convexity or concavity of the land surface. We distinguish between vertical, tangential, or horizontal curvatures. All are metrics for identifying areas where water and sediment fluxes are concentrated on the land surface. Mapping where how much sediment resides in mountainous terrain is an important step in locating potential sources of future sediment-laden floods and debris flows, and largely depends on appropriately resolved digital elevation models. The concavity of the upper flanks of the most regularly shaped stratovolcanoes may hold clues to their dominant eruptive style or even the time since their last eruption or collapse. Magmas responsible for rectilinear

volcanic slopes are characterized mostly by higher SiO$_2$ or higher H$_2$O contents and thus the likelihood of explosive eruptions, whereas concave-up slopes of stratovolcanoes are indicative of more effusive eruptions (Karátson et al. 2010). Former debris-avalanche scars may stand out as undulations or indentations in otherwise concentric elevation contours and thus distinct changes in hillslope curvature (Grosse et al. 2012).

Contributing catchment area [m^2] is a metric that hinges on a connected network of steepest flow paths, and is a widely used proxy for cumulative discharge and runoff in humid river catchments, assuming other factors such as groundwater flow or evaporation are negligible. Catchment area is also an important proxy for flood discharge and bulk sediment loads. Other catchment characteristics such as drainage density [m m^{-2}] and topographic relief [m] provide constraints on hillslope length and topographic stresses that contribute to slope instability. Brardinoni et al. (2009) analysed patterns of landsliding in mountain catchments of coastal British Columbia and argued that the distribution of relict glacial landforms controlled sediment yield from landslides over a 70-year period. Their study revealed that unchannelled topography was conducive to the highest sediment yields from landslides, whereas low-order tributaries and hanging valleys had lesser landslide sediment yields. They also showed that larger catchments are more likely to have sediments stored on broad valley floors and thus overall lower sediment yields.

Combinations of metrics such as slope gradient or topographic surface roughness can be useful for automatically detecting and delineating avalanches or landslides from DEMs (McKean and Roering 2004). Recent landslide deposits appear less smooth than the surrounding terrain on high-resolution DEMs from which vegetation cover has been removed. However, these surface roughness metrics rarely perform well in unmapped terrain with an unknown number of landslides, and thus

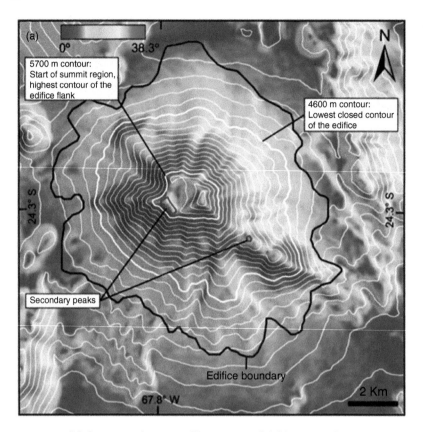

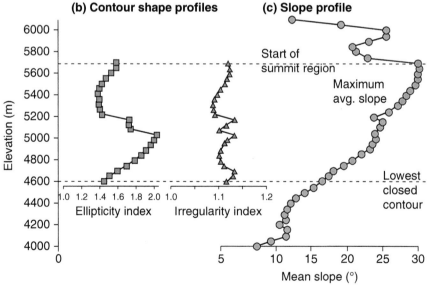

Figure 5.3 (a) Shaded slope map of Aracar volcano, Argentina (100 m contour interval). (b) Ellipticity and irregularity indexes of the main closed contours on the edifice flank plotted as a function of elevation. (c) Mean average slope plotted against elevation (elevation interval 50 m). From Grosse et al. (2012).

researchers have experimented with other methods, especially given the many proposed definitions or interpretations of topographic 'roughness' (Smith 2014). Methods from spatial statistics or time series analysis find increasing use in unravelling new information from DEMs. Wavelet transform, for example, can be a useful alternative to surface roughness for detecting the geomorphic footprints of landslides (Kalbermatten et al. 2012). In essence, we can characterize to first order many causes and effects of natural hazards and disasters with one or several topographic metrics.

5.3.2 Regression Models

In many hazard studies we wish to learn how a continuous target variable y responds to one or several predictor variables $\mathbf{x}$. For example, we might be interested in learning how storm-surge height increases with maximum sustained wind speed in a tropical cyclone, how the physical impact of a debris flow changes with peak discharge, or how the total number, area or volume of landslides triggered by seismic shaking relates to earthquake magnitude. In mathematical terms, these and many other problem can be expressed as:

$$y = \mathbf{w}^{\mathrm{T}}\mathbf{x} + \epsilon \qquad (5.1)$$

where $\mathbf{w}$ is a vector of weights or regression coefficients and ϵ is a random noise term that mainly reflects measurement errors in the response variable. This noise is often modelled by a Gaussian distribution. If measurement noise is negligible, ϵ can be interpreted alternatively as an observed variability of y; combining noise and variability is also possible. Most scientists would consider it to be good scientific practice to include information about the fitting errors (residuals) and prediction errors when reporting regression results. When dealing with natural hazards – especially when it comes to prediction – communicating this information is essential.

Equation (5.1) is a form of the generalized linear model, which is one of the most flexible and thus widely used in statistical analyses in natural hazards research; even state-of-the-art models of deep neural networks are based on thousands of stacked and interconnected forms of linear models. Most geomorphological studies use variants of the generalized linear model for highlighting possible connections between variables. Many geoscientific data, especially measurements of volume and area, are highly asymmetrical or skewed and may need some logarithmic transformation before being analyzed. Spatial and time series data also introduce the problem of autocorrelation, that is data points that are geographically closer together are more correlated than those farther apart. Ignoring autocorrelation in models may lead to incorrect variances, and thus over- or underconfident predictions. The field of geostatistics deals with this effect in interpolating and predicting spatial data. For example, geographically weighted regression is a form of the generalized linear model that attaches weights to each predictor to represent the effects of spatial autocorrelation, so that locations that are closer together have more similar values. More robust and flexible methods involve kriging, which allows specifying or even inferring the length scales, amplitudes, and directions of correlation between data.

Many simple empirical models collapse into their parameters much more complicated physical relationships of variables that can be time consuming or prohibitive to measure in the field. For example, the concept of downstream hydraulic geometry of rivers states that channel width W, depth d, slope S, and average flow velocity v scale with discharge Q as:

$$W \propto Q^{c}, d \propto Q^{f}, S \propto Q^{k}, v \propto Q^{m} \qquad (5.2)$$

where c, f, k, and m are scaling exponents that we can learn from regression (Griffiths 2003).

Similarly, predicting the maximum runout $L_{\max}$ [m] of a debris flow on a fan depends on many factors that elude direct measurement. Thus, it can be convenient to approximate some of these variables in a regression model that relates $L_{\max}$ to a predictor that we can readily measure, say the length of the fan L_f [m] (Hürlimann et al. 2008). We can then fit a mathematical function to the data to approximate the average changes in debris-flow runout on fans of different sizes. A worldwide data set of 154 debris flows shows that L_{max} can be estimated as:

$$L_{max} = \alpha V^b H^d \qquad (5.3)$$

where V is debris-flow volume [m^3], H is the vertical travel distance [m], $\alpha = 1.9$ [m^{1-3b+d}] is the coefficient, and $b = 0.16$ and $d = 0.83$ are the exponents obtained by multivariate regression.

In a similar spirit, the LAHARZ model (Iverson et al. 1998) estimates potential inundation zones of volcanic debris flows from digital elevation data. This widely used model relies on empirical relationships between deposit volume V [m^3], the cross-sectional channel area A [m^2], and the footprint area of the flow B [m^2]:

$$A = 0.05 \, V^{2/3} \qquad (5.4)$$

$$B = 200 \, V^{2/3} \qquad (5.5)$$

Given a user-specified flow volume, LAHARZ solves Eqs (5.4) and (5.5) pixel-by-pixel in a digital elevation model. The result is an objective and reproducible map of potential inundation zones. Like all model results, thus identified zones require close scrutiny and validation to see if these results are reliable.

Many important relationships in natural hazards research and geomorphology originated from, or sometimes even hinge on, findings from empirical regression. Another example is studies that aim to estimate the volume V of a landform or water body from its footprint area A. Volumetric information is often hard to come by, whereas planform geometry can be rapidly and widely mapped from satellite data or air photos. One popular regression model is:

$$V = \alpha A \qquad (5.6)$$

where α is a coefficient [L$^{3-2\beta}$] and β a dimensionless scaling exponent. Power relationships of this form are available for estimating

volumes of lakes, landslides, lava flows, coastal sediment bars, and many other landforms. Grinsted (2013) compiled 18 published relationships of this form for glaciers, for example, and reported fitted values of the scaling exponent of $1.12 < \beta < 2.9$. This example demonstrates that the convenience of estimating volumes with mathematically simple and appealing equations comes at the cost of a high scatter. This scatter – especially if prevalent in the scaling exponent – compromises accurate estimates for hazard appraisals and predictions. Hence, published volumetric estimates can differ widely for the same set of data. For example, estimated landslide volumes based on Eq (5.6) can differ measurably when considering shallower soil or deeper bedrock failures (Larsen et al. 2010).

Power relationships are also frequently used for inferring the average rates of processes. For example, Walder and O'Connor (1997) reviewed methods of estimating the average peak discharge Q_p [m^{3}s^{-1}] following the sudden failure of natural dams via regression models and reported that predictions differed by orders of magnitude, largely depending on the choice of input parameters such as dam height or volume of water released. Such high variance is common in empirical regression models that have few parameters, as these few parameters only partly represent more complex process interactions.

The generalized linear model can be extended to many predictors and is thus suitable for hazards that may depend on a number of controls. Landslide susceptibility studies routinely use up to dozens of potential predictors and use models to identify the most successful predictors of slope instability. Kritikos et al. (2015) developed a GIS-based method for predicting the distribution of relative susceptibility to landsliding due to a scenario earthquake. They used earlier landslide distributions mapped following the Northridge and Wenchuan earthquakes to calibrate a multivariate model that, when applied to the Chichi earthquake, gave notably accurate results. Robinson et al. (2016) subsequently developed a model to predict the numbers and volumes of earthquake-triggered landslides, though without pinning down their specific locations.

Predictions of natural disaster risk can also build on regression models if individual factors of the risk equation are beyond practical measurements. Scientists have long been searching to derive hazard indices from measured field data, combined with categorized data on the vulnerability of elements at risk. Gao et al. (2014) used this approach to model the windstorm risk to coastal cities in China from the observed frequency of landfalling tropic cyclones and the maximum elevation of their associated storm surges (see Fig. 2.6). They modelled vulnerability as a function of socioeconomic, land use, ecological, and resilience indices, and reported a distinct risk gradient from north (less risk) to south (more risk) irrespective of several uncertainties in the underlying data. These and other simple regression-based hazard and risk models can (at least in principle) be extended to full risk models by including the projected policy response, based on how much a community wishes to invest in flood-risk protection against future storm surges (Aerts et al. 2014).

Machine Learning, Artificial Intelligence, Data Science These and several related terms have begun to dominate many aspects of modern science. Data sets available for scientific study are becoming larger and computer algorithms more diverse and complex. Many useful and powerful methods have arisen at the interface between applied statistics and computational sciences, and machine learning has long attracted the attention of researchers interested in predicting natural hazards. Many algorithms build on individual or entire stacks of statistical models, although not all algorithms were designed with formal statistical inference in mind. Instead, algorithms such as random forests or artificial neural networks emphasize more the detection of (ideally previously unobserved) patterns in the data. In essence, the seemingly endless variety of models can be broken down into a few major classes, for example regression, classification, data reduction, or clustering methods. With more and more studies embracing these computational methods, we emphasize that a sound understanding of these tools is essential to avoid training overconfident models.

5.3.3 Classification Models

Models of classification are concerned with predicting a category. In natural hazards and risk research, we can often express our model prediction in terms of categories such as 'ground liquefaction is likely', 'storm surges can be higher than 2 m' or 'susceptible to debris flow'. The categories that these models predict can be linked to a threshold (e.g. the 2-m storm-surge height mentioned above), but the categories must be mutually exclusive. Let us stick with the debris-flow example: bridges on mountain roads must be designed to withstand floods and, ideally, also debris flows. However, discharge data from systematic monitoring are rare, while it is costly and generally impractical to survey in detail all channel deposits. Hence building models that identify likely debris-flow channels is a first step in discriminating those catchments that may need further attention through a field campaign. Several studies proposed that debris flows occur in distinct topographic domains that can be delimited using two or several terrain metrics. Finding out whether a densely settled alluvial fan is subject to debris flows is important. Metrics of the fans and their catchments, including planform area, steepness, topographic ruggedness, fan length, and fan gradient, can offer simple means to roughly separate fluvial fans from debris-flow fans (de Scally and Owens 2004; Welsh and Davies 2011). Most studies are in agreement that the gradient of the fan and the ruggedness of its catchment are important criteria for discerning debris-flow activity, although this approach hinges on how well we can separate 'flood-dominated' and 'debris flow-dominated' fans in the first place (Wilford et al. 2004): often this distinction rests on geomorphic evidence of the last few recent events that may have reshaped the fan. This may include rarer mechanisms such as catastrophic outbursts from naturally dammed streams that can also rapidly form new fans (Benn et al. 2006). Pronounced winnowing, reworking or pronounced weathering of surface clasts to the point of disintegration may also give a false visual impression of a mainly fluvial fan surface that was originally shaped by debris flows (de Haas et al. 2014).

Distinguishing between flood-dominated and debris-flow fans is a statistical classification problem in which we wish to predict a nominal attribute (or class) from continuous data such as topographic metrics. Similar classifications involve nominal states such as 'landslide' versus 'absent landslide', or, where relevant for operational early warning systems, 'tsunami' versus 'absent tsunami'. This assumption rests on a clear definition of what 'absent' means, however. Does absent mean that we have not seen the phenomenon in question? Does it mean that the phenomenon might have happened at the study site some time in the past but is too old and faded to be considered further? Clearly, the original class labels may carry some uncertainty. Some of the most popular methods applied to this classification problem include discriminant analysis, logistic regression, weights-of-evidence, and decision trees.

Logistic regression is a method of classifying data, and is widely used for assessing whether a phenomenon of interest – such as landsliding – is likely to occur given a set of environmental characteristics, for example, slope gradient, rock type, mean annual rainfall, and so on

(Ayalew and Yamagishi 2005). Logistic regression estimates the probability that target data y belong to one out of two (or several) classes, say class $c = 1$:

$$P(y = 1 | \mathbf{x}, \mathbf{w}) = \frac{1}{1 + \exp(-\mathbf{w}^{\mathrm{T}}\mathbf{x})} \quad (5.7)$$

where $\mathbf{w}$ is a vector of weights or regression coefficients, and $\mathbf{x}$ is a vector containing the predictors. The scalar product $\mathbf{w}^{\mathrm{T}}\mathbf{x}$ is called the 'log odds ratio'; it is positive if the probability of belonging to class 1 is higher, and negative otherwise. For a data point that belongs with equal probability to either class 0 or 1, the log odds ratio is zero. This zero value is also known as the 'decision boundary'.

Hundreds of scientific studies throughout the world tell us that landslide and debris-flow susceptibility largely depends on a few common terrain variables (Figure 5.4).

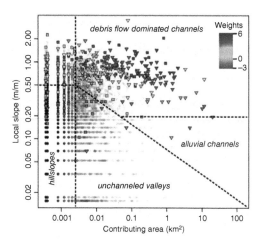

Figure 5.4 Slope-area plot for distinguishing geomorphic processes in Norwegian mountain catchments. The plot shows study area sample points of local surface inclination and contributing catchment area (grey) and documented debris-flow initiation points. Symbols specify debris-flow source locations: squares = open-slope, triangles = channelized debris flows. Color ramp indicates logistic regression weights estimated for debris flow locations: a positive weight indicates a >50% probability of being classified as debris flow; a negative weight indicates <50% probability of being classified as a debris flow. The steeper portions of the study area seems to be most debris-flow-prone. From Meyer et al. (2014).

Simple models using topographic metrics such as local slope gradient and curvature alone may offer predictions that are as reliable as those from more sophisticated models with more parameters (Meyer et al. 2014), especially if these parameters strongly correlate with each other. Logistic regression has seen many similar uses, for example, as a real-time warning model of rainfall-triggered lahars (Jones et al. 2017), or of the susceptibility of moraine dams to failure. For example, McKillop and Clague (2007) used this classification method to infer which of some 175 surveyed moraine-dammed lakes in the mountains of British Columbia are most prone to sudden failure, given a set of topographic and hydrographic characteristics of all lakes, and especially those that had known outbursts in the past.

Regression models offer alternative decision boundaries such as envelope curves fitted to a conditional quantile (one of several equal-sized fractions) of the data. For example, the critical or threshold values of antecedent rainfall at which landslides occur can be inferred by studying trends in the highest values. The assumption is that, if sufficiently constrained, rainfall measurements can be used to issue forecasts about slope failures during rainy periods. Here the combination of measured rainfall intensity I [mm h^{-1}] and duration D [h] seems to produce the most consistent results. One picks from a plot of I versus D an envelope line termed the rainfall intensity-duration threshold that separates rainfalls events with reported landslides from events without landslides. Yet the global conclusions that we can draw from the envelope approach are few, because rainfall intensity-duration thresholds vary between region and even between storms, often by orders of magnitude (Guzzetti et al. 2007). This scatter reflects the combined influence of differing land cover, antecedent soil moisture, and soil infiltration rates, rainfall interception by vegetation, and errors in interpolating rainfall from point measurements.

5.4 Probabilistic Models

We have seen that the term 'hazard' can be expressed as the probability that a process operating below, at or above the Earth's surface has the potential to impact human lives and assets during a specified period. Probabilistic models are widely used in natural hazard assessments. Some of these models provide satisfactory predictions, even though heir physics may remain obscure. Why use probability? Several points motivate this choice. Probability is a numerical and objective measure of uncertainty and thus approachable in a mathematically consistent way. Probability theory enables calculations and statistical inferences to be made regardless of whether we interpret them as measures of randomness or subjective uncertainties. Probability quantitatively also expresses the physical randomness that we see, but fail to fully explain, in many processes. Finally, probabilities are an integral part of our everyday lives, although we rarely appreciate this. Many of our questions and decisions such as 'Is it going to rain this afternoon?' or 'What happens if I miss that bus?' implicitly draw on the concept of probability.

5.4.1 Probability Expresses Uncertainty

Probability can be seen as a metric of subjective uncertainty. If used this way, it is helpful to distinguish between two types of uncertainty. Epistemic uncertainty arises from a lack of knowledge about a given phenomenon, often because we have insufficient data or observations. We can reduce epistemic uncertainty by carrying out further research or experiments that deliver more data and insight into the phenomenon in question and that reduce the uncertainty of measurements. Aleatoric uncertainty, on the other hand, refers to intrinsic randomness tied to a phenomenon. We can cast this randomness in numbers, but never eliminate it fully: collecting more data is unlikely to reduce the aleatoric uncertainty. In quantum physics, for example, Heisenberg's uncertainty principle states that we cannot accurately determine both the instantaneous position and momentum of a quantum particle. This principle exemplifies aleatoric uncertainty, because knowing more accurately the location of a particle is unavoidably tied to knowing less about its momentum. Dynamical chaos is another source of aleatoric uncertainty that is well known in fields such as fluid dynamics, where such uncertainty enters predictions of wind fields, precipitation patterns, or ocean currents as a probabilistic error term. While such numerical models do an increasingly good job of forecasting, they will always have an element of intrinsic randomness.

Probability naturally lends itself to measuring and linking epistemic and aleatoric uncertainties. The distinction between the two types of uncertainty is important because it is associated with two different interpretations of probability. If we toss a coin, we expect obtaining heads and tails with equal probability $\theta = 0.5$. But what if we wish to learn whether the coin does indeed yield this equal probability of heads and tails? By repeatedly tossing the coin, we could simply count the fraction of heads. In the long run, the observed frequency of heads then serves as a surrogate, or point estimate, of the probability of heads. The more often we toss the coin, the closer the observed frequency should come to the true probability. More data also increase statistical robustness, meaning that our estimate of θ will change only minutely as we add more data. This is why classical statistics are termed 'frequentist'. Tossing the coin only once is of limited informative value, as we would have to conclude that θ is either 0 or 1. Thus our estimate could be far from the true θ, which frequentist statistics argue is fixed. A different way of looking at the problem, though, is to argue that our knowledge about the true θ is uncertain. Hence, we can use probability to encode our belief about whether the coin is fair or not. By recording the tosses of a coin we can gradually update our initial subjective belief

regarding the coin's bias. With more and more data coming in, we can modify and reassign our uncertainty about θ. One difference to the frequentist approach is that we arrive at a richer and detailed result of whether the coin is biased or not, because we go beyond a single point estimate of θ.

Suppose we are interested in learning the probability of an M >9 earthquake occurring at a certain subduction zone next year. We duly analyse past seismic records and derive an estimate of how frequently earthquakes above a given magnitude happen. Assuming a(n almost) complete record of past earthquakes, including rare events such as those with M >9, we might use the frequencies of past events as proxies for the probability of future ones. But what if our record fails to capture such rare and strong earthquakes? Missing a rare event is similar to playing without winning the lottery regularly for some time. The observed frequency of wins is zero, but we would hardly maintain that the probability of winning is zero as well. You might concede that the probability is quite low, but indeterminable if we replaced probability by frequency. Frequentist statistics maintain that you simply need to play lottery often enough – in theory infinitely long – to arrive at an estimate of this probability. In contrast, the view of 'Bayesian' statistics is concerned with updating our uncertainty about the unknown probability of winning, using the data that we have. Using probability to express this subjective uncertainty is at the core of Bayesian statistics (Kruschke 2010). Coming back to our coin flipping example, the frequentist view would hold that θ is fixed and our data of coin tosses are random realizations. In the Bayesian view the data are fixed and θ a probability distribution describing our uncertainty. The foundation to the latter view is Bayes' Rule, which follows directly from the definition of conditional probabilities. For example, let $P(E)$ be the probability of the occurrence of an M >9 submarine earthquake within an ocean basin during a given time period. By analogy we consider $P(T)$ to be the probability that a tsunami of at least a given size (say with 5 m runup) occurs along a stretch of coastline bordering the ocean. Now the conditional probability of observing a tsunami given that we recorded an earthquake is:

$$P(T|E) = \frac{P(T,E)}{P(E)} \qquad (5.8)$$

where $P(T,E) = P(E,T)$ is the joint probability of observing both the earthquake and the tsunami. We can rearrange Eq (5.8) and solve for the probability that an earthquake had happened given that we observed a tsunami:

$$P(E|T) = \frac{P(T|E)P(E)}{P(T)} \qquad (5.9)$$

Note that $P(E|T) \neq P(T|E)$ in general – the probability of observing an earthquake given knowledge about a tsunami is something different than the probability of a tsunami given knowledge about an earthquake. Conditional probabilities are free of implications (or requirements) about any physical causes or effects between the states of E and T, neither do conditional probabilities reveal anything about order. For example, $P(E|T)$ is neutral about whether the tsunami had happened before the earthquake and also neutral about whether the tsunami caused the earthquake, which would make little sense physically, as far as we know. After more data become available about $P(T|E)$, we can update our initial belief, or scientific judgment, about the conditional earthquake probability and compute a new estimate of $P(E|T)$, which we can then take as the new prior probability $P(E)$. For more advanced applications of Bayes' Rule (Eq (5.9)), we replace the probabilities with full probability distributions. The things we thus can learn from a set of observed data include predictions of unobserved data, the parameters of models of, or hypotheses about, the data-generating processes. More generally, we can state

$$p(\theta|\mathscr{D}) = \frac{p(\mathscr{D}|\theta)p(\theta)}{p(\mathscr{D})} \qquad (5.10)$$

where θ are the model parameters and $\mathscr{D}$ are the data. We term $p(\theta|\mathscr{D})$ the posterior probability distribution, $p(\mathscr{D}|\theta)$ the likelihood function, $p(\theta)$ the prior distribution, and $p(\mathscr{D})$ the evidence or marginal likelihood. Note that we use in our notation upper-case P for probabilities and lower-case p for probability densities.

Regardless of which model or interpretation of probability we choose, probabilistic modelling of natural hazards must address how sensitive model results are to changes in input data and parameters. Instead of relying on a single solution from a physically based equation or empirical relationship, we assume that one, several, or all input parameters have their own probability density distributions. Assigning randomly drawn values from these distributions using pseudorandom number generators opens the door to assessing model results from thousands of model runs, each of which uses a slightly different combination of input parameter values. This approach is called a Monte Carlo simulation, and allows researchers to express their uncertainty about a given set of model input parameters. The method also provides insights into a model's sensitivity, especially at the boundaries of the parameter space, and also tracks the propagated errors within the model simulations. Modern slope-stability design, for example, relies on Monte Carlo methods, and many scientific studies have used this approach to predict the conditions under which landslides occur (Watts 2004).

Probability theory helps to objectively measure perceptions or cognitive biases related to natural hazards and disasters. In one branch of cognitive psychology, scientists are conducting systematic experiments of reproducible errors of human reasoning (Yudkowsky 2008). Some results from their experiments apply to natural hazards, risks, and disasters. Humans use methods of thought that are called heuristics, which in everyday circumstances return agreeable results that match the demands of life. These heuristics are prone to systematic errors called biases. We can appreciate biases when considering how reports on natural hazards, risks, and disasters routinely enumerate losses. Rarely do we read about a disaster without being informed about the fatalities or damage involved. Numbers may make us believe that we better comprehend what happened or how big the disaster was. But do we truly grasp the full meaning and implications of 10 000 lives lost or financial losses of several hundred million US$ during a single disaster? This is a case of scope neglect, a situation in which the human brain is limited to make us feel an emotion that would be several orders of magnitude higher than we would feel, for example, for a randomly selected single person's loss (Yudkowsky 2008). Small increases in our willingness to pay correspond to exponential increases in scope and require some reflection when dealing with data on natural hazards, risks, and disasters. These data are highly sensitive even if objectively derived by some of the methods that we present in this chapter. Communicating sensitive data should inform users such as the public, stakeholders, planners or decision makers, and avoid scaring them or making them oblivious to hazards.

Cognitive biases Subjective views strongly influence us when confronted with natural hazards and risks. Our emotional response undermines our objectiveness and gives rise to uncertainties. Dealing explicitly with uncertainties is one of the most essential parts of communicating hazards and risks. Illustrating such uncertainties to stakeholders, decision makers, and the public is a major task in the information age of the twenty-first century (Spiegelhalter et al. 2011). Even seemingly accurate scientific results are received differently. To this end, risk aversion has become an explicit factor in quantitative risk assessments and entails measuring personal biases and perceptions. Risk aversion influences how we make decisions under uncertainty,

depending on the different kinds of cognitive bias (Yudkowsky 2008), and can involve:

- *Anchoring bias*, the tendency to build decisions on the first piece of evidence. Persons relikely interpret any subsequent information in the light of initial data.
- *Affect heuristic* describes how subjective perceptions of bad and good might influence and bias our judgement, unintentionally favouring more rapid and nondeliberate decisions.
- *Availability bias* refers to our inclination to assess the frequency or probability of an event from data that we can obtain most readily or conveniently. Events that we can imagine more easily seem to us more frequent, whereas inconceivable events seem rare or even impossible to us.
- *Bystander apathy* arises when our feeling of responsibility diffuses because we hope that someone else will deal with an unwanted situation. This hope multiplies such that eventually nobody will take action.
- *Calibration and overconfidence* mean that we can be highly mistaken about very small probability events and give them more weight than their probability warrants.
- *Confirmation bias* occurs when we prefer supporting instead of falsifying our hypothesis about some evidence. If we tend to rationalize a given piece of evidence on rare disasters, we might be prone to a form of confirmation bias.
- *Conjunction fallacy* is our tendency to judge probability statements before thinking carefully about them. It applies in particular to a joint probability $P(A, B)$, which by definition is less than each of its components $P(A)$ and $P(B)$. For example, we overestimate whether all of three floods with an exceedance probability of 90% each will occur, whereas we

underestimate whether at least one of three floods will happen with probability of 10%.

- *Hindsight bias* occurs when our retrospective estimate of how predictable a given disaster was has measurably increased in comparison to the views of others who try to predict the disaster without any advance knowledge: 'We knew it all the time, right?'
- *Scope neglect* refers to small linear increases in our willingness to pay, which cause exponential increases in scope. Put simply, our brains cannot comprehend the scope of large natural disasters, although the amount of damage and the number of fatalities do seem to understandable to us.

Sutherland et al. (2013) suggested '20 tips' for dealing with factual claims in science, especially from the perspective of laypersons confronted with scientific findings. Much of their advice is based on detecting and, if possible, avoiding biases in perceptions and in the way that scientists handle their data analyses. This should always remind us that 'data can be dredged or cherry picked', and that the way we analyse data can also be misleading, poorly communicated, or simply inappropriate. Frequently encountered examples include relying too heavily on regression to the mean; confusing statistical correlation with a cause–effect relationship; unreflected extrapolation of trends in data; or putting too much faith in statistical significance.

5.4.2 Probability Is More than Frequency

Probability grounds our expectations about how large and frequent future harmful events might be. The systematic, if scattered, relationship between the magnitude and frequency of a hazardous process derives directly from a probability distribution (Figure 5.5). This distribution weighs how much each event contributes to the distribution through its size, magnitude, or intensity. Magnitude–frequency relationships are useful for selecting design criteria and lifetimes for infrastructure. Built structures need to withstand a predefined level of process intensity. In many building codes, for example, bridges must have enough freeboard to pass the 100-year flood without being damaged. This 100-year flood refers to the discharge that is being exceeded at an average return period $T = 100$ yr and thus has a probability of $p = 1/100 = 0.01$ of occurring in any one year. One misconception is that every century sees exactly one 100-year flood. Assuming that all the largest floods in each year are independent and drawn from the same distribution, we can calculate the probability of observing at least one 100-year flood in N years:

$$P_N = 1 - (1 - p)^N \qquad (5.11)$$

Recall that the probability of at observing least one event is equal to the inverse probability of observing zero events. Setting $N = 100$ in Eq (5.11) gives $P_{100} = 0.63$. Hence, the probability of observing a 100-year flood in 100 years is 63%, less than the 100% one might perhaps expect. Put differently, we have a probability of $1 - P_{100}$, or 37% that a 100-year long series of largest annual floods is devoid of the 100-year flood. Equation (5.11) also reveals that it is possible to have more than one 100-year flood in a century. Note how probability differs from observed frequency; nonetheless, using the frequency of past floods might be the only way to get first estimates of their future probabilities. Rare, large, and potentially destructive events often limit the reliability of such probabilistic forecasts by virtue of leaving behind fewer supporting data. The rarer the natural hazard of interest, the longer the period of time required to capture a sufficient number of past impacts, and to anticipate and quantitatively estimate future occurrences.

The analysis of seismic catalogues to estimate the recurrence of earthquakes in a

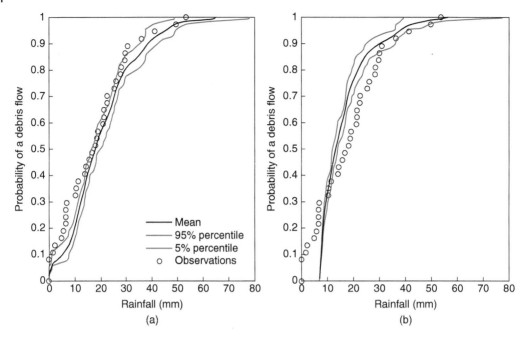

Figure 5.5 A probabilistic rainfall-driven sediment cascade model developed to predict debris flows at Illgraben in the Swiss Alps. (a) Modeled probability distribution of daily rainfall on 36 days with debris flows (with mean and 90% confidence bounds). (b) Modeled probability distribution with rainfall triggering only, i.e. bypassing the hydrological model component (snow accumulation, snowmelt, evapotranspiration, and soil water storage). From Bennett et al. (2014).

given area and interval is a classic example of magnitude–frequency studies and has led to the well-known Gutenberg–Richter relationship, sometimes also termed the 'Gutenberg law'. This relationship states that the observed frequency of an earthquake is inversely related to its magnitude (Turcotte and Schubert 2002):

$$\log \dot{N} = -bM + \log \dot{a} \qquad (5.12)$$

where $\dot{N}$ is the annual number of earthquakes greater than magnitude M, and b and $\dot{a}$ are empirically derived regression coefficients. The dots in Eq (5.12) denote rates or, more correctly, observed frequencies, thus $\dot{N}$ is the number of earthquakes per unit time, usually one year. Equation (5.12) shows that large earthquakes are systematically less frequent than smaller ones, conveniently condensed in an inverse power-law relation. This statistical model provides a mathematically straightforward way of predicting earthquakes.

Yet the model has several pitfalls that illustrate some general issues when dealing with frequency–magnitude data.

Firstly, we need to make sure that the data are sufficiently complete to capture the full distribution of earthquakes over a long period, as values of b and $\dot{a}$ tend to stabilize only for large sample sizes, often several thousands. For example, imagine that we have recorded only ten years of earthquake data. Some of the very large events, say M >8.5, may have occurred outside that particular period. It is easy to see how we could underestimate the probability for larger events from this short time series. This is what happened in the case of the 2011 Great Tohoku earthquake east of Japan. Seismologists were anticipating M~8 earthquakes offshore based on detailed historic and palaeoseismological evidence. Although potential locations of epicentres were known, the earthquake magnitude was underestimated, with devastating consequences.

Secondly, the power-law parameters differ according to the method by which they are estimated, and small differences in the exponent in Eq (5.12) can lead to large errors in estimates of earthquake probability. Thirdly, longer seismic records may contain only larger earthquakes, whereas with time the global network of seismic stations has become increasingly dense and is now sensitive to small earthquakes. Many earthquake catalogues have thus acquired a higher proportion of more frequent earthquakes over the years. In essence, we can use Eq (5.12) to estimate the average number of future earthquakes per year, assuming that seismic activity has remained, and will remain, unchanged over the interval we wish to make predictions. Constraining this relationship with robust data is an important step in better estimating seismic hazard and a blueprint for many other natural hazards for which we estimate their probability density function.

Finally, we often have to predict into a small population of large events that will occur during a practical planning period, for example several decades to a century. Such predictions are prone to substantial imprecisions even if we have a very large database of events (Davies and Davies 2018).

In general, we can write the probability density function of a power-law distribution of a continuous variable x as:

$$p(x|\alpha, x_{min}) = \frac{\alpha - 1}{x_{min}} \left(\frac{x}{x_{min}} \right)^{-\alpha} \quad (5.13)$$

where $\alpha > 0$ is the scaling exponent and $0 < x_{min} < x$ is an observed or arbitrarily set minimum value of x (Clauset et al. 2009). This minimum value is needed to ensure that the integral below the power-law function converges. Note that the Gutenberg-Richter relationship (Eq (5.12)) uses the complementary cumulative distribution function of x:

$$P(X > x) = \int_x^\infty p(x)dx = \left(\frac{x}{x_{min}} \right)^{1-\alpha} \quad (5.14)$$

where $P(X > x)$ is the probability that an observed value X exceeds the value of x.

Equations (5.13) and (5.14) have proven useful for characterizing the size distributions of floods (Malamud and Turcotte 2006), landslides (Pelletier et al. 1996), rock falls (Dussauge et al. 2003), snow avalanches (Birkeland and Landry 2002), wildfires (Malamud et al. 1998), the duration of lava-dome eruptions (Wolpert et al. 2016), and erosional and depositional events (Ganti et al. 2011). The inverse relationship between frequency and magnitude appears to be a common, and perhaps universal, model for hazardous natural processes (Malinverno 1997). Obtaining, comparing, and interpreting the parameters of these assumed power-law relationships has been motivating scientists for several decades.

A popular method for estimating power-law parameters involves binning the data in logarithmic space and fitting linear regression models to the transformed data because power laws plot as straight lines in log-log space. However, the regression parameters estimated in this way depend on the chosen bin size and are biased if derived by standard least squares regression. Similar problems arise from small sample sizes or insufficient knowledge about the largest events recorded, as x is unbounded in Eq (5.13). Thus a small number of bins with insufficiently recorded large events may further bias the fit parameters (Korup et al. 2012). Further issues with estimating power-law scaling parameters include extrapolations to two or more dimensions, the underestimation of smaller poorly resolved events, and the choice of the study area, which might incompletely cover larger events (Bonnet et al. 2001). One robust approach for estimating power-law exponents follows the principle of maximum likelihood:

$$\hat{\alpha} = 1 + n \left[\sum_{i=1}^{n} \ln \frac{x_i}{x_{min}} \right]^{-1} \quad (5.15)$$

where n is the number of samples. The estimate $\hat{\alpha}$ rests on the assumption that all the data were independently generated from an identical power-law distribution. Whether this assumption is valid remains an open field of study. For example, large rock avalanches occur every few years to decades in seismically active mountains. With time, erosion and deposition obliterate the traces of these catastrophic rock-slope failures. This erosional censoring ensures that geomorphic evidence of rock avalanches has different conditions of being preserved (Whitehouse and Griffiths 1983). Hence, deriving statistics for long time periods hinges on the assumption that the causes and triggers of rock avalanching have remained stationary: this is a strong assumption in the light of past climatic and environmental changes (Montanari and Koutsoyiannis 2014).

Power laws and natural hazards Power-law equations are simple and attractive mathematical models that are widely used in geomorphic and natural hazards research. Power laws characterize many empirical scaling relationships, such as downstream hydraulic geometry in rivers, and physically based equations, such as aerodynamic drag on suspended aeolian sediment. Estimating the volumes of lakes, glaciers, or landslides involves the application of a power-law equation to their footprint areas. Further, when cast as a probability distribution, the power law is a useful tool for characterizing the trade-off between the size and frequency of natural hazards. Its most famous example is the Gutenberg–Richter relationship of earthquake magnitudes. Researchers from outside seismology have adapted this model for wildfires, landslides, or floods. However the simplicity of power-law models and the seeming ease with which they can be fit to empirical data are countered by several limitations in practical applications. Establishing whether observed data are indeed power-law distributed requires rigorous statistical testing that goes beyond inferring a straight line from log-log plots or using standard regression techniques based on least squares. Statistical distributions such as the lognormal or exponential may also generate data that seemingly follow a straight line in log-log space. Small numerical differences in scaling exponents can lead to large differences in predictions. Such predictions differ strongly for small samples and also depend on the minimum and maximum observed events. More reliable predictions can be gained by applying methods such as cross-validation, which divides the data set into training and testing subsets. We build our predictions on the training set and test their success independently with the remaining data.

Researchers have also proposed alternative models to capture the observation that larger events occur more rarely than smaller ones. Candidates include the log-normal, Weibull, or inverse gamma distributions. For example, ten Brink et al. (2009) argued that the log-normal distribution characterizes the size distribution of submarine and terrestrial landslides better than a power law does. The log-normal distribution is rarely used to characterize the frequency and magnitude of terrestrial landslides. Although we can use several metrics to decide which particular distribution fits our data best, the ultimate goal is to decide which distribution is the best at predicting unobserved data. This task is known as model selection and today relies on either cross-validation techniques or Bayes' Rule. Cross-validation is a method that uses randomly selected and equal-sized fractions of the data as training input for the model and uses the remaining (holdout) fraction for testing, that is assessing how the trained model predicts these unobserved data.

The mathematical convenience of describing size distributions of natural hazards with only a few parameters remains a large incentive for analysing past events. The physical interpretations of thus inferred magnitude–

frequency relationships, however, do vary or remain at a level that few laypersons understand. The concept of self-organized criticality (SOC) derives from complex dynamic system theory (Turcotte et al. 2002), and offers one explanation of power-law behaviour in natural systems. Other scientists have stressed the scale invariance or fractal characteristics implicit in power-law distributions in the geological sciences (Bonnet et al. 2001). Scale invariance is the reason field photographs of rock outcrops or drainage networks, for example, require a scale bar to uniquely identify their size. True scale invariance occurs only where a power-law relationship holds over at least several orders of magnitude.

5.4.3 Extreme-value Statistics

For many hazard applications, we are interested mainly in the recurrence of larger and potentially more destructive events. The field of extreme-value statistics focuses on a specified fraction of the largest observed data, and discards all other observations. At the outset, we need to define what characterizes an extreme event. Given a time series of data, one way to select extreme values is to consider only the maxima in a given interval, for example, the largest flood discharges or storm-surge levels in each year or the highest daily rainfall. This approach therefore identifies what we call 'block maxima'; according to extreme-value theory these maxima should follow a Generalized Extreme-Value (GEV) Distribution, which in cumulative form can be written as:

$$P(x|\mu, \sigma, \alpha) = \exp\left[-\left[1 + \alpha\left[\frac{x-\mu}{\sigma}\right]\right]^{-1/\alpha}\right]$$

$$(5.16)$$

where α, β, and μ are the shape, scale, and location parameters, respectively. Note that we require that $1 + \alpha\left[\frac{x-\mu}{\sigma}\right] > 0$, and that $\sigma > 0$. We assume that the data are sampled independently from the same distribution, a condition statisticians call *i.i.d.* (identically and independently distributed). The shape parameter

α determines the type of extreme-value distribution generated from the GEV distribution. We call the distribution *heavy-tailed* if $\alpha > 0$. For $\alpha = 0$ the GEV distribution reduces to the Gumbel or Type I extreme-value distribution, which in cumulative form is:

$$P(x|\mu, \sigma) = \exp\left[-\exp\left[-\frac{x-\mu}{\sigma}\right]\right]$$

$$(5.17)$$

We obtain this result by taking the limit $\alpha \rightarrow 0$ in the GEV distribution. The Gumbel distribution has been used for modelling the recurrence of extreme rock avalanches and extreme floods (Katz et al. 2001; Whitehouse and Griffiths 1983).

Alternatively, we can define extreme events as those data that exceed an arbitrarily high threshold. For example, we could use the flood level or wind speed that is exceeded in only 5% of all records. This strategy is useful if we have only few data that would otherwise forbid a robust statistical analysis of block maxima. Using such a peak-over-threshold (POT) approach requires that these peaks are uncorrelated. Sticking with the flood example, all peaks are assumed to belong to different floods. In the case of POT samples, extreme-value theory predicts that the data follow a Generalized Pareto Distribution (GPD) (Bernardara et al. 2011), which is a more flexible version of the power-law distribution introduced in Equation (5.13):

$$P(x|\mu, \sigma, \alpha) = 1 - \left[1 + \left[\frac{\alpha(x-\mu)}{\sigma}\right]\right]^{-1/\alpha}.$$

$$(5.18)$$

Power-law and exponential distributions are closely related; the Generalized Pareto Distribution simplifies to the exponential distribution for $\alpha = 0$ and in this case is known as GPD Type I (Figure 5.6):

$$P(x|\mu, \sigma) = 1 - \exp\left[\frac{x-\mu}{\sigma}\right].$$

$$(5.19)$$

Extreme-value statistics have been central to estimating and predicting rare flood events and are widely used to determine the 100-year

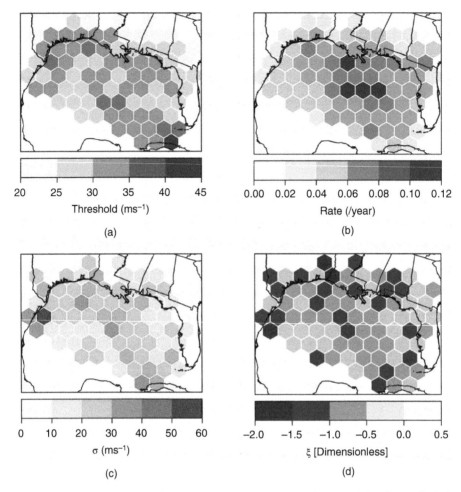

Figure 5.6 Extreme-value parameters for a model fitted to hurricane wind-speed data in the Gulf of Mexico. The parameter estimates of this extreme-value model include (a) the minimum wind speed threshold, (b) the average annual rate of exceeding this threshold, (c) the scale, and (d) the shape of the distribution. The latter two parameters describe the location and steepness of the model curve. From Trepanier et al. (2015).

flood. This 100-year return period is an important and traditional, though arbitrary, design criterion for many structures (Katz et al. 2001). Different hydrological regimes may require different statistical models, and hence different values of extreme-value parameters. For example, the variability of floods in rivers draining the Australian seaboard east of the Great Dividing Range changes markedly from north to south. Rivers in the seasonal tropics have lower flood variability than those farther south. Many catchments deviate strongly from this regional trend, and this deviation emphasizes the need for a tailored strategy

of determining 100-year floods from stream gauge data (Rustomji et al. 2009). How the shape of extreme-value distributions changes with environmental conditions such as topography, seasonal climate or seismicity, is an open field of research. In many streams in the United States, for example, the preferred model for predicting extreme flood discharges is based on the Pearson III or three-parameter inverse Gamma distribution:

$$p(x|\mu, \sigma, \alpha) = \frac{1}{\sigma \Gamma(\alpha)} \left[\frac{x - \mu}{\sigma}\right]^{\alpha - 1}$$
$$\times \exp\left[-\frac{x - \mu}{\sigma}\right], \quad (5.20)$$

where $\Gamma(\alpha)$ is the Gamma function of α, and $x > \mu$; note that $\mu, \sigma, \alpha > 0$. Many flood-frequency studies use the log-Pearson Type III distribution, such that $\log(x)$ is Pearson Type III distributed. Inspecting Eq (5.20) reveals that this extreme-value distribution combines an exponential relationship with a power-law tail for the larger observations. One frequent question that we face concerning 'extreme' natural hazards and disasters is whether their number and severity have been increasing, especially where models appear to have underestimated the resulting impacts. Modern extreme-value studies thus increasingly embrace the idea of dynamic hazards that change with time (Irish et al. 2011). One useful output of such nonstationary models is how long we can rely on a given hazard assessment: we thus can learn something about how long our predictions can be valid before needing revision.

5.4.4 Stochastic Processes

Studying magnitude–frequency relationships is only part of the story. We can similarly ask whether past events have a specific pattern in time, regardless of their size. This question is part of the study of time series or stochastic processes. Earthquake swarms or aftershock sequences are examples of hazards that can be clustered in time. Tropical cyclones are tied largely to seasonal ocean surface temperatures and hence also recur in distinct intervals each year. Some hazards are cyclic, while others may have a long-term 'memory' of previous events. For example, some plate-boundary faults appear to generate large earthquakes somewhat regularly (Berryman et al. 2012). The concept is vigorously debated, mainly because many models featuring characteristic earthquakes, seismic gaps or seismic cycles failed rigorous statistical tests, or were partly influenced by researchers' expectations (Kagan et al. 2012). The 2011 Great Tohoku earthquake rekindled the question of whether large subduction earthquakes occur in discrete clusters instead of occurring randomly in time. An answer to this question may depend largely on the choice of statistics (Kerr 2011). Historic records feature only five M >9 earthquakes since the beginning of the twentieth century, although these five earthquakes accounted for the bulk of total seismic moment during that period. Three of these great earthquakes – Kamtchatka 1952, Chile 1960, and Alaska 1964 – occurred closely together in time. Similarly, the Sumatra 2004 and Japan 2011 great earthquakes happened less than a decade apart after a gap of four decades without any comparably large events. Does that mean that there might be a higher probability of yet another mega-earthquake in the next few years? From our discussion of power-law statistics we would argue that the sample size of M >9 earthquakes is too small, hence predictions will be highly uncertain. Yet several statistical approaches are available to estimate and characterize a given sequence of discrete events. The concept of stochastic processes describes how these events are distributed in time (or in space). One can envisage a stochastic process as a random variable that depends on an indexed or sorted independent variable describing time, space, or both.

One of the most widely used stochastic processes in natural hazards is the Poisson process. It is a 'memory-less' process, in which the counted events occur randomly and uncorrelated with any previous or subsequent events, regardless of when or for how long we measure. We can derive the Poisson process from the Poisson distribution:

$$P(X = k|\lambda) = \frac{e^{-\lambda}\lambda^k}{k!} \tag{5.21}$$

where $P(X = k)$ is the probability of counting exactly k events with a rate λ. One convenient characteristic of the Poisson distribution is that its rate λ is also its mean. We can recast Eq (5.21) for an interval τ:

$$P(n(t + \tau) - n(t) = k) = \frac{e^{-\lambda t}(\lambda t)^k}{k!} \tag{5.22}$$

where n is the number of events counted and k is the number of events counted in τ. Suppose we are interested in the probability of having a measurement interval of length τ without any event, so that $k = 0$. We can then simplify Eq (5.22) to:

$$P(n(t + \tau) - n(t) = 0) = e^{-\lambda t} \quad (5.23)$$

which is the exponential distribution. Hence the time spans, or inter-arrival times, between any two subsequent events are exponentially distributed and independent of each other. We can begin measuring at any time since each event is 'without memory' of previous ones. We can also expect that longer waiting times for the next event are much rarer than shorter ones. One way to test whether a time series follows a Poisson process is to compare the observed exceedance probabilities p_i^{obs} with the theoretical, exponentially distributed ones, p_i^{theo}:

$$p_i^{\text{obs}} = \frac{R_i}{1 + n} \quad (5.24)$$

where R_i is the rank number i of the sorted inter-arrival times (with $i = 1$ being the highest), and n is the number of data. The theoretical exceedance probabilities are:

$$p_i^{\text{theo}} = e^{-t_i/T} \quad (5.25)$$

where t_i is the inter-arrival time for the ith rank and T is the mean inter-arrival time. Plots of p_i^{obs} versus p_i^{theo} will return data points close to the 1:1 line if the data follow a Poisson process. The more the data deviate from this line, the less likely is it that they follow a Poisson process. Many scientists model the times between subsequent earthquakes or large explosive eruptions with a Poisson process (Gusev et al. 2003).

Several alternative statistical models estimate the distribution of average waiting or interarrival times between consecutive events. Estimates of volcanic repose intervals, for example, rest on times between subsequent eruptions as identified from seismic records. The Weibull distribution has been used in studying time series of past eruptions, and this distribution also approximates the failure rates of materials:

$$p(x|k, \lambda) = \frac{k}{\lambda} \left(\frac{x}{\lambda} \right)^{k-1} e^{-(x/\lambda)^k} \quad (5.26)$$

where k is a shape parameter and λ is a scale parameter; the Weibull distribution is defined for $x \geq 0$. The corresponding cumulative distribution function is:

$$F_W(x|k, \lambda) = 1 - e^{-(x/\lambda)^k} \quad (5.27)$$

Applied to volcanic unrest, the Weibull distribution can model how the probability of an eruption increases exponentially with time since the last eruption. In a study of repose intervals of Soufrière Hills volcano on the island of Montserrat, Connor et al. (2003) found that the Weibull distribution was adequate only for predicting shorter waiting times between successive eruptions (Figure 5.7). The authors favoured a log logistic model for the intervals between larger, rarer eruptions, and argued that this model reflected better how processes such as pressurization and gas exsolution competed with each other. One important lessen from this and other similar studies is that single statistical distribution may be inadequate to capture a natural processes. Sometimes a combination of distributions offers more suitable interpretations (Watt et al. 2007).

5.4.5 Hazard Cascades, Event Trees, and Network Models

Many natural disasters involve a chain of adverse events that can form the building blocks for worst-case scenarios. A growing body of research is concerned with assessing these hazard and risk chains, which are sometimes termed secondary, subsequent, or cascading hazards, hazard or risk cascades: these are interconnected as opposed to a collection of unconnected individual hazards or risks, for which the terms 'multihazard' and 'multirisk' are more appropriate (Schmidt et al. 2011). It is useful to distinguish between

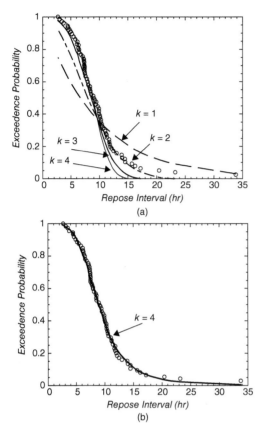

Figure 5.7 Statistical models of the probability of exceeding a specified volcanic repose interval, Soufrière Hills Volcano, Montserrat. This model is based on fitting the Weibull distribution to observed data for a variety of shape parameters k. Note how shorter repose intervals are more likely than longer ones. From Connor et al. (2003).

multiple natural hazards that are independent of each other and causally linked natural hazards that occur sequentially in time (Gill and Malamud 2014). Many volcanic hazards, for example, arise from the coupling of magmatic, hydrogeological, and atmospheric processes. Simply lumping all these processes into predictions about the expected size of a future volcanic eruption would incorrectly estimate the possible consequences of the processes involved: does a larger eruption size mean that lahars are more mobile or that collapse of an eruption column is more likely? Clearly a single index or metric is unable to capture

these and many other questions. An alternative approach is to focus on plausible scenarios that portray how these processes might be linked in hazard chains.

Tree-based models offer an intuitive structure for this kind of composite prediction (Figure 5.8). Decision trees, for example, can be either empirical or probabilistic models that specify mutually exclusive outcomes of how processes may combine, and are thus suited for simulating hazard or exposure scenarios (Marzocchi et al. 2004). A tree consists of nodes and branches, and often assumes a directional sense from the root (the node of origin) to various leaves (the nodes of termination). Each node defines a system state from which multiple possible outcomes or options can arise, and each of those outcomes is represented by a branch. By assigning mutually independent probabilities to each branch, we require that all probabilities emerging from a given node add up to unity. If interpreted as a sequence of possible events, such event trees can illustrate the range of possible outcomes in a hazard chain (Newhall and Hoblitt 2001), including possible decisions that people or computers make in the progress. Modern tsunami warning systems, for example, depend partly on pre-computed scenarios of how waves propagate from a given source; choosing the most likely or nearest source is an essential step in such decision trees (Strunz et al. 2011).

We can also assign different weights to the branches of a tree, thus incorporating expected benefits or, alternatively, costs or losses from a given course of action or outcome. This combination of decision theory with probabilistic reasoning can appeal more to decision makers than similarly complex, although purely physical or mathematical, models (Marzocchi and Woo 2007). Event trees are helpful when assessing natural hazards, such as natural dam breaks, that require a combination of processes that block rivers and make dams fail. Emmer and Vilímek (2014), for example, proposed a tree-based recipe for assessing whether glacier lakes in the Andes are susceptible to sudden

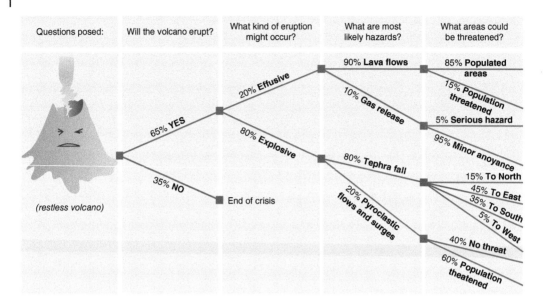

Figure 5.8 Event tree showing elements of possible process chains during a volcanic crisis. From Lockwood and Hazlett (2010).

failure. The main limitation is that event trees can only have parallel outcomes from a given node. Another item to consider is that in a probabilistic decision tree the errors on the probabilities need to be shown. These errors accumulate very rapidly along the branches, so that the ultimate outcomes at the leaves are likely to be less well specified than those emerging from parent nodes higher up.

We can also derive decision trees from data. In this case the branches define local decision boundaries between labelled groups of data. In the simplest form, decision trees separate data into discrete groups. Model trees classify data, and can be extended with local regression models for each data class. Trees are thus a very flexible class of models. A major drawback is that decision trees are prone to overfitting and thus need careful treatment such as pruning (reducing the number of excess branches) and cross-validation.

Any given tree model is a specific type of network, and can be interpreted as a probabilistic model rooted in mathematical graph theory. Networks are highly flexible and allow expression of the dependence or independence

of random variables or probabilities, the structure of their connectedness, the hierarchy of statistical models, or how interlinked systems respond to catastrophic disturbances (Buzna et al. 2006). Network models can identify the weakest link in a supply chain, information web, or connected early warning system that is affected by a natural disturbance. For example, Meyer et al. (2015) used graph theory to estimate increased travel times caused by road closures because of debris flows in southern Norway. They used a graph to represent the regional road network and investigated the effects of local traffic disruption assuming that a given road segment was blocked or unusable for some time. They chose to express the resulting risk in terms of excess travel times, higher vehicle loads, and additional fuel costs because of required detours.

5.5 Prediction and Model Selection

The growing bandwidth of statistical models available to study and better understand natural hazards might seem overwhelming. How do we find the 'right' or 'best' model for a given problem or data set? How well does the model reflect our observations? Is it compatible with our assumptions? Could a less well fitting model be physically more plausible or even more desirable in practice? For example, recall that repose intervals between subsequent volcanic eruptions seem to follow a power law, Weibull, or log-logistic distribution equally well (Connor et al. 2003). But which of these three models would we recommend for a reliable prediction? These questions are concerned with the important task of *model selection*.

When fitting a model to observed data, we commonly attempt to minimize errors resulting from measurement errors and inevitable simplifications of reality. We can express these errors as functions. One of the most widely used error functions, for example, is the summed squared differences between the observed data and the modelled fits; this is also known as *ordinary least squares*. Such fits differ from predictions, however, and need to meet the test of new or unobserved data that were excluded from the analysis. Randomly selecting a subset of the observed data as a training set to fit the model, and using the remaining or holdout data for testing the fitted model, is a standard strategy to estimate uncertainties tied to prediction. In using this strategy, we commonly we find that the error tied to prediction outweighs the error tied to training. Models tend to perform less reliably for data other than those they were trained on.

Few geoscientific data are normally distributed and so require adequate choice of loss functions. Data outliers in particular may represent extreme events rather than statistical artefacts and compromise ordinary least-squares fitting. In this case we need to resort to statistically more robust methods that are less sensitive to outliers. Transforming variables, for example into a logarithmic form, is another common remedy, although this means that we fit linear models in log space.

Regardless of model, we also need to consider that more complex models are always able to achieve a better fit than simpler ones. However, more parameters, also called 'degrees of freedom', make models more prone to fit noise in the data than the overall trend. The resulting overfitting promises higher precision for the training data, but poor accuracy for predicting testing data. Here accuracy means how far our model is from reality, whereas precision refers to the spread of the model predictions. Polynomial models fitted to bivariate data illustrate how increasing the degree of the polynomial yields steady increases in the goodness-of-fit; ultimately the model curve replicates the location of each data point. The same models, however, perform poorly when new data are added. By analogy, buying a brand new dress or suit that fits perfectly while you stand still (= the training stage) is only a good idea when you also test how the garment fits while walking, kneeling, or sitting down (= the testing stage). When we proceed from fitting a model to prediction, we need to deal explicitly and rigorously with these uncertainties. We can avoid overfitting by techniques such as cross-validation, by which we randomly subsample data sets and use portions of the data to train the models. The remainder of the data are used as test data to check how well the models do the job of predicting.

Simple models have many benefits over complex ones, and statisticians have proposed several criteria to assist us in this regard. A widely used metric is Akaike's Information Criterion (AIC) that penalizes the goodness-of-fit of a model by the number of model parameters (Forster and Sober 1994):

$$\text{AIC} = \frac{1}{n\hat{\sigma}^2}(\text{RSS} + 2d\hat{\sigma}^2) + K \qquad (5.28)$$

where n is the number of data points, d is the number of model parameters, $\hat{\sigma}^2$ is the estimated variance, RSS is the residual sum of squares, and K is a constant. Several alternative metrics are based on the same philosophy of penalizing model complexity.

A prediction can go beyond choosing a specific model, and can instead involve combinations of output from many models. These ensemble models take advantage of many different contributing models and base their eventual predictions on the overall mean, majority, or some other kind of summary vote. Many state-of-the-art predictions, whether they are concerned with tomorrow's likelihood of rainfall or mid-twenty-first century global average temperature, build on ensembles of numerical models. On the one hand, this approach pools a diverse range of approaches; on the other hand, it moderates the distorting effects of outlier simulations. The 'rainfall risk' that your weather app shows you is often just the fraction of models that forecast rain, based on the set of meteorological conditions that are most similar to those for the day that the forecast is made for. Simply modelling tomorrow's weather is too time-consuming and costly in computational resources. An alternative is to choose from detailed catalogues of weather conditions that resemble most closely the current ones. This combination of empirical and numerical modelling also aids the prediction and operational real-time warning of tropical cyclones. In this task, communicating clearly the underlying uncertainties in such predictions is essential (Vecchi and Villarini 2015). False predictions are more than model errors that we need to acknowledge, but also opportunities to identify where models systematically veer off real trends in the timing and locations of tropical cyclones.

Whatever the selected model or models, an important step is their testing or what some researchers call model validation or verification. Oreskes et al. (1994) argued that model validation may be impossible given that natural systems are always open and model results are always non-unique. Even if we test our model with data that were excluded from generating the model, we are unable to fully guarantee that these testing data come from the same probability distribution as the data used to generate the model. Prediction is always tied to some error. Good practice means to report this error as objectively and reproducibly as possible, so that decisions made under that uncertainty are informed at the best possible level.

5.6 Deterministic Models

Our last class of models rests on fundamental laws of physics and chemistry, and therefore almost exclusively uses deterministic formulations of natural processes that are found in modern text books on geomorphology. The underlying principles of these models include the conservation of mass, energy, and momentum that govern the motion and deformation of solids and fluids. However, some models include probabilities to acknowledge aleatoric dynamics of natural processes, such as turbulence in water flows. Deterministic models use formulations that capture the essential materials, masses, forces, and moments at play. We roughly divide these models into static and dynamic ones.

5.6.1 Static Stability Models

Static models assume negligible changes in forces or moments acting on the object of interest. Most stability models are of this type, and can be applied with some modification to hillslopes, natural dams, deep-sea sediments containing frozen gas hydrates, or any other structures with known geometry and material properties, including individual clasts (Nandasena et al. 2011). For example, we have seen that these models assume a static force balance to estimate slope stability from the ratio of shear strength to shear stresses on a given hillslope (Eq 8.4). The main inputs to these models are the geometry of the hillslope and its potential failure surfaces, the soil-water conditions, the mechanical properties of the soil and rock, and the location and friction of the sliding interface. (Figure 5.9). Dynamic short-lived loads during earthquake shaking,

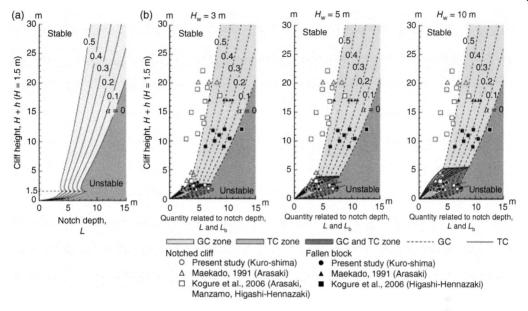

Figure 5.9 Threshold height of coastal cliffs for tsunami-induced collapse due to tsunami; this theoretical analysis holds for coral limestone cliffs of the Ryukyu Islands, Japan. (a) Relation between critical notch depth and cliff height for gravity-induced collapse. Colored panels: Relation between cliff height or block height, $H+h$, and a parameter characterizing notch depth, L. Open circles are values of $H+h$ and L for cliffs, and solid circles are $H+h$ and L-values for fallen blocks. The dashed curves show the relation between critical cliff height and notch depth for which gravity-induced collapse (GC) occurs at differing values of α, and the solid curves show the relationship for wave-induced collapse (TC). From Kogure and Matsukura (2012).

for example, can also be integrated easily (Li et al. 2009). Inverting the model set-up, we can also determine the set of critical parameter values needed to initiate failure and depict these thresholds on maps. The predictive success of these models depends largely on adequate calibration. Some parameter values may require tuning so that the output faithfully reproduces observations. In the case of models for rainfall-triggered slope failures, predicted sites of instability often deviate from landslide locations. Casadei et al. (2003) observed that their model overpredicted slope instability triggered by rainstorms mainly because rainfall, soil, and topographic data were were either missing, too coarsely resolved, or inaccurate. They also emphasized that previous landslides may have changed these inputs into the static stability model. Understanding better these precursory effects of landslides from a mechanistic viewpoint remains a largely open

research question. For example, Strenk and Wartman (2011) studied the reliability of the popular Newmark approach for predicting earthquake-triggered landslides, and found that the hillslopes that were identified as most prone to failure were also those with the highest model uncertainty. Other sources of uncertainty concern the direction(s) of ground acceleration, and topographic amplification of shaking along ridge crests. Numerical models of rock-slope stability during simulated seismic loads show that local contrasts in hillslope materials and the pattern of bedrock fracturing may be more important than hillslope geometry or position (Gischig et al. 2015).

5.6.2 Dynamic Models

Dynamic models involve single or coupled differential equations that describe the rates of change in the physical and chemical quantities of interest, from the motion of sand grains in a windstorm to the flow of a glacier. Solving these equations requires appropriate initial and boundary conditions such as parameter values at starting times and surrounding topography. Solutions of many coupled differential equations are rarely available in analytical form, so we need to compute numerically approximate solutions. One strategy is to discretize the model domain in space and time into a grid or mesh and to write computer code that iteratively evaluates local solutions of the differential equations at each node at each time step. Neighbouring grid nodes inform the local solution, and modellers can choose from various numerical schemes or solvers for achieving a compromise between computational efficiency, numerical precision, and physical accuracy (Perron 2011). Finite or discrete element codes similarly break down the model domain into simple discrete geometric forms that enable computation of the local static and dynamic stresses and strains for specified material properties. Returning to our slope stability example, the results of finite-element models eventually highlight regions of highest shear stresses or strain, thus revealing potential failure surfaces. A similar strategy applies to crustal blocks for models of seismic strains and stresses; when coupled to processes of erosion and sedimentation at the surface, these crustal models can help to predict how fault-slip rates may change with the changing amount of loading at the surface (Maniatis, et al. 2009).

Choosing field-calibrated initial and boundary conditions is key to successful numerical modelling, but introduces subjective and empirical elements to the simulation. Nonetheless, numerical modelling of atmospheric, hydrological or Earth surface processes is a powerful means of predicting physically plausible impacts from natural hazards. Computer simulations can reduce our uncertainty about the likely range of sizes, durations, and impacts of natural disasters, assuming that the models explore a broad enough range of scenarios. For example, modern models can now simulate realistic ground-motion scenarios during strong earthquakes (Barbot et al. 2012). Along similar lines, the Virtual Earthquake Approach (VEA) models long-period strong ground motions using Green's functions that are derived from measurements of the ambient seismic field (Denolle et al. 2014). The outputs of these and many other models can inform the choice of input values for models of subsequent hazards such as earthquake-triggered soil liquefaction or landslides.

Most deterministic models are concerned with how potentially damaging phenomena propagate from a specified source, for example how pyroclastic eruption columns form and collapse, how flood waves build up and attenuate downstream, or how tsunami waves spread from their point of origin. We sometimes call these simulations 'routing' or 'trajectory' models. With routing models we can generate or test various scenarios and assess the impacts of various types of mass flows on the landscape, built structures (Crosta et al. 2006), or types of land use (Nussbaumer et al. 2014). Many routing models depend on mathematical formulation of flow processes, which can be described by the Navier–Stokes equations or their simplified forms (Mangeney-Castelnau et al. 2005; George and Iverson 2014). Routing models shed light on how parameters such as flow geometry, velocity field, runout length, or physical impact forces are distributed and how they may affect the overall damage potential (Schneider et al. 2010). We can use them to calculate the ballistic curves of pyroclastic bombs or the runout of slope failures (Kelfoun and Druitt 2005). Routing models can also combine empirical and probabilistic inputs, allowing predictions, for example, of the flow direction of lava flows in real time. Infrared imagery obtained by satellites can provide data about eruption rates, which can then be used to drive lava-flow models such as MAGFLOW, which is based on a cellular automaton. The model solves for discrete pixels the steady-state solution of the Navier–Stokes equations assuming a Bingham fluid; these equations are tied to a simple heat-transfer model to estimate at each time step the spread of lava through the model domain (Vicari et al. 2011).

Sand box, flume, and natural hazards
Analogue models use fluids and solids with known properties to scale geomorphic processes down to sizes from several tens of meters to several metres (a sand box). Models of this kind might consist, for example, of a pile of rice or beans to simulate landslides (Densmore et al. 1997). Other experiments sprinkle granular materials with artificial rain to simulate how a topography resembling that of a mountain range evolves (Babault et al. 2005). High-resolution laser scanning can track how the surface morphology changes in these analogue models so that we can validate geomorphic transport laws in a controlled environment. In erosion boxes scientists can simulate tectonic uplift at prescribed rates and, together with measured rates of erosion, derive important data about processes that in real landscapes may take millennia or longer to alter landscapes noticeably (Lague et al. 2003). Physical models also mimic the dynamics of topography under gravitational stresses and show how hillslope relief and layers of weakness control the type and size of slope failure (Bachmann et al. 2006; Katz and Aharonov 2005) or even segments of active fault zones.

Flumes are small-scale artificial river channels used to reproduce bedforms or meanders, sediment pulses, debris flows, the build-up of alluvial fans, channel avulsions, the formation and failure of natural dams, or the effects of flow obstacles and vegetation. Flume experiments allow for reproducible conditions and have been used, for example, to simulate the geometry and the hydraulic conductivity of landslide dams, which together with inflow rates into the lake and local channel-bed hydraulics, are key controls on the lifespan and failure mode of the dam (Chen et al. 2015).

Sand-box and flume experiments reproduce many phenomena in real landscapes despite large differences in the governing dimensionless numbers. These numbers allow us to transfer small-scale models under controlled conditions to larger real-world settings. The surprising consistency between the sandbox and the real world possibly arises from a characteristic scaling observed in many natural systems (Paola et al. 2009). Keep in mind that some of these realistic outcomes from simple sandbox models remain only partly understood (Davies et al. 2003a; Beagley et al. 2020), so that their outcomes must be applied to field situations with great care.

The computer-based solution of wave equations is central to simulating how storm surges or tsunamis propagate. Some of the initial conditions depend on whether the tsunamis are triggered by earthquakes, landslides, or meteorite impacts. The solution to the differential equations is the water-surface elevation at a specified time and location. In its simplest form, we can derive a one-dimensional wave equation for deep water basins from two coupled expressions. The first expression specifies the change in water surface elevation η [L] with time t [T]:

$$\frac{\partial \eta}{\partial t} = -d\frac{\partial u}{\partial x} \tag{5.29}$$

where d is water depth [L], u is the current velocity [L T^{-1}], and x is a spatial coordinate [L]. The second expression relates the flow acceleration to the water surface slope:

$$\frac{\partial u}{\partial t} = -g\frac{\partial \eta}{\partial x} \tag{5.30}$$

where g is the acceleration due to gravity [L T^{-2}]. Combining Eqs (5.29) and (5.30) gives:

$$\frac{\partial^2 \eta}{\partial t^2} = C^2\frac{\partial^2 \eta}{\partial x^2} \tag{5.31}$$

where $C = \sqrt{gd}$ is the wave celerity that is the speed of the wave front [L T^{-1}]. We will again encounter the wave celerity in Eqs (9.1) and (9.2).

Solving more sophisticated versions of Eq (5.31) with a known bathymetry as the boundary condition can be used to estimate the travel times of tsunami waves across a body of water. One use of this approach is for transoceanic distances that take tsunamis several to up to some twenty hours to travel across. The high computational requirements of these simulations make it more practical to use precalculated scenarios in early warning systems. The case of an earthquake or other wave maker thus requires informed decisions about which of these scenarios is most appropriate. For coasts along the Pacific Ocean, for example, numerous precomputed model scenarios are based on epicentres of large earthquakes that are evenly spaced along the active subduction zones (Figure 5.10). In the case of an earthquake, the scenario with the conditions matching closest can be used to inform early warning systems. Making such predictions for landslide-triggered tsunami that can originate from continental shelves, the heads of submarine canyons, or steep rock walls of fjords requires considering many more local scenarios because these waves can reach populated coasts within a matter of minutes (Kelfoun et al. 2010). Another complicating factor is that the range of initial and runout conditions, including landslide volume, detachment location, and velocity and fragmentation of the failed mass, is too diverse to allow comprehensive preparatory modelling. Variants of the wave equation and the Navier–Stokes flow equations are central to many simulations of coastal waves, including tsunamis (Maza et al. 2015).

Numerical modelling is also useful for predicting more gradual processes, for example, how flood waves move through drainage networks or over alluvial fans (Pelletier and Mayer 2005), or how river beds build up following a sudden input of sediment and perhaps subsequently recover by incising once the sediment load drops. By linking equations of the conservation of mass and energy with hydraulic and empirical relationships we can predict how fast a river bed at a given location adjusts to incoming and outgoing water and sediment (Van De Wiel et al. 2011). This type of prediction is particularly useful if the response times of river channels are of the order of several months to decades, so that planners can respond accordingly. Picture, for example, a sufficiently large body of sediment added to a river. Field studies and flume experiments predict that the sediment will slowly move and spread downstream, reshaping the channel, and perhaps also the floodplain. The volume and calibre of sediment, together with the flow hydraulics, geometry, and ambient load of the channel, will determine how long it takes for the sediment pulse to disperse, raising the channel bed, swamping infrastructure, destroying bridges, and filling reservoirs. Engineering responses or countermeasures may require the rebuilding of bridges and roads on disturbed or newly formed valley floor. Ignoring the possibility that a slow sediment wave is still passing through the affected channel reach may leave new bridges swamped (stranded) once the river begins to deposit (incise) at the leading (trailing) edge of the sediment wave. Numerical models allow prediction of the outcomes of such anticipated channel changes given a set of engineering options. The models also inform us roughly how long it takes for the impacted channel bed to return to, or come close to, its elevation before the disturbance. In practice, it is prudent to monitor the field situation to detect as early as possible any departure of the real from the modelled situation.

Even simple models can reproduce how channel longitudinal profiles respond to sudden changes in base level, for example those induced by sea-level rise or earthquake faulting. For example, some of these models predict that downstream aggradation, and hence

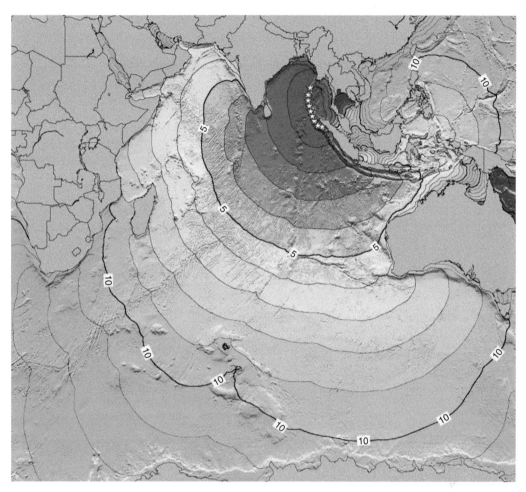

Figure 5.10 Computer simulation of travel times (in hours) of the leading wave of the 2004 Indian Ocean tsunami triggered by the Mw 9.3 Sumatra earthquake.
(https://www.ngdc.noaa.gov/hazard/tsu_travel_time_events.shtml#Simulate image source is
https://www.ngdc.noaa.gov/hazard/data/icons/2004_1226.jpg)

base-level rise, rather than changes in channel width, controls how much and how rapidly the longitudinal profile can adjust (Doyle and Harbor 2003). This finding contests the many studies that used channel width as a key diagnostic of river perturbation. Modern models now incorporate entire process cascades of hillslope erosion by combining in various forms soil creep, landsliding, fluvial transport, intermittent sediment storage, and the resulting sediment yield from a given catchment (Mouri et al. 2011). Such holistic approaches generally reduce the complexity of each of

the processes involved, but maintain as a key result the overall direction of the geomorphic response to disturbance, for example, whether channels are bound to incise or aggrade (Benda and Dunne 1997) or whether sediment yields are likely to increase or decrease (Ono and Yamaguchi 2011).

Similar 'reduced-complexity' models of fluvial sediment transport have produced some counterintuitive results that guide new research questions. For example, Van De Wiel and Coulthard (2010) showed that these models predicted sediment yields that are roughly

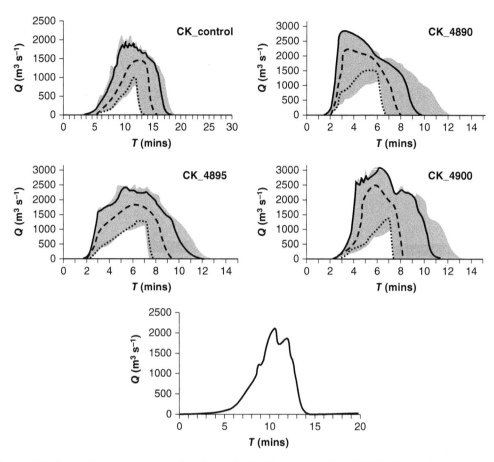

Figure 5.11 Percentile hydrographs derived from glacial lake outburst flood (GLOF) dam-breach simulations in the Himalaya. Dotted lines = 5th percentile; dashed lines = 50th percentile; solid black lines = 95th percentile. The hydrograph envelope for each scenario is displayed in grey. Bottom plot shows the modelled beach outflow hydrograph used as input for two-dimensional hydrodynamic modelling. From Westoby et al. (2014).

power-law distributed, with higher yields occurring less frequently than lower ones. Such model outputs motivate a more critical thinking about whether changing, if not highly variable, sediment yields can indicate whether changing environmental controls are ultimately responsible (Vanmaercke et al. 2011). Moreover, power-law distributed sediment yields lack any characteristic measure of central tendency: so that mean values are unsuitable for representing the distribution of recorded data.

Predicting river-channel changes with numerical models largely relies on the choice of equations to model sediment transport (Figure 5.11). How alluvial rivers respond to local damming or other disruptions can depend on the characteristic type of sediment transport, sediment calibre, and discharge regime. Gravel- and cobble-bed rivers that transport sediment mostly as bedload respond to changes in discharge and sediment supply in a way that maintains the relationship between discharge, channel slope, width, and grain size, and hence, transport capacity. Regulated or impounded rivers can respond to altered water and sediment inputs in multiple ways, although this response may sometimes be unrelated to the degree of sediment impoundment (Dade et al. 2011).

Many of these and similar insights are confounded by the ongoing search for widely applicable bed-load transport equations in steep rivers carrying a mixture of fine and coarse particles. Most proposed sediment transport equations come from laboratory- or field-based measurements and express sediment flux as a function of bed shear stress and particle diameter, but neglect important effects such as stress changes due to mostly immobile clasts or whether mobile sediment is available at all (Yager et al. 2007). Theory and field measurements argue that the rate of bedload transport should be a power-law function of water discharge, although in some cases this relationship can be nearly linear (Mueller and Pitlick 2005). Detailed time series of tagged gravel motion in a small experimental gravel-bed stream showed that both the transport distance and the volume of material moved scale approximately as power laws with peak stream power and the cumulative energy of hydrologic events. In essence, the views on how to parameterize sediment transport differ widely (Cui et al. 2005), and at least a dozen equations have been proposed for estimating sediment transport in ungauged channels.

A suite of numerical methods is now available for modelling various aspects of fluvial sediment transports, including the movement of sediment waves in channels (Cui et al. 2003a; 2003b), or the growth and decay of river meanders on floodplains via processes such as bed erosion and bank erosion (Figure 5.12) (Lauer and Parker 2008; Duan and Julien 2010). We need to understand how floodplains evolve if we wish to predict where floodplain deposits are most stable and long-lived. Age distributions of vegetation growing on floodplains, for example, reveal that the residence time of the underlying sediment is exponentially distributed, so that all sediment on a floodplain is equally likely to be eroded in a given interval. Computer simulations show, however, that meandering rivers that construct their own floodplains by lateral accretion

preferentially erode younger floodplains, thus producing a heavy-tailed distribution of sediment residence times (Bradley and Tucker 2013). This observation is consistent with flume experiments that show that the fraction of unreworked floodplain decays exponentially with time (Wickert et al. 2013). These insights offer estimates of how long a given portion of floodplain remains in place on average before being eroded again (Figure 5.13). Recall that floodplains sustain a large share of the Earth's human population, including many of the world's largest cities. Accordingly, most of these floodplains' rivers are heavily regulated, and efforts to re-establish natural floodplains or build ecosystem-based flood defenses might therefore benefit from numerical predictions of floodplain dynamics.

As with every other model, we need to think carefully about the assumptions and limitations of deterministic models. Many models concerned with fluvial geomorphology, for example, build largely on assumptions of steady flow and use empirical sediment transport and scour equations. The result is that these models inadequately capture processes such as local channel scour, high sediment loads or dam breach without additional tuning (Westoby et al. 2014).

Predicting the amplitude of landslide-triggered tsunamis The sudden displacement of water by landslides may trigger large displacement waves or landslide tsunamis. A number of deterministic models is now available for making predictions of this phenomenon. These physics-motivated or empirical equations attempt to predict the tsunami amplitude or wave height from characteristics of the wavemaker, but produce a broad range of results depending on the required parameters (Table 5.1). Essentially all of these equations predict that the maximum landslide-tsunami amplitude A is a power-law function (or a power law mixed with an exponential function) of

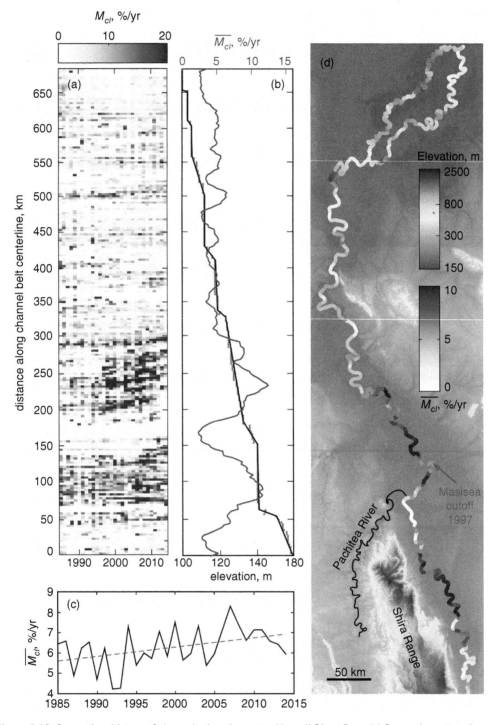

Figure 5.12 Space-time history of channel migration rates, Ucayali River, Peru. (a) Space-time plot of centerline-migrated areas normalized by channel areas ($\overline{M_{cl}}$). (b) Migration rates averaged over 30 years with M_{cl} showing spatial differences in migration rates (blue line) and SRTM-derived elevation profile (black line). (c) Time series and linear trend line (dashed red line) of average migration rates. (d) 2015 centerline plotted on top of an elevation map and colored by average migration rate for the 1985–2015 period. From Schwenk et al. (2016).

Floodplain Age (thousands of years)

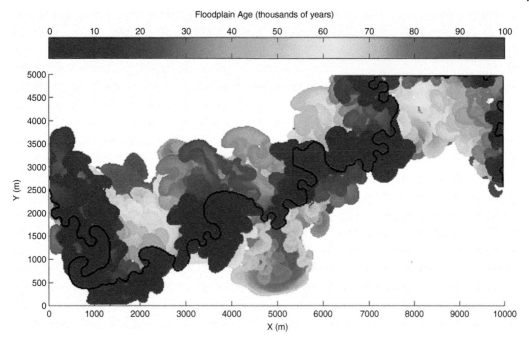

Figure 5.13 Floodplain age at the end of a numerical meander simulation of 100 000 model years. Flow direction is from left to right. Cool colors indicate young sediment, and warm colors older sediment. The channel location is colored black. Areas never visited by the channel are white. From Bradley and Tucker (2013).

parameters that describe the landslide size, its entrance velocity, the detachment slope angle, and the local water depth. Thus derived estimates of A differ considerably for a given landslide size, reflecting the many uncertainties that the various equations mask or collapse into a few parameters. Hence, we might find it more informative and reliable to predict landslide tsunami amplitude with more than a single model and use some weighted output, such as the mean (and standard deviation) of the model results. This simple strategy expresses much better the uncertainties tied to an ensemble of models instead of relying or preferring any single model.

Predicting processes on sedimentary coasts follows the same physical principles of conservation of mass, momentum, and energy and the numerical recipes to approximate local solutions of how surface elevations change with time (Roelvink et al. 2009). Models incorporate components that simulate erosion as a function of shoreline geometry and local sediment budget, and can include storm patterns, wind and wave regimes, tidal and sea-level changes, processes of dune erosion and migration, dynamics of tidal prism and overwash events, as well as gradual barrier migration (McNamara and Werner 2008). Recent additions to the coastal modelling toolkit include simulations of erosion and deposition by tsunamis (Yamashita et al. 2016).

Table 5.1 Empirical equations for predicting the near-field amplitude of landslide tsunamis (modified from Watt et al. 2012).

Slide model	Predictive equation
Subaerial	
2D empirical block	$A = 0.5d \sqrt{\dfrac{u}{\sqrt{gt}}} (t/d)^3$
2D empirical block	$A = 1.32d \left(4.5 \dfrac{wd^2}{V} \sqrt{l/d} \right)^{-0.68}$
2D empirical granular	$A = \dfrac{4}{9}d \left(u \sqrt{\dfrac{l}{ghd}} \left(\dfrac{v\rho_s}{\rho_w w d^2} \right)^{0.25} \sqrt{\cos(\tfrac{6}{7}\theta)} \right)^{0.8}$
3D empirical block	$A_{3D} \approx 0.035d \left[0.43 \left(\dfrac{ut}{d^2} \right)^{-1.27} \left(\dfrac{u}{\sqrt{gd}} \right)^{-0.66} \right]^{-0.45} e^{0.6 \cos\gamma} (\sin\theta)^{-0.286} \left(\dfrac{r}{d} \right)^{-0.44}$
Submarine	
2D hydrodynamic	$A = \dfrac{\pi t^2}{4h}$
2D energy scaling	$A = \left(\dfrac{\mu(\rho_s - \rho_w) t l \Delta z}{\rho_w \lambda} \right)^{0.5}$
2D empirical block	$A = 0.5\pi t \left(\dfrac{\rho_s}{\rho_w} + 1 \right) (0.0574 - 0.0431 \sin\theta) \left(\dfrac{l \sin\theta}{h} \right)^{1.25} \left(1 - e^{-2.2 \left(\frac{\rho_w}{\rho_s} - 1 \right)} \right)$
3D empirical block	$A = A_{2D} \dfrac{w}{w + 2\lambda}$

Parameters: A is maximum landslide-tsunami amplitude [m]; g is gravitational acceleration [m s^{-2}]; d is water depth at the base of the landslide slope [m]; h is the depth of submergence of the initial centre of mass of the landslide [m]; l is landslide length [m]; r is the distance of the point of maximum wave amplitude from the shoreline [m]; t is landslide (frontal) thickness [m]; u is entrance velocity [m s^{-1}]; V is landslide volume [m^3] estimated as $V = lwt$; w is landslide width [m]; Δz is the vertical drop in the landslide centre of mass [m]; γ is the radial angle of motion along the slide axis [°]; λ is the characteristic tsunami wavelength [m]; μ is the efficiency of energy conversion [1]; ρ_s is the density of the landslide [kg m^{-3}]; ρ_w is the density of sea water [kg m^{-3}]; and θ is the slope angle [°].

References

Aerts JCJH, Botzen WJW, Emanuel K, et al. 2014 Evaluating flood resilience strategies for coastal megacities. *Science* **344**(6183), 473–475.

Ayalew L and Yamagishi H 2005 The application of GIS-based logistic regression for landslide susceptibility mapping in the Kakuda-Yahiko Mountains, Central Japan. *Geomorphology* **65**(1-2), 15–31.

Babault J, Bonnet S, Crave A, and Van Den Driessche J 2005 Influence of piedmont sedimentation on erosion dynamics of an uplifting landscape: An experimental approach. *Geology* **33**(4), 301–304.

Bachmann D, Bouissou S, and Chemenda A 2006 Influence of large scale topography on gravitational rock mass movements: New insights from physical modeling. *Geophysical Research Letters* **33**(21), 1–4.

Barbot S, Lapusta N, and Avouac JP 2012 Under the hood of the earthquake machine: toward predictive modeling of the seismic cycle. *Science* **336**(6082), 707–710.

Beagley R, Davies T and Eaton B 2020 Past, present and future behaviour of the Waiho River, Westland, New Zealand: a new perspective. *Journal of Hydrology (NZ)* **59**(1), 41–61.

Benda L and Dunne T 1997 Stochastic forcing of sediment routing and storage in channel networks. *Water Resources Research* **33**(12), 2865–2880.

Benn DI, Owen LA, Finkel RC, and Clemmens S 2006 Pleistocene lake outburst floods and fan formation along the eastern Sierra Nevada, California: implications for the interpretation of intermontane lacustrine records. *Quaternary Science Reviews* **25**(21-22), 2729–2748.

Bennett GL, Molnar P, McArdell BW, and Burlando P 2014 A probabilistic sediment cascade model of sediment transfer in the Illgraben. *Water Resources Research* **50**, 1225–1244.

Bernardara P, Andreewsky M, and Benoit M 2011 Application of regional frequency analysis to the estimation of extreme storm surges. *Journal of Geophysical Research: Oceans* **116**(C2), 1–11.

Berryman KR, Cochran UA, Clark KJ, et al. 2012 Major earthquakes occur regularly on an isolated plate boundary fault. *Science* **336**(6089), 1690–1693.

Birkeland KW and Landry CC 2002 Power-laws and snow avalanches. *Geophysical Research Letters* **29**(11), 1–3.

Bonnet E, Bour O, Odling NE, et al. 2001 Scaling of fracture systems in geological media. *Reviews of Geophysics* **39**(3), 347–383.

Bradley DN and Tucker GE 2013 The storage time, age, and erosion hazard of laterally accreted sediment on the floodplain of a simulated meandering river. *Journal of Geophysical Research: Earth Surface* **118**(3), 1308–1319.

Brardinoni F, Hassan MA, Rollerson T, and Maynard D 2009 Colluvial sediment dynamics in mountain drainage basins. *Earth and Planetary Science Letters* **284**(3-4), 310–319.

Breiman L 2001 Statistical modeling: The two cultures (with comments and a rejoinder by the author). *Statistical Science* **16**(3), 199–231.

Buzna L, Peters K, and Helbing D 2006 Modelling the dynamics of disaster spreading in networks. *Physica A: Statistical Mechanics and its Applications* **363**(1), 132–140.

Casadei M, Dietrich WE, and Miller NL 2003 Testing a model for predicting the timing and location of shallow landslide initiation in soil-mantled landscapes. *Earth Surface Processes and Landforms* **28**(9), 925–950.

Chen SC, Lin TW, and Chen CY 2015 Modeling of natural dam failure modes and downstream riverbed morphological changes with different dam materials in a flume test. *Engineering Geology* **188**(C), 148–158.

Clauset A, Shalizi CR, and Newman MEJ 2009 Power-law distributions in empirical data. *SIAM Review* **51**(4), 661–703.

Connor CB, Sparks RSJ, Mason RM, et al. 2003 Exploring links between physical and probabilistic models of volcanic eruptions: The Soufriere Hills Volcano, Montserrat. *Geophysical Research Letters* **30**(13), 1–4.

Crosta GB, Chen H, and Frattini P 2006 Forecasting hazard scenarios and implications for the evaluation of countermeasure efficiency for large debris avalanches. *Engineering Geology* **83**(1), 236–253.

Cui Y, Parker G, Lisle TE, et al. 2003a Sediment pulses in mountain rivers: 1. Experiments. *Water Resources Research* **39**(9), 1–12.

Cui Y, Parker G, Pizzuto, JE, and Lisle TE 2003b Sediment pulses in mountain rivers: 2. Comparison between experiments and numerical predictions. *Water Resources Research* **39**(9), 1–11.

Cui Y, Parker G, Lisle TE, et al. 2005 More on the evolution of bed material waves in alluvial rivers. *Earth Surface Processes and Landforms* **30**(1), 107–114.

Dade WB, Renshaw CE, and Magilligan FJ 2011 Sediment transport constraints on river response to regulation. *Geomorphology* **126**(1-2), 245–251.

Davies TRH, McSaveney MJ, and Clarkson PJ 2003a Anthropic aggradation of the Waiho River, Westland, New Zealand: microscale modelling. *Earth Surface Processes and Landforms* **28**(2), 209–218.

Davies TRH, Smart CC, and Turnbull JM 2003b Water and sediment outbursts from advanced Franz Josef Glacier, New Zealand. *Earth Surface Processes and Landforms* **28**(10), 1081–1096.

Davies TRH and Davies AJ 2018 Increasing communities' resilience to disasters: An impact-based approach. *International journal of disaster risk reduction* **31**, 742–749.

de Haas T, Ventra D, Carbonneau PE, and Kleinhans MG 2014 Debris-flow dominance of alluvial fans masked by runoff reworking and weathering. *Geomorphology* **217**(C), 165–181.

de Scally FA and Owens IF 2004 Morphometric controls and geomorphic responses on fans in the Southern Alps, New Zealand. *Earth Surface Processes and Landforms* **29**(3), 311–322.

Denolle MA, Dunham EM, Prieto GA, and Beroza GC 2014 Strong ground motion prediction using virtual earthquakes. *Science* **343**(6169), 399–403.

Densmore AL, Anderson RS, McAdoo BG, and Ellis MA 1997 Hillslope evolution by bedrock landslides. *Science* **275**, 369–372.

Doyle MW and Harbor JM 2003 Modelling the effect of form and profile adjustments on channel equilibrium timescales. *Earth Surface Processes and Landforms* **28**(12), 1271–1287.

Duan JG and Julien PY 2010 Numerical simulation of meandering evolution. *Journal of Hydrology* **391**(1-2), 34–46.

Dussauge C, Grasso JR, and Helmstetter A 2003 Statistical analysis of rockfall volume distributions: Implications for rockfall dynamics. *Journal of Geophysical Research: Solid Earth* **108**, 1–11.

Emmer A and Vilímek V 2014 New method for assessing the susceptibility of glacial lakes to outburst floods in the Cordillera Blanca, Peru. *Hydrology and Earth System Sciences* **18**(9), 3461–3479.

Forster M and Sober E 1994 How to tell when simpler, more unified, or less ad hoc theories will provide more accurate predictions. *The British Journal for the Philosophy of Science* **45**(1), 1–35.

Ganti V, Straub KM, Foufoula-Georgiou E, and Paola C 2011 Space-time dynamics of depositional systems: Experimental evidence and theoretical modeling of heavy-tailed statistics. *Journal of Geophysical Research: Earth Surface* **116**(F2), 1–17.

Gao Y, Wang H, Liu GM, et al. 2014 Risk assessment of tropical storm surges for coastal regions of China. *Journal of Geophysical Research: Atmospheres* **119**(9), 5364–5374.

George DL and Iverson RM 2014 A depth-averaged debris-flow model that includes the effects of evolving dilatancy. II. Numerical predictions and experimental tests. *Proceedings of the Royal Society A:*

Mathematical, Physical and Engineering Sciences **470**(2170), 20130820–20130820.

Gill JC and Malamud BD 2014 Reviewing and visualizing the interactions of natural hazards. *Reviews of Geophysics* **52**(4), 680–722.

Gischig VS, Eberhardt E, Moore JR, and Hungr O 2015 On the seismic response of deep-seated rock slope instabilities — Insights from numerical modeling. *Engineering Geology* **193**(C), 1–18.

Goodman JA, Purkis SJ, and Phinn SR 2013 *Coral Reef Remote Sensing*. Springer.

Griffiths GA 2003 Downstream hydraulic geometry and hydraulic similitude. *Water Resources Research* **39**(4), 1–6.

Grinsted A 2013 An estimate of global glacier volume. *The Cryosphere* **7**(1), 141–151.

Grosse P, van Wyk de Vries B, Euillades PA, et al. 2012 Systematic morphometric characterization of volcanic edifices using digital elevation models. *Geomorphology* **136**(1), 114–131.

Guan M, Wright NG, Sleigh PA, and Carrivick JL 2015 Assessment of hydro-morphodynamic modelling and geomorphological impacts of a sediment-charged jökulhlaup, at Solheimajokull, Iceland. *Journal of Hydrology* **530**, 336–349.

Gusev AA, Ponomareva VV, Braitseva OA, et al. 2003 Great explosive eruptions on Kamchatka during the last 10,000 years: Self-similar irregularity of the output of volcanic products. *Journal of Geophysical Research: Solid Earth* **108**(B2), 1–18.

Guzzetti F, Peruccacci S, Rossi M, and Stark CP 2007 The rainfall intensity–duration control of shallow landslides and debris flows: an update. *Landslides* **5**(1), 3–17.

Hengl T and Reuter HI (Eds) 2009 *Geomorphometry: Concepts, Software, Applications*, Developments in Soil Science series Vol 33. Elsevier.

Hooke JM 2007 Complexity, self-organisation and variation in behaviour in meandering rivers. *Geomorphology* **91**(3-4), 236–258.

Hürlimann M, Rickenmann D, Medina V, and Bateman A 2008 Evaluation of approaches to calculate debris-flow parameters for hazard assessment. *Engineering Geology* **102**(3-4), 152–163.

Irish JL, Resio DT, and Divoky D 2011 Statistical properties of hurricane surge along a coast. *Journal of Geophysical Research* **116**(C10), 345.

Iverson RM, Schilling SP, and Vallance JW 1998 Objective delineation of lahar-inundation hazard zones. *Geological Society of America Bulletin* **110**(8), 972–984.

Jones R, Manville V, Peakall J, et al. 2017 Real-time prediction of rain-triggered lahars: incorporating seasonality and catchment recovery. *Natural Hazards and Earth System Sciences* **17**(12), 2301–2312.

Kagan YY, Jackson DD, and Geller RJ 2012 Characteristic earthquake model, 1884-2011, R.I.P.. *Seismological Research Letters* **83**(6), 951–953.

Kalbermatten M, Van De Ville D, Turberg P, et al. 2012 Multiscale analysis of geomorphological and geological features in high resolution digital elevation models using the wavelet transform. *Geomorphology* **138**(1), 352–363.

Karátson D, Favalli M, Tarquini S, et al. 2010 The regular shape of stratovolcanoes: A DEM-based morphometrical approach. *Journal of Volcanology and Geothermal Research* **193**(3-4), 171–181.

Katz O and Aharonov E 2005 Landslides in vibrating sand box: What controls types of slope failure and frequency magnitude relations?. *Earth and Planetary Science Letters* **247**(3), 280–294.

Katz RW, Parlange MB, and Naveau P 2001 Statistics of extremes in hydrology. *Advances in Water Resources* **25**(8), 1287–1304.

Kelfoun K and Druitt TH 2005 Numerical modeling of the emplacement of Socompa rock avalanche, Chile. *Journal of Geophysical Research: Solid Earth* **110**(B12), 1–13.

Kelfoun K, Giachetti T, and Labazuy P 2010 Landslide-generated tsunamis at Réunion Island. *Journal of Geophysical Research: Earth Surface (2003–2012)* **115**(F4), 1–17.

Kerr R 2011 More megaquakes on the way? That depends on your statistics. *Science* **332**(6028), 411.

Kogure T and Matsukura Y 2012 Threshold height of coastal cliffs for collapse due to tsunami; theoretical analysis of the coral limestone cliffs of the Ryukyu Islands, Japan. *Marine Geology* 323–325, 14–23.

Korup O, Gorum T, and Hayakawa Y 2012 Without power? Landslide inventories in the face of climate change. *Earth Surface Processes and Landforms* **37**(1), 92–99.

Kritikos T, Robinson TR, and Davies TRH 2015 Regional coseismic landslide hazard assessment without historical landslide inventories: A new approach. *Journal of Geophysical Research: Earth Surface* **120**(4), 711–729.

Kruschke JK 2010 What to believe: Bayesian methods for data analysis. *Trends in Cognitive Sciences* **14**(7), 293–300.

Lague D, Crave A, and Davy P 2003 Laboratory experiments simulating the geomorphic response to tectonic uplift. *Journal of Geophysical Research: Solid Earth* **108**(B1), 1–20.

Larsen IJ, Montgomery DR, and Korup O 2010 Landslide erosion controlled by hillslope material. *Nature Geoscience* **3**(4), 247–251.

Lauer JW and Parker G 2008 Modeling framework for sediment deposition, storage, and evacuation in the floodplain of a meandering river: Theory. *Water Resources Research* **44**(4), 1–16.

Li AJ, Lyamin AV, and Merifield RS 2009 Seismic rock slope stability charts based on limit analysis methods. *Computers and Geotechnics* **36**(1-2), 135–148.

Lockwood JP and Hazlett RlW 2010 Volcanoes. *Global Perspectives*. Wiley Blackwell.

Malamud BD and Turcotte DL 2006 The applicability of power-law frequency statistics to floods. *Journal of Hydrology* **322**(1-4), 168–180.

Malamud BD, Morein G, and Turcotte DL 1998 Forest fires: an example of self-organized critical behavior. *Science* **281**(5384), 1840–1842.

Malinverno A 1997 On the power law size distribution of turbidite beds. *Basin Research* **9**, 263–274.

Mangeney-Castelnau A, Bouchut F, Vilotte JP, et al. 2005 On the use of Saint Venant equations to simulate the spreading of a granular mass. *Journal of Geophysical Research: Solid Earth* **110**(B9), 1–17.

Maniatis G, KurfeÃŸ D, Hampel A, and Heidbach O 2009 Slip acceleration on normal faults due to erosion and sedimentation." Results from a new three-dimensional numerical model coupling tectonics and landscape evolution. *Earth and Planetary Science Letters* **284**(3-4), 570–582.

Marzocchi W and Woo G 2007 Probabilistic eruption forecasting and the call for an evacuation. *Geophysical Research Letters* **34**(22), 1–4.

Marzocchi W, Sandri L, Gasparini P, et al. 2004 Quantifying probabilities of volcanic events: the example of volcanic hazard at Mount Vesuvius. *Journal of Geophysical Research: Solid Earth* **109**, 1–18.

Maza M, Lara JL, and Losada IJ 2015 Tsunami wave interaction with mangrove forests: A 3-D numerical approach. *Coastal Engineering* **98**, 33–54.

McKean J and Roering J 2004 Objective landslide detection and surface morphology mapping using high-resolution airborne laser altimetry. *Geomorphology* **57**(3-4), 331–351.

McKillop RJ and Clague JJ 2007 Statistical, remote sensing-based approach for estimating the probability of catastrophic drainage from moraine-dammed lakes in southwestern British Columbia. *Global and Planetary Change* **56**(1-2), 153–171.

McNamara DE and Werner BT 2008 Coupled barrier island–resort model: 1. Emergent instabilities induced by strong human-landscape interactions. *Journal of Geophysical Research: Earth Surface* **113**, 1–10.

Meyer NK, Schwanghart W, Korup O, and Nadim F 2015 Roads at risk: traffic detours from debris flows in southern Norway. *Natural Hazards and Earth System Sciences* **15**(5), 985–995.

Meyer NK, Schwanghart W, Korup O, et al. 2014 Estimating the topographic predictability of debris flows. *Geomorphology* **207**(C), 114–125.

Montanari A and Koutsoyiannis D 2014 Modeling and mitigating natural hazards: Stationarity is immortal!. *Water Resources Research* **50**(12), 9748–9756.

Mouri G, Shiiba M, Hori T, and Oki T 2011 Modeling shallow landslides and river bed variation associated with extreme rainfall-runoff events in a granitoid mountainous forested catchment in Japan. *Geomorphology* **125**(2), 282–292.

Mueller ER and Pitlick J 2005 Morphologically based model of bed load transport capacity in a headwater stream. *Journal of Geophysical Research: Earth Surface* **110**, 1–14.

Nandasena NAK, Paris R, and Tanaka N 2011 Numerical assessment of boulder transport by the 2004 Indian ocean tsunami in Lhok Nga, West Banda Aceh (Sumatra, Indonesia). *Computers & Geosciences* **37**(9), 1391–1399.

Newhall C and Hoblitt R 2001 Constructing event trees for volcanic crises. *Bulletin of Volcanology* **64**(1), 3–20.

Nussbaumer S, Schaub Y, Huggel C, and Walz A 2014 Risk estimation for future glacier lake outburst floods based on local land-use changes. *Natural Hazards and Earth System Sciences* **14**(6), 1611–1624.

Ono K and Yamaguchi A 2011 Distributed specific sediment yield estimations in Japan attributed to extreme-rainfall-induced slope failures under a changing climate. *Hydrology and Earth System Sciences* **15**(1), 197–207.

Oreskes N, Shrader-Frechette K, and Belitz K 1994 Verification, validation, and confirmation of numerical models in the earth sciences. *Science* **263**(5147), 641–646.

Paola C, Straub K, Mohrig D, and Reinhardt L 2009 The "unreasonable effectiveness" of stratigraphic and geomorphic experiments. *Earth-Science Reviews* **97**(1-4), 1–43.

Pelletier J and Mayer L 2005 An integrated approach to flood hazard assessment on alluvial fans using numerical modeling, field mapping, and remote sensing. … *Society of America* ….

Pelletier JD, Malamud BD, Blodgett T, and Turcotte DL 1996 Scale-invariance of soil moisture variability and its implications for the frequency-size distribution of landslides. *Engineering Geology* **48**(3), 255–268.

Perron JT 2011 Numerical methods for nonlinear hillslope transport laws. *Journal of Geophysical Research: Earth Surface* **116**, 1–13.

Robinson TR, Davies TRH, Wilson TM, and Orchiston C 2016 Coseismic landsliding estimates for an Alpine Fault earthquake and the consequences for erosion of the Southern Alps, New Zealand. *Geomorphology* **263**, 71–86.

Roelvink D, Reniers A, van Dongeren A, et al. 2009 Modelling storm impacts on beaches, dunes and barrier islands. *Coastal Engineering* **56**(11-12), 1133–1152.

Roering JJ, Perron JT, and Kirchner JW 2007 Functional relationships between denudation and hillslope form and relief. *Earth and Planetary Science Letters* **264**(1-2), 245–258.

Rustomji P, Bennett N, and Chiew F 2009 Flood variability east of Australia's Great Dividing Range. *Journal of Hydrology* **374**(3-4), 196–208.

Schmidt J, Matcham I, Reese S, et al. 2011 Quantitative multi-risk analysis for natural hazards: a framework for multi-risk modelling. *Natural Hazards* **58**(3), 1169–1192.

Schneider D, Bartelt P, Caplan-Auerbach J, et al. 2010 Insights into rock-ice avalanche dynamics by combined analysis of seismic recordings and a numerical avalanche model. *Journal of Geophysical Research: Earth Surface* **115**, 1–20.

Schwenk J, Khandelwal A, Fratkin M, et al. 2016 High spatiotemporal resolution of river planform dynamics from Landsat: The RivMAP toolbox and results from the Ucayali River. *Earth and Space Science* **4**, 46–75.

Smith MW 2014 Roughness in the Earth Sciences. *Earth-Science Reviews* **136**(C), 202–225.

Spiegelhalter D, Pearson M, and Short I 2011 Visualizing uncertainty about the future. *Science* **333**(6048), 1393–1400.

Strenk PM and Wartman J 2011 Uncertainty in seismic slope deformation model predictions. *Engineering Geology* **122**(1-2), 61–72.

Strunz G, Post J, Zosseder K, et al. 2011 Tsunami risk assessment in Indonesia. *Natural Hazards and Earth System Sciences* **11**(1), 67–82.

Sutherland WJ, Spiegelhalter D, and Burgman MA 2013 Twenty tips for interpreting scientific claims. *Nature* **503**, 335–337.

ten Brink US, Barkan R, Andrews BD, and Chaytor JD 2009 Size distributions and failure initiation of submarine and subaerial landslides. *Earth and Planetary Science Letters* **287**(1-2), 31–42.

Trepanier JC, Ellis KN, and Tucker CS 2015 Hurricane risk variability along the Gulf of Mexico coastline. *PloS ONE* **10**, e0118196.

Turcotte DL and Schubert G 2002 *Geodynamics* 2nd edn. Cambridge University Press.

Turcotte DL, Malamud BD, Guzzetti F, and Reichenbach P 2002 Self-organization, the cascade model, and natural hazards. *Proceedings of the National Academy of Sciences* **99** (Suppl 1), 2530–2537.

Van De Wiel MJ and Coulthard TJ 2010 Self-organized criticality in river basins: Challenging sedimentary records of environmental change. *Geology* **38**(1), 87–90.

Van De Wiel MJ, Coulthard TJ, Macklin MG, and Lewin J 2011 Modelling the response of river systems to environmental change: Progress, problems and prospects for palaeo-environmental reconstructions. *Earth-Science Reviews* **104**(1-3), 167–185.

Vanmaercke M, Poesen J, Broeckx J, and Nyssen J 2014 Sediment yield in Africa. *Earth-Science Reviews* **136**(C), 350–368.

Vanmaercke M, Poesen J, Maetens W, et al. 2011 Sediment yield as a desertification risk indicator. *Science of the Total Environment, The* **409**(9), 1715–1725.

Vecchi GA and Villarini G 2015 Next season's hurricanes. *Science* **343**, 618–619.

Vicari A, Ganci G, Behncke B, et al. 2011 Near-real-time forecasting of lava flow hazards during the 12-13 January 2011 Etna eruption. *Geophysical Research Letters* **38**(13), 1–7.

Walder J and O'Connor J 1997 Methods for predicting peak discharge of floods caused by failure of natural and constructed earthen dams. *Water Resources Research* **33**(10), 2337–2348.

Watt SFL, Mather TA, and Pyle DM 2007 Vulcanian explosion cycles: Patterns and predictability. *Geology* **35**(9), 839–842.

Watt SFL, Talling PJ, Vardy ME, et al. 2012 Combinations of volcanic-flank and seafloor-sediment failure offshore Montserrat, and their implications for tsunami generation. *Earth and Planetary Science Letters* **319**-**320**, 228–240.

Watts P 2004 Probabilistic predictions of landslide tsunamis off Southern California. *Marine Geology* **203**(3-4), 281–301.

Welsh A and Davies, TRH 2011 Identification of alluvial fans susceptible to debris-flow hazards. *Landslides* **8**(2), 183–194.

Westoby MJ, Glasser NF, Hambrey MJ, et al. 2014 Reconstructing historic glacial lake outburst floods through numerical modelling and geomorphological assessment: Extreme events in the Himalaya. *Earth Surface Processes and Landforms* **39**(12), 1675–1692.

Whitehouse IE and Griffiths GA 1983 Frequency and hazard of large rock avalanches in the central Southern Alps, New Zealand. *Geology* **11**, 331–334.

Wickert AD, Martin JM, Tal M, et al. 2013 River channel lateral mobility: metrics, time scales, and controls. *Journal of Geophysical Research: Earth Surface* **118**(2), 396–412.

Wilford DJ, Sakals ME, Innes JL, et al. 2004 Recognition of debris flow, debris flood and flood hazard through watershed morphometrics. *Landslides* **1**(1), 61–66.

Wolpert RL, Ogburn SE, and Calder ES 2016 The longevity of lava dome eruptions. *Journal of Geophysical Research: Solid Earth* **121**, 676–686.

Yager EM, Kirchner JW, and Dietrich WE 2007 Calculating bed load transport in steep boulder bed channels. *Water Resources Research* **43**(7), 1–24.

Yamashita K, Sugawara D, Takahashi T, et al. 2016 Numerical simulations of large-scale sediment transport caused by the 2011 Tohoku earthquake tsunami in Hirota Bay, southern Sanriku coast. *Coastal Engineering Journal* **58**(04), 1640015.

Yudkowsky E 2008 Cognitive biases potentially affecting judgement of global risks. In *Global Catastrophic Risks* (ed. Bostrom N and Cirkovic MM), Oxford University Press, pp. 91–119.

6

Earthquake Hazards

Plate tectonic movements generate stress and accumulating elastic strain energy in the Earth's crust; when the crust eventually fractures the released elastic strain energy transforms into seismic waves. These seismic waves travel at speeds of many kilometres per second through the Earth's crust and interior. Slower shear waves, in particular, cause much of the damage to houses and other built structures. We cannot predict the timing and magnitude of future earthquakes, but the underlying physics of fault rupture are fairly well known. The list of processes and phenomena that interact with each other during fault rupture is long and contains, among others, the thermal pressurization of pore fluids; flash heating of micro-asperity contacts; fragmenting and melting of rocks and gouge; variation in normal stress due to material impurities across the fault surface; changes in the chemical environment; gouge microseismic (or acoustic) fluidization; effects of humidity; fault roughness; history of previous slip episodes; and coseismic triggering of mineral dehydration reactions (Bizzarri 2011). The release of elastic strain energy creates seismic waveforms that cause short-lived accelerations, and hence forces, on the Earth surface. These transient loads can cause structural damage, alter pore–water pressures, and destabilize hillslopes to create landslides (Figure 6.1).

6.1 Frequency and Magnitude of Earthquakes

About 90% of all earthquakes occur along the subduction zones of the Pacific Ring of Fire, which together form the longest continuous set of active plate boundaries on the planet. These subduction zones have also given rise to the strongest earthquakes ever recorded (M >9), and account for a disproportionately large fraction of the total seismic moment of all earthquakes, and therefore energy release and associated fault slip. Instrumental records now highlight the broad range of earthquake magnitudes and recurrence intervals at subduction zones, but also point out the gaps in knowledge. Geomorphic and geological evidence fills some of these gaps toward the ultimate aim of predicting future earthquakes (Satake and Atwater 2007). Tectonic structures, landforms, and sedimentary layers can affect the potential for generating large earthquakes. For example, the role of subducting submarine volcanoes known as seamounts continues to puzzle scientists (Watts et al. 2010). The topographic roughness of subducting seamounts may add frictional stresses to the plate boundary, contributing to its coupling, and thus priming future fault rupture.

Most earthquakes are clustered at subduction zones and other active plate boundaries, so that we know comparably less about of strong earthquakes in intraplate settings far

Geomorphology and Natural Hazards: Understanding Landscape Change for Disaster Mitigation, Advanced Textbook Series, First Edition. Tim R. Davies, Oliver Korup, and John J. Clague.
© 2021 John Wiley & Sons Ltd. Co-published 2021 by the American Geophysical Union and John Wiley & Sons Ltd.

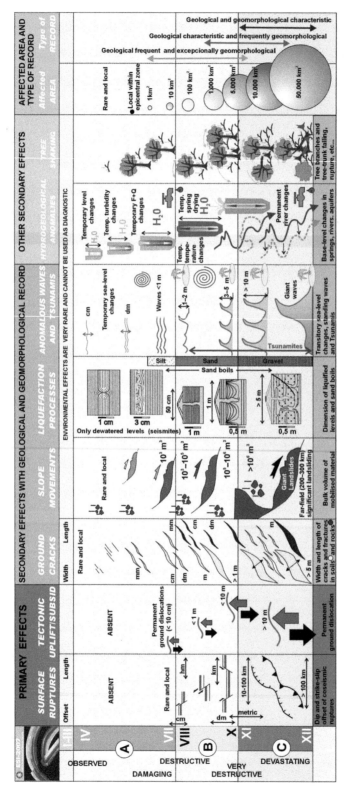

Figure 6.1 The INQUA Environmental Seismic Intensity Scale highlights several of the many cascading effects and subsequent hazards that strong earthquakes may trigger (https://en.wikipedia.org/wiki/Environmental_Seismic_Intensity_scale#/media/File:Graphic_representation_of_the_ESI_2007_intensity_degrees.tif).

away from any plate boundary (England and Jackson 2011). Because large intraplate quakes are infrequent, people are generally unprepared for them and buildings are rarely built to withstand strong shaking. A classic example is two M 8 intraplate earthquakes that occurred in the winter of 1811–1812 in the central Mississippi Valley in the United States. The earthquakes destroyed the town of New Madrid, Missouri, and were felt in nearly every city of eastern North America from New Orleans to Quebec City, an area of more than 2.5 million km^2. Rare moderate to large intraplate earthquakes also occur in southern Ontario and Quebec, Canada. Most of those quakes are associated with the ancient rifted edge of the North American continent. Strain accumulates along ancient faults along this rifted margin because of ocean-floor spreading on the Mid-Atlantic Ridge. The sparse records of large past intraplate earthquakes such as the New Madrid events result in considerably less reliable statistical estimates of the return periods of future events than those at plate boundaries.

In 1935, Charles Richter, working at the California Institute of Technology, developed the first quantitative scale for measuring the energy released by an earthquake. The Richter scale refers to the magnitude of local (California) earthquakes as the logarithm to the base 10 of the maximum signal wave amplitude recorded on a then-standard seismogram at a distance of 100 km from the epicentre. Although some news reports still refer to the Richter scale, seismologists today use other magnitude scales. The most commonly used measure of earthquake size is seismic moment, M_0. It is determined from an estimate of the area that ruptured along a fault plane during the quake, the amount of movement or slippage along the fault, and the rigidity of the rocks near the source. Seismic moment is a key measure of the size of an earthquake as a function of material properties and strain released when a rock mass ruptures:

$$M_0 = GA\Delta w \qquad (6.1)$$

where G is the shear modulus of the rocks that slip [N m^{-2}], A is the rupture area of the fault [m^2], and Δw is the average displacement across the fault during the earthquake [m]. This relationship is important, as it allows scientists to infer the approximate size of former earthquakes from offsets measured along fault scarps (Turcotte and Schubert 2002). It is important to understand that the moment magnitude scale, like all magnitude scales, is logarithmic. An increase from one whole number to the next higher one represents a ten-fold increase in the amount of shaking and a 32-fold increase in the amount of energy released. For example, the amount of ground motion from an M 7 earthquake is ten times that of an M 6 earthquake, but the amount of energy released is 32 times as much.

Dating methods that offer resolution down to a given year are useful for establishing the frequency and magnitude of past earthquakes that preceded recording by seismic stations. The measurement of lichen diameters and the thickness of weathering rinds on rockfall blocks have been promising in this regard. These low cost methods allow collection of a large number of ages that are statistically robust to be used for palaeoseismological analyses (Bull and Brandon 1998). Lichenometry must be based on reliable growth curves because environmental parameters such as ground temperature, exposure, windchill, rainfall, and substrate mineralogy can affect the local growth rate and diameter of a given lichen species. A few researchers have tried to document and estimate the uncertainties tied to lichenometric dating of rock surfaces (Naveau et al. 2007), although the user friendliness of the necessary statistical techniques has been questioned (Bradwell 2009), and others have called into question the entire technique (Osborn et al. 2015).

Figure 6.2 Observed ground deformation from the 2016 Kaikoura, New Zealand, earthquake. (a) and (b) Coastal uplift of 2–3 m associated with the Papatea block (labeled in (c). Red lines in inset show locations of known active faults. The black box indicates the Marlborough fault system. (c) Three-dimensional displacement field derived from satellite radar data. Vectors in (c) represent horizontal displacements, and the coloured background shows the vertical displacements. From Hamling et al. (2017).

Big earthquake on small faults The M 7.8 Kaikoura earthquake in New Zealand in 2016 was the best example to date of a major earthquake that took place not on a major fault, but on fault system that was less known (Figures 6.2 and 6.3). It ruptured about 20 smaller faults sequentially, about half of which were known prior to the event (Hamling et al. 2017). The ruptured faults cross the major Hope Fault, but only a very minor segment of the latter shows signs of movement. This earthquake shows that in the future we should discard the idea that major earthquakes are confined to major (and therefore known) faults. Instead, we should expect major earthquakes to occur on sequences of small faults, many of which will be unknown. The implications for seismic hazard assessment are serious; instead of basing seismic hazard models only on known faults, and thus delineating low hazard where faults are unknown, we should admit that many faults are unknown and delineate commensurately larger areas of potentially high seismic hazard.

6.2 Geomorphic Impacts of Earthquakes

6.2.1 The Seismic Hazard Cascade

Earthquakes are commonly the source of hazard cascades. Large earthquakes are followed

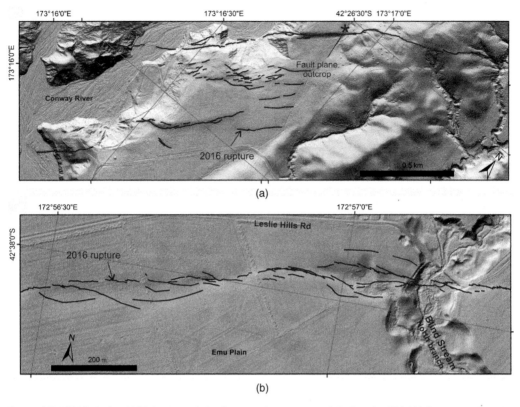

Figure 6.3 LiDAR-derived hillshade models of two surfaces ruptured during the 2016 Kaikoura earthquake, New Zealand: (a) Conway–Charwell fault and (b) The Humps fault. Surface ruptures (black lines) were interpreted from a combination of field mapping, LiDAR, and orthophotos. From Nicol et al. (2018).

by a prolonged series of aftershocks that may worsen the geomorphic impacts of the main shock. The clustering of earthquakes in space and time without any clear triggering mainshock is known as an earthquake swarm, which is manifest as an increase in the rate of seismicity above the background rate. Earthquake swarms can occur frequently at plate boundaries, mostly subduction zones, but also near volcanic centres (Holtkamp and Brudzinski 2011).

Earthquakes readjust crustal stresses by releasing accumulated strain along faults that cut blocks of the Earth's crust. Earthquakes may cause nearly instantaneous surface uplift or subsidence, depending on the type and geometry of fault rupture. The amount of ground deformation may be spectacular in

the case of stronger earthquakes, particularly where faults rupture the ground surface. Fault offsets may be several metres vertically and tens of metres horizontally for large earthquakes. But this deformation appears well beyond the corridors along fault ruptures. Large subduction earthquakes release accumulated strain along the plate-boundary interface over areas of up to several tens of thousands of square kilometres. The sudden coseismic subsidence of overriding plates in subduction zones makes coastal zones more vulnerable to incoming tsunamis, as the subsided land surface will be inundated by a greater depth of seawater that reaches farther inland.

Strong earthquakes have been responsible for some of the most deadly disasters in history. Yet seismic ground shaking is only one of the reasons for these dramatic losses. Earthquake-triggered soil liquefaction, tsunamis, landslides, and fires are responsible for additional losses, often greater than those caused by the ground motions themselves. However, liquefaction, landslides, and tsunamis occur directly with or shortly after ground shaking, so that the term 'coseismic' is more appropriate. Earthquake-induced landslides, liquefaction, tsunamis, and fire were responsible for fatalities in ~22% of all earthquakes between 1968 and 2008 (Marano et al. 2009). Disasters where the death toll from tsunamis exceeds that from the earthquake shaking are rare, but include the 2004 Sumatra earthquake. Removing the 2004 tsunami from the data leaves landslides as the cause of nearly 75% of all deaths from nonshaking effects of earthquakes – the 2011 Japan earthquake and tsunami are another exception, but occurred after this particular study had been conducted.

Earthquakes may trigger dozens to tens of thousands of landslides in a matter of minutes, and offer insights into the local site conditions affected by earthquake shaking (Keefer 2002; Fan et al. 2019). Topography tends to amplify seismic waves such that higher elevations, especially peaks or hill crests, are prone to more shaking than lower areas (Meunier et al. 2008; Buech et al. 2010). Hillslope inclination is another important element in models predicting local seismic site conditions (Allen and Wald 2009). The rationale is that slope steepness changes the local horizontal ground acceleration that, when considering the accelerated mass, raises shear stresses and thus steers the force balance of hillslopes towards more unstable conditions. Slopes that are stable under static conditions may thus fail under such dynamic loading. Numerical modelling, however, suggests that other factors that amplify seismic waves, such as material contrasts and rock-mass fractures, may be more important for triggering landslides during earthquakes than topographic location alone (Gischig et al. 2015).

Earthquake magnitude alone is rarely a good predictor of coseismic slope failure (Gorum et al. 2011). Local ground acceleration, the type and geometry of fault rupture, rock type and structure can affect the distribution of landslides. The directivity of incoming seismic waves can also trigger different geomorphic responses of hillslopes according to their aspect (Del Gaudio and Wasowski 2007). Quaternary sediments and Neogene rocks were the sources of ~80% of all terrestrial coseismic landslides triggered by the 2011 Tohoku earthquake; older rocks were more resistant to seismic shaking (Wartman et al. 2013). Previous earthquakes may weaken hillslopes and prime them for failure during the next quake, thus distorting the pattern of coseismic landslides predicted by simple slope-stability models (Alfaro et al. 2012). Detailed field investigations have shown that earthquake shaking can mobilize large landslides without much displacement, creating ground cracks and incipient head scarps that allow enhanced water infiltration and slow-moving slope deformation long after the main shock. Analyses of high-resolution air photographs taken before and after the 1999 M 7.6 Chi-Chi earthquake, Taiwan, with particle image velocimetry (PIV) revealed in detail how large slope failures can also respond noncatastrophically to seismic shaking. Tseng et al. (2009) calculated an average 24-m horizontal displacement for a 1.2-km^2 rockslide in the western foothills of central Taiwan during the earthquake. The landslide occurred in the same massif as the catastrophic and destructive Chiufengershan rock avalanches that were triggered by the same earthquake, highlighting topographic and lithologic controls on seismically induced slope instability. The geomorphic role of earthquakes thus extends to mechanically weakening rock masses by creating cracks for water infiltration, and thus preparing hillslopes for subsequent failure.

Studies involving mapping of hundreds to tens of thousands of earthquake-triggered landslides consistently show that most landslides cluster in corridors along fault ruptures (Massey et al. 2018). Also, in most recorded cases the decay of landslide density [km^{-2}] away from the fault is exponential. However, some isolated slope failures may occur at much larger distances from the earthquake epicentre than expected from such a general trend. These far-field landslides happen because of local seismic site effects, weak lithology, or antecedent rainfall (Delgado et al. 2011). Several researchers have also found that the number of coseismic landslides per unit area also increases nonlinearly with hillslope inclination. Field and aerial photo mapping of landslides triggered by the 2007 Niigata Chuetsu-Oki earthquake, Japan, revealed that slope failures were more than one hundred times more common per unit area on 70° slopes than they were on 35° slopes (Collins et al. 2012). Even between major earthquakes the seismically shattered and weakened rocks of fault zones may give rise to increased slope instability. For example, hundreds of seasonally active earthflows straddle the San Andreas fault zone in California (Scheingross et al. 2013).

aSoil liquefaction results from a transient increase in pore-water pressures, and hence reduced stability in saturated silty to sandy sediments. These sediments dominate river floodplains, sandy shorelines, deltas, some alluvial fans, deltas, and even landfills. These mostly flat to very gently inclined surfaces are areas that people preferentially develop, inhabit, and have often created artificially to obtain new parcels of land. Water-logged sedimentary bodies in coastal settings are especially prone to coseismic liquefaction. Frequently reported effects of liquefaction include the loss of the ground's bearing strength, lateral displacements of near-surface materials, ground cracking, sand boils, and flotation and settlement of built rigid structures. Even major buildings may topple on liquefied ground without losing their overall structural integrity, although damage to houses, streets, bridges, and other built structures can be widespread. The 2011 Christchurch, New Zealand, earthquake caused widespread liquefaction in the city of Christchurch, swamping many streets with thick sheets of sand expelled from the extensive Pleistocene and Holocene river sediments on which the city is built. Soil cracks and lateral ground displacements due to liquefaction were up to three metres, though limited mostly to within several hundred metres of streams and rivers (Cubrinovski et al. 2012). This and other field observations raised an intriguing question, because these fine deposits had previously been interpreted as tsunami deposits. However, the latter would have contained remains of marine microfauna, which liquefaction deposits would not, even though the liquefied soils may have been deposited originally in a near-coast environment.

Lake and sea floors also record the impact of earthquake shaking. Strong offshore earthquakes give rise to turbidity currents, which are rapid submarine gravity flows of sediment. Turbidity currents originate from deltas, continental shelves, and the heads of submarine canyons hosting large stacks of sediment. Coseismic turbidity currents form thick beds of layered subaquatic sediments known as turbidites. Occasionally, turbidity currents evolving from earthquake-triggered subaquatic mass movements may also trigger tsunamis. The M 7.2 Grand Banks earthquake off Newfoundland in 1929 was an early hint at the occurrence of such catastrophic sediment flows under water. The earthquake triggered a landslide that scraped ~2 × 10^{11} m^3 of sediments from the edge of the Atlantic continental shelf off Newfoundland, Canada. The landslide transformed into a turbidity current, which travelled some 1000 km eastward at speeds of between 60 and 100 km h^{-1}, judging from the times at which the flow successively severed twelve submarine telegraph cables during its runout (Fine et al. 2005). The landslide itself triggered a tsunami that claimed 27 lives in coastal communities in Newfoundland.

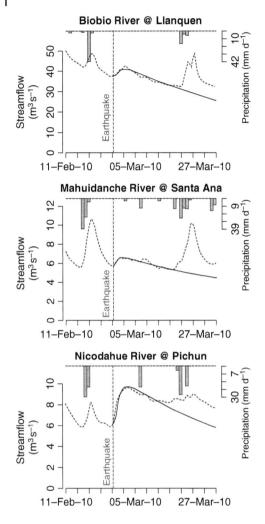

Figure 6.4 Stream flow response to the 2010 Maule, Chile, megathrust earthquake. Sample hydrographs show postseismic increase in river discharge. Black curve is fit from a seismically-triggered groundwater flow model; blue dashed line is measured stream flow; histogram shows daily precipitation data. Modified from Mohr et al. (2017).

6.2.2 Postseismic and Interseismic Impacts

Much of the damage and geomorphic impact of earthquakes occurs in the days, weeks, years, and even decades after the event (Figure 6.4). Inter- and postseismic ground deformation can have long-lasting consequences, particularly in coastal regions. Prince William Sound was the site of the 1964 Great Alaska earthquake,

the second largest earthquake of the twentieth century. The earthquake caused up to two metres of land subsidence, submerging coastlines and creating new estuaries. Other parts of southern Alaska experienced metres of uplift during the earthquake. Coseismic uplift and subsidence directly affect the locations and rates of river erosion and sedimentation. Coastal river mouths raised during the 1960 Great Chile earthquake induced fluvial sedimentation in shallow estuarine lakes, and high rates of postseismic sedimentation promoted ongoing channel instability for several decades (Reinhardt et al. 2010). Observations from earthquake-impacted rivers in Taiwan suggest that the amount of landslide sediment entering the fluvial system and the proximity to the earthquake source may eventually control whether river incision into bedrock is dampened or enhanced (Yanites et al. 2010).

Earthquake shaking can also contribute substantially to longer-term sediment budgets. Soil, sediment, and biomass mobilised by large earthquakes move slowly through the landscape by processes of surface runoff, landslides, debris flows, and fluvial transport. Estimated residence times of coseismic landslide debris in the drainage network range from several years to nearly a century and depend on rates of sediment transport and the storage capacity of river valleys. Debris generated by the 1923 Kanto earthquake still contributes the very high contemporary sediment yields in parts of the Japanese Alps, keeping channel beds elevated, filling in sediment retention dams, and providing sources for debris flows (Koi et al. 2008). Managing such delayed sediment pulses can be an intermediate to long-term task for hydraulic engineers and river managers who may only have access to contemporary water and sediment discharge data.

Fluvial suspended sediment loads fed by landslides triggered by the 1999 Chi-Chi earthquake, Taiwan, peaked sharply before dropping back to pre-earthquake levels some six years later (Dadson et al. 2004; Hovius et al. 2011). The mobility of such sediment pulses depends heavily on rainfall to rework and deliver the landslide debris from hillslopes to river channels. In the case of Taiwan, frequent tropical cyclones ensure ample heavy rainfall for flushing unconsolidated landslide debris from hillslopes and river floods are the key drivers of such sediment pulses or waves (see Chapter 10.2). The episodic reworking of coseismic landslide debris hampers efforts to plan and create new infrastructure and to safeguard valley floors from floods.

A continuing serious hazard resulting from the 2008 Wenchuan, China, earthquake is rainfall-triggered debris flows that are reworking coseismic landslide material. Yu et al. (2012) reported a single debris flow that moved 3.1 Mm^3 of landslide debris following heavy rain in the Wenjia Gully, Mianyuan River, Sichuan, in 2010. During the earthquake, a rock slide-debris avalanche of 30 Mm^3 occurred in the headwaters of this 'gully', which spans an elevation range of >1500 m over a catchment area of only 7.8 km^2. Five debris flows entrained as much as 14% of the coseismic landslide debris in 29 months following the earthquake, giving rise to very high sediment yields (0.46–0.5 Mt km^{-2} yr^{-1}). Check-dams designed to protect the downstream reaches were rapidly infilled or washed out. Given the high amounts of crushed landslide material in the headwaters, more sediment is bound to leave Wenjia Gully in the future. Heavy rainfall in tributaries of the nearby Min River on 10 July 2013 triggered more than 240 debris flows with peak discharges between 330 and 2400 m^3 s^{-1}. The debris flows were responsible for 29 fatalities and the loss of more than 22 000 houses, 44 factories, and two reservoirs (Ge et al. 2015). They also temporarily dammed 36 lakes along some 120 km of the river, submerging at least 2000 more houses. The total financial losses were estimated at US$ 633 million. The average amount of landscape lowering caused by this disaster ranged from 2–68 mm in the source catchments, and the sediment raised trunk-river channel beds by 2–11 m. Gauging station data recorded that, in the two years following the earthquake, the flux of sediment in the upper Min River increased 75% above the average of the preceding 27 years (Ding et al. 2014).

Where coseismic hillslope material is flushed directly into rivers, it may form landslide dams that impound large amounts of water and sediment (Figure 6.5). The 2008 Wenchuan earthquake generated more than 820 full and partial river impoundments; more than 90% of them were gone within one year of the earthquake (Fan et al. 2012). Landslide-dammed lakes may impede access into earthquake-struck regions and thus compromise immediate relief or rehabilitation efforts. Bridges and major traffic arteries may be destroyed or become partly flooded, and the impounded lakes may drown settlements, industrial facilities, or agricultural lands, contaminating water supplies and thus increasing the risk of diseases. Hydropower schemes are also prone to temporary inundation, especially because much sediment is involved.

The case of the Tangjiashan landslide dam, which formed during the 2008 Wenchuan earthquake, illustrates the many problems Chinese authorities encountered with coseismic river blockage. The landslide formed an 82-m high dam on the Jiang River, posing a hazard to 1.2 million citizens of Beichuan and Mianyang City downstream. More than 200 000 people were evacuated ten days before the dam breached via an artificial diversion channel that engineers completed 19 days after the earthquake (Peng and Zhang 2012). The breach released an estimated 161 Mm^3 of water at a peak discharge of 6500 m s^{-3}. The outbreak flood might have been much larger and more destructive had the dam incized much faster than only 0.83 mm s^{-1} on average, which is in the lower range of reported breach rates for landslide dams worldwide (O'Connor and Beebee 2009).

Postseismic sediment pulses may construct new landforms if they reach lake or ocean shores. Large coastal dune ridges may form at river mouths in response to large volumes of river-borne sediments that are transferred by longshore drift. Sets of parallel dune ridges

Figure 6.5 Remnants of former lakes impounded behind landslides that were triggered by the 2008 Wenchuan earthquake. (a) Upstream view of one of hundreds of smaller lakes that was artificially drained, leaving a pool flanked by dissected backwater sediments. (b) An earthquake-damaged building partly buried beneath backwater sediments. (c) Boulder (~ 3 m high) emplaced at the upstream side of a bridge by catastrophic landslide-dam outburst flood. (d) Floor of drained landslide-dammed lake covered with anthropogenic debris. From Korup and Wang (2015).

may contain sedimentary signals of past earthquakes (Goff et al. 2008), and represent the coastal end member of a suite of landforms that could be termed a 'seismic staircase'. In the model of Goff and Sugawara (2014), earthquake-triggered landslides increase sediment availability, thus inducing river aggradation and coastal dune building. The model predicts that multiple dune ridges along tectonically active coasts record sediment flux driven by river discharge, sea level, and climatic changes, and thus bear an imprint of past seismic disturbances. Evidence suggests that parallel dune ridges on the Sendai Plain, which was struck by the 2011 Great Tohoku earthquake, could record known major earthquakes in this area in the past. The establishment of vegetation can post-date the trigger earthquake by several decades (Goff et al. 2008), and this protracted postseismic river adjustment could be a serious concern for authorities planning post-earthquake recovery.

6.3 Geomorphic Tools for Reconstructing Past Earthquakes

Palaeoseismology is the science concerned with reconstructing the magnitude and frequency of earthquakes that predated systematic instrumental recording (Figure 6.6). It has helped to reveal many large prehistoric or poorly documented historic earthquakes throughout the world (Washburn et al. 2001). The magnitudes of past poorly documented earthquakes have been estimated from detailed analysis of observed damage. The Modified Mercalli Scale ranks observable earthquake effects, such as people's reaction or the degree of damage to buildings, to the local intensity of seismic ground shaking. Many geomorphic and sedimentary footprints of earthquakes offer additional useful proxies for reconstructing strong earthquakes in the past.

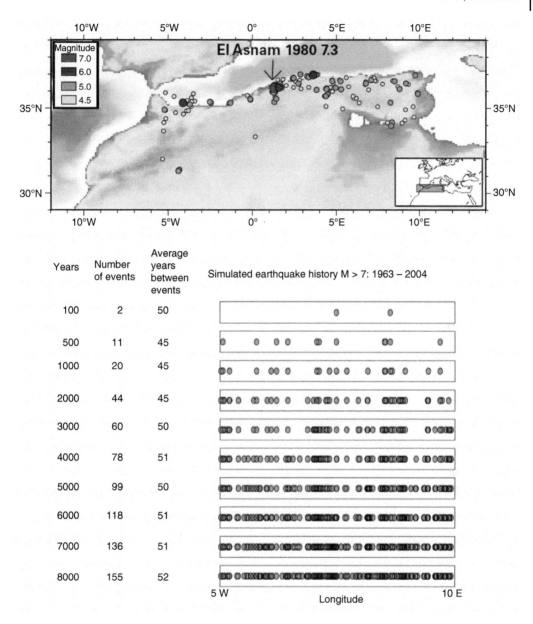

Figure 6.6 How many years of earthquake records do we need to estimate robustly the seismic hazard in a region with low to moderate activity? For the North Africa plate boundary, model simulations based on the magnitude and frequency of earthquakes indicate that about 8000 years of record are needed to avoid apparent gaps; compare this to the historic seismicity on the map. From Stein and Stein (2014).

6.3.1 Offset Landforms

Ironically, surface evidence of earthquakes in tectonically active landscapes can be removed through erosion and deposition. As a result fewer traces of active faults might be visible in active landscapes than we would expect (Cox et al. 2012). This natural censoring is problematic because many palaeoseismological proxies are largely based on landforms such as river channels, terraces, or fans that are truncated and offset along fault traces (De Pascale et al.

Table 6.1 Methods and proxies of palaeoseismology

Investigation method	Aim	Tools
Surface geological mapping and reconnaissance	• Locate faults • Identify and map landforms offset by fault trace • Mapping of landslide deposits • Mapping of precariously balanced boulders • Detect earthquake signals in caves • Survey uplifted marine landforms	• Aerial photography • Digital terrain analysis • Geological mapping • Back-calculation • Cave survey • LiDAR scanning
Geophysical methods	• Locate blind faults • Identify subaquatic mass-movement deposits	• Microgravity, resistivity • Seismic reflection
Fault trenching	• Measure and date fault displacements	• Logging, digital photography • Soil analysis, sedimentology
Drilling	• Recover lake and submarine sediments • Sample landslide deposits, speleothems	• Kullenberg, Livingstone • UWITEC, ram cores, drilling
Laboratory methods	• Estimate liquefaction potential • Correlate drill cores • Analyse deformation structures • Analyse sediments	• Particle size analysis, CPT • Magnetic susceptibility • X-ray tomography • Smear analysis, shear strength
Absolute age dating	• Date geological and biological samples	• ^{14}C, U series, OSL • Pollen, tephra, varves

Source: Modified from Becker et al. (2005).

(2014); Table 6.1). Nevertheless, such offsets, if preserved and accurately dated, constrain long-term slip rates along faults ranging from local to subcontinental scale (Cowgill 2007). Staircases of terraces or alluvial fans are particularly well suited for studying the relation between episodic fault slip and local channel aggradation and incision (Hubert-Ferrari et al. 2005), thus revealing the potential geomorphic response of rivers to seismic shaking. The slip rates that researchers estimate from such offset landforms have errors arising from topographic measurements and the uncertainties associated with the absolute age dating methods such as radiocarbon and OSL (Mason et al. 2006). The geometry of tilted or folded fluvial terraces can also yield valuable information on fault

slip rates (Thompson et al. 2002). River terraces can also provide evidence of pulsed long-term interseismic deformation. The arrangement of some terraces flanking the rivers of the New Zealand Southern Alps is consistent with a cycle in which several hundred years of accelerated river incision may have preceded major earthquakes (Campbell et al. 2003).

Palaeoseismologists have used many proxies to resolve recurrent coseimic uplift events along tectonically active coasts (Berryman et al. 2010). Confidence is increased when more than one proxy points to the same conclusion. Consistent results derived from independent methods are likely to yield the most robust estimates of former earthquake magnitudes and dates. In coastal environments uplifted staircase-like beach ridges (McSaveney et al. 2006), marine terraces, and coral reefs offer insights into relative sea-level changes driven by tectonic uplift. Sessile organisms such as calcareous tube worms and intertidal corals live on rock surfaces close to mean sea level, where they leave their mark. Vertical uplift of such organisms provides a benchmark for assessing coseismic uplift at a resolution of several decimetres (Shishikura et al. 2009). Marine terraces and uplifted coral reefs may, however, record processes other than coseismic uplift. Numerical modelling of coral growth including erosion, deposition, sea-level changes, and coseismic uplift suggests that uplifted coastal landforms may also contain signals of rapid uplift from sources other than large earthquakes, including volcanic activity or regional tectonic uplift. In the Ryukyu Islands of southwestern Japan, for example, GPS stations have recorded rapid coastal uplift of the order of several mm yr^{-1}, while Holocene reef uplift had occurred since 6.3 ka. Modelling confirmed that only part of this uplift was coseismic and associated with large earthquakes. The implication is that reconstruction of past earthquakes from uplifted marine landforms may potentially overestimate the magnitude of large seismic events (Shikakura 2014). This finding confounds previous interpretations that marine terraces along the east coast of Taiwan are the geomorphic outcome solely of coseismic uplift during megathrust earthquakes along the Ryukyu Trench (Hsu et al. 2012). Sedimentary markers of past earthquakes also include beds of freshwater peat that accumulated in coastal wetlands and abruptly overlie tidal mud, a composition that is characteristic of episodic and rapid uplift during large earthquakes. On the coast of eastern Hokkaido, Japan, such peat beds record abrupt coastal uplift, on average, every 500 years; the peat beds are preserved due to postseismic deep slip on the down-dip extension of the seismogenic plate boundary (Kelsey et al. 2006).

Sea-floor geomorphology also offers valuable insights into the history of past earthquakes, especially where coastal or terrestrial evidence is sparse (Henstock et al. 2006). Blind faults are those without any obvious surface expression or faults that have been buried by sediment. They are a common problem in palaeoseismology in that they leave gaps in the earthquake record. McHugh et al. (2011) used multibeam bathymetry and cores to identify a 50 Mm^3 turbidite in the Canal de Sud basin, offshore Haiti. The tell-tale excess amount of ^{234}Th, which is a radioactive isotope with a half life of 24 days, in the turbidite revealed that it was triggered during the destructive M 7.0 Haiti earthquake in 2010. The cross-bedded and normally graded turbidite deposit overlies similar older deposits, and can be seen as evidence that similar earthquakes happened before the 2010 disaster.

An offshore earthquake proxy that has been extensively exploited is the record of deep-sea turbidites from the Cascadia Basin west of Washington and Oregon. Adams (1990) first attributed Holocene turbidites in the basin to great plate-boundary earthquakes at the Cascadia subduction zone, which extends along the Pacific coast from Northern California to Vancouver Island, British Columbia. He used what has become known as the 'confluence test', counts of turbidites upstream and downstream of junctions between deep-sea channels, to make the case for earthquake triggering. Adams (1990) found the same number of turbidites upstream and downstream of junctions, suggesting that turbidity currents from different submarine canyons reached the junction at about the same time, as would be expected if the turbidity currents had been triggered by widespread seismic triggering. He documented 13 correlative turbidites younger than the Mazama tephra layer, which is 7700 years old. Importantly, he argued that the same number of turbidites in multiple channels of the same submarine canyons and at widely separated sites in the basin showed that the turbidity currents were synchronously generated by subduction zone earthquakes that affected the entire Cascadia margin. He further argued that other possible mechanisms, such as severe storms, could hardly explain synchronous turbidity currents over such a large length of the margin. Goldfinger et al. (2007) tested Adams' hypothesis with data from more drill cores, AMS radiocarbon ages, visible and X-ray imagery, and stratigraphic correlation using continuous physical property measurements to extend the turbidite record in space and time to the earliest Holocene. They developed a method that tests for synchronicity of the turbidite events and evaluated potential triggers against the time, space, and physical requirements imposed by the faults that might be active at the Cascadia margin and different triggering mechanisms.

6.3.2 Fault Trenching

Detailed logging of offset sedimentary layers in trenches dug across fault traces is a widely used palaeoseismological method for reconstructing the magnitude and frequency of prehistoric earthquakes. Dateable material contained in soil layers, fluvial beds, or colluvial wedges helps to reconstruct the ages of prehistoric earthquakes, whereas the geometry of the offset constrains the type and amount of slip, and hence the approximate magnitude of each major ground-shaking episode (Figures 6.7 and 6.8). Absolute age dating of risers of fault traces with cosmogenic nuclides such as ^{10}Be is another method to determine when earthquakes happened. The assumption here is that the riser surface has escaped any subsequent reworking by erosion or sedimentation following the earthquakes, thus offering a minimum age of fault rupture (Ritz et al. 2003).

Some natural exposures in steep river banks also offer valuable insight into fault geometry and history. Studies in the Lesser and Sub-Himalayas have provided a detailed picture of the past seismic activity and future seismic risk in Earth's highest mountain belt (Sapkota et al. 2012). Much of the active deformation in this region is located along the Himalayas Main Frontal Thrust (MFT), where population density is also among the highest in the mountain belt (Kumar et al. 2001). The MFT is a broad fault zone comprising several structures rather than a single fault, thus palaeoseismological reconstructions need to take into account the offsets along each fault strand. Offset river terraces and alluvial fans are prime diagnostic landforms that document this evidence along the trace of the MFT in the Himalayan foreland (Yeats and Thakur 2008). Deformed Holocene river terraces in areas of active tectonic folding such as the Siwalik Hills of central Nepal offer additional indirect evidence of strain accommodation and shortening along the MFT. Using this evidence, Lavé and Avouac (2000) inferred that the MFT has accommodated an average of 21 ± 1.5 mm yr^{-1} of north–south shortening during the Holocene, a finding that is consistent with decadal averages of GPS measurements. Radiocarbon dating of soil liquefaction features such as injected sand dykes has helped to link these features to historically documented earthquakes in the eastern Himalaya, while adding several undocumented older events to the palaeoearthquake catalogue (Reddy et al. 2009). In the Higher Himalayas, the forefields of Bhutanese glaciers host evidence of fault traces that offset moraine-fringed outwash surfaces. Together with liquefaction features and pervasively cracked gravel in kame terraces, these faults attest to very recent but prehistoric seismic activity, most likely at magnitudes higher than recorded earthquakes (Meyer et al. 2006). Seismic shaking in this heavily glaciated, although sparsely populated, area repeatedly triggers destructive landslides and avalanches that could displace meltwater stored in some of the thousands of glacial lakes in the mountain belts, causing catastrophic lake bursts and sudden flood waves.

We recall that a given trench may fail to record the occurrence of an earthquake: 'absence of evidence is not evidence of absence' should be a general principle. As a final note

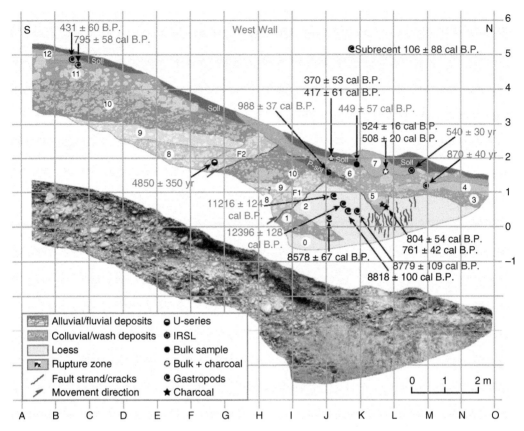

Figure 6.7 Stratigraphic interpretation and photolog of a trench across the Issyk-Ata Fault near Bishkek, Kyrgyzstan. Fluvial deposits (units 8–11) in the hanging wall have been thrust over loess (unit 2). At least three earthquakes ruptured two fault zones (F1 and F2). The footwall comprises earthquake-related colluvial deposits interbedded with wash deposits and a coseismic fissure surrounded by cracks extending downward. From Patyniak et al. (2017).

on palaeoseismology, the 2016 M 7.8 Kaikoura earthquake in New Zealand's South Island occurred near the Hope Fault, one of the nation's known active major faults. The earthquake involved the cascading rupture of about 20 minor faults about half of which were unknown prior to this event. Had the earthquake occurred before historical records, palaeoseismologic research might have confused it with multiple smaller ruptures. Researchers originally thought that all major earthquakes occurred on first-order faults. This may need an update: effectively, a major earthquake can be expected to occur anywhere in South Island.

Megathrust earthquakes in the Himalayas
The Himalayas are the Earth's highest mountain belt and result from the ongoing collision of the Indian and Eurasian tectonic plates. The steep dissected topography reflects the dynamic interplay between rapid uplift and monsoon-driven erosion. Interseismic rates of surface uplift in the Nepal Himalayas measured from interferometric synthetic aperture radar (InSAR) between 2003 and 2010 average ~7 mm yr^{-1} at the foot of the Higher Himalayas and are compatible with 18–21 mm yr^{-1} of slip on the deep shallow-dipping portion of the Main

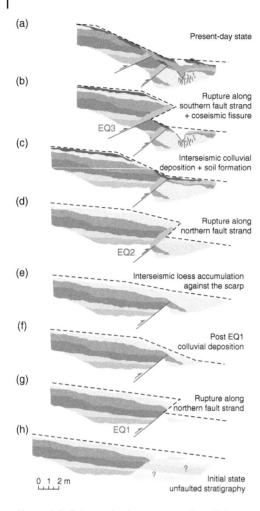

(a) Present-day state

(b) Rupture along southern fault strand + coseismic fissure

EQ3

(c) Interseismic colluvial deposition + soil formation

(d) Rupture along northern fault strand

EQ2

(e) Interseismic loess accumulation against the scarp

(f) Post EQ1 colluvial deposition

(g) Rupture along northern fault strand

(h) EQ1

0 1 2 m

Initial state unfaulted stratigraphy

Figure 6.8 Palaeoseismic reconstruction of the trench across the Kyrgyzstan (previous figure). From Patyniak et al. (2017).

Himalayan Thrust (Grandin et al. 2012). The Himalayas are home to tens of million of people, and many more live in the adjacent foreland. Strong earthquakes are part of the history of this region and have had devastating results (Figure 6.9). Three large earthquakes struck the Himalayas in the twentieth century without leaving any traces of surface rupture (Kumar et al. 2010). Nepal's capital city of Kathmandu was struck by strong earthquakes in 1934 and 2015, and sits on top of lake sediments with deformed layers that are testaments to past events of comparable magnitude (Mugnier et al. 2011). Along with the 1950 Assam and 2005 Kashmir earthquakes, these were some of the most destructive natural disasters in the Himalayas. Yet little is known about great Himalayan earthquakes that happened prior to the eighteenth century (Bilham et al. 2001). Palaeoseismological work finds more and more evidence of strong earthquakes in Medieval to prehistoric times, using historic evidence and fault trenches, although some uncertainty about the timing and extent of fault ruptures remains (Bollinger et al. 2014).

Reviews of field evidence of large Himalayan earthquakes in the past 2000 years reveal a regionally consistent picture (Kumar et al. 2001). A late Holocene record of fault offsets in eastern Nepal holds that large earthquakes occur there every 750–870 years (Bollinger et al. 2014). Many detailed studies of natural exposures in fault zones have provided evidence of the few large historic earthquakes in this region (Lavé et al. 2005), while adding previously unrecognized faults away from the MFT to the list of seismogenic sources (Munier et al. 2013; Meade 2010; Cummins 2007). The records illustrate a conspicuous clustering of large earthquakes during Medieval times (Mugnier et al. 2013). Geodetically constrained modelling of interseismic rock uplift and channel erosion in bedrock suggests that convex river longitudinal profiles should result if interseismic deformation is greater than coseismic deformation. This prediction corresponds well with the contemporary river channel geometry, and raises questions about the nature of earthquake cycles in the Himalayas (Meade 2010). Megathrust earthquakes may also lurk along parts of the active subduction zone that is partly buried beneath the Ganges-Brahmaputra fan delta (Cummins 2007).

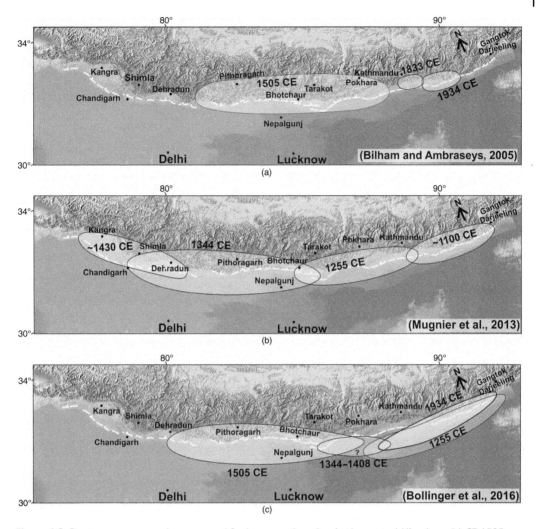

Figure 6.9 Rupture zone scenarios proposed for large earthquakes in the central Himalaya: (a) CE 1505, 1833, and 1934 (Bilham); (b) CE 1430, 1344, 1255, and 1100 (Mugnier); and (c) CE 1505, 1255, and 1934 (Bollinger). From Rajendran et al. (2019).

6.3.3 Coseismic Deposits

Land surfaces that subside or uplift during large earthquakes can form useful markers of past geomorphic change, if correctly identified and dated. This rationale has attracted a lot of research to subduction zones around the world to unravel the chronology of past plate-boundary ruptures. While the rationale is simple and appealing, a lot of scientific discussion revolves around recognizing with certainty deposits formed or disturbed by past earthquakes. For example, buried wetland peats resulting from episodic sudden subsidence are present at more than 20 estuaries on the Pacific coast of North America between Northern California and Vancouver Island (Figure 6.10) (Clague 1997). Each peat layer is abruptly overlain by intertidal mud; the sharp contact between the peat and mud records lowering of the wetland surface due to coseismic subsidence. How well these tsunami deposits may survive in sedimentary archives depends also on the rate of sea-level rise along

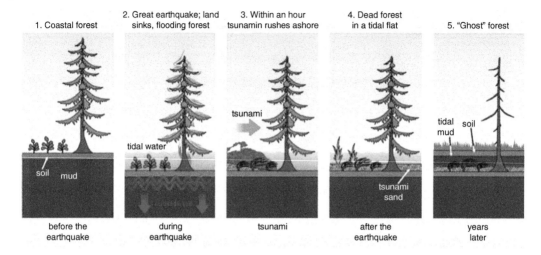

Figure 6.10 Cartoons showing the sequence of events responsible for the diagnostic stratigraphy accompanying a great earthquake at the Cascadia subduction zone in coastal tidal marshes in the Pacific Northwest. Pre-earthquake tidal marsh (panel 1). Submergence of the marsh and adjacent forest margin during the earthquake (panel 2). Deposition of sand on the submerged marsh during the tsunami following the earthquake (panels 3 and 4). Burial of the submerged marsh by tidal mud and subsequent recolonization of the emergent surface by marsh plants (panel 5). From Clague et al. (2007).

a given coast. Kelsey et al. (2015) argued that the Banda Aceh coastal plain, one of the locations hit most badly by the 2004 Indian Ocean tsunami, was essentially made up by a stack of tsunami sediments. In this interpretation, about half a dozen of these catastrophically emplaced deposits were mostly responsible for accreting the coastal plain over the past 3800 years, a period during which sea level remained stable.

Radiocarbon ages of intertidal plants in growth position at the top of the peat layers approximate the times of past earthquakes and allow researchers to correlate events over distances of hundreds of kilometres with a high degree of confidence. At a few wetland sites, old conifers that died when they were lowered from the forest fringe into the intertidal zone during the most recent great earthquake at the Cascadia subduction zone remain standing as ghostly sentinels to a past disaster. Cross-dating of the outer rings of these 'ghost' trees with very old living trees in the region narrowed the age of the earthquake to a few years around 1700 AD (Jacoby 1997). Subsequently the tsunami that was spawned

by this earthquake was identified in Japanese written records dated 27 January 1700. One can back-calculate the time it took the tsunami to travel from the west coast of North America to the coast of Honshu, and infer that the event happened in the evening of 26 January 1700 Pacific Standard Time (Satake and Atwater 2007)! In Chapter 9.3 we deal in more detail with the geomorphic evidence of tsunami, particularly those triggered by earthquakes.

The dating of deposits of coseismic landslides is another option to estimate the ages of past earthquakes. Establishing an earthquake trigger for landslides also requires that the landslide be close to an active fault, and that other triggers such as rainstorms can be excluded (Crozier et al. 1995); one clue may lie in the landslide source area, which in the case of a coseismic landslide is likely to be bowl-shaped because of resonant shaking of the edifice generates deep-seated failure surfaces. Such constraints require ample testing and excluding alternative scenarios, although several studies have proposed a link between large rockslides and earthquakes (Schulz et al. 2012). Radiometric dating of landslide deposits can be highly time- and resource-intensive and requires a good understanding of the stratigraphy of a landslide deposit, which is rarely described in textbooks. Small numbers of landslides that can be accurately dated rarely allow us to infer reliably that the landslides were triggered by an earthquake. The unavoidable limitations on the accuracy of age dates can only give us a probabilistic measure that we use to ascribe a seismic trigger to landslides. For example, several large (>1 Mm^3) landslides have occurred in the Southern Alps of New Zealand since 1991. If this sequence had occurred, say, some 200 years ago before the first written records, there could have been a strong temptation to infer, incorrectly, a widespread seismic trigger.

Precariously balanced boulders can be useful geomorphic proxies of past earthquake magnitude. These are large and freestanding rock clasts resting on narrow pedestals that would be prone to toppling during moderate seismic shaking. In seismically active areas, the detailed geometry of these perched boulders can be mathematically converted to the forces and hence ground acceleration necessary to topple them. Together with cosmogenic nuclide exposure dating (Gosse and Phillips 2000), precariously balanced boulders may contribute valuable information independent of fault trenching to earthquake chronologies (Balco et al. 2011).

Widespread river aggradation following historic earthquake-induced landslides is well documented in the literature, but remains somewhat elusive in the recent geological record (Figure 6.11). Sedimentary landforms bear testimony to past aggradation events, but the ages of these landforms must be consistent with that of past earthquakes. For example, fan terraces several tens of metres above current channel beds line many bedrock rivers of the Southern Alps of New Zealand and the Central Range of Taiwan (Hsieh and Chyi 2010). Despite several attempts to date these former valley floors, radiocarbon ages have yielded seemingly inconclusive results, at least as far as linkages to past earthquakes are concerned.

Lakes in mountain valleys or at mountain fronts can also host biogenic or sedimentary evidence of postseismic sediment pulses (Figure 6.12). Radiocarbon ages of trees drowned in Lake Poerua in the piedmont of the western Southern Alps of New Zealand are consistent with times of large Alpine Fault earthquakes, suggesting that aggradation by several metres in the wake of such earthquakes could have affected lake-level changes in the foreland, and thus contributed to drowning forest stands (Langridge et al. 2012). Discrete layers of turbiditic sediments found in a lake close to the Alpine Fault provide a detailed record of episodic postseismic sedimentation from small rivers draining the nearby range front of the Southern Alps. These sediment pulses were sustained for several years to decades; compared with modern sediment fluxes, the dated lake sediments could have involved about a quarter of the estimated background erosion on average (Howarth et al. 2012).

Slump blocks supporting conifer trees in growth position rest on the floor of Lake Washington in the Pacific Northwest of the United States. Dating of the trees showed that the slumps happened, and the trees were drowned, during a large earthquake on the Seattle Fault about 1000 years ago (Jacoby et al. 1992). The earthquake produced a maximum vertical displacement of about seven metres on one of the strands of the fault, which in turn triggered a tsunami in Puget Sound that is recorded as a landward-thinning and -fining sand sheet along its shores (Atwater and Moore 1992).

Barnes et al. (2013) investigated submarine turbidites from sedimentary basins less than ten kilometres offshore of Fiordland, New Zealand. These deposits yielded a more than 2000-yr record of large-magnitude paleoearthquakes, assumed to be associated mainly with the Alpine fault and the Fiordland subduction zone. The records include two recent interplate thrust earthquakes in 2003

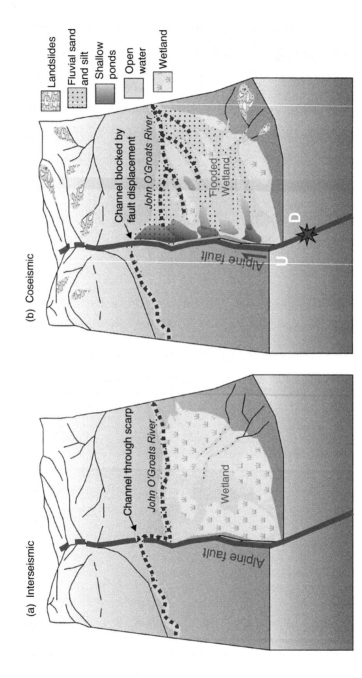

Figure 6.11 Depositional model for an earthquake record, John O'Groats River, New Zealand. (a) In an interseismic period, the river flows freely via a channel across the fault scarp and a wetland is accumulating peat and organic silt. (b) Following an earthquake, the river is blocked by vertical and horizontal displacement along the fault and the wetland is flooded and covered by silt and sand delivered from the catchment. From Cochran et al. (2017).

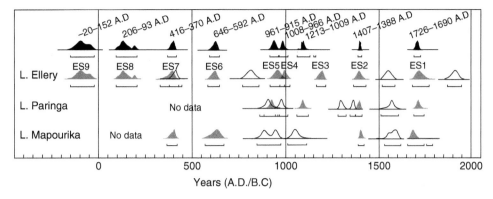

Figure 6.12 A 2000-year rupture history for the Alpine Fault derived from Lake Ellery, South Island, New Zealand. This figure shows probability density functions (PDFs) for ages of subaqueous mass wasting and high sediment-flux events (grey PDFs) and subaqueous mass wasting (open PDFs) in lakes Ellery, Paringa, and Mapourika, New Zealand. PDFs for ages of regional subaqueous mass wasting and increased sediment-flux events (black PDFs) were determined using the 'combine function' in OxCal 4.2. From Howarth et al. (2016).

and 2009, and probably the 1826 AD Fiordland earthquake and the well-dated 1717 AD Alpine fault earthquake. The recurrence intervals are shorter than recently published recurrence data from the Alpine fault on land, reflecting various fault sources and a potentially higher segmentation of the Alpine fault offshore. De Pascale et al. (2014) reviewed data from fault trenches, dated tree cohorts, dune sequences and rockfall deposits, lacustrine turbidites, and LiDAR-scanned offset landforms as independent proxies of past earthquakes on the Alpine Fault, and pointed out a mismatch between thus reconstructed offsets, timing, and slip rates, at least for the past 1100 years. One explanation to reconcile those divergent proxies is that some of the recorded seismic shaking may have happened during earthquakes on faults in the Southern Alps rather than on the range-front Alpine Fault (Briggs et al. 2018). Evidence from lake sediments featuring largely coeval sequences of turbidites topped by beds presumably delivered by rivers fed by coseismic and postseismic landslides, however, is consistent with very strong ground shaking at several locations along the Alpine Fault that none of the known active faults in the Southern Alps could have produced simultaneously (Howarth et al. 2016).

Seismic shaking may also deform water-saturated lake sediments, generating dyke-like sand injections, loading structures, or undulating contacts, collectively known as seismites or homogenites. Fluidized and horizontally sheared layers are other proxies of a sudden disturbance of lake sediments (Mugnier et al. 2011). Together with deposits of hyperpycnal flows, subaquatic slumps, and turbidites, these features are often used as diagnostic tools of former earthquakes that destabilized lacustrine sediments. Large subaquatic slope failures preserved in Pleistocene lake clays may occur on slopes <1°, and in some cases attest to seismic triggering (Garcia-Tortosa et al. 2011). Such landslides, however, may also occur on nearly flat surfaces in response to nonseismic disturbances. The occurrence of such features depends on the instability of lake sediments in the basin, and sediment supply rates from rivers draining into the lake (Bertrand et al. 2008). These constraints aside, palaeoseismological proxies are especially useful in areas with only patchy chronologies of strong earthquakes such as the central European Alps (Beck 2009).

Deformed lake sediments provide ambiguous evidence for former earthquakes, as subaquatic slope failures may entrain large amounts of sediment that liquefy during runout and deform and entrain underlying layers. Large rockslides entering Aysén Fjord, Chile, during the 2007 Aysén M 6.2 earthquake bulldozed and deformed sediments on the fjord floor (Van Daele et al. 2013), but non-seismically triggered rockslides might do the same thing. A definitive distinction between earthquake-induced and landslide-induced deformation structures in sediments might seem elusive, given that landslides may arise from processes such heavy rainfall or overloading of delta slopes where sedimentation rates are high. The sudden collapse of lake-delta sediments, for example, may occur without any apparent trigger, as in the case of the 1996 delta collapse in Lake Brienz, Swiss Alps (Girardclos et al. 2007). A comprehensive study of more than a hundred sediment cores in 17 Chilean lakes bearing impacts by the 1960 Valdivia and 2010 Maule megathrust earthquakes offers a systematic approach to better identifying the sedimentary signatures of seismic shaking: Van Daele et al. (2015) recognized five major types of 'event deposits' tied to the two earthquakes. These deposits included mass-transport deposits; structures that were deformed in situ; megaturbidites tied to lake seiches; and two types of lacustrine turbidites differing in the their compositions with respect to that of the 'background' lake sediments. The turbidites with more terrestrial contributions in their sedimentology are likely linked to local mass wasting along lake shores, delta collapse, and landslides or fluvially reworked landslide debris entering the lake. The more lacustrine turbidites are mostly attributed to shallow mass wasting on lake beds. Van Daele et al. (2015) argued that it is those turbidites and their relative area that unambiguously record the sedimentary legacy of past earthquakes together with their approximate local shaking intensity.

6.3.4 Buildings and Trees

The interdisciplinary field of archaeoseismology makes use of archaeological evidence for reconstructing the times and local intensities of former earthquakes with a high level of accuracy. Scientists interpret archaeological layers as earthquake-derived and attribute them to damage or collapse of buildings, where preserved (Galadini et al. 2006). Detailed LiDAR scans of damaged buildings, for example, offer insights into the local forces ancient buildings had experienced, and hence ground accelerations. This young multidisciplinary field has shifted from a first phase dedicated to extending the palaeoseismic records to a second phase that now aims at better unraveling the importance of site effects during strong earthquakes (Sintubin 2011). These local site effects are essential for seismic microzonation efforts, and also allow detailed reconstruction of the localized geomorphic response to earthquakes.

Trees are another source of palaeo-earthquake information, and dendrogeomorphic studies have taken advantage of the direct physical impacts of seismic shaking on forest vegetation to more than 50 km from the epicentre (Wells et al. 2001). Tree rings afford highly accurate dating compared to most other methods and also offer insight into annual, even seasonal growth conditions (Jacoby 1997). These conditions may be influenced by both geomorphic and hydrological changes linked to strong earthquakes. For example, consistent outer growth rings in even-aged tree cohorts may bear signs of forest damage by a past earthquake if other causes such as wildfires and storms can be ruled out. Detailed studies of forest stands in areas affected by historic earthquakes argue that both direct shaking and impact or burial by landslides, debris flows or river sediments may kill entire cohorts of trees (Vittoz et al. 2001). Tree-ring dating of alluvial surfaces has shown strong links with the dates of known earthquakes in Westland, New Zealand (Yetton and Wells 2010). Wells and Goff (2007) used a combination of fault trenching, anomalies in tree-ring widths, ^{14}C dating of tree cohorts that established on young, postseismic aggradation surfaces and landslide deposits, and dune coastal sequences to reconstruct an earthquake history of the central Alpine Fault, New Zealand, from 1200 AD to the present. They showed that coastal dune formation followed decades after the occurrence of major earthquakes along the Alpine Fault, mainly because of enhanced sediment delivery from rivers overloaded by coseismic landslide debris from the nearby Southern Alps. Detailed work on cyclic sedimentation in a pond dammed by the same fault yielded evidence for 24 surface ruptures over the past 8000 years (Berryman et al. 2012).

References

Adams J 1990 Palaeoseismicity of the Cascadia Subduction Zone: evidence from turbidites off the Oregon-Washington Margin. *Tectonics* **9**, 569–583.

Alfaro P, Delgado J, Garcia-Tortosa FJ, et al. 2012 Widespread landslides induced by the Mw 5.1 earthquake of 11 May 2011 in Lorca, SE Spain. *Engineering Geology* **137-138**(C), 40–52.

Allen TI and Wald DJ 2009 On the use of high-resolution topographic data as a proxy for seismic site conditions. *Bulletin of the Seismological Society of America* **99**(2A), 935–943.

Atwater BF and Moore AL 1992 A tsunami about 1000 years ago in Puget Sound. *Science* **258**, 1614–1617.

Balco G, Purvance MD, and Rood DH 2011 Exposure dating of precariously balanced rocks. *Quaternary Geochronology* **6**(3-4), 295–303.

Barnes PM, Bostock HC, Neil HL, et al. 2013 A 2300-year paleoearthquake record of the Southern Alpine Fault and Fiordland Subduction Zone, New Zealand, based on stacked turbidites. *Bulletin of the Seismological Society of America* **103**(4), 2424–2446.

Beck JL 2009 Late Quaternary lacustrine paleo-seismic archives in north-western Alps: Examples of earthquake-origin assessment of sedimentary disturbances. *Earth-Science Reviews* **96**(4), 327–344.

Becker A, Ferry M, Monecke K, et al. 2005 Multiarchive paleoseismic record of late Pleistocene and Holocene strong earthquakes in Switzerland. *Tectonophysics* **400**, 153–177.

Berryman K, Litchfield N, Cochran U, and Little T 2010 Evaluating the coastal deformation mechanisms of the Raukumara Peninsula, northern Hikurangi subduction margin, New Zealand and insights into forearc uplift processes. *New Zealand Journal of Geology and Geophysics* **53**(4), 341–358.

Berryman KR, Cochran UA, Clark KJ, et al. 2012 Major earthquakes occur regularly on an isolated plate boundary fault. *Science* **336**(6089), 1690–1693.

Bertrand S, Charlet F, Chapron E, et al. 2008 Reconstruction of the Holocene seismotectonic activity of the Southern Andes from seismites recorded in Lago Icalma, Chile, 39 °S. *Palaeogeography, Palaeoclimatology, Palaeoecology* **259**(2-3), 301–322.

Bilham R and Ambraseys N 2005 Apparent Himalayan slip deficit from the summation of seismic moments for Himalayan earthquakes, 1500–2000. *Current Science* **80**, 1658–1663.

Bilham R, Gaur VK, and Molnar P 2001 Himalayan seismic hazard. *Science* **293**(5534), 1442–1444.

Bizzarri A 2011 On the deterministic description of earthquakes. *Reviews of Geophysics* **49**, 1–32.

Bollinger L, Sapkota SN, Tapponnier P, et al. 2014 Estimating the return times of great Himalayan earthquakes in eastern Nepal: Evidence from the Patu and Bardibas strands of the Main Frontal Thrust. *Journal of Geophysical Research: Earth Surface* **120**, 1623–1641.

Bollinger L, Tapponnier P, Sapkota SN, and Klinger Y 2016 Slip deficit in central Nepal: Omen for a repeat of the 1344 AD earthquake? *Earth Planets and Space* **68**, 12.

Bradwell T 2009 Lichenometric dating: A commentary, in the light of some recent statistical studies. *Geografiska Annaler: Series A, Physical Geography* **91**, 61–69.

Briggs J, Robinson T, and Davies T, 2018 Investigating the source of the c. AD 1620 West Coast earthquake: implications for seismic hazards. *New Zealand Journal of Geology and Geophysics,* **61**(3), 376–388.

Buech F, Davies TR, and Pettinga JR 2010 The little red hill seismic experimental study: topographic effects on ground motion at a bedrock-dominated mountain edifice. *Bulletin of the Seismological Society of America,* **100**(5A), 2219–2229.

Bull WB and Brandon MT 1998 Lichen dating of earthquake-generated regional rockfall events, Southern Alps, New Zealand. *Geological Society of America Bulletin* **110**(1), 60–84.

Campbell JK, Nicol A, and Howard ME 2003 Long-term changes to river regimes prior to late Holocene coseismic faulting, Canterbury, New Zealand. *Journal of Geodynamics* **36**(1-2), 147–168.

Clague JJ 1997 Evidence for large earthquakes at the Cascadia subduction zone. *Reviews of Geophysics* **35**(4), 439–460.

Clague J, Yorath C, Franklin R, and Turner B 2007 *At Risk; Earthquakes and Tsunamis on the West Coast of Canada.* Tricouni Press.

Cochran UA, Clark KJ, Howarth JD, et al. 2017 A plate boundary earthquake record from a wetland adjacent to the Alpine Fault in New Zealand refines hazard estimates. *Earth and Planetary Science Letters* **464**, 175–188.

Collins BD, Kayen R, and Tanaka Y 2012 Spatial distribution of landslides triggered from the 2007 Niigata Chuetsu–Oki Japan Earthquake. *Engineering Geology* **127**(C), 14–26.

Cowgill E 2007 Impact of riser reconstructions on estimation of secular variation in rates of strike–slip faulting: Revisiting the Cherchen River site along the Altyn Tagh Fault, NW China. *Earth and Planetary Science Letters* **254**(3-4), 239–255.

Cox SC, Stirling MW, Herman F, et al. 2012 Potentially active faults in the rapidly eroding landscape adjacent to the Alpine Fault, central Southern Alps, New Zealand. *Tectonics* **31**(2), 1–24.

Crozier MJ, Deimel MS, and Simon JS 1995 Investigation of earthquake triggering for deep-seated landslides, Taranaki, New Zealand. *Quaternary International* **25**, 65–73.

Cubrinovski M, Robinson K, Taylor M, et al. 2012 Lateral spreading and its impacts in urban areas in the 2010–2011 Christchurch earthquakes. *New Zealand Journal of Geology and Geophysics* **55**(3), 255–269.

Cummins PR 2007 The potential for giant tsunamigenic earthquakes in the northern Bay of Bengal. *Nature* **449**(7158), 75–78.

Dadson SJ, Hovius N, Chen H, et al. 2004 Earthquake-triggered increase in sediment delivery from an active mountain belt. *Geology* **32**(8), 733.

De Pascale GP, Quigley MC, and Davies TRH 2014 Lidar reveals uniform Alpine fault offsets and bimodal plate boundary rupture behavior, New Zealand. *Geology* **42**(5), 411–414.

Del Gaudio V and Wasowski J 2007 Directivity of slope dynamic response to seismic shaking. *Geophysical Research Letters* **34**(12), 1–8.

Delgado J, Garrido J, López-Casado C, et al. 2011 On far field occurrence of seismically induced landslides. *Engineering Geology* **123**(3), 204–213.

Ding H, Li Y, Ni S, et al. 2014 Increased sediment discharge driven by heavy rainfall after Wenchuan earthquake: A case study in the upper reaches of the Min River, Sichuan, China. *Quaternary International* **333**(C), 122–129.

England P and Jackson J 2011 Uncharted seismic risk. *Nature Geoscience* **4**(6), 348–349.

Fan X, van Westen CJ, Korup O, et al. 2012 Transient water and sediment storage of the decaying landslide dams induced by the 2008 Wenchuan earthquake, China. *Geomorphology* **171-172**(C), 58–68.

Fan X, Scaringi G, Korup O, et al. 2019 Earthquake-induced chains of geologic hazards: Patterns, mechanisms, and impacts. *Reviews of Geophysics* **57**(2), 421–503.

Fine IV, Rabinovich AB, Bornhold BD, et al. 2005 The Grand Banks landslide-generated tsunami of November 18, 1929: Preliminary analysis and numerical modeling. *Marine Geology* **215**(1-2), 45–57.

Galadini F, Hinzen KG, and Stiros S 2006 Archaeoseismology: Methodological issues and procedure. *Journal of Seismology* **10**(4), 395–414.

Garcia-Tortosa FJ, Alfaro P, Gibert L, and Scott 2011 Seismically induced slump on an extremely gentle slope (¡1 °) of the Pleistocene Tecopa paleolake (California). *Geology* **39**(11), 1055–1058.

Ge Y, Cui P, Zhang J, et al. 2015 Catastrophic debris flows on July 10th 2013 along the Min River in areas seriously-hit by the Wenchuan earthquake. *Journal of Mountain Science* **12**(1), 186–206.

Girardclos S, Schmidt OT, Sturm M, et al. 2007 The 1996 AD delta collapse and large turbidite in Lake Brienz. *Marine Geology* **241**(1-4), 137–154.

Gischig VS, Eberhardt E, Moore JR, and Hungr O 2015 On the seismic response of deep-seated rock slope instabilities – Insights from numerical modeling. *Engineering Geology* **193**(C), 1–18.

Goff J and Sugawara D 2014 Seismic-driving of sand beach ridge formation in northern Honshu, Japan?. *Marine Geology* **358**(C), 138–149.

Goff J, McFadgen B, Wells A, and Hicks M 2008 Seismic signals in coastal dune systems. *Earth-Science Reviews* **89**(1-2), 73–77.

Goldfinger C, Morey AE, Nelson CH, et al. 2007 Rupture lengths and temporal history of significant earthquakes on the offshore and north coast segments of the Northern San Andreas Fault based on turbidite stratigraphy. *Earth and Planetary Science Letters* **254**(1-2), 9–27.

Gorum T, Fan X, van Westen CJ, et al. 2011 Distribution pattern of earthquake-induced landslides triggered by the 12 May 2008 Wenchuan earthquake. *Geomorphology* **133**(3-4), 152–167.

Gosse JC and Phillips FM 2000 Terrestrial in situ cosmogenic nuclides: Theory and application. *Quaternary Science Reviews* **20**(14), 1475–1560.

Grandin R, Doin MP, Bollinger L, et al. 2012 Long-term growth of the Himalaya inferred from interseismic InSAR measurement. *Geology* **40**(12), 1059–1062.

Hamling IJ, Hreinsdóttir S, Clark K, et al. 2017 Complex multifault rupture during the 2016 M_w7.8 Kaikōura earthquake, New Zealand. *Science* **356**(6334), eaam7194.

Henstock TJ, McNeill LC, and Tappin DR 2006 Seafloor morphology of the Sumatran subduction zone: Surface rupture during megathrust earthquakes?. *Geology* **34**(6), 485–488.

Holtkamp SG and Brudzinski MR 2011 Earthquake swarms in circum-Pacific subduction zones. *Earth and Planetary Science Letters* **305**(1-2), 215–225.

Hovius N, Meunier P, Ching-Weei L, et al. 2011 Prolonged seismically induced erosion and the mass balance of a large earthquake. *Earth and Planetary Science Letters* **304**(3-4), 347–355.

Howarth JD, Fitzsimons SJ, Norris RJ, and Jacobsen GE 2012 Lake sediments record cycles of sediment flux driven by large earthquakes on the Alpine fault, New Zealand. *Geology* **40**(12), 1091–1094.

Howarth JD, Fitzsimons SJ, Norris RJ, et al. 2016 A 2000 yr rupture history for the Alpine fault derived from Lake Ellery, South Island, New Zealand. *Geological Society of America Bulletin* **128**(3-4), 627–643.

Hsieh ML and Chyi SJ 2010 Late Quaternary mass-wasting records and formation of fan terraces in the Chen-yeo-lan and Lao-nung catchments, central-southern Taiwan. *Quaternary Science Reviews* **29**(11-12), 1399–1418.

Hsu YJ, Ando M, Yu SB, et al. 2012 The potential for a great earthquake along the southernmost Ryukyu subduction zone. *Geophysical Research Letters* **39**(14), 1–5.

Hubert-Ferrari A, Suppe J, Van der Woerd J, et al. 2005 Irregular earthquake cycle along the southern Tianshan front, Aksu area, China. *Journal of Geophysical Research: Solid Earth* **110**, 1–8.

Jacoby GC 1997 Application of tree ring analysis to paleoseismology. *Reviews of Geophysics* **35**(2), 109–124.

Jacoby GC, Williams PL, and Buckley BM 1992 Tree ring correlation between prehistoric landslides and abrupt tectonic events. *Science* **258**, 1621–1623.

Keefer DK 2002 Investigating landslides caused by earthquakes – A historical review. *Surveys in Geophysics* **23**(6), 473–510.

Kelsey H, Satake K, Sawai Y, et al. 2006 Recurrence of postseismic coastal uplift, Kuril subduction zone, Japan. *Geophysical Research Letters* **33**(13), 1–5.

Kelsey HM, Engelhart SE, Pilarczyk JE, et al. 2015 Accommodation space, relative sea level, and the archiving of paleo-earthquakes along subduction zones. *Geology* **43**(8), 675–678.

Koi T, Hotta N, Ishigaki I, et al. 2008 Prolonged impact of earthquake-induced landslides on sediment yield in a mountain watershed: The Tanzawa region, Japan. *Geomorphology* **101**(4), 692–702.

Korup O and Wang G 2015 Mutliple landslide-damming episodes. In *Landslide Hazards, Risks, and Disasters* (ed Davies TR). Elsevier, 241–261.

Kumar S, Wesnousky SG, Jayangondaperumal R, et al. 2010 Paleoseismological evidence of surface faulting along the northeastern Himalayan front, India: Timing, size, and spatial extent of great earthquakes. *Journal of Geophysical Research: Solid Earth* **115**(B12), 1–20.

Kumar S, Wesnousky SG, Rockwell TK, et al. 2001 Earthquake recurrence and rupture dynamics of Himalayan Frontal Thrust, India. *Science* **294**(5550), 2328–2331.

Langridge RM, Basili R, Basher L, and Wells AP 2012 Late Holocene landscape change history related to the Alpine Fault determined from drowned forests in Lake Poerua, Westland, New Zealand. *Natural Hazards and Earth System Sciences* **12**(6), 2051–2064.

Lavé J and Avouac JP 2000 Active folding of fluvial terraces across the Siwaliks Hills, Himalayas of central Nepal. *Journal of*

Geophysical Research: Earth Surface (2003–2012) **105**(B3), 5735–5770.

Lavé J, Yule D, Sapkota S, et al. 2005 Evidence for a great Medieval earthquake (1100 AD) in the central Himalayas, Nepal. *Science* **307**(5713), 1302–1305.

Marano KD, Wald DJ, and Allen TI 2009 Global earthquake casualties due to secondary effects: a quantitative analysis for improving rapid loss analyses. *Natural Hazards* **52**(2), 319–328.

Mason DPM, Little TA, and Van Dissen RJ 2006 Rates of active faulting during late Quaternary fluvial terrace formation at Saxton River, Awatere fault, New Zealand. *Geological Society of America Bulletin* **118**(11-12), 1431–1446.

Massey C, Townsend D, Rathje E, et al. 2018 Landslides triggered by the 14 November 2016 Mw 7.8 Kaikōura earthquake, New Zealand. *Bulletin of the Seismological Society of America* **108**(3B), pp.1630–1648.

McHugh CM, Seeber L, Braudy N, et al. 2011 Offshore sedimentary effects of the 12 January 2010 Haiti earthquake. *Geology* **39**(8), 723–726.

McSaveney MJ, Graham IJ, Begg JG, et al. 2006 Late Holocene uplift of beach ridges at Turakirae Head, south Wellington coast, New Zealand. *New Zealand Journal of Geology and Geophysics* **49**(3), 337–358.

Meade BJ 2010 The signature of an unbalanced earthquake cycle in Himalayan topography?. *Geology* **38**(11), 987–990.

Meunier P, Hovius N, and Haines JA 2008 Topographic site effects and the location of earthquake induced landslides. *Earth and Planetary Science Letters* **275**(3-4), 221–232.

Meyer MC, Wiesmayr G, Brauner M, et al. 2006 Active tectonics in Eastern Lunana (NW Bhutan): Implications for the seismic and glacial hazard potential of the Bhutan Himalaya. *Tectonics* **25**(3), 1–21.

Mohr CH, Manga M, Wang C-Y, and Korup O 2017 Regional changes in streamflow after a megathrust earthquake. *Earth and Planetary Science Letters* **458**, 418–428.

Mugnier JL, Huyghe P, Gajurel AP, et al. 2011 Seismites in the Kathmandu basin and seismic hazard in central Himalaya. *Tectonophysics* **509**(1-2), 33–49.

Mugnier JL, Gajurel A, Huyghe P, et al. 2013 Structural interpretation of the great earthquakes of the last millennium in the central Himalaya. *Earth-Science Reviews* **127**(C), 30–47.

Murphy MA, Taylor MH, Gosse J, et al. 2014 Limit of strain partitioning in the Himalaya marked by large earthquakes in western Nepal. *Nature Geoscience* **7**(1), 38–42.

Naveau P, Jomelli V, Cooley D, et al. 2007 Modeling uncertainties in lichenometry studies. *Arctic, Antarctic, and Alpine Research* **39**(2), 277–285.

Nicol A, Khajavi N, Pettinga JR, et al. 2018 Preliminary geometry, displacement, and kinematics of fault ruptures in the epicentral region of the 2016 Mw 7.8 Kaikoura, New Zealand, earthquake. *Bulletin of the Seismological Society of America* **108**, 1521–1539.

O'Connor JE and Beebee RA 2009 Floods from natural rock-material dams. In *Megaflooding on Earth and Mars* (ed. Burr DM, Carling PA, and Baker VR) Cambridge University Press pp. 128–171.

Osborn G, McCarthy D, LaBrie A, and Burke R 2015 Lichenometric dating; science or pseudo-science?. *Quaternary Research* **83**(1), 1–12.

Patyniak M, Landgraf A, Dzhumabaeva A, et al. 2017 Paleoseismic record of three Holocene earthquakes rupturing the Issyk-Ata fault near Bishkek, North Kyrgyzstan. *Bulletin of the Seismological Society of America* **107**, 2721–2737.

Peng M and Zhang LM 2012 Analysis of human risks due to dam break floods – part 2: application to Tangjiashan landslide dam failure. *Natural Hazards* **64**(2), 1899–1923.

Reddy DV, Nagabhushanam P, Kumar D, et al. 2009 The great 1950 Assam Earthquake revisited: Field evidences of liquefaction and search for paleoseismic events. *Tectonophysics* **474**(3-4), 463–472.

Rajendran CP, Sanwal J, John B, et al. 2019 Footprints of an elusive mid-14th century earthquake in the central Himalaya; consilience of evidence from Nepal and India. *Geological Journal* **54**, 2829–2846.

Reinhardt EG, Nairn RB, and Lopez G 2010 Recovery estimates for the Río Cruces after the May 1960 Chilean earthquake. *Marine Geology* **269**(1-2), 18–33.

Ritz JF, Bourlès D, Brown ET, et al. 2003 Late Pleistocene to Holocene slip rates for the Gurvan Bulag thrust fault (GobiâAltay, Mongolia) estimated with 10Be dates. *Journal of Geophysical Research: Solid Earth* **108**, 1–16.

Sapkota SN, Bollinger L, Klinger Y, et al. 2012 Primary surface ruptures of the great Himalayan earthquakes in 1934 and 1255. *Nature Geoscience* **6**(1), 71–76.

Satake K and Atwater BF 2007 Long-term perspectives on giant earthquakes and tsunamis at subduction zones. *Annual Review of Earth and Planetary Sciences* **35**(1), 349–374.

Scheingross JS, Minchew BM, Mackey B, et al. 2013 Fault-zone controls on the spatial distribution of slow-moving landslides. *Geological Society of America Bulletin* **125**(3-4), 473–489.

Schulz WH, Galloway SL, and Higgins JD 2012 Evidence for earthquake triggering of large landslides in coastal Oregon, USA. *Geomorphology* **141-142**(C), 88–98.

Shikakura Y 2014 Marine terraces caused by fast steady uplift and small coseismic uplift and the time-predictable model: Case of Kikai Island, Ryukyu Islands, Japan. *Earth and Planetary Science Letters* **404**(C), 232–237.

Shishikura M, Echigo T, and Namegaya Y 2009 Evidence for coseismic and aseismic uplift in the last 1000 years in the focal area of a shallow thrust earthquake on the Noto Peninsula, west-central Japan. *Geophysical Research Letters* **36**(2), 1–5.

Sintubin M 2011 Archaeoseismology: Past, present and future. *Quaternary International* **242**(1), 4–10.

Stein S and Stein J 2014 *Playing Against Nature: Integrating Science and Economics to Mitigate*

Natural Hazards in an Uncertain World. AGU/Wiley.

Thompson SC, Weldon RJ, Abdrakhmatov K, et al. 2002 Late Quaternary slip rates across the central Tien Shan, Kyrgyzstan, central Asia. *Journal of Geophysical Research: Solid Earth* **107**, 1–32.

Tseng CH, Hu JC, Chan YC, et al. 2009 Non-catastrophic landslides induced by the M_w 7.6 Chi-Chi earthquake in central Taiwan as revealed by PIV analysis. *Tectonophysics* **466**(3-4), 427–437.

Turcotte DL and Schubert G 2002 *Geodynamics* 2nd edn. Cambridge University Press.

Van Daele M, Versteeg W, Pino M, et al. 2013 Widespread deformation of basin-plain sediments in Aysén fjord (Chile) due to impact by earthquake-triggered, onshore-generated mass movements. *Marine Geology* **337**(C), 67–79.

Van Daele M, Moernaut J, Doom L, et al. 2015 A comparison of the sedimentary records of the 1960 and 2010 great Chilean earthquakes in 17 lakes: Implications for quantitative lacustrine palaeoseismology. *Sedimentology* **62**(5), 1466–1496.

Vittoz P, Stewart GH, and Duncan RP 2001 Earthquake impacts in old-growth *Nothofagus* forests in New Zealand. *Journal of Vegetation Science* **12**(3), 417–426.

Wartman J, Dunham L, Tiwari B, and Pradel D 2013 Landslides in Eastern Honshu induced by the 2011 off the Pacific Coast of Tohoku earthquake. *Bulletin of the Seismological Society of America* **103**(2B), 1503–1521.

Washburn Z, Arrowsmith J, Forman S, et al. 2001 Late Holocene earthquake history of the central Altyn Tagh fault, China. *Geology* **29**(11), 1051–1054.

Watts AB, Koppers AAP, and Robinson DP 2010 Seamount subduction and earthquakes. *Oceanography* **23**(1), 166–173.

Wells A and Goff J 2007 Coastal dunes in Westland, New Zealand, provide a record of paleoseismic activity on the Alpine fault. *Geology* **35**(8), 731–734.

Wells A, Duncan RP, and Stewart GH 2001 Forest dynamics in Westland, New Zealand: the importance of large, infrequent earthquake-induced disturbance. *Journal of Ecology* **89**(6), 1006–1018.

Yanites BJ, Tucker GE- Mueller KJ, and Chen YG 2010 How rivers react to large earthquakes: Evidence from central Taiwan. *Geology* **38**(7), 639–642.

Yeats RS and Thakur VC 2008 Active faulting south of the Himalayan Front: Establishing a new plate boundary. *Tectonophysics* **453**(1), 63–73.

Yetton MD and Wells A 2010 Earthquake rupture history of the Alpine fault over the last 500 years. In Williams AL, Pinches GM, Chin CY, et al. (eds) *Geologically Active*, Proceedings of the 11th IAEG Congress, Auckland, New Zealand, 5–10 September, Taylor and Francis Group.

Yu B, Ma Y, and Wu Y 2012 Case study of a giant debris flow in the Wenjia Gully, Sichuan Province, China. *Natural Hazards* **65**(1), 835–849.

7

Volcanic Hazards

Volcanoes offer some of the most spectacular displays of Earth's endogenic and exogenic processes. The chemical composition of ascending magma, which combines solid, liquid, and gaseous phases, largely determines the type of volcanic eruption, which involves the spectrum from effusive to explosive processes. Volcanic hazards arise from the coupling of magmatic, hydrogeological, and atmospheric processes, and offer a high diversity of potentially damaging processes ranging from lava flows and pyroclastic flows to lahars, catastrophic edifice collapses, caldera-lake outbursts, acid rains, and massive deposition of tephra over large areas of the landscape. Some volcanic hazards can be tied to eruptions directly, whereas others may occur many years to millennia after a volcano has become extinct. Volcanic islands offer some of the highest submarine relief on Earth and may release correspondingly large debris avalanches that can trigger large tsunami without any seismic trigger. Ice-covered volcanoes are the source of destructive outburst flows triggered by the sudden release of subglacial meltwater. The volumes of material released by a given volcano differ by many orders of magnitude, and can vary from the growth of local lava domes to landscape-altering sheets of ignimbrites or flood basalts. These latter impacts are often invoked in connection with some of

the major mass extinctions of species on our planet.

7.1 Frequency and Magnitude of Volcanic Eruptions

Measuring and comparing the sizes of volcanic eruptions rely on well documented case studies (Figure 7.1). The Global Volcanism Program (GVP), operated under the auspices of the Smithsonian Institution National Museum of Natural History (https://volcano.si.edu), lists 1556 volcanoes that either erupted at least once during the Holocene or that are currently at unrest. The global database on large explosive volcanic eruptions (LaMEVE, www.bgs.ac.uk/vogripa) features information on 3000 volcanoes and more than 1800 eruptions during the Quaternary (Crosweller et al. 2012). Many currently active volcanoes are monitored to provide early warning of impending eruptions and thus prevent future disasters. However, Wilson et al. (2010) pointed out that:

> Globally, there is very limited information on how affected communities recover following large volcanic eruptions. Typically, there is substantial scientific, political and media interest when impacts are at their worst, but there is little in the way of ongoing monitoring of recovery

Geomorphology and Natural Hazards: Understanding Landscape Change for Disaster Mitigation, Advanced Textbook Series,
First Edition. Tim R. Davies, Oliver Korup, and John J. Clague.
© 2021 John Wiley & Sons Ltd. Co-published 2021 by the American Geophysical Union and John Wiley & Sons Ltd.

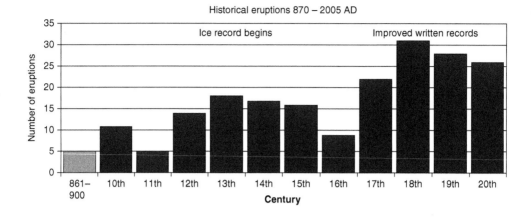

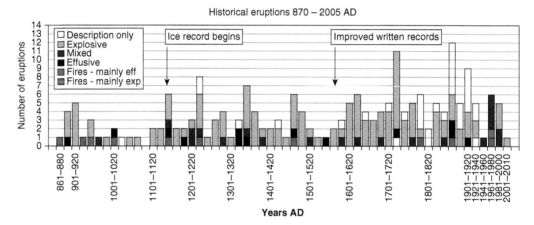

Figure 7.1 Frequency of known volcanic events in Iceland since the Norse settlement around CE 870. Data are binned per (a) 100 years and (b) 20 years. Thirteen events are labelled 'Fires', which designate annual to decade-long episodes with 2–11 eruptions. The number of eruptions in 'Fires' predating AD 1700 is generally unknown. From Thorardson and Larsen (2007).

processes despite this information being crucial for long-term volcanic risk management.

This statement holds for other natural hazards as well, but in this case it is important to remember that, according to the GVP, about 757 million people live within 100 km of a volcano that has erupted during the Holocene (https://volcano.si.edu).

One of the most important metrics for comparing eruptive events is the total mass of material that leaves the volcano during the eruption, allowing us to specify the size of an eruption in a physically meaningful way

(Table 7.1). The mass produced by documented volcanic eruptions ranges over many orders of magnitude. The largest among these are super-eruptions, defined as those yielding more that 1000 Gt of magma (Bryan et al. 2010). Decades of detailed geological mapping have revealed that at least 42 eruptions of that size have occurred during the past 36 million years, making them very rare, but also very destructive events (Mason et al. 2004). The most violent of these eruptions had peak intensities – the maximum rate of magma discharge per unit time – of 1.1 Mt s^{-1}, which is enough to fill a couple of the world's largest oil tankers in a mere second.

Table 7.1 Size and characteristics of selected explosive eruptions.

Volcano	Date	Magma type	Max. column height (km)	Total DRE (km³)	VEI
Toba, Indonesia	74 ka	Dacite, rhyolite	80	2750 ± 250	8
Mangakino, New Zealand	1.01 Ma	Rhyolite	Unknown	1200 ± 680	8
Yellowstone, USA	2.13 Ma	Rhyolite	Unknown	~900	8
Long Valley, USA	760 ka	Rhyolite	Unknown	~600	8
Taupo, New Zealand	27.1 ka	Rhyolite	30	~530	8
Yellowstone, USA	1.27 Ma	Rhyolite	Unknown	>280	8
Aso, Japan	87 ka	Rhyolite	Unknown	>200	<8
Campi Flegrei, Italy	39.3 ka	Trachyte	44	158 ± 53	6
Kurile Lake, Russia	8.4 ka	Rhyolite	Unknown	75 ± 5	7
Crater Lake, USA	7.7 ka	Dacite	55	75 ± 9	7
Yellowstone, USA	173 ka	Rhyolite	Unknown	50?	7
Tambora, Indonesia	1815 AD	Trachyandesite	43	32 ± 2	7
Santorini, Greece	3.6 ka	Dacite, rhyolite	36	32 ± 1.5	6
Novarupta, USA	1912 AD	Dacite	26	13 ± 3	6
Krakatau, Indonesia	1883 AD	Dacite	50	12.4	6
Pinatubo, Philippines	1991 AD	Dacite	40	5.4 ± 0.6	6
St. Helens, USA	1980 AD	Dacite	Unknown	0.44	5
Soufrière, West Indies	1902 AD	Basalt, andesite	16	0.15	4
Eyjafjöll, Iceland	2010 AD	Unknown	6.4	>0.04	4

DRE = Dense rock equivalent; VEI = Volcanic Explosivity Index; Source: All data compiled from the catalogue of the Volcano Global Risk Identification and Analysis Project (VOGRIPA, http://www.bgs.ac.uk/vogripa/index.cfm).

The volume of material delivered by volcanoes may also seem a reasonable measure of eruption size. However, the density of volcanic materials can differ greatly. Rhyolitic magmas are less dense than basaltic ones, and thus require larger volumes (~410 km³ versus ~360 km³) to qualify for a super-eruption, although such volumes can often overlap given the large error bars. In the geological past, many of these eruptions had landscape-changing impacts. During the formation of the trap basalts of India's Deccan Plateau 64.8 million years ago more than 9000 km³ of magma erupted. The Miocene basalts of the Columbia River flood basalt province (~15 Ma), United States, are a composite of more than 300 different flows with

average volumes of 500–600 km³ each (Bryan et al. 2010).

Newhall and Self (1982) proposed a Volcanic Explosivity Index (VEI), a logarithmic scale based on the total erupted mass (Figure 7.2). The VEI is widely used for comparing and ranking different eruptions. They also defined volcanic vigour as a metric for the activity of volcanoes in different arc settings. Volcanic vigour may be expressed in three ways: the number of volcanoes in an arc active during the Holocene, the total duration of their eruptive phases, or the total number of eruptions above an arbitrarily set threshold based on the VEI. All three metrics should be normalized to the length of the volcanic arc. Based on these metrics, the Cascades and Northeast Japan volcanic arcs are the most vigorous and

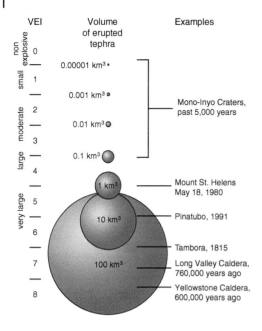

Figure 7.2 The Volcanic Explosivity Index (VEI). From Lockwood and Hazlett (2010).

the most productive, with an average eruptive volume of 5–6 km^3 km^{-1} over the past 1.8 Myr (Lockwood and Hazlett 2010). The Taupo volcanic zone in New Zealand is by far the most productive rhyolite volcanic zone on the planet, having produced 2 m^3 s^{-1} of fresh magma on average over the past 65 kyr (Wilson et al. 1995).

Extreme events and the denudation of tropical volcanic islands Tropical volcanic islands offer opportunities for assessing the role of extreme events such as large eruptions or flank collapse in the overall long-term history of volcanic landscapes. Erosion rates can be high on the rain-drenched flanks of tropical volcanoes, which are underlain by weak and geochemically altered rocks. The chronologies of debris avalanches that can carry many cubic kilometres from many volcanic islands afford first-order estimates of the history of catastrophic failures, as well as erosional budgets of tropical volcanic islands.

Detailed studies of other types of erosional processes operating on tropical volcanoes are growing and provide useful benchmarks for comparison (Ferrier et al. 2013). Erosion estimates reported in these studies can be compared to long-term denudation rates inferred by reconstructing the initial shape of dissected volcanic edifices (Salvany et al. 2012). The main assumption behind this approach is that volcanoes will be approximately cone-shaped before subsequent erosion takes place. If the age of the original edifice is known, it is possible then to estimate average rates of subsequent erosion by fluvial and mass-wasting processes over intervals of 10 000–100 000 years. Large catastrophic flank collapses involving up to 100 km^3 of rock (Hildenbrand et al. 2006) are sufficiently infrequent that a direct evaluation of the role of extreme denudation events should be possible.

On the island of Réunion in the Indian Ocean, torrential rain brought by tropical cyclones can efficiently entrain and rework landslide-derived sediment, and raise river beds where flushed to channels. A major rock-wall collapse in 1965 introduced large quantities of coarse sediment into the Remparts River. The river aggraded as a consequence, burying a 15-m high terrace, and raising the channel bed by nearly 40 m in only five years (Garcin et al. 2005). Such catastrophic sediment pulses are accompanied by average sediment yields of the order of 10^4 t km^{-2} yr^{-1}, aided by lahars and hyperconcentrated flows. Smaller rainfall-triggered landslides dislodged as much as 60 000 t of sediment per square kilometre of study area on Fiji's island of Viti Levu (Terry 2007) during Tropical Cyclone Wally in 1980. Most landscapes require decades to erode this mass per unit area.

Although making up less than 5% of the total denudation, chemical erosion driven by high rainfall can be high on tropical volcanic islands (Rad et al. 2006). The average of 1000 t km^{-2} yr^{-1} on Réunion Island is five and nearly 20 times higher than in the Amazon and the Congo river catchments, respectively (Louvat and Allègre 1997).

Emphasizing these extreme events in shaping volcanic edifices could lead one to overlook the geomorphic efficacy of less dramatic, but more frequent erosion and transport processes. Volcanic islands may shed sediment rapidly even without rare catastrophic flank collapse. Deep-sea fans off Réunion Island, Indian Ocean, for example, contain layers of hyperpycnal flows caused by tropical cyclones (Saint-Ange et al. 2011). Headward erosion along deeply cut canyons off the volcano Piton des Neiges appears to be the main mechanism of sediment export from this island, at rates of >1.2 km^3 kyr^{-1} over the past 180 kyr. Such gradual removal of mass reduces the material available for major slope instability, and may put a limit on the number of Pleistocene catastrophic collapses (Salvany et al. 2012).

7.2 Geomorphic Impacts of Volcanic Eruptions

7.2.1 The Volcanic Hazard Cascade

Volcanic eruptions are among the most spectacular of Earth's natural processes. They remind us of its fiery and dynamic interior. The range of hazards associated with volcanic eruptions is large and arises from the dynamic coupling of magmatic, hydrologic, and atmospheric processes. Only some of these hazards are linked directly to eruptions. Many historic volcanic disasters have happened long after an eruption or between eruptive phases due to catastrophic reworking of large amounts of volcaniclastic material, the sudden collapse of volcanic dams, or failure of the flank of volcanic edifices.

In the volcanic hazard cascade, large eruptions may indirectly affect nonvolcanic hazards such as floods: increased concentrations of aerosols following large explosive eruptions reflect more sunlight and cause cooling, which can depress rainfall and eventually cause streamflows to decrease over long distances. Iles and Hegerl (2015) showed that large rivers such as the Amazon, Congo, Nile, or Yenisey had reduced discharges in the first few years following major eruptions. Whether these altered streamflows are sufficient for modifying floods or sediment transport remains to be explored. Though far-reaching, these hydrological effects largely step behind the more immediate and rapid response of landscapes to being buried by volcanic products.

7.2.2 Geomorphic Impacts During Eruption

Material that falls and flows during volcanic eruptions generates most hazards. Pyroclastic particles <2 mm in diameter are referred to as volcanic ash, and may travel as airborne sediment for up to several thousands of kilometres. The possibility of these fine shards of volcanic glass interfering with jet turbines was the main the reason why many airplanes remained on the ground in Europe and North America during the moderate (VEI 4) 2010 eruption of Eyjafjalla volcano in Iceland (Figure 7.3). Thick piles of ash also induce loads that can be great enough to collapse the roofs of poorly constructed houses. Ash sheets may bury areas of tens of thousands to hundreds of thousands of square kilometres of the landscape, depending on the explosivity of the eruption, the degree of fragmentation, the height of the eruption column, and prevailing wind directions (Oppenheimer 2011). The ash sheets may suffocate living vegetation, severely affect agriculture, and create hydrophobic layers that reduce the amount of rainfall that enters the ground.

Pyroclastic flows, also known as 'nuées ardentes', 'glowing clouds' or 'pyroclastic

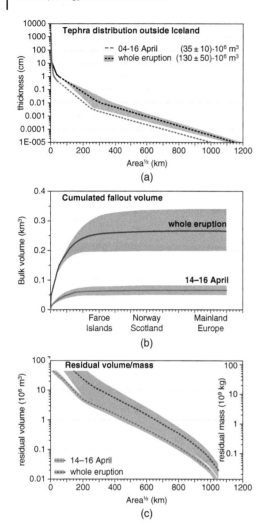

Figure 7.3 Distribution of tephra from the April–May 2010 eruptions of Eyjafjallajökull, Iceland, and the volume of tephra erupted. Tephra erupted from 14 to 16 April is estimated separately because it was responsible for the first ash transported to Europe. (a) Tephra thickness plotted as a function of the square root of area within each isopach. (b) Estimated cumulative fallout volume and its distribution with distance. (c) Estimate of volume (left axis) and mass (right axis) transported beyond a specified area. From Gudmundsson et al. (2011).

density currents', have been the most deadly of volcanic processes in history. The worst such disaster in the twentieth century was the eruption of Mount Pelée on the island of Martinique in 1902. The eruption released a

pyroclastic flow that killed nearly 29 000 people (Witham 2005). Pyroclastic flows involve the rapid motion of fine particles suspended in superheated gases of up to 1000°C. They travel at speeds of up to 700 km h^{-1}, so that running away from them is futile. Even buildings offer little protection, as temperatures remain lethally high in the distal zones of pyroclastic flows. This was the fate of residents of Pompeii and Herculaneum during the famous 79 AD eruption of Vesuvius, Italy (Mastrolorenzo et al. 2010).

Pyroclastic flows can be divided in three types, based on their grain size and mobility: firstly, block-and-ash flows involve less than 1 km^3 of pyroclastic material and feature abundant angular and rounded blocks with little, if any, pumice. Ash flows, the second type, are much more mobile, can bury areas of up to tens of thousands of square kilometres, and have volumes of up to 1000 km^3; in some cases ash flows may climb uphill or travel over water. The largest ash flows form landscape-mantling deposits termed ignimbrites that cover several hundreds to thousands of square kilometres. The eruption of Taupo volcano, New Zealand, in 1850 BP, buried some 20 000 km^2 of densely forested terrain beneath thick layers of ignimbrite; in an even larger area affected by tephra fall wildfires broke out and lasted for at least a century (Wilmshurst and McGlone 1996). Understanding the flow properties and runout dynamics of pyroclastic flows is essential for hazard zonations around volcanoes. The degree of dissection of hillslopes and valleys fringing a volcano can partly control the lateral mobility and tendency of block-and-ash flows to flow overbank or create avulsions (Cronin et al. 2013). Laboratory experiments using dilute mixtures of warm talcum powder suspended in air suggest that the height of topographic obstacles may play a role in diverting or even partly reflecting flows if they are at least 1.5 times higher than the flow depth (Andrews and Manga 2011). The third type of pyroclastic flow involves extremely rapid ice-slurry flows or 'mixed avalanches' (Pierson

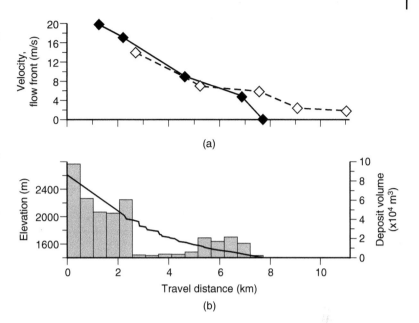

Figure 7.4 (a) Frontal velocities of two highly mobile ice-slurry flows (E1 and E2) from Ruapehu volcano, New Zealand, in September 2005 (filled diamonds: E1; open diamonds: E2). (b) Elevation and deposit distribution for E1. Modified from Lube et al. (2009).

and Janda 1994) that form when pyroclastic flows rapidly melt snow and ice (Figure 7.4). Deposits of ice-slurry flows are nearly invisible in the geological record, thus masking a potentially very destructive process (Lube et al. 2009).

'Lahar' is an Indonesian word for rapidly flowing mixtures of water and volcaniclastic sediment that may occur during volcanic eruptions (primary lahars) or after volcanic activity has ceased (secondary lahars). Abundant water, steep slopes, and sufficient amounts of loose volcaniclastic sediments are prerequisites for lahars (Figure 7.5). Common triggers include contact and mixing of hot pyroclastic material with ice or snow, rain-induced failure of tephra, and the catastrophic burst of caldera lakes. Lahars lie along a continuum of water–sediment mixtures. One end member of this continuum is normal or 'clear-water' floods that carry a suspended load and bedload as part of the most frequent sediment transport in rivers. Higher amounts of suspended and bed load characterize hyperconcentrated flows with volumetric sediment concentrations of ~20–50%, and bulk densities of 1.3–1.8 t m^{-3}. A high concentration of suspended load alters the mechanics of bedload transport and causes

debris flows that carry all coarse material available. These debris flows lie at the other end of the continuum and have solid concentrations of more than ~50% and bulk densities of 1.8–2.3 t m^{-3} (Iverson 1997). Lahars, which are hyperconcentrated flows or debris flows, produce much higher physical impact forces for a given discharge than clear-water floods because of their higher bulk density. Lahars are very mobile and destructive on the flanks of volcanoes and in river channels draining terrain mantled by pyroclastic materials (Figure 7.6).

The Lusi mud volcano, east Java Volcanoes can also form by erupting mud, and the largest and fastest growing example is the Sidoarjo mud flow (also known as the Lusi mud volcano), on the Indonesian island of Java. The mud volcano began its life on 29 May 2006 and was highly productive during its early stages, with a peak eruption rate of 0.12 Mm^3 of hot mud per day. The mud flow flooded the surrounding flat terrain and displaced more than 30 000 people, destroyed more than 10 000 homes, and also ponded a lake. Scientists proposed various hypotheses concerning

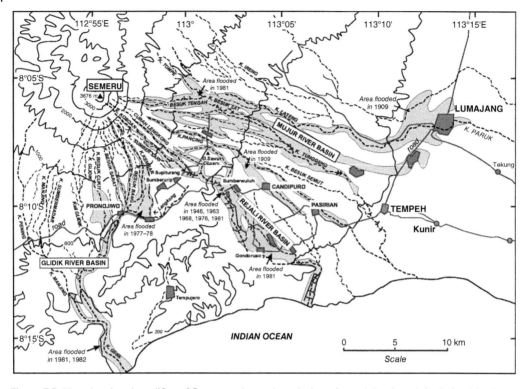

Figure 7.5 Map showing the edifice of Semeru volcano, Java, Indonesia, and the ring plain drained by the Mujur, Rejali, and Glidik rivers. Grey areas outline valleys and plains flooded by lahars during the twentieth century. From Thouret et al. (2007).

the trigger of the eruption, including the reactivation of an old fault during rupture of a nearby fault, geothermal processes, and a man-made disturbance because of nearby drilling for gas exploration that may have caused fracturing or fault reactivation (Davies et al. 2008). Regardless of the trigger, the Lusi mud volcano exemplifies how rare and less well studied volcanic processes can cause unexpected damage for people and the environment.

7.2.3 Impacts on the Atmosphere

Large volcanic eruptions can also change climate in the short term, and therefore affect weather patterns that influence other natural hazards such as temperature anomalies or droughts. Only recently have scientists discovered that the role of volcanic eruptions in triggering regional and global cooling episodes may have been underestimated, both in terms of frequency and duration (Sigl et al. 2015). The effects of a volcanic eruption on climate depend on the location, season, style, magnitude, and chemistry of the eruption, and have lasted for up to a decade for the strongest eruptions in the past 2500 years. The effects of volcanic eruptions on atmospheric temperatures are now routinely measured by satellite sensors, although these measurements may have underestimated the contribution of volcanic eruptions to cooling, at least in the twenty-first century (Ridley et al. 2014). Detailed analyses of historic evidence and palaeoenvironmental proxies of past large volcanic eruptions show that the temperature changes they effect might have been severe enough to drive the decline of entire communities, or even civilizations (Oppenheimer 2011).

Figure 7.6 Photographs of (a) the market (José Oseguera) and (b) the church in the town of Atenquique during a normal day prior to a lahar on 15 October 1955 (José Oseguera). (c) Photograph of Atenquique during the 1955 lahar, showing the remains of the market and church (Manuel Ponce). (d) Close-up of the church and the bouldery surface of the deposit (the largest blocks are 3–4 m in size) (Manuel Ponce). From Saucedo et al. (2008).

Glaciers in Arctic Canada and Iceland advanced between 1275 and 1300 AD, and especially from 1430 to 1455 AD, and some scientists link these advances with explosive eruptions rich in sulfur dioxide (SO_2) that may have cooled climate early during the Little Ice Age (Miller et al. 2012). The coincidence of reduced summer temperatures and some of the strongest volcanic eruptions of the past millennium seems compelling, but requires that volcanic aerosols remain in the atmosphere long enough to cool climate for decades to centuries. An alternative explanation is that the eruptions kicked-started cooling, which subsequently caused changes in the extent of Arctic sea ice, albedo, and ocean temperatures. Such feedbacks might be responsible for decades of lower summer temperatures long after the volcanic aerosol concentration had returned to background levels (Miller et al. 2012). In comparison, the historic eruption of Pinatubo,

Philippines, in 1991 depressed global temperatures only for a few years, mainly because SO_2 was injected into the lower stratosphere only. Other volcanic impacts on climate include changes to incoming solar radiation due to aerosols introduced by eruptions.

7.2.4 Geomorphic Impacts Following an Eruption

The fallout from explosive eruptions can rapidly increase the availability of loose sediment to a given landscape; the magnitude of the eruption, the distance from the crater, and the specific pyroclastic processes largely determine how landscapes and drainage basins respond in terms of erosion and sedimentation (Figure 7.7) (Pierson and Major 2014). Surface runoff increases when landscapes are freshly blanketed by ash. Widespread gullying of thick fine-grained tephra sets in motion a cascade

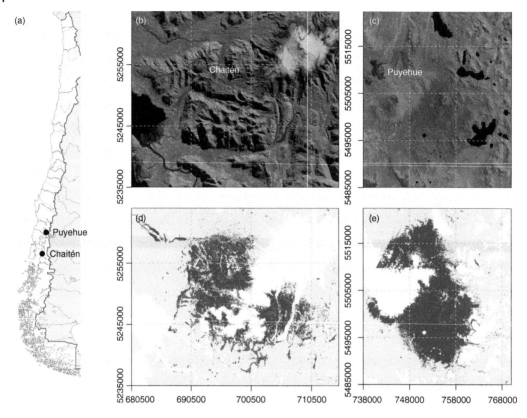

Figure 7.7 Post-eruptive landslides in tephra-covered rainforest, south-central Chile. (a) Location of Chaitén and Puyehue volcanoes. (b) Satellite image of Chaitén volcano (red triangle). (c) Satellite image of the Puyehue-Cordón Caulle volcanic complex. The red triangle shows the vent of the 2011 volcanic eruption. (d) and (e) Forest losses and landslides in the Chaitén and Puyehue-Cordon Caulle areas, respectively: forests with >90% tree cover in 2000 (light grey), forest losses between 2000 and 2014 (dark grey), and posteruptive landslides (red). White areas include bare rock surfaces, water, ice, grass, and shrub cover. From Korup et al. (2019).

of erosion, reworking, and deposition. The excess sediment enters river channels, spawning hyperconcentrated flows and lahars that cause widespread aggradation in channels and on floodplains (Figure 7.8). Three destructive rainfall-triggered lahars happened in Italy, Mexico, and Nicaragua in 1998 alone, each mobilizing and depositing >1 Mm^3 of sediment along populated river valleys. In 1955, heavy rainfall triggered catastrophic runoff from the flanks of the Nevado de Colima volcano in Mexico. The floodwaters entrained pyroclastic sediment and initiated a lahar that rapidly moved downstream, and increased in its size and speed before overrunning an artificial water reservoir en route. The lahar had a flow depth of 8–9 m in the village of Atenquique some 25 km from the volcano, where it killed more than 23 people, taking away houses and parts of the local railway, and deposited 3.2 Mm^3 within only 10–15 minutes. Most of the village was buried beneath more than two metres of bouldery deposits (Fig. 7.6; Saucedo et al. 2008).

Some of the highest sediment yields in river channels, recorded over periods of several months to years, result from reworking of fine pyroclastic materials; yields are up to 1 Mt km^{-2}, more than two orders of magnitude higher than undisturbed fluvial sediment

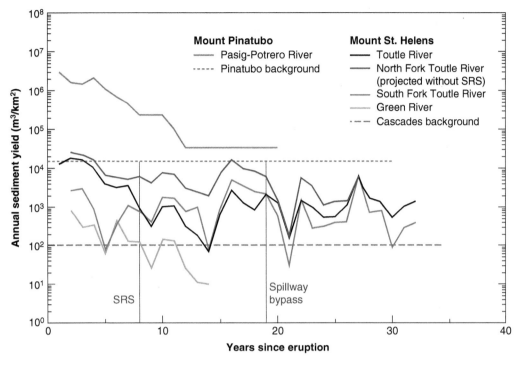

Figure 7.8 Comparison of sediment yield from severely disturbed basins at Mt St Helens and Mt Pinatubo as functions of time since eruption. SRS represents the time that the US Army Corps of Engineers completed a large sediment retention structure on the North Fork Toutle River, Washington, which trapped both bedload and suspended load. About ten years after completion of the sediment retention structure. Mt St Helens sediment yields are based on suspended-sediment data alone, whereas Mt Pinatubo data come mainly from measurements of accumulated deposits and thus represent both suspended load and bedload sediment. Note the rapid decline of sediment yield within the first five to ten years after each eruption. From Pierson and Major (2014).

yields. Tephra can remain highly mobile where vegetation cover offers insufficient roughness for trapping particles. For example, semi-arid and windy areas such as Patagonia are at risk from ash storms of reworked tephra sourced from volcanoes of the Andean chain. Such storms may bury farmland beneath dune-like deposits, bury or abrade vegetation, and contaminate feed supplies for livestock (Wilson et al. 2010), necessitating the clean up of wind-blown material. Deposition of reworked tephra in irrigation canals and reservoirs may compromise agricultural production and human health, especially in drier landscapes that are largely devoid of trees. Vehicular traffic that resuspends fine volcanic ash articles can also

worsen respiratory health problems (Martin et al. 2009).

Volcanologists have long recognized that large portions of oceanic volcanic islands are composed of deposits dumped by catastrophic flank collapse (Figure 7.9). The same holds for many volcanoes on continents. Many volcanologists consider the catastrophic collapse of volcanic edifices an integral stage in their growth and decay. Such destruction may form calderas or deposits of large debris avalanches. A global database of calderas that formed in the past ~90 Myr shows that they range in size from 0.01 to 4700 km^2, although most are smaller than 500 km^2 (Geyer and Martí 2008). Most of these documented calderas are associated with stratovolcanoes located

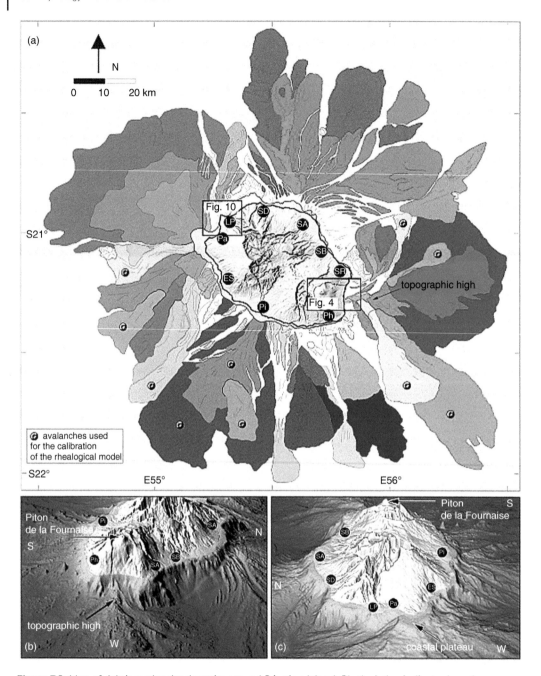

Figure 7.9 Map of debris avalanche deposits around Réunion Island. Black circles indicate densely populated regions: SD, St Denis; LP, Le Port; Pa, St Paul; ES, Etang Salé; Pi, St Pierre; Ph, St Philippe; SR, Ste Rose; SB, St Benoît; SA, St André. Many volcanic islands have such aprons, made up by their own debris. From Kelfoun et al. (2010).

920 m

0 m

−1000 m

2 km

Figure 7.10 Digital terrain and marine model of Stromboli volcano, Italy, draped with an aerial photograph. Grey shades denote areas below sea level. The deep scar at the lower left is the location of a slope failure that triggered a tsunami in December 2002. From Chiocci et al. (2008).

in subduction or continental rift settings, and some now contain major lakes. Outbursts from such lakes have produced catastrophic floods that were surpassed in discharge and water volumes only by the Quaternary megafloods originating from major glacial meltwater lakes. Manville (2010) estimated that five of the nearly 30 known floods with peak discharges >0.1 Mm^3 s^{-1} occurred during the Quaternary. Some calderas have given rise to repeated outburst floods that may have resulted from episodic, or even periodic, interaction between lake filling and volcanic processes. Other forms of sudden instability in calderas can involve catastrophic landslides. In 2004, for example, some 0.2 km^3 of rock detached from Bawakaraeng volcano on the Indonesian island of Sulawesi, forming a long-runout debris avalanche that travelled for seven kilometres down the Jeneberang valley. Tsuchiya et al. (2009) argued on the basis of slope-stability modelling that rising groundwater levels could

have triggered this sudden slope failure in an area where annual rainfall exceeds 4000 mm.

Catastrophic flank or sector collapse may occur either during or between eruptions, or even long after a volcano has become dormant (Figure 7.10). Large debris avalanches occur somewhere on Earth about once every 25 years on average (Siebert 1992). Triggers of catastrophic edifice collapse include: changes in the volcano's slope due to magma chamber inflation or deflation; subvolcanic basement spreading irrespective of ongoing magmatic activity (van Wyk de Vries and Francis 1997); increases in fissure-bound gas pressures in lava domes due to rainfall (Elsworth et al. 2004); and hydrothermal pressurization (Reid 2004). Submarine deep-seated gravitational slope deformation may explain the gradual downward sliding of some volcanic flanks at rates of several centimetres per year, for example at Etna volcano, Italy (Chiocci et al. 2011). Debris avalanches and related types of volcanic flank collapse mostly detach along extensive weak layers in shield volcanoes, formed by marine strata at depth, hyaloclastites, hydrothermally altered rocks, or volcaniclastic deltas. Analogue models using sand and plaster show at a small scale that deep-seated gravitational spreading of shield volcanoes on submarine strata can lead to catastrophic sliding if the thickness of the sedimentary layer is less than 10% of the volcanic edifice height (Oehler et al. 2005).

Debris avalanches produce a diverse set of geomorphic landforms and sedimentary facies (Shea and van Wyk de Vries 2008). Most studies emphasize the hummocky mounds dotting the surface of debris-avalanche deposits and the chaotic and poorly sorted nature of the sediments themselves, rich in megaclasts. Distal debris-flow facies might show where the debris avalanche entered river channels or standing water bodies (Figure 7.11) (Capra et al. 2002). Toreva blocks are large intact pieces of volcanic rock that have been rafted a short distance to proximal positions during these extreme mass movements (Roa 2003). Multiple aprons of debris avalanches cover the roots of volcanic islands throughout the world,

Figure 7.11 Photographs showing the effects of the displacement wave in Lago Cabrera, Chile, following a landslide from the summit of Yate in 1965. (a) Southwest corner of the lake, showing bare ground overran by the displacement wave and the site of a settlement several hundred meters from the lakeshore. Yate, the source of the landslide, is visible in the distance. (b) Southwest corner of Lago Cabrera showing the displacement wave trimline up to 30 m above present lake level. (c) Aerial photograph of the southwest corner of the lake showing directions the displacement wave travelled. Points x, y, and z are locations shown in images (a) and (b). (d) Rip-ups of soil in mud deposited at the southwest corner of the lake following the tsunami and preserved beneath a felled tree. Selected clasts are outlined for clarity. (e) Outflow from Lago Cabrera at the edge of mud deposited by the displacement wave at the southwest corner of the lake. Outflow is escaping beneath volcanic flows. (f) Clast embedded in a felled tree; scale in centimeters. (g) Block of volcanic rock from the northeast summit Yale carried about 12 km in a debris flow in 1870 or 1896. Source: Kench et al. (2006).

and attest to the catastrophic failure of volcano flanks between eruptions (Boudon et al. 2007). The sudden collapse of Ritter Island in 1888 in the Bismarck archipelago near Papua New Guinea was the largest documented edifice failure in historical time. Sudden slope failure removed ~5 km^3 of volcanic rock – most of the island's formerly visible topography – and caused a destructive tsunami tens of metres high in the Bismarck Sea. The debris avalanche transformed into a debris flow that ran out at least 20 km on the sea floor (Silver et al. 2005).

Volcanoes, like any other mountains, are subject to the forces of seismicity, tectonic uplift, weathering, and erosion. Aggressive headward erosion along canyons cut into the flanks of volcanoes can induce oversteepening and eventual catastrophic collapse of caldera rims by preferentially exploiting weaknesses in the rock mass developed along ring fault systems (Merle et al. 2008). A detailed study of 47 catastrophic flank collapses on volcanoes of the Caribbean arc showed that smaller (<1 km^3) and more frequent debris avalanches had sources on northern volcanoes with intense hydrothermally altered rocks and pervasively fractured summits, whereas larger (>10 km^3) and rarer collapses mainly detached from oversteepened western flanks of the southern volcanoes (Boudon et al. 2007). The study also showed that large debris avalanches were quite frequent in the Lesser Antilles, occurring on average every 800 years over the past 12 kyr.

Catastrophic unroofing of volcanic edifices by mass wasting causes rapid drops of overburden pressure and can promote magma ascent from shallow reservoirs, potentially spawning new eruptive activity. Manconi et al. (2009) computed that large landslides can reduce overburden pressures ranging from 10^{-2} to 10^1 MPa, and that decompression scales as a power law with landslide volumes ranging from 1 to 1000 km^3. The Icod debris avalanche, emplaced by a large sector collapse of Teide volcano on Tenerife, Canary Islands, unloaded the underlying magma chamber, and prompted a large explosive eruption 175 ± 3 ka ago (Boulesteix et al. 2012). Rapid extrusion of lavas at eruptive rates as high as 8 km^3 yr^{-1} built a new volcanic cone at the head of the landslide during the following 120 kyr. As a consequence the volcanic and geomorphic evolution of Teide volcano has been largely conditioned by catastrophic flank collapse.

The eruption of Mount St. Helens, Washington, in 1980 gave important insights into the long-lasting hydrologic and geomorphic impacts of an explosive volcanic eruption. A catastrophic debris avalanche buried 60 km^2 of valley floors, and a lateral blast and pyroclastic flows obliterated 550 km^2 of mature forest (Figure 7.12). The landscape was covered with several decimetres of pumiceous silt, sand, and gravel, while debris flows eroded riparian corridors and filled channels with up to several metres of lahar deposits (Major and Mark 2006). The ensuing sediment loads of rivers draining the volcano climbed dramatically, and prompted short-lived changes in the frequency and magnitude of autumn and winter flows. Twenty years after the eruption, suspended sediment yields remained about a hundred times higher than yields before the eruption (Major et al. 2000).

Pre-eruption channel courses can change rapidly following sudden sediment inputs, and thus displaced river channels may incise rapidly into previously protected bedrock spurs and lips along valley floors (Whipple et al. 2000). Pyroclastic deposits are easily reworked, spawning hyperconcentrated flows and lahars that raise sediment yields in rivers to some of the highest ever reported. Detailed studies of sedimentation in rivers impacted by the 1991 eruption of Pinatubo volcano, Philippines, showed that sediment yields remained high even two decades later, mainly because of channels continued to be unstable and widen (Gran et al. 2011). Similarly high sediment yields (>0.1 Mt km^{-2} yr^{-1}) arose from lahars that descended from the flanks of Merapi volcano, Indonesia, in the first decade following its 1994 eruption (Lavigne 2004). Lahars are a

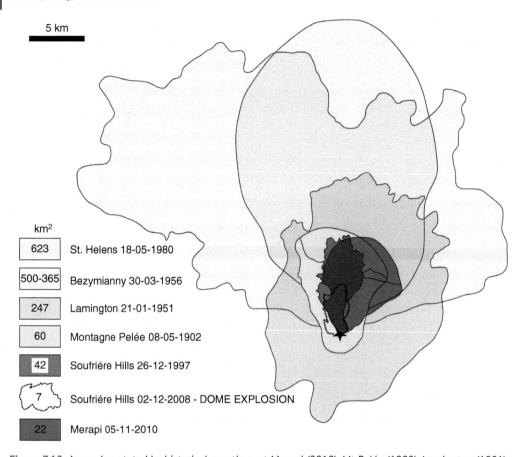

5 km

km²	
623	St. Helens 18-05-1980
500-365	Bezymianny 30-03-1956
247	Lamington 21-01-1951
60	Montagne Pelée 08-05-1902
42	Soufriére Hills 26-12-1997
7	Soufriére Hills 02-12-2008 - DOME EXPLOSION
22	Merapi 05-11-2010

Figure 7.12 Areas devastated by historical eruptions at Merapi (2010), Mt Pelée (1902), Lamington (1951), Bezymianny (1956), Mt St Helens, Soufrière Hills (1997), and a smaller directed explosion at Soufrière Hills (2010). From Komorowski et al. (2013).

permanent hazard on many volcanoes simply because of the high availability of pyroclastic material in steep terrain that is prone to heavy rainfall.

The long-anticipated collapse of a tephra dam at Crater Lake, Ruapehu Volcano, New Zealand, on 18 March 2007, provided an opportunity to document in detail the dynamics and geomorphic impact of a lahar generated by a catastrophic outburst from a lake (Figure 7.13). Detailed laser scanning revealed a complex and alternating pattern of erosion and deposition along the lahar track. The 1.3 Mm^3 of water released from the lake entrained 2.5–3.6 Mm^3 of volcaniclastic sediment from the tephra dam, and added channel materials over the first five kilometres of the path, attaining a total volume of 4.4 Mm^3 (Procter et al. 2010). Volcanic materials transported by rivers may remain stored in the landscape for long periods. For example, many alluvial plains on the Japanese islands contain, or have even formed from, large amounts of volcaniclastic sediment reworked by Pleistocene and Holocene volcanic lake outbursts (Kataoka et al. 2009).

7.3 Geomorphic Tools for Reconstructing Past Volcanic Impacts

7.3.1 Effusive Eruptions

Magma is a three-phase system with solid, liquid, and gaseous components. The chemistry

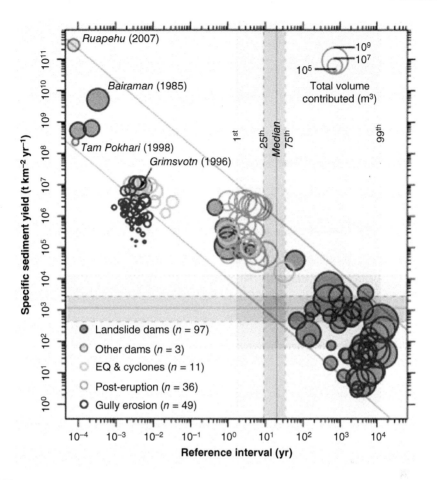

Figure 7.13 Average specific fluvial sediment yields following failures of natural dams (brown and blue circles), regional landslides caused by earthquakes and tropical cyclones (green circles), pyroclastic eruptions (orange circles), and rainstorm-driven gully erosion in loess terrain (violet circles). Labels show the largest estimated instantaneous yields for historic volcanic, landslide, moraine, and glacier dam failures: Ruapehu, New Zealand; Bairaman, New Britain; Tam Pokhari, Nepal; and Grimsvotn, Iceland, respectively. Landslide-driven sediment yields are from the 1999 Chi-Chi earthquake and several subsequent tropical cyclones in Taiwan. Gully erosion data are from the Loess Plateau, China. The disturbance-driven yields differ by eight orders of magnitude, and by three orders of magnitude for a given interpolation period. The red lines show time trajectories that describe the declining contributions of high, although short-lived, sediment yields with increasing reference intervals. The lower red line defines extreme sediment yields involved $>10^4$ t km^{-2} yr^{-1} averaged over a single year. Grey shaded areas are percentiles of sediment yields reported over 600 mountain rivers worldwide. Source: Korup O, Earth's portfolio of extreme sediment transport events. Earth-Science Reviews 112, 115–125. © 2012, Elsevier.

and percentages of each of these phases largely determine the character of the volcanic eruption. Effusive eruptions generally involve the extrusion and flow of lava from volcanic craters or fissures. The mobility of a lava flow depends mainly on its temperature, effusion rate, crystallinity, density, viscosity, and yield strength.

A rule of thumb holds that lava starts to flow at temperatures of $>700°$C (Lockwood and Hazlett 2010). Flowing lava can be modelled as a viscoplastic Bingham fluid, which requires a critical shear stress τ_c [N m^{-2}] or yield strength to set the fluid in motion on a slope β [°]:

$$\tau_c = h_c \rho_l g \sin \beta \qquad (7.1)$$

where g is gravitational acceleration [m s^{-2}], h_c is the minimum or critical thickness [m], and ρ_l is the density [kg m^{-3}] of the lava flow with slope β [°]. The yield strength thus defines the point at which lava begins to flow, although it may stop flowing under different conditions. Knowing the flow depth and the steepness of the terrain across which the lava flows thus helps estimating its flow parameters. The bulk viscosity η of the lava [kg m^{-1} s^{-1}] is a measure of its internal resistance to flow, and is defined as the local gradient of the applied shear stress τ to the strain rate $\dot{\epsilon}$ [s^{-1}]:

$$\tau = \tau_c + \eta \dot{\epsilon}^m \tag{7.2}$$

Note that Eq (7.2) only describes Bingham fluids with the nondimensional scaling exponent $m = 1$. Fluids for which $m = 1$ and $\tau_c = 0$, such as water, are called Newtonian fluids. Fluids in which η varies with shear stress exhibit either dilatant (shear-thickening, $m > 1$) or pseudoplastic (shear-thinning, $m < 1$) behaviour. Think of a bottle of ketchup. It helps to shake the bottle before pouring, because the shaking increases the shear stress while reducing the viscosity. At room temperature ketchup can be modelled as a shear-thinning, or Bingham pseudoplastic, fluid.

The viscosity of lava flows depends on several other factors, including temperature and the content of silica, water, and gas bubbles. Hotter lava flows are less viscous, but become more viscous as silica content increases. Strong temperature gradients in lavas alter the flow dynamics and therefore necessitate more complex models that account for differing rheologies within the flow.

For steady state conditions, however, we can neglect changes in rates. Using this assumption, we can compute the velocity of a lava flow $u(z)$ from its viscosity and channel geometry:

$$u(z) = \frac{\rho g \sin\theta (t^2 - z^2)}{4\eta} \tag{7.3}$$

where z is the flow depth or the distance from the bank [m], t is the channel depth or radius [m], respectively, ρ is the density of the lava

corrected for the volumetric content of gas vesicles [kg m^{-3}], η is the bulk viscosity of the lava [kg m^{-1} s^{-1}] and θ [°] is the slope of the underlying surface (Harris 2013). Equation (7.3) predicts a parabolic velocity profile across the lava channel, and a maximum velocity at the flow surface, that partly depend on the local geomorphology. Lava flows are faster when contained in channels or tubes (> 10 m s^{-1}), where heat loss is minimized. Unconfined flows across rocky terrain, however, can be up to two orders of magnitude slower.

Lava flows leave conspicuous footprints in the landscape, and those with sufficient effusion rates and durations can dominate landscapes. The literature on volcanic geomorphology highlights the many landforms produced by volcanoes and lava flows. The largest volcanic eruptions on Earth involved >1000 km^3 of low-viscosity and thus highly mobile trap basalt flows and silicic ignimbrites that covered areas of tens to hundreds of thousand square kilometres. Volcanologists estimate the volumes of eruptions in terms of 'dense-rock equivalent' (DRE) to account in particular for the low density of pyroclastic materials; DRE is estimated as the volume of magma just before it erupts. Several 'mega'-eruptions are known to have occurred briefly before, or at, times of major biotic change, so that some researchers argued that these peaks of volcanic activity may have triggered some of the mass extinctions on our planet (Bryan et al. 2010). Some mega-eruptions may have themselves been triggered by meteorite or bolide impacts such as during the Cretaceous–Tertiary (or 'K/T') boundary, which marks the extinction of the dinosaurs about 65 million years ago.

Most lava flows move slowly enough that people can escape their paths. However, lava flows can burn or destroy housing and other infrastructure, trigger wildfires, and cause anoxia through the release of carbon dioxide. Some lava flows of low or moderate mobility can be diverted by artificial structures called lava dams (Scifoni et al. 2010).

Another measure that successfully stopped a slow-moving lava flow on the island of Heimaey, Iceland, in 1973 (Mattsson and Höskuldsson 2003), was to apply water. Initially, the edges and surface of the flow were cooled with water discharged from fire hoses. Then a large water pipe with small holes was placed in the path of the flow to accelerate cooling.

Lava flows can dam rivers and thus affect valley geometry or alter drainage directions. Examples of late Quaternary lava-dammed lakes can today be found along the Colorado River in the Grand Canyon and the Owyhee River in Oregon, USA (Fenton et al. 2006). Lavas flowed down and infilled river valleys in the Blue Mountains of southeastern Australia during Eocene to Miocene times and remain a marker that geologists can use to trace river-channel incision and adjustment to a new equilibrium over millions of years. Reconstructions suggest that these deeply carved and impressive sandstone canyons formed at average incision rates of 14–40 m Myr^{-1} since the lava flows were emplaced (van der Beek et al. 2001).

7.3.2 Explosive Eruptions

Explosive eruptions involve large amounts of gas that propel hot and highly fragmented rock and ashy volcanic ejecta into the atmosphere. The peak intensity of an explosive eruption, in terms of mass erupting per unit time, scales nonlinearly with its magnitude, because larger eruptions necessitate larger magma chambers with higher pressures. The relationship between the size of a volcanic vent and the peak intensity can be modelled in one dimension as:

$$\dot{m} = \frac{\pi r^4 g \rho_m (\rho_{cr} - \rho_m)}{8\eta} \quad (7.4)$$

where $\dot{m}$ is the peak mass eruption rate [$kg\,s^{-1}$], r is the vent radius [m], g is gravitational acceleration [$m\,s^{-2}$], ρ_m is the bulk density of magma [$kg\,m^{-3}$], ρ_{cr} is the bulk density of the crust [$kg\,m^{-3}$], and η is bulk magma viscosity

[$kg\,m^{-1}\,s^{-1}$] (Wilson et al. 1980), providing another direct link between geomorphology (vent size) and volcanic activity (eruption intensity).

The peak mass eruption rate further controls how high the eruptive plume or column rises into the atmosphere. This rise height H [m] can be estimated as:

$$H = \lambda \left(\frac{\delta\rho}{\rho_0} g \dot{V} \right)^{1/4} \left(-\frac{g}{\rho_0} \frac{\partial \rho_a}{\partial z} \right)^{-3/8} \quad (7.5)$$

where $\lambda (\approx 5)$ is an experimentally derived constant, $\delta\rho$ is the difference between the bulk density of the eruptive column and the density of the surrounding ambient air (ρ_0) at the level of the vent [$kg\,m^{-3}$], $\dot{V}$ is the volumetric flux from the vent [$m^3\,s^{-1}$], ρ_a is the density of the ambient air at local height [$kg\,m^{-3}$], and z is the local height coordinate [m] (Woods 2013). Equation (7.5) predicts that the height of the volcanic plume scales as a power law with volumetric flux $\dot{V}$, and thus also with peak mass eruption rate $\dot{m}$. Hence larger and more vigorous eruptions are more likely to inject pyroclastic material to higher atmospheric levels, even as far as the lower stratosphere, where particles may take weeks to several years to disperse. The height of the eruption column also determines the mode and timing of its subsequent collapse that produces pyroclastic flows. The intensity of the collapse may be expressed by the fraction of the total mass flux that eventually descends down the edifice as pyroclastic flows (Carazzo et al. 2015).

Airborne pyroclastic material, termed tephra, ranges widely in grain size. Coarse tephras are known as lapilli or bombs. The trajectories of bombs can be modelled using a ballistic approach based on Newton's second law of motion, which states that force is equal to the product of mass and acceleration (Clarke 2013). We solve for the horizontal acceleration dv_x/dt [$m\,s^{-2}$] by dividing the force resulting from air drag, which resists the clast's motion, by its mass:

$$\frac{dv_x}{dt} = -\frac{v \cos\theta \rho_a v' C_d}{2m} \quad (7.6)$$

where v is the speed of the clast [m s^{-1}], θ is the angle of motion relative to the horizontal [°], ρ_a is atmospheric density [kg m^{-3}], v' is the relative speed of the clast with respect to the ambient fluid (which is air in most cases), C_d is a dimensionless drag coefficient with values ranging between 0.06 and 1, and m is the mass of the clast [kg]. We derive the vertical acceleration dv_y/dt [m s^{-1}] correspondingly, but adjust for gravitational acceleration g [m s^{-2}], and the density contrast:

$$\frac{dv_y}{dt} = -\frac{v \sin \theta \rho_a v' C_d}{2m} - g\frac{\rho_r - \rho_a}{\rho_r} \quad (7.7)$$

where ρ_r is the density of the clast [kg m^{-3}]. Note that the acceleration of the clast scales with the square of its velocity if the surrounding fluid is immobile ($v = v'$). Note, however, that empirical data indicate that the spatial distribution of ejecta size is sometimes affected by initial entrainment of even large clasts in a high-velocity ascending ash cloud. Equations (7.6) and (7.7) thus predict mass-dependent parabolic trajectories that result in sorting of tephra by grain size as a function of distance from the vent.

Reconstructing the geomorphic impact of volcanic eruptions and volcanic mass movements depends largely on deposits and landforms that can be identified in the geological record (Thouret 1999). Few regions on Earth, however, offer complete detailed records of volcanic eruptions. An exception is the Icelandic sagas, which contain a wealth of information about Medieval eruptions and support a detailed record spanning some 1100 years (Thorardson and Larsen 2007). In spite of the rich historic record of volcanic eruptions in the Mediterranean, several reported eruptions remain elusive in the geological record. The Plinian eruption of Santorini volcano, Greece, sometime between 1650 and 1620 BC during the Minoan period, is among the large events that have attracted many efforts to track and date in sedimentary archives both onshore and offshore (Walsh 2014). Researchers have arrived at different conclusions about the environmental impacts of this eruption on Bronze Age people in the eastern Mediterranean, and whether or not the collapse of Minoan society following that time was a result the eruption.

Tephrochronology is one of the main tools used to reconstruct both the times and extents of former volcanic eruptions. Volcanic tephras have distinct mineralogical and geochemical fingerprints. Tephras from a single volcano may change from eruption to eruption, and generally are sufficiently different from one another to distinguish between their sources and times of deposition. Absolute age dating of tephra layers enables researchers to build local to regional chronologies of past eruptions. Icelandic terrestrial soils that formed during the Holocene host nearly 100 different silicic tephras (Larsen and Eiríksson 2008). Basaltic tephras are more than eight times more abundant than silicic ones, but even they provide only a partial catalogue of Icelandic eruptions during the Holocene because the source volcanoes are commonly situated beneath glaciers, in areas of high groundwater, or are offshore. In the last case the eruptive products can be poorly preserved in terrestrial settings. In some instances experts are able to distinguish tephra beds from field evidence, although in most cases detailed laboratory analyses are needed. Assuming that tephras of different ages are equally well preserved, one can infer important details about the frequency and magnitude of past eruptions. Chronologies are now available for many volcanic centres, with as many as several dozens to hundreds of eruptions documented in a given region (Watt et al. 2011). Traces of major volcanic eruptions are also found in ice-core records, where they offer valuable insights into the regional to global spread of volcanic aerosols as well as their associated climate impacts.

The type, magnitude, and peak intensity of past eruptions can be estimated from tephra deposits in the field. By mapping the thickness and degree of fragmentation of air-fall deposits, one can infer the type of eruption and its approximate magnitude using the Walker classification (Lockwood and Hazlett 2010). This method relies on isopach maps showing the thickness of volcaniclastic deposits attributed to a specific eruption. Field measurements have shown that the local average thickness T [m] of tephra deposits scales as a power law with their area, such that detailed field mapping of well preserved deposits allows estimates to be made of the original eruption volumes:

$$T = T_0 \left(\frac{A_0}{A} \right)^{m/2} \tag{7.8}$$

where T_0 is the maximum thickness of tephra deposits [m], A is the area enclosed within a given isopach contour [m^2], A_0 is the area enclosed by the isopach of maximum thickness [m^2], and $m > 2$ is the scaling exponent (Bonadonna and Costa 2013). From Equation (7.8) we can derive the total erupted volume V [m^3]:

$$V = \frac{2}{2 - m} T_0 A_0 \left(\left(\frac{A_{dist}}{A_0} \right)^{1-m/2} - 1 \right) \tag{7.9}$$

where A_{dist} is the area enclosed by the isopach of zero thickness [m^2], which can be approximated as the maximum downwind aerial extent of the deposit. If the substrate topography is adequately known, isopach maps can also provide first-order estimates of the deposit volume, which when multiplied by mean density provides an estimate of the total mass involved. This method is best suited for younger explosive eruptions – those that happened in the past 100 000 years – because erosion quickly alters the sedimentary evidence of past eruptions. To infer the peak intensity of a past eruption, volcanologists commonly map the maximum diameters of intact ejecta to produce an isopleth map showing contours of maximum particle size. This information can then be used in theoretical and empirical models of eruption columns to estimate peak intensity.

Kataoka et al. (2009) pointed out that the products of explosive eruptions may so overload rivers that channels and floodplains can be buried by sediment hundreds of kilometres from the volcanic centre. Thus volcanic hazard assessments based solely on mapped distributions of pyroclastic material may severely underestimate the risk to rivers. Manville et al. (2009) reconstructed how some 15 major rivers responded to the explosive 1.8 ka eruption of Taupo volcano, New Zealand. This eruption emplaced $\sim$30 km^3 of nonwelded ignimbrite up to 10 m thick over an area of $\sim$20 000 km^2. Following the eruption, hyperconcentrated sheet flows began to rework the pyroclastic deposits in the headwaters; rivers also began to re-incise, occasionally boosted by flood waves following the sudden failures of several ignimbrite-dammed water bodies. Yet between one-third and two-thirds of the material remains as it was deposited in the larger impacted catchments. Most of the average specific sediment yields attributable to this post-eruptive flushing of volcaniclastic material are remarkably similar, at 200–300 t km^{-2} yr^{-1}.

Sudden or sustained inputs of volcanic sediment input may continue to force channel adjustments more than 100 yr after an eruption. The Sandy River aggraded a confined bedrock gorge >20 m after having been impacted by a lahar that travelled for more than 90 km following the 1781–1791 AD eruption of Mount Hood, Oregon (Pierson et al. 2010). Peak sedimentation rates were >2.3 m yr^{-1}, although very short-lived. Still, the river has yet to incise to its original pre-eruption level.

Lahars can be modelled using the same principles as lava flows. We can reformulate Eq (7.2) by expressing the yield strength τ_c as a combination of Coulomb friction and effective fluid pore pressure:

$$\tau = c' + (\sigma - u_f) \tan \phi + \eta \dot{\varepsilon}^m, \tag{7.10}$$

where c' is the apparent cohesive strength of the lahar material [kPa], σ is the normal stress [kPa], u_f is the fluid pore pressure [kPa], ϕ [°]

is the internal angle of friction of the material and η is the dynamic viscosity of the lava [kg m^{-1} s^{-1}] and $\acute{\epsilon}$ is the shear strain rate; $m = 1$ for a Herschel-Bulkey model (Manville et al. 2013). This simple Coulomb-viscous model is strictly valid only for a single-phase flow, and ignores the broad and unsorted mix of particles ranging from clays to boulders that characterize lahars. Collision and momentum transfer between particles also requires including a dispersive stress term in rheological models.

Detailed field mapping soon after deposition is helpful for accurately reconstructing the flow dynamics of lahars. The geometry of the fresh deposits reveals some insights, for the peak discharge Q_{pl} of a lahar [m^3 s^{-1}] appears to scale with its volume V_l [m^3]:

$$Q_{pl} = aV_l^b \qquad (7.11)$$

where reported values of a range from 0.003 to 0.006 [s^{-1}] and b ranges from 0.83 to 1.01 (Bovis and Jakob 1999). The lahar volume V_l [m^3] can in turn be estimated from the undisturbed planimetric deposit area A_l [m^2]:

$$A_l = \alpha V_l^\gamma \qquad (7.12)$$

where $\alpha = 200$ [m] and $\gamma = 2/3$ (Iverson et al. 1998). Slightly different coefficient and exponent values extend these simple relationships to nonvolcanic debris flows and emphasize the close relationship between flow processes in volcanic and nonvolcanic terrains.

Lakes may also hold valuable archives of past eruptions or volcaniclastic sediment pulses. The Mount Meager volcanic massif occupies only 2% of the area of the Lillooet River basin, British Columbia, but may well be Canada's most landslide-prone region. At least 25 slope failures larger than 0.5 Mm^3 have removed >2.4 km^3 of volcanic rock from the massif over the past 10 kyr (Guthrie et al. 2012) and much of that sediment has ended up in Lillooet Lake near the mouth of the catchment. Huge amounts of volcaniclastic sediment accumulated in the lake immediately after the last eruption of the volcano ~2.4 ka ago; at that time background rates of sedimentation

and delta progradation increased by a factor of 25 (Friele et al. 2005). But even without detailed information of sedimentation rates lakes offer valuable archives of past eruptions. Moreno et al. (2015) found 26 tephra layers in Lago Teo, a small lake close to Chaitèn volcano, south-central Chile. The volcano erupted unexpectedly in 2008 after having been seemingly inactive during the Holocene. However, dating of the tephra layers in the lake revealed that explosive eruptions had occurred once every 200–300 years on average during that period, making the 2008 eruption rather unexceptional.

Archaeologists, anthropologists, and volcanologists have teamed up to reconstruct volcanic disasters in historic and prehistoric times. A key question these research teams attempt to answer is whether some ancient civilizations suddenly collapsed or underwent major changes because of volcanic eruptions (Newhall et al. 2000). We need to ask whether environmental changes or disasters alone caused the cultural collapse or whether they just worsen socioeconomic conditions such as 'institutional incompetence or corruption, civil strife and insecurity, invasion, or pandemics' (Butzer 2012). Detailed reconstructions of past disasters such as the Avellino Plinian eruption of Vesuvius (3780 yr BP), which impacted the Campanian Bronze age culture, may serve as instructive analogues for future disasters (Mastrolorenzo et al. 2006). Yet these multidisciplinary investigations have also revealed some surprising past success stories. Neall et al. (2008) examined a sequence of 22 tephra beds on the Willaumez Isthmus, New Britain, Papua New Guinea, to elucidate how volcanic eruptions had impacted human settlement during the past 40 kyr. The team concluded that people successfully coped with these eruptions by maintaining mobile and flexible settlement patterns, deliberate and planned use of natural resources, social exchange, and more intense landscape management.

Lowe et al. (2012) examined cryptotephra – volcanic ash layers invisible to the naked eye – from the ~40 ka Campagnian Ignimbrite eruption (Fitzsimmons et al. 2013) at about a dozen archaeological sites in Italy, Greece, and northern Africa. This eruption was the most explosive in Europe in the past 200 kyr and coincided with the arrival of anatomically modern humans in Europe, and the gradual decline of the Neanderthals. The researchers concluded that the combined effects of the eruption and climate cooling had less grave consequences to Neanderthals than the spread of modern humans. Further, they argued that Palaeolithic cultures were remarkably resilient to this large volcanic eruption, which deposited tephra as far away as the Russian Plain, the eastern Mediterranean, and northern Africa. The eruption of Toba volcano, Indonesia, ~78 ka ago has been the focus of a major debate about the resilience of Palaeolithic people to great volcanic eruptions. While some researchers have argued that this super-eruption nearly annihilated modern humans by causing a bottleneck in their evolutionary history, others have argued for less apocalyptic, though still catastrophic, impacts (Oppenheimer 2002).

Sedimentary and landform evidence below sea level has become indispensable for reconstructing volcanic flank collapses and large-scale landsliding. The recognition that many volcanic islands are mantled by their own debris reflects the importance of large catastrophic flank collapses and debris avalanches that mark brief, but important, stages in the evolution of these islands. Deposits of giant landslides, each covering hundreds of square kilometres of sea floor, have been mapped around Hawaii (Morgan et al. 2007), the Caribbean (Boudon et al. 2007), the Canary Islands (Boulesteix et al. 2012), the Aleutians (Coombs et al. 2007), Tahiti (Hildenbrand et al. 2006), and many other oceanic islands.

Terrestrial records with similar detail are available from places like the Trans-Mexican volcanic belt (Capra et al. 2002) and Kamtchatka (Ponomareva et al. 2006).

Inventories of such collapses feature hundreds of large volcanic debris avalanches. Yet many of the data fields in these inventories remain empty because the necessary parameters can only be crudely estimated. Determining the runout and impacted area of such events requires knowledge of provenance and mode of transport. For example, catastrophic collapse of Asama volcano, Japan, about 24 ka ago produced a large debris avalanche containing ~5 km^3 of sediment that was later reworked by catastrophic debris flows, some of which reached nearly 90 km from the source (Yoshida and Sugai 2007). The late Pleistocene Rio Teno debris avalanche with a source at Planchon volcano, Chile, ran down the Claro and Teno rivers, entraining sediment en route to grow from 3 to 10 km^3. The debris avalanche deposited material as far as 95 km from the volcano over an area of some 370 km^2 on the Andean foreland (Tormey 2010). The catastrophic collapse of volcanic islands or volcanic edifices close to the sea may also cause tsunamis (see Chapter 9).

Discrete element modelling of the stability of volcanic slopes provides insights as to how the shape of volcanoes reflects something about their mechanical properties and material discontinuities. In these models, volcanic flanks tend to approach their angle of repose when subject to frequent debris avalanches. In contrast, concave-up slopes may develop where basal volcanic spreading is favoured by weak décollements (Morgan and McGovern 2005).

7.4 Climate-Driven Changes in Crustal Loads

The short-term impacts that volcanic eruptions may impose on the atmosphere point in one direction of a two-way coupling. McGuire (2013) summarized how the solid Earth responds to climate change. He identified volcanic landscapes, together with ocean basins and margins, high-latitude permafrost areas, and mountainous terrain as particularly susceptible in this respect, mainly because water and ice in these areas may undergo rapid phase changes. Some of these pathways are based on systematic responses of the lithosphere to past climate changes, notably glacial-interglacial transitions when global-scale changes in the ratio of water to ice induced major shifts of crustal loads. These changing loads were of the order of 10^0–10^1 MPa and would have affected patterns of volcanism, large-scale strain, seismicity, and erosion, partly through the commensurate triggering of surface mass movements. The rapidity at which these changes in crustal loads occurred may have also triggered responses from the solid Earth; sea levels in the early Holocene, for example, may have risen as fast as 45 mm yr^{-1}, annually adding a crustal load of roughly 1 kPa (McGuire 2013), equivalent to 1 MPa per millennium.

Weather and climate affect crustal deformation The eruptive character of some volcanoes can respond to changes in the crustal stress field. Ice loading and unloading is an important mechanism determining the shape and composition of volcanoes in cold environments. But weather and climate change can also cause crustal stresses to oscillate. Rainfall and snow loads may change shallow crustal pore pressures and eventually modulate the patterns of local seismicity (Hainzl et al. 2006; Husen et al. 2007). Detailed measurements of a decade's worth of data from borehole strainmeters together with time series of rainfall, nontidal sea-level variations, and surface air pressure during and after the passage of tropical cyclones in Taiwan revealed how these storms cause a short-lived deformation of the crust (Mouyen et al. 2017). The low pressure at the eye of the storms caused a dilatation in the crust, similar to the inverse barometer effect that tropical cyclones have on the open ocean, causing transient increases in

local sea level. After the storms moved on, the strainmeters in the crust recorded a compression, which would largely reflect the added loading by the storm rainfall that concentrated as runoff in river valleys. Changes in air pressure due to atmospheric tides can cause air and water in sediment pores to flow vertically, and thus contribute to altering the frictional stress along shear surfaces in landslides, and thus introducing a periodic signal in slip rates (Schulz et al. 2009). These and other interactions and feedbacks can also contribute potentially unwanted noise when inferring deformation signals. Hsu et al. (2014) reported that nearly a third of the continuous GPS stations in the Central Range of Taiwan ride on deep-seated slow-moving landslides that systematically accelerate their movement following heavy rains or the wet season in general, thus 'contaminating' the deformation field of interseismic motion.

Quaternary geological and geomorphological studies support the idea that climate change indirectly affects volcanic activity by altering the crustal overburden pressures of ice or water. Melting of glaciers on volcanoes unloads the underlying edifices and can decompress magma chambers sufficiently to trigger eruptions (Figure 7.14). Researchers have reported notable increases in volcanic activity and seismicity during and following widespread deglaciation at the end of the Pleistocene (Watt et al. 2013; Mörner et al. 2000). The loss of thick ice cover around 10.3–9.4 ka BP caused rapid isostatic rebound of Iceland's oceanic crust at rates of at least 20 mm yr^{-1}, judging from radiocarbon-dated shells in uplifted marine sediments and remnant peat beds now below sea level (Ingólfsson et al. 2008). The eruptive activity also increased around that time, and the style of eruptive activity also changed, with a growing contribution of subaerial, as opposed to subglacial, eruptions. Similarly, the glacial history of the Kamchatka volcanic

Figure 7.14 Landslide-induced decompression calculated for different volcanic systems and three magma reservoir depths (5, 10, and 20 km). Horizontal log-scale axes show decompression (bottom) and flank-collapse volume (top). Modified from Manconi et al. (2009).

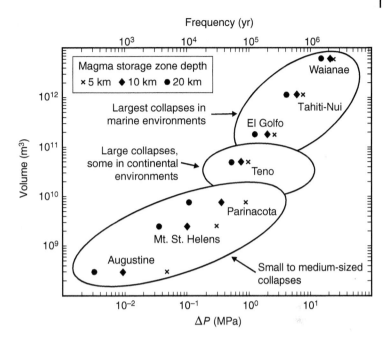

arc appears to be linked closely to phases of eruptive activity; the abundance of Pleistocene ice-rafted debris found in marine drill cores in the western North Pacific increased just before and during the deposition of volcanic ash layers at around 40 ka BP. Comparable volcanic activity was absent in the region since then (Bigg et al. 2008). Trajectory models of icebergs support the idea that most of the ice-rafted debris came from the area of Kamchatka, which may have hosted a major ice sheet at the time.

References

Andrews BJ and Manga M 2011 Effects of topography on pyroclastic density current runout and formation of coignimbrites. *Geology* **39**(12), 1099–1102.

Bigg GR, Clark CD, and Hughes ALC 2008 A last glacial ice sheet on the Pacific Russian coast and catastrophic change arising from coupled ice–volcanic interaction. *Earth and Planetary Science Letters* **265**(3-4), 559–570.

Bonadonna C and Costa A 2013 Modeling tephra sedimentation from volcanic plumes. In *Modeling Volcanic Processes* (eds Fagents S, Gregg TKP, and Lopes RMC) Cambridge University Press pp. 173–202.

Boudon G, Le Friant A, Komorowski JC, et al. 2007 Volcano flank instability in the Lesser Antilles Arc: Diversity of scale, processes, and temporal recurrence. *Journal of Geophysical Research: Solid Earth* **112**, 1–28.

Boulesteix T, Hildenbrand A, Gillot PY, and Soler V 2012 Eruptive response of oceanic islands to giant landslides: New insights from the geomorphologic evolution of the Teide–Pico Viejo volcanic complex (Tenerife, Canary). *Geomorphology* **138**(1), 61–73.

Bovis MJ and Jakob M 1999 The role of debris supply conditions in predicting debris flow activity. *Earth Surface Processes and Landforms* **24**, 1039–1054.

Bryan SE, Peate IU, Peate DW, et al. 2010 The largest volcanic eruptions on Earth. *Earth-Science Reviews* **102**(3-4), 207–229.

Butzer KW 2012 Collapse, environment, and society. *Proceedings of the National Academy of Sciences* **109**, 3632–3639.

Capra L, Macías JL, Scott KM, et al. 2002 Debris avalanches and debris flows transformed from collapses in the Trans-Mexican Volcanic Belt, Mexico – behavior, and implications for hazard assessment. *Journal of Volcanology and Geothermal Research* **113**(1), 81–110.

Carazzo G, Kaminski E, and Tait S 2015 The timing and intensity of column collapse during explosive volcanic eruptions. *Earth and Planetary Science Letters* **411**(C), 208–217.

Chiocci FL, Romagnoli C, Tommasi P, and Bosman A 2008 The Stromboli 2002 tsunamigenic submarine slide: Characteristics and possible failure mechanisms. *Journal of Geophysical Research – Solid Earth* **113**,B10102.

Chiocci FL, Coltelli M, Bosman A, and Cavallaro D 2011 Continental margin large-scale instability controlling the flank sliding of Etna volcano. *Earth and Planetary Science Letters* **305**(1-2), 57–64.

Clarke AB 2013 Unsteady explosive activity: vulcanian eruptions. In *Modeling Volcanic Processes* (ed Fagents SA, Gregg TKP, and Lopes RMC) Cambridge University Press pp. 129–152.

Coombs ML, White SM, and Scholl DW 2007 Massive edifice failure at Aleutian arc volcanoes. *Earth and Planetary Science Letters* **256**(3-4), 403–418.

Cronin SJ, Lube G, Dayudi DS, et al. 2013 Insights into the October–November 2010 Gunung Merapi eruption (Central Java, Indonesia) from the stratigraphy, volume and characteristics of its pyroclastic deposits. *Journal of Volcanology and Geothermal Research* **261**(C), 244–259.

Crosweller HS, Arora B, Brown SK, et al. 2012 Global database on large magnitude explosive volcanic eruptions (LaMEVE). *Journal of Applied Volcanology* **1**(4), 1–13.

Davies RJ, Brumm M, Manga M, et al. 2008 The East Java mud volcano (2006 to present): An earthquake or drilling trigger?. *Earth and Planetary Science Letters* **272**(3-4), 627–638.

Elsworth D, Voight B, Thompson G, and Young SR 2004 Thermal-hydrologic mechanism for rainfall-triggered collapse of lava domes. *Geology* **32**(11), 969–972.

Fenton CR, Webb RH and Cerling TE 2006 Peak discharge of a Pleistocene lava-dam outburst flood in Grand Canyon, Arizona, USA. *Quaternary Research* **65**(2), 324–335.

Ferrier KL, Perron JT, Mukhopadhyay S, et al. 2013 Covariation of climate and long-term erosion rates across a steep rainfall gradient on the Hawaiian island of Kaua'i. *Geological Society of America Bulletin* **125**(7-8), 1146–1163.

Fitzsimmons KE, Hambach U, Veres D, and Iovita R 2013 The Campanian ignimbrite eruption: New data on volcanic ash dispersal and its potential impact on human evolution. *PLoS ONE* **8**(6), e65839.

Friele PA, Clague JJ, Simpson K, and Stasiuk M 2005 Impact of a Quaternary volcano on Holocene sedimentation in Lillooet River valley, British Columbia. *Sedimentary Geology* **176**(3-4), 305–322.

Garcin M, Poisson B, and Pouget R 2005 High rates of geomorphological processes in a tropical area: the Remparts River case study (Réunion Island, Indian Ocean). *Geomorphology* **67**(3-4), 335–350.

Geyer A and Martí J 2008 The new worldwide collapse caldera database (CCDB): A tool for studying and understanding caldera processes. *Journal of Volcanology and Geothermal Research* **175**(3), 334–354.

Gran KB, Montgomery DR, and Halbur JC 2011 Long-term elevated post-eruption sedimentation at Mount Pinatubo, Philippines. *Geology* **39**(4), 367–370.

Guthrie RH, Friele P, Allstadt K, et al. 2012 The 6 August 2010 Mount Meager rock slide-debris flow, Coast Mountains, British Columbia: characteristics, dynamics, and implications for hazard and risk assessment. *Natural Hazards and Earth System Sciences* **12**(5), 1277–1294.

Hainzl S, Kraft T, Wassermann J, et al. 2006 Evidence for rainfall-triggered earthquake activity. *Geophysical Research Letters* **33**(19), 1–5.

Harris AJL 2013 Lava flows. In *Modeling Volcanic Processes* (eds Fagents S, Gregg TKP, and Lopes RMC), Cambridge University Press, pp. 85–106.

Hildenbrand A, Gillot PY, and Bonneville A 2006 Offshore evidence for a huge landslide of the northern flank of Tahiti-Nui (French Polynesia). *Geochemistry Geophysics Geosystems* **7**(3), 1–12.

Hsu YJ, Chen RF, Lin CW, et al. 2014 Seasonal, long-term, and short-term deformation in the Central Range of Taiwan induced by landslides. *Geology* **42**(11), 991–994.

Husen S, Bachmann C, and Giardini D 2007 Locally triggered seismicity in the central Swiss Alps following the large rainfall event of August 2005. *Geophysical Journal International* **171**(3), 1126–1134.

Iles CE and Hegerl GC 2015 Systematic change in global patterns of streamflow following volcanic eruptions. *Nature Geoscience* **8**(11), 838–842.

Ingólfsson Ó, Norddahl H, and Haflidason H 2008 Rapid isostatic rebound in southwestern Iceland at the end of the last glaciation. *Boreas* **24**(3), 245–259.

Iverson RM 1997 The physics of debris flows. *Reviews of Geophysics* **35**(3), 245–296.

Iverson RM, Schilling SP, and Vallance JW 1998 Objective delineation of lahar-inundation hazard zones. *Geological Society of America Bulletin* **110**(8), 972–984.

Kataoka KS, Manville V, and Yamaguchi A 2009 Impacts of explosive volcanism on distal alluvial sedimentation: Examples from the Pliocene–Holocene volcaniclastic successions of Japan. *Sedimentary Geology* **220**(3-4), 306–317.

Kelfoun K, Giachetti T, and Labazuy P 2010 Landslide-generated tsunamis at Reunion Island. *Journal of Geophysical Research* **115**(F4), F04012.

Komorowski J-C, Jenkins S, Baxter PJ, et al. 2013. Paroxysmal dome explosion during the Merapi 2010 eruption; processes and facies relationships of associated high-energy pyroclastic density currents. *Journal of Volcanology and Geothermal Research* **261**, 260–294.

Korup O 2012 Earth's portfolio of extreme sediment transport events. *Earth-Science Reviews* **112**, 115–125.

Korup O, Seidemann J, and Mohr CH 2019 Increased landslide activity on forested hillslopes following two recent volcanic eruptions in Chile. *Nature Geoscience* **12**, 284–289.

Larsen G and Eiríksson J 2008 Late Quaternary terrestrial tephrochronology of Iceland – frequency of explosive eruptions, type and volume of tephra deposits. *Journal of Quaternary Science* **23**(2), 109–120.

Lavigne F 2004 Rate of sediment yield following small-scale volcanic eruptions: a quantitative assessment at the Merapi and Semeru stratovolcanoes, Java, Indonesia. *Earth Surface Processes and Landforms* **29**(8), 1045–1058.

Lockwood JP and Hazlett RW 2010 *Volcanoes. Global Perspectives.* Wiley Blackwell.

Louvat P and Allègre CJ 1997 Present denudation rates on the island of Reunion determind by river geochemistry: Basalt weathering and mass budget between chemical and mechanical erosions. *Geochimica et Cosmochimica Acta* **61**, 3645–3669.

Lowe J, Barton N, Blockley S, et al. 2012 Volcanic ash layers illuminate the resilience of Neanderthals and early modern humans to natural hazards. *Proceedings of the National Academy of Sciences* **109**(34), 13532–13537.

Lube G, Cronin SJ, and Procter JN 2009 Explaining the extreme mobility of volcanic ice-slurry flows, Ruapehu volcano, New Zealand. *Geology* **37**(1), 15–18.

Major J and Mark L 2006 Peak flow responses to landscape disturbances caused by the cataclysmic 1980 eruption of Mount St.

Helens, Washington. *Geological Society of America Bulletin* **118**(7-8), 938–958.

Major J, Pierson T, Dinehart R, and Costa JE 2000 Sediment yield following severe volcanic disturbance – a two-decade perspective from Mount St. Helens. *Geology* **28**(9), 819–822.

Manconi A, Longpre MA, Walter TR, et al. 2009 The effects of flank collapses on volcano plumbing systems. *Geology* **37**(12), 1099–1102.

Manville V 2010 An overview of break-out floods from intracaldera lakes. *Global and Planetary Change* **70**(1-4), 14–23.

Manville V, Segschneider B, Newton E, et al. 2009 Environmental impact of the 1.8 ka Taupo eruption, New Zealand: Landscape responses to a large-scale explosive rhyolite eruption. *Sedimentary Geology* **220**(3-4), 318–336.

Manville V, Major JJ, and Fagents SA 2013 Modeling lahar behaviour and hazards. In *Modeling Volcanic Processes* (eds Fagents S, Gregg TKP, and Lopes RMC), Cambridge University Press, pp. 300–330.

Martin RS, Watt SFL, Pyle DM, et al. 2009 Environmental effects of ashfall in Argentina from the 2008 Chaitén volcanic eruption. *Journal of Volcanology and Geothermal Research* **184**(3-4), 462–472.

Mason BG, Pyle DM, and Oppenheimer C 2004 The size and frequency of the largest explosive eruptions on Earth. *Bulletin of Volcanology* **66**(8), 735–748.

Mastrolorenzo G, Petrone P, Pappalardo L, and Sheridan MF 2006 The Avellino 3780-yr-B.P. catastrophe as a worst-case scenario for a future eruption at Vesuvius. *Proceedings of the National Academy of Sciences* **103**(12), 4366–4370.

Mastrolorenzo G, Petrone P, Pappalardo L, and Guarino FM 2010 Lethal thermal impact at periphery of pyroclastic ssurges: Evidences at Pompeii. *PLoS ONE* **5**(6),e11127.

Mattsson H and Höskuldsson Á 2003 Geology of the Heimaey volcanic centre, south Iceland: early evolution of a central volcano in a propagating rift?. *Journal of Volcanology and Geothermal Research* **127**(1-2), 55–71.

McGuire B 2013 Hazardous responses of the solid Earth to a changing climate. In *Climate forcing of geological hazards* (eds McGuire B and Maslin M), Cambridge University Press, pp. 1–33.

Merle O, Michon L, and Bachèlery P 2008 Caldera rim collapse: A hidden volcanic hazard. *Journal of Volcanology and Geothermal Research* **177**(2), 525–530.

Miller GH, Geirsdóttir Á, Zhong Y, et al. 2012 Abrupt onset of the Little Ice Age triggered by volcanism and sustained by sea-ice/ocean feedbacks. *Geophysical Research Letters* **39**(2), 1–5.

Moreno P, Alloway BV, Villarosa G, et al. 2015 A past-millennium maximum in postglacial activity from Volcan Chaiten, southern Chile. *Geology* **43**(1), 47–50.

Morgan JK and McGovern PJ 2005 Discrete element simulations of gravitational volcanic deformation: 1. Deformation structures and geometries. *Journal of Geophysical Research: Solid Earth* **110**, 1–22.

Morgan JK, Clague DA, Borchers DC, et al. 2007 Mauna Loa's submarine western flank: Landsliding, deep volcanic spreading, and hydrothermal alteration. *Geochemistry Geophysics Geosystems* **8**(5), 1–42.

Mörner NA 1991 Intense earthquakes and seismotectonics as a function of glacial isostasy. *Tectonophysics* **188**, 407–410.

Mörner NA, Tröften PE, Sjöberg R, et al. 2000 Deglacial paleoseismicity in Sweden: the 9663 BP Iggesund event. *Quaternary Science Reviews* **19**, 1461–1468.

Mouyen M, Canitano A, Chao BF, et al. 2017 Typhoon-induced ground deformation. *Geophysical Research Letters* **44**(21), 11,004–11,011.

Neall VE, Wallace RC, and Torrence R 2008 The volcanic environment for 40,000 years of human occupation on the Willaumez Isthmus, West New Britain, Papua New Guinea. *Journal of Volcanology and Geothermal Research* **176**(3), 330–343.

Newhall CG and Self S 1982 The volcanic explosivity index (VEI) an estimate of explosive magnitude for historical volcanism. *Journal of Geophysical Research: Atmospheres* **87**, 1231–1238.

Newhall CG, Bronto S, Alloway B, et al. 2000 10,000 Years of explosive eruptions of Merapi Volcano, Central Java: archaeological and modern implications. *Journal of Volcanology and Geothermal Research* **100**, 9–50.

Oehler JF, van Wyk de Vries B, and Labazuy P 2005 Landslides and spreading of oceanic hot-spot and arc shield volcanoes on Low Strength Layers (LSLs): an analogue modeling approach. *Journal of Volcanology and Geothermal Research* **144**(1), 169–189.

Oppenheimer C 2002 Limited global change due to the largest known Quaternary eruption, Toba 74 kyr BP?. *Quaternary Science Reviews* **21**, 1563–1609.

Oppenheimer C 2011 *Eruptions that Shook the World*. Cambridge University Press.

Pierson TC and Janda RJ 1994 Volcanic mixed avalanches: a distinct eruption-triggered mass-flow process at snow-clad volcanoes. *Geological Society of America Bulletin* **106**(10), 1351–1358.

Pierson TC and Major JJ 2014 Hydrogeomorphic effects of explosive volcanic eruptions on drainage basins. *Annual Review of Earth and Planetary Sciences* **42**(1), 469–507.

Pierson TC, Pringle PT, and Cameron KA 2010 Magnitude and timing of downstream channel aggradation and degradation in response to a dome-building eruption at Mount Hood, Oregon. *Geological Society of America Bulletin* **123**(1-2), 3–20.

Ponomareva VV, Melekestsev IV, and Dirksen OV 2006 Sector collapses and large landslides on Late Pleistocene–Holocene volcanoes in Kamchatka, Russia. *Journal of Volcanology and Geothermal Research* **158**(1-2), 117–138.

Procter J, Cronin SJ, Fuller IC, et al. 2010 Quantifying the geomorphic impacts of a lake-breakout lahar, Mount Ruapehu, New Zealand. *Geology* **38**(1), 67–70.

Rad SD, Allègre CJ, and Louvat P 2006 Hidden erosion on volcanic islands. *Earth and Planetary Science Letters* **262**(1), 109–124.

Reid ME 2004 Massive collapse of volcano edifices triggered by hydrothermal pressurization. *Geology* **32**(5), 373.

Ridley DA, Solomon S, Barnes JE, et al. 2014 Total volcanic stratospheric aerosol optical depths and implications for global climate change. *Geophysical Research Letters* **41**(22), 7763–7769.

Roa K 2003 Nature and origin of toreva remnants and volcaniclastics from La Palma, Canary Islands. *Journal of Volcanology and Geothermal Research* **125**(3-4), 191–214.

Saint-Ange F, Savoye B, Michon L, et al. 2011 A volcaniclastic deep-sea fan off La Reunion Island (Indian Ocean): Gradualism versus catastrophism. *Geology* **39**(3), 271–274.

Salvany T, Lahitte P, Nativel P, and Gillot PY 2012 Geomorphic evolution of the Piton des Neiges volcano (Réunion Island, Indian Ocean): Competition between volcanic construction and erosion since 1.4 Ma. *Geomorphology* **136**(1), 132–147.

Saucedo R, Macías JL, Sarocchi D, and Bursik M 2008 The rain-triggered Atenquique volcaniclastic debris flow of October 16, 1955 at Nevado de Colima Volcano, Mexico. *Journal of Volcanology and Geothermal Research* **173**, 69–83.

Schulz WH, Kean JW, and Wang G 2009 Landslide movement in southwest Colorado triggered by atmospheric tides. *Nature Geoscience* **2**(12), 863–866.

Scifoni S, Coltelli M, Marsella M, et al. 2010 Mitigation of lava flow invasion hazard through optimized barrier configuration aided by numerical simulation: The case of the 2001 Etna eruption. *Journal of Volcanology and Geothermal Research* **192**(1-2), 16–26.

Shea T and van Wyk de Vries B 2008 Structural analysis and analogue modeling of the kinematics and dynamics of rockslide avalanches. *Geosphere* **4**(4), 657.

Siebert L 1992 Threats from debris avalanches. *Nature* **356**, 658–659.

Sigl M, Winstrup M, McConnell JR, et al. 2015 Timing and climate forcing of volcanic eruptions for the past 2,500 years. *Nature* **523**(7562), 543–549.

Silver E, Day S, Ward S, et al. 2005 Island arc debris avalanches and tsunami generation. *Eos, Transactions American Geophysical Union* **86**(47), 485–496.

Terry JP 2007 *Tropical Cyclones. Climatology and Impacts in the South Pacific.* Springer.

Thorardson T and Larsen G 2007 Volcanism in Iceland in historical time: Volcano types, eruption styles and eruptive history. *Journal of Geodynamics* **43**, 118–152.

Thouret JC 1999 Volcanic geomorphology – an overview. *Earth-Science Reviews* **47**, 95–131.

Thouret J-C, Oehler JFS, Gupta A, et al. 2007 Erosion and aggradation on persistently active volcanoes; a case study from Semeru Volcano, Indonesia. *Bulletin of Volcanology* **70**, 221–244.

Tormey D 2010 Managing the effects of accelerated glacial melting on volcanic collapse and debris flows: Planchon–Peteroa Volcano, Southern Andes. *Global and Planetary Change* **74**(2), 82–90.

Tsuchiya S, Sasahara K, Shuin S, and Ozono S 2009 The large-scale landslide on the flank of caldera in South Sulawesi, Indonesia. *Landslides* **6**(1), 83–88.

van der Beek P, Pulford A, and Braun J 2001 Cenozoic landscape development in the Blue Mountains (SE Australia): lithological and tectonic controls on rifted margin morphology. *The Journal of Geology* **109**(1), 35–56.

van Wyk de Vries BVW and Francis PW 1997 Catastrophic collapse at stratovolcanoes induced by gradual volcano spreading. *Nature* **387**(6631), 387–390.

Walsh K 2014 *The Archaeology of Mediterranean Landscapes.* Cambridge University Press.

Watt SFL, Pyle DM, and Mather TA 2013 The volcanic response to deglaciation; evidence from glaciated arcs and a reassessment of global eruption records. *Earth-Science Reviews* **122**(C), 77–102.

Watt SFL, Pyle DM, Naranjo JA, et al. 2011 Holocene tephrochronology of the Hualaihue region (Andean southern volcanic zone,

42 ° S), southern Chile. *Quaternary International* **246**(1-2), 324–343.

Watt SFL, Pyle DM, Naranjo JA, and Mather TA 2008 Landslide and tsunami hazard at Yate volcano, Chile as an example of edifice destruction on strike-slip fault zones. *Bulletin of Volcanology* **71**, 559–574.

Whipple KX, Snyder NP, and Dollenmayer K 2000 Rates and processes of bedrock incision by the Upper Ukak River since the 1912 Novarupta ash flow in the Valley of Ten Thousand Smokes, Alaska. *Geology* **28**, 835–838.

Wilmshurst JM and McGlone MS 1996 Forest disturbance in the central North Island, New Zealand, following the 1850 BP Taupo eruption. *The Holocene* **6**(4), 399–411.

Wilson CJN, Houghton BF, McWilliams MO, et al. 1995 Volcanic and structural evolution of Taupo Volcanic Zone, New Zealand: A review. *Journal of Volcanology and Geothermal Research* **68**, 1–28.

Wilson L, Sparks RSJ, and Walker GPL 1980 Explosive volcanic eruptions – IV. The control of magma properties and conduit geometry on eruption column behaviour. *Geophysical Journal Royal Astronomical Society* **63**, 117–148.

Wilson TM, Cole JW, Stewart C, et al. 2010 Ash storms: impacts of wind-remobilised volcanic ash on rural communities and agriculture following the 1991 Hudson eruption, southern Patagonia, Chile. *Bulletin of Volcanology* **73**(3), 223–239.

Witham CS 2005 Volcanic disasters and incidents: A new database. *Journal of Volcanology and Geothermal Research* **148**(3-4), 191–233.

Woods AW 2013 Sustained explosive activity: volcanic eruption columns and hawaiian fountains. In *Modeling Volcanic Processes* (eds Fagents S, Gregg TKP, and Lopes RMC), Cambridge University Press, pp. 153–172.

Yoshida H and Sugai T 2007 Magnitude of the sediment transport event due to the Late Pleistocene sector collapse of Asama volcano, central Japan. *Geomorphology* **86**(1-2), 61–72.

8

Landslides and Slope Instability

Landslides are the downward movement of slope materials under the influence of gravity. Water is involved in most cases, at least to some degree, as completely dry hillslopes are rare in nature, failure of massive rock slopes being the main exception. Slopes fail when shear forces acting on slope materials exceed the shear strength of the materials. 'Landslide' is a generic term for a wide range of gravitational mass movements that we classify by their dominant material (earth, debris, rock), type of motion (fall, slide, slump, flow, topple), rapidity of motion (slow, rapid, extremely rapid), and state of activity (active, inactive, fossil). Many classifications are available, though most combine type of movement and materials involved in landslides. Hungr et al. (2013) modified the definition of landslide-forming materials to be compatible with geotechnical and geological terminology for rock and soil materials. Landslides may occur as single events or as a series of failures separated by minutes to days. Landslide episodes (Crozier 2005) are a consequence of strong earthquakes, intense rainstorms, sudden lake drainage, or rapid snow melt. Submarine landslides along continental margins and offshore of volcanic islands may be several orders of magnitude larger than their terrestrial cousins. Triggers of submarine landslides include strong earthquakes, volcanic eruptions, strong wave action during storms, loading due to rapid sedimentation, dissociation of clathrate layers, gas-rich fluid migration to planes of weaknesses in marine sediments, and weak layer contacts.

8.1 Frequency and Magnitude of Landslides

Losses from landslides are large and often underestimated. An inventory of fatal slope failures between 2004 and 2010 listed more than 32 000 deaths, which is an order of magnitude higher than fatalities recorded in landslide data sets for other periods (Petley 2012; Froude and Petley 2018). This mismatch points to a tendency to underestimate the damage associated with landslides because the relevant information is spread through unpublished newspaper and media articles and engineering reports, and because deaths and damage are attributed to hazards of other types (Figure 8.1). A global inventory of 213 fatal debris flows between 1950 and 2011 lists nearly 78 000 deaths, with developing countries having four times the median per-event mortality rate of more advanced countries (Dowling and Santi 2013). About one-third of these lethal debris flows were triggered by processes other than rainfall, and in those cases the median per-event mortality rate is more than 50 times higher than that of rainfall-triggered debris flows.

Geomorphology and Natural Hazards: Understanding Landscape Change for Disaster Mitigation, Advanced Textbook Series,
First Edition. Tim R. Davies, Oliver Korup, and John J. Clague.
© 2021 John Wiley & Sons Ltd. Co-published 2021 by the American Geophysical Union and John Wiley & Sons Ltd.

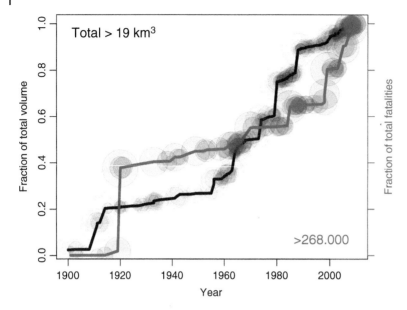

Figure 8.1 Cumulative volumes and impacts of 242 large (> 10^6 m³) landslides reported since 1900. These landslides claimed more than 250 000 lives. Bubbles are scaled to individual volumes and number of fatalities. Adapted from Korup et al. (2007).

Most of Earth's largest known landslides have happened under water, including submarine debris avalanches and sediment slumps with volumes of thousands of cubic kilometres. Submarine slope failures can spawn turbidity currents that run out over distances of several hundreds of kilometres, reaching deep-sea basins, where they deposit layers of fine sediment (Talling et al. 2007, 2013). Thousands of large submarine slope failures dot the rims of all major ocean basins, although they are much better preserved along passive continental margins than along subduction zones (Twichell et al. 2008). The Mediterranean alone hosts at least 700 submarine landslides, and about 250 of these have deposit volumes of more than 1 km³. From size statistics we can estimate that landslides of this size happen, on average, once every 40 years in this basin (Urgeles and Camerlenghi 2013).

Terrestrial landslides originate preferentially on slopes with low strength, but can also occur in stronger rocks during earthquake shaking. Rocks along fault zones that are mechanically weakened and geochemically altered are prime candidates for slope failures, even where the rocks are otherwise competent (Osmundsen et al. 2009). Where the bedrock has decayed

into clayey fault gouge and cataclasite, unstable gully systems may develop within and at the margins of catastrophic slope failures (Korup 2004). Clusters of large, slow-moving earthflows along fault zones lend support to the notion that tectonically crunched and seismically shattered rocks are conducive to slope failure (Figure 8.2) (Scheingross et al. 2013).

Popular belief holds that landslides are limited to steep or high hillslopes. Yet slope failures also occur in areas of modest relief, supporting the notion that landslides are prominent processes of denuding landscapes (Figure 8.3). Some of the oldest known terrestrial landslide deposits are in dry and largely flat places. The hyperarid climate along the margin of Chile's Atacama Desert may be responsible for preserving the deposits of large landslides that date back to the Neogene, while dozens of deposits of large landslides also dot the arid coastal ranges of Chile (Mather et al. 2014). One of the largest terrestrial mass movements is the Baga Bogd landslide in the Gobi Desert of southern Mongolia. This giant earth-block slide contains about 50 km³ of material that had moved catastrophically along a basal shear plane inclined at only about 3°. Luminescence dating of lacustrine sediments

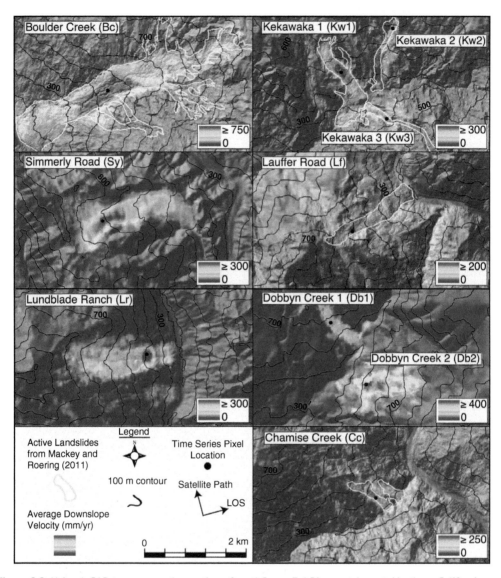

Figure 8.2 Using InSAR to measure the motion of earthflows, Eel River catchment, Northern California. Time-averaged (2007–2011) downslope velocity draped over shaded relief maps derived from LiDAR (1 m grid) and 10 m grid. Line-of-sight velocities are projected downslope for each slide using the local slope and azimuth. Colour scales were chosen to highlight kinematic zones instead of the full range of velocity (i.e. red pixels are the maximum listed value). From Handwerger et al. (2013).

capping the landslide deposit suggests that the failure happened shortly before 140 ± 16 ka, just when the Penultimate Glaciation began to give way to the Last Interglaciation. At that time climate became warmer and wetter, possibly allowing fluidisation of gypsiferous clays along which failure occurred. The landslide was most likely triggered by a strong

earthquake on a nearby fault (Balescu et al. 2007). The Zagros Mountains of Iran host most of the large landslides in the country, including one of the world's largest rock avalanches (38 km³) at Seimareh (Shoaei 2014). The semi-arid Columbia River basin in eastern Oregon has more than 400 landslides >0.1 km² in size in terrain with local topographic relief that is

Figure 8.3 Examples of landslides. (a) The 2014 Oso landslide, which killed 43 people in Washington (Washington State Department of Transportation). (b) Coloured LiDAR terrain model of the Oso landslide (Jason Stoker, US Geological Survey). (c) The 1993 Lemieux landslide in Ontario, Canada, a quick clay failure in Leda Clay (Steve G. Evans; Geological Survey of Canada). (d) A slump in the valley of Beatton River near Fort St John, British Columbia. This landslide interrupted traffic between British Columbia and Alberta (John Clague).

mostly below 100 m (Safran et al. 2011). Busche (2001) noted that large landslides in the Sahara Desert are far from just a 'decorative element of some escarpment and inselberg slopes'. He described dozens of kilometre-scale rotational landslides along the 100-m-high eastern rim of Djado plateau in Niger.

Giant landslides can also happen in gentle landscapes in humid areas. In northern Ukraine slope failures moved volumes of several cubic kilometres each (Pánek et al. 2008). Recognizing and mapping such extensive slope failures in forested landscapes is easier done using satellite images than on the ground, especially where they have issued from the fringes of mountain ranges. Very large landslides also generally have long runouts, up to ~20 km. Even gently sloping forelands composed of large alluvial fans may be impacted episodically by rock avalanches travelling beyond mountain-range fronts (Hubert-Ferrari

et al. 2005). A particular class of very large, low-angle landslides involve the sliding of largely intact blocks of rock many kilometres over gently-inclined substrates. The largest of these is at Heart Mountain, USA (Beutner and Gerbi 2005), with a volume of about 3500 cubic km, that travelled 45 km on a basal slope of two degrees. The dynamics of frictional motion at such low slopes are a remaining challenge.

Despite the large size and widespread geomorphic impacts of both submarine and terrestrial landslides, large ones are rare events. Smaller and much more frequent landslides have received most of the attention of researchers and practitioners (Figure 8.4). Nevertheless, it is the larger events that are responsible for most of the damage and fatalities. In eastern Gansu province on the Chinese Loess Plateau, for example, more than 1000 large landslides killed at least 2000 people between 1965 and 1979. The seven

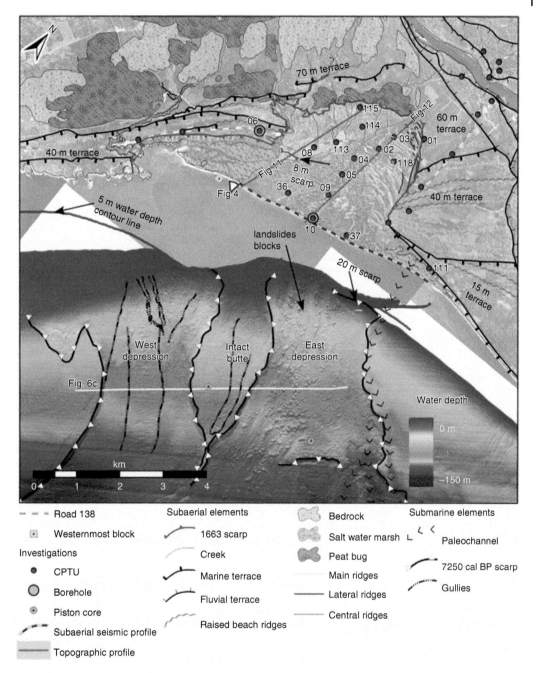

Figure 8.4 Geomorphologic interpretation of the Betsiamites landslide complex, Quebec, Canada. Landslides occur on land and the seafloor off the shoreline of the Gulf of St Lawrence. From Cauchon-Voyer et al. (2011).

largest of these landslides caused about half of the fatalities, destroyed some 17 500 houses, and buried 225 km² of farmland (Derbyshire 2000). Rapid and destructive loess flows in this terrain can detach on slopes less than 15° steep and travel for several kilometres. In terms of the ratio of their drop height to travel distance H/L, these loess flows are much more mobile than most landslides elsewhere around the world (Zhang and Wang 2006). Laboratory tests on the shear strength of the loess suggest that an increase in pore-water pressures resulting from a rapid collapse of the soil structure may explain these highly mobile flows. A final low-slope landslide category, also caused by soil structure change, involves so-called 'quick clay', which is found where, for example, postglacial coastal uplift has elevated marine (saline) clays to form coastal lowlands. These clays have been deposited under the ocean with their platelets in edge-to-face configuration that is stable in saline conditions. When exposed to subaerial rainfall, the change to freshwater environmental chemistry eventually renders the edge-to-face platelet configuration unstable, and a sufficient trigger can cause local reversion to face-to-face configuration. This has much lower structural strength and higher pore-water pressure, and if the reversion is large enough it can destabilize adjacent materials, leading to progressive failure. Because coastal lowlands are desirable for residence and agriculture, these 'quick clay' slides can have catastrophic outcomes; the 1978 Rissa event in Norway initially affected 30 ha but extended to about 330 ha; one person died and a dozen homes were destroyed (Gregerson 1981).

Landslide velocity and mobility (which determines deposit area) are of prime concern in hazard assessments. The speeds of extremely rapid landslides can exceed 100 m s⁻¹. The maximum velocity v_{max} [m s⁻¹] of a landslide can be roughly estimated from its run-up (or 'swash') on a topographic obstacle or a valley flank, assuming friction can be neglected:

$$v_{max} = \sqrt{2gh_r} \tag{8.1}$$

where g is gravitational acceleration [m s²], and h_r is the run-up height [m] (Figure 8.5). Note that Eq (8.1) can also be used to estimate the speed of pyroclastic density currents or other types of rapid flows; the underlying physical principle, conservation of energy, is the same. Rapidly moving landslides also leave lateral trimlines along valley flanks. The trimlines may be superelevated at outer valley or channel bends, providing data from which velocities can be estimated:

$$u = \sqrt{\frac{k_e g \cos S \Delta e R_c}{w}} \tag{8.2}$$

where u if the flow velocity [m s⁻¹], k_e is an empirical coefficient ranging between 0.1 and 1, S is the channel gradient [m m⁻¹],

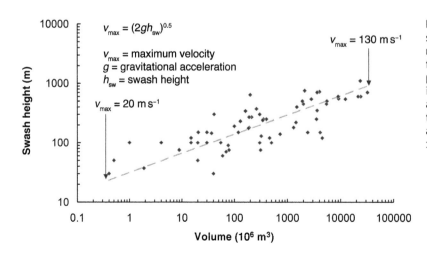

Figure 8.5 The swash height, or run-up, of a fast-moving landslide provides a rough indication of velocity at those points. The fastest landslides can attain speeds of > 100 m s⁻¹.

Δe [m] is the elevation difference in the flow across the channel, R_c is the radius of curvature of the channel bend [m], and w is the flow width [m]. Equation (8.2) can also be used to reconstruct the velocity of debris flows and mudflows (Manville et al. 2013), but applies only to Newtonian flows and will thus yield incorrect results for sediment-rich flows such as landslides and debris flows.

The stability of hillslopes can be estimated by using a force-equilibrium approach that relates the summed measures of hillslope shear strength $\sum_{F_r}$ to the driving forces $\sum_{F_d}$ that can destabilize the slope. This ratio is commonly expressed as the so-called Factor of Safety FS:

$$FS = \frac{\sum_{F_r}}{\sum_{F_d}} \qquad (8.3)$$

The hillslope is unstable when $FS < 1$, and stable otherwise. The metastable state of $FS = 1$ is termed 'limit equilibrium'. A simple case to consider is that of a potentially unstable rectangular block on a planar soil-mantled hillslope of infinite length:

$$FS = \frac{c' + (W \cos \beta - u) \tan \phi}{W \sin \beta} \qquad (8.4)$$

where c' is the apparent cohesive strength along the failure plane [kPa or kN m^{-2}], W is the weight force of the potentially unstable block [kPa], β is the slope angle [°], u is the pore-water pressure [kPa], and ϕ is the angle of internal friction [°] (Sidle and Ochiai 2006). Equation (8.4) is one of the simplest forms of the 'infinite slope model', and states that the stability of a hillslope depends on its material, water content, and geometry. The infinite slope model uses the concept of Coulomb friction, for which shear stress τ [kPa] changes linearly with the effective normal stress $\sigma' = \sigma - u$ [kPa], where σ [kPa] is normal stress in the dry state. We refer to this as a linear failure criterion:

$$\tau = c' + (\sigma - u) \tan \phi \qquad (8.5)$$

Compare this equation to Eq (7.10), in which Coulomb friction is part of a simple model for lahars and debris flows.

The weight force W per unit area [kPa] of the potential sliding block can be computed as:

$$W = (\gamma_t (H - h) + \gamma_{sat} h) \cos \beta \qquad (8.6)$$

where γ_t is the unit (or specific) weight of partly water-saturated soil [kPa m^{-1} or kN m^{-3}], H is the thickness of the potentially unstable block [m], which we take to be orders of magnitude smaller than the length of the slope (hence 'infinite slope model'), h is the thickness of the fully saturated soil [m], and γ_{sat} is the unit weight of fully water-saturated soil [kPa m^{-1}]. The pore-water pressure u can be derived from:

$$u = \gamma_w h \cos^2 \beta \qquad (8.7)$$

where γ_w is the unit weight of water [kPa m^{-1}]. We can rearrange Eq (8.4) in a form that better highlights some basics about slope stability:

$$FS = \frac{c'}{W \sin \beta} + \frac{\tan \phi}{\tan \beta} - \frac{u \tan \phi}{W \sin \beta} \qquad (8.8)$$

Now consider a completely dry and cohesionless hillslope made of sand or gravel, with zero porewater pressures ($u = 0$), and $c' = 0$. These constraints allow us to eliminate the first and third terms in Eq (8.8), and we are left with the fundamental insight that the limiting slope in this case is defined by the internal angle of friction ϕ. This means that a pile of dry granular material cannot be inclined more steeply than its internal angle of friction.

A slope can fail at very low rates of deformation, and giant landslides involving the slow movement of several cubic kilometres of rock are common in many mountain belts. For example, the European Alps contain at least 900 such 'deep-seated gravitational slope deformations' (DSGSDs). They move at rates as low as several mm yr^{-1}, although some accelerate and may fail catastrophically as rock slides and rock avalanches (Agliardi et al. 2013). Some researchers have argued that the main distinguishing characteristic of DSGSDs is their lack of a distinct failure surface at depth, whereas others believe that they represent a transitional phenomenon between gravitationally and tectonically induced slope deformation (Agliardi et al. 2013).

Detailed field monitoring of slow-moving landslides is largely limited to hillslopes adjacent to settlements, highways or railway infrastructure, or mines. One example is the Yakatabari mudslide complex in the Porgera River catchment, Papua New Guinea, which is home to one of the world's largest gold mines. Repeat surveys showed that this 16 Mm^3 landslide moves at average rates of 50 mm per month, and may have been doing so for the past several thousand years (Blong and Goldsmith 1993). Mining and other human activities may reactivate slow-moving landslides such as this and in extreme cases cause them to fail catastrophically. Sluicing for gold likely undermined the toe of Yakatabari landslide, triggering a rapid failure of at least 4 Mm^3 in 1974.

Modern remote sensing techniques such as interferometric synthetic aperture radar (InSAR) (Colesanti and Wasowski 2005; Wasowski and Bovenga 2015) and high-resolution LiDAR scanning have revealed that large portions of mountain rock slopes undergo creep deformation. These slow (or periodic) movements may prime hillslopes for large-scale catastrophic failure during intense rainstorms (Chigira et al. 2013) or earthquakes. Researchers studying slow-moving landslides have long used a multidisciplinary approach combining engineering geology, geomorphology, near-surface geophysics, and hydrology. Ground measurements and remote sensing have provided insights into patterns of movement of slow-moving landslides, especially in response to heavy rainfall, seismicity, and fluvial erosion at their toes (Handwerger et al. 2013; Hilley et al. 2004; Massey et al. 2016; Roering et al. 2009).

8.2 Geomorphic Impacts of Landslides

8.2.1 Landslides in the Hazard Cascade

The geomorphic coupling of landslides and river channels produces a diverse array of disturbances that alter landscapes; in particular

it affects river-channel geometry as well as water and sediment discharge. Catastrophic rock slides and rock avalanches can bury large reaches of river channels beneath sediment, damming stream flow and creating lakes. In some places such blockages can divert or even reverse river courses (Costa and Schuster 1988). Whether a landslide is sufficiently large to block a river for an extended period depends largely on the characteristics of the material forming the dam, the height of the dam, the volume of the lake, and the upstream contributing catchment area. To date researchers have largely relied on empirical relationships between these controls (Ermini and Casagli 2003; Dong et al. 2011), and physically more rigorous models wait to be developed.

Landslides introduce to rivers sediment that often differs in size from that transported by the river. Together with the added volume of sediment, the altered calibre of the ambient fluvial load initiates downstream sediment dispersal and the formation of coarse lag deposits (Brummer and Montgomery 2006), hydraulic flow obstacles, and persistent knickpoints (Korup 2006). Clusters of boulders emplaced by a catastrophic rock avalanche from the range front of the Southern Alps, New Zealand, have remained in place for more than 700 years despite being located in the channel of one of the major gravel-bed rivers flowing from the Alps (Chevalier et al. 2009). In mountainous headwater catchments, debris delivered to streams by catastrophic landslides may raise background sediment yields by up to two orders of magnitude, attaining peak rates of $\sim$0.2 Mt km^{-2} yr^{-1} (Mikoš et al. 2006). Once in the channel, some or all of the landslide debris begins migrating downstream in sediment pulses, which are slow-moving waves of bedload material. Field observations show that such sediment waves translate downstream at average rates of 0.1–5 km yr^{-1}, depending, among other things, on the sequence of effective discharge events, the channel longitudinal profile, the local sediment storage capacity of the channel, and any major constrictions imposed by resistant bedrock outcrops (Kasai et al. 2004). Dated deposits of landslide-derived sediment waves on floodplains or in river terraces show that shifting channels may remain unstable for several decades to centuries (Clague et al. 2003).

The implications for artificial reservoirs or hydropower stations can involve higher costs of dredging, structural damage or even inoperability, especially if the volume of landslide debris is of the same order of magnitude as that of the capacity of the reservoir. The sudden flank collapse of Bawakaraeng volcano (2830 m) on Sulawesi, Indonesia, in 2004 produced a debris avalanche of ~0.2 km^3 that obliterated the upper seven kilometres of the Jenebarang River. Reworking of the debris-avalanche deposit triggered a sediment wave that raised the channel bed and widened the channel by >200% in places. Sediments from the landslide continue to infill the Bili-bili reservoir downstream, which has a capacity of 0.35 km^3 (Tsuchiya et al. 2009). Landslides can directly affect water reservoirs. In late 2005, creeping slope deformation heralded the catastrophic failure of 2.5 Mm^3 of Palaeogene flysch rocks, which entered and blocked the Siriu Reservoir in the Romanian Carpathians (Micu and Bălteanu 2013). Such landslides may generate displacement waves that may overtop the dam or cause it to fail. The Vajont Slide, which happened in northeast Italy on 9 October 1963, is one of the most famous natural disasters in modern history (Crosta et al. 2015). Approximately 250 Mm^3 of rock slid catastrophically from Monte Toc into the then newly created Vajont Reservoir, generating a 140-m-high displacement wave that overtopped the Vajont Dam and flooded villages both upstream and downstream of the landslide. The landslide itself did no damage to the dam, but the sudden displacement wave that overtopped the dam claimed 1910 lives. Repeated raising and lowering of the reservoir level in response to accelerated creep of the Vaiont slide mass preceded the catastrophic collapse.

We emphasize the feedback between slopes and river processes. For example, a river can cause a slope failure by undercutting the slope toe. In response the sudden input of a large volume of sediment might alter the behaviour of the river so as to either increase or decrease undercutting. If the former, a positive feedback might lead to further landsliding that would cause a major change in the system. Overall, the geomorphic impacts of landslides depend on the rates at which landslides happen and how efficiently rivers dissect and remove their deposits. Generally, larger landslide deposits have a higher chance of remaining in the landscape for longer times, especially if rates of rainfall and runoff are low (Korup and Clague 2009). Blocky deposits of large landslides may resist erosion and persist on valley floors for thousands of years, even in mountain belts receiving several metres of rainfall annually such as the fjords of Norway and southwest New Zealand (Hancox and Perrin 2009).

Unstable fronts of mountain ranges Given the large populations on alluvial fans close to steep hillslopes and large active rivers, the fronts of mountain ranges are sites of a range of natural hazards. Range fronts commonly coincide with large active faults, where repeated seismic shaking, surface rupture, rock shattering, and hydrothermal alteration can weaken the rock mass more than in the interiors of the mountain belt. Range fronts also intercept moist airstreams, causing intensive rainfalls. Hence many forms of slope instability ranging from localized gully erosion to catastrophic rock-slope failures develop along these fronts, shedding large amounts of sediment onto large alluvial fans. Some of these fans may have formed from countless debris flows, whereas others record catastrophic fluvial aggradation from nearby landslides (Davies and Korup 2007). Rock avalanches from Central Asian mountain ranges have buried tens of kilometres of foreland alluvial fan surfaces with hummocky deposits that have been mistaken for moraines (Paguican et al. 2012; Robinson et al. 2014). Large landslides close to a range front may have a greater impact on the size of the ensuing sediment pulse (Acharya et al. 2009) than larger landslides in the interior of the range. In coastal settings landslides may increase rates of fan-delta progradation to several metres per year (Gallousi and Koukouvelas 2007).

We infer that incised or trenched fan heads may record little evidence of tectonic or climatic forcing (Nicholas and Quine 2007). Rather, trenched fan heads in actively aggrading settings might result from pulsed sediment supply followed by incision. In contrast, aggradation at a mountain front could migrate headward and infill lower mountain valleys with sediment. Numerical modelling suggests that such foreland deposition controls the local base level of fans and thus potentially also the rate of denudation within a mountain belt (Pelletier 2004). Sustained gully erosion along mountain fronts also feeds the growth of new fans and mixes sediments from the mountain belt with local material. The magnitude and frequency of debris flows entering and reshaping the fan may largely govern the rate and extent of channel avulsions and, ultimately, local flooding frequencies (de Haas et al. 2018).

8.2.2 Landslides on Glaciers

Landslides falling onto glaciers are special in that the underlying ice appears to favour longer runout. Landslide debris on top of glaciers can alter their mass balance by suppressing ice-surface ablation (Reznichenko et al. 2010). In some cases, the debris cover is sufficient to cause the glacier to advance and produce a moraine that bears the imprint of a landslide rather than regional climate oscillations (Reznichenko et al. 2016). In such cases, making inferences about former climates solely based on moraine positions may turn out to be flawed, while the hazard of catastrophic rock-slope failures may be underestimated. Whether landslides cause moraines to form and thus contaminate what glaciologists and palaeoclimatologists traditionally have viewed as a purely climatic proxy has caused some debate. Central to this debate is how a glacier responds to abrupt but partial covering by rockslide or rock-avalanche debris. Numerical models of glacier motion under sediment load show that any advance caused by emplacement of rock-avalanche debris on a glacier is likely to cause eventual stagnation of the advanced ice lobe. The result would be distributed, hummocky deposits quite different from the single moraine ridges that are traditionally associated with climate-driven glacier advances (Vacco et al. 2010). The results of this model questioned the original concept of Tovar et al. (2008), who had proposed that New Zealand's Waiho Loop moraine was the outcome of a supraglacial rock avalanche. This landslide would have thickened the glacier terminus dramatically, eventually creating a large terminal dump moraine (Alexander et al. 2014).

Menounos et al. (2013) described a late Holocene (~2.7 ka) advance of the Tiedemann Glacier in the Coast Mountains of British Columbia that reached well beyond the maximum extent of the glacier during the Little Ice Age, forming a single moraine ridge. This advance seems at odds with the regional glacial chronology, unless a rock avalanche blanketed the glacier and provoked an advance at that time. By coupling an algorithm that describes the motion of rock debris on the glacier surface and an ice dynamics model, the authors found that even a moderately sized rock avalanche of 10 Mm^3 delivered to the top of the ablation zone could cause the glacier to thicken and advance far beyond its Little Ice Age limit. Work on contemporary glacier response to landslides seems to corroborate some of these findings. For example, a detailed satellite-based mass-balance and ice velocity study of Black Rapids Glacier, Alaska, which was blanketed by debris from earthquake-triggered rock avalanches in 2002, revealed distinct phases of acceleration and deceleration that soon faded to the background levels of large seasonal and yearly variations (Shugar et al. 2012). More instructive sedimentological studies capable of reliably separating landslide from glacier processes might resolve these issues. Jamieson et al. (2015) reported an intriguing and unequivocal example of human-driven glacier advance following dumping of mine waste onto a glacier ablation zone. Over a 15-year period, two glaciers in Kyrgyzstan advanced by 1.2 and 3.2 km respectively at a rate of up to 350 m yr^{-1}, associated with dumping of spoil of up to 180 m thick on large parts of these valley glaciers.

8.2.3 Submarine Landslides

Detailed studies of submarine landslides broaden our insights into the overall hazard of slope failure and its contribution to shaping Earth's surface. Quantitative studies suggest that submarine landslides represent the lion's share of sediment moved along passive continental margins (Taylor et al. 2001). Turbidity currents along the northwest African continental margin have transported 10–100 km^3 of sediment to abyssal basins over the past 22 Myr, which amounts to some 20% of the total sediment delivered during that time (Weaver 2003). Deposits of seven giant landslides with volumes ranging from 3 to 500 km^3 originated from submarine slopes as low as 1–2° off the mouth of the Nile River, and constitute 40% of the total Quaternary sediment pile on the sea floor, while in some locations the sediment column contains up to 90% of mass-transport deposits (Garziglia et al. 2008). Scientists find many fewer deposits of large submarine landslides along active margins, where deposits are being recycled by subduction, and frequent seismicity causes submarine sediments to strengthen by settling and dewatering. High sedimentation rates fed by influx from glacierized onshore sources may

offset this effect and build up pore pressures in sea-floor sediments, thus making them more unstable. This seems to be the case for offshore southern Alaska, where Sawyer et al. (2017) found that the submarine Surveyor Fan, which accretes at about 3 mm yr^{-1} with fine glacial outwash from the St. Elias Mountains, had much lower sediment strengths than marine geologists would expect in this active subduction zone, whereas sea-floor portions off the fan had much stronger sediments. This result underlines some of the practical value of studying in detail the flux of sediment from terrestrial sources to submarine sinks to inform hazard appraisals for submarine landslides and tsunamis. Local uplift and steepening of the continental shelf add to the potential causes for giant slope failures along active margins; Geersen et al. (2011) inferred these causes for three large Middle Pleistocene slope failures involving 250–370 km^3 of sediment in the active trench margin of southern Chile. Finally, large submarine landslides in active margins triggered during earthquakes may modulate tsunami potential, so that wave characteristics combine those of seismogenic and landslide sources (Tappin et al. 2014).

8.3 Geomorphic Tools for Reconstructing Landslides

8.3.1 Landslide Inventories

The foci of much contemporary landslide research are highly diverse, but generally concerned with predicting the locations and impacts of slope failures and understanding their dynamics. Such research mostly relies on detailed engineering geological, sedimentary, hydrometeorological, and geomorphological field data. Landslide hazard studies aim at estimating the frequency and magnitude of slope failures for a given area and period. Yet another research focus is the regional distribution of landslides, with the aim of identifying locations that are most susceptible to slope failure. This approach builds on a combination of satellite images, aerial photograph interpretation, elevation data, and ground observations. Deciphering landscapes and landforms shaped by landslides from remote sensing data is prone to several sources of error and amibuigity, and even experts may differ about the exact location and boundaries of a given landslide (Van Den Eeckhaut et al. 2005). Still, these methods have produced landslide inventories that cover study areas from river catchments to whole nations, and contain the location and planform shape of several dozens to more than 100 000 landslides (Guzzetti et al. 2012).

Few of these catalogues contain detailed information about the timing of landslides. If mapped following large earthquakes, landslides that fell in aftershocks may be indistinguishable from those triggered by the mainshock (Bommer and Rodríguez 2001). Similar problems of temporal separation apply to landslides that happened during a close sequence of earthquakes and rainstorms (Yamagishi and Iwahashi 2007). Seismic waveforms emitted by detaching and moving slope masses offer a means to pinpoint both timing and location, and have helped to detect many large rock-slope failures in remote terrain (Ekstrom and Stark 2013). Such seismic tracking reveals much about the energy, forces, and eventually mass, of single slope failures or even precursory signals of fracturing (Walter and Schwaderer 2012), but the method reaches its limits if the signals of many landslides are superimposed, or mixed with earthquake signals (Figure 8.6). Most landslide inventories include only approximate times of landslides that are constrained by the timestamps of the aerial photographs or satellite images on which they were identified (Reid 1998). Nonetheless, these data allow statements about average landslide frequency over the period between two sequential images. Bear in mind that such average landslide frequencies are snapshots of a long-term picture; average landslide frequencies may be affected by recent disturbances such as earthquakes, rainstorms, or snowmelt that trigger a large number of slope failures,

outnumbering those that happen during more quiescent periods. Databases of prehistoric landslides (Prager et al. 2008) have poorer age control, thus inferences about their causes and triggers are commonly speculative.

The spatial pattern of landsliding lends itself to comparison with parameters such as lithology, discontinuities in rock masses, topography, land cover, and road or river networks. The ability to derive these data from a rapidly growing number of digital data sets has given rise to hundreds to tens of thousands of landslide susceptibility studies. Systematic comparisons of different methods of landslide mapping and inventory compilation are rare (Galli et al. 2008), and so is the comparison of landslide maps prepared by different scientists. Van Den Eeckhaut et al. (2005) asked several experts to map large, deep-seated landslides in partly forested terrain of the Flemish Ardennes from digital terrain models. One key outcome was that the locations and sizes of purported landslides differed quite markedly, though partly depended on previous experience with landslide mapping in that particular area.

Studies like these provide new grounds for moving the field forward toward establishing a common standard, especially when summarizing those data for landslide hazard and risk assessments (Corominas et al. 2014). Comparing how well different methods predict the occurrence of landslides or debris flows is considerably more advanced. Many studies routinely use a selection of several methods to produce maps of landslide susceptibility. However, only a handful of these methods are reliable and robust (Carrara et al. 2008). The literature is teeming with landslide susceptibility and hazard studies, but few studies appear to have practical applicability, partly because of the many limitations involved in using landslide inventories for hazard assessments (Van Westen et al. 2005). The same holds for studies of landslide risk that require detailed and reliable data on vulnerability and elements at risk (Dai et al. 2002).

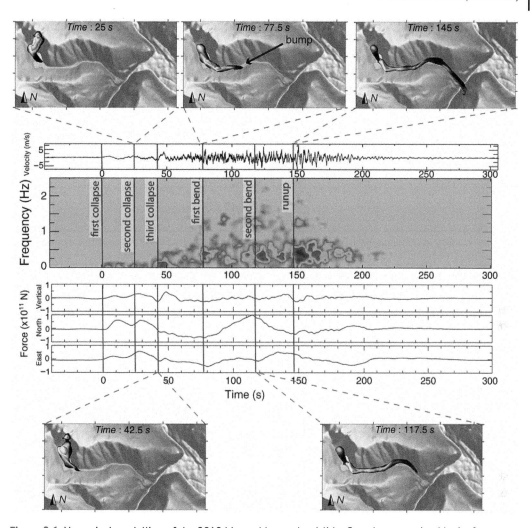

Figure 8.6 Numerical modelling of the 2010 Mount Meager landslide, Canada, constrained by its force history derived from seismic data. Seismograph and spectrogram of the vertical component of the signal recorded at the closest seismic station and inverted force–time function. Snapshots of simulation of the best scenario at the times corresponding to each seismic pulse are shown. From Moretti et al. (2015).

Many landslide susceptibility studies promise high success rates, but fail when it comes to rigorous tests of their performance. A review of some 150 studies concerned with landslide susceptibility over the past decade showed that estimated success rates in prediction of landslide locations remained unchanged regardless of the number of landslides studies or parameters used in the models. Nonetheless most studies reported success rates between 71% and 98%(!). These optimistic results were largely based on models using five to 15 geological, topographic, or other environmental variables. Among these variables, local hillslope inclination is most prevalent and often also most relevant (Korup and Stolle 2014).

8.3.2 Reconstructing Slope Failures

Recognizing and reconstructing the geomorphic footprints of former landslides often captures a tiny sample of the full spectrum of mass wasting occurring in a given landscape. Evidence of larger failures survives longer in the landscape and its sedimentary archives (Korup and Clague 2009). Erosion and sedimentation commonly rapidly obliterate evidence of smaller landslides. Even larger landslide deposits may be invisible to the untrained eye if dissected or covered by younger sediments. Although all landslide deposits are destined for eventual removal by fluvial, glacial, or aeolian erosion, some deposits may linger in the landscape for hundreds of millennia (Hewitt et al. 2008).

Once identified, the age of landslide deposits may be of interest, and a wide range of dating techniques is available (Pánek 2015). Cosmogenic exposure dating is now commonly used for dating deposits that are at least several hundreds of years old. This method requires rock surfaces that have remained stable, thus warranting constant exposure angles to incoming cosmic rays: thus finding large immobile boulders or exposed slip surfaces is essential (Ivy-Ochs et al. 2009). Other methods include radiocarbon dating of soils buried under landslides, or organic peat beds in landslide-dammed sediment wedges. Radiocarbon dating is limited to events less than about 50 000 years old. Tephrochronology is useful for dating landslide deposits in volcanically active areas (Hermanns et al. 2000). Dendrochronology or tree-ring dating can provide annually resolved ages for damage to trees caused by impacts of landslide debris (Šilhán et al. 2012), but the technique is generally limited to slope failures that are less than about 1000 years old. Lichenometry (Bull and Brandon 1998) allows rapid and cheap collection of rock surface-exposure ages, and can yield statistical data to allow event age estimation. Other surface exposure techniques include weathering-rind depth analysis (Fruaenfelder et al. 2005) and Schmidt-hammer dating (Shakesby et al. 2006).

Deciding whether a given landslide deposit is a product of single or multiple failures has concerned surprisingly few studies (Orwin 1998). The surface morphology of larger landslides allows the researcher to identify multiple failures at a given site over periods of decades to millennia (Oehler et al. 2009). As a consequence, studies of former landslides only roughly approximate the frequency and magnitude of slope instability in a given study area. It follows that detecting fluctuations in past landslide activity that is above the noise of measurement errors requires even more care.

Long-lived landslide dams offer opportunities for palaeoenvironmental reconstruction. Stacks of landslide-dammed lake sediments >100 m thick are nested in the steep and deeply dissected river valleys at the margins of the Tibetan Plateau, such as the upper Indus, Shyok, Yarlung Tsangpo, and Yangtze rivers. Absolute age constraints on these sediments show that many of the lakes have persisted for several thousands of years (Wang et al. 2011) despite aggressive erosion, rapid sedimentation ($\sim 10^1$ mm yr^{-1}), high rainfall, and episodic strong earthquakes. Some dams of prehistoric landslides are used for hydropower projects such as at Lake Waikaremoana, New Zealand (Allan et al. 2002). The 300 MW Baspa II dam, India's largest private hydroelectric facility, was built on top of rock avalanche-dammed lake sediments of early Holocene age. The landslide dam remained stable and the lake that formed behind it completely filled with sediment over some 3000 years. At least five stratigraphic levels of deformed lake sediments have been attributed to earthquakes that failed to breach the landslide dam (Draganits et al. 2014).

Lake sediments can record catastrophic subaerial landslides if other processes such as flood-driven hyperpycnal flows or subaqueous mass failures can be excluded. For example, organic-rich lacustrine layers of terrestrial origin have supported detailed landslide chronologies in the Dolomites of Italy as far back as the early Holocene (Borgatti et al. 2007). The use of different dating methods on different landslide deposits further complicates straightforward comparisons. Targets for inferring the age of landslides include surface exposure dating of head scarps or large and

presumably immobile boulders on the deposit surface, or radiocarbon dating of sediments trapped upstream of, or directly on, the landslide body. Organic material enclosed or buried by landslides can also be radiocarbon dated, but these ages are only maxima for the age of the event (Wood et al. 2011). K-Ar-dating of lava flows or lava domes that were emplaced following volcanic flank collapse provide minimum ages for volcanic debris avalanches associated with the collapse (Samper et al. 2007).

Reconstruction of larger terrestrial landslides depends on the correct identification of diagnostic sedimentary and geomorphic evidence. The potential for confusing landslide deposits with those of other processes is high in mountain areas, where many different processes can lay down coarse deposits. In such settings, deposits of large rock avalanches have been confused with moraines and vice versa (McColl and Davies 2011). In the Karakoram Range of Pakistan, large parts of the Quaternary glacial history will have to be rewritten given that many 'moraines' have turned out to be deposits of catastrophic rock avalanches rather than glaciers (Hewitt 1999). In a similar spirit, Jarman et al. (2011) have argued that several conspicuously large fans in the Italian Alps could have resulted from single catastrophic long-runout landslides instead of prolonged phases of alluvial processes. This re-interpretation uproots some long-held notions in geomorphology that alluvial fans faithfully record tectonic and climatic perturbations over long periods of time. But landslide features may also resemble other geological features, and Hart et al. (2012) report several examples of engineering geological studies that had confused scarps and tension cracks of landslides with geological fault zones of Pleistocene to Holocene age.

Rare exposures of basal landslide facies containing up to decimetre-thick volcanic-glass like melt bands termed 'frictionite' or 'pseudotachylite' have helped to identify deposits of catastrophic rock-slope failures of at least several millions of cubic metres (Weidinger and Korup 2009). The family of diagnostic sedimentary features of large catastrophic rock-slope failures now includes pervasively fragmented and interlocked jigsaw-cracked rock masses, basal mélange containing rip-up clasts and phantom blocks, bands of microbreccia, and thin bands of basal frictionite. However, fracturing is so pervasive that, even at the submicrometre scale, the grain-size distributions of landslide deposits are nearly indistinguishable from those of geological fault zones or meteorite impact breccias. Microscopic signatures characteristic of rapid, high-stress comminution have also been proposed as diagnostic indicators of landslides (Reznichenko et al. 2012). All these sedimentary features can be useful fingerprints of dissected landslide debris that may be barely detectable otherwise (Weidinger et al. 2014). Nonetheless, the sedimentology of terrestrial landslides is still in need of a better systematics, and we may yet learn much from detailed studies of the stratigraphy of volcanic debris avalanches, and submarine landslides.

Passive continental margins off Norway and Svalbard, the eastern United States, the Northwest African shelf, and the Amazon fan are sites of some of the largest submarine landslides known. Three landslides at the edge of Norway's continental shelf are among the largest known landslides. The debris from these landslides has a total volume of about 3500 km^3, equivalent to an area the size of Iceland buried to a depth of 34 m. The 'Storegga landslides' triggered a very large tsunami in the North Atlantic Ocean, with recognizable deposits along the northeastern Scottish coastline. Based on carbon dating of plant material recovered from these tsunami deposits, the landslides occurred about 8200 years ago. On the Amazon fan, catastrophic mass-transport deposits (a generic term that marine geologists use for landslides) that are on average 200 m thick cover an area of ~15 000 km^2 (Maslin et al. 2005; Silva et al. 2010). These deposits involve as much as 5000 Gt and document infrequent catastrophic mass movements and their potential for triggering tsunamis. These rare but giant slope failures anchor the extreme end of events for statistical hazard models.

Sequences of turbidites provide valuable time series of local to regional catastrophic sediment transport, and may in some cases augment and extend palaeoseismological chronologies (Goldfinger et al. 2007; Barnes et al. 2013). However, only in some cases does strong ground motion during large subduction earthquakes trigger turbidity currents. Investigations of bathymetry data after the M 8.8 Maule, Chile, earthquake in 2010 revealed many minor submarine landslides, but fewer than would be expected from an earthquake of that magnitude. Völker et al. (2011) argued that frequent earthquakes in any area may and exhaust sediment supplies and thus favour small slope failures at the expense of larger and rarer ones.

Debris flows in urban areas Debris flows form where large sediment volumes can mix with runoff in steep terrain (Jakob et al. 2005). This prerequisite supports the view that debris flows mostly occur in remote mountainous terrain. However, debris flows have also struck many urban areas around the world. In December 1999, the coastal state of Vargas, Venezuela, was hit by ~1000-year rainfall amounts that triggered countless landslides and debris flows that travelled onto densely settled fan deltas, killing an estimated 19 000 people and destroying much of the urban infrastructure, with an estimated total damage of $1.9 billion across the region (https://pubs .usgs.gov/of/2001/ofr-01-0144/). Terraces consisting of debris-flow deposits along some of the impacted streams record similar events in prehistoric times. Highly seasonal rainfall is an important factor determining the hazard from debris flows in cities built partly adjacent to steep terrain. For example, many regions with a Mediterranean climate of hot summers and rainy winters have flashy creeks and rivers that occasionally also generate debris flows. In November 2011, rainstorms that had an estimated 200-year return period triggered debris flows that claimed more than 800 lives in the city of Algiers, Algeria (Machane et al. 2008).

8.4 Other Forms of Slope Instability: Soil Erosion and Land Subsidence

Earth's land surface may become unstable through processes other than landsliding. Lal (2003) estimated that the global land area impacted by water and wind erosion may be as much as 10.9 million km^2, and 3.0 million km^2, respectively. Soil and gully erosion are widespread and damaging processes, and have both natural and human causes. Agricultural lands lack a protective cover of vegetation during parts of the year and are among the areas with some of the highest rates of soil erosion and sediment yield on our planet. Human history is rich with accounts of soil erosion, but soil conservation is more of a modern invention and began to expand and diversify in the late nineteenth and early twentieth centuries (Dotterweich 2013). The natural processes of erosion, transport, and sedimentation change the land and, eventually, affect human lives and assets, particularly in areas with seasonal precipitation and sparse vegetation cover. Some of the recommendations that García-Ruiz et al. (2013) distilled from a detailed review of erosion in Mediterranean landscapes are to improve land management by more appropriate guidelines for agriculture and reducing wildfires, to improve erosion monitoring methods and coverage, to understand badland dynamics better, and to conserve culturally important landscape elements.

Gully erosion is a distinctly linear erosion process and an important indicator of land-use and environmental change. Eroding gullies may produce peak sediment yields of more than 5000 t km^{-2} yr^{-1} (Poesen et al. 2003). Entrainment of sediment in gullies may commence after some rainfall intensity or duration threshold is crossed. Gullies may expand into badlands in poorly consolidated and easily erodible sedimentary rocks. Detailed measurements of water and sediment discharge in the Isábena River in the central Pyrenees, however, showed that only 1% of the catchment area, which is prone to badland erosion in Eocene marls and sandstones, was feeding its high rates of sediment yield (López-Tarazón et al. 2012).

Solifluction and creep processes also gradually move sediment from hillslopes to valley floors. These processes are active in soil-mantled mountainous terrain, especially in permafrost regions. Lobes of soliflucted debris are ubiquitous on mountain slopes that support permafrost. During summer the active layer on such slopes may fail, producing thaw flows that, in some cases, reach the valley floor. Differential thaw of permafrost in low-relief, high-latitude landscapes can produce irregularities in the land surface. Depressions may become occupied by 'thermokarst' lakes that expand over time.

Subsurface dissolution of soluble rocks is another important process for changing the terrestrial land surface, often producing very characteristic landforms. The German term 'karst' refers to processes and landforms arising from the solution of calcareous rocks by carbon dioxide dissolved in rainwater. Spectacular caves and sink holes or dolines, dry valleys, and underground rivers are well known karst phenomena and characterize landscapes such as the Dinaric Alps in southeastern Europe and South China, part of which has been a UNESCO World Heritage site since 2007. However, rainfall and runoff can also dissolve rocks other than calcareous; it can also affect gypsum and other evaporatic rocks ones and even silicate rocks such as sandstones. In these cases the term 'parakarst' can be used (Sponholz 1994). 'Pseudokarst' refers to underground cavities created by physical forces rather than chemical solution processes.

Measuring rates of soil erosion in a given area relies on detailed, although often small-scale, measurements, stratigraphic analyses of reworked soil and colluvial deposits, and catchment-scale to regional-scale modelling. The Universal Soil Loss Equation (USLE) – or its refined form (RUSLE) – is one of the most popular and widely applied models, especially on agricultural lands. Application of this equation in more mountainous terrain is ill-advised, given that many of its parameters are calibrated to low-relief landscapes. Lazzari et al. (2015) tested the RUSLE model together with several other methods of estimating catchment-wide sediment yields, including landslide mapping, and compared the estimates to yields computed from sediment trapped in an artificial reservoir in the Basilicata region of southern Italy. Their results suggest that other models perform better than RUSLE for predicting material fluxes in mountainous catchments. The historic records in their study also demonstrated how a four-year phase of increased landslide activity raised the background sediment yields by a factor of three on average, thus largely outweighing the contributions of other soil-erosion processes.

Other models offer a larger range of input parameters for processes such as sheet, rill, gully, and bank erosion; landslides; and connectivity to simulate catchment-wide losses of sediment (de Vente and Poesen 2005). Many of these methods are designed for measuring soil erosion and sediment yields from the plot to the catchment scale, and the required level of instrumentation limits the monitoring periods to a few years to decades. Most metrics that have been suggested for sediment delivery from hillslopes are based on the distance from, or travel time to, the river network below. Some metrics also seek to include effects of particle size, local terrain inclination, or even rainfall and runoff characteristics during hillslope sediment transport (Vigiak et al. 2012).

Measuring the concentration of cosmogenic nuclides such as ^{10}Be in river sands can provide catchment-averaged values of mean denudation rates mostly over centuries to tens of millennia, against which the impact of anthropogenic soil erosion can be compared. Hewawasam et al. (2003) applied this method in the tropical highlands of central Sri Lanka and found that contemporary sediment yields ranged from 130 to 2100 t km^{-1} yr^{-1}, which is ten to nearly one hundred times natural background rates. Whether such case studies can be generalized to the regional scale remains an important question that is pertinent for future strategies aimed at mitigating soil erosion. Vanacker et al. (2007) studied erosion in catchments of the southern Andes of Ecuador and concluded that widespread deforestation has increased rates of erosion and sediment yield up to two orders of magnitude above natural background rates inferred from concentrations of cosmogenic ^{10}Be in river sands. The extent of vegetation cover influenced local erosion rates greatly; catchments with dense and near-natural forest stands were among those with the lowest rates of soil loss.

In some cases the long-term sediment flux inferred from ^{10}Be-derived denudation rates greatly exceeds contemporary rates. This imbalance may hint at possible effects of sediment storage within sampled catchments, or pulsed sediment fluxes (Kirchner et al. 2001). These explanations might be the case for an analysis of up to 27-year long records of daily sediment yield recorded at gauging stations in southwest China and southeast Tibet (Schmidt et al. 2011). This study revealed unchanged sediment yields over the period of record that appeared to be independent of the fraction of cropland determined from satellite imagery. Annual sediment yields were highly variable but less so in the larger catchments, so that the commensurately higher sediment storage in these catchments could act as a buffer. Sediment storage likely masks the anthropogenic impacts on sediment yields in the larger and more densely populated catchments.

Despite the widely recognized and well publicized protective role of vegetation in mitigating erosion, several studies maintain that it is topography and runoff that are the principal controls on catchment-wide denudation. Carretier et al. (2013) argued that slope and runoff outweigh the role of vegetation cover in controlling ^{10}Be-derived denudation rates, mimicking the strong bioclimatic gradient along the Chilean Andes. A confounding issue is that vegetation type and cover generally strongly correlate with topographic and climatic variables (think of the tree line, for example), so that the direct effect of vegetation on erosion rates may remain obscured. If the long-term rate of sediment yield is limited by the rate at which bedrock is weathered into erodible material, changes in vegetation will affect the maximum soil depth that plants can stabilize. Thus deforestation will release soil beyond the stabilizing effects of tree roots, and reforestation will cause weathered soil to build up. The former will increase erosion rates, whereas the latter will reduce it. In the long term, however, the sediment delivery rate under both types of vegetation will be similar; however forests will yield infrequent larger amounts of sediment while grass will yield more frequent smaller amounts (Bloomberg and Davies, 2012).

8.5 Climate Change and Landslides

In the debate about the relationship between climate change and landslide occurrence, some have argued that global warming will alter the frequency and magnitude of landslides (Keiler et al. 2010). Yet few inventories are able to support this notion, reasonable as it may seem. Huggel et al. (2012) listed some of the main mechanisms that could alter landslide magnitude and frequency, and thus hazard, under warming conditions, including: (i) positive feedbacks acting on mass movement processes that, after an initial climatic stimulus, may evolve independently of climate change; (ii) threshold behavior and tipping points in geomorphic systems; and (iii) storage of sediment and ice involving important lag-time effects. However, landslides also occur under cold and wet conditions, as shown by the unusually high incidence of slope failures during the exceptionally cold and wet winters between 2008 and 2010 on the Mediterranean island of Majorca (Mateos et al. 2012). To avoid drawing false conclusions from such synoptic to seasonal deviations from an overall warming trend, we need longer archives of landslides and associated climate data. A study that linked climate records to landslides mapped from sequential air photos in California suggests that debris-slide frequency changed from ~1.6 to 8.3 km^{-2} yr^{-1} during the late 1930s mostly because of the increased frequency of high-intensity rainstorms (Reid 1998).

One approach is to check for statistically tangible differences in the magnitude-frequency distribution of landslides over time. Such differences might reflect variations in climate or seismicity. If indeed valid, this strategy could be a valuable tool for detecting a climatic change signal in landslide databases. However, it also seems that the shapes and parameters of landslide size distributions reflect to a first-order other nonseismic and nonclimatic controls on slope stability, including topography (ten Brink et al. 2009) and material strength (Frattini and Crosta 2013). Assuming that these controls can be disentangled, landslide inventory statistics might offer a way to detect the effects of climatic or environmental changes on slope stability. The number of landslide inventories is growing, and some of them include several hundred to tens of thousands of landslides. Unfortunately many data catalogues suffer from under-sampling, and only the most recent events can be used for representative data analysis (Kirschbaum et al. 2009). Measured size distributions of landslides from large inventories can be statistically robust, and in some cases reveal little, if any, apparent changes regardless of differing trigger mechanisms and varying environmental boundary conditions (Malamud et al. 2004). In the Japanese islands, for example, the frequency and magnitude of landslides differ little across different precipitation zones. Examining systematic simulations of measurement noise in landslide databases, however, Korup et al. (2012) showed that frequency–magnitude relationships based on inverse power laws may be too insensitive to reveal changes that might be caused by climate or global environmental change.

Precipitation and runoff are key controls on erosion rates in terrestrial settings and thus affect the delivery of sediment to continental margins. The frequency and magnitude of submarine landslides may thus respond to changes in the rates at which sediment is delivered to continental shelves. High-latitude regions appear to be most sensitive to changes in sediment input, and marine geological studies have shown that this input was higher in glacial periods than in interglacial periods. For example, Dimakis et al. (2000) reported very high Pleistocene sedimentation rates (~600 mm yr^{-1}) near the former ice-sheet margin along the Svalbard–Barents Sea margin. The submarine Bear Island fan contains about as much late Cenozoic sediment as the Amazon or Mississippi fans located at lower latitudes. The researchers proposed that the high mass input during glacial periods favoured the rapid build up of large sediment piles, which spawned huge debris flows involving involving 10–30 km^3 of sediment. High rates of sediment production and delivery to continental shelves and submarine fans during glacial periods might also provide a climate driver for submarine slope failures along Arctic passive margins (Tappin 2010).

Sea-level rise during deglaciation increases the load of water on the sea floor, while also gradually raising bottom-water temperatures. If the effects of this warming outweigh those of the added loads by the water column, the probability increases that frozen methane hydrates (also known as clathrates), which occur as layers in sea-floor sediments, would

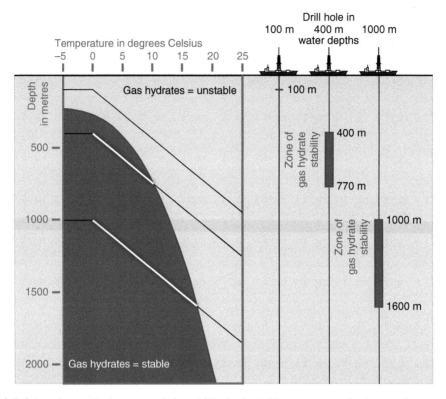

Figure 8.7 Submarine gas hydrates and their stability (indicated by purple domain characterized by sufficient hydrostatic pressure from the overlying water column and sufficiently cold bottom-water temperatures) play an important role in submarine slope stability. From World Ocean Review (2010, https://worldoceanreview.com/wp-content/downloads/wor1/WOR1-en.pdf)

sublimate, thus destabilizing sediments and leading to sudden slope failures (Figure 8.7). It is currently estimated that clathrates contain between 1600 and 2000 Gt C stored in ocean sediments (Maslin et al. 2010). The sudden release of large volumes of methane into the atmosphere could easily outweigh the climatic forcing effects of CO_2. Archer (2007) noted that the reservoir of frozen methane hydrates in submarine sediments and arctic permafrost is so large that if 10% of the methane were released to the atmosphere within a few years, it would have an impact on the Earth's radiation budget equivalent to a factor of 10 increase in atmospheric CO_2.

If mobilized, submarine methane hydrates could reduce the stability of large tracts of sea floor, potentially triggering large (>1 Mm^3) sediment failures and tsunamis (Vogt and Jung 2002).

Marine sedimentary sequences record potential climate-driven phases of submarine slope instability. For example, causal links between sea-level change and turbidite frequency have been proposed for the Ulleung Basin between South Korea and Japan. Turbidity currents were three times more frequent in this basin between 19 and 29 ka than after this period (Lee et al. 2009). An explanation for the more frequent turbidity currents during the earlier period is that lower hydrostatic pressures due to the lower sea levels at this time promoted dissolution of frozen gas hydrates, which in turn destabilised the seafloor. This interpretation follows the opposite argument to that made for clathrate stability earlier and needs to be adjusted for the likelihood that sedimentation rates were higher during glacial periods. The interpretation also hinges on the assumption that the cored turbidite stacks are complete and without erosional hiatuses.

Although the frequency of turbidites may change during times of lower sea level, their size may show a different trend. Two very large turbidite 'megabeds', each containing 3–6 × 10^{11} m^3 of sediment, underlie the Balearic and Herodotus abyssal plains of the western Mediterranean. They were deposited during sea-level lowstands during the last glacial maximum (Rothwell et al. 2000). The two megabeds are among the most important sedimentation events of the past 120 000 years and support the idea that increasing terrestrial input at times of low sea levels are likely responsible for some events of this type. The timing of emplacement of these enormous turbidites also appears to be consistent with the possible release of seafloor methane hydrates following decompression when the sea level was much lower than today. The rarity of these large events, however, is problematic for inferring a statistically robust relationship.

Higher temperatures and more extreme weather conditions also affect the occurrence of terrestrial slope failures. Scientists have argued that atmospheric warming causes more frequent landslides due to more frequent and intense rainstorms, ongoing deglaciation, and degrading mountain permafrost (Huggel et al. 2012). Although this argument seems intuitive, available data to support it are scarce and rarely robust enough for statistical testing (Allen et al. 2010). These tests are especially important in areas where humans have intensively modified the landscape, and thus altered the boundary conditions of slope stability. Land-use changes such as widespread deforestation before or during the early stages of industrialization may have compromised the stability of hillslopes eroded in weak rocks

or soils. The resulting pattern of landsliding may reflect a lagged adjustment to the loss of protective vegetation cover rather than to contemporary climate change (Crozier 2010), especially as short-term time series might mask long-term trends.

Not surprisingly, the bulk of available data applies to changes in either the frequency or magnitude of landslides, although rarely both. Geertsema et al. (2006), for example, noted that catastrophic landslides with volumes >0.5 Mm^3 or running out >1 km in northern British Columbia were becoming more frequent, increasing from 1.4 to 2.3 yr^{-1} over a 30-year period. These large landslides are rare compared to smaller ones, so we may need to invest more in assessing whether the observed increase is long-term trend or simply a manifestation of the medium term variance in their occurrence. This and many similar observations call for detailed studies of how, and which types of, landslides are sensitive to climate forcing.

Lake sediments trapped behind former landslide dams may offer detailed archives of palaeoclimate changes and their possible connections to landslides. Most studies of this kind have unravelled chronologies covering several millennia that are seemingly well beyond those during which anthropogenic warming has happened. Nonetheless, these archives illuminate some of the natural links between climate change and slope stability. For example, sediments from landslide-dammed lakes in the eastern Argentinian Cordillera may preserve climatic signals in their varves. The power spectra of these varves include contributions that are compatible with ENSO phases and possibly also those of oscillating sea-surface temperatures of the tropical Atlantic. The study concluded that the landslide-dammed lakes formed during distinctly wetter phases between 35 and 25 ka BP, and after 5 ka BP (Trauth et al. 2000). Whether this also means that landslides were more frequent during those wet periods is debatable and requires detailed time series of past landslides, because missing evidence of slope failure during more dry periods hardly supports the inference of a period with more stable hillslopes.

Numerical climate models are another means of simulating changes in rainfall characteristics and thus anticipating how landslides respond. For example, Jakob and Lambert (2009) used 19 different climate models as predictors of 6% and 10% increases in, respectively, short-term precipitation and antecedent precipitation in arguing for higher landslide frequencies in the future in the North Shore Mountains near Vancouver, Canada. This forecast requires the assumption that changes in rainfall intensity and duration, together with antecedent moisture availability, directly affect the dynamics of pore-water pressures in hillslopes irrespective of any indirect effects on vegetation or land cover, which in turn govern infiltration rates of soil water.

Empirical models are useful for pinpointing critical rainfall intensity–duration thresholds above which shallow landslides and debris flows are likely to occur (Baum and Godt 2010). The idea is to collect data on the rainfall intensity and duration together with reported occurrences of landslides to record the minimum rainfall characteristics required for slope failure. A global compilation of 2626 landslide-triggering rainfall events by Guzzetti et al. (2007) showed that with increasing rainfall duration the minimum average rainfall intensity required to trigger shallow slope failures decreases nonlinearly. Satellite-based sensors such as those of the Tropical Rainfall Measurement Mission (TRMM) allowed detailed monitoring of global rainfall patterns between latitudes 40° N and 40° S, and provide valuable early warnings of areas that might be impacted by rainfall-triggered landslides (Hong et al. 2006). In areas with strong seasonal rainfall regimes such as the monsoon-dominated Himalayas, other rainfall thresholds for triggering landslides may come into play. Antecedent soil moisture lowers thresholds for slope failure and thus must be considered in warnings based on rainfall amounts and intensities

alone. Gabet et al. (2004) argued on the basis of a three-year time series that both accumulated rainfall totals over the monsoon season and daily totals were critical for triggering landslides in the Annapurna region of Nepal. These and similar estimates may be controlled by local site conditions that differ from conditions elsewhere in the mountain belt. Dahal and Hasegawa (2008) noted that reliable rainfall measurements are available for only 193 of the 677 recorded landslides in the Nepal Himalaya between 1951 and 2006. Nevertheless, these 193 landslides all happened when daily precipitation exceeded 144 mm, which largely occurs during the monsoon season.

The regional picture of rainfall thresholds for landslides can be more variable, as Saito et al. (2014) demonstrated in their analysis of >4500 rainfall-triggered landslides throughout Japan between 2001 and 2011. They concluded that it was neither the most frequent nor the most extreme values of rainfall intensity or duration that triggered most of the total landslide volume. A possible reason for the poor relationship is previous landscape disturbance. A strong earthquake, for example, may drastically reduce the precipitation intensity–duration thresholds for debris flows because more unconsolidated and readily erodible debris is available in the landscape after seismic disturbance (Yu et al. 2012). Snowfall may also delay and distribute moisture to hillslopes and could be an important factor for triggering debris flows in cold environments (Decaulne and Sæmundsson 2007). These findings show that empirical models based on rainfall duration-intensity curves must be calibrated for the region of interest to demonstrate that they are reliable.

Exemplifying a completely different effect of climate change on landsliding, McColl et al. (2012) modelled the effect of the presence of a high ice surface in a mountain range suppressing the response of edifices to seismic shaking. Because ice acts as a strong rigid solid at seismic shaking frequencies, the ice surrounding edifices prevents them from acquiring large amplitudes during earthquakes, effectively suppressing topographic amplification. Hence, coseismic landsliding is reduced during large glaciations and deglaciation is accompanied by enhanced coseismic landsliding, perhaps increased further by deglaciation-induced seismicity (Mörner 1991).

References

Acharya G, Cochrane TA, Davies T, and Bowman E 2009 The influence of shallow landslides on sediment supply: A flume-based investigation using sandy soil. *Engineering Geology* **109**(3-4), 161–169.

Agliardi F, Crosta GB, Frattini P, and Malusá MG 2013 Giant non-catastrophic landslides and the long-term exhumation of the European Alps. *Earth and Planetary Science Letters* **365**(C), 263–274.

Allan JC, Stephenson WJ, Kirk RM, and Taylor A 2002 Lacustrine shore platforms at Lake Waikaremoana, North Island, New Zealand. *Earth Surface Processes and Landforms* **27**(2), 207–220.

Allen SK, Cox SC, and Owens IF 2010 Rock avalanches and other landslides in the central Southern Alps of New Zealand: a regional study considering possible climate change impacts. *Landslides* **8**(1), 33–48.

Archer D 2007 Methane hydrate stability and anthropogenic climate change. *Biogeosciences* **5**, 521–544.

Balescu S, Ritz JF, Lamothe M, et al. 2007 Luminescence dating of a gigantic palaeolandslide in the Gobi-Altay mountains, Mongolia. *Quaternary Geochronology* **2**(1-4), 290–295.

Barnes PM, Bostock HC, Neil HL, et al. 2013 A 2300-year paleoearthquake record of the southern Alpine Fault and Fiordland Subduction Zone, New Zealand, based on stacked turbidites. *Bulletin of the Seismological Society of America* **103**(4), 2424–2446.

Baum RL and Godt JW 2010 Early warning of rainfall-induced shallow landslides and debris flows in the USA. *Landslides* **7**(3), 259–272.

Beutner EC and Gerbi GP 2005 Catastrophic emplacement of the Heart Mountain block

slide, Wyoming and Montana, USA. *Geological Society of America Bulletin* **117**(5-6), 724–735.

Blong RJ and Goldsmith RCM 1993 Activity of the Yakatabari mudslide complex, Porgera, Papua New Guinea. *Engineering Geology* **35**, 1–17.

Bloomberg M and Davies TRH 2012. Do forests reduce *erosion*? The answer is not as simple as you may think. *New Zealand Journal of Forestry* **56**(4), 16–20.

Bommer JJ and Rodríguez CE 2001 Earthquake-induced landslides in Central America. *Engineering Geology* **63**(3), 189–220.

Borgatti L, Ravazzi C, Donegana M, et al. 2007 A lacustrine record of early Holocene watershed events and vegetation history, Corvara in Badia, Dolomites (Italy). *Journal of Quaternary Science* **22**(2), 173–189.

Brummer CJ and Montgomery DR 2006 Influence of coarse lag formation on the mechanics of sediment pulse dispersion in a mountain stream, Squire Creek, North Cascades, Washington, United States. *Water Resources Research* **42**(7), 1–16.

Bull WB and Brandon MT 1998 Lichen dating of earthquake-generated regional rockfall events, Southern Alps, New Zealand. *Geological Society of America Bulletin* **110**(1), 60–84.

Busche D 2001 Early Quaternary landslides of the Sahara and their significance for geomorphic and climatic history. *Journal of Arid Environments* **49**(3), 429–448.

Carrara A, Crosta G, and Frattini P 2008 Comparing models of debris-flow susceptibility in the alpine environment. *Geomorphology* **94**(3-4), 353–378.

Carretier S, Regard V, Vassallo R, et al. 2013 Slope and climate variability control of erosion in the Andes of central Chile. *Geology* **41**(2), 195–198.

Cauchon-Voyer G. 2011 Large-scale subaerial and submarine Holocene and recent mass movements in the Betsiamites area, Quebec, Canada. *Engineering Geology* **121**(1), 28–45.

Chevalier G, Davies T, and McSaveney M 2009 The prehistoric Mt Wilberg rock avalanche, Westland, New Zealand. *Landslides* **6**(3), 253–262.

Chigira M, Tsou CY, Matsushi Y, et al. 2013 Topographic precursors and geological structures of deep-seated catastrophic landslides caused by Typhoon Talas. *Geomorphology* **201**(C), 479–493.

Clague JJ, Turner RJW, and Reyes AV 2003 Record of recent river channel instability, Cheakamus Valley, British Columbia. *Geomorphology* **53**(3-4), 317–332.

Colesanti C and Wasowski J 2005 Investigating landslides with space-borne Synthetic Aperture Radar (SAR) interferometry. *Engineering Geology* **88**(3), 173–199.

Corominas J, van Westen C, Frattini P, et al. 2014 Recommendations for the quantitative analysis of landslide risk. *Bulletin of Engineering Geology and the Environment* **9**(3), 209–263.

Costa JE and Schuster RL 1988 The formation and failure of natural dams. *Geological Society of America Bulletin* **100**, 1054–1068.

Crosta GB, Imposimato S, and Roddeman D 2015 Landslide spreading, impulse water waves and modelling of the Vajont rockslide. *Rock Mechanics and Rock Engineering* **49**(6), 2413–2436.

Crozier MJ 2005 Multiple-occurrence regional landslide events in New Zealand: Hazard management issues. *Landslides* **2**(4), 247–256.

Crozier MJ 2010 Deciphering the effect of climate change on landslide activity: A review. *Geomorphology* **124**(3-4), 260–267.

Dahal RK and Hasegawa S 2008 Representative rainfall thresholds for landslides in the Nepal Himalaya. *Geomorphology* **100**(3-4), 429–443.

Dai F, Lee C, and Ngai YY 2002 Landslide risk assessment and management: an overview. *Engineering Geology* **64**, 65–87.

Davies TRH and Korup O 2007 Persistent alluvial fanhead trenching resulting from large, infrequent sediment inputs. *Earth Surface Processes and Landforms* **32**(5), 725–742.

de Haas T, Densmore AL, Stoffel M, et al. 2018 Avulsions and the spatio-temporal evolution of debris-flow fans. *Earth-Science Reviews* **177**, 53–75.

de Vente J and Poesen J 2005 Predicting soil erosion and sediment yield at the basin scale: Scale issues and semi-quantitative models. *Earth-Science Reviews* **71**(1-2), 95–125.

Decaulne A and Sæmundsson T 2007 Spatial and temporal diversity for debris-flow meteorological control in subarctic oceanic periglacial environments in Iceland. *Earth Surface Processes and Landforms* **32**(13), 1971–1983.

Derbyshire E 2000 Geological hazards in loess terrain, with particular reference to the loess regions of China. *Earth-Science Reviews* **54**(1), 231–260.

Dimakis P, Elverhoi A, Hoeg K, et al. 2000 Submarine slope stability on high-latitude glaciated Svalbard-Barents Sea margin. *Marine Geology* **162**, 303–316.

Dong JJ, Tung YH, Chen CC, et al. 2011 Logistic regression model for predicting the failure probability of a landslide dam. *Engineering Geology* **117**(1–2), 52–61.

Dotterweich M 2013 The history of human-induced soil erosion: Geomorphic legacies, early descriptions and research, and the development of soil conservation – A global synopsis. *Geomorphology* **201**(C), 1–34.

Dowling CA and Santi PM 2013 Debris flows and their toll on human life: a global analysis of debris-flow fatalities from 1950 to 2011. *Natural Hazards* **71**(1), 203–227.

Draganits E, Grasemann B, Janda C, et al. 2014 300 MW Baspa II – India's largest private hydroelectric facility on top of a rock avalanche-dammed palaeo-lake (NW Himalaya): Regional geology, tectonic setting and seismicity. *Engineering Geology* **169**(C), 14–29.

Ekstrom G and Stark CP 2013 Simple scaling of catastrophic landslide dynamics. *Science* **339**(6126), 1416–1419.

Ermini L and Casagli N 2003 Prediction of the behaviour of landslide dams using a geomorphological dimensionless index. *Earth Surface Processes and Landforms* **28**(1), 31–47.

Frattini P and Crosta GB 2013 The role of material properties and landscape morphology on landslide size distributions. *Earth and Planetary Science Letters* **361**(C), 310–319.

Frauenfelder R, Laustela M, and Kääb A 2005 Relative age dating of Alpine rockglacier surfaces. *Zeitschrift für Geomorphologie* **49**(2), 145–166.

Froude MJ and Petley D 2018 Global fatal landslide occurrence from 2004 to 2016. *Natural Hazards and Earth System Sciences* **18**, 2161–2181.

Gabet EJ, Burbank DW, Putkonen JK, et al. 2004 Rainfall thresholds for landsliding in the Himalayas of Nepal. *Geomorphology* **63**(3-4), 131–143.

Galli M, Ardizzone F, Cardinali M, et al. 2008 Comparing landslide inventory maps. *Geomorphology* **94**(3-4), 268–289.

Gallousi C and Koukouvelas IK 2007 Quantifying geomorphic evolution of earthquake-triggered landslides and their relation to active normal faults. An example from the Gulf of Corinth, Greece. *Tectonophysics* **440**, 85–104.

García-Ruiz JM, Nadal-Romero E, Lana-Renault N, and Beguería S 2013 Erosion in Mediterranean landscapes: Changes and future challenges. *Geomorphology* **198**(C), 20–36.

Garziglia S, Migeon S, Ducassou E, et al. 2008 Mass-transport deposits on the Rosetta province (NW Nile deep-sea turbidite system, Egyptian margin): Characteristics, distribution, and potential causal processes. *Marine Geology* **250**(3-4), 180–198.

Geersen J, Volker D, Behrmann JH, et al. 2011 Pleistocene giant slope failures offshore Arauco Peninsula, Southern Chile. *Journal of the Geological Society* **168**(6), 1237–1248.

Geertsema M, Clague JJ, Schwab JW, and Evans SG 2006 An overview of recent large catastrophic landslides in northern British Columbia, Canada. *Engineering Geology* **83**(1-3), 120–143.

Goldfinger C, Morey AE, Nelson CH, et al. 2007 Rupture lengths and temporal history of significant earthquakes on the offshore and north coast segments of the Northern San Andreas Fault based on turbidite stratigraphy. *Earth and Planetary Science Letters* **254**(1-2), 9–27.

Gregersen O 1981 The quick clay landslide in Rissa, Norway. *Norwegian Geotechnical Institute Publication* **135**, 1–6.

Guzzetti F, Peruccacci S, Rossi M, and Stark CP 2007 The rainfall intensity–duration control of shallow landslides and debris flows: an update. *Landslides* **5**(1), 3–17.

Guzzetti F, Mondini AC, Cardinali M, et al. 2012 Landslide inventory maps: New tools for an old problem. *Earth-Science Reviews* **112**(1-2), 42–66.

Hancox GT and Perrin ND 2009 Green Lake Landslide and other giant and very large postglacial landslides in Fiordland, New Zealand. *Quaternary Science Reviews* **28**(11-12), 1020–1036.

Handwerger AL, Roering JJ, and Schmidt DA 2013 Controls on the seasonal deformation of slow-moving landslides. *Earth and Planetary Science Letters* **377-378**(C), 239–247.

Hart MW, Shaller PJ, and Farrand GT 2012 When landslides are misinterpreted as faults: Case studies from the Western United States. *Environmental & Engineering Geoscience* **18**, 313–325.

Hermanns R, Trauth M, and Niedermann S 2000 Tephrochronologic constraints on temporal distribution of large landslides in northwest Argentina. *The Journal of Geology* **108**, 35–52.

Hewawasam T, Von Blanckenburg F, Schaller M, and Kubik P 2003 Increase of human over natural erosion rates in tropical highlands constrained by cosmogenic nuclides. *Geology* **31**(7), 597–600.

Hewitt K 1999 Quaternary moraines vs catastrophic rock avalanches in the Karakoram Himalaya, Northern Pakistan. *Quaternary Research* **51**, 220–237.

Hewitt K, Clague JJ, and Orwin JF 2008 Legacies of catastrophic rock slope failures in mountain landscapes. *Earth-Science Reviews* **87**(1-2), 1–38.

Hilley G, Bürgmann R, Ferretti A, et al. 2004 Dynamics of slow-moving landslides from permanent scatterer analysis. *Science* **304**, 1952–1954.

Hong Y, Adler R, and Huffman G 2006 Evaluation of the potential of NASA multi-satellite precipitation analysis in global landslide hazard assessment. *Geophysical Research Letters* **33**(22), 1–5.

Hubert-Ferrari A, Suppe J, Van der Woerd J, et al. 2005 Irregular earthquake cycle along the southern Tianshan front, Aksu area, China. *Journal of Geophysical Research: Solid Earth* **110**, 1–8.

Huggel C, Clague JJ, and Korup O 2012 Is climate change responsible for changing landslide activity in high mountains?. *Earth Surface Processes and Landforms* **37**(1), 77–91.

Hungr O, Leroueil S, and Picarelli L 2013 The Varnes classification of landslide types, an update. *Landslides* **11**(2), 167–194.

Ivy-Ochs S, von Poschinger A, Synal HA, and Maisch M 2009 Surface exposure dating of the Flims landslide, Graubünden, Switzerland. *Geomorphology* **103**(1), 104–112.

Jakob M and Lambert S 2009 Climate change effects on landslides along the southwest coast of British Columbia. *Geomorphology* **107**(3-4), 275–284.

Jakob M, Bovis M, and Oden M 2005 The significance of channel recharge rates for estimating debris-flow magnitude and frequency. *Earth Surface Processes and Landforms* **30**(6), 755–766.

Jamieson SSR, Ewertowski MW, and Evans DJA 2015 Rapid advance of two mountain glaciers in response to mine-related debris loading.

Journal of Geophysical Research: Earth Surface **120**, 1418–1435.

Jarman D, Agliardi F, and Crosta GB 2011 Megafans and outsize fans from catastrophic slope failures in Alpine glacial troughs: the Malser Haide and the Val Venosta cluster, Italy. *Geological Society, London, Special Publications* **351**(1), 253–277.

Kasai M, Marutani T, and Brierley GJ 2004 Patterns of sediment slug translation and dispersion following typhoon-induced disturbance, Oyabu Creek, Kyushu, Japan. *Earth Surface Processes and Landforms* **29**(1), 59–76.

Keiler M, Knight J, and Harrison S 2010 Climate change and geomorphological hazards in the eastern European Alps. *Philosophical Transactions of the Royal Society A: Mathematical, Physical and Engineering Sciences* **368**(1919), 2461–2479.

Kirchner JW, Finkel RC, Riebe CS, et al. 2001 Mountain erosion over 10 yr, 10 k.y., and 10 m.y. time scales. *Geology* **29**(7), 591–594.

Kirschbaum DB, Adler R, Hong Y, et al. 2009 A global landslide catalog for hazard applications: method, results, and limitations. *Natural Hazards* **52**(3), 561–575.

Korup O 2004 Geomorphic implications of fault zone weakening: slope instability along the Alpine Fault, South Westland to Fiordland. *New Zealand Journal of Geology and Geophysics* **47**, 257–267.

Korup O 2006 Rock-slope failure and the river long profile. *Geology* **34**(1), 45.

Korup O and Clague JJ 2009 Natural hazards, extreme events, and mountain topography. *Quaternary Science Reviews* **28**(11-12), 977–990.

Korup O and Stolle A 2014 Landslide prediction from machine learning. *GSA Today* **30**(1), 26–33.

Korup O, Gorum T, and Hayakawa Y 2012 Without power? Landslide inventories in the face of climate change. *Earth Surface Processes and Landforms* **37**(1), 92–99.

Korup O, Clague JJ, Hermanns RL, et al. 2007 Giant landsldies, topography, and erosion.

Earth and Planetary Science Letters **261**, 578–589.

Lal R 2003 Soil erosion and the global carbon budget. *Environment International* **29**(4), 437–450.

Lazzari M, Gioia D, Piccarreta M, et al. 2015 Sediment yield and erosion rate estimation in the mountain catchments of the Camastra artificial reservoir (Southern Italy): A comparison between different empirical methods. *Catena* **127**(C), 323–339.

Lee SH, Bahk JJ, Kim HJ, et al. 2009 Changes in the frequency, scale, and failing areas of latest Quaternary (¡29.4 cal. ka B.P.) slope failures along the SW Ulleung Basin, East Sea (Japan Sea), inferred from depositional characters of densely dated turbidite successions. *Geo-Marine Letters* **30**(2), 133–142.

López-Tarazón JA, Batalla RJ, Vericat D, and Francke T 2012 The sediment budget of a highly dynamic mesoscale catchment: The River Isábena. *Geomorphology* **138**(1), 15–28.

Machane D, Bouhadad Y, Cheikhlounis G, et al. 2008 Examples of geomorphologic and geological hazards in Algeria. *Natural Hazards* **45**(2), 295–308.

Malamud BD, Turcotte DL, Guzzetti F, and Reichenbach P 2004 Landslide inventories and their statistical properties. *Earth Surface Processes and Landforms* **29**(6), 687–711.

Manville V, Major JJ, and Fagents SA 2013 Modeling lahar behaviour and hazards. In *Modeling Volcanic Processes* (eds Fagents S, Gregg TKP, and Lopes RMC), Cambridge University Press, pp. 300–330.

Maslin M, Owen M, Betts R, et al. 2010 Gas hydrates: past and future geohazard?. *Philosophical Transactions of the Royal Society A: Mathematical, Physical and Engineering Sciences* **368**(1919), 2369–2393.

Maslin M, Vilela C, Mikkelsen N, and Grootes P 2005 Causes of catastrophic sediment failures of the Amazon Fan. *Quaternary Science Reviews* **24**(20-21), 2180–2193.

Massey CI, Petley DN, McSaveney MJ, and Archibald G 2016 Basal sliding and plastic deformation of a slow, reactivated landslide in New Zealand. *Engineering Geology* **208**, 11–28.

Mateos RM, García-Moreno I, and Azañón JM 2012 Freeze–thaw cycles and rainfall as triggering factors of mass movements in a warm Mediterranean region: the case of the Tramuntana Range (Majorca, Spain). *Landslides* **9**(3), 417–432.

Mather AE, Hartley AJ, and Griffiths JS 2014 The giant coastal landslides of Northern Chile: Tectonic and climate interactions on a classic convergent plate margin. *Earth and Planetary Science Letters* **388**(C), 249–256.

McColl ST and Davies TR 2011 Evidence for a rock-avalanche origin for 'The Hillocks' "moraine", Otago, New Zealand. *Geomorphology* **127**(3-4), 216–224.

McColl ST, Davies TRH, and McSaveney MJ 2012 The effect of glaciation on the intensity of seismic ground motion. *Earth Surface Processes and Landforms* **37**, 1290–1301.

Menounos B, Clague JJ, Clarke GKC, et al. 2013 Did rock avalanche deposits modulate the late Holocene advance of Tiedemann Glacier, southern Coast Mountains, British Columbia, Canada?. *Earth and Planetary Science Letters* **384**(C), 154–164.

Micu M and Bălteanu D 2013 A deep-seated landslide dam in the Siriu Reservoir (Curvature Carpathians, Romania). *Landslides* **10**(3), 323–329.

Mikoš M, Fazarinc R, and Ribičič M 2006 Sediment production and delivery from recent large landslides and earthquake-induced rock falls in the Upper Soča River Valley, Slovenia. *Engineering Geology* **86**(2-3), 198–210.

Moretti L, Allstadt K, Mangeney A, et al. 2015 Numerical modeling of the Mount Meager landslide constrained by its force history derived from seismic data. *Journal of Geophysical Research Solid Earth* **120**, 2579–2599.

Nicholas AP and Quine TA 2007 Modeling alluvial landform change in the absence of external environmental forcing. *Geology* **35**(6), 527–530.

Oehler JF, Lénat JF, and Labazuy P 2009 Successive Holocene rock avalanches at Lake Coleridge, Canterbury, New Zealand. *Landslides* **70**(4), 287–297.

Orwin JF 1998 The application and implications of rock weathering-rind dating to a large rock avalanche, Craigieburn Range, Canterbury, New Zealand. *New Zealand Journal of Geology and Geophysics* **41**(3), 219–223.

Osmundsen PT, Henderson I, Lauknes TR, et al. 2009 Active normal fault control on landscape and rock-slope failure in northern Norway. *Geology* **37**(2), 135–138.

Paguican EMR, Van Wyk de Vries B, and Lagmay AMF 2012 Hummocks: how they form and how they evolve in rockslide-debris avalanches. *Landslides* **11**(1), 67–80.

Pánek T 2015 Recent progress in landslide dating. *Progress in Physical Geography* **39**(2), 168–198.

Pánek T, Hradecký J, Smolková V, and šilhán K 2008 Gigantic low-gradient landslides in the northern periphery of the Crimean Mountains (Ukraine). *Geomorphology* **95**(3-4), 449–473.

Pelletier JD 2004 The influence of piedmont deposition on the time scale of mountain-belt denudation. *Geophysical Research Letters* **31**(15), 1–4.

Petley D 2012 Global patterns of loss of life from landslides. *Geology* **40**(10), 927–930.

Poesen J, Nachtergaele J, Verstraeten G, and Valentin C 2003 Gully erosion and environmental change: importance and research needs. *Catena* **50**, 91–133.

Prager C, Zangerl C, Patzelt G, and Brandner R 2008 Age distribution of fossil landslides in the Tyrol (Austria) and its surrounding areas. *Natural Hazards and Earth System Sciences* **8**, 377–407.

Reid L 1998 Calculation of average landslide frequency using climatic records. *Water Resources Research* **34**(4), 869–877.

Reznichenko NV, Davies TRH, Shulmeister J, and Larsen SH 2012 A new technique for identifying rock avalanche-sourced sediment in moraines and some paleoclimatic implications. *Geology* **40**(4), 319–322.

Reznichenko N, Davies T, Shulmeister J, and McSaveney M 2010 Effects of debris on ice-surface melting rates: an experimental study. *Journal of Glaciology* **56**(197), 384–394.

Reznichenko NV, Davies TR, and Winkler S 2016 Revised palaeoclimatic significance of Mueller Glacier moraines, Southern Alps, New Zealand. *Earth Surface Processes and Landforms* **41**(2), 196–207.

Robinson TR, Davies TRH, Reznichenko NV, and De Pascale GP 2014 The extremely long-runout Komansu rock avalanche in the Trans Alai range, Pamir Mountains, southern Kyrgyzstan. *Landslides* **12**(3), 523–535.

Roering JJ, Stimely LL, Mackey BH, and Schmidt DA 2009 Using DInSAR, airborne LiDAR, and archival air photos to quantify landsliding and sediment transport. *Geophysical Research Letters* **36**(19), 1–5.

Rothwell RG, Reeder MS, Anastasakis G, et al. 2000 Low sea-level stand emplacement of megaturbidites in the western and eastern Mediterranean Sea. *Sedimentary Geology* **135**, 75–88.

Safran EB, Anderson SW, Mills-Novoa M, et al. 2011 Controls on large landslide distribution and implications for the geomorphic evolution of the southern interior Columbia River basin. *Geological Society of America Bulletin* **123**(9-10), 1851–1862.

Saito H, Korup O, Uchida T, et al. 2014 Rainfall conditions, typhoon frequency, and contemporary landslide erosion in Japan. *Geology* **42**(11), 999–1002.

Samper A, Quidelleur X, Boudon G, et al. 2007 Radiometric dating of three large volume flank collapses in the Lesser Antilles Arc. *Journal of Volcanology and Geothermal Research* **176**(4), 485–492.

Sawyer DE, Reece RS, Gulick SPS, and Lenz BL 2017 Submarine landslide and tsunami hazards offshore southern Alaska: Seismic strengthening versus rapid sedimentation. *Geophysical Research Letters* **44**(16), 8435–8442.

Scheingross JS, Minchew BM, Mackey B, et al. 2013 Fault-zone controls on the spatial distribution of slow-moving landslides. *Geological Society of America Bulletin* **125**(3-4), 473–489.

Schmidt AH, Montgomery DR, Huntington KW, and Liang C 2011 The question of communist land degradation: New evidence from local erosion and basin-wide sediment yield in southwest China and southeast Tibet. *Annals of the Association of American Geographers* **101**(3), 477–496.

Shakesby RA, Matthews JA, and Owen G 2006 The Schmidt hammer as a relative-age dating tool and its potential for calibrated-age dating in Holocene glaciated environments. *Quaternary Science Reviews* **25**(21–22), 2846–2867.

Shoaei Z 2014 Mechanism of the giant Seimareh Landslide, Iran, and the longevity of its landslide dams. *Environmental Earth Sciences* **72**(7), 2411–2422.

Shugar DH, Rabus BT, Clague JJ, and Capps DM 2012 The response of Black Rapids Glacier, Alaska, to the Denali earthquake rock avalanches. *Journal of Geophysical Research: Earth Surface (2003–2012)* **117**, 1–14.

Sidle RC and Ochiai H 2006 *Landslides. Processes, Prediction, and Land Use*, Water Resources Monograph 18. American Geophysical Union.

šilhán K, Pánek T, and Hradecký J 2012 Tree-ring analysis in the reconstruction of slope instabilities associated with earthquakes and precipitation (the Crimean Mountains, Ukraine). *Geomorphology* **173-174**(C), 174–184.

Silva C, Araújo E, Reis A, et al. 2010 Megaslides in the Foz do Amazonas Basin, Brazilian Equatorial Margin. In *Submarine Mass Movements and Their Consequences* (eds Mosher DC, Shipp RC, Moscardelli M, et al.), Springer, pp. 581–591.

Sponholz B 1994 Silicate karst associated with lateritic formations (examples from eastern Niger). *Catena* **21**(2), 269–278.

Talling PJ, Paull CK, and Piper D 2013 How are subaqueous sediment density flows triggered, what is their internal structure and how does it evolve? Direct observations from monitoring of active flows. *Earth-Science Reviews* **125**, 244–287.

Talling PJ, Wynn RB, Masson DG, et al. 2007 Onset of submarine debris flow deposition far from original giant landslide. *Nature* **450**(7169), 541–544.

Tappin DR 2010 Submarine mass failures as tsunami sources: their climate control. *Philosophical Transactions of the Royal Society A: Mathematical, Physical and Engineering Sciences* **368**(1919), 2417–2434.

Tappin DR, Grilli ST, Harris JC, et al. 2014 Did a submarine landslide contribute to the 2011 Tohoku tsunami?. *Marine Geology* **357**(C), 344–361.

Taylor J, Dowdeswell JA, and Siegert MJ 2001 Late Weichselian depositional processes, fluxes, and sediment volumes on the margins of the Norwegian Sea (62–75 °N). *Marine Geology* **188**(1), 61–77.

ten Brink US, Barkan R, Andrews BD, and Chaytor JD 2009 Size distributions and failure initiation of submarine and subaerial landslides. *Earth and Planetary Science Letters* **287**(1-2), 31–42.

Tovar SD, Shulmeister J, and Davies TR 2008 Evidence for a landslide origin of New Zealand's Waiho Loop moraine. *Nature Geoscience* **1**(8), 524–526.

Trauth MH, Alonso RA, Haselton KR, et al. 2000 Climate change and mass movements in the NW Argentine Andes. *Earth and Planetary Science Letters* **179**(2), 243–256.

Tsuchiya S, Sasahara K, Shuin S, and Ozono S 2009 The large-scale landslide on the flank of caldera in South Sulawesi, Indonesia. *Landslides* **6**(1), 83–88.

Twichell DC, Chaytor JD, ten Brink US, and Buczkowski B 2008 Morphology of late Quaternary submarine landslides along the U.S. Atlantic continental margin. *Marine Geology* **264**(1), 4–15.

Urgeles R and Camerlenghi A 2013 Submarine landslides of the Mediterranean Sea: Trigger mechanisms, dynamics, and frequency-magnitude distribution. *Journal of Geophysical Research: Earth Surface* **118**(4), 2600–2618.

Vacco DA, Alley RB, and Pollard D 2010 Glacial advance and stagnation caused by rock avalanches. *Earth and Planetary Science Letters* **294**(1-2), 123–130.

Van Den Eeckhaut M, Poesen J, Verstraeten G, et al. 2005 The effectiveness of hillshade maps and expert knowledge in mapping old deep-seated landslides. *Geomorphology* **67**(3-4), 351–363.

Van Westen CJ, Asch TWJ, and Soeters R 2005 Landslide hazard and risk zonation – why is it still so difficult?. *Bulletin of Engineering Geology and the Environment* **65**(2), 167–184.

Vanacker V, von Blanckenburg F, Govers G, et al. 2007 Restoring dense vegetation can slow mountain erosion to near natural benchmark levels. *Geology* **35**(4),303.

Vigiak O, Borselli L, Newham LTH, et al. 2012 Comparison of conceptual landscape metrics to define hillslope-scale sediment delivery ratio. *Geomorphology* **138**(1), 74–88.

Vogt PR and Jung WY 2002 Holocene mass wasting on upper nonPolar continental slopes – due to postGlacial ocean warming and hydrate dissociation?. *Geophysical Research Letters* **29**(9), 1–4.

Völker D, Scholz F, and Geersen J 2011 Analysis of submarine landsliding in the rupture area of the 27 February 2010 Maule earthquake, Central Chile. *Marine Geology* **288**(1-4), 79–89.

Walter M and Schwaderer U 2012 Seismic monitoring of precursory fracture signals from a destructive rockfall in the Vorarlberg Alps, Austria. *Natural Hazards and Earth System Sciences* **12**, 3545–3555.

Wang P, Zhang B, Qiu W, and Wang J 2011 Soft-sediment deformation structures from the Diexi paleo-dammed lakes in the upper reaches of the Minjiang River, east Tibet. *Journal of Asian Earth Sciences* **40**(4), 865–872.

Wasowski J and Bovenga F 2015 Remote sensing of landslide motion with emphasis on satellite multitemporal interferometry applications: An overview. In *Landslide Hazards, Risks and Disasters* (eds Shroder JF and Davies TR). Elsevier, pp. 345–403.

Weaver PPE 2003 Northwest African continental margin: History of sediment accumulation, landslide deposits, and hiatuses as revealed by drilling the Madeira Abyssal Plain. *Paleoceanography* **18**(1), 1–13.

Weidinger JT and Korup O 2009 Frictionite as evidence for a large Late Quaternary rockslide near Kanchenjunga, Sikkim Himalayas, India – Implications for extreme events in mountain relief destruction. *Geomorphology* **103**(1), 57–65.

Weidinger JT, Korup O, Munack H, et al. 2014 Giant rockslides from the inside. *Earth and Planetary Science Letters* **389**, 62–73.

Wood JR, Wilmshurst JM, and Rawlence NJ 2011 Radiocarbon-dated faunal remains correlate very large rock avalanche deposit with prehistoric Alpine Fault rupture. *New Zealand Journal of Geology and Geophysics* **54**(4), 431–434.

Yamagishi H and Iwahashi J 2007 Comparison between the two triggered landslides in Mid-Niigata, Japan by July 13 heavy rainfall and October 23 intensive earthquakes in 2004. *Landslides* **4**(4), 389–397.

Yu B, Ma Y, and Wu Y 2012 Case study of a giant debris flow in the Wenjia Gully, Sichuan Province, China. *Natural Hazards* **65**(1), 835–849.

Zhang D and Wang G 2006 Study of the 1920 Haiyuan earthquake-induced landslides in loess (China). *Engineering Geology* **94**(1), 76–88.

9

Tsunami Hazards

'Tsunami' is a Japanese word meaning 'harbour wave'. This word entered the common lexicon following the catastrophic tsunami in the Indian Ocean that killed more than 250 000 people on 26 December 2004. Since this catastrophe, scientific attention to the phenomenon has skyrocketed, bolstering appraisals of tsunami hazard and risk (Løvholt et al. 2012). The tsunami triggered by the 2011 Great Tohoku earthquake, Japan, further spurred these research efforts and also prompted refinements to tsunami risk projections for Japan's coastlines (Cyranoski 2012). Tsunamis are trains of transient impulse waves in oceans or lakes that result from the sudden displacement of the water column by submarine earthquakes, landslides, volcanic eruptions, or meteorite impacts. The process is never a stand-alone one; tsunamis are always triggered by another process. Meteotsunamis are oceanic waves that form because of rapidly propagating changes in air pressure. They are surface waves, whereas 'real' tsunamis are body waves and thus are fundamentally different. They are rarely as destructive as their geologically triggered counterparts, but may cause water-level fluctuations in harbours of up to several metres (Renault et al. 2011).

9.1 Frequency and Magnitude of Tsunamis

For most natural hazards, the amount of detail available for documented events decreases rapidly as we go back in time, and this holds particularly for tsunamis. One of the largest databases on past tsunamis is maintained by the NOAA National Center for Environmental Information and features more than 26 000 run-up measurements that were recorded for nearly 2600 tsunamis throughout the world, ranging from Paatuut, Greenland, to Scott Base, Antarctica, and going back some 4000 years (ngdc.noaa.gov/hazard/hazards.shtml,). Yet half of these records relate to tsunamis that occurred after 1996, and 24% of all entries (more than 6000) were on the 2011 Tohoku tsunami. Japan is prominently represented in this database, featuring 44% of all data locations, closely followed by the United States, Chile, and Indonesia. Hence the censoring of older events is accompanied by a strong geographic bias of past tsunamis.

Despite this bias, the data allow some conservative estimates and demonstrate that very high local run-ups have occurred: at least six past tsunamis resulted in local run-ups of more than 100 m above sea level. The highest tsunami run-up in historic time happened in

Geomorphology and Natural Hazards: Understanding Landscape Change for Disaster Mitigation, Advanced Textbook Series, First Edition. Tim R. Davies, Oliver Korup, and John J. Clague.
© 2021 John Wiley & Sons Ltd. Co-published 2021 by the American Geophysical Union and John Wiley & Sons Ltd.

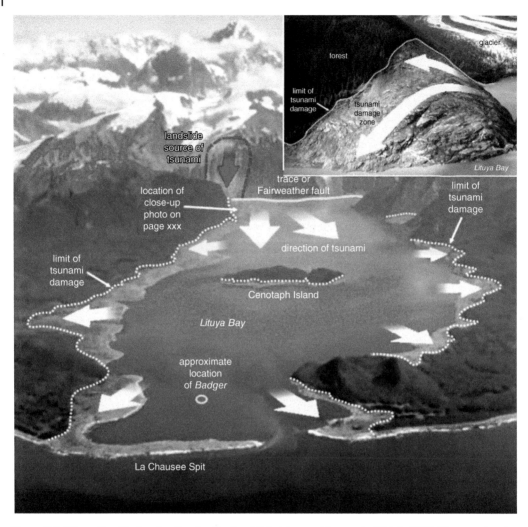

Figure 9.1 Lituya Bay, Alaska, shortly after the landslide-triggered tsunami of 7 July 1958. The prominent trimline delineates the upper limit of the tsunami. Forest below this line was obliterated by the surging waters. The upper photograph shows the limit of removal of forest (ca. 565 m above sea level) on the bedrock slope opposite the source of the landslide. The lower photograph shows the location of a fishing boat (*Badger*) that was anchored at the head of the bay when the landslide happened. The boat was ripped from its anchor and carried over La Chausee Spit into the open Pacific Ocean. The husband and wife on board the boat were uninjured and rescued shortly after the event. From Clague et al. (2007).

1958 in Lituya Bay, Alaska (Figure 9.1), when a large rockslide entered an inlet and displaced water that rose up to about 525 m above sea level on the opposite mountain ridge, stripping off mature conifer forest (Fritz et al. 2009). Two other tsunamis with run-ups >100 m happened earlier in Lituya Bay, in 1853 and 1936. Similarly, a large landslide dropped without any observable trigger into Taan Fjord,

Alaska, in 2015, displacing water by up to 190 m above sea level. The data set further reveals that tsunamis with run-ups higher than 50 m have occurred, on average, at least once in every 20 years since the late seventeenth century. The sources of these tsunamis were nearby, meaning less than 500 km away, and prominent examples include the 2004 Indian Ocean and the 2011 Tohoku tsunamis.

Local run-up decreases as the distance to the tsunami source increases, as we would expect from attenuating waves, although tsunamis that crossed the Pacific, for example, were still capable of generating local waves that were several metres high. While rocky cliffs increase local run-ups, gently sloping sedimentary coasts and estuaries allow the water masses to invade deeply inland. At least 31 of the documented tsunamis inundated places more than 1 km inland from the shoreline, and four of them have flooded locations more than 5 km inland. Keeping in mind the caveats about the completeness of information about past tsunamis, any global damage assessment is likely to be an underestimate: reports account for at least 860 000 fatalities, with more of a quarter of those attributable to the 2004 Indian Ocean tsunami. A global study that combined inundation models with estimates of average population density projected that about 19 million people in 76 countries were exposed to tsunamis in 2007 (Løvholt et al. 2012). The estimated return period of this inundation level is roughly 475 years, following the convention for global earthquake return periods in the Global Seismic Hazard Assessment program. These numbers imply that roughly 40 000 people, with the majority living in Indonesia and Japan, would be hit by tsunamis worldwide each year, if all susceptible areas were inundated with equal probability. This estimate must be treated with care, and may seem unduly high compared to the average mortality rate of ~2600 people per year since the beginning of the twentieth century. The projected annual monetary risk is of the order of $US 392 million worldwide, and currently mostly aggregated in the tsunami-prone coastal areas of Japan (Løvholt et al. 2012).

What matters most for such projections are reliable estimates of the timing and magnitude of run-up and inundation areas for a given coastal spot. Average return periods for very large and destructive tsunami such as those in the Indian Ocean in 2004 or off the Pacific Coast of Japan in 2011 are beyond the length of detailed instrumented records, and likely of the order of several hundred years at least. More detailed time series of tsunami occurrences at a given location have led some researchers to propose that run-up heights approximately follow an inverse power-law distribution with an upper size limit (Burroughs and Tebbens 2005). This model may also hold for the waiting time between two successive tsunami arrivals, which is a key metric for hazard assessments (Løvholt et al. 2012), though rarely known reliably for larger tracts of coastline. Even local studies that can afford detailed information about wave regimes, coastline topography, and nearshore bathymetry may eventually suffer from poor constraints about the return periods of the mechanisms that trigger tsunamis, particularly for rare events. Short historic records with few data leave scientists several choices of appropriate distributions of run-up frequency and magnitude. Geist and Parsons (2008) argued that only part of the global and local data could be explained by a Poisson process (see Chapter 5.4.4), for which interevent times are exponentially distributed, independent of each other, and also independent of the time when measurements start. This distribution is characteristic of earthquakes as the most common triggers, though Geist and Parsons (2008) found for a global record of tsunamigenic earthquakes that shorter interevent times were more abundant than this model predicted, thus revealing some clustering in time. The researchers thus envisaged a mixture of two components in tsunami interevent times, interpreted as both a long-term clustering of earthquakes in time and a decay associated with earthquake aftershocks. This interpretation assumes that other tsunami triggers such as landslides or volcanic eruptions are negligible, even when they occur together with earthquakes. Indeed, the global NOAA database records earthquakes as exclusive triggers in 72% of all cases, whereas volcanic eruptions are listed for only 5%.

Nonetheless, tsunami hazard assessments face the problem of accumulated uncertainty;

by design they must include the summed errors relating to modelling both triggering process and tsunami propagation. In estimating the tsunami hazard from earthquake-triggered submarine landslides along the eastern seaboard of the United States, for example, ten Brink et al. (2009) noted the high uncertainties regarding the choice of the spectral acceleration period, amplifying effects of soft sediments, sediment thickness and liquefaction potential, or earthquake return periods, let alone the use of empirical relationships between seismic shaking and slope instability. Some of these uncertainties can be explored using Monte Carlo simulations of these parameters, assuming that their range of values can be constrained plausibly. Grilli et al. (2009) used this strategy to estimate that, for the upper east coast of the United States, the run-up hazard of tsunamis triggered by submarine landslides with return periods of between 100 and 500 years was below that of 100-year storm surges. Again, the authors emphasized that we should view such rough estimates from the perspective of orders of magnitude. Running many different models with probabilistic inputs and using their joint outcomes for a given region could be one way forward for regional tsunami hazard appraisals.

9.2 Geomorphic Impacts of Tsunamis

9.2.1 Tsunamis in the Hazard Cascade

Two key characteristics of tsunamis are their long wavelengths (10^4–10^5 m) and small wave heights or amplitudes (10^{-1}–10^0 m) in deep open water bodies. The average speed C [m s^{-1}] of the advancing wave, also known as wave celerity, in the open ocean can be expressed in a simplified form as a function of gravitational acceleration g [m s^{-2}], wave period T [s], and wavelength λ [m]:

$$C_{\text{deep}} = \frac{(g\lambda)^{0.5}}{(2\pi)} \tag{9.1}$$

Equation (9.1) applies to deep-water waves with wavelengths that are much smaller than water depth d [m]. Tsunami wavelengths are several hundred kilometres, thus shallow-water wave equations are commonly used to calculate the speed of the wave (Conley, 2014):

$$C_{\text{shallow}} = \sqrt{gd} \tag{9.2}$$

Equation (9.2) predicts speeds of >210 m s^{-1} (or >750 km h^{-1}) for tsunami waves in water depths of 4500 m, which are representative of the open ocean. Tsunamis thus cross oceans at the speed of jet planes but are hardly noticed by ships because of the low heights and long periods of the waves. However, when the waves move into shallower water they slow to 10–20 m s^{-1} due to friction of the sea bed and increasing wave height, producing much higher waves and turbulent surges of water that, in extreme cases, can run up to elevations of several tens of metres above sea level. These coastal waves create sufficient basal shear stress to entrain marine sediments up to boulder size from depths of 20–30 m below sea level.

Different triggers produce different types of tsunami waveforms: earthquakes produce fast (10^1–10^2 m s^{-1}), low-frequency (10^{-4}–10^{-3} s^{-1}) waves that propagate over long distances without much attenuation. The directionality of these waves is determined by the orientation of the rupturing sea-floor fault. Seismically triggered tsunami wave fronts are thus roughly linear or beam-shaped and can be hundreds of kilometres long. As a consequence the run-up along coasts can be similarly widespread. In contrast, tsunamis triggered by landslides and pyroclastic flows have point sources and their wave fronts are radial, and they attenuate more rapidly. Nevertheless, landslide tsunamis were responsible for the highest recorded run-up heights in history. The wave heights at the source exceed the average thickness of the failed mass when failure is rapid, but attenuate fast as they travel away from the source. Numerical experiments reveal that the amplitudes of landslide tsunamis are roughly proportional to the volume of the water-displacing landslide mass, and that landslides originating onshore before entering the sea or a lake generally have a higher tsunamogenic potential than submarine or lake-floor landslides (Watts, 2004). How efficiently a landslide can generate a tsunami wave also depends on a special type of the Froude number Fr, which relates the maximum landslide speed to that of the shallow water wave Eq (9.2). Terrestrial landslides that enter a large water body produce the largest tsunamis when $Fr > 1$, whereas submarine landslides require $Fr \sim 1$ (Watt et al. 2012). Catastrophic landslides that dislodge entire flanks of volcanic islands, for example, often involve both terrestrial and submarine phases, and thus require more elaborate assessments of their tsunami hazard (Table 5.1). Overall, NOAA's tsunami database attributes only some 3% of all recorded events to landslide triggers, although this fraction may vary locally by an order of magnitude (Hornbach et al. 2010). Nonetheless, several hundreds of large submarine landslide deposits are now documented on the world's ocean floors, but the number of known associated tsunami deposits is surprisingly miniscule (Tappin, 2010).

Based on computer simulations of tsunami run-up in shallow coastal waters, also known as the near field, Okal and Synolakis (2004) proposed that landslide-triggered tsunamis had characteristic aspect ratios of maximum amplitude to lateral extent along the coast of $>10^{-4}$, whereas the ratios for earthquake-triggered tsunamis remained below this value. These characteristics may yield important clues when reconstructing the triggers of tsunamis from their deposits in coastal areas, particularly where earthquakes, volcanic eruptions, and landslides might all be candidate triggers.

Tsunamis triggered by large meteorite impacts can be larger than the largest tsunamis produced by other processes. Those that penetrate to the deep ocean floor have wave heights that approach the depth of the water. However, such impacts create wave trains with many periods and wavelengths that attenuate much more rapidly than earthquake-generated tsunamis (Figure 9.2) (Dawson and Stewart, 2007); fortunately, they are very rare.

9.2.2 The Role of Coastal Geomorphology

Nearshore bathymetry is another important control on the height and run-up of tsunami waves. An example is harbour resonance, where amplification of tsunami waves in bays, harbours, and other coastal embayments can be estimated using the following equation:

$$T_* = \frac{4l}{\sqrt{gd_m}} \qquad (9.3)$$

where T_* is harbour resonance [s], l is the length of a rectangular harbour [m], g is gravitational acceleration [m s^{-2}], and d_m is the water depth at the mouth of the bay [m]. Equation (9.3) will need to be adjusted for different types of coastlines. A rule of thumb is that the shape of a coastline will amplify waves if T_* exceeds the wave period of the incoming tsunami. Equation (9.3) predicts that shallow river mouths and estuaries are particularly prone to higher wave heights and are thus more vulnerable to tsunamis than steeper or rocky coasts. The equation also implies that harbour resonance is more sensitive to water depth than to harbour length. For example, in Banda Aceh, Indonesia, the region most severely hit by the 2004 Indian Ocean tsunami, the earthquake-triggered water waves invaded several river mouths by more than 8 km (Umitsu et al. 2007). Similarly, the local coastal geomorphology played a large role in explaining much of the run-up, inundation and damage pattern caused by the tsunami of the 2011 Tohoku earthquake. Post-disaster field surveys were unprecedented in detail and the most extensive ever undertaken following a tsunami, reporting local run-up of more than 40 m especially in narrow bays

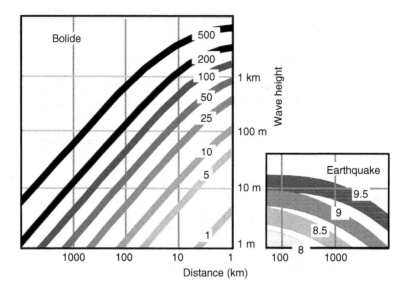

Figure 9.2 Computed peak open-ocean tsunami height versus distance from source for bolide impacts (left) and earthquakes (right). Bolide impact attenuation curves are based on radii between 1 and 500 m. Earthquake attenuation curves are for magnitudes 8.0–9.5, with grey bands including an allowance for anomalous events. In both plots the ocean depth is 4000 m and shoaling amplification factors are excluded. Impact-derived waves may be as high as the ocean is deep at the impact point, but the resulting complex tsunami-wave trains attenuate rapidly. Earthquake-induced tsunamis may be less high but suffer less attenuation. From Dawson and Stewart (2007).

Table 9.1 Selected coastal geomorphic impacts of tsunamis.

Location	Year	Net erosion or deposition ($m^3\ m^{-1}$)	Material	Trigger	Reference
Papua New Guinea	1998	−14 to +36	Beach sand	EQ	Gelfenbaum and Jaffe (2003)
Lhok Nga, Indonesia	2004	−16 to +160	Beach sand	EQ	Paris et al. (2010)
Kuril Is., Russia	2006	−200 to +6.3*	Beach sand, gravel	EQ	MacInnes et al. (2009)

Net erosion and deposition are averaged per metre of coastline and given in negative and positive values, respectively; *per metre of surveyed profile length; EQ = earthquake. Compare these values with those in Table 10.1.

and drowned river valleys that funnelled the approaching waves (Mori et al. 2011). Terrestrial and airborne laser-scanning surveys in a small valley of northeastern Japan's ria (drowned valley) coastline demonstrated that the 2011 Tohoku tsunami scraped all soil cover from the bedrock valley sidewalls, also removing several millimetres to decimetres of bedrock in places; a pronounced knickpoint in the longitudinal profile of the small stream

may record the cumulated erosive action of multiple tsunamis in the past (Hayakawa et al. 2015). Despite these and many other documented local geomorphic impacts, (Table 9.1) (Figure 9.3), Kench et al. (2006) suggested that the long-term net effect of tsunamis may be overrated. In their study of coastline changes in the Maldives following the 2004 Indian Ocean tsunami, they documented local erosion of exposed island scarps by several metres,

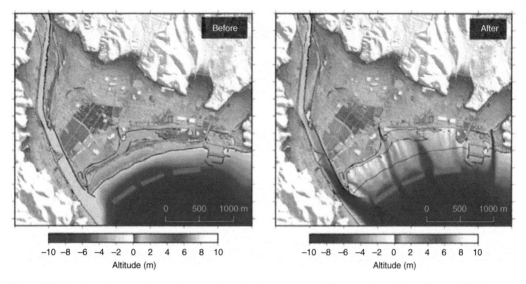

Figure 9.3 Results of tsunami shoreline erosion for Rikuzentakata City and Hirota Bay after the 2011 Tohoku tsunami, Japan. Maps show the topography and bathymetry before and after the tsunami. From Yamashita et al. (2016).

but also underlined similar amounts of net deposition, so that the overall island shape and area remained largely unchanged. They noted that, in the Maldives at least, the seasonal turnover of sediment during monsoon storms was higher than that during much rarer tsunamis such as the one in 2004, concluding that 'far-field tsunamis are unlikely to be important mechanisms of atoll island destruction or formation and that any permanent addition of tsunami sediment to island surfaces is unlikely to be distinguishable in the stratigraphic record' (Figure 9.4).

Tsunamis have sufficient momentum that, upon reaching the shoreline, they can entrain particles ranging up to boulder size and may deposit sand containing marine fossils and debris far inland. Once the waves reach their upper limit, they recede as sediment-laden backwash flows. These flows may be funnelled along topographic lows and can enter submarine canyons and generate turbidity currents that transport fine material offshore into deep-sea basins. Studies that measure in detail the erosional power and geomorphic work of tsunamis are surprisingly few among

the plethora of scientific reports dealing with tsunamis in the past decade. Gelfenbaum and Jaffe (2003) reported that the 1998 Papua New Guinea tsunami removed between 50 and 150 m of beaches, berms, and soils, equal to an average of about 14 m³ m⁻¹ (per meter of coastline) of sediment. In places, the tsunami also dumped a sheet of sand that was 8 cm thick on average, with a specific volume of up to 36 m³ m⁻¹. About two-thirds of this sediment came from offshore sources. By chance, MacInnes et al. (2009) surveyed in detail a sandy coastline shortly before and after the 2006 Kuriles tsunami, adding much needed quantitative detail to the many mostly qualitative observations elsewhere. They found that erosion of onshore material was greater than sedimentation along the measured sections. The tsunami also scoured most of this material where it ran up highest, especially where more than 15 m; erosion was most likely aided by a steep topography favouring powerful backwash flows. They also noted that the landward limit of the tsunami deposit coincided closely with the limit of inundation and run-up. Local sediment budgets, however, revealed

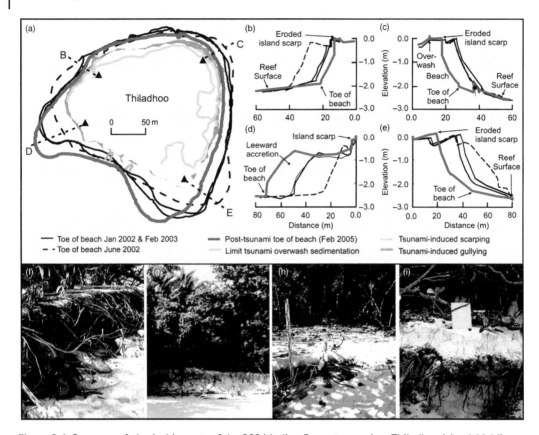

Figure 9.4 Summary of physical impacts of the 2004 Indian Ocean tsunami on Thiladhoo Island, Maldives. (a) Thiladhoo Island showing pre- and post-tsunami vegetated shoreline and toe of beach positions. Note erosion of the vegetated shoreline along the northwest and northeast corners of the island and extension of the shoreline due to deposition on the southwest. (b)–(e) Island-to-reef topographic surveys showing landward movement of vegetated scarp (b, c, e) and significant deposition along the southwest shore (d). (f) Tsunami-induced scarping. (g) Example of pre-tsunami scarping. (h) Overwash sand sheet extending 15 m landward and covering old eroded scarp. (i) Overwash sand sheet (0.2 m thick) exposed at island edge overlying older soil (dark), which has undergone post-event scarping. From Kench et al. (2006).

little about flow roughness, vegetation cover, or distance from the shore. Numerical simulations of erosion and sediment transport during tsunamis backed by detailed field measurements confirm that the first wave, and especially its associated return flow, can have high erosional power: the 2011 Tohoku tsunami on Rikuzentakata City and Hirota Bay, Japan, locally eroded and deposited sediment stacks up to several metres thick, causing a net loss of some 2 Mm^3 (Yamashita et al. 2016; Figure 9.3).

More studies of this kind would provide vital data for managing shorelines that are vulnerable to tsunamis, and establish net material gains or losses, where this material is coming from, and where it is deposited eventually (Figure 9.5). Better understanding of changes in sediment concentration is crucial in this regard. In contrast to run-up waves, backwash and offshore flows of receding tsunami waves are aided by gravity and the local slope, so that they may pick up more sediment. The lower reaches of small coastal rivers might channelize such backwash flows and focus erosion (Umitsu et al. 2007).

Theory predicts that the sediment budget of tsunami-disturbed beaches will gradually

Figure 9.5 Erosion of dunes at Lampuuk, West Banda Ache, Sumatra, by the 2004 Indian Ocean tsunami. The secondary scarp on the flank of the dune, below the maximum inundation height and pre-tsunami surface, probably reflects the erosion by the backwash. From Paris et al. (2009).

regain a new equilibrium according to the principles of sediment continuity. How long it takes for a given size and type of coast to recover remains largely unknown, though highly desirable for designing immediate repair as well as long-term coastal management. Determining in detail the tsunami-induced net losses and gains of beach sediment may thus eventually inform numerical models of how the coast might adjust later when the wave regime returns to a more normal state.

9.3 Geomorphic Tools for Reconstructing Past Tsunamis

Fresh traces of destructive tsunami are hard to overlook and damaged buildings, snapped trees, stranded debris, or eroded beach profiles provide ample markers for inferring inundation depths and local run-up heights. In the aftermath of the 2004 Indian Ocean tsunami, scientists collected over 1500 measurements of local run-up levels, and over 500 estimates of how far the waves reached inland; the numbers of comparable measurements for the 2011 Tohoku tsunami, Japan, were four and ten times higher, respectively. This increase in data availability also partly reflects the growing interest in understanding the geomorphic consequences of tsunami, an issue that had been largely overlooked in research on tsunami. In some places, detailed historical archives can expand the chronology of past tsunamis. For example, written accounts helped to identify some 17 tsunamis in the past 700 years along the Apulian coast of Italy (Pignatelli et al. 2009).

Reconstructions of geomorphic impacts of prehistoric or poorly documented tsunamis, however, face the problem that tsunamis selectively leave telltale sedimentary traces that are preserved long after the event. Even strong submarine earthquakes may leave only sparse geomorphic footprints of tsunamis along nearby coastlines, and frequent storm-wave action might soon erase a lot of the traces of former tsunamis. We thus cannot expect to find a complete sedimentary archive of past tsunamis onshore (Clark et al., 2011). Hence, researchers were very interested in elucidating in detail the geomorphic and sedimentary legacy of the destructive Indian Ocean and Tohoku tsunamis. Both occurred mainly along actively accreting sedimentary coastlines along active margins, and the available array of proxies has grown thanks to detailed studies of these and other recent events. For example, the experience of the 2011 earthquake in Christchurch, New Zealand, has shown that onshore liquefaction of sediments by strong earthquakes can generate deposits of sand that can be confused with tsunami deposits. Techniques such as magnetic susceptibility (Wassmer et al. 2010) and detailed stratigraphical analysis are needed to distinguish with certainty between tsunami deposits and coastal deposits of cyclone or liquefaction origin. Chagué-Goff et al. (2011) compiled the diverse array of proxies for identifying deposits of past tsunamis, including:

- sediments that are generally normally graded and that fine inland, although inversely graded subunits are possible;
- basal contacts that are unconformable or erosional;
- rip-up clasts of natural and anthropogenic material;
- shell, wood and less dense debris 'floating' in upper deposit layers;
- anisotropic magnetic susceptibility that, combined with grain size analysis, reveals hydrodynamic conditions;

- an abundance of marine to brackish diatoms and more reworked terrestrial diatoms in upper deposit layers;
- marked changes in foraminifera and other marine microfossil assemblages;
- lower pollen concentrations or high fraction of coastal pollen;
- increased concentrations of sodium, sulfur, and chlorine due to former saltwater intrusions;
- both water-worn and intact shells and shell-rich units;
- buried vascular plant material or buried soils;
- archaeological remains such as middens;
- partly eroded coastal dunes, catastrophically widened channel mouths, and coseismically uplifted or subsided areas.

Sheets of sand rich in marine fauna and trapped in, or alternating with, buried soils or terrestrial organic wetland deposits are telltale indicators of large wave run-up. Marine diatoms and nannoliths (mixtures of biogenic carbonate particles of silt to clay size) are among the most tell-tale indicators in these sediments (Paris et al. 2010). Locations inland of beach berms, spits, and beach dunes, and also coastal lakes and lagoons form good sediment traps to archive potential incursions by large waves, especially if providing additional evidence of rapid coseismic subsidence (Nichol et al. 2007). Coastal lagoons and lakes offer complementary proxies for establishing former environmental conditions, especially if the water was brackish or fresh, and how rapidly these conditions may have changed (Kitamura et al. 2013). Depending on the local topography, such sand sheets can extend up to a kilometre or more inland, contain vegetation debris and rip-up clasts, and have distinct erosional contacts at the base. Some tsunami deposits contain rip-up clasts made of soft undeformed sediment or shells without any signs of wear, a characteristic of high lift forces in a basal flow layer but little shear in the upper flow. Alternating layers of darker marsh and lighter sand are important markers of prehistoric tsunamis (Cisternas et al. 2005), and sometimes distinct layers of coral rubble or local imbricated boulder berms and beds also mark the reach of former wave action. The generic term 'tsunamiite' refers to all of these deposits, although Shiki et al. (2011) point out that this term also encompasses sediments laid down by processes other than only tsunami run-up onto the shore: these processes include backwash currents that may also rework or destabilize large amounts of sediment off-shore, potentially triggering submarine mass wasting. Tsunamiites also occur in the geological record, and prominent examples occur as sedimentary rocks interpreted to have formed by giant waves stirred by an extraterrestrial impactor at the transition between Cretaceous and Tertiary times (Dawson and Stewart, 2007).

By dating and measuring the onshore extent of diagnostic deposits, the tsunami scientist can establish a frequency–magnitude relationship for past tsunamis for a coastal section of interest. Once corrected for potential effects of vertical crustal movement and sea-level change, the height of these sediments above sea level is a measure of both tsunami run-up and inundation. Generally, tsunami deposits become thinner farther inland; however, where steep topography such as cliffs or terrace risers stops the advancing tsunami, the flow rapidly loses most of its sediment load, leaving behind thick sheets that may lack a tell-tale landward thinning (Hori et al. 2007). Along low-lying coasts, the grain size in these deposits gradually decreases in the direction of flow. At active subduction zones, palaeotsunami studies based on coastal deposits are an ideal tool to support efforts to reconstruct, or independently test, the chronologies of past megathrust earthquakes.

The 2004 Indian Ocean tsunami motivated many scientists to elucidate the past history of large waves in the region. Detailed and independent investigations of soil pits and sections derived from auger holes along the two coastal locations in Thailand and Indonesia agreed that the last tsunami in the Indian Ocean with run-ups at these locations comparable to the one in 2004 occurred sometime between 1300 and 1450 AD, and 1290 and 1400 AD respectively (Figure 9.6) (Jankaew et al. 2008; Monecke et al. 2008). Both age intervals are consistent within the errors of radiocarbon dating so that the researchers believe that they have captured evidence of the same event. Another independent study of rapidly drowned mangrove swamps, uplifted coral terraces, traces of liquefaction, and soils sealed by sand and coral rubble sheets on the Nicobar and Andaman islands, managed to trace some five large tsunamis that had occurred in the past 2000 years. The medieval event detected in Thailand and Indonesia also appears in the Nicobar and Andaman chronology, although a tsunami sometime in the eighth to eleventh century AD is the strongest contender for having had similarly severe impacts to the one in 2004 (Rajendran et al. 2013). Stratigraphic evidence from the coastal plain of Aceh province, which was strongly impacted by the 2004 tsunami, reveals even older traces of sudden coseismic land subsidence similar to that experienced in the 2004 earthquake. Radiocarbon ages between 6500 and 7000 cal yr BP show that marine flooding of this part of the coast is rare (Pre et al. 2012). More field evidence of large Holocene tsunamis in the Indian Ocean is accumulating and slowly extending the chronology of palaeotsunami in this region (Rhodes et al. 2011). Coastal lagoon sediments in Sri Lanka may hold sandy sediments of perhaps three more large tsunamis between 1600 and 5500 years ago.

The west coast of North America has one of the most complete palaeotsunami records. The Pacific coast of northern California, Oregon, Washington, and southern British Columbia lies along the Cascadia subduction zone, which has been the source of about 20 M 8–9+ earthquakes in the past 10 000 years (Clague et al., 2000). The tsunami of the most recent of these earthquakes, which happened in 1700 AD, is

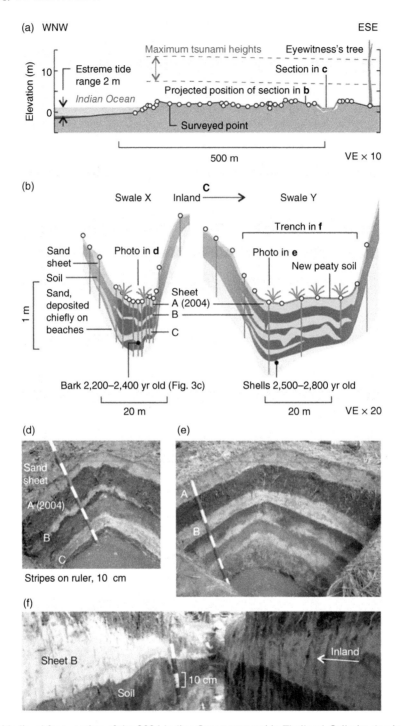

Figure 9.6 Medieval forewarning of the 2004 Indian Ocean tsunami in Thailand. Soil pits dug in a coastal back-barrier beach on Phra Thong Island, 125 km north of Phuket, Thailand, reveal pale sandy layers alternating with darker, organic-rich salt marsh deposits. (a) Topographic profile across beach (vertical exaggeration (VE) × 10). (b) and (c) Cross-sections in (b) swale X and (c) swale Y, based on pits and auger borings (vertical grey lines). Note soil peaty layers in swales (dark brown) and sand layers on ridges (light brown). Sand layers are inferred to result from infrequent catastrophic marine inundation during tsunamis. (d) and (e) Sand sheets alternate with dark peaty soils on the walls of pits in (d) swale X and (e) swale Y. (f) Lateral continuity of sand sheet B exposed in trench. From Jankaew et al. (2008).

particularly well documented. It left a sheet of sand in coastal wetlands and low-lying coastal lakes along the entire 1,100-km length of the subduction zone and also caused damaging run-up on the east coast of Honshu, Japan, almost 7500 km from the source (Satake and Atwater, 2007). Japanese written records of this damage mention that the earthquake happened on the evening of 26 January 1700.

The west coast of South America hosts another large subduction zone that has spawned many tsunamis in recent times. Cisternas et al. (2005) identified and dated buried soils and sand layers in south-central Chile to reconstruct a 2000-year history of coastal subsidence and attendant tsunamis. They concluded that the great 1960 Chile earthquake, which is the largest earthquake ever recorded, may have had similar-sized predecessors with average recurrence intervals of several centuries. The same methods and diagnostic sediments led Nanayama et al. (2003) to deduce that large tsunamis struck the coast of Hokkaido, Japan, in the mid- to late Holocene about every 500 years on average, with waves much higher than those triggered by M~8 earthquakes recorded in the Kuril-Japan subduction zone in the nineteenth and twentieth centuries.

Historically documented tsunamis offer independent confirmation of dates inferred from geomorphic and sedimentary evidence. Detailed fieldwork in both Japan and the Pacific Northwest of North America has brought to light the evidence of the 1700 AD tsunami triggered by an earthquake at the Cascadia subduction zone (Clague et al., 2000). The Pacific coast of Kamtchatka has an excellent sedimentary record of past tsunamis, anchored by dated tephra layers from nearby volcanoes. Pinegina et al. (2003) identified 41 sediment sheets left by tsunamis that ran up by more than five metres on the shores of Kamchatka in the past 4000 years.

Active plate boundaries in the Mediterranean Sea are also sources of destructive tsunami. Radiocarbon ages on corals and bryozoa from a raised shoreline in western Crete are consistent with rapid coseismic uplift of nearly ten metres during the 21 July 365 AD earthquake. This earthquake generated a huge tsunami that destroyed cities and killed thousands of people between the Nile Delta and the modern-day city of Dubrovnik (Shaw et al. 2008), partly according to the vivid details captured by historian Ammianus Marcellinus. Polonia et al. (2013) proposed that this tsunami triggered a large megaturbidite bed in the Ionian Sea, regionally known as 'Homogenite', a largely structureless and fine-grained deposit with mixed pelagic (Type A) and terrestrial (Type B) sources originally attributed to the collapse of Santorini volcano in Minoan times around 3500 yr BP. A sequence of megaturbidite beds south of Santorini Island contains reworked volcaniclastic material that may have been associated with eruptions of the volcano (Anastasakis, 2007). In any case, older and similar-sized deposits in the Ionian Sea point to another episode of catastrophic seafloor sedimentation after 15 kyr BP. Return periods of catastrophic tsunami in the eastern Mediterranean may be of the order of several thousand years. Such rare events seem to have exceptional outcomes, as the deposits from these prominent earthquake-triggered turbidity currents make up more than 90% of the volume of all Late Pleistocene to Holocene sediment on the Ionian Sea floor (Polonia et al. 2013).

Finding evidence of big waves is still a step away from proving that they were of tsunami origin given that strong storms, storm surges, and meteotsunamis can drive waves and sediment far inland. Some debate has revolved around the possibility of confusing tsunamiites with 'tempestites', which are storm-derived sediments. While sedimentary archives may faithfully record the deposits by extreme waves or flows, the contrasting inferences of the underlying process cause a lot of debate. Large wave-transported boulders or megaclasts, 'boulderites' or 'cliff-top boulders' stranded well above the tidal range on gently sloping coasts or marine terraces have become popular research objects because they reflect key characteristics of large waves, particularly minimum wave heights and flow velocities needed to set these boulders in motion (Nott, 2003). In the northern British Isles, mostly the Orkney and Shetland Islands, such partly imbricated cliff-top deposits occur up to 50 m above sea level, where they cover shell-rich sands or peat-bog deposits, offering markers of past wave heights (Hansom and Hall, 2009). Boulders composed of coral from nearby reefs and dumped onto the shore attest to both the erosional and depositional work of tsunamis. The threshold of motion for such boulders depends on their size, shape, and orientation to the incoming flow, whether they are partly or completely submerged or bounded by joints or other hydraulic flow obstacles, as well as the sediment concentration of the flow. Much of the argument centres on the different waveforms involved. Tsunami waves have a much longer wavelength than storm waves and thus a different momentum per unit length. Tsunami waves can also attain higher flow velocities and flow depths at much higher wave periods, so scientists have been searching for corresponding characteristic thresholds in the geometry and setting of megaclasts that can separate storm from tsunami transport. The common approach is to use wave-competence equations that are based on a hydrodynamic force, or momentum, equilibrium approach and approximate the ability of a given wave to dislodge and transport boulders. In practice, these equations predict that storm waves have to be four times higher than tsunami waves to entrain the same boulder; for example, according to Scheffers (2008), storm waves of up to 15 m high are needed to move a boulder of up to 10 t if in a joint-bounded position (Table 9.2).

However, given that the underlying flow equations simplify a lot of physical processes, researchers have some doubt whether this is the right approach to use for assessing exclusive boulder transport by tsunamis, as some boulders could have also moved during storms

Table 9.2 Characteristics of coastal megaclasts moved by tsunamis and storms.

Location	Boulder mass (t)	Height a.s.l. (m)	Distance (m)	Process	Max. wave height (m)	Reference
Mallorca, Spain	2.5			Storm	9	Bryant (2004)
Bonaire Is., Antilles	>5			Storm	12	Scheffers (2008)
Niue, Samoa	8-10	23		Storm		Terry et al. (2013)
Mallorca, Spain	40		<0.1	Storm	9	Scheffers et al. (2008)
Tuamotu, French Polynesia	60-75			Storm		Terry et al. (2013)
Sydney, Australia	200		50	Storm		Scheffers et al. (2008)
Phi Phi Don Is., Thailand	40	5	200	Tsunami		Scheffers (2008)
Lhok Nga, Indonesia	11		900	Tsunami		Paris et al. (2010)
Apulia, Italy	70	1.8	40	Tsunami?		Pignatelli et al. (2009)
Lhok Nga, Indonesia	85		<5	Tsunami		Paris et al. (2010)
Ryukyu Is., Japan	~250	30	2500	Tsunami		Scheffers (2008)
Long Is., Bahamas	350	17	150	Tsunami?		Scheffers (2008)
Krakatau, Indonesia	~600		100	Tsunami	36	Frohlich et al. (2009)
Japan	>1,000	>7		Tsunami	14*	Scheffers (2008)
Tongatapu Is., Tonga	~1,800	10	130	Tsunami?		Frohlich et al. (2009)

Height given above sea level (a.s.l.); distance refers to the total landward transport distance during the event; *refers to run-up height. See Scheffers and Kelletat (2003) for more data.

(Paris et al. 2010). In a detailed review, Lorang (2011) called for better consideration of how much sediment is available and how the coastal topography modulates the advancing waves when using a wave-competence approach for discriminating storm- from tsunami-deposited boulders. He further pointed out that mega-clasts are useful for determining the wave period T, thus providing a better distinction between storm and tsunami waves:

$$T = \frac{2h_c}{gd_i} \frac{\rho_w}{\rho_s - \rho_w} \frac{C_d}{S} u_{\max} \qquad (9.4)$$

where h_c is the height to which the megaclast was transported above an arbitrary datum [m], g is gravitational acceleration [m s^{-2}], d_i is the intermediate diameter of the megaclast [m], ρ_w and ρ_s are the seawater and sediment densities [kg m^{-3}], respectively, $C_d = 2$ is the drag coefficient, S is the local slope [m m^{-1}], and $u_{\max}$ is the maximum swash velocity [m s^{-1}]. Storm waves have periods of several tens of seconds, whereas tsunami wave have periods of several tens of minutes to hours; hence Eq (9.4) should offer a way to infer the type of transporting process from a boulder's relative location and size; at the same time, the many parameter values that need to be either measured accurately or approximated can quickly introduce large error margins and possibly faulty interpretations (Lorang, 2011). Flow acceleration, drag coefficient, boulder geometry, and boulder exposure to flow are some examples of parameters that these models treat as arbitrary or constant, although they can vary considerably during wave swash and backwash. Dewey and Ryan (2017) argued that boulder size alone may be insufficient for distinguishing tsunami from storm waves as the original transport mode, motivated by the results from a numerical model using local oceanographic real-time data to test whether storm conditions could pluck and move cliff-top boulders.

In a similar spirit, Nandasena et al. (2011b) cautioned that the widely used hydrodynamic equations derived by Nott (2003) only reflect the initial threshold of motion. In revising these equations for inertia forces, they derived the minimum flow velocities $u_{\min}$ [m s^{-1}] required to move coastal megaclasts. For example, rolling or overturning of boulders without joint contacts occurs at:

$$u_{\min} = \left(\frac{2(\rho_s/\rho_w - 1)gc(\cos\theta + (c/b)\sin\theta)}{C_d(c^2/b^2) + C_l} \right)^{1/2} \qquad (9.5)$$

where b and c are the intermediate and short axes of the boulder [m] and oriented parallel to the flow direction, θ [°] is the slope angle, and C_l is the lift coefficient [1]. For at least four field settings, however, the velocities predicted by Eq (9.5) were less than half of those originally proposed. Nandasena et al. (2011b) also noted that the equations revealed little about the dynamics of boulder transport. To meet this shortcoming, they used a numerical model to simulate in one dimension the transport of rectangular boulders sliding or saltating (except for rolling) across a uniform and nonerodible bed (Nandasena et al. 2011a). Some of their simulations showed that boulders were deposited near the highest run-up or subsequently during backwash flows, but also overestimated most transport distances observed in the field. Again, the many parameters that are part of this model make it easy for errors or unwarranted assumptions to propagate. Weiss (2012) emphasized that rough surfaces could impede the incipient motion of boulders when hit by large waves, and cause entrainment to be more selective than if merely dictated by boulder size. He proposed a stability model for cylinder-shaped megaclasts, combining drag and lift forces derived from linear wave theory with their moment arms from a geometrical force balance. This model expresses the Factor of Safety FS as the ratio of resisting over destabilizing moments:

$$\mathrm{FS} = \frac{l_2 \mathbf{W}_s \cos\theta}{l_1 \mathbf{W}_s \sin\theta + l_3 \mathbf{F}_d + l_2 \mathbf{F}_l} \qquad (9.6)$$

where l_1 and l_2 are the moment arms [m] of the downslope and normal vector components of the boulder's weight force W_s [N], respectively, l_3 is the moment arm [m] of the drag force F_d [N], and F_l is the lift force [N]. Here, l_2 and $l_3 - l_1$ are interpreted as the bed roughness and the exposure of the boulder, respectively. Weiss (2012) showed that, for this model, cylindrical boulders remain stable against wave action if the bed roughness is larger than 30% of their total height. This potential importance of bed roughness calls for more detailed measurements in the field, however, as previous roughness elements such as cobbles or small boulders may have been eradicated by large waves. The model also allows specifying wavelengths and water depths via the drag and lift forces, and predicts that both tsunamis and storms of similar amplitudes may move similar-sized boulders. This insight may further confound the discussion about whether we can learn something about the transporting process from the size of coastal megaclasts unless taking into account more data on the waves themselves, particularly their wave period. On this point, Weiss (2012) remarked that tsunamis largely leave random boulder deposits, whereas only storms were capable

of organizing boulders in lines, clusters, and ridges, mainly because of the higher number of waves and the higher total energy involved.

Studying imbricated coastal boulder deposits in areas near the equator where the Coriolis force, and hence one ingredient for tropical cyclones, is low may be one way to constrain better the assumptions about the mechanisms that generated prehistoric tsunami (Rajendran et al. 2013), although a systematic investigation of such locations remains to be done. Similarly, systematic attempts to date the timing of boulder deposition are rare (Scheffers et al. 2008), but could fill important gaps in our knowledge about past tsunamis. Paris et al. (2010) broadened the scope of megaclast studies by also considering side-scan sonar images of over 300 boulders submerged at 15–10 m below sea level at Lhok Nga, Banda Aceh province, Indonesia. They argued that boulders transported by the 2004 Indian Ocean tsunami in that region were more angular and coated by fine sediment as opposed to bouldery blocks from the underlying coastal platform (Figure 9.7).

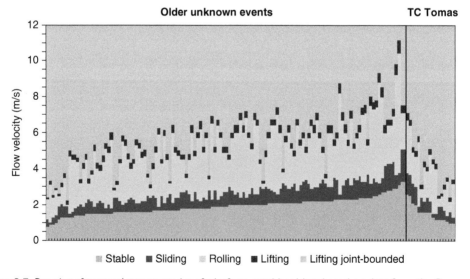

Figure 9.7 Bar plot of tsunami-transported reef-platform coral boulders based on data from the Bouma and Levena boulder fields on Taveuni Island, northern Fiji. Individual boulders are plotted side-by-side along the horizontal axis. Fresh boulders deposited by Tropical Cyclone (TC) Tomas in 2010 are grouped on the right-hand side of the diagram; boulders deposited by older cyclones are grouped on the left. From Terry et al. (2013).

In volcanic settings, the lithology of tsunami sediments offers clues about their origin, if containing tephra with distinct macroscopic characteristics or geochemical fingerprints. Carey et al. (2001) attributed stranded pumice rafts containing large amounts of coral debris to a tsunami triggered by the 1883 eruption of Krakatau, Indonesia. The conspicuous rounding of the pumice clasts and the fractal characteristics of particle shapes point to reworking by a tsunami rather than direct deposition from airfall. May et al. (2015) studied in detail the sediments and landforms associated with washover deposits of the delta of the Ashburton River in northwestern Australia. Using detailed sedimentological fingerprinting based on heavy minerals and X-ray diffraction together with OSL dating, the researchers established a record of coastal flooding for the past 150 years, and distinguished with high confidence deposits laid down by historic tsunami (including that from the Krakatau eruption in 1883) from those deposited by tropical cyclones.

Boulders on sedimentary coasts: left by storms or tsunamis? Large boulders on beaches far from any rock cliffs, or even lodged on top of cliffs, provide clues about past wave regimes, if other processes such as weathering or mass movement can be excluded (Figure 9.8). The size, shape, and orientation of boulders to incoming waves are part of a set of variables that constitute wave-competence equations. The idea rests on a quantitative relationship between boulder size and wave height, all other things being equal. These equations are based on the equilibrium of drag forces F_d, lift forces F_l, inertia forces F_i, resisting (gravity) forces F_r, and frictional force F_μ (Paris et al. 2010). From this equilibrium one can compute the forces at the threshold of motion for submerged boulders:

$$F_d + F_l = F_r \qquad (9.7)$$

In the case of subaerial boulders, the inertia force must also be considered, and the limit equilibrium at which the boulder is just about to move is:

$$F_d + F_l + F_i = F_r \qquad (9.8)$$

Equations (9.7) and (9.8) can be solved for the minimum tsunami wave height H_t [m] required to mobilize boulders above the tidal range following the assumptions initially made by Nott (2003), and later revised by Nandasena et al. (2011b):

$$H_t = \frac{(\rho_s/\rho_w - 1)(2a - C_m a\,\ddot{v}/bg)}{4(Cd(ac/b^2) + C_l)} \qquad (9.9)$$

and a similar solution can be obtained for the minimum storm wave height required to move boulders H_s [m]:

$$H_s = \frac{(\rho_s/\rho_w - 1)(2a - 4C_m a\,\ddot{v}/bg)}{4(Cd(ac/b^2) + C_l)} \qquad (9.10)$$

where ρ_s is the density of sediment [t m^{-3}], ρ_w is the density of seawater [t m^{-3}], a, b, and c are the long, intermediate, and short boulder axis lengths [m], respectively, $C_m = 1$ is the mass or inertia coefficient (which becomes higher for submerged boulders), $C_d = 2$ is the drag coefficient, $C_l = 0.178$ is the lift coefficient, $\ddot{v}$ is the instantaneous flow acceleration [m s^{-2}], and g is gravitational acceleration [m s^{-2}]. Similar expressions can be derived for subaerial boulders or those nested in coastal rock platforms and bounded by joints. Equations (9.9) and (9.10) predict that storm wave heights need to be four times higher than tsunami wave heights to move a given boulder.

These equations rely on simplifying assumptions but have been used widely to infer wave heights of prehistoric tsunamis and storms. Numerical modelling of tsunamis has increasingly been used as an independent means of checking the validity of coastal boulders as diagnostic

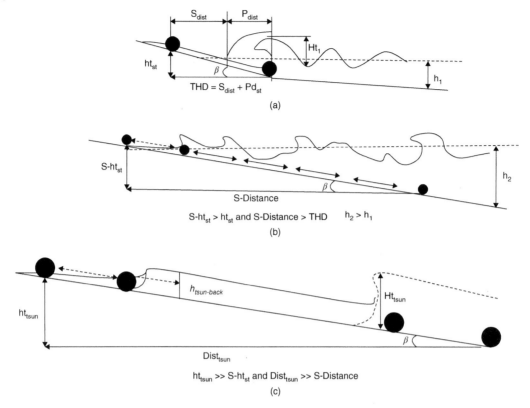

Figure 9.8 Storm versus tsunami processes: (a) A conceptual diagram showing wave breaking on a boulder beach occurring over some plunge and horizontal swash (run-up). The wave entrains, transports, and deposits a boulder to some elevation above its original position. (b) Storm-wave driven process of boulder entrainment, transport and shoreward deposition on a smooth inclined plane due to storm surge and multiple breaking waves. (c) Boulder entrainment on a smooth inclined plane due to tsunami-generated waves moving as giant swash bores that deposit the boulder at some elevation up the slope. From Lorang (2011).

proxies of past tsunami wave heights (Nandasena et al. 2011b). However, until these models are tested in the field or in flumes they should be used with full consideration of their possible inaccuracies. Strategies to estimate such uncertainties include Monte Carlo simulations that use plausible probability distributions for the input parameters instead of fixed values.

The problem of confidently distinguishing storm-wave from tsunami-wave impacts also applies to the case of high-level boulder and marine deposits on the flanks of oceanic islands or along fjords. Coseismic submarine

landslides may also generate exceptional tsunami run-ups along coastlines (Fritz et al. 2007). Attributing tsunamis to landslides commonly requires numerical model experiments to demonstrate that other triggers can be precluded (Figure 9.9). Such models may identify situations in which earthquake-driven displacement of the sea floor is insufficient to create the wave trains responsible for the observed pattern of run-up (Figure 9.10) (Fryer et al. 2004). Prehistoric examples are found in Hawaii, where mysterious – but seemingly wave-transported – trains of boulders have been found several hundred metres above sea level (McMurtry et al. 2004). The origins of these deposits, however, remain disputed given

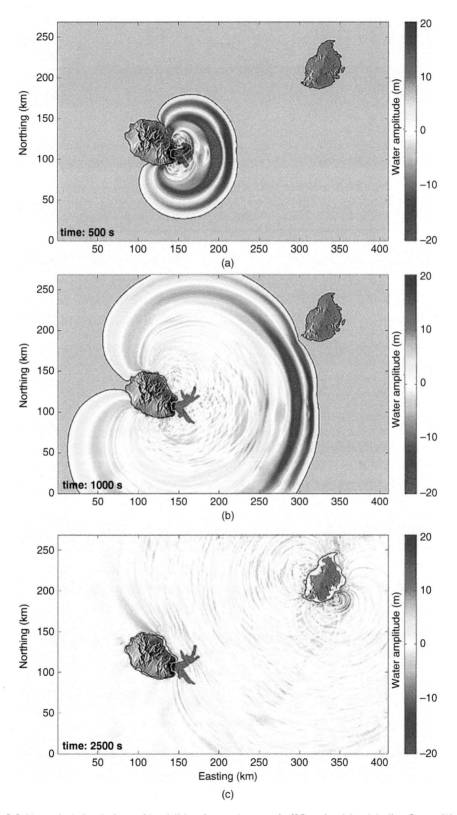

Figure 9.9 Numerical simulations of landslide-triggered tsunami off Reunion Island, Indian Ocean. Water amplitude (meters) generated by a 10 km^3 landslide at (a) t = 500 s, (b) t = 1000 s, (c) t = 2500 s. Landslide deposits appear in dark. From Kelfoun et al. (2010).

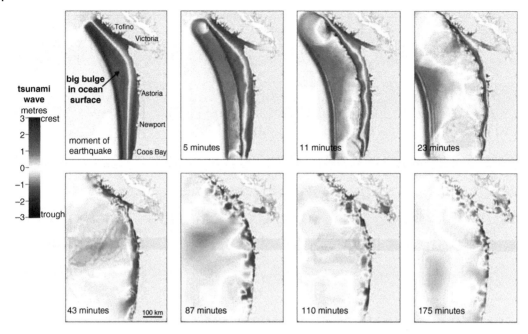

Figure 9.10 Numerical simulation of tsunami amplitudes on southern Vancouver Island, British Columbia, resulting from a great earthquake at the Cascadia subduction zone. From Clague et al. (2007).

evidence about their distinct stratigraphic order supported by uranium-thorium dates that reflect a deposition by multiple events during two different interglacials (Rubin et al. 2000). Pérez-Torrado et al. (2006) argued that Pleistocene marine conglomerates containing broken marine fossils dislodged from their original growth positions up to ~190 m above sea level on the Canary Islands were emplaced by tsunamis triggered by catastrophic collapses of the flanks of volcanoes on these islands.

Among these and many other recent advances in reconstructing prehistoric tsunamis, scientists have also begun to pay more attention to identifying evidence of large displacement waves in lakes. For example, tsunami-borne boulder ridges also occur along the shores of Lake Tahoe, California, where Moore et al. (2006) linked them to the catastrophic McKinney Bay landslide deposit (10 km³). When this gigantic failure took place is only roughly known. Available age estimates range anywhere between 15 and 7 ka, and demonstrate the need to understand better the

timing of past disturbances of lakes, which like ocean shores, attract high numbers of people and infrastructure.

9.4 Future Tsunami Hazards

Can climate change also affect tsunami hazard? Geological evidence points to this possibility. Many of Earth's volcanic islands are made up of their own debris, emplaced by large catastrophic collapses of entire flanks of the volcanic edifices. The sudden impact of such debris avalanches has caused many large tsunamis in the past. Some researchers have proposed a climate trigger for these catastrophic collapses of island volcanoes, pointing out the clustering of large edifice failures at the end of Pleistocene glacial stages. One proposition is that increased warming following glaciations leads to sea-level rise, which in turn increases coastal erosion, and eventually triggers large-scale collapse of volcanic islands (Quidelleur et al. 2008). We can use a similar line of reasoning for the case of a possible climatic cause of the large submarine landslides that frequently initiate tsunamis. Catastrophically destabilized methane hydrates or overpressurized pore fluids can trigger submarine slope failures, but convincing links to changing climatic conditions remain elusive. Submarine methane hydrates appear to have been stable over the past glacial–interglacial cycles, and many dated submarine landslides have occurred during times of rising sea levels, which tend to stabilize methane hydrates through higher hydrostatic pressures on the sea floor, although part of this may be countered by warmer ocean waters (Tappin, 2010).

References

Anastasakis G 2007 The anatomy and provenance of thick volcaniclastic flows in the Cretan Basin, South Aegean Sea. *Marine Geology* **240**(1-4), 113–135.

Bryant E 2004 *Natural Hazards* 2nd edn. Cambridge University Press.

Burroughs SM and Tebbens SF 2005 Power-law scaling and probabilistic forecasting of tsunami runup heights. *Pure and Applied Geophysics* **162**(2), 331–342.

Carey S, Morelli D, Sigurdsson H, and Bronto S 2001 Tsunami deposits from major explosive eruptions: An example from the 1883 eruption of Krakatau. *Geology* **29**(4), 347–350.

Chagué-Goff C, Schneider JL, et al. 2011 Expanding the proxy toolkit to help identify past events — Lessons from the 2004 Indian Ocean Tsunami and the 2009 South Pacific Tsunami. *Earth-Science Reviews* **107**(1-2), 107–122.

Cisternas M, Atwater BF, Torrejón F, et al. 2005 Predecessors of the giant 1960 Chile earthquake. *Nature* **437**(7057), 404–407.

Clague JJ, Bobrowsky PT, and Hutchinson I 2000 A review of geological records of large tsunamis at Vancouver Island, British Columbia, and implications for hazard. *Quaternary Science Reviews* **19**(9), 849–863.

Clague J, Yorath C, Franklin R, and Turner B 2007 *At Risk; Earthquakes and Tsunamis on the West Coast of Canada*. Tricouni Press.

Clark KJ, Johnson PN, Turnbull IM, and Litchfield NJ 2011 The 2009 M$_w$ 7.8 earthquake on the Puysegur subduction zone produced minimal geological effects around Dusky Sound, New Zealand. *New Zealand Journal of Geology and Geophysics* **54**(2), 237–247.

Conley DC 2014 Drivers: Waves and tides. In *Coastal Environments & Global Change* (eds Masselink G and Gehrels R), AGU Wiley, pp. 79–102.

Cyranoski D 2012 Tsunami simulations scare Japan. *Nature* **484**(7394), 296–297.

Dawson AG and Stewart I 2007 Tsunami deposits in the geological record. *Sedimentary Geology* **200**(3), 166–183.

Dewey JF and Ryan PD 2017 Storm, rogue wave, or tsunami origin for megaclast deposits in western Ireland and North Island, New Zealand?. *Proceedings of the National Academy of Sciences* **114**(50), E10639–E10647.

Fritz HM, Kongko W, Moore A, et al. 2007 Extreme runup from the 17 July 2006 Java tsunami. *Geophysical Research Letters* **34**(12), 1–5.

Fritz HM, Mohammed F, and Yoo J 2009 Lituya Bay Landslide impact generated mega-tsunami 50th anniversary. *Pure and Applied Geophysics* **166**(1-2), 153–175.

Frohlich C, Hornbach MJ, Taylor FW, et al. 2009 Huge erratic boulders in Tonga deposited by a prehistoric tsunami. *Geology* **37**(2), 131–134.

Fryer GJ, Watts P, and Pratson LF 2004 Source of the great tsunami of 1 April 1946: a landslide in the upper Aleutian forearc. *Marine Geology* **203**(3-4), 201–218.

Geist EL and Parsons T 2008 Distribution of tsunami interevent times. *Geophysical Research Letters* **35**(2), 1–6.

Gelfenbaum G and Jaffe B 2003 Erosion and sedimentation from the 17 July, 1998 Papua New Guinea tsunami. *Pure and Applied Geophysics* **160**(10-11), 1969–1999.

Grilli ST, Taylor ODS, Baxter CDP, and Maretzki S 2009 A probabilistic approach for determining submarine landslide tsunami hazard along the upper east coast of the United States. *Marine Geology* **264**(1-2), 74–97.

Hansom JD and Hall AM 2009 Magnitude and frequency of extra-tropical North Atlantic cyclones: A chronology from cliff-top storm deposits. *Quaternary International* **195**(1-2), 42–52.

Hayakawa YS, Oguchi T, Saito H, et al. 2015 Geomorphic imprints of repeated tsunami waves in a coastal valley in northeastern Japan. *Geomorphology* **242**(C), 3–10.

Hori K, Kuzumoto R, Hirouchi D, et al. 2007 Horizontal and vertical variation of 2004 Indian tsunami deposits: An example of two transects along the western coast of Thailand. *Marine Geology* **239**(3-4), 163–172.

Hornbach MJ, Braudy N, Briggs RW, et al. 2010 High tsunami frequency as a result of combinedstrike-slip faulting and coastal landslides. *Nature Geoscience* **3**(11), 783–788.

Jankaew K, Atwater BF, Sawai Y, et al. 2008 Medieval forewarning of the 2004 Indian Ocean tsunami in Thailand. *Nature* **455**(7217), 1228–1231.

Kelfoun K, Giachetti T, and Labazuy P 2010 Landslide-generated tsunamis at Reunion Island. *Journal of Geophysical Research* **115**(F4), F04012.

Kench PS, McLean RF, Brander RW, et al. 2006 Geological effects of tsunami on mid-ocean atoll islands: The Maldives before and after the Sumatran tsunami. *Geology* **34**(3), 177–180.

Kitamura A, Fujiwara O, Shinohara K, et al. 2013 Identifying possible tsunami deposits on the Shizuoka Plain, Japan and their correlation with earthquake activity over the past 4000 years. *The Holocene* **23**(12), 1684–1698.

Lorang MS 2011 A wave-competence approach to distinguish between boulder and megaclast deposits due to storm waves versus tsunamis. *Marine Geology* **283**(1-4), 90–97.

Løvholt F, Glimsdal S, Harbitz CB, et al. 2012 Tsunami hazard and exposure on the global scale. *Earth-Science Reviews* **110**(1-4), 58–73.

MacInnes BT, Bourgeois J, Pinegina TK, and Kravchunovskaya EA 2009 Tsunami geomorphology: Erosion and deposition from the 15 November 2006 Kuril Island tsunami. *Geology* **37**, 995–998.

May SM, Brill D, Engel M, et al. 2015 Traces of historical tropical cyclones and tsunamis in the Ashburton Delta (north-west Australia). *Sedimentology* **62**(6), 1546–1572.

McMurtry GM, Watts P, Fryer GJ, et al. 2004 Giant landslides, mega-tsunamis, and paleo-sea level in the Hawaiian Islands. *Marine Geology* **203**(3-4), 219–233.

Monecke K, Finger W, Klarer D, et al. 2008 A 1,000-year sediment record of tsunami recurrence in northern Sumatra. *Nature* **455**(7217), 1232–1234.

Moore JG, Schweickert RA, Robinson JE, et al. 2006 Tsunami-generated boulder ridges in Lake Tahoe, California-Nevada. *Geology* **34**(11), 965–968.

Mori N, Takahashi T, Yasuda T, and Yanagisawa H 2011 Survey of 2011 Tohoku earthquake tsunami inundation and run-up. *Geophysical Research Letters* **38**, 1–6.

Nanayama F, Satake K, Furukawa R, et al. 2003 Unusually large earthquakes inferred from tsunami deposits along the Kuril trench. *Nature* **424**(6949), 660–663.

Nandasena NAK, Paris R, and Tanaka N 2011a Numerical assessment of boulder transport by the 2004 Indian ocean tsunami in Lhok Nga, West Banda Aceh (Sumatra, Indonesia). *Computers & Geosciences* **37**(9), 1391–1399.

Nandasena NAK, Paris R, and Tanaka N 2011b Reassessment of hydrodynamic equations: Minimum flow velocity to initiate boulder transport by high energy events (storms, tsunamis). *Marine Geology* **281**(1-4), 70–84.

Nichol SL, Goff JR, Devoy RJN, Chagué-Goff C, Hayward B and James I 2007 Lagoon subsidence and tsunami on the west coast of

New Zealand. *Sedimentary Geology* **200**(3-4), 248–262.

Nott J 2003 Waves, coastal boulder deposits and the importance of the pre-transport setting. *Earth and Planetary Science Letters* **210**(1-2), 269–276.

Okal EA and Synolakis CE 2004 Source discriminants for near-field tsunamis. *Geophysical Journal International* **158**(3), 899–912.

Paris R, Wassmer P, Sartohadi J, et al. 2009 Tsunamis as geomorphic crises; lessons from the December 26, 2004 tsunami in Lhok Nga, west Banda Aceh (Sumatra, Indonesia). *Geomorphology* **104**, 59–72.

Paris R, Fournier J, Poizot E, et al. 2010 Boulder and fine sediment transport and deposition by the 2004 tsunami in Lhok Nga (western Banda Aceh, Sumatra, Indonesia): A coupled offshore–onshore model. *Marine Geology* **268**(1-4), 43–54.

Pérez-Torrado FJ, Paris R, Cabrera MC, et al. 2006 Tsunami deposits related to flank collapse in oceanic volcanoes: The Agaete Valley evidence, Gran Canaria, Canary Islands. *Marine Geology* **227**(1-2), 135–149.

Pignatelli C, Sansó P, and Mastronuzzi G 2009 Evaluation of tsunami flooding using geomorphologic evidence. *Marine Geology* **260**(1-4), 6–18.

Pinegina TK, Bourgeois J, Bazanova LI, et al. 2003 A millennial-scale record of Holocene tsunamis on the Kronotskiy Bay coast, Kamchatka, Russia. *Quaternary Research* **59**(1), 36–47.

Polonia A, Bonatti E, Camerlenghi A, et al. 2013 Mediterranean megaturbidite triggered by the AD 365 Crete earthquake and tsunami. *Scientific Reports* **3**, 1–12.

Pre CAG, Horton BP, Kelsey HM, et al. 2012 Stratigraphic evidence for an early Holocene earthquake in Aceh, Indonesia. *Quaternary Science Reviews* **54**(C), 142–151.

Quidelleur X, Hildenbrand A, and Samper A 2008 Causal link between Quaternary paleoclimatic changes and volcanic islands evolution. *Geophysical Research Letters* **35**(2), 1–5.

Rajendran CP, Rajendran K, Andrade V, and Srinivasalu S 2013 Ages and relative sizes of pre-2004 tsunamis in the Bay of Bengal inferred from geologic evidence in the Andaman and Nicobar Islands. *Journal of Geophysical Research: Solid Earth* **118**(4), 1345–1362.

Renault L, Vizoso G, Jansá A, et al. 2011 Toward the predictability of meteotsunamis in the Balearic Sea using regional nested atmosphere and ocean models. *Geophysical Research Letters* **38**(10), 1–7.

Rhodes BP, Kirby ME, Jankaew K, and Choowong M 2011 Evidence for a mid-Holocene tsunami deposit along the Andaman coast of Thailand preserved in a mangrove environment. *Marine Geology* **282**(3-4), 255–267.

Rubin KH, Fletcher CH, and Sherman C 2000 Fossiliferous Lana'i deposits formed by multiple events rather than a single giant tsunami. *Nature* **408**(6813), 675–681.

Satake K and Atwater BF 2007 Long-term perspectives on giant earthquakes and tsunamis at subduction zones. *Annual Review of Earth and Planetary Sciences* **35**(1), 349–374.

Scheffers A 2008 Tsunami boulder deposits. In *Tsunamiites. Features and Implications* (eds Shiki T, Tsuji Y, Yamazaki T and Minoura K), Elsevier, pp. 299–317.

Scheffers A and Kelletat D 2003 Sedimentologic and geomorphologic tsunami imprints worldwide—a review. *Earth-Science Reviews* **63**(1-2), 83–92.

Scheffers SR, Scheffers A, Kelletat D, and Bryant EA 2008 The Holocene paleo-tsunami history of West Australia. *Earth and Planetary Science Letters* **270**(1-2), 137–146.

Shaw B, Ambraseys NN, England PC, et al. 2008 Eastern Mediterranean tectonics and tsunami hazard inferred from the AD 365 earthquake. *Nature Geoscience* **1**(4), 268–276.

Shiki T, Tsuji Y, Yamazaki T, and Minoura K (eds) 2011 *Tsunamiites. Features and Implications*. Elsevier.

Tappin DR 2010 Submarine mass failures as tsunami sources: their climate control. *Philosophical Transactions of the Royal Society A: Mathematical, Physical and Engineering Sciences* **368**(1919), 2417–2434.

ten Brink US, Lee HJ, Geist EL, and Twichell D 2009 Assessment of tsunami hazard to the U.S. East Coast using relationships between submarine landslides and earthquakes. *Marine Geology* **264**(1-2), 65–73.

Terry JP, Lau AA, and Etienne S 2013 *Reef-Platform Coral Boulders: Evidence for High-Energy Marine Inundation Events on Tropical Coastlines.* Springer Science & Business Media.

Umitsu M, Tanavud C, and Patanakanog B 2007 Effects of landforms on tsunami flow in the plains of Banda Aceh, Indonesia, and Nam Khem, Thailand. *Marine Geology* **242**(1-3), 141–153.

Wassmer P, Schneider JL, Fonfrége AV, et al. 2010 Use of anisotropy of magnetic susceptibility (AMS) in the study of tsunami deposits: Application to the 2004 deposits on the eastern coast of Banda Aceh, North Sumatra, Indonesia. *Marine Geology* **275**(1-4), 255–272.

Watt SFL, Talling PJ, Vardy ME, et al. 2012 Combinations of volcanic-flank and seafloor-sediment failure offshore Montserrat, and their implications for tsunami generation. *Earth and Planetary Science Letters* **319-320**, 228–240.

Watts P 2004 Probabilistic predictions of landslide tsunamis off Southern California. *Marine Geology* **203**(3-4), 281–301.

Weiss R 2012 The mystery of boulders moved by tsunamis and storms. *Marine Geology* **295-298**(C), 28–33.

Yamashita K, Sugawara D, Takahashi T, et al. 2016 Numerical simulations of large-scale sediment transport caused by the 2011 Tohoku earthquake tsunami in Hirota Bay, Southern Sanriku Coast. *Coastal Engineering Journal* **58** (04), 1640015.

10

Storm Hazards

According to the World Meteorological Organization, 90% of all natural disasters are meteorological in origin. Meteorological hazards range from local-scale whirlwinds, dust devils, and tornadoes, to more regional-scale cyclones. Hemispheric-scale atmosphere-ocean phenomena such as the El Niño–Southern Oscillation (ENSO) can provide the necessary preconditions for forming storms and anomalous rainfall. Storms dissipate thermal and mechanical energy in atmospheric systems. The Earth's wind field is mainly driven by gravity, i.e. the thermally induced density contrasts and pressure gradients in air masses between equatorial and polar regions owing to the differences in solar radiation and the planet's rotation. The distribution of land masses and water bodies affect the general circulation pattern, as does topography in the form of mountain belts.

We distinguish between cyclonic and convective storms. Cyclonic storms are largely driven by advection, which is the horizontal motion of air masses, whereas convective storms largely feed from convection, which is the upward motion of air masses. Tropical cyclones are large rotating storms that contain cloud-free and calm cores and form over seas with high surface temperatures in the tropics and subtropics. The strongest of tropical cyclones, also known as typhoons in the Pacific, and hurricanes in the Atlantic and the Caribbean, can have wind speeds >300 km h^{-1}. Tornadoes are much smaller, but highly destructive winds that occur in association with severe convective storms (Figure 10.1).

10.1 Frequency and Magnitude of Storms

10.1.1 Tropical Storms

Tropical cyclones are the most destructive hazards on Earth, at least if judged from their annual recurrence. An annual average of some 22 million people directly affected and an estimated US$ 29 billion spent over the past two decades motivate detailed studies of how the risk from tropical cyclones has evolved and how it will change in the future (Figure 10.2) (Geiger et al. 2018). These storms develop in tropical waters with sea surface temperature (SST) above 26° C and sufficient potential for generating rotating wind fields away from the equator. Typhoon Haiyan in 2013 holds the current record for the strongest gusts (up to 378 km h^{-1}) measured during a tropical cyclone. The storm obliterated the coastal city of Tacloban in the Philippines, killing at least 6300 people. The strong winds and torrential rain of tropical cyclones have caused more fatalities and incurred more insured losses than any other natural disaster, even in high-income countries such as the United States, Japan, and

Geomorphology and Natural Hazards: Understanding Landscape Change for Disaster Mitigation, Advanced Textbook Series, First Edition. Tim R. Davies, Oliver Korup, and John J. Clague.
© 2021 John Wiley & Sons Ltd. Co-published 2021 by the American Geophysical Union and John Wiley & Sons Ltd.

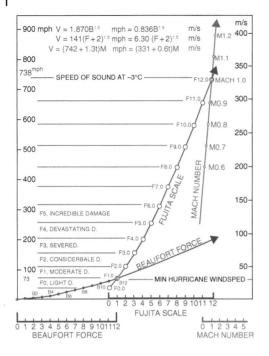

Figure 10.1 Wind speed scales. www.spc.noaa .gov/efscale/

Taiwan. Recent record-breaking storms seem to have occurred in close sequence. Hurricane Katrina, which inundated large parts of New Orleans in 2005, had been the most destructive tropical cyclone to hit the United States ever. Only twelve years later, however, Hurricane Harvey brought torrential rains that flooded large parts of Houston, causing an estimated US$ 150–180 million of damage, thus breaking Katrina's record. Similarly, Hurricane Sandy in 2012 demonstrated the vulnerability of more northern cities, including New York City, to these storms. Typhoon Morakot, which struck Taiwan in 2009, was the nation's most destructive tropical cyclone in 50 years; it caused 619 deaths and damage in excess of US$ 5 billion according to the National Disasters Prevention and Protection Commission. This typhoon currently holds the record for rainfall during a single storm event – 2749 mm, some 1000 mm more than the previous record-breaking storm, Typhoon Herb in 1996.

The average year sees about 80–90 tropical cyclones worldwide, and some 20% of them

make landfall at hurricane intensity, i.e. at wind speeds of >33 m s^{-1} or >118 km h^{-1} (Woodruff et al. 2014). Tropical cyclones are unevenly distributed on Earth; between 1945 and 2006 nearly a third of them occurred in the 'typhoon alley' of the western Pacific (Figure 10.3). Within that area, the Philippines are struck by 8–9 typhoons per year on average; Taiwan about seven, and the Japanese islands about two. Although only 5% of all tropical cyclones worldwide occur in the Indian Ocean, 65% of all fatalities from tropical cyclones during the period 1945–2006 were in the Bay of Bengal. In contrast, hurricanes in the western North Atlantic, mostly along the United States eastern and southeastern seaboards, comprise only 10% of all tropical cyclones worldwide, but are responsible for 60% of the total insured losses (Woodruff et al. 2014). A worldwide prediction of the risk from tropical cyclones holds that the potential for loss of life depends mainly on measures of tropical cyclone intensity, exposure, poverty, and governance, and trends project that the mortality risk from tropical cyclones is increasing (Peduzzi et al. 2012).

Wind speed is an intuitive measure of storm strength. Meteorologists use the Accumulated Cyclone Energy (ACE) value, which is the square of a tropical storm's maximum sustained wind speed, v_{max}^2, summed over four standard six-hour intervals, i.e. 0000, 0600, 1200, and 1800 Coordinated Universal Time (UTC). The minimum wind speed to qualify for a tropical storm, and thus to enter the ACE summation, is 35 knots or ~65 km h^{-1}. Systematic data on tropical wind speeds such as the historical worldwide 'best-track' archives (www.ncdc.noaa.gov/ibtracs) extend back to 1950 or even earlier but are subject to uncertainties and inaccuracies (Kossin et al. 2015). Data obtained since 1990 provide a more reliable global assessment of tropical cyclone activity, particularly around the time the storm reaches its maximum intensity. Between 1981 and 2010, the northwestern Pacific Ocean experienced some of the world's highest and most extensive ACE values topping 0.5×10^4 m^2 s^{-2}

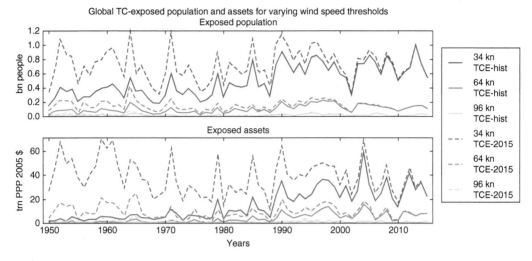

Figure 10.2 Annual global tropical cyclone (TC) exposure for different thresholds of wind speed given in knots (kn). Dashed lines are estimates based on fixed patterns of population in billions (bn) and assets for 2015 (TCE-2015) in purchasing power parity (PPP). Solid lines are estimates based on the historical evolution of population and assets (TCE-hist). From Geiger et al. (2018).

in places. The post-1990 peak in ACE in the North Atlantic was in 2005, the year that Hurricane Katrina struck the United States. During that tropical cyclone season more than 3900 people lost their lives and the estimated (insured) damages exceeded US$159 billion. The trend in ACE for the post-1990 period in the North Atlantic is inconclusive, but there has been a decline in the eastern Pacific, spurring further debate on the role of global warming in tropical cyclone activity. Related metrics of cyclone energy, such as the simplified power dissipation index, use v_{max}^3 over the life time of a tropical cyclone and yield similar conclusions, but show a stronger correlation with observed storm damage (Emanuel 2005). Sea-surface temperature has a strong effect on the maximum wind speed (also known as potential intensity) of tropical cyclones, although other influences such as regional wind fields can alter their strength (Woodruff et al. 2014).

Another measure that roughly expresses the strength of storms in coastal settings is the 'significant wave height'. It is a useful metric for characterizing the wave climate and the likelihood of coastal erosion or sedimentation.

Engineers and planners use it to establish design criteria for coastal infrastructure. Empirically calibrated models relate significant wave height to wave swash run-up levels, and are of practical use for managing coastal erosion with engineering structures (Komar et al. 2013). A related measure is the expected maximum wave height, which one can determine from regular wave records (You and Nielsen 2013). A common approach is to use extreme-value distributions that are fitted to the time series of measured wave heights at a given station. The underlying idea is that the maximum wave heights that were recorded at various stations are part of a statistical distribution that we can estimate from station data.

10.1.2 Extratropical Storms

Extratropical cyclones are the characteristic storm type in the temperate climate zone. These storms have an asymmetric footprint, and are frontal systems that arise from the mixing of contrasting dry cold subpolar and warm moist subtropical air masses. Extratropical storms mainly occur in the spring

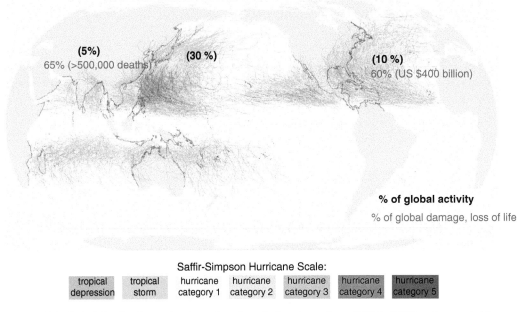

% of global activity

% of global damage, loss of life

Saffir-Simpson Hurricane Scale:

| tropical depression | tropical storm | hurricane category 1 | hurricane category 2 | hurricane category 3 | hurricane category 4 | hurricane category 5 |

Figure 10.3 Global distribution and trajectories of tropical cyclones from 1945 to 2006, colour-coded by the type of storm on the Saffir–Simpson Hurricane Scale. The reported impacts are highly skewed, with only 5% of all tropical cyclones being responsible for 65% of all fatalities, particularly in the Indian Ocean. In contrast, 10% of all tropical cyclones caused 60% of the insured damage in the North Atlantic. Figure modified from Woodruff et al. (2013). Map background: http://en.wikipedia.org/wiki/File:Tropical_cyclones_1945_2006_wikicolor.png

and autumn seasons when the potential for atmospheric instability is greatest and can produce wind speeds in excess of 200 km h^{-1} with heavy precipitation. The frontal character of these storms favours rapid changes in temperature and humidity, as well as lightning and thunder. Wind speeds of storms in the temperate latitudes are routinely measured by the Beaufort wind force scale, which is number that relates wind speed to observed conditions. Wind speed [m s^{-1}] on the Beaufort scale is is based on the empirical relationship:

$$v = 0.836 \, B^{3/2} \qquad (10.1)$$

where v is the equivalent wind speed at ten metres above the sea surface and B is the Beaufort scale number. The Beaufort Scale only applies up to Force 12. Tropical cyclones can have much higher wind speeds, but they are defined according to the Saffir–Simpson Hurricane Scale (Figure 10.1).

Convective storms are much smaller than tropical or extratropical cyclones and arise from the thermal updraft of converging and moist air masses. Commonly known as thunderstorms, these short-lived storms can occur very locally in response to differential heating of the land surface, frontal activity, or topographic obstacles. Every day sees some 40 000 thunderstorms on Earth, with wind gusts, lightning, rain, and hail. In the worst of cases, thunderstorms generate slowly rotating super cells that give rise to tornadoes (also named 'twisters' in North America), which are ground-touching funnel clouds. Conditions required to form tornadoes include the mixing of cold dry and warm moist air masses at falling air pressure and a highly variable wind field. The Fujita tornado intensity scale and its enhanced version categorize tornadoes by maximum wind speeds, which in exceptional circumstances can exceed

500 km h^{-1}. In contrast, the ground travel speed of tornadoes is mostly between 15 and 30 km h^{-1}, though unpredictable bursts may exceed 100 km h^{-1}. Tornadoes can form corridors of destruction several hundred metres wide, in which only the most sturdy of buildings remain intact. Funnel clouds commonly touch down and lose contact with the ground repeatedly, and tornadoes can leave tracks of destruction over distances of tens to hundreds of kilometres.

Blizzards are another form of extratropical storm in North America and are especially common where low-relief terrain allows cold arctic air to sweep down to mid latitudes and rapidly reduce ground surface temperatures to below −5° C, sometimes delivering dumps of snow that can be more than five metres thick. Such snow loads can bury roads and cause buildings to collapse. Wind speeds may be exceed 60 km h^{-1}, thus increasing wind chill, and reducing visibility to less than a few hundred metres.

10.2 Geomorphic Impacts of Storms

10.2.1 The Coastal Storm-Hazards Cascade

Storms bring strong winds that may also be accompanied by high amounts of precipitation. Storms, therefore, can both build and destroy landforms, and set in motion a cascade of damaging processes that depend on whether the storm travels over water or land. Most of the detailed work on the geomorphic impact of storms has focused on selected river catchments or strips of coastline, while fewer studies have looked into the response of entire landscapes.

In coastal settings, shallow water depths along gently sloping sedimentary coasts are particularly prone to elevated water levels during storms (Masselink and van Heteren 2014), as waves and surges can inundate greater areas

than along steep coasts (Figure 10.4). These coasts tend to be higher and more rocky and prone to sudden failure during storms, however. Nonetheless, parts of river deltas have also been reported to collapse during storms because of transient wave-induced pressures, while steeper sections of coral reefs can shed avalanches of rubble triggered by strong waves. High yields of suspended sediment, and possibly also contaminants, from flooded coastal rivers also compromise the health of coral reefs, and can induce partial reef decay, which in turn may influence the pattern and intensity of waves moderated by the reef and hitting the coastline. Numerical experiments show that wave run-up along coasts fringed by coral reefs depends largely on the width and slope of the fore reef; degraded coral with lower roughness can also promote higher wave heights hitting the coast (Quataert et al. 2015). Averaged over decades to centuries, the sediment budget, and hence stability, of some tropical coasts fringed by coral reefs may thus greatly depend on how quickly and thoroughly reefs recover from episodic storms (Terry et al. 2013).

Strong winds and low atmospheric pressure temporarily elevate sea level and promote storm surges. These storm surges build on background tidal levels to form storm tides. The magnitude of a storm surge is controlled by the dominant wind directions and velocities and the barometric effect of low pressures in storm systems. The Coriolis force due to the Earth's rotation affects the local pattern of storm surges on top of local controls such as the planform shape of the coastline, the approach angle of the storm, the near-shore water depth, instantaneous river discharge to the sea, and the tidal level at the time of landfall. Low atmospheric pressures associated with many storms lead to local doming of the sea surface such that storm-surge heights are highest near the location of lowest atmospheric pressure. This phenomenon is known as the 'inverted barometer effect'. For a pressure of 870 mbar – the lowest ever recorded in the eye of a tropical cyclone – the corresponding

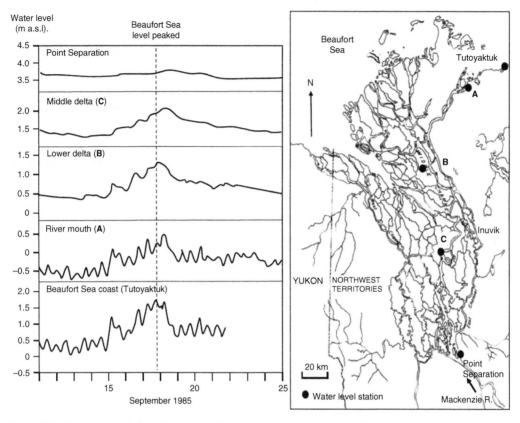

Figure 10.4 Storm-surge-induced water-level rise along the east channel of Mackenzie River at the south end of Mackenzie Delta to the river mouth on the north. From Woo (2012).

rise in the water column would be as much as 6.6 m (Bryant 2004). Tropical cyclones hitting the eastern seaboard of the United States from a southern direction, for example, rotate counterclockwise, with winds piling up higher storm waves along the eastern, seaward flanks of the storm, whereas the western flanks have lower waves and storm surges thanks to dominant offshore winds upon landfall. Extratropical cyclones on oceans can also produce far-reaching swell waves. Hoeke et al. (2013) reported that swell – wind waves with long wavelengths – from several mid-latitude storms in the North Pacific contributed to a regional episode of coastal flooding that hit Micronesia, the Marshall Islands, Kiribati, Papua New Guinea, and the Solomon Islands in December 2008, adding to already high local sea levels driven by pronounced La Niña

conditions at the time. The authors also noted that tide gauges in the affected island nations largely underestimated the locally damaging inundation levels, partly because the tide gauges are situated in sheltered bays and harbours, and thus unable to capture fully the wave set-up due to incoming swell.

Storm surges pushed onshore by tropical cyclones have caused major damage and loss of life on densely populated coasts. Between 1990 and 2010, an average of 11.5 million people per year were affected by storm surges associated with tropical cyclones in eastern China (Gao et al. 2014). Yet disentangling the processes responsible for storm-surge damage is rarely the focus of these and similar reports, though highly desirable for planning effective countermeasures. Tropical cyclone Nargis in 2008 was one of the rare storms to hit the low-latitude Ayeyarwady (Irrawaddy) delta in Myanmar (Burma) in historic times. The Category IV storm cost more than 138 000 people their lives and ranked as one of the most deadly and most destructive tropical cyclones in history. Wave heights in coastal areas were up to seven metres due to a combination of a 5-m storm surge and 2-m storm waves. Vertical erosion was more than one metre in places, causing coastal retreat of up to 100 m inland. The delta was inundated as far as 50 km from the shore, particularly in low-lying areas, destroying agricultural land and with it the livelihood of hundreds of thousands of people (Fritz et al. 2009). Detailed measurements showed that the delta coast retreated by 47 m on average when the tropical cyclone made landfall and, further, this retreat nearly tripled in the following two years (Besset et al. 2017).

Many similar examples illustrate how storm waves or surges have modified gently sloping sedimentary coastlines such as deltas, estuaries, and barrier-beach systems on top of seasonal variations (Table 10.1). Yet net erosion and deposition following storms can differ by at least an order of magnitude along a given coastline, and depend mainly on its local topography, but also on the storm-wave heights and on water levels (Mahabot et al. 2017). The major estuaries along the Wadden Sea coast in the Netherlands and northern Germany are legacies of Medieval storm surges that inundated extensive tracts of land close to or partly below sea level. The 14 December 1287 Lucia flood, for example, inundated the Zuidersee, Dollart, and Jade Bay, which today are large coastal embayments. The 1570 AD All Saints Day flood rose more than four metres above the land, and remains the most devastating storm surge in western Europe, killing an estimated 400 000 people. The 'Grote Manndränke' in January 1632 was one of the higher storm surges along the German North Sea coast, eroding large parts of Nordfriesland up to 15 km inland, and destroying of about half of its agricultural land. These and previous disasters prompted people to construct dykes along the coast as early as the eleventh century to protect the lands lying at or below sea level. For nearly 1300 years prior to the building of dykes, coastal inhabitants had built artificial hillocks or dwelling mounds called 'wurten' or 'warften' to provide refuges for people on elevated dry ground during storm surges. Archaeological excavations have revealed how coastal inhabitants kept on stacking layer on

Table 10.1 Selected estimates of coastal erosion and deposition during tropical cyclones.

Location	Year	Net erosion or deposition ($m^3\ m^{-1}$)	Material	Event	Reference
Lifuka Is., Tonga	1982	−28	Beach sand	TC Isaac	Terry (2007)
West coast, Reunion Is.	2014	$−42.5 \pm 1.6$	Beach sand	TC Bejisa	Mahabot et al. (2017)
Galveston Is., USA	2008	5.4–5.6	Beach sand	TC Ike	Hawkes and Horton (2012)
Galveston Is., USA	1983	6.2*	Beach sand	TC Alicia	Hawkes and Horton (2012)
Upolu Is., Samoa	1990	15–60	Coral rubble	TC Ofa	Terry (2007)
Ontong Java, Salomon Is.	1967	20–60	Coral rubble	TC Annie	Terry (2007)
Texas coast, USA	1961	24–52*	Beach sand	TC Carla	Hawkes and Horton (2012)
Ocracoke Is., USA	2003	24	Beach sand	TC Isabel	Conery et al. (2018)
Funafati, Tuvalu	1972	78	Coral rubble	TC Bebe	Terry (2007)
Jaluit, Marshall Is.	1958	≤150	Coral rubble	TC Ophelia	Terry (2007)

Net erosion and deposition are averaged per metre of coastline and have negative and positive signs, respectively; TC = Tropical Cyclone; *Washover deposits. Compare these values with those in Table 9.1.

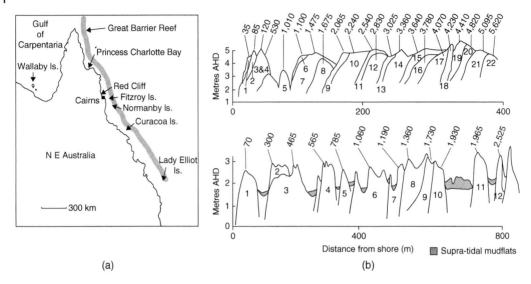

Figure 10.5 High frequency of 'super-cyclones' along the Great Barrier Reef over the past 5000 years. (a) Location map of study sites. (b) Stratigraphic relationship of storm deposit/ridges on Curacoa island (top) and Princess Charlotte Bay (bottom). Successive storm deposits are numbered. Mean reservoir-corrected radiocarbon age (in yr BP) for each ridge is shown. Note progressive increase in age with distance inland. AHD is Australian Height Datum. From Nott and Hayne (2001).

layer of sediment over the centuries to be safe from gradually rising sea levels, choosing the highest locations on the salt marshes that were nevertheless subjected to recurring floods each year (Bazelmans et al. 2012).

Storm waves dump mostly sands in the landward portions of barrier beaches or build successive parallel beach ridges, thus nourishing seaward-advancing coastlines (Figure 10.5) (Woodruff et al. 2014). High sediment yields supplied by flooded rivers aid this processes. Locally also known as cheniers, many of these sandy beach ridges lie on higher tidal flats or supratidal flats, though some may have also formed mostly during high spring tides (Morales et al. 2014). Low-lying marshland and ponds between beach ridges can trap storm-derived sediments and provide instructive chronologies of large storms and waves in the past (Donnelly et al. 2015). In turn, the sediment that storms strip from beaches is often replenished during long fair-weather periods. The more detailed chronologies of beach-ridge accretion cover several thousands of years, and may show more overarching

links with oscillations in sea-level rise and sediment supply than a one-to-one correspondence to individual storms alone (Lampe and Lampe 2018). Coastal geomorphologists distinguish overwash from washover. Overwash is the process by which sediment is deposited inland of a beach crest by overtopping waves, whereas washover refers to the resulting deposits (Switzer 2014). Storm waves can also entrain and redistribute sediments coarser than sand residing offshore or in barrier reefs. These sediments only make their way into the coastal zone episodically during extreme storms, where they are deposited as gravel ridges, sheets or ramparts that can rapidly raise the foreshore by several metres. For example, in 1972 the strong waves of Tropical Cyclone Bebe pushed some 2.8 Mt of coral rubble above sea level, forming an 18-km long and 3.5-m high natural embankment along Funafuti atoll in the South Pacific (Terry 2007). Storms, however, also erode coastlines, enlarging or creating new tidal inlets and pushing much of the sediment removed from the foreshore to back-barrier settings. For example, a systematic inspection of air photographs of Ocracoke Island, North Carolina, showed that this barrier island lost 40% of its width between 1949 and 2006, and that a single hurricane was responsible for 23% of this net erosion (Conery et al. 2018). However, the storm also dumped about 0.6 Mm^3 of material eroded from the foreshore back onto landward parts of the islands. This cannibalization and resulting rollover of sediments stored along barrier coasts eventually causes coastlines to retreat. Partly eroded barriers offer less protection from storm waves and make coasts more susceptible to flooding and erosion by waves. Coarser storm-derived gravel deposits on coral-reef atolls undergo a similar evolution, becoming less steep and thick as less extreme waves redistribute the deposits.

To characterize this cascade of geomorphic consequences of storms for barrier islands, Sallenger (2000) proposed an impact scale based on four distinct regimes, in which processes and coastal landforms are in close feedback with each other. The least of impacts occur in Level 1, termed the 'swash' regime, during which waves run up and erode the foreshore, though the sediment lost is replenished between storms so that the net geomorphic change is zero. In the 'collision' regime (Level 2) waves overrun the base of the foredune ridge, causing net losses of material. Still higher waves fall into the 'overwash' regime (Level 3), which features net sediment deposition in a landward direction associated with the barrier also moving into that direction. Finally, Level 4 describes the 'inundation' regime, during which storm waves are large enough to submerge the barrier island, dumping large amounts of sediment at the landward side. These four impact levels derive from several threshold ratios of the lower- and uppermost elevations of swashing waves and the elevations of the dune base and crest (Sallenger 2000). Coupled numerical models are capable of bringing together these different regimes in physics-based equations of wave regimes and wave-breaking mechanisms with equations on sediment transport from undercut collapsing dunes, overwash, and channel formation after a barrier has been breached. The outputs of these models are useful for informing planners and decision makers on the costs and benefits of potential coastal protection measures, as the models highlight how the coast might evolve with or without such measures (Roelvink et al. 2009).

Few of these models, however, directly connect with the effects of economic development. Humans continue to modify and rework sedimentary coasts, often to a degree that could be problematic or even misleading to ignore. A comparison of numerical storm-surge models for the area of what is now New York City for the periods 850–1800 AD and 1970–2005 AD shows that rising sea level was mostly responsible for mean flood heights that increased by more than 1.2 m (Reed et al. 2015). These models used downscaled results from Coupled Model Intercomparison Project Phase 5 (CMIP5) models to generate synthetic centennial time series of tropical cyclones, and suggest that flood heights have increased for a given return period. McNamara and Werner (2008) offered a model that links the natural dynamics of barrier islands with impacts of resort development, hazard mitigation, and policy decisions. They show that lower-lying islands are preferred for development, and are thus prone to be more exposed to storm surges and flooding than when in their natural state. This increased exposure results partly from human-made changes in the sediment budget, but also partly from protecting the growing assets in developed barrier islands. McNamara and Werner (2008) go as far as to argue that coastal areas that have benefitted recently from major protection measures have accumulated a greater overall risk, and will also experience greater damage, even if discounting for effects of rising sea levels or climate change. This parallels the increase in flood hazard caused by the construction of levees on rivers (Criss and Shock 2001). Such hypotheses await further testing but open new insights into the response of coastal systems heavily modified by human actions.

By damaging or killing vegetation, storms also indirectly determine the locations and rates of coastal erosion in the years after. Strong winds have the power to topple trees or kill them via sustained salt spray. Storm surges inundate and smother less salt-tolerant vegetation, and contaminate shallow freshwater aquifers. Nikitina et al. (2014) estimated that salt marshes that were eroded during land-falling tropical cyclones along the New Jersey coast, United States, had completely recovered after several decades to a few centuries; this protracted recovery points to the use of eroded salt marshes as possible proxies for past storms.

10.2.2 The Inland Storm-Hazard Cascade

Some geomorphic impacts of storms arise directly from the shearing effect of strong winds on the ground surface, which erodes soil and deposits it somewhere else. In semi-arid and arid areas this erosion may cause large sand and dust storms, which, apart from their geomorphic impacts, also involve hazards to human health, as well as ground and air traffic (Baddock et al. 2013). The vast loess deposits of central and eastern Asia are testimony to prolonged phases of dust transport from glacial and periglacial land-scapes in Pleistocene glacial periods. A rule of thumb holds that areas with annual rainfall greater than 1000 mm experience less than one dust storm per year. Dust-storm frequency is highest where annual rainfall is 100–200 mm (Figure 10.6).

Contemporary source areas of dust identified from regional monitoring campaigns spanning several decades include the Tibetan Plateau, Mongolia, and Inner Mongolia (Zhang et al. 2003). The Gobi Desert, in particular, is a hot spot of dust production and emits an estimated 190 $\mu g\ m^{-2}s^{-1}$ during the peak dust-storm months in spring and early summer, corresponding to a net loss of

unconsolidated surface materials, including soil, of ~1000 t km^{-2} during that time (Shao and Dong 2006). In Australia, the world's driest continent after Antarctia, estimated average dust yields are 107–122 t $km^{-2}\ yr^{-1}$, which is roughly three times higher than sediment yields of all the continent's rivers, that is 32–45 t $km^{-2}\ yr^{-1}$ (McTainsh and Strong 2007). This comparison stresses the importance of wind erosion and transport in dry regions, but also motivates more research on how and why the underlying fluxes vary. Bear in mind that dust storms occur beyond hot and dry climates. The vast glacial outwash plains of Iceland store large amounts of fine sediments, and gave rise to an average of nearly 17 dust days per year from 1949 to 2011, which is more than the recorded frequencies in places like Utah or Iran, but comparable to other Arctic outwash plains such as in the Yukon Territory of Canada, or in western Greenland (Dagsson-Waldhauserova et al. 2013). Globally, the contemporary dust flux is estimated to be 0.13–1.8 Gt yr^{-1} (Mahowald et al. 2005); the WMO arrives at higher estimates of the yearly dust flux of 1.0–2.15 Gt yr^{-1} (www.wmo.int).

Asian regions downwind of the Gobi Desert regularly receive high dumps of dust. Winds deposit dust at an average rate of 50 t km^{-2} per month during the peak of the dust season on the Chinese Loess Plateau, and high loads also reach more eastern parts of the country, Korea, and Japan (Shao and Dong 2006). A single dust storm originating from the Chinese Loess Plateau in April 1980 deposited up to 24 t km^{-2} per day in Beijing some 1000 km to the east (Derbyshire 2000). Dust storms carry soil and sediment, but also pathogens from dried-out lake beds such as the Aral Sea. Hence, dust storms can transport and possibly worsen diseases such as asthma, adding to the list of negative health effects of high suspended particle loads in the atmosphere.

Shifting of sand dunes can gradually, although relentlessly, bury settlements and infrastructure, and exhume those buried long ago. Wind-transported debris may impact

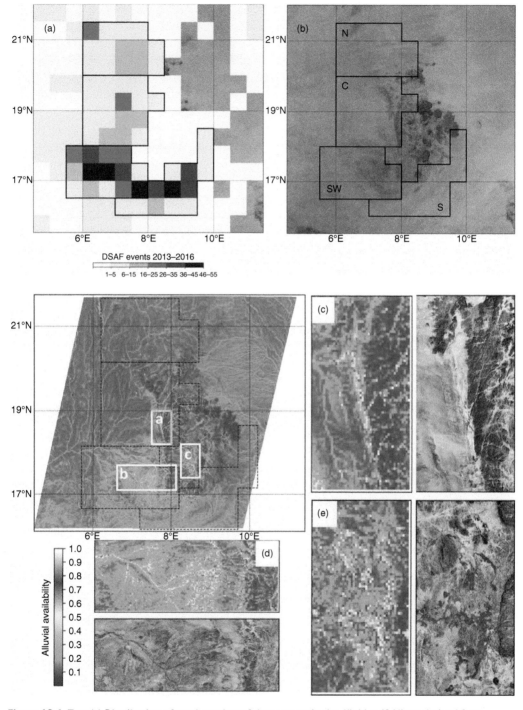

Figure 10.6 Top: (a) Distribution of total number of dust events in the Aïr Massif, Niger, derived from observations of Meteosat Second Generation Desert-Dust-RGB images from 2013 to 2016; DSAF = dust source activation frequency. (b) True color image of the study area showing four dust hot-spot zones. Bottom: Alluvial source sediment maps (AFMs) and Sentinel-2 images of three subset areas. (c), (d) and (e) show subsets of the AFM map within the study area. Values close to 1 represent high availability of alluvial sediments as dust sources. From Feuerstein and Schepanski (2019).

and abrade rock surfaces, forming ventifacts (wind-abraded rock clasts) and yardangs (wind-abraded streamlined and partly aerodynamic landforms). Strong winds may topple trees and thus compromise slope stability, down powerlines, block roads, and damage buildings. In rugged mountain terrain, tree fall resulting from prevailing winds may be responsible for the preferential occurrence of landslides on windward slopes (Rulli et al. 2007).

Floods and landslides count among the most frequently reported geomorphic consequences of storms. Storms can deliver large amounts of precipitation in the form of rain,

hail, or snow, causing flooding. Hydrostatic loading and cyclic wave pumping are two possible mechanisms for multiple slope failures that occurred in a submarine canyon of the Ganges-Brahmaputra river delta after it was hit by Cyclone Sidr in 2007 (Rogers and Goodbred 2010). Heavy rainfall associated with storms such as tropical cyclones triggers dozens to thousands of landslides, mainly through increases of pore-water pressures in hillslope materials (Figure 10.7). Rainstorms are especially effective in denuding hillslopes where previous disturbances have made large amounts of material available. A study of

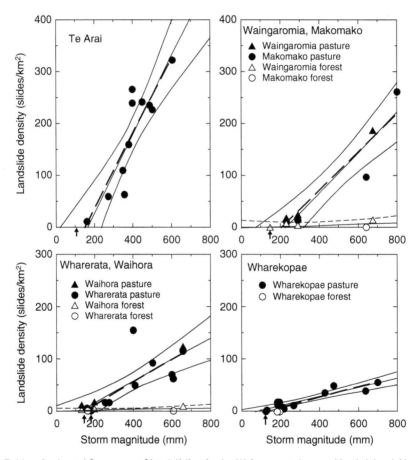

Figure 10.7 Magnitude and frequency of landsliding in the Waipaoa catchment, North Island, New Zealand. The figures show the relationship between flood magnitude and areal landslide density for forest and pasture areas within four landslide-prone land systems in the catchment. Curved lines indicate the 95% confidence interval for unconstrained regressions (solid lines), and bold dashed lines indicate regressions constrained to fit observations of threshold magnitudes for landslide generation. Maximum magnitudes for which landslides were not generated are indicated by arrows on the x-axis. From Reid and Page (2003).

typhoon-triggered landslides in the Tachia River catchment, Taiwan, pointed to the frequent renewal and enlargement of landslide scars that were created previously during the 1999 Chi-Chi earthquake. Typhoons Toraji and Mindulle reactivated 59–66% of the coseismic landslide areas in the catchment in the first five years after the earthquake (Chuang et al. 2009). Landslides often cluster most densely in the areas of highest rainfall intensity during such storms. Examples of this spatial coincidence can also be found in areas outside the reach of tropical storms. In the central Appalachians of Virginia, average catchment denudation during historic storms correlates with rainfall totals recorded at climate stations (Eaton et al. 2003). In that case, rainfall-triggered debris flows were responsible for the bulk of the transported sediment. Eaton et al. (2003) inferred on the basis of radiocarbon dating of prehistoric debris-flow deposits that such events occurred only every 2000–4000 years on average, most likely during rare and large storms.

Storm floods can catastrophically widen and flatten river channels by causing bank erosion and redeposition of coarse materials in channels and on floodplains or even low-lying terraces. Tropical cyclones have caused sedimentation of up to several metres in places along the coast. Where the average return periods of such storms are known, we can extrapolate to obtain mean sedimentation rates that range mostly between 10^0 and 10^2 mm yr^{-1} in episodically disturbed channel and floodplains reaches.

Hobley et al. (2012) inferred the size of the unusual 2010 South Asian summer monsoon storms in the semiarid high mountain desert of Ladakh, India, from the geomorphic signature of destructive debris flows triggered by the storms. The team mapped channels impacted by debris flows and used hydraulic and hydrologic back-calculations to arrive at local precipitation estimates from climate stations and satellite observations. Stolle et al. (2015) studied the same storm event and concluded that such reconstructions could be subject to large errors that depend largely on the choice of the hydraulic model used to reconstruct discharge from debris flows, but also on inaccuracies involved in reconstructing the impacted channel geometry from field evidence.

Lightning strikes give rise to wildfires that remove protective vegetation and increase the likelihood of subsequent erosion and landslides (see Chapter 13). Wildfires affect soils differently depending on the vegetation type, the type and moisture content of the soil and the duration and intensity of the fire. The amount and intensity of precipitation after a fire also influence how a wildfire affects soil. Hot fires that scorch dry coarse soil may leave a near-surface, water-repellent layer called a hydrophobic layer. Water repellency is caused by the accumulation of chemicals derived from burning vegetation (e.g. the 'chapparal' vegetation of the San Gabriel Mountains, USA). The water-repellent layer increases surface runoff and erosion because the burned surface lacks vegetation to hold the loose soil above it. Soil above the hydrophobic layer quickly becomes saturated during rains and may wash downslope. Soil erosion and debris flows are common following wildfires, although the effects are variable. Areas that are susceptible to erosion commonly experience increased erosion for a few years after a fire, whereas areas that normally experience little erosion may show little effect. Erosion rates and landslide frequencies are higher on steep slopes charred by a severe burn than on gentler, less severely burned ones. Heavy rains further increase the incidence of erosion and landslides in burned areas (see Chapter 13).

10.3 Geomorphic Tools for Reconstructing Past Storms

The science of reconstructing the magnitude and frequency of prehistoric or only partly instrumented storms from geomorphic and sedimentary evidence is termed palaeotempestology (Nott 2004). The main focus has been on the geomorphic legacy of storms in coastal areas and therefore differs from studies that infer the times and frequency of atmospheric dust transport and storminess from marine sediment or ice cores (Goudie and Middleton 2001). The concept that:

(i) fair-weather periods are dominated by alongshore sediment reworking and coastal sand accumulation; and (ii) monsoon-driven storm periods are characterized by increased wave-energy and offshore-directed downwelling storm flow that occur simultaneously with peak fluvial discharge caused by storm precipitation ('storm-floods')

has also been embraced in longer-term depositional models of continental margins (Collins et al. 2017). These models emphasize the sedimentary diagnostics of sequences of storms in the geological record.

10.3.1 Coastal Settings

Storms reshape coastlines by removing and adding sediment (Figure 10.8). Tropical cyclones leave tell-tale sediments and landforms, but they must be carefully documented and interpreted to avoid misinterpretations and inaccurate assessments, particularly when it comes to relating the strength of these storms to local flood stages (Otvos 2011). Deposits of past tropical cyclones have been dubbed tempestites, and most of them have been studied in coastal settings (Nott 2004). Parallel dune ridges, beach berms, and coastal barriers are characteristic sedimentary legacies of storms along coasts. Stacked beach ridges made of coral rubble provide detailed archives of prehistoric tropical cyclones (Cunningham et al. 2011). For example, Nott and Hayne (2001) reported absolute ages of 36 beach ridges composed of coral rubble thought to have been washed ashore by tropical cyclones at six different study sites along Australia's

Great Barrier Reef. The radiocarbon ages of the ridges become systematically older away from the shore, attesting to episodic accretion, and their relative heights above sea level yield information about wave run-up and the potential pressure field in each of the cyclones.

Other landforms created during tropical cyclones include clusters of megablocks, gravel sheets, and rubble banks (or ramparts); these are mostly composed of coral detritus that large waves dislodged from nearby reefs. These landforms record the reach of physical impact by extreme waves hitting the coast. For example, large sea swells during Tropical Cyclone Heta in 2004 dumped coral boulders of several metres in diameter on a marine terrace located 23 m above sea level on Niue, a raised coral atoll in the South Pacific (Terry et al. 2013).

Beach ridges, washover fans, marine sediment sheets, coastal dunes, coral rubble, and young aggradation terraces provide information about the precipitation and wind speed of storms, data that are routinely measured today for most storms. Reconstructing palaeostorms, however, requires that one exclude all other potential processes that might form similar geomorphic and sedimentary evidence. Even agreeing on a systematic and unambiguous terminology for some coastal landforms, especially those found along barrier coasts, could be a step forward: we still have much to learn about how these coasts form and develop over time (Otvos 2012). Low-lying depressions in coastal dune fields can be ideal sites for trapping marine sediment delivered by high waves. Cunningham et al. (2011) combined sedimentological logging of dune outcrops with ground-penetrating radar (GPR) to trace former storm-surge deposits along a 1-km reach of coast near Heemskerk, The Netherlands. The team used optically stimulated luminescence (OSL) dating to infer that the deposits were probably laid down in AD 1775/1776 during a time of increased storminess in northwestern Europe. Where successive storms episodically stack sand sheets on top of each other, the

Figure 10.8 Erosional impacts of three 2017 tropical cyclones on Saint-Martin Island, Lesser Antilles. (a) Exhumed beachrock slabs at Longue Bay. The red arrow shows the 5 m-wide inner part of the beachrock slab that was exhumed by waves during the cyclones. (b) Beach lowering indicated by exhumation of the previously buried lower part of a retaining wall at Red Bay. Note destruction of vegetation in front of the wall. (c) Cyclone-driven sand-dune retreat exposing underlying soil. (d) and (e) Cyclone-generated trenches extending from the upper beach to the inner land area at Orient Bay North and Embouchure Bay, respectively. (f) Marked soil scouring over a distance of 20 m at the southern end of Guana Bay. Before the cyclone, this area was entirely covered by dense vegetation. From Duvat et al. (2019).

resulting layers can be read like an open book. For example, a sequence of 32 coarse washover deposits with occasional rip-up clasts in a brackish coastal pond near Cape Cod, Massachusetts, offers a chronology of hurricanes in the western North Atlantic over the past 2000 years, constrained by ^{14}C and ^{137}Cs ages and lead pollution layers (Figure 10.9) (Donnelly et al. 2015). This detailed storm archive records that hurricanes were making landfall

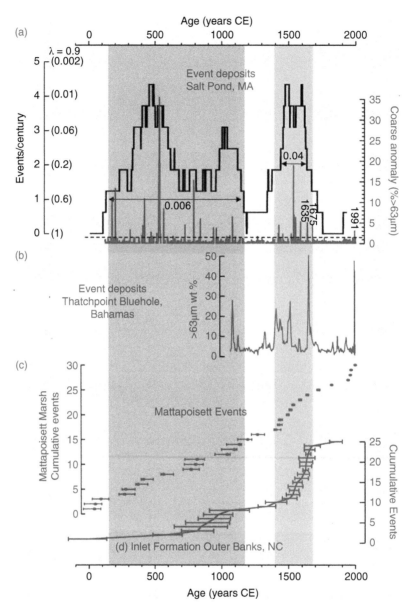

Figure 10.9 Comparison of hurricane proxy records from the east coast of North America. (a) Coarse anomaly plot from Salt Pond with event bed threshold of 1.34% coarse material shown as a dashed line. Historical hurricane strikes attributed to event beds are noted. Grey is event frequency with associated Poisson probabilities of occurrence assuming 0.9 events per century (i.e. modern conditions). Arrows are continuous centuries with more than one event per century and two events per century and their associated probabilities under modern conditions. (b) Event beds from a sediment core from Thatchpoint blue hole in the Bahamas. (c) Cumulative event frequency of overwash events preserved in Mattapoisett Marsh, MA. Shading is intervals CE 150–1150 and 1400–1675, when Salt Pond records heightened intense hurricane-related event beds. From Donnelly et al. (2015).

at the site most frequently between 1400 and 1675 AD. European sedimentary coasts might similarly record evidence of extreme storm surges, but systematic studies are few. Swindles et al. (2018) reported traces of the 1953 North Sea storm surge, one of the most disastrous to strike central European coasts in the twentieth century, recorded as a distinct sand layer in a salt marsh on the eastern British coast, and called for reconstructing in more detail the record of other such extreme events (Figure 10.10). Jackson et al. (2019) noted that many transgressive coastal dunes along European coasts were highly active during the Little Ice Age, though mostly between AD 1400 and 1900, partly encroaching on settlements and agricultural lands. Rather than pointing out a single mechanism, they largely identified the combination of enhanced storminess, effects of sea-level rise, and human disturbance of coastal dunes as the main drivers of the increased dune mobility.

These and other sedimentary archives illuminate past storm chronologies. However, May et al. (2013) pointed out that, judging from the contemporary frequency of large tropical cyclones, over 100 000 tropical cyclones would have occurred during the past 8000 years. Detailed geomorphic and sedimentary studies are patchy at best in the light of this potentially overwhelming legacy of Holocene storms. Their review of studies devoted to extracting storm data from washover deposits and beach ridges shows in the United States and Australia that surveys rarely reconstructed more than 20 storm deposits each, with average inferred storm frequencies ranging between five and 400 years. Hence, our current knowledge of the time series of prehistoric storms is far from complete, and a large mismatch remains between historically recorded storms and those inferred from geomorphic and sedimentary evidence.

Coastal boulder deposits can also provide information on former storms. Entire nests of boulders, for example, fringe cliff tops and beaches on the south coast of the Reykjanes

Peninsula of Iceland. The largest boulders, up to 70 t in mass, are as far as 65 m inland, and smaller boulders lie some 200 m inland (Etienne and Paris 2010). Assuming that this coast very rarely experiences tsunamis, makes these boulder deposits a potentially useful proxy of the wave regime during the largest Atlantic storms. Nonetheless, a debate about whether wave-transported coastal boulders or 'megaclasts' in Iceland or elsewhere have a storm or tsunami origin continues. In the case of Iceland, either hypothesis seems equally testworthy at a given coastal site (see Chapter 9.3), given the country's very active seismic and volcanic setting.

10.3.2 Inland Settings

Fewer studies have attempted to track prehistoric storms in inland settings, given that detailed time series on wind speeds and precipitation are the basis for creating robust assessments of the frequency and magnitude of storms. Finding geomorphic evidence of prehistoric storms in mountainous terrain is often a matter of coincidence or inference because climate stations are sparse, and because the high rates of erosion and deposition can swiftly obliterate sedimentary and geomorphic evidence.

One novel method for augmenting or independently testing the palaeostorm record derived from geomorphic and sedimentary evidence takes advantage of the isotopic composition of speleothems. Haig et al. (2015) measured the $^{18}O/^{16}O$ ratio of seasonally accreting carbonate layers of actively growing stalagmites from Queensland and Western Australia. Rainfall brought by tropical cyclones making landfall in Australia is depleted in ^{18}O by >0.6% compared to monsoonal rainfall, mainly because of water recycling and condensation efficiency in large and long-lived storms. The ratio of $\delta^{18}O$ in the dark calcites in the stalagmites reflects wet season storms and depends on the lifespan, intensity, and source region of each tropical cyclone. According

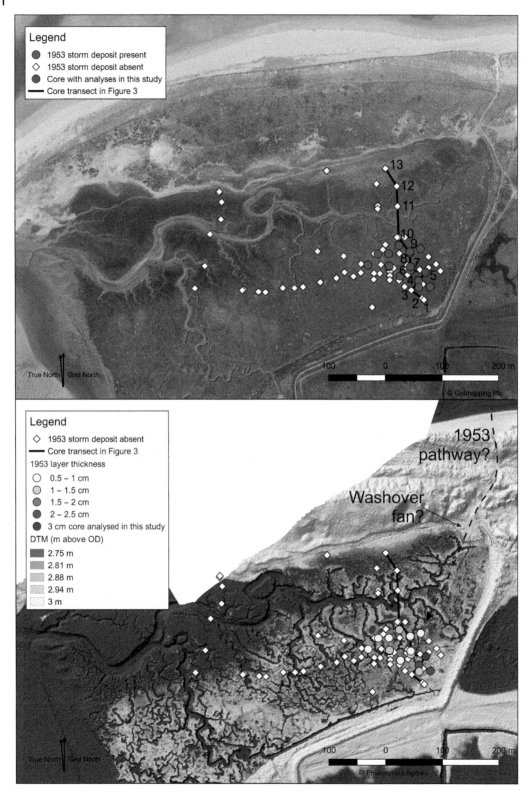

Figure 10.10 Maps showing the spatial distribution of the 1953 storm deposit at Holkham, UK. Top: Aerial photograph of the salt marsh showing the distribution of the 1953 storm sand layer based on acquired cores. Bottom: Thickness of the 1953 storm sand within the cores. The suggested pathway of the storm wave is indicated by a dashed arrow. From Swindles et al. (2018).

to this geochemical proxy, Australia may be currently experiencing the lowest tropical cyclone activity of the past 500 to 1500 years.

Lake sediments can also store important information about past storms, judging from a detailed investigation of a 27-m core from landslide-dammed Lake Tutira in the rolling hill country of New Zealand's North Island. The core contains 1400 layers of mostly terrestrial sediment that Page et al. (2010) interpreted to be storm-derived deposits. The lake was dammed by a landslide ~7200 cal yr BP and offers a detailed archive of the changing magnitude and frequency of past storms over that period, including a phase of nearly twice the frequency of contemporary storm activity ~2000 years ago. The high average sedimentation rate of 3.3 mm yr^{-1} in the lake attests to the geomorphic work of storms. About 55 of these 1400 storms may have caused distinct sediment pulses, perhaps similar to that unleashed by Cyclone Bola that struck North Island in 1988, triggering tens of thousands of landslides in largely deforested pastoral hill terrain underlain by weak mudstones and sandstones and easily erodible regolith (Carter et al. 2010).

10.4 Naturally Oscillating Climate and Increasing Storminess

While single storms are weather-related, long sequences of storms can be part of climate oscillations that happened long before humans started to influence the atmosphere's chemical composition and temperature field detectably. Climate change is a natural process that has occurred throughout our planet's history and is still happening, even if betraying an increasing distinct human fingerprint.

El Niño–Southern Oscillation (ENSO) refers to natural fluctuations in the tropical atmosphere and ocean over phases of up to several years. The ENSO phenomenon arises from the dynamic coupling of wind fields, sea-surface temperature, and sea level that alternate between two end-member states. The warm phase of ENSO is called El Niño, and its counterpart cold phase is called La Niña. Intermediate conditions rarely persist for long as the system swings between the two extreme phases. We may view ENSO as a quasideterministic see-saw between these different states. It is deterministic because the physical processes and equations behind the phenomenon are fairly well understood. The prefix 'quasi-' emphasizes that ENSO phases can be predicted for a few seasons in advance at best. One measure of ENSO strength is the Southern Oscillation Index (SOI), which is the standardized air pressure difference between Papeete, Tahiti, and Darwin, Australia, measured at sea level. The dynamics of ENSO are tied to the Walker circulation of surface air masses from east to west, and episodically changing sea levels and precipitation patterns in the equatorial Pacific.

The El Niño–Southern Oscillation (ENSO) describes one of Earth's largest natural climate fluctuations. ENSO strongly influences climate in large parts of the tropical Pacific Ocean, Indonesia, Australasia, tropical South America, and even parts of the North Pacific. It causes precipitation anomalies with a periodicity of between about two and seven years. The characteristics of the warm El Niño phase are warming of surface and near-surface waters in the eastern tropical Pacific, collapse of the trade winds, and thunderstorm activity migrating eastward from Indonesia to the central Pacific. El Niño events tend to bring abnormally dry conditions in Northern Australia, Indonesia, and the Philippines. Drier-than-normal conditions may also prevail in southeast Africa and northern Brazil. The most severe drought in the Amazon basin occurred in 2010 during the early stages of an El Niño when many river

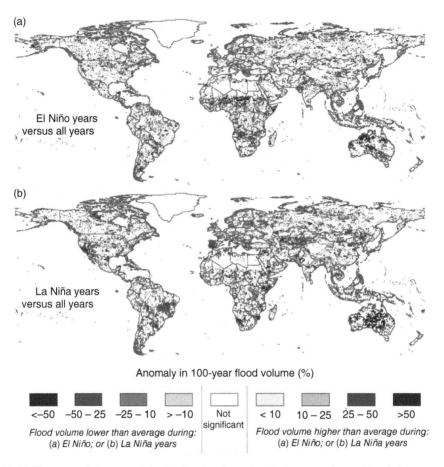

Figure 10.11 The strong influence of the El Niño Southern Oscillation on estimated flood risk around the world. Percentage anomaly in flood volumes with return periods of 100 years during (a) El Niño years and (b) La Niña years (compared with all years). From Ward et al. (2014).

levels dropped to unprecedented levels, cutting off whole communities that live on the floodplains of the Amazon and depend on rivers for daily transport (Marengo et al. 2011). Conversely, the arid to hyperarid coasts of Peru and northern Chile receive heavy rainfall during El Niño phases, triggering flash floods and debris flows in the most extreme cases (Vargas et al. 2006); during those phases river discharge and sediment fluxes can be multiples of their long-term average values (Tote et al. 2011).

The number and strength of ENSO teleconnections, or the long-distance effects, decrease towards the poles; nevertheless these effects underscore the potential global reach of this phenomenon. In a global analysis, Ward et al.

(2014) showed that the likelihood of flooding differed between El Niño and La Niña phases over more than one-third of the Earth's land surface (Figure 10.11). They used worldwide estimates of the exposed population, gross domestic product, and urban damage from floods to compute ENSO-driven anomalies in flood risk and found that the nationwide annual damage deviated by more than 50% during warm or cold phases of the oscillation. During the Northern Hemisphere summer, Indian monsoon rainfall tends to be lower than normal during El Niño years, especially in the northwestern parts of the subcontinent and East Asia. River discharge and sediment transport in major East Asian rivers such as the Yangtze and the Mekong systematically vary with ENSO and are also lower during El Niño years (Wang et al. 2011). In contrast, wetter-than-normal conditions prevail along the west coast of tropical South America, the North American Gulf Coast, and from southern Brazil to central Argentina. Winters are warmer in the north-central United States, but cooler in the southeastern and southwestern parts of the country (Viles and Goudie 2002). Some 300 km of southern California coastline retreated 10 m landward on average because of higher wave erosion during the 2015–2016 El Niño, and comparable coastal losses were recorded during the 2009–2010 El Niño (Young et al. 2018). Tropical cyclones occur more frequently in the Atlantic Ocean during La Niña years than they do during El Niño years, such that the likelihood that a hurricane will strike North America is about three times higher. The opposite holds for the western Pacific, where tropical cyclones are more frequent during El Niño phases (McPhaden et al. 2006). ENSO also has an effect on the interannual variability of the global carbon cycle. Drought and wildfires during El Niño years noticeably increase CO_2 emissions (see Chapter 17.1).

El Niño is also responsible for diverse indirect or intangible damages. Examples include a marked decrease in primary productivity in the tropical Pacific and high animal mortality along western South American coastlines (McPhaden et al. 2006). The El Niño in 1997/1998 was the strongest ever recorded and was accompanied by equatorial sea-surface temperatures 3–5° C above normal, especially in the tropical Indian Ocean. This heating resulted in up to 90% mortality of shallow reef corals in Sri Lanka, the Maldives, India, Kenya, Tanzania, and the Seychelles. Reef mortality was still 50% in waters >20 m deep, and an estimated 16% of reef-building corals were killed worldwide (Wilkinson et al. 1999). But even weaker El Niño phases have had distinct teleconnections. In Italy, for example, the intensity of torrential rainfall (>128 mm d^{-1}) increased fourfold during the

period 1951–1995, with peaks coinciding with El Niño years (Alpert et al. 2002). Indeed, an ENSO signal has been detectable over much of Europe in the past few centuries, although other climate controls, such as the North Atlantic Oscillation (NAO) or the atmospheric effects of volcanic eruptions, also had strong contributions (Brönnimann et al. 2006).

Heavy rainfall and strong wave action during El Niños accelerate erosion of coastal cliffs in California, mainly through landsliding (Hapke and Green 2006). Dryland areas also suffer impacts of this natural climate oscillation. In Egypt about 25% of the variability in the flow of the Nile River may be linked to ENSO phases (Eltahir 1996). In eastern Australia, La Niña phases are accompanied by higher precipitation, higher river discharge, and flooding of its internally drained lakes (Kiem et al. 2003). Higher flood risks in New South Wales have also been tied to La Niña years, although another regional climate oscillation – the Interdecadal Pacific Oscillation (IPO) – also plays a role in determining these impacts (Kiem, Franks and Kuczera 2003). The extensive Fitzroy River floods in the austral summer of 2010–2011 have been linked to La Niña. The flooding damaged 22 cities including parts of Brisbane, Australia's third largest city, where 200 000 people were affected. The total area covered by floodwaters equalled the area of Germany and France combined. In southeastern Australia, these floods were followed by a nine-year drought that was likely the most severe since 1783 (van Dijk et al. 2013).

In essence, some researchers consider ENSO to be the most important global-scale climate oscillation driving year-to-year weather variability and extreme events (McPhaden et al. 2006). Hundreds of millions of people are exposed to disasters precipitated by ENSO events. It is the onset and early stages of El Niño phases that have affected people the most (Bouma et al. 1997).

Most climate-change scenarios for the twenty-first century predict more pronounced weather extremes, and many climatological time series support this prediction. Extreme monsoon-related rainfall increased significantly from 1951 to 2000, whereas more moderate rainfall events have declined, thus masking changes in average trends (Goswami et al. 2006). The occurrence of tornadoes in the United States has changed dramatically since the early 1970s. The number of days with at least one tornado has dropped by one-third, while the probability of several tornadoes happening during any given tornado day has increased, and these storms have become more clustered in their occurrence (Figure 10.12) (Brooks et al. 2014). Japan's Ministry of Land, Infrastructure and Transport (MLIT, www .mlit.go.jp) reported a gradual increase in extreme rainfall over the past few decades (Figure 10.13). According to this study, rainfall events with intensities higher than 50 mm h^{-1} have increased by about 50%. Whereas between 1977 and 1986, an average of 200 such events happened each year, the number rose to an average of 313 per year in the decade from 1997 to 2006. The occurrence of intense rainfall surpassing 100 mm h^{-1} has more than doubled between these two periods. In Japan, these rainfall intensities commonly give rise to damaging landslides; a detailed inventory records over 4700 such incidences in minute detail between 2001 and 2011 (Figure 10.14) (Saito et al. 2014).

With warmer temperatures, the atmosphere will be able to hold more water vapour (about 7% per 1 K of warming), thus supplying storms with more moisture and favouring more extreme thunderstorms and tropical and extratropical cyclones (Trenberth 2011). Some debate has centred on the potential impacts of atmospheric warming on the frequency and magnitude of tropical cyclones. Using the summed maximum sustained wind speed measured in a tropical cyclone as a simplified metric of energy dissipation, Emanuel (2005) argued that tropical cyclones have become

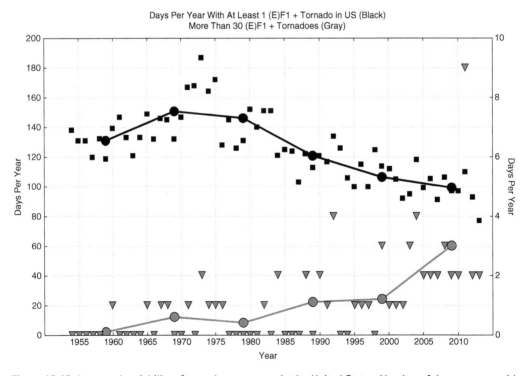

Figure 10.12 Increased variability of tornado occurrence in the United States. Number of days per year with at least one and more than 30 (E)F1+ tornadoes (on the Enhanced Fujita scale; see Figure 10.1). Black squares indicate one (E)F1+ tornado, and gray triangles indicate more than 30. Large dots and lines are decadal means. Modified from Brooks et al. (2014).

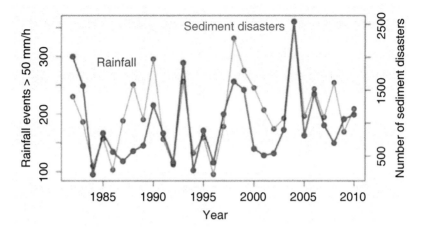

Figure 10.13 Reported rainfall events with intensities of > 50 mm/h and sediment disasters in Japan. Data from http://www.mlit.go.jp/english/white-paper/2011.pdf

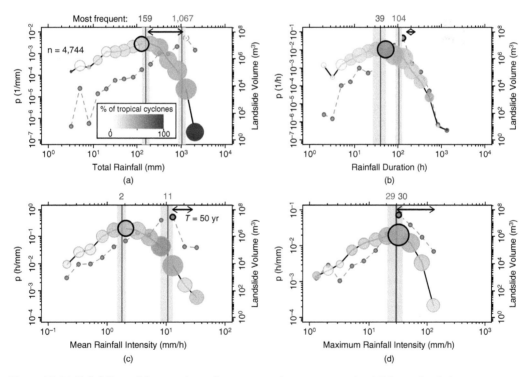

Figure 10.14 Rainfall conditions, typhoon frequency, and contemporary landslide erosion in Japan. Log-binned probability density estimates of rainfall parameters (blue solid curves) and total landslide volume (orange dashed curves for (a) rainfall totals, (b) rainfall duration, (c) rainfall intensity, and (d) maximum intensity. Vertical lines and grey boxes are means and ±2 sigma error bars (based on multiple bin widths) of rainfall parameters that were most frequent (blue), and associated with the highest total landslide volumes from 2001 to 2011 (orange). Black circles outline rainfall parameter values most frequently associated with reported landslides. Bubble size is scaled to total landslide volume per bin; bubble brightness represents percentage of tropical cyclones out of the total number of landslide rainfall events; darker tones indicate higher fractions. Black arrows span the range of 50-year rainfall events computed for two stations (Abashiri and Owase) that are in regions of the lowest and highest frequencies of heavy rainfall in Japan, respectively. From Saito et al. (2014).

stronger since the first available measurements in the 1930s and that this trend correlates with increases in sea-surface temperature. However, part of this trend may be poorly constrained because reliable measurements of wind speeds and storm locations have become available with improved satellite monitoring only since the early 1990s. Potential biases in older data must be dealt with accordingly. Some of the statistical treatment to remove potential bias in older data may be questionable and thus undermine such findings (Knutson et al. 2010). The assertion that hurricanes have become more destructive has met criticism on the basis of data on hurricane-induced financial losses normalized by trends in inflation, population, and wealth. The data document an average per-storm loss of US\$ 9.3 billion for 40 large storms – defined here as a storm causing at least US\$ 1 billion damage – between 1900 and 1950. The subsequent five decades had 46 large storms with an average damage of US\$ 7.0 billion, though without any statistical evidence of change (Pielke 2005).

Measured increases in sea-surface temperatures (SSTs) of ~0.2–0.4 K between 1986 and 2005 might be linked to more intense and longer-lived tropical cyclones in the North Atlantic, but fewer and less energetic tropical cyclones in the Pacific (Klotzbach 2006). Yet for nearly the same observation period, Elsner et al. (2008) argued that only the highest wind speeds of tropical cyclones increased measurably through time, supporting the idea that warmer oceans provide more energy for generating stronger storms. In contrast, a statistical analysis of tropical cyclones in the Gulf of Mexico, showed a decline in frequency with higher coastal sea-surface temperatures (Trepanier et al. 2015), pointing out that we cannot necessarily expect more landfalling cyclones because of near-shore waters that are warming locally. Instead, rising sea-surface temperatures matter more in areas where tropical cyclones form in the first place.

Atmospheric conditions also come into play. For example, a sixfold increase in anthropogenic aerosol emissions since the 1930s may have played a role in lowering the vertical wind shear over the Arabian Sea, thus raising the potential for tropical cyclone formation. Increases in the premonsoon intensity of tropical cyclones in that area are consistent with a concomitant increase in black carbon and sulfate emissions, suggesting that growing air pollution adds to higher tropical storm intensities over the Arabian Sea (Evan et al. 2011). Tropical cyclones also appear to have moved poleward at rates of 5.3 and 6.2 km yr^{-1} from 1982 to 2012 in the northern and southern hemispheres, respectively (Kossin et al. 2015). In contrast, satellite data of cloud-top temperatures obtained over a 30-year period showed an unchanged tropical cyclone energy. Global circulation models (GCMs) predict a 30% decrease of tropical cyclone activity in the western Pacific, with lowered maximum wind speeds, but the opposite for the North Atlantic (Oouchi et al. 2006). In eastern Asia, the centres of tropical cyclones seem to have shifted closer toward the densely populated coasts of China, Korea, and Japan between 1977 and 2010 (Park et al 2014). The distribution of land and sea in that area is likely to complicate any similar trends of rising rainfall intensities.

References

Alpert P, Ben-Gai T, Baharad A, et al. 2002 The paradoxical increase of Mediterranean extreme daily rainfall in spite of decrease in total values. *Geophysical Research Letters* **29** (10), 1–4.

Baddock MC, Strong CL, Murray PS, and McTainsh GH 2013 Aeolian dust as a transport hazard. *Atmospheric Environment* **71** (C), 7–14.

Bazelmans J, Meier D, Nieuwhof A, et al. 2012 Understanding the cultural historical value of the Wadden Sea region. The co-evolution of environment and society in the Wadden Sea area in the Holocene up until early modern times (11,700 BC-1800 AD): An outline. *Ocean & Coastal Management* **68** (C), 114–126.

Besset M, Anthony EJ, Dussouillez P, and Goichot M 2017 The impact of Cyclone Nargis on the Ayeyarwady (Irrawaddy) River delta shoreline and nearshore zone (Myanmar): Towards degraded delta resilience?. *Comptes Rendus Geoscience* **349** (6-7), 238–247.

Bouma MJ, Kovats RS, Goubet SA, et al. 1997 Global assessment of El Nino's disaster burden. *The Lancet* **350** (9089), 1435–1438.

Brönnimann S, Xoplaki E, Casty C, et al. 2006 ENSO influence on Europe during the last centuries. *Climate Dynamics* **28** (2-3), 181–197.

Brooks HE, Carbin GW, and Marsh PT 2014 Increased variability of tornado occurrence in the United States. *Science* **346** (6207), 349–352.

Bryant E 2004 *Natural Hazards*, 2nd edn. Cambridge University Press.

Carter L, Orpin AR, and Kuehl SA 2010 From mountain source to ocean sink – the passage of sediment across an active margin, Waipaoa Sedimentary System, New Zealand. *Marine Geology* **270** (1-4), 1–10.

Chuang SC, Chen H, Lin GW, et al. 2009 Increase in basin sediment yield from landslides in storms following major seismic disturbance. *Engineering Geology* **103** (1-2), 59–65.

Collins DS, Johnson HD, Allison PA, et al. 2017 Coupled 'storm-flood' depositional model: Application to the Miocene-Modern Baram Delta Province, north-west Borneo. *Sedimentology* **64** (5), 1203–1235.

Conery I, Walsh JP, and Corbett DR 2018 Hurricane overwash and decadal-scale evolution of a narrowing barrier island, Ocracoke Island, NC. *Estuaries and Coasts* **41**, 1626–1642.

Criss RE and Shock EL 2001 Flood enhancement through flood control. *Geology* **10**, 875–878.

Cunningham AC, Bakker MAJ, van Heteren S, et al. 2011 Extracting storm-surge data from coastal dunes for improved assessment of flood risk. *Geology* **39** (11), 1063–1066.

Dagsson-Waldhauserova P, Arnalds O. and Olafsson H 2013 Long-term frequency and characteristics of dust storm events in Northeast Iceland (1949–2011). *Atmospheric Environment* **77** (C), 117–127.

Derbyshire E 2000 Geological hazards in loess terrain, with particular reference to the loess regions of China. *Earth-Science Reviews* **54** (1), 231–260.

Donnelly JP, Hawkes AD, Lane P, et al. 2015 Climate forcing of unprecedented intense-hurricane activity in the last 2000 years. *Earth's Future* **3**, 49–65.

Duvat V, Pillet V, Volto N, et al. 2019 High human influence on beach response to tropical cyclones in small islands: Saint-Martin Island, Lesser Antilles. *Geomorphology* **325**, 70–91.

Eaton LS, Morgan BA, Kochel RC, and Howard AD 2003 Role of debris flows in long-term landscape denudation in the central Appalachians of Virginia. *Geology* **31** (4), 339–342.

Elsner JB, Kossin JP, and Jagger TH 2008 The increasing intensity of the strongest tropical cyclones. *Nature* **455** (7209), 92–95.

Eltahir EAB 1996 El Niño and the natural variability in the flow of the Nile River. *Water Resources Research* **32** (1), 131–137.

Emanuel K 2005 Increasing destructiveness of tropical cyclones over the past 30 years. *Nature* **436** (7051), 686–688.

Etienne S and Paris R 2010 Boulder accumulations related to storms on the south coast of the Reykjanes Peninsula (Iceland). *Geomorphology* **114** (1-2), 55–70.

Evan AT, Kossin JP, Chung CE, and Ramanathan V 2011 Arabian Sea tropical cyclones intensified by emissions of black carbon and other aerosols. *Nature* **479** (7371), 94–97.

Feuerstein S and Schepanski K 2019 Identification of dust sources in a Saharan dust hot-spot and their implementation in a dust-emission model. *Remote Sensing* **11**, 1–24.

Fritz HM, Blount CD, Thwin S, et al. 2009 Cyclone Nargis storm surge in Myanmar. *Nature Geoscience* **2** (7), 448–449.

Gao Y, Wang H, Liu GM, et al. 2014 Risk assessment of tropical storm surges for coastal regions of China. *Journal of Geophysical Research: Atmospheres* **119** (9), 5364–5374.

Geiger T, Frieler K and Bresch DN 2018 A global historical data set of tropical cyclone exposure (TCE-DAT). *Earth System Science Data* **10**, 185–194.

Goswami BN, Venugopal V, Sengupta D, et al. 2006 Increasing trend of extreme rain events over India in a warming environment. *Science* **314** (5804), 1442–1445.

Goudie AS and Middleton NJ 2001 Saharan dust storms: nature and consequences. *Earth-Science Reviews* **56** (1), 179–204.

Haig J, Nott J, and Reichart GJ 2015 Australian tropical cyclone activity lower than at any time over the past 550–1,500 years. *Nature* **505** (7485), 667–671.

Hapke CJ and Green KR 2006 Coastal landslide material loss rates associated with severe climatic events. *Geology* **34** (12), 1077.

Hawkes AD and Horton BP 2012 Sedimentary record of storm deposits from Hurricane Ike, Galveston and San Luis Islands, Texas. *Geomorphology* **171-172** (C), 180–189.

Hobley DEJ, Sinclair HD, and Mudd SM 2012 Reconstruction of a major storm event from its geomorphic signature: The Ladakh floods, 6 August 2010. *Geology* **40** (6), 483–486.

Hoeke RK, McInnes KL, Kruger JC, et al. 2013 Widespread inundation of Pacific islands triggered by distant-source wind-waves. *Global and Planetary Change* **108** (C), 128–138.

Jackson DWT, Costas S, and Guisado-Pintado E 2019 Large-scale transgressive coastal dune behaviour in Europe during the Little Ice Age. *Global and Planetary Change* **175**, 82–91.

Kiem AS, Franks SW, and Kuczera G 2003 Multi-decadal variability of flood risk. *Geophysical Research Letters* **30** (2), 1–4.

Klotzbach PJ 2006 Trends in global tropical cyclone activity over the past twenty years (1986–2005). *Geophysical Research Letters* **33** (10), 1–4.

Knutson TR, McBride JL, Chan J, et al. 2010 Tropical cyclones and climate change. *Nature Geoscience* **3** (3), 157–163.

Komar PD, Allan JC, and Ruggiero P 2013 U.S. Pacific Northwest coastal hazards: Tectonic and climatic controls. In *Coastal Hazards* (ed Finkl CW), Springer Science & Business Media, pp. 587–674.

Kossin JP, Emanuel KA, and Vecchi GA 2015 The poleward migration of the location of tropical cyclone maximum intensity. *Nature* **509** (7500), 349–352.

Lampe M and Lampe R 2018 Evolution of a large Baltic beach ridge plain (Neudarss, NE Germany): A continuous record of sea-level and wind-field variation since the Homeric Minimum. *Earth Surface Processes and Landforms* **43** (15), 3042–3056.

Mahabot MM, Pennober G, Suanez S, et al. 2017 Effect of tropical cyclones on short-term evolution of carbonate sandy beaches on Reunion Island, Indian Ocean. *Journal of Coastal Research* **33** (4), 839–853.

Mahowald NM, Baker AR, Bergametti G, et al. 2005 Atmospheric global dust cycle and iron inputs to the ocean. *Global Biogeochemical Cycles* **19** (4), 1–15.

Marengo JA, Tomasella J, Alves LM, et al. 2011 The drought of 2010 in the context of historical droughts in the Amazon region. *Geophysical Research Letters* **38** (12), 1–5.

Masselink G and van Heteren S 2014 Response of wave-dominated and mixed-energy barriers to storms. *Marine Geology* **352** (C), 321–347.

May SM, Engel M, Brill D, et al. 2013 Coastal hazards from tropical cyclones and extratropical winter storms based on holocene storm chronologies. In *Coastal Hazards* (ed Finkl CW), Springer Science & Business Media, pp. 557–585.

McNamara DE and Werner BT 2008 Coupled barrier island–resort model: 1. Emergent instabilities induced by strong human-landscape interactions. *Journal of Geophysical Research: Earth Surface* **113**, 1–10.

McPhaden MJ, Zebiak SE, and Glantz MH 2006 ENSO as an integrating concept in earth science. *Science* **314** (5806), 1740–1745.

McTainsh G and Strong C 2007 The role of aeolian dust in ecosystems. *Geomorphology* **89** (1-2), 39–54.

Morales JA, Borrego J, and Davis, Jr RA 2014 A new mechanism for chenier development and a facies model of the Saltés Island chenier plain (SW Spain). *Geomorphology* **204** (C), 265–276.

Nikitina DL, Kemp AC, Horton BP, et al. 2014 Storm erosion during the past 2000 years along the north shore of Delaware Bay, USA. *Geomorphology* **208** (C), 160–172.

Nott J 2004 Palaeotempestology: the study of prehistoric tropical cyclones—a review and implications for hazard assessment. *Environment International* **30** (3), 433–447.

Nott J and Hayne M 2001 High frequency of 'super-cyclones' along the Great Barrier Reef over the past 5,000 years. *Nature* **413** (6855), 508–512.

Oouchi K, Yoshimura J, Yoshimura H, et al. 2006 Tropical cyclone climatology in a global-warming climate as simulated in a 20 km-mesh global atmospheric model: Frequency and wind intensity analyses. *Journal of the Meteorological Society of Japan* **84** (2), 259.

Otvos EG 2011 Hurricane signatures and landforms—toward improved interpretations and global storm climate chronology. *Sedimentary Geology* **239** (1-2), 10–22.

Otvos EG 2012 Coastal barriers – Nomenclature, processes, and classification issues. *Geomorphology* **139-140** (C), 39–52.

Page MJ, Trustrum NA, Orpin AR, et al. 2010 Storm frequency and magnitude in response to Holocene climate variability, Lake Tutira, North-Eastern New Zealand. *Marine Geology* **270** (1-4), 30–44.

Park DSR, Ho CH, and Kim JH 2014 Growing threat of intense tropical cyclones to East Asia over the period 1977–2010. *Environmental Research Letters* **9** (1), 014008.

Peduzzi P, Chatenoux B, Dao H, et al. 2012 Global trends in tropical cyclone risk. *Nature Climate Change* **2** (4), 289–294.

Pielke RA 2005 Meteorology: Are there trends in hurricane destruction?. *Nature* **438** (7071), E11–E11.

Quataert E, Storlazzi C, van Rooijen A, et al. 2015 The influence of coral reefs and climate change on wave-driven flooding of tropical coastlines. *Geophysical Research Letters* **42**, 6407–6415.

Reed AJ, Mann ME, Emanuel KA, et al. 2015 Increased threat of tropical cyclones and coastal flooding to New York City during the anthropogenic era. *Proceedings of the National Academy of Sciences* **112** (41), 12610–12615.

Reid LM and Page MJ 2003 Magnitude and frequency of landsliding in a large New Zealand catchment. *Geomorphology* **49**, 71–88.

Roelvink D, Reniers A, van Dongeren A, et al. 2009 Modelling storm impacts on beaches, dunes and barrier islands. *Coastal Engineering* **56** (11-12), 1133–1152.

Rogers KG and Goodbred SL 2010 Mass failures associated with the passage of a large tropical cyclone over the Swatch of No Ground submarine canyon (Bay of Bengal). *Geology* **38** (11), 1051–1054.

Rulli MC, Meneguzzo F, and Rosso R 2007 Wind control of storm-triggered shallow landslides. *Geophysical Research Letters* **34** (3), 1–5.

Saito H, Korup O, Uchida T, et al. 2014 Rainfall conditions, typhoon frequency, and contemporary landslide erosion in Japan. *Geology* **42** (11), 999–1002.

Sallenger, Jr AH 2000 Storm impact scale for barrier islands. *Journal of Coastal Research* **16** (3), 890–895.

Shao Y and Dong CH 2006 A review on East Asian dust storm climate, modelling and monitoring. *Global and Planetary Change* **52** (1-4), 1–22.

Stolle A, Langer M, Blöthe JH, and Korup O 2015 On predicting debris flows in arid mountain belts. *Global and Planetary Change* **126** (C), 1–13.

Swindles GT, Galloway JM, Macumber AL, et al. 2018 Sedimentary records of coastal storm surges: Evidence of the 1953 North Sea event. *Marine Geology* **403**, 262–270.

Switzer AD 2014 Coastal hazards: Storms and tsunamis. In *Coastal Environments & Global Change* (eds Masselink G and Gehrels R), Wiley Blackwell, pp. 104–127.

Terry JP 2007 *Tropical Cyclones. Climatology and Impacts in the South Pacific*. Springer.

Terry JP, Lau AA, and Etienne S 2013 *Reef-Platform Coral Boulders: Evidence for High-Energy Marine Inundation Events on Tropical Coastlines*. Springer Science & Business Media.

Tote C, Govers G, Van Kerckhoven S, et al. 2011 Effect of ENSO events on sediment production in a large coastal basin in northern Peru. *Earth Surface Processes and Landforms* **36** (13), 1776–1788.

Trenberth KE 2011 Changes in precipitation with climate change. *Climate Research* **47** (1), 123–138.

Trepanier JC, Ellis KN, and Tucker CS 2015 Hurricane risk variability along the Gulf of Mexico coastline. *PLoS ONE* **10** (3), e0118196.

van Dijk AIJM, Beck HE, Crosbie RS, et al. 2013 The Millennium Drought in southeast Australia (2001-2009): Natural and human causes and implications for water resources, ecosystems, economy, and society. *Water Resources Research* **49** (2), 1040–1057.

Vargas G, Rutllant J, and Ortlieb L 2006 ENSO tropical–extratropical climate teleconnections and mechanisms for Holocene debris flows along the hyperarid coast of western South America (17 °–24 °S). *Earth and Planetary Science Letters* **249** (3-4), 467–483.

Viles HA and Goudie AS 2002 Interannual, decadal and multidecadal scale climatic variability and geomorphology. *Earth-Science Reviews* **61** (1), 105–131.

Wang H, Saito Y, Zhang Y, et al. 2011 Recent changes of sediment flux to the western Pacific Ocean from major rivers in East and Southeast Asia. *Earth-Science Reviews* **108** (1-2), 80–100.

Ward PJ, Jongman B, Kummu M, et al. 2014 Strong influence of El Niño Southern Oscillation on flood risk around the world. *Proceedings of the National Academy of Sciences* **111** (44), 15659–15664.

Wilkinson C, Linden O, Cesar H, et al. 1999 Ecological and socioeconomic impacts of 1998 coral mortality in the Indian Ocean: An ENSO impact and a warning of future change? *Ambio* **28**, 188–196.

Woo M-K 2012 *Permafrost Hydrology*. Springer.

Woodruff JD, Irish JL, and Camargo SJ 2013 Coastal flooding by tropical cyclones and sea-level rise. *Nature* **504**, 44–52.

Woodruff JD, Irish JL, and Camargo SJ 2014 Recent shifts in coastline change and shoreline stabilization linked to storm climate change. *Nature* **504** (7478), 569–585.

You ZJ and Nielsen T 2013 Extreme coastal waves, ocean surges and wave runup. In *Coastal Hazards* (ed Finkl CW), Springer Science & Business Media, pp. 677–733.

Young AP, Flick RE, Gallien TW, et al. 2018 Southern California coastal response to the 2015-2016 El Niño. *Journal of Geophysical Research: Earth Surface* **123** (11), 3069–3083.

Zhang XY, Gong SL, Zhao TL, and Arimoto R 2003 Sources of Asian dust and role of climate change versus desertification in Asian dust emission. *Geophysical Research Letters* **30** (24), 1–4.

11

Flood Hazards

Rivers are found almost everywhere on the Earth's land surface and are the prime conduits for the transport of water, sediment, and dissolved solids from catchments into basins inland and the oceans. Understanding the dynamics of rivers requires an understanding of their behaviour in flood, because most changes in channel morphology occur when flow rates are high and thus sediment motion is rapid. We may define a flood as the discharge that exceeds the carrying capacity of the channel, referred to as bank-full discharge, when water spills over onto the floodplain. Yet this definition restricts our understanding of river dynamics to an anthropogenic viewpoint. In geomorphic terms out-of-channel flows are simply part of the continuum of river dynamics. Discerning between active channels and floodplains may be problematic for rivers that convey their discharge highly episodically or along multiple channel threads, such as the large anabranching channels of interior Australia, the extensive braidplains in New Zealand and Alaska, or large inland deltas such as those of the Okavango River in Botswana and the Peace-Athabasca River in Canada. The simple concept of overbank flow causing flooding is problematic to apply in these settings.

Running water is a key agent of moving solid material on Earth. It exerts shear stresses

on the surface over which it flows, while transporting sediment that may abrade and scour bedrock, and it entrains and deposits sediment. The size of river-transported sediment ranges from large boulders to submicron particles. Running water also exerts shear stresses that undermine river banks and hillslopes, thus making more sediment available for downstream transport and enabling the channel to migrate laterally.

The movement of large masses of water across the land surface leads to measurable changes in its gravity field. Satellite sensors measure minute changes in Earth's gravity field and provide information about regional-scale changes in the hydrological cycle. The annual discharge of major Himalayan rivers, for example, involves several cubic kilometres of water that flow through the mountain belt seasonally owing to the South Asian summer monsoon. This seasonal transfer of water can be accompanied by measurable increases in microseismicity due to crustal loading (Bollinger et al. 2007).

Water is also a solvent and rivers transport large amounts of cations and anions. Clay-size particles suspended in river water may absorb chemical species including nutrients and pathogens. At the other end of the grain-size spectrum, bedload may account for up to 35% of the total sediment recorded in databases. Fluvial sediment transport is a classic field of study in quantitative geomorphology and among the key tasks for both hydraulic

Geomorphology and Natural Hazards: Understanding Landscape Change for Disaster Mitigation, Advanced Textbook Series,
First Edition. Tim R. Davies, Oliver Korup, and John J. Clague.
© 2021 John Wiley & Sons Ltd. Co-published 2021 by the American Geophysical Union and John Wiley & Sons Ltd.

engineers and water resource managers. Measuring, estimating, and predicting the concentration and flux of sediment in rivers uses methods ranging from detailed site sampling to modelling and statistical approaches.

We can distinguish between static and dynamic flooding. Static flooding refers to low-velocity overbank flows that inundate floodplains. The areas impacted by floods can be enormous. The great 1998 floods on the Yangtze River, China, inundated 210 000 km² of land, destroyed five million houses, and caused losses of some US$ 20 billion (Piao et al. 2010). Most agricultural land is located in natural or embanked floodplains. Yet the large losses of agricultural areas or crop yields that result from floods hardly make it into disaster databases, at least if comparing nation by nation (Lesk et al. 2016). In contrast, dynamic flooding involves flow speeds of >1 m s^{-1}, a velocity at which hydrodynamic pressure and the physical impact of flood debris contributes strongly to changes in river channels and floodplains. In extreme cases, turbulent runoff forms flash floods in small and steep catchments with specific discharges (meaning that these are normalised by unit catchment area) of up to 100 m³ s^{-1} km^{-2}. Flash floods are especially common in mountain regions with semiarid or Mediterranean climates, highly seasonal precipitation, sparse vegetation, and high land-use pressure (Gaume et al. 2009).

11.1 Frequency and Magnitude of Floods

Floods and their effects on channels and floodplains have been a backbone of geomorphic research for many decades. Much of this work is based on case studies that have investigated how and where floods entrain, transport, and deposit sediment. Many studies used the resulting net effects of erosion and accretion to characterize how effective floods are in shaping channels, floodplains, and hillslopes. These studies sometimes identify specific

floods, but mostly stress the cumulative effect of multiple floods.

Some fluvial geomorphologists argue that water discharge, the quantity and calibre of sediment carried by rivers, and average valley slopes mainly determine the form of alluvial channels (Eaton et al. 2004). In particular, the concept of a channel-forming or 'dominant discharge' has attracted both enquiry and debate for decades (Pickup and Warner 1975). The concept states that only part of the range of discharges passing through a channel is effective in shaping its geometry. Early studies proposed that floods with average return periods close to the mean annual flood – the discharge that a river (statistically) exceeds every $2\frac{1}{3}$ years on average – are responsible for creating channel forms that best accommodate these flows, so that form and process condition each other. This feedback offers little information on how channels may have looked before having adjusted to the dominant discharge.

Water flow in many rivers is distinctly seasonal, so that average annual flows may only poorly represent the discharge regime. For example, Henck et al. (2010) argued that the mean monsoonal flow was a better contender than individual storms for characterizing the effective channel-shaping discharge of rivers draining the southeastern Tibetan Plateau. There, the monsoon season lasts about four months each year, and during that time rivers carried more than 80% of the annual suspended sediment.

The debate over whether dominant discharge is a useful concept includes the questions about erosional thresholds, that is the minimum discharge needed to entrain sediment and thus perform geomorphic work. The debate also includes questions about rarer floods that may erode or deposit more sediment, but may recur insufficiently frequently to do so in a way that would form the channel in the long term. These rarer and commensurately larger floods may, however, completely reshape large tracts of channels and floodplains. Some stream channels may

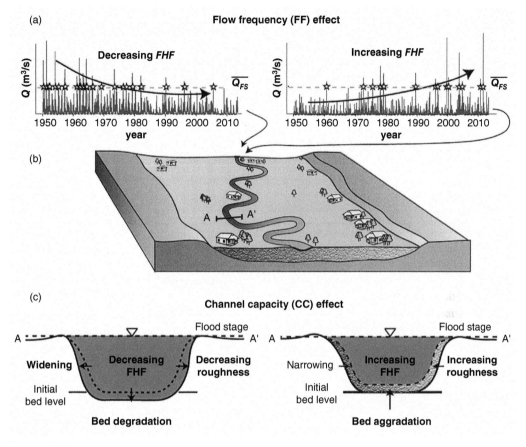

Figure 11.1 Schematic of flow frequency and channel capacity effects on flood hazard frequency. (a) Time series showing decreasing and increasing flow frequency effects on flood hazard frequency (FHF; stars are days with discharge equaling or exceeding the average discharge at flood stage marked by dashed line). (b) Flood hazard frequency at channel-floodplain cross section A–A′ may change due to altered flow frequency or to altered channel capacity. (c) Changes in channel capacity change flood hazard frequency at A–A′ through shifts in cross-sectional area or roughness. From Slater et al. (2015).

adjust their geometry to frequent lower floods, but studies from semiarid environments show that the channel geometry of seasonal rivers in particular may reflect rare and large floods (Kale 2007). However, rare and large floods may be exceptions to the rule (see Chapter 11.2.2).

Whatever the dominant or geomorphically most effective discharge (range), any change in channel geometry will effect its capability for carrying flood waters, and hence flood frequency. Slater et al. (2015) systematically studied flow records of 400 gauging stations in the United States, and found that nearly 60%

of the stations had trends of average discharge that varied with time because the channel cross-sections also changed because of local erosion and sedimentation (Figure 11.1). This geometric effect of increasing or reducing local channel capacity altered the local flood frequencies. In contrast, the inferred changes in flood frequencies arising from hydrological controls were higher, but occurred at fewer stations.

Even heavily disturbed catchments, such as those that received high loads of volcaniclastic sediment following the 1980 eruption of Mount St Helens, have channel forms that mainly reflect moderate-sized floods. A 20-year record from these catchments elucidates the geomorphic work of floods following the eruption: although large floods transported up to half of the annual suspended sediment loads in a single day, more moderate-sized floods with average return periods of less than two years were responsible for transporting 60–95% of the annual suspended sediment load over periods of 1–3 weeks per year (Major 2004). It is tempting to generalize and apply these findings to other catchments. However, similarly detailed measurements have shown that extreme floods carry the bulk of water and sediment in much shorter times than smaller floods. For example, a 17-year bedload measuring campaign in a boulder-littered headwater stream in the Italian Alps highlighted that a single extreme flood abruptly raised the ratio of bedload volume to effective runoff by more than tenfold; it took about a decade for the stream to regain the ambient ratio (Lenzi et al. 2004). While such measurements underline how single floods may perturb or even control the subsequent transport of sediment in channels, many time series are too short to establish the 'ambient' water and sediment fluxes or how they vary without disturbances. What defines a long enough interval depends in part on how frequently disturbing floods occur. This definition is important because we can also observe similarly pulsed and unbalanced water and sediment discharge when measuring over much shorter intervals. For example, three-minute time series of water and suspended sediment transport in forested headwater streams in south-central Chile showed that about 80% of the total sediment was flushed by floods that in total took only 5% of the monitored time (Mohr et al. 2014).

Bear in mind that such measurements are always tied to the geometry of a given channel cross-section, and how sensitive that geometry is to changes by local erosion or sedimentation.

The geomorphic results of floods vary strongly along the channel network, and can influence local measurements. For example, terrestrial laser scanning of channel sections before and after a 2000-year flood in the upper Lockyer catchment in the subtropical part of Australia highlighted how bedrock-confined reaches had incised, whereas less constricted reaches had aggraded by nearly eight metres. These differing effects on local channel-bed elevation were partly tied to varying stream power that the authors estimated to have been 2–3 times lower in the unconfined reaches (Thompson and Croke 2013).

Extreme floods have completely reshaped river channels, and 'river metamorphosis' (Schumm 2005) refers to this abrupt and flood-driven rearrangement of channel planform and cross-section (Figure 11.2). Sudden channel changes can also affect tectonically inactive terrain in which rivers may appear to be sluggish most of the time. Macdonald River north of Sydney, Australia, is an example. The river dissects a forested sandstone plateau and has partly infilled an estuary during the course of the Holocene. The lower reaches of the river channel were navigable until the late nineteenth century. A series of floods in the mid-twentieth century with estimated return periods of well above 100 years reshaped this meandering and largely bedrock-confined sand-bed river, widening the channel, and raised the bed by 2–3 m with sandy sediments. The rapidly aggrading channel blocked several tributaries and formed shallow floodplain lakes at the confluences. The bed aggradation also pushed the tidal interface some 7 km downstream, or nearly a third of the total distance that the river mouth prograded downstream in the past 6 kyr (Rustomji 2008). Most of this excess sediment was eroded during the floods from older floodplain deposits in upstream reaches, possibly augmented by minor contributions of sandy material from nearby hillslopes.

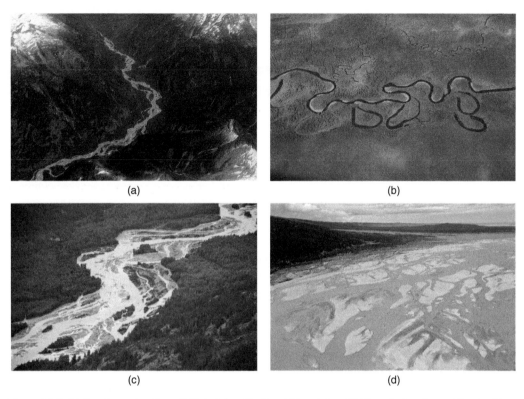

Figure 11.2 (a) Anastomosing river, British Columbia Coast Mountains. (b) Meandering river with cut-offs, central Alaska. (c) Braided river planforms, Lillooet River, British Columbia. (d) Braidplain of the Delta River, Alaska (all photos by John Clague).

11.2 Geomorphic Impacts of Floods

11.2.1 Floods in the Hazard Cascade

We distinguish between meteorological floods triggered by rainfall and runoff, and nonmeteorological floods triggered by the sudden release of impounded water masses. Meteorological floods require a storm, snowmelt or periods of sustained precipitation, whereas nonmeteorological floods require an unstable natural dam to let the impounded water escape and therefore always occur as a follow-up process in the hazard cascade.

Topography, soils, vegetation, and land use influence runoff generation that lead to weather-related floods. Larger catchments generally produce less flashy runoff than smaller ones, though different land-use and land-cover histories complicate this relation, and the hydrological literature has produced several ways of estimating these effects and taking them into account when appraising flood hazards.

From a geomorphic perspective, most floods following rainstorms also trigger at least soil erosion, bank collapse, debris flows, or landslides. Hence the geomorphic fingerprint of flooding alone – that is without additional inputs of hillslope sediments from mass wasting – may be hard to pin down. The resulting input of sediment that moves downstream dynamically changes the local channel geometry and thus also the properties of channel flow. Indeed, much of the economic impact of floods is related to the damage by inundation and the sediment that floodwaters carry. Using a 40-year database of costs incurred by floods and mass movements in Switzerland, Badoux et al. (2014) estimated that coarse sediment transport by rivers in the Valais and Ticino cantons of the Swiss Alps caused annual average losses of 215–230 CHF per capita. Most of the damage recorded in this study happened during summer floods following local convective rainstorms.

Sediment rating curves relate instantaneous measurements of sediment transport to water discharge, and are widely used for quickly estimating bulk sediment transport by extrapolating these data across the range of discharges (Figure 11.3). Most rating curves are based on simple power-law relationships between suspended sediment concentrations and water discharge (Morehead et al. 2003). However, as appealing as this method might be, it suffers from at least two major shortcomings. The first characterizes almost all sediment-transporting flows and is defined by time lags between the peaks in sediment transport and water discharge, an effect known as hysteresis. This effect can result from an increase in sediment entrainment during the rising limb of a flood when erosion of the channel bed and banks as well as hillslopes is more effective, because discharge is increasing so water surface slope is steeper than during later stages when the sediment supply is exhausted. Waning stages of floods in turn cause sediment to drop out of suspension, resulting in deposition of sediment. One consequence is that suspended sediment concentrations often vary by orders of magnitude for a given discharge, and that measurements of annual sediment yields may also vary more than 100-fold, even across neighbouring catchments (Warrick et al. 2015). The second shortcoming arises from the popular use of ordinary least-squares regression with log-transformed data on sediment concentration, sediment flux, and water discharge. This approach returns biased equations with slope exponents that consistently underestimate the correlation parameters (Xiao et al. 2011). The use of thus biased slope exponents results in predictions of lower sediment concentrations for the highest of discharges.

A sudden increase in the supply of sediment to rivers may result in the gradual downstream motion of sediment waves or pulses. These slow-moving slugs of bedload sediment alter the local channel geometry, flow conditions, and habitats as they propagate according to the principles of mass conservation and sediment continuity (Eq (3.1)). Sediment pulses can widen active channels, raise their beds, and cause water to spill over onto floodplains, thus reorganizing channel form and hydraulic roughness, and altering stream-flow conditions (Madej 2001). The time it takes a channel to recover from a sediment wave is termed the 'recovery time', and depends on what defines the state of the channel before the sediment input. Points of reference could be the original channel-bed elevation or average

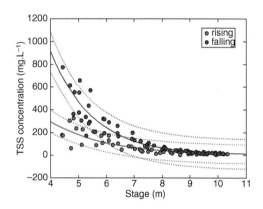

Figure 11.3 Sediment rating curves describing the average relation between suspended sediment concentration and stage in the floodplain of the lower Amazon River near Óbidos, Brazil. Separate curves were fitted (solid lines; dotted lines indicate 95% confidence level limits) to samples taken during the rising and falling limb of the hydrograph to account for a hysteresis effect. Sediment concentration was sampled about every 10 days between the years 2000 and 2003. Water level for the sampling dates was simulated with the LISFLOOD-FP model. From Rudorff et al. (2017).

sediment flux. The recovery time of mountainous rivers can be shorter than the average return period of sediment inputs that cause further disturbance, especially where landslides abound or where soil and gully erosion is prolific. Diagnostic in-channel landforms that point to stabilized conditions may be suitable indicators of recovery if we can be sure that they are long-lived features.

For many river management applications, recovery time (like dominant discharge) appears to be an attractive concept. When examined in detail, however, it often eludes direct and objective measurement because water and sediment inputs to a river reach vary continuously. Nevertheless, if recovery times are long enough, the cumulative effect of many perturbations may lead to a highly diverse set of valley-floor landforms. How this plays out over the Holocene, or possible even longer, is exemplified by the 'disturbance regime landscape' of the upper Indus basin, where hundreds of large and river-blocking rock slides and rock avalanches have created a natural cascade of sedimentary basins interspersed by steep bedrock gorges (Hewitt 2009).

Floods in tropical rivers Rivers are the arteries of life for people living in the tropics. These rivers drain roughly a third of the land surface, and many are known for their seasonal flood levels that can attain several metres. Yet studies focusing on the hazards of tropical rivers in flood are rare compared to those focusing on temperate regions. The rapidly growing population and expanding forest clearing in the Amazon River basin illustrate the increasing pressure on tropical rainforests and their rivers. The population in the Amazon basin increased more than fourfold from 1960 to 2010, now amounting to more than 25 million, while some 20% of the rainforest area was lost to agricultural and other uses (Davidson et al. 2012). More than 60% of the original Brazilian savannas, termed *cerrado*, have been fragmented by

deforestation and converted to agriculture. This loss of forest canopy increased water discharge, and caused high rates of soil erosion that feed aggrading and shifting channels on the Amazon river. One example is the Araguaia, which is the largest river in the seasonal tropics of Brazil, with a catchment area of 375 000 km^2. Between 1965 and 1998, following widespread forest losses, the bedload carried by the Araguaia increased by 30%, accompanied by dumping >0.39 Mm^3 of sediment on average on its floodplain for each kilometre of channel length (Latrubesse et al. 2009).

Many rivers in the seasonal tropics debouch onto huge alluvial fans – sometimes called megafans – that cover >1000 km^2 (Leier et al. 2005). These fans, like all alluvial fans, are prone to catastrophic channel changes and attendant flooding through major avulsions. Highly pulsed flushing of sediments and nutrients mostly occurs during tropical storms and cyclones (Figure 11.4). However, tropical rivers seem to carry low amounts of sediment in general because most of their headwaters are densely forested and have low relief (Syvitski et al. 2014). However, the sediment and nutrients that some tropical rivers do deliver to coasts also nourish tidal flats and mangrove forests, and decreased sediment fluxes reaching the coasts could impact mangrove populations sufficiently to expose coastlines to higher wave erosion during storms and tsunamis. Yet some studies suggested that the high chemical weathering rates and water availability in many tropical river catchments might dampen, or at least delay, this degradation of coastal ecosystems (Jennerjahn 2012).

11.2.2 Natural Dam-break Floods

Rainstorms may be the most frequent cause of flooding, but many of the world's larger floods resulted instead from the sudden failure of natural dams. Most larger 'megafloods' happened

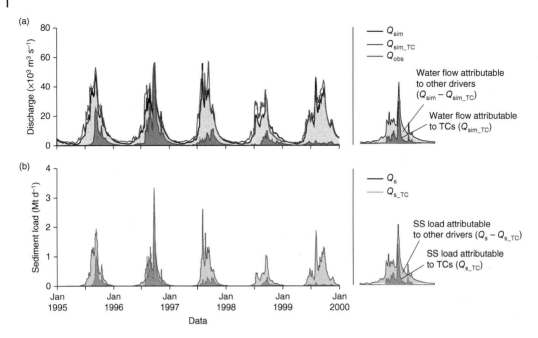

Figure 11.4 Daily flow discharge and suspended solids load at Kratie, Mekong River delta, from 1 January 1995 to 31 December 1999. a, Daily simulated (Q_{sim}) and observed (Q_{obs}) water flows, along with the daily water flows attributable to tropical cyclones (Q_{sim_TC}). b, Daily total suspended solids (SS) load (Q_s; in Mt per day) and daily suspended solids load attributable to TCs ($Q_{s_{TC}}$; also in Mt per day). Note that the period 1995 to 1999 encompasses the years during the 1981–2005 study period that are the most (1996) and least (1999) strongly affected by TCs. Jan, January. From Darby et al. (2016).

tens of thousands to millions of years ago, but some of their geomorphic legacy is still visible today (see Section 11.4). Natural dams are formed by glaciers, moraines, landslides, lava, and sand dunes (Costa and Schuster 1988) that block river channels and impound water bodies for less than one day to tens of thousands of years (Figure 11.5). The largest of these naturally dammed lakes may store thousands of cubic kilometres of water, and their sudden drainages may have transported the largest volumes of sediment in the shortest amount of time (Korup 2012). When we mention outburst 'floods', we acknowledge that these processes rather entail the full spectrum from clear-water floods to hyperconcentrated flows to debris flows. Sediment yields in rivers affected by natural dam breaks can exceed 10 000 t km^2 yr^{-1} for years, and can thus be on a par with the high and short-lived sediment yields produced by volcanic eruptions,

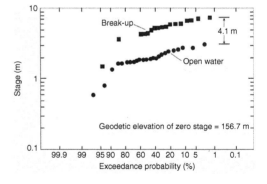

Figure 11.5 Flood stages and the probabilities of their exceedance for Hay River, NWT, Canada. Circles mark normal floods, whereas squares mark ice-jam floods. From Beltaos and Prowse (2009).

earthquakes, or rainstorms (Figure 7.1). Accordingly, reworking and evacuating the sediment from natural dam failures can take several years to centuries, or longer. Laboratory flume experiments show that river flows incise rapidly into sediments stored behind failed dams, causing the channel to first incise and scour sediment from steep and unstable side walls (Cantelli et al. 2004). The incised channel migrates upstream while widening, delivering sediment to sections downstream of the failed dam. Field studies show that the over-spilling lake waters can cut new channels into bedrock lows that the higher lake level has made accessible, thus bypassing the natural dam and incising epigenetic gorges (Hewitt 1999). The drastic and potentially long-lasting effects of natural dam-break floods on river erosion and sedimentation add to the discussion of whether more frequent or less frequent discharges contribute more to controlling the geometry of channels or even entire valley floors (Korup and Tweed 2007).

Another largely unresolved research mandate concerns obtaining more data on, and mechanistic insights into, the failure and breaching processes of natural dams. The dynamics and rates of dam breach are a first-order control on peak discharge and hence potential flood power. Reconstructions of prehistoric dam breaks and their floodways partly rely on rare direct observations of much smaller dam breaks in historic times (Lamb and Fonstad 2010). Systematic collections of natural dam data thus have many gaps about specific properties of the dam, such as its origin, age or size. The same goes for details about the outburst process, such as the volume of water and sediment released, the duration of outburst flooding, and the resulting peak discharge. For example, a database recognizing more than 1200 landslide dams throughout the world includes only 52 cases that offer some data on breach parameters (Peng and Zhang 2012). Some types of natural dams seem to have enjoyed more enquiry than others. Failures of volcanic dams, for example, are rarely documented compared to failures of landslide, glacier, or moraine dams. Yet outbursts from volcanic dams can be equally destructive: pyroclastic surges and flows from the 1982 eruptions of the El Chichón

volcano, Mexico, dammed the Magdalena River and impounded a lake that stored some 48 Mm^3 of water at near the boiling point. The sudden collapse of the dam released a hot hyperconcentrated flow that travelled 35 km downstream before being contained by an artificial reservoir. Thanks to early evacuation, all residents were saved (Macías et al. 2004). This and many other similar accounts might give the impression that such destructive – and geomorphically effective – floods can be measured in a straightforward manner. In fact, most reports of natural dam failures feature parameters that can be hazardous or simply impossible to measure directly. Many quantitative estimates are biased towards outburst floods that caused damaged or that happened in populated valleys. Large uncertainties may arise as a consequence. On 9 April 2000, a 0.3-km^3 rock slide-debris avalanche dammed the Yigong River, a tributary of the Yarlung Tsangpo in southeast Tibet, forming a lake several kilometres long. Two months later, the dam breached, with a reported release of some 3 km^3 of water and a peak discharge of 120 000 $m^3 s^{-1}$ (Shang et al. 2003). This figure was based on the observed flood level that was 32 m above a bridge 17 km downstream from the dam. Xu et al. (2012) re-examined this event and reported a flood level ten metres higher at the same location; however, they inferred a peak discharge of only about 12 000 $m^3 s^{-1}$ instead. The error margins associated with estimates like these require that more than the few available historic case studies be reanalysed to improve planning for future floods. We might also wish to refine the tools for inferring peak discharge, as these tools mostly rest on empirical relationships that are poorly calibrated. Placing instruments to measure natural dam breaks in the field is often out of the question because time is short and resources to mobilize are tight. However, the decommissioning of artificial dams can provide more foreseeable opportunities for monitoring in detail the processes and rates of dam break (Schneider et al. 2014).

River erosion and slope instability, Buller River, New Zealand Closure of mountain roads because of landslides means extra driving distances and extra accidents. State Highway 6 runs through the Buller Gorge in northwest South Island, New Zealand. Earthquakes damaged the highway in 1929 and 1968, when ground rupture and rockfalls caused lengthy closures. Since the mid-1990s, the road has remained prone to erosion along a steep rock slope known as 'Newmans'. Cutting the road farther back into the slope to maintain a wide enough road is costly. The cause of the ongoing slope erosion is undercutting by the river, and a groyne was constructed to deflect the river away from the toe of the eroding slope.

For most of its course through Buller Gorge the river is gravel-bedded, narrow, and deeply incised. Its channel bends have a consistent shape, reflecting water and sediment discharge and past incision episodes, except for the eroding bend at Newmans that is much sharper and nearly right-angled at the undercut slope. Aerial photographs taken in 1965 show that Newmans bend had a comparable form to other bends along the gorge. Between 1965 and 1976, the Newmans bends sharpened as the river shifted sideways just upstream, by about 70 m, eroding an alluvial fan from a tributary. An investigation into the cause of the erosion eventually linked it to a large landslide known as 'Earthquake Slip' that fell into the river in the 1968 Inangahua earthquake about 3 km upstream of Newmans. The landslide blocked the river for three days before it overtopped and breached the barrier, generating a flood of $\sim$3000 m^3 s^{-1}. This is the ultimate cause of the road problem decades later – the dam-break flood, carrying a high concentration of sediment from the landslide site, forced the Newmans bend to sharpen, as the river cut deeply into the slope. It took a decade or so before the erosion at the toe of the cliff propagated upslope to the road.

Left to itself, Buller River will further alter the shape of the Newmans bend. The focus of intense erosion would move downstream, subjecting other stretches of road to undercutting. The river bank is in broken rock, so that the bend will migrate only slowly. The groyne installed to deflect the river from the eroding toe of the slope may affect this future development of the bend. If the groyne remains in place, it will cause severe erosion farther downstream. If the groyne is unmaintained and eventually eroded, the new bend will revert to its natural location. In either case, further troublesome slope erosion appears to be inevitable along this stretch of the highway in the future.

Hazard appraisals should consider that single natural dams can produce more than one flood. Some river-blocking landslides have released multiple outburst floods. The earthquake-triggered Tombi debris avalanche in 1858, with an estimated volume of 100–120 Mm^3, is such an example. This landslide descended from the slopes of Tateyama Caldera, Japan, and blocked two tributaries of the Joganji River, impounding an estimated 37 Mm^3 of water (Inoue et al. 2010). Historic records provide illustrative accounts of two subsequent outburst floods:

> Kaga-Han clan records state that debris and flood flows from the first outburst flood damaged 66 towns and villages, mainly on the east side of the Joganji River, drowning five persons, sweeping away or destroying over 250 houses and 78 storehouses and barns, and rendering agricultural land with a recognized yield of over 5,236 koku (about 150 kg) barren. The second outburst flood hit 74 towns and villages, again mainly on the

east side of the Joganji River, drowning 135, adversely affecting the lives of 7,350, sweeping away or destroying over 1,360 houses and 808 storehouses and barns, and rendering agricultural land yielding over 20,560 koku barren. Records of the second dam failure also show that 18 towns and villages were adversely affected and agricultural land with a yield of over 7,360 koku was rendered barren in the Toyama-Han fief east of the Itachi River (Inoue et al. 2010).

Other types of natural dams may form and fail in response to major climatic changes, so that the hazard of outburst is likely to vary with time. The catastrophic failure of moraine dams generally follows extended periods of cool climate when large lateral and end moraines are built at stable glacier margins. Subsequent warming causes glaciers to downwaste and retreat, trapping meltwater in the newly formed basin behind these moraines, thus slowly elevating the potential for catastrophic emptying, as more meltwater and precipitation input continues to raise lake levels (Richardson and Reynolds 1999). Landslides, ice-falls, and avalanches into the lakes, groundwater sapping, or gradual settling of ice-cored dams are among the most common causes of moraine-dam failures. Unstable moraine walls fringing meltwater lakes can release landslides capable of producing large displacement waves (Klimeš et al. 2016). Heavy rainfall may also trigger overtopping and breaching, subsequent incision of moraine dams, and rapid release of impounded meltwater. This was the case during an exceptional rainstorm in the headwaters of the Alaknanda River, Indian Himalaya, where a moraine-dammed lake catastrophically emptied after it was overtopped by runoff in June 2013. The resulting flood swept through the pilgrimage town of Kedarnath located only 1 km downstream from the lake. The flood destroyed 44% of the buildings in the town, and left only a quarter

of all built structures in Kedarnath intact or slightly affected (Das et al. 2015). Warmer temperatures may also compromise the stability of moraine dams. The Nyenchen Tanglha mountains of monsoonal southeastern Tibet host some of the most active temperate glaciers in Asia. A gauging station at Guxiang Glacier recorded 95 debris flows between 1964 and 1965 from this single source, although only 25 of them had peak discharges >50 m^3 s^{-1}. Assuming that all these debris flows resulted from glacial lake outbursts, Liu et al. (2013) argued that more debris flows occurred during periods of high temperatures and rainfall. Some of these lake outbursts eventually formed new lakes downstream after choking tributary mouths with excess sediment, and one of the larger of these debris flows from Guxiang gully delivered 11 Mm^3 into the trunk stream of the Parlung Tsangpo in 1953, forming a 5-km long lake that remains to this day.

Detailed satellite monitoring of ice-surface changes contributes to documenting contemporary mass balances of glaciers in the Himalayas and other mountain regions, and to assessing the potential impacts of glacial lake outburst floods (Ives et al. 2010). Satellite imagery enables rapid and systematic mapping of inaccessible mountain lakes and allows researchers to track the formation, growth, and disappearance of these lakes. The Khumbu region of the Nepal Himalaya, for example, had at least 377 glacial lakes between the 1960s and 1970s (Bajracharya et al. 2007). A more recent survey in October 2008 gave a count of 624 lakes, of which 437 had formed directly on glaciers (Salerno et al. 2012). An accurate count may be elusive because smaller lakes, especially those on glaciers, may elude detection and form or disappear rapidly. Steep mountainous terrain prone to frequent cloud cover and shadows compromises even modern satellite sensors, so that results from different studies may be inconsistent.

Being prepared for the physical impacts of glacial lake outburst floods (GLOFs) requires at least some crude knowledge about their possible sources, the highest possible peak discharge at the failing dam, and how the resulting flood wave travels downstream. Hydraulic and hydrodynamic models can capture some of the basic physics operating during dam breach and constrain the associated discharge for a specified geometry of both dam and lake (Westoby et al. 2014). The size and orientation of boulders deposited by GLOFs can provide clues on local water depths or flow velocities, and these can be used to run hydraulic step-backwater models that offer estimates of discharge at a given channel cross-section (Cenderelli and Wohl 2001). The underlying assumption of these models, however, is that the flow is steady, uniform, and without sediment; this assumption is rarely accurate. The step-backwater modelling technique requires a series of adjacent channel cross-section geometries together with field-mapped or modelled flood levels, and reconstructs the former water surface, velocity, and discharge from these geomorphic constraints. Sediment and wood transport are rarely factored into the calculations, simply because so few data are available and because subsequent stream flow alters the sedimentary signature of GLOFs.

Repeat satellite imagery has paved the way for routinely deriving digital elevation models (DEMs) of metre-resolution or finer. If these DEMs are taken before and after an outburst flood and if they are adequately tied into a common coordinate system, it is straightforward to capture volumetric changes in the runout track (Figure 11.6). Jacquet et al. (2017) used such DEMs-of-difference (DoD) to capture channel and floodplain changes following a sequence of nearly two dozen GLOFs from Lago Cachet Dos on Rio Baker, Chilean Patagonia. They inferred that the floods transported as much as 25 Mm^3 of sediment through a network of rapidly shifting braided channels, and that some of these channels incised

by up to 40 m in places. This method also highlighted nicely the active dynamics of the braided river on the valley train with channel braids forming and disappearing after each flood. Guan et al. (2016) used thus acquired changes in channel-bed elevation as an input to a hydraulic flood model to estimate the effects on flood hazard in Jökulsá River, Iceland, in response to a series of outburst floods. They noted that channel incision and the commensurate change in local floodwater storage was the main cause for altered flood frequencies.

DEMs-of-difference allow detailed assessments of how the accommodation space for floodwaters and sediment change in a river reach after each major flood, and thus provide information that is essential to update flood management regularly. The level of detail captured by these DEMs aids tracking the movement of large wood debris that GLOFs recruit from forested valley floors. Large amounts of coarse wood form log jams that affect the dynamics of river flow by causing local erosion and deposition. One hypothesis holds that the distribution of log jams deposited by GLOFs differs distinctly from that produced by meteorological floods. If correct, the distribution of clusters of large woody debris could be used as a diagnostic of former GLOFs in forested reaches of river catchments (Oswald and Wohl 2008).

Finally, the aspect that eludes most studies of natural dams is their interior composition. This holds clues to their structural integrity and stability against hydrostatic pressures and seepage erosion. Noninvasive near-surface geophysical investigations can shed some light on these conditions, apart from situations where only breached dams provide direct exposures of the dam interior (Casagli et al. 2003). This factor is particularly important for dams formed by rock avalanches, whose interior is likely to be highly fragmented and thus easily erodible once the surface carapace of coarse boulders has been incised by overtopping flow.

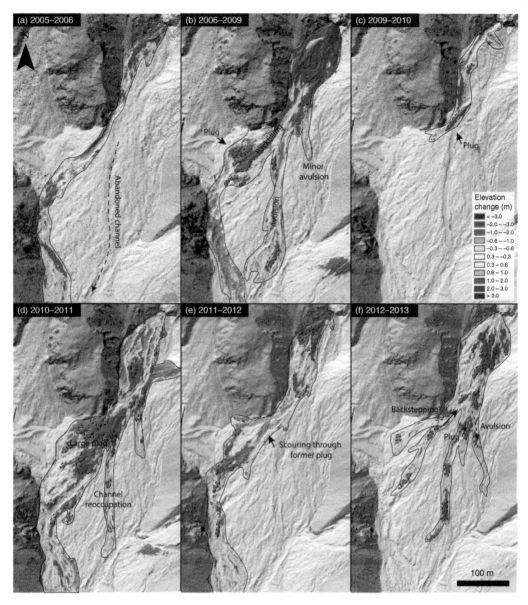

Figure 11.6 DEMs-of-difference for elevation of the Ohya, Japan, debris-flow fan over six periods. Differences: (a) 2005–2006 (four debris flows). (b) 2006–2009 (four events). (c) 2009–2010 (three events). (d) 2010–2011 (five events). (e) 2011–2012 (five events). (f) 2012–2013 (two events). Solid lines indicate areas affected by debris flows; dashed lines indicate distinct debris flow deposits. From de Haas et al. (2018).

11.2.3 Channel Avulsion

Sedimentation in river channels may reach the point that a given cross section becomes clogged and water is forced out of the channel and onto the floodplain, regardless of the cause of flooding. The term 'avulsion' refers to the sudden lateral shifting of a river channel on its floodplain (Slingerland and Smith 2004). This shifting differs from gradual cut-and-fill processes or smaller cut-offs in that the intervening parts of floodplains remain unaffected during avulsions. Avulsions mainly take place in rapidly aggrading river systems with laterally unconfined floodplains, on alluvial and subaquatic fans, and on deltas. Many large rivers have well documented histories of avulsions, including the Yellow River (China), the lower Rhine (Netherlands), and the Po (Italy).

Fluvial megafans are also sites prone to frequent avulsions. These fans cover areas of 1000–100 000 km^2, have very low surface slopes (1–4°), and lie mostly in forelands of mountain belts with highly seasonal discharge regimes (Leier et al. 2005). The Kosi fluvial megafan in the Himalayan foreland is a classic example of a system of distributary channels subject to recurring avulsions triggered by large floods. Over the past two centuries, the Kosi River channel migrated more than 110 km through random avulsions, putting the livelihoods of millions of people at risk. Aggradation rates in the active channels have been very high (>50 mm yr^{-1}) and sustained by the high sediment loads from the Himalayas (Chakraborty et al. 2010). Engineered channels abound on the fan to protect settlements and agriculture. The latest avulsion in 2008 followed an embankment breach induced by a moderate flood that overtopped and breached the bank of the confined channel at the fan head, presumably due to rapid bed aggradation. The resulting avulsion displaced the main channel by nearly 60 km and affected roughly three million people. Neighbouring megafans compete for the tributaries of the Kosi megafan and incorporate channels that issue from the mountainous hinterland. If these tributary channels are cut off from their active sediment sources, they start to incise and rework sediments of the megafans (Chakraborty and Ghosh 2010), potentially setting the stage for future channel instabilities and avulsions.

Large channel avulsions may take years to complete. A ~150-km long channel segment on the Taquari megafan in the Pantanal wetland of west-central Brazil established a new course over a period of ten years (1988–1998).

The change began with gradually establishing crevasse splays, which are smaller fans forming on the downstream side of breached channel levees, and terminated with a complete abandonment of the former channel (Assine 2005). Several large floods during wetter-than-average La Niña conditions triggered an avulsion on the fan of the Rio Pastaza in the Ecuadorian piedmont of the Andes, maintaining two adjusting channel branches for more than 30 years (Bernal et al. 2012), highlighting the gradual adjustment of rivers to these channel shifts. Observations from flume experiments on model fans show that the frequency of avulsions increases with higher sedimentation rates, such that lesser amounts of vertical accretion are needed to trigger an avulsion. At the same time the required bed-elevation changes occur more readily with higher bedload (Bryant et al. 1995). Field data to support this observation are still rare, but it is clear and supported by numerical models that sediment flux controls changes in avulsion-dominated channels. Frequent channel avulsions may be a key factor in forming branching rivers (Jerolmack and Mohrig 2007). Anabranching rivers are examples of multichannel forms that are particularly prone to avulsions. These rivers are characterized by actively aggrading channels that gradually become elevated above the valley floor. Levees or alluvial ridges gradually also build up in response to the high sedimentation rates, but when breached cause new avulsions (Carling et al. 2013).

11.3 Geomorphic Tools for Reconstructing Past Floods

The systematic reconstruction of floods and their geomorphic impacts has a long research tradition and has relied largely on data from stream gauges, detailed historic records, preserved or documented flood marks, and the study of fluvial sediments. The field of 'palaeoflood hydrology' is concerned with studying floods that happened before the instrumental period in particular (Baker 2006). Only the legacy of larger floods remains in people's memories and in geological archives. Nevertheless, documenting these extreme events is often the only way to extend gauging records, as these measurements only provide robust estimates of short return-period floods (Figure 11.7).

Floodplains are sedimentary archives that can preserve information on the times and magnitudes of previous floods, and on lateral migration and reworking rates of channels (Figure 11.8). Yet much research in fluvial geomorphology has focused on changes in river-channel geometry in response to changing water discharge and sediment supply. Floodplains have attracted less attention despite being both natural archives and retention spaces for floodwaters and sediment. The rates at which floodplains store and release river sediments vary widely. Rowland et al. (2005), for example, determined the age of OSL samples from sand layers in tie channels on the floodplains of three different meandering rivers – the Mississippi, United States; the Middle Fly River, Papua New Guinea; and Birch Creek, Yukon River, Alaska – to

test the applicability of this method for dating rapidly aggrading valley floors. On river floodplains, tie channels convey sediment-laden water between the main channel and adjacent floodplain waterbodies. The OSL ages, which were supported by historic maps, air photos, and hydrographic surveys, were consistent with sedimentation rates ranging from 2.5 to 58 mm yr^{-1} over periods of decades to almost a millennium. Several studies of floodplain accretion in tropical rivers have reported similarly high rates. As with many other time-averaged rates, these become lower with longer interpolation periods as the fraction of geomorphically inactive phases increases (Table 11.1). The rates of vertical floodplain growth in rivers in temperate climates can be similarly high, especially in lowland catchments with a high fraction of agricultural land use. For example, Belyaev et al. (2012) used spiked concentrations of ^{137}Cs in the radioactive fallout from the 1986 Chernobyl nuclear accident as a marker of floodplain sedimentation in the Plava River of central European Russia. This method shows that the floodplain accreted vertically at rates of up to 14 ± 2.8 mm yr^{-1} after the accident.

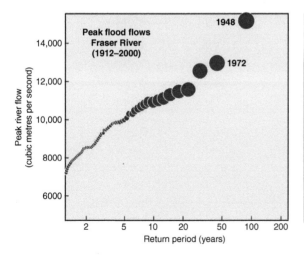

Figure 11.7 Left: Return periods for historic Fraser River floods. Right: Levels reached by historic Fraser River floods on a cabin the Fraser Lowland east of Vancouver, British Columbia. The flood of record occurred in 1884 (top line on cabin wall) (British Columbia Ministry of Environment, Lands and Parks). From Clague and Turner (2003).

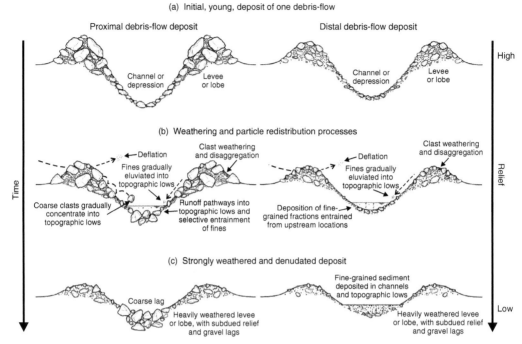

Figure 11.8 Conceptual model of the effects of weathering and fluvial reworking on alluvial fan-surface texture and morphology. The fresh surface (a) represents fans or fan sectors dominated by primary processes of deposition, whereas the weathered surface (b and c) represents fans or fan sectors dominated by secondary processes, including local disintegration, winnowing, and downslope transport of fines. Modified from de Haas et al. (2014).

Table 11.1 Historic rates of vertical accretion on tropical river floodplains.

River	Location	Accretion rate (mm yr^{-1})	Interval (yr)	Reference
Dumbea, Grande Terre	New Caledonia	<100	TC Brenda, 1968	Terry (2007)
Guadalcanal	Solomon Islands	300–500	TC Namu, 1986	Terry (2007)
Jurua	Brazil	<600	Flood 1985/86	Terry (2007)
Amazon	Brazil	270–420	Floods	Terry et al. (2002)
Brahmaputra-Jamuna	Bangladesh	6.7–11.5	50	Terry et al. (2002)
Brahmaputra	Bangladesh	1.6	50–100	Terry (2007)
Brahmaputra	Bangladesh	32–55	1	Rogers and Overeem (2017)
Hanalei	Hawaii	8.2–30.9	100	Terry et al. (2002)
Wainimala, Viti Levu	Fiji	32	45	Terry et al. (2002)
Felefa, Upolu	Samoa	41	40	Terry (2007)
Jourdain, Santo	Vanuatu	48	40	Terry (2007)
Labasa, Vanua Levu	Fiji	10–58	40	Terry (2007)
Krishna, Cauvery	India	3.5–11	100	Terry et al. (2002)

'Interval' refers to the approximate time difference these rates have been averaged over, or simply the flood that triggered rapid vertical accretion.

Such high sedimentation rates mean that a lot of material is trapped on floodplains, but eventually this also increases the amount of material available to future channel erosion. For example, balancing up gauging data from the Ganges–Brahmaputra River that drains a large portion of the Himalayas, Islam et al. (1999) found that the floodplains of this river retained nearly 30% of the total annual suspended sediment load. Another ~20% of the total load was deposited in channels, raising their beds at average rates of up to nearly 40 mm yr^{-1}. The river system owes much of its mobility and floodplain turnover to its high sediment loads: in Assam, northeast India, the Brahmaputra River alone transports an average of 0.4 Gt of suspended load per year. Its shifting channels have migrated up to several hundred metres per year and undercut some 868 km^2 of banks and floodplains in the twentieth century (Sarma 2005). Hence, some fraction of the sediment stored in floodplains is also bound to be recycled by wandering channels.

A similarly high trapping efficiency and reworking characterizes the tropical Strickland River floodplain, Papua New Guinea, which retained between 17% and 27% of the transported suspended sediment over some 65 years, judging from floodplain sediment cores dated by ^{210}Pb (Aalto et al. 2008). These newly formed floodplains can be completely cannibalized by migrating cut-bank erosion over a period of about 1000 years. Up to half of the contemporary sediment flux of the Strickland River may thus arise from floodplain material that is recycled by shifting channels. Similar times for floodplain sediment recycling might be applicable to the Amazon River, or at least to its tributaries that rework their floodplains equally rapidly (Figure 11.9). Nonetheless, Gautier et al. (2007) pointed out that estimated recycling times differed by an order of magnitude between tributaries. The rate of meander migration and cut-off in alluvial lowland rivers such as the Amazon also seems to depend on the sediment delivered from the more mountainous headwaters. Looking at a 30-year record of channel pattern changes mapped from LANDSAT images, Constantine et al. (2014) found that rivers with higher sediment loads had more mobile meanders than those with lower loads.

From the perspective of planning and preparing for natural hazards, we are concerned with the stability of floodplains, especially if they are meant to provide foundation for buildings, bridges, or infrastructure. Therefore, we need to know both average and flood-related rates of bank erosion and channel migration to work out the average residence time of floodplain sediments. From the perspective of engineering, structures and buildings need to be flood-proofed and secured against the natural tendency of channels to migrate. River dykes or stopbanks are a common engineering solution to this problem, but constrain the channel's natural tendency to shift laterally on floodplains, thus squeezing the accommodation space for sediment. Rogers and Overeem (2017) report that nearly 6000 km of dykes enclose more than 11 000 km^2 on the active Ganges–Brahmaputra delta, thus reducing coastal flooding at the cost of decreasing areas of sedimentation. They also noticed the scarcity of data on sedimentation rates despite the many predictions that large parts of the delta would be (at least more frequently) flooded because of rapidly rising sea levels. Local measurements of sedimentation in traps and trenches following a monsoon season on the lower delta, however, showed that rates of floodplain accretion exceeded those of local sea-level rise by far. More detailed and extensive study will be needed to provide more robust evidence of whether these short-term rates are representative, so that we can predict better the future of these coastal lowlands and the livelihoods that they sustain (Rogers and Overeem 2017).

Geoarchaeology is another field closely linked to fluvial geomorphology and has contributed rich evidence of former floods and their impact on human settlements and land

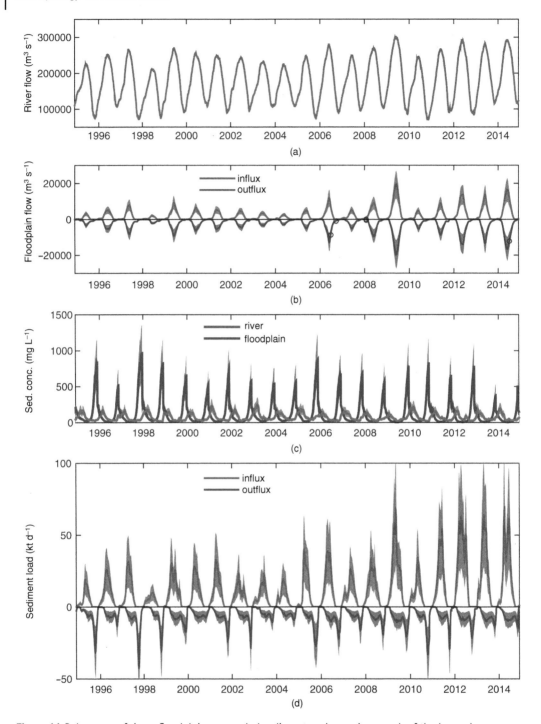

Figure 11.9 Increase of river–floodplain suspended sediment exchange in a reach of the lower Amazon River from December 1994 to November 2014. (a) River discharge gauged at Óbidos, Brazil. (b) Overbank flow discharge into and out of the floodplain. (c) Total suspended sediment concentrations in the main channel and floodplain. (d) Total suspended sediment discharge in and out of the floodplain. Light shades indicate standard errors obtained through an error propagation analysis of estimates of sediment concentration and routing of flow. From Rudorff et al. (2018).

use. Such work can also highlight early means of engineering that people have used to manage floods. In this spirit, geoarchaeological studies have also begun to embrace more diversified models of past interactions between humans and river systems instead of relying on the more traditional interpretations that past cultures or civilisations mostly vanished because of social and economic conflicts (Bintliff 2002). For example, Macklin and Lewin (2015) argued that channel abandonment driven by avulsion during major floods caused major problems to ancient civilisations, such as that along the Nile in Egypt, but also cautioned that inferring across-the-board hydrological changes from declines of ancient societies may be misleading. In ancient Rome, archaeological remains of major drainage structures such as the *Cloaca Maxima* date back to the third century BC; these were designed to route flood waters away from the city. Even older, although still functioning, hydraulic engineering structures include a large diversion dam and drainage canal built to protect the Bronze Age city of Tiryns in Greece from floods (Walsh 2014). Roman ships encased in sandy floodplain sediments of the Arno River near Pisa, Italy, document how sedimentation and lateral channel shifts during high-magnitude floods obliterated the harbour originally built by the Etruscans in the sixth century BC and later used by the Romans (Benvenuti et al. 2006). The early Etruscan harbour palisades are now buried beneath two metres of sediment that accumulated at an average rate of 0.8 mm yr^{-1} over the past 2600 years, largely driven by coastal subsidence.

These and many other studies nicely demonstrate the net effects of floods over decades to millennia, and how people have dealt or failed to deal with them. Studying the sedimentary evidence related to single floods or entire sequences of channel and floodplain environments has been the traditional focus of fluvial sedimentology. Different sedimentary facies can reveal much information about flow types, dynamics, and directions, as well as the general depositional environment such as channels, levees, floodplains or backswamps. Imbricated and (partly) rounded megaclasts offer insight into the erosive power and transport capacity of outburst floods. Large flood-transported boulders give a rough indication of the forces available for entrainment during floods. Minimum flow velocities and local stream power can be estimated by introducing the size of the largest of these boulders into equations of transport competence. This approach relies on the simplistic assumption that the fluid moving the boulder was a clear-water flood (and thus had negligible sediment concentrations) with steady flow. In many cases rheological flow models that more suitably characterize hyperconcentrated flows and debris flows may be more appropriate (Davies et al. 2003). Such turbulent, nonsteady flows have high and rapidly changing flow velocities with multiple surges that can cause transient forces possibly being responsible for entraining boulders, such that the traditional use of empirical equations may end up delivering crude overestimates of flow velocities or discharge (Alexander and Cooker 2016).

Near the other extreme of the grain-size spectrum are slackwater deposits: these are fine, commonly silt-sized sediments that rapidly drop out of suspension when flow velocities decrease. They are best preserved in rock niches, alcoves, and confluences of bedrock gorges. Pioneering studies used slackwater deposits to reconstruct prehistoric floods in arid climates, which were assumed to offer optimal conditions for preserving these sediments. Recent work has demonstrated that slackwater deposits can also provide suitable palaeoflood proxies in wetter settings. Huang et al. (2010) reported slackwater deposits intercalated with loess layers along the Jinghe River, a tributary of the middle Yellow River, China. The fine-grained flood deposits blanket a late Neolithic (4.3–4.0 kyr BP) site and thus offer tight stratigraphic constraints for floods thought to have had peak discharges between 19 500 and 22 000 m^3 s^{-1}, hence much larger than historic floods. The Neolithic floods coincided with a period of severe drought in monsoon-dominated areas of China, and seem to be consistent with anecdotal evidence of great floods during that time. Wasson et al. (2013) used slackwater and floodplain deposits to obtain a chronology of extreme floods along the Alaknanda River in the headwaters of the Ganges, Indian Himalaya. There, large floods were common between 1100 and 1300 AD, and several unusually high discharge estimates point to catastrophic dam breaks as the cause rather than meteorological extremes. This study is one of the few that called attention to a potential mix of dam-burst and rainfall-driven events in palaeoflood chronologies. Overall, the use of slackwater deposits effectively extends known flood records. Sheffer et al. (2008) reported evidence for at least five large palaeofloods during the past 500 years on the Gardon River, southern France. All five floods exceeded the historic flood of record in 2002, which claimed 23 lives and caused €1.2 billion of damage to several thousand houses along the river. The palaeoflood record puts this destructive event into a longer-term perspective, and extends flood hazard assessments by information on inundation levels that preceded instrumental records.

Other tell-tale palaeoflood indicators include rhythmites with an erosional base, and an upward decrease of particle size (Froese et al. 2003). Palaeocurrent directions that face upstream also are evidence of high-energy floods that entered tributaries and deposited sediment there (Stolle et al. 2017). Catastrophic floods may also form alluvial lakes at river junctions, where excess sediment deposited during the flood blocks the river branch with lesser transport capacity. The relevance of these proxies is that, if undisturbed, they freeze the minimum position of former flood levels. Such palaeo-stage indicators also provide valuable input for numerical models that are used to back-calculate cross-sectional discharge and flood routing. Assuming that the

channel geometry and the input hydrograph are sufficiently accurate, such models can generate detailed insights into inundation heights, flow velocities, and flood durations (Denlinger and O'Connell 2010; Carling et al. 2010). Adequately calibrated numerical models identify locations of extreme energy expenditure and stream power where bedrock erosion might be especially pronounced (Alho et al. 2010). Once again, the lack of detailed data on how much sediment was transported by which processes during these exceptional floods renders modelling results largely indicative.

Absolute age dating of palaeo-stage indicators can reveal the timing of prehistoric floods. For example, the concentration of cosmogenic nuclides in the exposed surface of a flood-transported boulder can tell us something about the time it was last moved. This method measures the amount of cosmogenic [10]Be that has accumulated in the surface rind of boulders as a proxy for the duration of their exposure to cosmic radiation. With this method, Reuther et al. (2006) determined that the most recent cataclysmic flood from Pleistocene ice-dammed lakes in the Russian Altai Mountains, Siberia, swept across the landscape at 15.8 ± 1.8 ka. Users of the method assume that the sampled boulders have remained in place and unweathered since they were deposited. Fujioka et al. (2014) proposed a numerical method that simultaneously solves for the predicted nuclide concentrations from the exposed and hidden surfaces of boulders potentially flipped by floods to calculate the time since they were overturned. They applied this method to imbricated boulders of the Durack River in the Kimberley Ranges of Australia, and inferred that catastrophic floods happened there at 10.3 ± 1.9 ka and 5.6 ± 1.0 ka. Based on these results, the authors argued that these extreme floods had produced long-lived evidence in the bedrock channels, partly aided by the arid climate. These and similar reconstructions can refine assessments of flood return periods based on contemporary gauging data, especially with respect to changes in climate from drier to wetter phases and vice versa.

A detailed Pleistocene radiocarbon chronology of flood and debris-flow deposits on alluvial fans in south-coastal Peru provided evidence for recurring large floods associated with El Niño phases over the past 38 000 years. These floods appear to have had much stronger geomorphic impacts than historical ones (Keefer et al. 2003). Three plausible explanations that affect current flood-hazard assessments are 'mega-niños' with anomalously heavy rainfall along most or all of the central Andean coast, El Niño events that exported large amounts of debris generated by great earthquakes, and El Niño events that delivered anomalously heavy local rainfall.

Few studies have attempted to compile regional chronologies of past outburst floods following natural dam breaks. A combination of historic evidence, remote sensing, and geomorphic field study has revealed that moraine-dam failures in British Columbia occurred during distinct phases in the Holocene Epoch and especially during recent centuries, although this may reflect the better preservation of the youngest evidence. Glaciers built many moraines in British Columbia's high mountains during the Little Ice Age, and particularly during the 1700s and 1800s. The depressions behind these debris barriers started to collect large ponds of meltwater when climate warmed in the 1900s, and a spate of outburst floods followed (Clague and Evans 2000). These general cycles can aid the compilation of comprehensive chronologies of prehistoric outburst floods from naturally dammed lakes (Figure 11.10), but will nevertheless require detailed site studies to validate the size of the dams and the volumes of water and sediment flushed from formerly impounded lakes.

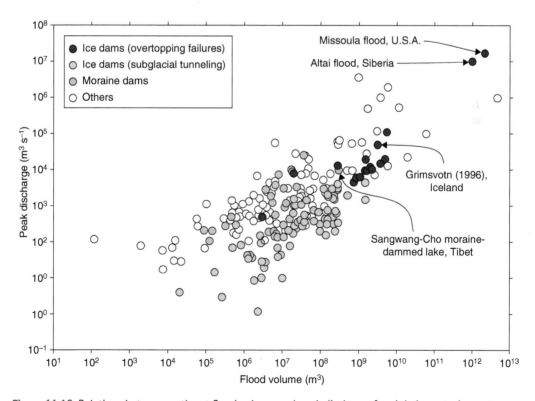

Figure 11.10 Relations between outburst flood volume and peak discharge for global events. Largest floods of each type are labelled. From O'Connor et al. (2013).

11.4 Lessons from Prehistoric Megafloods

Some of largest floods scientists are aware of happened thousands to millions of years ago (Figure 11.11). Consider the Messinian salinity crisis, a period of enhanced aridity in the Mediterranean basin (Garcia-Castellanos and Villaseñor 2011). During that period most of basin had much lower sea levels, and a bedrock sill at the Strait of Gibraltar separated the basin from the Atlantic Ocean. At the end of the Messinian 5.33 Myr ago sea level rose and overtopped the sill, causing rapid incision and catastrophic inflow into the Mediterranean basin. This sudden refilling of the nearly desiccated Mediterranean Sea may have involved a peak discharge of $0.1 \text{ km}^3 \text{ s}^{-1}$, three orders of magnitude larger than the largest reported flood on the Amazon River (Garcia-Castellanos et al. 2009); a numerical dam-breach model indicates that the sea level may have risen by several metres per day. Pleistocene dam-break floods also appear to have completely reshaped former valleys on the continental shelf in the region of what is now Dover Strait and the English Channel, thus contributing to separating the British Isles from the European mainland (Gupta et al. 2007). Megafloods can also arise as a consequence of volcanic eruptions: prehistoric lava dams along the Colorado River, United States, produced large megafloods with peak discharges in the same league as those caused by Pleistocene ice-dam outbursts (Fenton et al. 2006).

The largest terrestrial floods date to the late Pleistocene and when stagnating and downwasting continental ice sheets in North America, Europe, and Siberia impounded giant meltwater lakes at their margins. Sudden failure of local ice dams allowed floodwaters to escape at estimated peak discharges of the order of $1 \text{ Mm}^3 \text{ s}^{-1}$, an order of magnitude higher than any known rainfall-fed flood peaks (O'Connor and Costa 2004). These Pleistocene floods were so large that their discharge can be measured in Sverdrups [$\text{Mm}^3 \text{ s}^{-1}$], a unit that is normally used for ocean currents. Optically stimulated luminescence dates provide some evidence that major meltwater lakes formed in northern Eurasia at 90–80 ka and 60–50 ka.

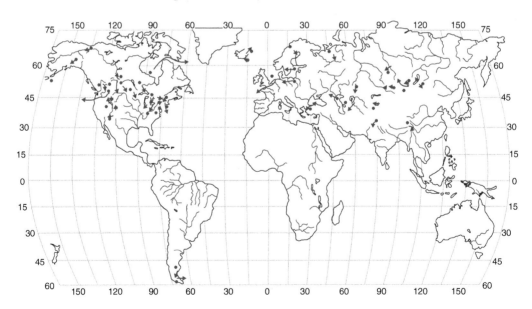

Figure 11.11 Locations for 41 regions where terrestrial megaflooding or related large-scale flooding has been documented. From Carling (2013).

Theory holds that, upon the collapse of the ice dams, some the lakes burst out catastrophically, rapidly flushing meltwater volumes of several hundreds to thousands of cubic kilometres along flood routes southward as far as the Aral Sea and the Black Sea. Several low-elevation corridors may have been spillways and connect these water bodies like a giant cascade (Mangerud et al. 2004).

Both flume and field studies have helped to compile a diagnostic suite of landforms and deposits that characterize outburst floods. Pleistocene meltwater floods have scoured entire landscapes to bedrock, created fluvially sculpted rock surfaces, formed multiple channels, created dry waterfall cliffs with plunge pools, scoured oversize potholes, and deposited megaripples. Catastrophic outburst floods can also produce canyons with amphitheatre-shaped heads. For decades researchers had attributed these landforms to slow and gradual processes such as groundwater erosion. Lamb et al. (2014) used cosmogenic ^{3}He dating to examine the ages of rock surfaces in such canyons at Malad Gorge, Idaho. They found that canyons with vertical or stubby head walls had closely clustered rock-surface ages indicative of catastrophic formation, whereas canyons with more pointed headwalls showed clear evidence of gradual knickpoint retreat consistent with groundwater and river erosion. Detailed studies of outburst flood sediments were also instrumental for reconstructing many late Quaternary megafloods. Complete stratigraphic sections of megafloods are hard to find, but many include, at least in parts, successions of '(i) basal, thick, coarse parallel-bedded units, (ii) large-scale clinoforms, (iii) horizontally-bedded thin laminated units, (iv) ripple and dune cross-beds, (v) silt-beds, and (vi) succession-capping debris flow deposits' (Carling 2013).

The Channeled Scablands in Washington State, United States, is emblematic of the geomorphic impacts of Pleistocene ice-dam floods that had estimated peak discharges of the order of 10 $Mm^3 s^{-1}$ (Baker 2009).

Perhaps even larger outburst floods have been identified in southern Siberia and Mongolia on the basis of very similar landforms and deposits (Rudoy 2002). Some megafloods have so altered landscapes that the overall impact is better seen from space than from the ground. New megaflood tracts in North America and Siberia are still being discovered on satellite imagery (Murton et al. 2010; Komatsu et al. 2009; Rayburn et al. 2005). Reconstructing these floods hinges on the size and shape that bedrock channels, gravel dunes, slackwater deposits, megaclasts, and other proxies have today, and scientists continue to identify new major flood paths of late Pleistocene megafloods (Murton et al. 2010). However, Larsen and Lamb (2016) pointed out that some of these proxies are consistent with the cumulative impacts of more gradual flooding and incision, and cautioned against potentially overestimating peak discharges of past megafloods. A numerical threshold shear stress model for simulating Pleistocene flood-water surfaces at Moses Coulee, eastern Washington, led the authors to conclude that discharges could have been five to tenfold smaller than previously thought.

The large areas that megafloods affected encourage the search for multiple independent evidence. For example, distinct freshwater inflows by megafloods are archived in offshore sediments (Blais-Stevens et al. 2003) at and after the Pleistocene–Holocene transition. The Younger Dryas and the 8.2 ka cold events, which are attributed to outburst floods from glacier-dammed lakes in North America (Clarke et al. 2003; Alley and Ágústsdóttir 2004), had far-reaching impacts, including changes in the oceanic thermohaline circulation and abrupt decreases in the strength of the South Asian summer monsoon (Dixit et al. 2014; Grant et al. 2014).

Sediments of former lakes that were sources of megafloods are another valuable archive. The Kathmandu Basin in the Nepal Himalayas is likely a remnant of a vast debris-flow-dammed lake that formed along the former Bagmati River between 1.07 and 0.97 Ma (Sakai et al. 2006). Coincident rapid tectonic uplift of the Mahabharat Ranges where the dam was located aided the trapping of more than 600 m of Pleistocene valley-fill sediments before the lake emptied rapidly sometime during the early Holocene, exposing the extensive lake floor on which Nepal's capital is located. Thick deposits of Holocene lake sediments in the large mountain valleys of southeastern Tibet record natural damming by glacier dams up to 700 m high, which upon their sudden failure may have released megafloods with estimated peak discharges of 1–5 $Mm^3 s^{-1}$ down the Tsangpo gorge (Montgomery et al. 2004). Detrital zircon U-Pb provenance data from correlative outburst flood deposits downstream of the gorge appear to support this hypothesis (Lang et al. 2013).

The interplay of active volcanism and glaciers or ice sheets produces large outburst floods. Jökulsá á Fjöllum, a large river in northeastern Iceland, shows evidence of Holocene outburst floods with estimated peak discharges as large as 0.9 $Mm^3 s^{-1}$, which inundated an estimated 1400 km^2 (Alho et al. 2005). Rapidly draining caldera lakes have produced megafloods with peak discharges up to ∼2 $Mm^3 s^{-1}$ (Manville 2010). These megaflood sources form irrespective of changes to ice sheets or glaciers linked to Quaternary glacial–interglacial cycles. Several alluvial plains and intramontane fans on the Japanese islands bear traces of former outburst floods from caldera lakes. Bedrock-confined upstream reaches feature diagnostic landforms, including fluvially sculpted hanging valleys well below Quaternary glacial limits, boulder bars, and dry valleys in regions without karst (Kataoka 2011). Large fill terraces contain abundant reworked pyroclastic deposits lacking any erosional features, which are indicative of large-magnitude and high-energy transport.

Despite the global importance of these prehistoric megafloods, we must take care in choosing which, if any, data on their frequency and magnitude we might wish to use in hazard models. At least Earth lacks continental ice sheets for feeding such large meltwater lakes today. Yet the legacy of these exceptional outburst floods may control the geometry of bedrock channels over many millennia. Baynes et al. (2015), for example, used surface exposure dating based on cosmogenic ^{3}He concentrations in river-cut bedrock surfaces to infer that three phases of outburst flooding dominated the entire Holocene history of incision of the Jökulsárgljúfur canyon in northeast Iceland. Bedrock knickpoints created by these rare outburst floods have remained largely in place despite high 'normal' flood discharges in this region. Several of these knickpoints were former flood-overspill locations, and now remain in the landscape as striking waterfalls popular with tourists.

11.5 Measures of Catchment Denudation

The concentration of cosmogenic ^{10}Be in quartz-rich river sands offers another opportunity to indirectly infer the geomorphic relevance of multiple floods, or put differently, the cumulative effect of catchment denudation processes that we infer from river sediment. Assumptions underlying this method are that rivers flush all the sediment from the catchment to the location of sampling; that detachment of material from hillslopes occurs largely by grain-by-grain erosion; that sediment storage is negligible; and that the sediment sample is perfectly mixed and hence representative of the denudation processes operating upstream. Thousands of catchment denudation rates have been inferred this way and are now available for study in a systematic documentation. Tomkins et al. (2007) used the method in the Blue Mountains west of Sydney, Australia, and found that erosion rates inferred from historic suspended sediment measurements were lower than the rates inferred from cosmogenic ^{10}Be concentrations. They also concluded that erosion following wildfires accounted for only 5% of the long-term rates, whereas catastrophic floods – including some palaeofloods that were much larger than historically recorded ones – accounted for about a third.

Similarly, Siame et al. (2011) noted the contrasting rates of [10]Be-derived denudation and modern suspended sediment loads in northeastern Taiwan. There, denudation rates of 2 ± 1 mm yr^{-1} averaged over several centuries were lower than contemporary rates of 5–7 mm yr^{-1} averaged over several decades. Frequent disturbances by earthquakes and typhoons may account for this difference by promoting highly intermittent storage and pulsed export of sediment. Bedload is also often unresolved in gauging data and might explain part of the discrepancy between the decadal and centennial estimates. Landslides in particular enrich river loads with sediments of highly differing concentrations of cosmogenic nuclides as the material comes from different depths. Landslides with failure surfaces extending more than 2 m below the surface (a depth at which most incoming cosmogenic radiation is largely attenuated) may add high contributions of material depleted in [10]Be to rivers (Yanites et al. 2009).

Studies that attempt to infer catchment erosion from historic gauging records have identified several potential shortcomings. For example, different erosional thresholds can further limit the sediment output from catchments, especially those nourished mainly by gully incision and landslides (Hicks et al. 2000). These sediment sources may only activate episodically and for short times during rainstorms or strong earthquake shaking. It is also possible that extrapolations based on suspended sediment loads underestimate the total load, especially where the bedload fraction is high (Figure 11.12). Gauging campaigns have shown that tropical rivers in particular can have bedload concentrations of 20–40% of the total load, or at least more than the ~10% that is commonly assumed (Ziegler et al. 2014). Average estimates of erosion or sediment transport also inevitably integrate both the geomorphically active and inactive phases of a given process, which results in systematically lower mean rates for longer observation periods (Sadler 1981; Schumer and Jerolmack 2009). Catchment erosion rates that are measured over years or decades may be of the same order of magnitude as rates of landscape lowering over millions of years

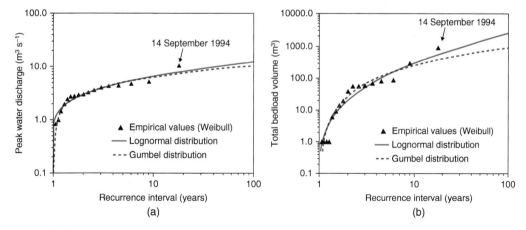

Figure 11.12 Magnitude and frequency of gauged water and bedload discharge (triangles) in Rio Cordon, Italian Alps. Left: Magnitude–frequency relationship for annual maximum peak discharge. Right: Magnitude–frequency relationship for annual maximum bedload volume. Red and dashed lines are two possible model fits to the data. From Lenzi et al. (2004).

(Peeters et al. 2008), but it is unclear whether this is merely a coincidence.

Sedimentary archives in lakes and oceans are another option to appreciate in an integrated manner the legacy of floods. Lakes trap incoming fluvial, hillslope, and aeolian sediments, as well as biogeochemical species, and can be very useful recorders of past environmental conditions. Modern studies employ a wide range of sedimentological and geochemical fingerprinting techniques on well-dated lake-sediment cores to unravel the timing and processes of sedimentation. Distinct event layers in these lake sediments often link up with inputs by catastrophic floods (Gomez et al. 2007), but may also instead have resulted from other processes such as local landslides or delta collapse (Girardclos et al. 2007). Some

of these archives can span several tens of thousands of years, allowing detailed study of the dominant frequencies of such event layers and their potential link to former climate time series (Schlolaut et al. 2014). Delta sediments can also trap coarse flood layers, and Schulte et al. (2015) used such beds to build a 2600-year chronology of floods entering Lake Brienz, Swiss Alps. Fed mainly by the Aare, a gravel-bed river, the delta had been aggrading at rates of 1–6 mm yr^{-1} over the past few millennia, conserving amongst fluvial sediments also traces of organic soils and peat beds away from the active channel (Figure 11.13). Aside from coarser grain size, the flood layers had a distinct geochemical signature in terms of their heavy mineral content, and the ratio of zirconium to titanium (Zr/Ti) seemed particularly

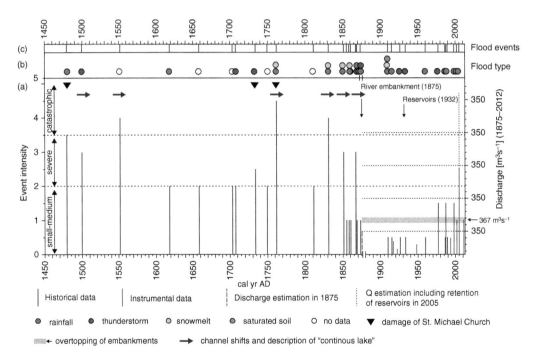

Figure 11.13 Flood chronology of the Aare River in the Hasli Valley, Switzerland, from AD 1480 to 2012. (a) Flood intensities reconstructed from documentary and geomorphological evidence (blue columns) and instrumental data (red columns). Event intensities (M, left scale) were estimated for the period AD 1480–2012, whereas maximum annual discharges >300 m^3 s^{-1} (right scale) recorded at the Brienzwiler gauging station apply the period AD 1908–2012. Discharges in AD 1875 were made by engineers of the Aare correction project. Measured discharges and event intensity levels since AD 1875 are influenced by construction of river embankments and changes in retention capacities of reservoirs. Triangles represent damage to the Sankt Michael church by floods and aggradation by the Alpbach River. (b) Hydro-climatological causes of floods. (c) Composite flood frequencies. From Schulte et al. (2015).

useful for independently distinguishing 12 out of 14 historically documented floods in the past 500 years.

The marine environment offshore of river mouths is a place of strong density and salinity contrasts. This interface is the breeding ground for hyperpycnal plumes and turbidity currents that happen when denser, sediment-rich river waters enter the ocean (Mulder et al. 2003). This process requires sediment concentrations >40 g l^{-1} that can occur on a yearly basis at the mouth of the Yellow River, China, and less frequently at the mouths of many smaller rivers (Lamb and Mohrig 2009). Even slight density contrasts in seawater alone might be sufficient to set in motion sediment in turbidity currents. Canals et al. (2006) called this process 'dense shelf water cascading', and argued that it is partly responsible for creating large erosional bedforms in submarine canyons. Clastic mud beds devoid of sedimentary structures but rich in organic matter in the deep-sea fan of the Nile River show characteristics of hyperpycnal plumes that Ducassou et al. (2008) directly attributed to larger-than-average river floods. Identifying and confirming similar linkages between deep-sea sedimentary records and terrestrial flood inputs could offer new tools for independently testing past flood chronologies.

However, the hypothesis that hyperpycnal flow deposits faithfully record terrestrial floods hinges on the assumption that the resulting turbidites preserve the rising and falling limbs of the flood. Otherwise the sedimentary signal would be lost and the turbidite sedimentology inconclusive about onshore sources. Lamb and Mohrig (2009) used a one-dimensional flow model to demonstrate that expected bed forms and sediment grading patterns in hyperpycnal flow deposits may record multiple flow accelerations and decelerations even during a simple single-peaked flood. This result could compromise a straightforward, one-to-one correspondence of marine turbidites and terrestrial floods.

These examples show, however, that the flux of river sediment continues to have geomorphic impacts even after leaving the fluvial setting. Large sediment plumes issuing from the Burdekin River in northeastern Australia only occasionally reached the inner fringes of the Great Barrier Reef from the mid-eighteenth to the mid-nineteenth century. However, after about 1870 the flux of river-borne sediment increased five to tenfold, especially after drought-breaking floods. Geochemical evidence of such palaeofloods reaching the Great Barrier Reef comes from barium-calcium (Ba/Ca) ratios in long-lived *Porites*, a coral species that is sensitive to changing sediment supply (McCulloch et al. 2003). The corals record an abrupt increase in sediment input from the Burdekin River in the late nineteenth century, immediately following the time when European settlers arrived, clearing forests and establishing cattle grazing. This practice seems to have degraded the land surface heavily, eroding large amounts of topsoil in the Burdekin River catchment. Overall, the total suspended sediment loads delivered to coastal waters fringing the Great Barrier Reef increased more than fivefold as a consequence (Brodie et al. 2012). Despite the promise of corals as offshore proxies of terrestrial sediment flux, they likely record only the most far-reaching of floods, as most suspended river sediment and the particulate nutrients that it contains settle rapidly within a few kilometres of river mouths. In these areas continuous fresh water discharge from the river inhibits coral growth, while longshore drift further incorporates coarser sediments, mainly into sand bars, beach ridges, and subaqueous dunes.

11.6 The Future of Flood Hazards

Much research has been concerned with whether floods will become more frequent or extreme in a warming world with more variable pattens of precipitation. A widespread method has been to investigate how flood frequency curves that relate contemporary discharge to average return periods might change in the future. Altered weather patterns and increases in droughts and storms will likely alter soil infiltration rates, groundwater flux, and surface runoff, all key variables controlling the generation and routing of floods.

Warming of the atmosphere and oceans involves changes in the transport of air masses and the precipitation they carry, thus directly affecting runoff and, indirectly, river discharge. Such changes affect entire regions so that analyses have adopted the strategy of estimating the frequency and magnitude of discharge by extrapolating data from many stations or even many catchments. Motivated by projections that hold that Arctic regions will warm most rapidly in the future, McClelland et al. (2004) reported that discharge from Siberia's six largest Arctic rivers increased by an average of 128 km^3 yr^{-1} from 1936 to 1999. They ran numerical experiments to show that artificial damming, permafrost thaw, and effects of wildfires were too small to explain this increased discharge. Instead, they concluded that the enhanced northward transport of moist air masses tied to atmospheric warming was the most likely cause of the greater water flows in the rivers, and thus changed flood frequencies.

Milly et al. (2002) argued that flood risk is also increasing as climate warms in Europe. 'Risk' in this study was defined as the 100-year flood stage in 29 worldwide river catchments that have areas >0.2 million km^2. The 100-year flood can be estimated from extreme-value statistics (see Chapter 5.4.3) and is widely used in flood studies, but we recall that it is more a measure of flood hazard than of risk. Milly et al. (2002) reported that about 75% of all 100-year floods in these large catchments happened in the second half of the twentieth century. Aided by numerical climate models they argued that this frequency is likely to increase in the future. This projection may need further testing for smaller catchments and regional variations, especially as many large rivers at low latitudes seemed to deviate from the modelled data.

Studies focusing on single catchments confirm the high variance in flood trends past and future. For example, a detailed analysis of extreme floods on the German Elbe and Oder rivers for the past 80–150 years, failed to find any rising trend in the rates of floods occurrence (Mudelsee et al. 2003). The authors grouped the floods in the historic records according to thresholds in discharge and distinguished between summer and winter floods. On average, the frequency of larger floods seemed to have remained constant or even declined in both river catchments, regardless of the season that the floods occurred in. The authors also argued that recorded precipitation extremes, defined as 25-year maxima, similarly appeared to have remained unchanged in central Europe in the twentieth century.

Regional and global assessments aggregate such uncertainties and possibly smooth out diverging trends in flood activity. Yet correlations between extreme discharges of major river catchments can show up even at these aggregated scales, and allow some extrapolation. Probabilistic projections hold that average flood losses across Europe could well double by 2050 at the current rates of atmospheric warming and socioeconomic development (Jongman et al. 2014). At the global scale, Jongman et al. (2015) proposed that the average vulnerability to floods decreased between 1980 and 2010, mostly because of the growing income per capita. They approximated vulnerability to floods using a dynamic high-resolution model of flood hazard and exposure, and backed their findings by the documented decrease of flood victims, losses normalized by population, and gross domestic product exposed to inundation. Ensembles that average the results from several climate, flood routing, and inundation models predict large regional differences in flood activity in the twenty-first century. Hirabayashi et al. (2013) offered a worldwide prognosis that the 100-year flood discharge in the twentieth century might have altered return periods anywhere between two and more than a thousand years in the twenty-first century (Figure 11.14). For example, rivers in most of Scandinavia, eastern Europe, and western Russia would exceed this flood discharge less often in the future, whereas rivers in India and southeast Asia would exceed it more frequently. This study interpolated gauging discharge with inputs from a hydrological model to estimate the 100-year flood in 15' × 15' grid cells worldwide. While this approach may mask a lot of uncertainty regarding local flood

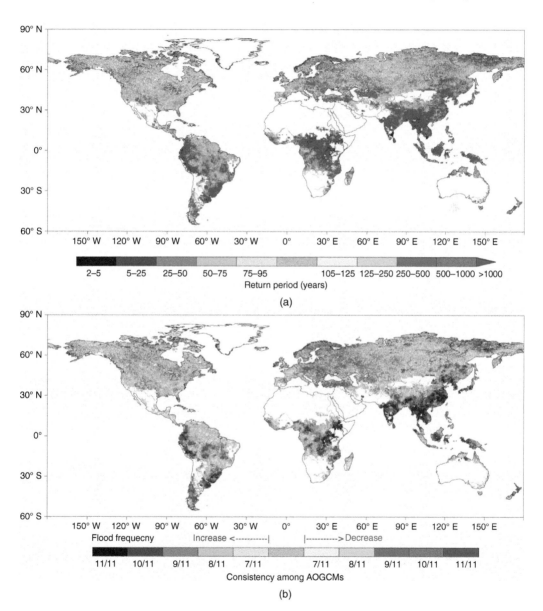

Figure 11.14 Projected change in flood frequency in a warming climate. (a) Multimodel median return periods (years) in the twenty-first century for discharge corresponding to the twentieth century 100-year flood. (b) Model consistency. Grid cells with mean annual discharge of a retrospective simulation for 1979–2010 of <0.01 mm/day are screened out. The case for the IGPCC RCP8.5 scenario is shown. From Hirabayashi et al. (2013).

levels, this and many other regional to global studies of changing flood hazards and risks allow testing of different appraisals of flood hazard, and also highlight regions or countries that stand out for some reason.

Predictions of flood frequency rarely consider changes in sediment transport or intermittent storage. In many tectonically active mountain ranges, sediment production, mobilization, and deposition is highly episodic and often tied to processes other than floods. The term 'sediment pollution' might characterize the problems associated with large volumes of river- and hillslope-derived material that are sluiced along steep mountain rivers, creating large fluvial and alluvial deposits in the wake of rainstorms. The process may be entirely natural, so giving it an anthropogenic label is possibly misleading. In any case, such sediment may remain stored in rivers for periods that are much longer than predictions of future flood hazards. This lag effect is at least one reason why any prediction about whether sediment yields from catchments will increase with atmospheric warming needs several simplifying assumptions. For example, increased occurrence of wildfires as a result of warmer and drier weather might favour higher net erosion rates assuming that hillslopes have sufficient amounts of sediment that can be mobilized after each fire. Changes to erosion rates also depend on how the frequency, magnitude, and the resulting spatial distribution of fires change and whether repeated fires increase weathering and thus sediment production. Goode et al. (2012) reviewed the geomorphic impacts of wildfires in semiarid catchments in Idaho, United States, with the idea that enhanced sediment yields could impact downstream reservoirs that were initially designed for lower historic sediment yields. One instructive conclusion from this study is that rare but destructive debris flows that dominate the postfire sediment flux are impractical to mitigate. The most practical mitigation option appears to be the reconstruction of damaged roads in an attempt to balance the negative impacts of climate change-induced increases in sediment supply. However, this strategy of maintaining roads might turn out to be a 'drop in the bucket', if rates of road-related erosion are smaller than those associated with increased wildfire activity.

Combining numerical climate models with measured rates of sediment production, transport, and commensurate reservoir sedimentation caused by rainstorms can be useful for anticipating the potential impacts of climate change on future sediment yields. Ono and Yamaguchi (2011) followed this approach to forecast sediment yields on the Japanese islands, and projected average increases of 20% in rainfall-induced sediment yields for the more likely climate scenarios. Independent studies offered largely consistent results for suspended sediment yields (Mouri et al. 2010), although the underlying assumption is that we can measure contemporary sediment yields within a variance that is less than the projected changes. This increase might seem small, but Japanese authorities go to great pains to maintain thousands of reservoirs and sediment retention basins with sediment yields of the order of 100–1,000 t km^{-2} yr^{-1}. Many of these artificial sediment traps, like most others elsewhere, require regular dredging or flushing to keep them operational. Of course these projections of future sediment yields depend on reliable and statistically robust trends of past sediment fluxes and their response to rainfall events, especially in mountainous areas with high and variable erosion rates. Data from 59 large Japanese dams suggest that extreme sedimentation events in mountain catchments impacted by rainstorms or earthquakes are frequent, with mean return periods of only 5–7 years (Ono and Yamaguchi 2011). Detailed sedimentation measurement campaigns of this kind are valuable for refining numerical models that estimate or predict future changes in water and sediment fluxes in rivers, although data on sedimentation in reservoirs often remain outside of the public domain or scientific publications.

References

Aalto R, Lauer JW, and Dietrich WE 2008 Spatial and temporal dynamics of sediment accumulation and exchange along Strickland River floodplains (Papua New Guinea) over decadal-to-centennial timescales. *Journal of Geophysical Research: Earth Surface* **113**, 1–22.

Alexander J and Cooker MJ 2016 Moving boulders in flash floods and estimating flow conditions using boulders in ancient deposits. *Sedimentology* **63**(6), 1582–1595.

Alho P, Baker VR, and Smith LN 2010 Paleohydraulic reconstruction of the largest Glacial Lake Missoula draining(s). *Quaternary Science Reviews* **29**(23-24), 3067–3078.

Alho P, Russell AJ, Carrivick JL, and Käyhkö J 2005 Reconstruction of the largest Holocene jökulhlaup within Jökulsá á Fjöllum, NE Iceland. *Quaternary Science Reviews* **24**(22), 2319–2334.

Alley RB and Ágústsdóttir AM 2004 The 8k event: Cause and consequences of a major Holocene abrupt climate change. *Quaternary Science Reviews* **24**(10), 1123–1149.

Assine ML 2005 River avulsions on the Taquari megafan, Pantanal wetland, Brazil. *Geomorphology* **70**(3-4), 357–371.

Badoux A, Andres N, and Turowski JM 2014 Damage costs due to bedload transport processes in Switzerland. *Natural Hazards and Earth System Sciences* **14**(2), 279–294.

Bajracharya B, Shrestha AB, and Rajbhandari L 2007 Glacial lake outburst floods in the Sagarmatha region. *Mountain Research and Development* **27**(4), 336–344.

Baker VR 2006 Palaeoflood hydrology in a global context. *Catena* **66**(1-2), 161–168.

Baker VR 2009 The channeled scabland: A retrospective. *Annual Review of Earth and Planetary Sciences* **37**(1), 393–411.

Baynes ERC, Attal M, Niedermann S, et al. 2015 Erosion during extreme flood events dominates Holocene canyon evolution in northeast Iceland. *Proceedings of the National Academy of Sciences* **112**(8), 2355–2360.

Beltaos S and Prowse T 2009 River-ice hydrology in a shrinking cryosphere *Hydrological Processes* **23**, 122–144.

Belyaev VR, Golosov VN, Markelov MV, et al. 2012 Using Chernobyl-derived ^{137}Cs to document recent sediment deposition rates on the River Plava floodplain (Central European Russia). *Hydrological Processes* **27**(6), 807–821.

Benvenuti M, Mariotti-Lippi M, Pallecchi P, and Sagri M 2006 Late-Holocene catastrophic floods in the terminal Arno River (Pisa, Central Italy) from the story of a Roman riverine harbour. *The Holocene* **16**(6), 863–876.

Bernal C, Christophoul F, Soula JC, et al. 2012 Gradual diversions of the Rio Pastaza in the Ecuadorian piedmont of the Andes from 1906 to 2008: Role of tectonics, alluvial fan aggradation, and ENSO events. *International Journal of Earth Sciences* **101**(7), 1913–1928.

Bintliff J 2002 Time, process and catastrophism in the study of Mediterranean alluvial history: A review. *World Archaeology* **33**(3), 417–435.

Blais-Stevens A, Clague JJ, Mathewes RW, et al. 2003 Record of large, Late Pleistocene outburst floods preserved in Saanich Inlet sediments, Vancouver Island, Canada. *Quaternary Science Reviews* **22**(21-22), 2327–2334.

Bollinger L, Perrier F, Avouac JP, et al. 2007 Seasonal modulation of seismicity in the Himalaya of Nepal. *Geophysical Research Letters* **34**(8), 1–5.

Brodie JE, Kroon FJ, Schaffelke B, et al. 2012 Terrestrial pollutant runoff to the Great Barrier Reef: An update of issues, priorities and management responses. *Marine Pollution Bulletin* **65**(4-9), 81–100.

Bryant M, Falk P, and Paola C 1995 Experimental study of avulsion frequency and rate of deposition. *Geology* **23**(4), 365–368.

Canals M, Puig P, de Madron XD, et al. 2006 Flushing submarine canyons. *Nature* **444**(7117), 354–357.

Cantelli A, Paola C, and Parker G 2004 Experiments on upstream-migrating erosional narrowing and widening of an incisional channel caused by dam removal. *Water Resources Research* **40**, 1–12.

Carling P, Jansen J, and Meshkova L 2013 Multichannel rivers: their definition and classification. *Earth Surface Processes and Landforms* **39**(1), 26–37.

Carling P, Villanueva I, Herget J, et al. 2010 Unsteady 1D and 2D hydraulic models with ice dam break for Quaternary megaflood, Altai Mountains, southern Siberia. *Global and Planetary Change* **70**(1-4), 24–34.

Carling PA 2013 Freshwater megaflood sedimentation: What can we learn about generic processes?. *Earth-Science Reviews* **125**(C), 87–113.

Casagli N, Ermini L, and Rosati G 2003 Determining grain size distribution of the material composing landslide dams in the Northern Apennines: Sampling and processing methods. *Engineering Geology* **69**(1-2), 83–97.

Cenderelli DA and Wohl EE 2001 Peak discharge estimates of glacial-lake outburst floods and "normal" climatic floods in the Mount Everest region, Nepal. *Geomorphology* **40**(1), 57–90.

Chakraborty T and Ghosh P 2010 The geomorphology and sedimentology of the Tista megafan, Darjeeling Himalaya: Implications for megafan building processes. *Geomorphology* **115**(3-4), 252–266.

Chakraborty T, Kar R, Ghosh P and Basu S 2010 Kosi megafan: Historical records, geomorphology and the recent avulsion of the Kosi River. *Quaternary International* **227**(2), 143–160.

Clague JJ and Evans SG 2000 A review of catastrophic drainage of moraine-dammed lakes in British Columbia. *Quaternary Science Reviews* **19**(17), 1763–1783.

Clague JJ and Turner RJ 2003 *Vancouver, City on the Edge; Living with a Dynamic Geological Landscape*. Tricouni Press.

Clarke G, Leverington D, Teller J, and Dyke A 2003 Superlakes, megafloods, and abrupt climate change. *Science* **301**(5635), 922–923.

Constantine JA, Dunne T, Ahmed J, et al. 2014 Sediment supply as a driver of river meandering and floodplain evolution in the Amazon Basin. *Nature Geoscience* **7**(12), 899–903.

Costa JE and Schuster RL 1988 The formation and failure of natural dams. *Geological Society of America Bulletin* **100**, 1054–1068.

Darby SE, Hackney CR, Leyland J, et al. 2016 Fluvial sediment supply to a mega-delta reduced by shifting tropical-cyclone activity. *Nature* **539**, 276–279.

Das S, Kar NS, and Bandyopadhyay S 2015 Glacial lake outburst flood at Kedarnath, Indian Himalaya: a study using digital elevation models and satellite images. *Natural Hazards* **77**(2), 769–786.

Davidson EA, de Araújo AC, Artaxo P, et al. 2012 The Amazon basin in transition. *Nature* **481**(7381), 321–328.

Davies TRH, Smart CC, and Turnbull JM 2003 Water and sediment outbursts from advanced Franz Josef Glacier, New Zealand. *Earth Surface Processes and Landforms* **28**(10), 1081–1096.

de Haas T, Ventra D, Carbonneau PE, and Kleinhans MG 2014 Debris flow dominance of alluvial fans masked by runoff reworking and weathering. *Geomorphology* **217**, 165–181.

de Haas T, Densmore AL, Stoffel M, et al. 2018 Avulsions and the spatio-temporal evolution of debris-flow fans. *Earth-Science Reviews* **177**, 53–75.

Denlinger RP and O'Connell DRH 2010 Simulations of cataclysmic outburst floods from Pleistocene Glacial Lake Missoula. *Geological Society of America Bulletin* **122**(5-6), 678–689.

Dixit Y, Hodell DA, Sinha R, and Petrie CA 2014 Abrupt weakening of the Indian summer monsoon at 8.2 kyr B.P. *Earth and Planetary Science Letters* **391**, 16–23.

Ducassou E, Mulder T, Migeon S, et al. 2008 Nile floods recorded in deep Mediterranean sediments. *Quaternary Research* **70**(3), 382–391.

Eaton BC, Church M, and Millar RG 2004 Rational regime model of alluvial channel morphology and response. *Earth Surface Processes and Landforms* **29**(4), 511–529.

Fenton CR, Webb RH, and Cerling TE 2006 Peak discharge of a Pleistocene lava-dam outburst flood in Grand Canyon, Arizona, USA. *Quaternary Research* **65**(2), 324–335.

Froese DG, Smith GD, Westgate JA, et al. 2003 Recurring middle Pleistocene outburst floods in east-central Alaska. *Quaternary Research* **60**(1), 50–62.

Fujioka T, Fink D, Nanson G, et al. 2014 Flood-flipped boulders: In-situ cosmogenic nuclide modeling of flood deposits in the monsoon tropics of Australia. *Geology* **43**(1), 43–46.

Garcia-Castellanos D and Villaseñor A 2011 Messinian salinity crisis regulated by competing tectonics and erosion at the Gibraltar arc. *Nature* **480**(7377), 359–363.

Garcia-Castellanos D, Estrada F, Jiménez-Munt I, et al. 2009 Catastrophic flood of the Mediterranean after the Messinian salinity crisis. *Nature* **462**(7274), 778–781.

Gaume E, Bain V, Bernardara P, et al. 2009 A compilation of data on European flash floods. *Journal of Hydrology* **367**(1-2), 70–78.

Gautier E, Brunstein D, Vauchel P, et al. 2007 Temporal relations between meander deformation, water discharge and sediment fluxes in the floodplain of the Rio Beni (Bolivian Amazonia). *Earth Surface Processes and Landforms* **32**(2), 230–248.

Girardclos S, Schmidt OT, Sturm M, et al. 2007 The 1996 AD delta collapse and large turbidite in Lake Brienz. *Marine Geology* **241**(1-4), 137–154.

Gomez B, Carter L, and Trustrum NA 2007 A 2400 yr record of natural events and anthropogenic impacts in intercorrelated terrestrial and marine sediment cores: Waipaoa sedimentary system, New Zealand. *Geological Society of America Bulletin* **119**(11-12), 1415–1432.

Goode JR, Luce CH, and Buffington JM 2012 Enhanced sediment delivery in a changing climate in semi-arid mountain basins: Implications for water resource management and aquatic habitat in the northern Rocky Mountains. *Geomorphology* **139-140**(C), 1–15.

Grant KM, Rohling EJ, Ramsey CB, et al. 2014 Sea-level variability over five glacial cycles. *Nature Communications* **5**, 1–9.

Guan M, Carrivick JL, Wright NG, et al. 2016 Quantifying the combined effects of multiple extreme floods on river channel geometry and on flood hazards. *Journal of Hydrology* **538**(C), 256–268.

Gupta S, Collier JS, Palmer-Felgate A, and Potter G 2007 Catastrophic flooding origin of shelf valley systems in the English Channel. *Nature* **448**(7151), 342–345.

Henck AC, Montgomery DR, Huntington KW, and Liang C 2010 Monsoon control of effective discharge, Yunnan and Tibet. *Geology* **38**(11), 975–978.

Hewitt K 1999 Quaternary moraines vs catastrophic rock avalanches in the Karakoram Himalaya, northern Pakistan. *Quaternary Research* **51**, 220–237.

Hewitt K 2009 Glacially conditioned rock-slope failures and disturbance-regime landscapes, Upper Indus Basin, northern Pakistan. *Geological Society, London, Special Publications* **320**(1), 235–255.

Hicks DM, Gomez B, and Trustrum NA 2000 Erosion thresholds and suspended sediment yields, Waipaoa River basin, New Zealand. *Water Resources Research* **36**(4), 1129–1142.

Hirabayashi Y, Mahendran R, Koirala S, et al. 2013 Global flood risk under climate change. *Nature Climate Change* **3**(9), 816–821.

Huang CC, Pang J, Zha X, et al. 2010 Extraordinary floods of 4100-4000 a BP recorded at the late neolithic ruins in the Jinghe River gorges, middle reach of the Yellow River, China. *Palaeogeography, Palaeoclimatology, Palaeoecology* **289**(1-4), 1–9.

Inoue K, Mizuyama T, and Sakatani Y 2010 The catastrophic Tombi landslide and accompanying landslide dams induced by the 1858 Hietsu earthquake. *Journal of Disaster Research* **5**(3), 245–256.

Islam MR, Begum SF, Yamaguchi Y, and Ogawa K 1999 The Ganges and Brahmaputra rivers in Bangladesh: Basin denudation and sedimentation. *Hydrological Processes* **13**, 2907–2923.

Ives J, Shrestha R, and Mool PK 2010 *Formation of glacial lakes in the Hindu Kush-Himalayas and GLOF risk assessment*. International Centre for Integrated Mountain Development (ICIMOD), Kathmandu.

Jacquet J, McCoy SW, McGrath D, et al. 2017 Hydrologic and geomorphic changes resulting from episodic glacial lake outburst floods: Rio Colonia, Patagonia, Chile. *Geophysical Research Letters* **44**(2), 854–864.

Jennerjahn TC 2012 Biogeochemical response of tropical coastal systems to present and past environmental change. *Earth-Science Reviews* **114**(1-2), 19–41.

Jerolmack DJ and Mohrig D 2007 Conditions for branching in depositional rivers. *Geology* **35**(5), 463.

Jongman B, Hochrainer-Stigler S, Feyen L, et al. 2014 Increasing stress on disaster-risk finance due to large floods. *Nature Climate Change* **4**(4), 264–268.

Jongman B, Winsemius HC, Aerts JCJH, et al. 2015 Declining vulnerability to river floods and the global benefits of adaptation. *Proceedings of the National Academy of Sciences* **112**(18), E2271–E2280.

Kale VS 2007 A half-a-century record of annual energy expenditure and geomorphic effectiveness of the monsoon-fed Narmada River, central India. *Catena* **75**(2), 154–163.

Kataoka KS 2011 Geomorphic and sedimentary evidence of a gigantic outburst flood from Towada caldera after the 15ka Towada–Hachinohe ignimbrite eruption, northeast Japan. *Geomorphology* **125**(1), 11–26.

Keefer DK, Moseley ME, and deFrance SD 2003 A 38 000-year record of floods and debris flows in the Ilo region of southern Peru and its relation to El Niño events and great earthquakes. *Palaeogeography,*

Palaeoclimatology, Palaeoecology **194**(1-3), 41–77.

Klimeš J, Novotný J, Novotná I, et al. 2016 Landslides in moraines as triggers of glacial lake outburst floods: example from PalcacochaLake (Cordillera Blanca, Peru). *Landslides* **13**, 1–17.

Komatsu G, Arzhannikov SG, Gillespie AR, et al. 2009 Quaternary paleolake formation and cataclysmic flooding along the upper Yenisei River. *Geomorphology* **104**(3-4), 143–164.

Korup O 2012 Earth's portfolio of extreme sediment transport events. *Earth-Science Reviews* **112**, 115–125.

Korup O and Tweed F 2007 Ice, moraine, and landslide dams in mountainous terrain. *Quaternary Science Reviews* **26**(25-28), 3406–3422.

Lamb MP and Fonstad MA 2010 Rapid formation of a modern bedrock canyon by a single flood event. *Nature Geoscience* **3**(7), 477–481.

Lamb MP and Mohrig D 2009 Do hyperpycnal-flow deposits record river-flood dynamics?. *Geology* **37**(12), 1067–1070.

Lamb MP, Mackey BH, and Farley KA 2014 Amphitheater-headed canyons formed by megaflooding at Malad Gorge, Idaho. *Proceedings of the National Academy of Sciences* **111**(1), 57–62.

Lang KA, Huntington KW, and Montgomery DR 2013 Erosion of the Tsangpo Gorge by megafloods, Eastern Himalaya. *Geology* **41**(9), 1003–1006.

Larsen IJ and Lamb MP 2016 Progressive incision of the Channeled Scablands by outburst floods. *Nature* **538**(7624), 229–232.

Latrubesse EM, Amsler ML, de Morais RP, and Aquino S 2009 The geomorphologic response of a large pristine alluvial river to tremendous deforestation in the South American tropics: The case of the Araguaia River. *Geomorphology* **113**(3-4), 239–252.

Leier AL, DeCelles PG, and Pelletier JD 2005 Mountains, monsoons, and megafans. *Geology* **33**(4), 289.

Lenzi MA, Mao L, and Comiti F 2004 Magnitude-frequency analysis of bed load data in an Alpine boulder bed stream. *Water Resources Research* **40**, 1–12.

Lesk C, Rowhani P, and Ramankutty N 2016 Influence of extreme weather disasters on global crop production. *Nature* **529**(7584), 84–87.

Liu J, Cheng Z, and Li Q 2013 Meteorological conditions for frequent debris flows from Guxiang Glacier, Mount Nyenchen Tanglha, China. *Mountain Research and Development* **33**(1), 95–102.

Macías JL, Capra L, Scott KM, et al. 2004 The 26 May 1982 breakout flows derived from failure of a volcanic dam at El Chichón, Chiapas, Mexico. *Geological Society of America Bulletin* **116**(1), 233.

Macklin MG and Lewin J 2015 The rivers of civilization. *Quaternary Science Reviews* **114**(C), 228–244.

Madej MA 2001 Development of channel organization and roughness following sediment pulses in single-thread, gravel bed rivers. *Water Resources Research* **37**(8), 2259–2272.

Major JJ 2004 Posteruption suspended sediment transport at Mount St. Helens: Decadal-scale relationships with landscape adjustments and river discharges. *Journal of Geophysical Research: Earth Surface* **109**, 1–22.

Mangerud J, Jakobsson M, Alexanderson H, et al. 2004 Ice-dammed lakes and rerouting of the drainage of northern Eurasia during the Last Glaciation. *Quaternary Science Reviews* **23**, 1313–1332.

Manville V 2010 An overview of break-out floods from intracaldera lakes. *Global and Planetary Change* **70**(1-4), 14–23.

McClelland JW, Holmes RM, Peterson BJ, and Stieglitz M 2004 Increasing river discharge in the Eurasian Arctic: Consideration of dams, permafrost thaw, and fires as potential agents of change. *Journal of Geophysical Research: Atmospheres* **109**, 1–12.

McCulloch M, Fallon S, Wyndham T, et al. 2003 Coral record of increased sediment flux to the inner Great Barrier Reef since European settlement. *Nature* **421**(6924), 727–730.

Milly PCD, Wetherald RT, Dunne KA, and Delworth TL 2002 Increasing risk of great floods in a changing climate. *Nature* **415**(6871), 514–517.

Mohr CH, Zimmermann A, Korup O, et al. 2014 Seasonal logging, process response, and geomorphic work. *Earth Surface Dynamics* **2**(1), 117–125.

Montgomery DR, Hallet B, Yuping L, et al. 2004 Evidence for Holocene megafloods down the Tsangpo River gorge, southeastern Tibet. *Quaternary Research* **62**(2), 201–207.

Morehead MD, Syvitski JP, Hutton EWH, and Peckham SD 2003 Modeling the temporal variability in the flux of sediment from ungauged river basins. *Global and Planetary Change* **39**(1-2), 95–110.

Mouri G, Golosov V, Chalov S, et al. 2010 Assessment of potential suspended sediment yield in Japan in the 21st century with reference to the general circulation model climate change scenarios. *Global and Planetary Change* **102**, 1–9.

Mudelsee M, Börngen M, Tetzlaff G, and Grünewald U 2003 No upward trends in the occurrence of extreme floods in central Europe. *Nature* **425**(6954), 166–169.

Mulder T, Syvitski JP, Migeon S, et al. 2003 Marine hyperpycnal flows: initiation, behavior and related deposits. A review. *Marine and Petroleum Geology* **20**(6), 861–882.

Murton JB, Bateman MD, Dallimore SR, et al. 2010 Identification of Younger Dryas outburst flood path from Lake Agassiz to the Arctic Ocean. *Nature* **464**(7289), 740–743.

O'Connor JE, Clague JJ, Walder JS, et al. 2013 Outburst floods. In *Treastise on Geomorphology* (ed Shroder JE), vol 9. Academic Press, pp. 475–510.

O'Connor JE and Costa JE 2004 The world's largest floods, past and present: Their causes and magnitudes. *U.S. Geological Survey Circular* **1254**, 1–13.

Ono K and Yamaguchi A 2011 Distributed specific sediment yield estimations in Japan attributed to extreme-rainfall-induced slope failures under a changing climate. *Hydrology and Earth System Sciences* **15**(1), 197–207.

Oswald EB and Wohl E 2008 Wood-mediated geomorphic effects of a jökulhlaup in the Wind River Mountains, Wyoming. *Geomorphology* **100**(3-4), 549–562.

Peeters I, Van Oost K, Govers G, et al. 2008 The compatibility of erosion data at different temporal scales. *Earth and Planetary Science Letters* **265**(1-2), 138–152.

Peng M and Zhang LM 2012 Breaching parameters of landslide dams. *Landslides* **9**(1), 13–31.

Piao S, Ciais P, Huang Y, et al. 2010 The impacts of climate change on water resources and agriculture in China. *Nature* **467**(7311), 43–51.

Pickup G and Warner RF 1975 Effects of hydrologic regime on magnitude and frequency of dominant discharge. *Journal of Hydrology* **29**(1), 51–75.

Rayburn JA, K Knuepfer PL, and Franzi DA 2005 A series of large, Late Wisconsinan meltwater floods through the Champlain and Hudson Valleys, New York State, USA. *Quaternary Science Reviews* **24**(22), 2410–2419.

Reuther AU, Herget J, Ivy-Ochs S, et al. 2006 Constraining the timing of the most recent cataclysmic flood event from ice-dammed lakes in the Russian Altai Mountains, Siberia, using cosmogenic in situ [10]Be. *Geology* **34**(11), 913.

Richardson SD and Reynolds JM 1999 An overview of glacial hazards in the Himalayas. *Quaternary International* **65**, 31–47.

Rogers KG and Overeem I 2017 Doomed to drown? Sediment dynamics in the human-controlled floodplains of the active Bengal Delta. *Elementa* **5**, 1–15.

Rowland JC, Lepper K, Dietrich WE, et al. 2005 Tie channel sedimentation rates, oxbow formation age and channel migration rate from optically stimulated luminescence (OSL) analysis of floodplain deposits. *Earth Surface Processes and Landforms* **30**(9), 1161–1179.

Rudorff CM, Dunne T, and Melack JM 2018 Recent increase of river-floodplain suspended sediment exchange in a reach of the lower Amazon River. *Earth Surface Processes and Landforms* **43**, 322–332.

Rudoy A 2002 Glacier-dammed lakes and geological work of glacial superfloods in the Late Pleistocene, Southern Siberia, Altai Mountains. *Quaternary International* **87**, 119–140.

Rustomji P 2008 A comparison of Holocene and historical channel change along the Macdonald River, Australia. *Geographical Research* **46**(1), 99–110.

Sadler P 1981 Sediment accumulation rates and the completeness of stratigraphic sections. *The Journal of Geology* **89**, 569–584.

Sakai H, Sakai H, Yahagi W, et al. 2006 Pleistocene rapid uplift of the Himalayan frontal ranges recorded in the Kathmandu and Siwalik basins. *Palaeogeography, Palaeoclimatology, Palaeoecology* **241**(1), 16–27.

Salerno F, Thakuri S, D'Agata C, Smiraglia C, et al. 2012 Glacial lake distribution in the Mount Everest region: Uncertainty of measurement and conditions of formation. *Global and Planetary Change* **92-93**(C), 30–39.

Sarma JN 2005 Fluvial process and morphology of the Brahmaputra River in Assam, India. *Geomorphology* **70**(3-4), 226–256.

Schlolaut G, Brauer A, Marshall MH, et al. 2014 Event layers in the Japanese Lake Suigetsu 'SG06' sediment core: Description, interpretation and climatic implications. *Quaternary Science Reviews* **83**(C), 157–170.

Schneider JM, Turowski JM, Rickenmann D, et al. 2014 Rapid reservoir erosion, hyperconcentrated flow, and downstream deposition triggered by breaching of 38 m tall Condit Dam, White Salmon River, Washington. *Journal of Geophysical Research: Earth Surface* **119**, 1376–1394.

Schulte L, Peña JC, Carvalho F, et al. 2015 A 2600-year history of floods in the Bernese Alps, Switzerland: frequencies, mechanisms and climate forcing. *Hydrology and Earth System Sciences* **19**(7), 3047–3072.

Schumer R and Jerolmack DJ 2009 Real and apparent changes in sediment deposition rates through time. *Journal of Geophysical Research: Earth Surface* **114**, 1–12.

Schumm SA 2005 *River Variability and Complexity*. Cambridge University Press.

Shang Y, Yang Z, Li L, et al. 2003 A super-large landslide in Tibet in 2000: background, occurrence, disaster, and origin. *Geomorphology* **54**(3-4), 225–243.

Sheffer NA, Rico M, Enzel Y, et al. 2008 The Palaeoflood record of the Gardon River, France: A comparison with the extreme 2002 flood event. *Geomorphology* **98**(1-2), 71–83.

Siame LL, Angelier J, Chen RF, et al. 2011 Erosion rates in an active orogen (NE-Taiwan): A confrontation of cosmogenic measurements with river suspended loads. *Quaternary Geochronology* **6**(2), 246–260.

Slater LJ, Singer MB, and Kirchner JW 2015 Hydrologic versus geomorphic drivers of trends in flood hazard. *Geophysical Research Letters* **42**, 370–376.

Slingerland R and Smith ND 2004 River avulsions and their deposits. *Annual Review of Earth and Planetary Sciences* **32**(1), 257–285.

Stolle A, Bernhardt A, Schwanghart W, et al. 2017 Catastrophic valley fills record large Himalayan earthquakes, Pokhara, Nepal. *Quaternary Science Reviews* **177**, 88–103.

Syvitski JPM, Cohen S, Kettner AJ, and Brakenridge GR 2014 How important and different are tropical rivers? — An overview. *Geomorphology* **227**(C), 5–17.

Terry JP 2007 *Tropical Cyclones. Climatology and Impacts in the South Pacific*. Springer.

Terry JP, Garimella S, and Kostaschuk RA 2002 Rates of floodplain accretion in a tropical island river system impacted by cyclones and large floods. *Geomorphology* **42**, 171–182.

Thompson C and Croke J 2013 Geomorphic effects, flood power, and channel competence of a catastrophic flood in confined and unconfined reaches of the upper Lockyer valley, southeast Queensland, Australia. *Geomorphology* **197**(C), 156–169.

Tomkins KM, Humphreys GS, Wilkinson MT, et al. 2007 Contemporary versus long-term denudation along a passive plate margin: the role of extreme events. *Earth Surface Processes and Landforms* **32**(7), 1013–1031.

Walsh K 2014 *The Archaeology of Mediterranean Landscapes*. Cambridge University Press.

Warrick JA, Melack JM, and Goodridge BM 2015 Sediment yields from small, steep coastal watersheds of California. *Journal of Hydrology: Regional Studies* **4**, 516–534.

Wasson RJ, Sundriyal YP, Chaudhary S, et al. 2013 A 1000-year history of large floods in the Upper Ganga catchment, central Himalaya, India. *Quaternary Science Reviews* **77**, 156–166.

Westoby MJ, Glasser NF, Hambrey MJ, et al. 2014 Reconstructing historic glacial lake outburst floods through numerical modelling and geomorphological assessment: Extreme events in the Himalaya. *Earth Surface Processes and Landforms* **39**(12), 1675–1692.

Xiao XX, White EPE, Hooten MBM, and Durham SLS 2011 On the use of log-transformation vs. nonlinear regression for analyzing biological power laws. *Ecology* **92**(10), 1887–1894.

Xu Q, Shang Y, van Asch T, et al. 2012 Observations from the large, rapid Yigong rock slide – debris avalanche, southeast Tibet. *Canadian Geotechnical Journal* **49**(5), 589–606.

Yanites BJ, Tucker GE, and Anderson RS 2009 Numerical and analytical models of cosmogenic radionuclide dynamics in landslide-dominated drainage basins. *Journal of Geophysical Research: Earth Surface* **114**(F1), 1–20.

Ziegler AD, Sidle RC, Phang VXH, et al. 2014 Bedload transport in SE Asian streams—Uncertainties and implications for reservoir management. *Geomorphology* **227**(C), 31–48.

12

Drought Hazards

Drought is a subtle and elusive natural hazard. It has a slow onset, may last several months to years, and has poorly defined, often shifting areal limits. Droughts affect more people globally than any other natural hazard. Droughts kill crops and nonagricultural vegetation, and also cause many indirect and intangible losses arising from reduced access to water, higher electricity prices, lowered air quality, or damage to road surfaces. Droughts can affect large parts of continents and can trigger severe geomorphic changes such as reduced river flows and lake levels, enhanced wildfire activity, wind-driven soil erosion, and dust storms. Human activities, including inappropriate farming practices, deforestation, overuse of water, and consequent erosion, worsen the effects of droughts. Hence, we are often confronted with deciding whether droughts are exclusively natural or rather partly human-made disasters. Droughts affect many parts of the planet, including those well outside semiarid regions prone to desertification. For example, drought conditions of the two highest categories – 'extreme' and 'exceptional' – have caused more than US$ 40 billion in damages, and claimed more than 200 lives in the United States since 2010 (Overpeck 2013). Heat waves also affect a large number of people, but they are of shorter duration than droughts. The very warm Northern Hemisphere summers of 2003, 2010, and 2014 were record breaking in terms of their temperature deviations from instrumentally recorded climate normals and were dubbed 'mega-heat waves'.

12.1 Frequency and Magnitude of Droughts

The exceptionally warm summer of 2003 caused an estimated 70 000 heat-related deaths and losses of more than €8.7 billion in western and central Europe. The hot summer of 2010 led to the death of at least 50 000 people, mostly in rural areas of Russia. High temperatures and dry vegetation favoured the burning of >10 000 km^2 of forests by wildfires in western Russia; the estimated national crop failure rate was ~25%, and temperature anomalies lasting from a fortnight to two months were recorded from the largest contiguous region in history (Barriopedro et al. 2011). Overall, these mega-heat waves of 2003 and 2010 involved unprecedented summer temperatures over nearly half of Europe, judging from the available seasonal temperature data for the past 500 years. Droughts were among the most costly natural disasters in the United States between 1980 and 2003, accounting for US$ 144 billion, or 41% of the estimated

Geomorphology and Natural Hazards: Understanding Landscape Change for Disaster Mitigation, Advanced Textbook Series,
First Edition. Tim R. Davies, Oliver Korup, and John J. Clague.
© 2021 John Wiley & Sons Ltd. Co-published 2021 by the American Geophysical Union and John Wiley & Sons Ltd.

US\$ 349 billion total cost of all weather-related disasters; average annual losses from drought amount to US\$ 6 to 8 billion each year (Mishra and Singh 2010).

The natural disaster database EM-DAT records more than 70 major droughts worldwide since the beginning of the twentieth century. These disasters killed nearly 12 million people, while affecting more than 2.7 billion (www.emdat.be). As a global average, droughts affect more people than any other natural hazard. High economic losses add to the list of worries. Droughts in the Canadian Prairies from 1999 to 2002, for example, caused losses in grain production of US\$ 3.6 billion. The direct impacts of droughts on water and food security have been tied to human development ever since human society began. Some ancient civilizations kept written accounts that allow rough estimates of how often droughts occurred in past centuries. From Chinese written records we can reconstruct that major droughts had hit the nation every 2–6 years on average during the past two millennia, but seem to be more frequent in recent times. This increase could well reflect more attention to detail in recording exceptionally dry periods, and higher population densities with commensurately more dependence on water and damage, but also awareness. Droughts were most prolific in the twentieth century, with more than three quarters of all years recording pronounced water shortages in the Chinese Loess Plateau (Derbyshire 2000). This striking regularity raises the question of whether such recurring drought damages are still compatible with the definition of natural disasters, or rather recurring losses from mainly societal processes.

12.1.1 Defining Drought

It is tempting to define a drought as an extended period of dryness. While this definition might seem appealing and intuitive, scientists have had many suggestions and disagreements as to what defines a drought.

This terminological diversity is perhaps one of the reasons that complicate strategies for mitigating drought because the necessary first stage depends on recognizing what qualifies as a drought (Mishra and Singh 2010). Suggestions as to how to classify droughts generally use meteorological, hydrological, agricultural or socioeconomic criteria. For example, the World Meteorological Organization (https://public.wmo.int/en) states that a drought 'is a prolonged dry period in the natural climate cycle that can occur anywhere in the world'. Note here the reference to a natural state in an oscillating climate system; an older definition still used by the Integrated Drought Management Programme (www.droughtmanagement.info) holds that drought is the:

> (1) Prolonged absence or marked deficiency of precipitation. (2) Period of abnormally dry weather sufficiently prolonged for the lack of precipitation to cause a serious hydrological imbalance.

The UN Convention to Combat Desertification (https://www2.unccd.int/) defines drought as the:

> naturally occurring phenomenon that exists when precipitation has been significantly below normal recorded levels, causing serious hydrological imbalances that adversely affect land resource production systems.

The Food and Agriculture Organization of the United Nations (www.fao.org) defines drought hazard as 'the percentage of years when crops fail from the lack of moisture'. Other sources mention prolonged periods of below-average rainfall relative or comparable to low-flow conditions in rivers or groundwater, though few definitions entail high temperatures. Mishra and Singh (2010) reviewed many of these and other definitions, and concluded that viable definitions should best use more than a single climatological or hydrological variable.

This lack of clarity about how to define drought is partly due to the many types of drought that are recognized in the literature. For example, meteorological droughts involve a lack of precipitation over a defined region for a period of time. Hydrological drought in turn refers to periods during which established water users have inadequate surface and subsurface water to meet their needs. We could argue, however, that this also meets most key criteria of a socioeconomic drought (see below), depending on how people are managing their water supplies. The type of agricultural droughts refers to periods with declining soil moisture and consequent crop failure without any reference to surface water resources, though with an explicit link to vegetation and land use. Socioeconomic drought implies the failure of water resource systems to meet water demands, and this type of drought is thus associated with the supply of and demand for a resource, in this case water. Van Loon et al. (2016) ventured yet a step further by proposing that any water shortage caused and modified by people be included when defining drought. These authors observed that analyses of drought monitoring and impacts had long ago adopted this practical approach without strictly distinguishing natural droughts from those made worse or even caused by human intervention.

12.1.2 Measuring Drought

The multitude of drought definitions reinforces the need for objective metrics that we can easily determine in any climate zone. The same goes for reconciling the many different or incompatible time series of past droughts if the aim is to learn something about their frequency and magnitude. To this end, simple measures are the most flexible. For example, the concept of Dust Event Days (DEDs) is meant to capture any observed and documented dust activity for a given calendar date and location (O'Loingsigh et al. 2014). Dust Event Days characterize how frequently wind erosion occurs at selected stations, but leaves open how intensive or severe this erosion is. The Dust Storm Index (DSI) tries to fill this gap by computing the weighted sum of three erosion-intensity levels across all stations that recorded the dust event within a set time period, usually a year. Compared to moderate dust storms, which have a weight of unity in the DSI, severe storms are given weights of five, and low dust storms a weight of 0.05. These simple definitions allow harmonizing of historic meteorological records from different stations and observers, and thus extending the time series of droughts or wind erosion beyond those having had a standard methodology for recording (O'Loingsigh et al. 2014).

The Palmer Drought Severity Index (PDSI) is a widely used, standardized measure for ranking droughts. Some time series allow reconstructing this index date back to the 1870s, when systematic climatic measurements began (Dai et al. 2004). The PDSI emphasizes meteorological characteristics of drought by taking into account data on both temperature and rainfall, and expressing anomalies in atmospheric moisture availability and demand at the Earth surface. Despite its popularity, this index has several limitations. For example, the interval over which observations are made is mostly applicable to agricultural impacts and may insufficiently characterize hydrological droughts. The PDSI is based on rainfall but it only partly captures the onset and termination of drought. The index is sensitive to time lags in both temperature and precipitation and, therefore, it is only partly suitable for a real-time assessment of drought conditions. Researchers have proposed several modifications of, and alternatives to, the PDSI to account for these shortcomings (Mishra and Singh 2010). One extension is the a self-calibrating PDSI (sc-PDSI), which replaces the use of constant monthly temperature and precipitation by time series of these data for a given climate station, and thus allows more consistent comparison across a region (Wells et al. 2004). For example, Pederson et al. (2014) calculated a sc-PDSI from the growth rings of more than a hundred trees in central Mongolia, and established a regional time series of drought for the past 1100 years. If adequately calibrated by and validated with instrumental data, such long-term records put the more recent droughts into perspective, and the authors argued that the early twenty-first-century drought that hit Mongolia was the hottest since the beginning of their tree-based chronology.

Bonsal et al. (2012) used the Standardized Precipitation Index (SPI) as another drought indicator besides the PDSI to evaluate the variability of summer drought duration and intensity over part of the Canadian Prairies during the pre-instrumental, spanning several hundred years, and the instrumental record (1901–2005). Based on these records, they forecast conditions during the twenty-first century using statistically downscaled climate variables from several global climate models with different emission scenarios. They found that twentieth century droughts were mild in comparison to those of the pre-instrumental period, but that the severe droughts of the latter period are likely to return, or even worsen, in the future due to warming of the atmosphere. However, they also found that future drought projections are distinctly different for the two indices. All PDSI-related model runs show greater drought frequency and severity, mainly due to increasing temperatures. Conversely, the precipitation-based SPI shows that, in the future, the frequency of summer drought might remain unchanged, although multiyear droughts may tend to be more persistent in central and southern portions of the Canadian Prairies.

Satellite remote sensing of Earth's gravity field during NASA's Gravity Recovery and Climate Experiment (GRACE) provided unprecedented insights into the regional pattern and mass balance of surface water. Changes in the loads that water exert on the Earth's crust trigger tiny fluctuations in local gravity that satellites can now measure. For example, the sensor measured a water deficit of 240 Gt during the severe 2013–2014 drought in the western United States. This mass loss induced an uplift of the land surface in the mountains of California of up to 15 mm (Borsa et al. 2014). This transient, but rapid, surface deformation is comparable to that following earthquakes or even to tectonic uplift in rising mountain ranges (Figure 12.1).

12.2 Geomorphic Impacts of Droughts

12.2.1 Droughts in the Hazard Cascade

The choice and difficulty of determining the start, duration, and end of droughts also applies directly to assessing the role that droughts play in the hazard cascade. Most studies on the impacts of droughts emphasize the hydrological, ecological, agricultural, and socioeconomic impacts (van Dijk et al. 2013), but largely fail to mention the geomorphic consequences and the damage that these cause. For example, droughts frequently promote the formation or growth of wind-blown sand sheets and dunes. The gradual onset and termination of droughts also means that effects on soil water, river flow, lake levels, wind erosion, and dune migration may set in less abruptly or in a delayed manner compared to other natural hazards. Similar to other geomorphic impacts of droughts, however, attributing an exclusively natural origin to many recently forming dunes is problematic, and often requires considering additional human interference. It can take effort to attribute a geomorphic footprint to a particular drought while singling out other effects and processes, especially in areas where a dry or seasonally dry climate may have modified landforms and processes over millennia to millions of years.

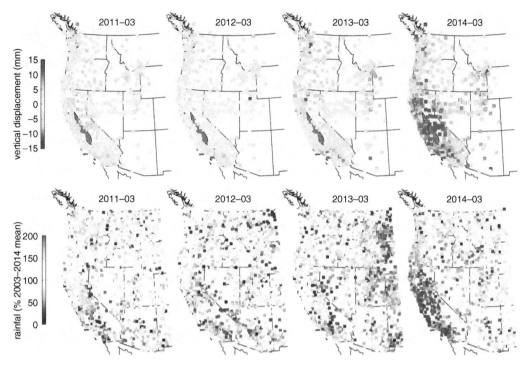

Figure 12.1 Drought-induced uplift in the Western United States. Upper: Maps of vertical GPS displacements from March 2011 to 2014. Uplift is indicated by yellow-red colors and subsidence by shades of blue. The grey region is where stations were excluded in the Central Valley of California. Lower: Maps of annual precipitation anomalies. Deviation of annual precipitation from the 2003–2013 mean at meteorological stations in NOAA's Global Historical Climatology Network for 2011–2104. The pattern of precipitation, in particular the surplus in California in 2011 and the deficit in 2014, mirrors the pattern of uplift seen in the GPS data. From Borsa et al. (2014).

12.2.2 Soil Erosion, Dust Storms, and Dune Building

Many of the geomorphic footprints of droughts arise from deficits of water and reduced vegetation cover. Hence one of the most prominent geomorphic impacts of droughts is soil erosion, largely because rates of soil loss depend on how much moisture is, and has been, available for a given site. Reduced or lost plant cover can lead to increased erosion rates, especially in semi-arid and arid regions. Dried out and indurated soils may seal the subsurface from water and increase the flashiness of surface runoff when drought-breaking rains occur. Wherever vegetation cover on poorly consolidated substrates succumbs to drought, wind erosion may increase and entrain dust and sand more frequently (Figure 12.2). In the worst cases,

roads, houses, and other infrastructure are buried by wind-blown sand and silt.

As with many natural hazards and disasters, much damage can arise from the coupling of adverse processes. The effects of water shortage can indirectly add to other processes that suppress vegetation growth (and thus increase the potential for soil erosion). For example, prolonged heat and water stresses on plants, together with high numbers of grazing snails, severely damaged nearly 1300 km^2 of marshes on the Louisiana coastal plain in the southeastern United States, between 1999 and 2001 (Silliman et al. 2005). In that area intact wetlands are vital for protecting the coast from storm surges and eroding waves. The drought, estimated to be a 100-year event, decreased soil moisture and made the soils more saline

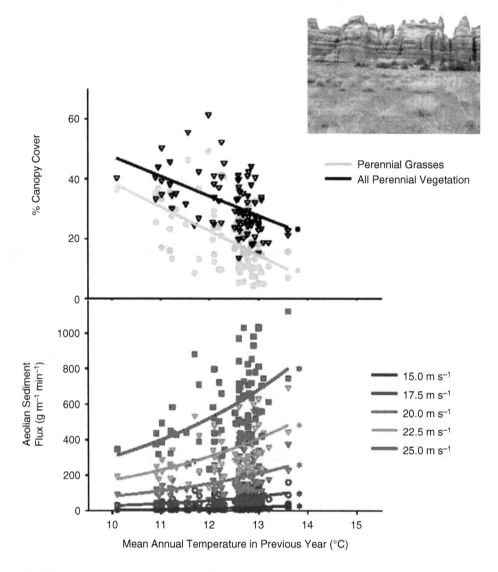

Figure 12.2 Vegetation and wind erosion as a function of antecedent temperature on the Colorado Plateau. Dominant plant species and functional type canopy cover (a) and modeled aeolian sediment flux (b) at five wind speeds in relation to mean annual temperature in the previous year in perennial grasslands and shrublands. From Munson et al. (2011).

and acidic. Grazing by a high number of snails amplified these stresses and caused a major die-back of the marsh vegetation.

Drought-induced changes to groundwater can also trigger changes in the land surface. Kaste et al. (2016) studied the pattern of geomorphic changes in the desert ecosystem of California's Owens Valley, using soil radionuclide inventories and groundwater data to trace changes in erosion over the past decades. They found that soil erosion was patchy, but highest where plant cover reduced to below 20%, and where the groundwater table dropped by several metres. Such drying and loss of protective root cohesion may make upper soil layers more susceptible to wind erosion.

Dust storms may occur more frequently during droughts, and entrain and transport millions of tons of topsoil over thousands to tens of thousands of square kilometres. Such impacts were felt in the central United States from 1933 to 1938. At that time huge dust storms produced conditions known as the 'Dust Bowl'. A combination of natural drought and poor agricultural practices during the Great Depression caused severe soil erosion in parts of five states. During this period, wind erosion affected 28 000 km^2 or some 43% of the then arable farm land. Bolles et al. (2017) summarized from previous sources that wind eroded between 0.4 and 5.5 Gt of topsoil during the Dust Bowl. Frequent, sometimes daily, dust storms destroyed crops and pastureland, building dunes, and burying buildings, fences, and farmlands. These drought conditions extended into Alberta and Saskatchewan, causing severe hardship for farmers and ranchers. This depletion of fertile soils is a characteristic of many droughts and adds to the indirect and intangible losses that continue for centuries, given that new soil forms at much slower rates than the old soil has been lost.

Yet a purely human cause of the 'Dust Bowl' is now debated, as part of the drought conditions seem to have been linked to variations in the surface temperature of the eastern tropical Pacific, and especially the Pacific Decadal Oscillation (Cook et al. 2008; Schubert et al. 2004). This anomalous airflow and sea-surface temperatures tied to La Niña conditions are now recognized as the main natural causes of the 'Dust Bowl'. The year 1934 was the region's driest in the past millennium (Cook et al. 2014). At the same time, however, agriculture was expanding rapidly throughout the Midwest during the early twentieth century, and this unprecedented spread would have promoted further soil erosion. In a study intended to detect geomorphic traces of the time, Cordova and Porter (2015) emphasized that dust storms and soil erosion had already affected the area affected by the 'Dust Bowl' much earlier, particularly in the late nineteenth century, partly because the spread of farming then paved the way for land degradation. They also noted that the dust storms were fed by the finer particles in bare soils, but also by blowout of fine materials entrained from older sand dunes (Figure 12.3). Some of these dunes were last active during megadroughts in past centuries or as early as the mid-Holocene. Compared to these dunes, active river beds appeared to contribute much less sediment to the storms, however. Several widened and drier river channels offered new sources of dust and sand, forming local channel-bordering dunes that have remained to the present day, although most sediment was delivered from older dunes. A study of playa deposits in the Mojave Desert of California and Nevada confirmed that a major sand sheet was deposited during a period of drought and a pronounced cool phase of the Pacific Decadal Oscillation in the twentieth century (Whitney et al. 2015). Subsequent wetter conditions promoted the erosion of the sand sheet and older yardangs, while remnants of a previous generation of such eroded forms dating to the nineteenth century indicate a likely cycle of sand-sheet deposition and reworking in response to major droughts (Figure 12.4).

Drawing mainly on historic sources of information on dust storms, Cattle (2016) argued that a similarly severe period of intensive wind erosion affected southeastern Australia between 1895 and 1945. A combination of natural and human impacts seems to have been responsible for this down-under version of the 'Dust Bowl', when drier-than-average rainfall conditions coincided with a shift of land-use practice to more intensive grazing. A developing rabbit plague enhanced vegetation and soil losses. Cattle (2016) tracked dozens of major dust storms in newspaper articles from the time. These clippings featured very graphic accounts of red rains that fell in major Australian cities such as Melbourne and Sydney, and across parts of New Zealand. One particularly strong dust storm on New Years' eve of 1927 dumped nearly ten tons of sediment per square kilometre on average onto a Melbourne suburb. Historic photos of Australian

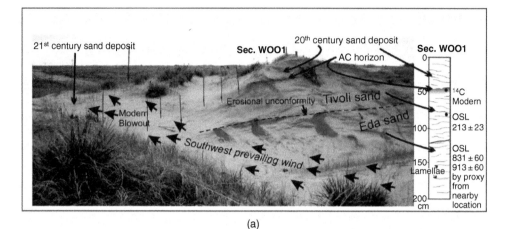

(a)

(b)

Figure 12.3 Examples of recent dynamics in aeolian and alluvial localities in northwest Oklahoma. (a) Sand dune showing several generations of aeolian activity at one place north of Woodward. (b) Upstream incision in Tesesquite Creek; photographs of the same location taken in 1968 (left) (Wilson) and 2004 (right). From Cordova and Porter (2015).

farmland document that local soil erosion and deposition were more than a meter in places. Estimates from the time hold that the storm carried 0.1–0.2 Mt of dust to New Zealand.

We can find many similar historic reports from areas under widespread agricultural use. For example, in 1928, dust storms in Ukraine entrained up to 15 Mm^3 of topsoil, causing an average loss of 0.25 m over the deflated area. What goes up must come down and the eroded Ukrainian soil was redeposited over an area of 6 million km^2 in Poland and Romania (Bryant 2004). Similar regional impacts have

been reported. Leys et al. (2011) studied dust concentrations and their consequences for respiratory health during the 3000-km long 'Red Dawn' dust storm on 22 and 23 September 2009, which was the largest on record in Sydney, Australia. During the storm, the amount of fine particles in the air temporarily exceeded the nation's air quality standard by a factor of nearly 50 in places. Overall, the storm carried about 2.5 Mt of dust from eastern Australia offshore, and remains the largest ever documented soil loss during a single storm.

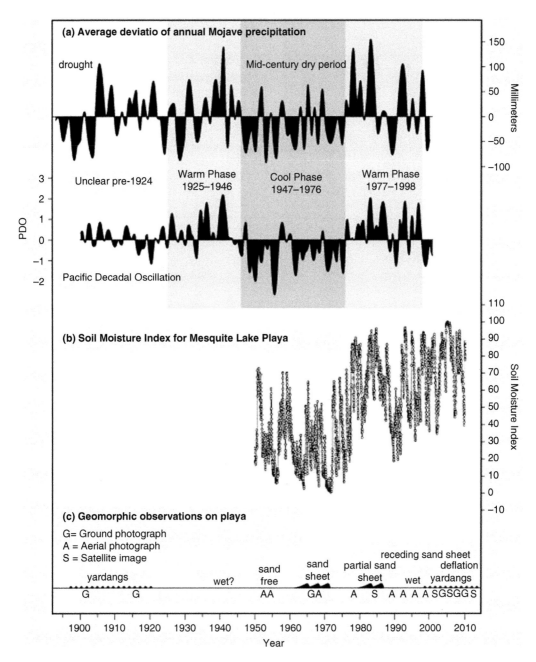

Figure 12.4 Aeolian responses to climate variability during the past century at Mesquite Lake Playa, Mojave Desert, USA. (a) Average deviation of annual Mojave Desert precipitation and Pacific Decadal Oscillation (PDDO phases). (b) Soil-moisture index for Mesquite Lake playa from 1950 to 2010. (c) Geomorphic observations on Mesquite Lake playa. From Whitney et al. (2015).

12.2.3 Surface Runoff and Rivers

Assessing the natural geomorphic impact of droughts while singling out any potential human interference offers many opportunities for future research. Here we need to distinguish between how people alter the natural processes that give rise to droughts, and how people become (less) susceptible to the geomorphic impacts arising from those droughts. In times of prolonged water stresses, it is essential to know how rivers and freshwater bodies respond to drought through changes in water and sediment fluxes, and how these changes alter geomorphic and ecological conditions. For example, the discharge curves of some of the largest rivers are well correlated with time series of the PDSI averaged over their catchment areas. This observation holds for most of the twentieth century and offers a straightforward means to reconstruct river response to drought and vice versa, even if discounting time lags between local runoff and streamflow (Dai et al. 2004).

In a study of land clearance and its hydrological and geomorphic consequences in the West African semiarid belt of the Sahel, Leblanc et al. (2008) collected evidence from air photos dating back to the 1950s, fieldwork observations, and groundwater data. They noted that removal of woody savannah vegetation for agricultural use and firewood caused increased surface runoff, local ponding, and drainage density. After several decades of lag time, these land-use changes eventually led to an average annual rise of the groundwater table of about 0.1 m. The study area had been hit by a severe and long-lasting drought in the 1970s and 1980s, but the authors had to conclude that the anthropogenic impacts of land clearance on the water balance were in total stronger than during that pronounced drought. Long periods of reduced water availability may also compromise the stability of earthen levees along river channels. The sustained reduction of pore water and drawdown can lead to mechanical

weakening as the soil material in levees softens, shrinks, and cracks during desiccation. These volumetric changes can alter internal water flow and erosion. In consequence, the levee surface can subside, erode, and expose soil organic carbon to microbial oxidation. In the worst cases, dried-out levees may fail catastrophically, while subsided levees that remain stable offer lesser flood protection because of lost freeboard (Robinson and Vahedifard 2016).

How much droughts change runoff and river discharge varies with many physical and biological characteristics of a given catchment. Identifying amongst these some useful predictors of low flows during drought can aid water and sediment management in drought-prone catchments. For example, Price et al. (2011) found that drainage density, topographic roughness, the amount of hillslope sediments, and the fraction of low-order streams were well correlated with the duration of drought-induced low flows in largely undisturbed rivers in the Blue Ridge of North Carolina (Figure 12.5). The degree of forest cover was much less correlated with low flows, in turn, although the authors emphasized that rates of infiltration and groundwater recharge were nourishing low flows during droughts.

Reduced streamflows can also curtail the delivery of nutrients from upstream sources and raise the salinity of river water. Both effects can alter biological productivity to the point at which channel and floodplain vegetation is dying, eventually affecting the pattern of eroding and accumulating channel sediments (Leigh et al. 2014). Other studies have come to quite different insights, and reported that extended periods of low flow or reduced flood activity can also encourage riparian vegetation to colonize and stabilize channel bars. Gaeuman et al. (2005) studied channel changes in the Duchesne River, Utah, during periods of drought and artificially reduced channel flows in the twentieth century. They reported that the decline in frequent flooding allowed plants to establish in the channel, thus causing it to

Figure 12.5 Stream low flows during severe drought conditions in the southern Blue Ridge Mountains, Georgia and North Carolina, USA. Seven day low-flow recurrence curves constructed from three long-term US Geological Survey gauges. From Price et al. (2011).

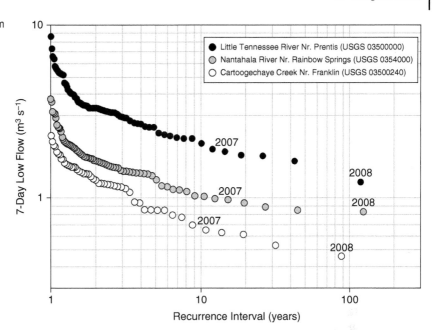

gradually narrow over several decades as a result of fewer disturbing floods.

Periods of drought may increase the potential for more flashy flood regimes, something that Persico and Meyer (2012) noted when studying how climatic changes affected beaver populations in the Greater Yellowstone Ecosystem. The authors used radiocarbon dating to establish the timing of up to 2-m thick beaver-dam deposits in the region's headwater rivers, and inferred that beavers built these impoundments mostly during more moist periods in the past 4000 years. During drought, beavers would have preferred trunk rivers with more reliable access to water and less of the flashy discharges that introduced more coarse deposits to the channels and floodplains and compromised the stability of beaver dams. This study illustrates how we need to consider many facets when trying to attribute geomorphic response to droughts. In this example, the role of ecosystem engineers such as beavers cannot be ignored, especially when concerned with finding templates for rehabilitating streams towards a more natural and perhaps even drought-resilient state.

The formation of in-channel bars in sand-bed rivers of southeastern Australia is one possible effect of prolonged periods of lesser rainfall and runoff, and has led to the idea of alternating drought- and flood-dominated regimes that largely control channel geometry in a quasicyclic manner (Erskine and Warner 1998). Rainfall data document that some of these cycles lasted for several decades. The overall concept has attracted some critique, because rivers in the region were also modified by people (Kirkup et al. 1998), so that channels and floodplains were also affected by soil erosion and the widespread and enhanced deposition of alluvial sediments after European settlers arrived. Floods and low flows in eastern Australian rivers also correlate strongly with several modes of natural climate, such as the El Niño Southern Oscillation or the Pacific Decadal Oscillation (van Dijk et al. 2013), and their contributions need to be taken into account. For example, between 2001 and 2006, during the time of the 'Millenium Drought', some 126 gauging stations in upland southeastern Australia had 82% less streamflow than their long-term average. Yet only about 45% of this reduction was due to

(a)

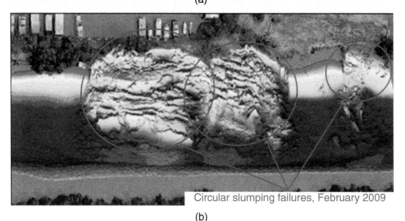

Circular slumping failures, February 2009

(b)

Figure 12.6 Drought-induced slumping of the banks of the Murray River, Australia, at the peak of the Millennium Drought. (a) The lower pool level of the Murray River induced desiccation cracks along the crest of the riverbank. These cracks were the primary weakening mechanisms that led to significant reductions in soil strength. (b) Multibeam bathymetry map of the slumps on the channel floor. The shallowest water is red; the deepest water is dark blue. From Robinson and Vahedifard (2016). (Prof. Tom Hubble and Elyssa De Carli).

lower rainfall, whereas the remainder was due to stream regulation such as impoundments or water extraction upstream. Measurements of erosion and sediment transport during this drought also show some major changes. Vigiak et al. (2016) found that stream-bank erosion in the Latrobe River in southeastern Australia became a more dominant source of river sediment than nearby hillslopes during the 'Millenium Drought', at least judging from the records spanning the period from 1997 to 2005 (Figure 12.6). On top of that, the sediment yield in the river dropped to about one fifth of that during the previous wetter conditions.

12.3 Geomorphic Tools for Reconstructing Past Drought Impacts

Looking back in time at archives of past climate, we can more easily discern pronounced periods of aridity than we can trace droughts in sedimentary and geomorphic records. Dust grains transported by storms during the 1930s 'Dust Bowl', for example, can be traced in well-dated Greenland ice cores. In many cases, however, we can only surmise that such drier periods included more frequent and pronounced droughts than did wetter periods. Dust storms, sand storms, and migrating sand dunes are characteristic of arid environments, but can also be episodic consequences of drought. Only where sedimentary and geomorphic records reveal that these processes operated for longer periods and over larger areas can we be certain to capture droughts as possible drivers instead of more local disturbances like grazing or wildfire.

The Great Plains of the United States are amongst the largest grasslands in the Northern Hemisphere, but also feature extensive deposits of former sand dunes, sand sheets, and loess mantles that cover several tens of thousands of square kilometres. Luminescence dates of sand sheets in erosion hollows (or blowouts) created during the 1930s 'Dust Bowl' group into several periods of activity over the past centuries that Cordova and Porter (2015) interpreted as the legacy of past megadroughts. Following a similar reasoning, Miao et al. (2007) proposed that extensive loess deposits such as those of the Great Plains could only accumulate if nourished by landscapes with extensive and actively shifting dunes, which in turn require suitably dry conditions. They used this reasoning in dating both sand dunes and loess layers by OSL at 21 locations throughout Nebraska, mostly in the Sand Hills area, and established a chronology of droughts spanning the past 10 kyr. The sand dunes and loess deposits yielded consistent ages, with several distinct clusters during the Holocene. The inferred drought periods occurred at 4.5–2.3 ka and 1.0–0.7 ka, assuming that new sand dunes formed in response to drier conditions without much delay. A few of the OSL ages are consistent with the interpretation that soils were forming while at least some dunes were active. The authors argued that these ages reflect local disturbances instead of regional drought. Forman et al. (2001) came to a similar conclusion and reported that times of

higher dune activity in the Great Plains in the past two millennia hardly matched the timing of about a dozen major droughts inferred from tree-ring records and lake sediments. In essence, the Nebraska Sand Hills is one of the best examples of pronounced dune activity during late Pleistocene and mid-Holocene times. Some of the moving sand dunes were over a hundred meters high and blocked rivers, forming more than a thousand highly alkaline lakes and raising local groundwater tables by up to 25 m (Loope et al. 1995).

This extent of formerly active dune fields reminds us that wind-driven erosion and sedimentation were more active then, and that prehistoric droughts were more severe than in historic times. However, both active and fossil sand dunes can occur in the same area and climatic conditions, and this motivates studying in more detail the role of wind erosion, precipitation rates, and land-use changes. A numerical model considering wind power, precipitation rate, and vegetation cover shows that sand dunes that are originally stabilized may become active or reactivated only once a critical threshold of wind power is exceeded. Similarly, once activated, these dunes may only stabilize after wind power has dropped again (Yizhaq et al. 2009). This study shows that drought conditions alone may be insufficient to explain the mobilization of dunes. Yan and Baas (2018) noted that lag times lie between the onset of dryer or wetter climate, changes in vegetation, and the activation or stabilization of sand dunes, and indicated that these lag times may span decades to almost a century. Thomas et al. (2005) used global climate models to predict that, by the end of the twenty-first century, large dune fields in the southern Kalahari desert of Africa would be reactivated and mobile if ongoing atmospheric warming continues. To measure these changes, they used a modified dune mobility index $A_{p,GCM}$:

$$A_{p,GCM} = \bar{u}^3 \left(\frac{P_{lag}}{E_{p,lag}} + \frac{P_{rain}}{E_{p,rain}} \right)^{-1} \quad (12.1)$$

where $\bar{u}$ is mean wind speed [m s^{-1}], P_{lag} is the mean precipitation of the previous and current month [mm], $E_{p,lag}$ is the mean potential evapotranspiration of the previous and current month [mm], P_{rain} is the running mean precipitation per month during the rainy season [mm], and $E_{p,rain}$ is the running mean potential evapotranspiration per month during the rainy season [mm]. This mobility index was specifically designed for model outputs for southern African climates, but helps to characterize the activity of present dune fields. For more flexible applications, this and similar dune-mobility indices need to take into account more variable time lags as vegetation cover adjusts to changing climate conditions. Coastal settings might have different dynamics driving active dune phases that may elude clear links to past droughts. For example, Zular et al. (2018) reported that transgressive dune fields blocked rivers in northeast Brazil in the early to mid-Holocene (mainly 11–6 ka), mainly driven by rising sea levels that made more sediment available, but also by stronger wind patterns during a generally wetter climate.

Tree rings, varved lake sediments, and cave speleothems can offer detailed and reliable time series of past droughts. The growing number of such archives can offer robust chronologies for testing independently whether past (mega)droughts caused detectable shifts in rates of erosion or sedimentation in rivers, lakes, or on hillslopes. Tree rings also contain important information about the conditions limiting or favouring tree growth, and the discipline of dendroclimatology has informed us about the past occurrence of droughts. Tree rings are among the most promising and widely used proxies of past droughts. For example, variations in tree-ring widths record several megadroughts in the South Asian summer monsoon circulation during the mid-fourteenth to fifteenth centuries. These dry phases seem to have been linked to the beginning of the Little Ice Age, but carried on after. Each of them lasted for several years to decades, though without comparable analogs in the instrumental record (Sinha et al. 2011). Indirect evidence comes from scattered historical and archaeological sources from the upper Sutlej valley, India, and western Tibet. There, famines and reduced access to water occurred because of lower groundwater levels during these megadroughts. During those periods, profound societal changes took place also in other parts of South and Southeast Asia, possibly involving – or at least contributing to – total collapse. Sinha et al. (2011) mentioned in this context the demise of the Yuan dynasty in China, the Rajarata civilization in Sri Lanka, and the Khmer civilization of Angkor Wat in Cambodia.

Cook et al. (2010) synthesized data from a regional network of tree-ring chronologies and compiled the Monsoon Asia Drought Atlas (MADA), which features a gridded map of Asian monsoon droughts and pluvials over the past millennium (Figure 12.7). The dendroclimatologic data provide seasonal resolution and identify many historically mentioned droughts in monsoonal Asia. The data in the MADA revealed previously unknown megadroughts and their relationship to changes in sea surface temperature linked to the El Niño Southern Oscillation. One the most pronounced events in the record is the 'Strange Parallels' drought that occurred between 1756 and 1768, and affected much of western India, southeast Asia, and western Siberia during a time of societal upheaval and political reorganization in these regions (Cook et al. 2010). The situation became much worse about a century later:

The late Victorian Great Drought of 1876 to 1878 occurred during one of the most severe El Niño events of the past 150 years. The effects of this devastating drought were felt across much of the tropics and were particularly acute in India. A revolt

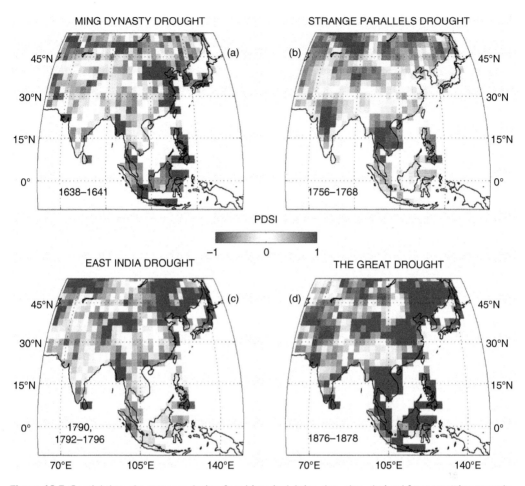

Figure 12.7 Spatial drought patterns during four historical Asian droughts, derived from tree-ring records of the Monsoon Asia Drought Atlas. Mean Palmer Drought Drought Severity Index (PDSI) for each of four regional droughts identified from the historical record. (a) The Ming Dynasty drought (1638–1641). (b) The Strange Parallels drought (1756–1768). (c) The East India drought of the late eighteenth century (1790 and 1792–1796). In 1791, much of India was slightly wet, except the region around Chennai where the drought persisted. (d) The late Victorian Great Drought (1876–1878). From Cook et al. (2010).

against the French in Vietnam also took place as a consequence of severe drought and famine at this time, and the drought was felt as far away as Jakarta, Borneo, and New Guinea. More than 30 million people are thought to have died from famine worldwide, and Colonial-era imperialism left regional societies ill-equipped to deal with the effects of drought (Cook et al. 2010).

Similar reconstructions of alternating wet and dry periods have become available for the North American continent. A network of tree-ring chronologies has enabled a reconstruction of droughts over the past 1000 years, partly taking in the work by Miao et al. (2007) and Cook et al. (2010). The reconstructions reveal the occurrence of megadroughts of greater severity and duration than any experienced during the past century, including that of the 1930s. Numerical simulations of major droughts in North America from the mid-1800s to the period 1998–2004, when a severe drought gripped the western United

States, revealed the dominant importance of tropical Pacific Ocean sea-surface temperatures (SSTs) in determining how much precipitation falls over large parts of North America (Cook et al. 2010). Development of cool SSTs in the eastern tropical Pacific during La Niña phases is thought to be linked to drought conditions in North America and may be driven by changes in radiative forcing over that region.

Pederson et al. (2014) noted that most studies of the impacts of climate change on people have been concerned with linking the downfall of past societies with drought, whereas fewer studies attempted to demonstrate the opposite, that wetter conditions favoured the rise of complex social groups or societies. The researchers noted that the warm-season water balance inferred from tree rings at a site in central Mongolia is correlated with proxies of the regional water balance and steppe productivity. Their reconstruction further posits that the Mongolian empire under Chinggis Khan rose during an exceptionally warm and wet phase in the thirteenth century. This phase was unprecedented in the past 1100 years, and may have favoured a thriving of the Mongolian grasslands, possibly facilitating the political and military conquests at the time. Radiocarbon-dated archaeological finds from south-central Siberia are consistent with the hypothesis that a shift toward more humid climate conditions at around 850 BC may have promoted the spread of the central Asian horse-riding Scythian culture into semiarid desert areas that had transformed into more productive steppes (van Geel et al. 2004).

While these and many other studies inform us about how people dealt with former droughts or wetter phases, many of the geomorphic consequences remain shrouded or implicit in these records. Soil erosion, for example, is a largely irreversible process, and can force people to migrate to areas with more favourable resources. Prehistoric soil erosion as a sole consequence of droughts can be hard to track down, but important nevertheless: McAuliffe et al. (2006) investigated the bared roots of pinyon pines (*Pinus edulis*) in the Colorado Plateau to obtain a history of past soil erosion that they attributed to alternating drier and wetter periods. The scientists measured the depth to which roots were exposed and linked the inferred amount of soil erosion to dated tree rings. The inferred average rate of soil erosion was 1.9 mm yr^{-1} over the past 400 years, but was especially high when droughts gave way to longer periods with above-average precipitation, as captured by distinct changes in the incremental growth of the tree rings.

Droughts can also leave geochemical signals in deep-sea sediments. Seasonal differences in the content of bulk titanium in a sediment core from the anoxic Cariaco Basin in the southern Caribbean may reflect variations in the regional hydrological cycle and sediment input from nearby rivers (Haug et al. 2003). The times of declining riverine input and pronounced droughts inferred from this geochemical proxy coincide closely with the established dates for the collapse of the Mayan culture in Central America. Past droughts and water shortages might have contributed to this demise. While these deep-sea proxies constrain the timing of major changes in rivers, we still require detailed studies in the affected catchments to learn whether these droughts had major traceable impacts on channels and floodplains.

Studies of past dust transport can complement attempts to reconstruct prehistoric droughts. Atmospheric dust appears to be an important forcing parameter in the Asian and African monsoonal system. Mulitza et al. (2010) established a record covering more than three millennia of terrestrial dust deposition on the sea floors just offshore from the Senegal River, Africa. The authors reported that the late Holocene dust flux correlated mostly with changes in precipitation regimes, but rapidly increased at the beginning of the nineteenth century to rates of up to five times higher, when commercial agriculture took hold in the Sahel zone. The information contained in dust deposits can provide valuable insights into environmental and geomorphic changes in the source regions, so far as these can be identified. Yet often the geomorphic consequences of droughts blend with those of dust storms and wildfires that potentially accompany droughts. In some regions the causes of wind erosion and dust transport may be elusive. Both climate change and human land use may be responsible for increased dust-storm activity in arid subtropical zones such as Inner Mongolia and northwestern China, which are the centres of global dust production (Zhang et al. 2003).

An intriguing suggestion from a geomorphic standpoint is to use sediment yield as a risk indicator for desertification (Vanmaercke et al. 2011). While the term 'desertification' encompasses various phenomena of land degradation in mostly dry areas, it is most often linked to droughts. For example, Asian deserts including the Gobi, Karakum, Lut, Taklimakan and Thar have been been changing their size, with decreases of ~10% in the 1990s, and increases of ~9% in the early twenty-first century. These cycles of expanding and contracting deserts can arise from local changes in temperature and precipitation, but can also reflect changes in moisture transport from far-away sources (Jeong et al. 2011).

Many studies of desertification focus on local effects of soil erosion, but often disregard the broader effects of thus mobilized soil (and sediment) that travels through – and eventually leaves – catchments, causing unwanted sediment deposition, flooding, and ecological disturbances along the way. Soil erosion measured at local plots rarely reflects what is going on in the catchment, so sediment yield can be a more integrative measure of desertification, and hence, indirectly, also droughts. Similarly,

gully erosion, another key process frequently associated with desertification, often develops and propagates along headwater reaches so that a static, local perspective on the problem is of little use. Vanmaercke et al. (2011) noted that most of some 70 previously proposed physical, ecological, and economic diagnostics of desertification focused on local effects of erosion. Only five indicators addressed the dynamics of entire catchments with respect to runoff, flooding, and sedimentation. In summary, methods to attribute the natural and anthropogenic drivers of geomorphic responses to desertification and droughts remain a field with a lot of scope for future research.

12.4 Towards More Megadroughts?

It seems reasonable to expect that increasing temperatures over the land surface will change the frequency and severity of droughts. Judging from long temperature records, we can see that contemporary droughts were unprecedented in the past decades to centuries, so that scientists have proposed new terms such as 'hot droughts', 'hotter droughts' or even 'megadroughts' to underline this seemingly growing frequency or intensity of dry and hot periods (Figure 12.8; Millar and Stephenson 2015). Model simulations based on climatic conditions with and without anthropogenic forcing reveal a strong human fingerprint in the occurrence of recent and continuing drought in California, for example. More drought years were experienced in California in the decades from 1990 to 2010 than in the entire preceding century (Diffenbaugh et al. 2015).

Ongoing atmospheric warming may also reduce air quality due to enhanced dust and associated contaminant transport. Central Asia and Inner Mongolia, China, are leading areas

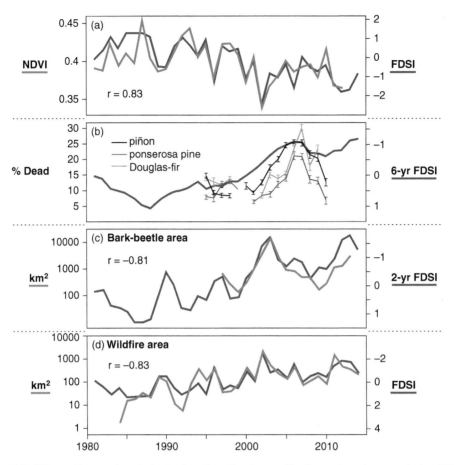

Figure 12.8 Effects of hotter drought in the American Southwest. The forest drought-stress index (FDSI) integrates the effects of warm-season water deficit (controlled mostly by high temperature) and cold-season precipitation. Declining values of FDSI correspond to increasing drought. In the Southwest, increasing drought has been accompanied by (a) declining vegetation greenness [normalized difference vegetation index (NDVI)], a remotely sensed index of greenness; (b) increasing mortality of the three most common conifers; (c) increasing area affected by barkbeetle outbreaks; and (d) increasing area affected by wildfires. From Millar and Stephenson (2015).

of dust production. Analysis of historic time series has shown that the incidence of dust storms in China has increased. Between 300 and 1949 AD dust storms had an average return period of about 31 years, whereas since the 1990s dust storms have happened almost every year (Liu and Diamond, 2005). From 1951 to 1955 some 3882 dust storms were recorded in Central Asia, with an average of 100–174 such storms per year in the Gobi Desert. Whether this is due entirely to atmospheric warming or includes effects of land-use change and desertification is a subject of debate. Numerical

simulations of particle-size dependent soil dust erosion and transport identified the Takli-makan and Badain Juran deserts of Mongolia and northwestern China as the main sources, supporting the notion that meteorological and climatic conditions are more important than desertification as a driver of historic dust emissions in Asia (Zhang et al. 2003).

Elsewhere, land-use changes may have a stronger impact. In the Amazon basin, for example, the large-scale pattern of present and future deforestation may contribute to raising the probability of drought on top

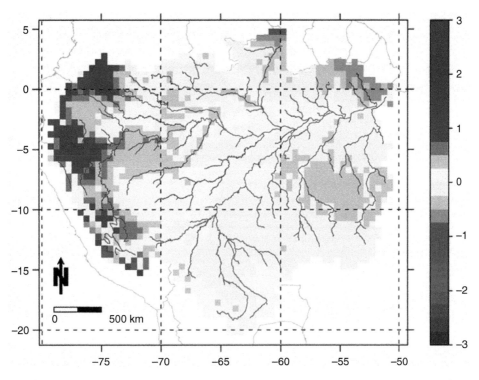

Figure 12.9 Lengthening (blue) and shortening (red) of the duration of inundation in the Amazon River basin in months (mean over 24 IPCC AR4 climate models) between the reference period (1961–1990) and the projection for 2070–2099. From Langerwisch et al. (2013).

of any regional projections of atmospheric warming driven solely by greenhouse gas emissions (Figure 12.9) (Malhi et al. 2008). However, it is unclear whether these models sufficiently disentangle the feedbacks between climate change and desertification, that is whether they treat both as independent variables. Hence, the use of ensembles of numerical climate models has become standard for expressing uncertainties related to predictions of future droughts (or climate conditions in general). The idea of ensemble modelling is that uncertainties can be expressed by the percentage of outputs that share common characteristics or trends.

The effects of future droughts and heatwaves are likely to be felt also in major cities. Changing atmospheric chemistry and rising temperatures also affect most large cities, where emissions are highest. Urban areas have higher surface temperature fields than surrounding rural areas, and increased heat stress ranks high amongst the natural hazards affecting city dwellers. The strength of such urban 'heat islands' depends on their aerodynamic roughness and background climate, so that urban areas may be prone to higher humidity and temperature gradients than rural areas (Zhao et al. 2015). Detailed analyses show that added heat stresses might have been responsible for 5% of all deaths in Berlin, for example, between 2001 and 2010, mostly affecting people aged 65 and over (Scherer et al. 2014). Heat stress also compromises vegetation growth, which in turn might affect the way that it stabilizes soils, river banks, or road embankments, for example. Yet detailed measurements are still needed to assess whether such effects are relevant enough to grossly change rates of sediment erosion, deposition, and transport in urban areas.

References

Barriopedro D, Fischer EM, Luterbacher J, et al. 2011 The hot summer of 2010: Redrawing the temperature record map of Europe. *Science* **332**(6026), 220–224.

Bolles K, Forman SL, and Sweeney M 2017 Eolian processes and heterogeneous dust emissivity during the 1930s Dust Bowl Drought and implications for projected 21st-century megadroughts. *The Holocene* **27**(10), 1578–1588.

Bonsal BR, Aider R, Gachon P, and Lapp S 2012 An assessment of Canadian prairie drought: past, present, and future. *Climate Dynamics* **41**(2), 501–516.

Borsa AA, Agnew DC, and Cayan DR 2014 Ongoing drought-induced uplift in the western United States. *Science* **345**(6204), 1587–1590.

Bryant E 2004 *Natural Hazards* 2nd edn. Cambridge University Press.

Cattle SR 2016 The case for a southeastern Australian Dust Bowl, 1895–1945. *Aeolian Research* **21**(C), 1–20.

Cook BI, Miller RL, and Seager R 2008 Dust and sea surface temperature forcing of the 1930s "Dust Bowl" drought. *Geophysical Research Letters* **35**(8), 1–5.

Cook BI, Seager R, and Smerdon JE 2014 The worst North American drought year of the last millennium: 1934. *Geophysical Research Letters* **41**, 7298–7305.

Cook ER, Anchukaitis KJ, Buckley BM, et al. 2010 Asian monsoon failure and megadrought during the last millennium. *Science* **328**(5977), 486–489.

Cordova C and Porter JC 2015 The 1930s Dust Bowl: Geoarchaeological lessons from a 20th century environmental crisis. *The Holocene* **25**(10), 1707–1720.

Dai A, Trenberth KE, and Qian T 2004 A global dataset of Palmer Drought Severity Index for 1870-2002: Relationship with soil moisture and effects of surface warming. *Journal of Hydrometeorology* **5**(6), 1117–1130.

Derbyshire E 2000 Geological hazards in loess terrain, with particular reference to the loess regions of China. *Earth-Science Reviews* **54**(1), 231–260.

Diffenbaugh NS, Swain DL, and Touma D 2015 Anthropogenic warming has increased drought risk in California. *Proceedings of the National Academy of Sciences* **112**(13), 3931–3936.

Erskine WD and Warner RF 1998 Further assessment of flood- and drought-dominated regimes in south-eastern Australia. *Australian Geographer* **29**(2), 257–261.

Forman S, Oglesby R, and Webb RS 2001 Temporal and spatial patterns of Holocene dune activity on the Great Plains of North America: megadroughts and climate links. *Global and Planetary Change* **29**, 1–29.

Gaeuman D, Schmidt JC, and Wilcock PR 2005 Complex channel responses to changes in stream flow and sediment supply on the lower Duchesne River, Utah. *Geomorphology* **64**(3-4), 185–206.

Haug GH, Günther D, Peterson LC, et al. 2003 Climate and the collapse of Maya civilization. *Science* **299**(5613), 1731–1735.

Jeong SJ, Ho CH, Brown ME, et al. 2011 Browning in desert boundaries in Asia in recent decades. *Journal of Geophysical Research: Atmospheres* **116**, 1–7.

Kaste JM, Elmore AJ, Vest KR, and Okin GS 2016 Groundwater controls on episodic soil erosion and dust emissions in a desert ecosystem. *Geology* **44**(9), 771–774.

Kirkup H, Brierley G, Brooks A, and Pitman A 1998 Temporal variability of climate in south-eastern Australia: a reassessment of flood- and drought-dominated regimes. *Australian Geographer* **29**(2), 241–255.

Langerwisch F, Rost S, Gerten D, et al. 2013 Potential effects of climate change on inundation patterns in the Amazon Basin. *Hydrology and Earth System Sciences* **17**(6), 2247.

Leblanc MJ, Favreau G, Massuel S, et al. 2008 Land clearance and hydrological change in the Sahel: SW Niger. *Global and Planetary Change* **61**(3-4), 135–150.

Leigh C, Bush A, Harrison ET, et al. 2014 Ecological effects of extreme climatic events on riverine ecosystems: insights from Australia. *Freshwater Biology* **60**(12), 2620–2638.

Leys JF, Heidenreich SK, Strong CL, et al. 2011 PM_{10} concentrations and mass transport during "Red Dawn" – Sydney 23 September 2009. *Aeolian Research* **3**, 327–342.

Liu J and Diamond J 2005 China's environment in a globalizing world. *Nature* **435**, 1179–1186.

Loope DB, Swinehart JB, and Mason JP 1995 Dune-dammed paleovalleys of the Nebraska Sand Hills: Intrinsic versus climatic controls on the accumulationof lake and marsh sediments. *Geological Society of America Bulletin* **107**(4), 396–406.

Malhi Y, Roberts JT, Betts RA, et al. 2008 Climate change, deforestation, and the fate of the Amazon. *Science* **319**(5860), 169–172.

McAuliffe JR, Scuderi LA, and McFadden LD 2006 Tree-ring record of hillslope erosion and valley floor dynamics: Landscape responses to climate variation during the last 400 yr in the Colorado Plateau, northeastern Arizona. *Global and Planetary Change* **50**(3-4), 184–201.

Miao X, Mason JA, Swinehart JB, et al. 2007 A 10,000 year record of dune activity, dust storms, and severe drought in the central Great Plains. *Geology* **35**(2), 119.

Millar CI and Stephenson NL 2015 Temperate forest health in an era of emerging megadisturbance. *Science* **349**, 823–826.

Mishra AK and Singh VP 2010 A review of drought concepts. *Journal of Hydrology* **391**(1-2), 202–216.

Mulitza S, Heslop D, Pittauerova D, et al. 2010 Increase in African dust flux at the onset of commercial agriculture in the Sahel region. *Nature* **466**(7303), 226–228.

Munson SM, Belnap J, and Okin GS 2011 Responses of wind erosion to climate-induced vegetation changes on the Colorado Plateau. *Proceedings of the National Academy of Sciences* **108**, 3854–3859.

O'Loingsigh T, McTainsh GH, Parsons K, et al. 2014 Using meteorological observer data to compare wind erosion during two great droughts in eastern Australia; the World War II Drought (1937-1946) and the Millennium Drought (2001-2010). *Earth Surface Processes and Landforms* **40**(1), 123–130.

Overpeck JT 2013 The challenge of hot drought. *Nature* **350**, 350–351.

Pederson N, Hessl AE, Baatarbileg N, et al. 2014 Pluvials, droughts, the Mongol Empire, and modern Mongolia. *Proceedings of the National Academy of Sciences* **111**(12), 4375–4379.

Persico L and Meyer G 2012 Natural and historical variability in fluvial processes, beaver activity, and climate in the Greater Yellowstone Ecosystem. *Earth Surface Processes and Landforms* **38**(7), 728–750.

Price K, Jackson CR, Parker AJ, et al. 2011 Effects of watershed land use and geomorphology on stream low flows during severe drought conditions in the southern Blue Ridge Mountains, Georgia and North Carolina, United States. *Water Resources Research* **47**(2), 1–19.

Robinson JD and Vahedifard F 2016 Weakening mechanisms imposed on California's levees under multiyear extreme drought. *Climatic Change* **137**, 1–14.

Scherer D, Fehrenbach U, Lakes T, et al. 2014 Quantification of heat-stress related mortality hazard, vulnerability and risk in Berlin, Germany. *Die Erde – Journal of the Geographical Society of Berlin* **144**(3-4), 238–259.

Schubert SD, Suarez MJ, Region PJ, et al. 2004 On the cause of the 1930s Dust Bowl. *Science* **303**, 1855–1859.

Silliman BR, Van de Koppel J, Bertness MD, et al. 2005 Drought, snails, and large-scale die-off of southern US salt marshes. *Science* **310**, 1803–1806.

Sinha A, Stott L, Berkelhammer M, et al. 2011 A global context for megadroughts in monsoon

Asia during the past millennium. *Quaternary Science Reviews* **30**(1-2), 47–62.

Thomas DSG, Knight M, and Wiggs GFS 2005 Remobilization of southern African desert dune systems by twenty-first century global warming. *Nature* **435**(7046), 1218–1221.

van Dijk AIJM, Beck HE, Crosbie RS, et al. 2013 The Millennium Drought in southeast Australia (2001-2009): Natural and human causes and implications for water resources, ecosystems, economy, and society. *Water Resources Research* **49**(2), 1040–1057.

van Geel B, Bokovenko NA, Burova ND, et al. 2004 Climate change and the expansion of the Scythian culture after 850 BC: a hypothesis. *Journal of Archaeological Science* **31**(12), 1735–1742.

Van Loon AF, Gleeson T, Clark J, et al. 2016 Drought in the Anthropocene. *Nature Geoscience* **9**(2), 89–91.

Vanmaercke M, Poesen J, Maetens W, et al. 2011 Sediment yield as a desertification risk indicator. *Science of the Total Environment, The* **409**(9), 1715–1725.

Vigiak O, Beverly C, Roberts A, et al. 2016 Detecting changes in sediment sources in drought periods: The Latrobe River case study. *Environmental Modelling & Software* **85**, 42–55.

Wells N, Goddard S, and Hayes MJ 2004 A self-calibrating Palmer Drought Severity Index. *Climate* **17**(12), 2335–2351.

Whitney JW, Breit GN, Buckingham SE, et al. 2015 Aeolian responses to climate variability during the past century on Mesquite Lake Playa, Mojave Desert. *Geomorphology* **230**(C), 13–25.

Yan N and Baas ACW 2018 Transformation of parabolic dunes into mobile barchans triggered by environmental change and anthropogenic disturbance. *Earth Surface Processes and Landforms* **43**(5), 1001–1018.

Yizhaq H, Ashkenazy Y, and Tsoar H 2009 Sand dune dynamics and climate change: A modeling approach. *Journal of Geophysical Research: Earth Surface* **114**, 1–11.

Zhang XY, Gong SL, Zhao TL, and Arimoto R 2003 Sources of Asian dust and role of climate change versus desertification in Asian dust emission. *Geophysical Research Letters* **30**(24), 1–4.

Zhao L, Lee X, Smith RB, and Oleson K 2015 Strong contributions of local background climate to urban heat islands. *Nature* **511**(7508), 216–219.

Zular A, Utida G, Cruz FW, et al. 2018 The effects of mid-Holocene fluvio-eolian interplay and coastal dynamics on the formation of dune-dammed lakes in NE Brazil. *Quaternary Science Reviews* **196**, 137–153.

13

Wildfire Hazards

Wildfire has been part of the Earth surface system for hundreds of millions of years. Its earliest traces – fossil charcoal – date to Silurian times when vegetation occupied the land; trees evolved and spread across the land at least 320 million years ago, but grasses appeared only about 20 million years ago, providing fuel (Bowman et al. 2009). Many coal seams of the early Carboniferous contain large amounts of charcoal that attest to wildfires during the time when plants began to diversify. Ecologists stress that fire is a natural and essential element of many ecosystems (Pausas et al. 2008). Fire requires a trigger, fuel, and oxygen (Bowman et al. 2009), whereas the frequency of ignition, the availability of fuel, and the humidity of the climate all set limits to fire. Topography partly governs the spread of wildfires (Scott et al. 2014). Steeper hillslopes cause tilting flames that can more easily reach fuels upslope of the advancing fire front. Slope aspect affects the local temperature regime and how quickly fuel dries as a consequence, and also determines the exposure to prevailing winds, which can aid the spread of wildfire. Narrow valleys, canyons, and ravines in particular can act as natural chimneys because of enhanced air flow, while the radiating heat can dry out the neighbouring valley flanks more rapidly. Landscape features with little fuel, such as lakes, bare rock surfaces, or landslide deposits, can slow down or impede advancing fires. The effect of elevation is superimposed on these local controls. High-lying mountainous areas that are colder and more humid than lower areas are likely to have less easily combustible fuel.

13.1 Frequency and Magnitude of Wildfires

About 8.7 Gt of terrestrial biomass burn every year on average (Neary et al. 1999). An estimated area of 3.5 million km^2, which is more than one-and-a-half times the area of Greenland, burned in 2000 alone. Fires in southeastern Australia burnt an area almost the size of Switzerland in the first decade of the twenty-first century (Nyman et al. 2011). The 1990s saw some 50 000 wildfires in countries bordering the Mediterranean, with an average annual area burned of ~6000 km^2 (Shakesby and Doerr 2006). More than four times this area burned each year in Canada during the same period (Stocks et al. 2002), highlighting that wildfires can hit areas without pronounced Mediterranean climates, but also occur where fuel, in this case boreal forest, is abundant. Boike et al. (2016), for example, mapped changes in surface water, vegetation, and fire occurrence in a 315 000 km^2 portion of the Lena River catchment, eastern Siberia,

Geomorphology and Natural Hazards: Understanding Landscape Change for Disaster Mitigation, Advanced Textbook Series, First Edition. Tim R. Davies, Oliver Korup, and John J. Clague.
© 2021 John Wiley & Sons Ltd. Co-published 2021 by the American Geophysical Union and John Wiley & Sons Ltd.

from satellite images between 2002 and 2009. During that period wildfires burnt some 17% of the study area, which is mostly underlain by continuous permafrost. Incidentally, the total area covered by lakes also increased by about 18% during that interval. However, most wildfires occurred in areas away from those where lakes expanded or new ones formed. Hence clear links between wildfire activity, possible melt of permafrost, and subsequent growth of lakes remained elusive.

Wildfires, be they natural or anthropogenic, cause indirect losses that can exceed the direct losses from burning alone. In 1998, prolonged burning of tropical rain forests in southern Mexico led the government to mandate reductions in industrial emissions to regain acceptable levels of air quality, leading to estimated losses of US$ 8 million per day. Fire-haze-related sickness and respiratory health problems caused by fires in Indonesia in 1997 resulted in a nationwide loss of labour of 2.5 million workdays (Cochrane 2009).

What all these studies have in common is that satellite monitoring has become indispensable for mapping fires. Regional to global inventories are the backbone for estimating the frequency and magnitude of wildfires. The size distributions of areas burnt by fires are heavy-tailed like those of landslides, so that scientists approximate the probability of fire occurrence with simple models such as power-law distributions (Malamud et al. 1998). On a global scale, variations in the form of this power-law distribution appear and can be interpreted as the result of a distinct human fingerprint, or even regionally differing patterns of fire management (Figure 13.1) (Hantson et al. 2014). If this assumption holds, fire mapping and size statistics could enable more refined estimates of wildfires in areas with a given land use.

Fire is a self-sustaining, rapid, high-temperature biochemical oxidation reaction that releases heat, light, carbon dioxide, and water vapour (Bowman et al. 2009); the prefix 'wild' identifies a fire ignited naturally and without any human trigger. The combustion process releases many chemical compounds in solid, liquid, and gaseous form. Common trace gases include nitrogen oxides, carbonyl sulfide, carbon monoxide, methyl chloride, and hydrocarbons such as methane. These gases, along with solid particles of ash and soot, form

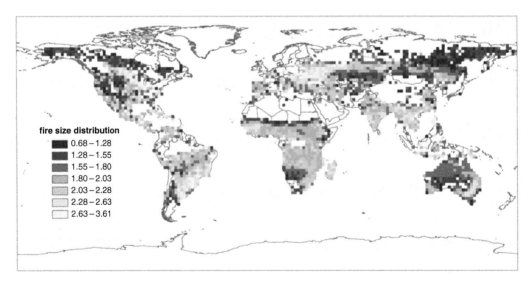

fire size distribution
0.68 – 1.28
1.28 – 1.55
1.55 – 1.80
1.80 – 2.03
2.03 – 2.28
2.28 – 2.63
2.63 – 3.61

Figure 13.1 Global map of the fire size distribution as estimated by β of an inverse power law, fitted to fire areas in each grid cell. Areas with low or no fire occurrence, where the power law could not be fitted, are marked in white. From Hantson et al. (2015).

the smoke observed during a wildfire. A trigger (e.g. a lightning strike or volcanic eruption) is required to initiate a wildfire, and three things are essential to maintain one: fuel, oxygen, and heat. If any of these three things is removed, the fire goes out. After a fire, colonizing plants become established on the burned landscape. The vegetation goes through a postfire cycle from early colonization to a mature ecosystem adapted to the climate at that particular location and particular time. This cycle, which operates today, is so ancient that some plants have evolved to rely on and use fire to their advantage. For example, oak and redwood trees have bark that resists fire damage, and some species of pine trees have seed cones that open only after a fire. Native eucalyptus trees in southeast Australia, which is regularly swept by bush fires in summer, sprout new leaves after the trees are charred. Often within a year or two, the 'bush' has grown back to such an extent that it can be hard to see where it had burned.

As with other natural hazards, we need a scale for measuring the frequency and magnitude of wildfires. An intuitive and simple measure is the total area burnt during a given event. Many detailed inventories give a good idea about the size distribution of areas burned by wildfires. These distributions are heavy-tailed, with a few very large areas representing the lion's share of the total area. In Canada, for example, only 3% of all fires reported between 1959 and 1997 were larger than 2 km², but these burned ~97% of the total burnt area (Stocks et al. 2002). The power-law distribution is a frequently used model to characterize this imbalance and offers a way to describe wildfires as a phenomenon of self-organized criticality (Malamud et al. 1998).

A more physical metric of wildfires describes the energy output from a wildfire as 'fireline intensity' I_B [kW m^{-1}]:

$$I_B = b_f L_f^{\beta} \tag{13.1}$$

where b_f is a coefficient ranging from 13 to 258 [kW$^{-(1+\beta)}$], L_f is the flame height [m], and $\beta = 2.17$ is the scaling exponent. I_B refers to the energy that is produced for each metre of a fire front for a given flame height. Fireline intensity thus measures the heat release of burning fuel per unit time per unit length of the fire front, irrespective of the fire front's width (Scott et al. 2014). Fireline intensity ranges widely from 10 to $> 10^5$ kW m^{-1}. Depending on fuel conditions, low-intensity fires (<500 kW m^{-1}) rarely have flames taller than two metres. High-intensity fires (>10^5 kW m^{-1}), in contrast, may have flames that are several tens of metres high (Table 13.1). Fireline intensity is directly proportional to the linear rate at which fire spreads R [m s^{-1}], the specific mass of fuel consumed w_a [kg m^{-2}], and the net low heat of combustion (or simply heat yield of the burning fuel) H [kJ kg^{-1}]:

$$I_B = R w_a H \tag{13.2}$$

Equations (13.1) and (13.2) permit us to relate the size and energy release of a fire to

Table 13.1 Rating scheme for fireline intensity and severity rating for eucalypt-dominated sclerophyll vegetation communities in southeastern Australia. (Source: Modified from Shakesby and Doerr 2006).

Min. fireline intensity (kW m^{-1})	Max. flame height (m)	Severity rating	Postfire vegetation characteristics
500	1.5	Low	Only ground fuel and shrubs <2 m high burnt
3000	5	Moderate	All ground fuel and shrub vegetation <4 m high consumed
7000	10	High	Same as above; lower tree canopy <10 m high scorched
70 000	10–30	Very high	Green vegetation and tree canopy <30 m high consumed
100 000	20–40	Extreme	All green and woody vegetation <10 mm diameter consumed

the amount of fuel destroyed and imply that the spreading rate of a fire is inversely related to its fuel consumption. For a given value of I_b, faster moving fires burn much less biomass and organic litter per unit area than slower moving ones.

Satellites capable of detecting and recording fires as small as 50 m² produce information that is vital to estimate robustly the return period of frequent burning in particular. The data show that subtropical savanna and shrubland currently have the highest number of fires per unit area globally, owing to an optimal combination of solar radiation, sufficient vegetation stands as fuel, high thunderstorm activity, and human land-use practices favouring fire. Burning affects a given patch of land every two to five years on average in the savanna (Scott et al. 2014).

13.2 Geomorphic Impacts of Wildfires

13.2.1 Wildfires in the Hazard Cascade

Wildfires occur in several compartments of the hazard cascade. On one hand, fires ignite frequently from lightning strikes by thunderstorms and thus frequently evolve from meteorological hazards. On the other hand, the loss of protective vegetation cover following wildfire promotes enhanced soil erosion and slope instability. Wildfires often have various and interlinking impacts on the hydrology, geomorphology, and ecology in burned areas and beyond.

13.2.2 Direct Fire Impacts

Wildfires can have temperatures of up to 1000° C and thus accelerate rates of bedrock weathering by causing rock fragments to spall off the surface. Rock flakes thus detaching from burnt surfaces can be several millimetres thick, and some outcrops shed several kilograms of flakes per square metre in a single wildfire (Shakesby and Doerr 2006).

Fire-induced rock weathering can be an important process in priming landscapes for subsequent erosion. Wildfires alter vegetation stands and thus affect the way that precipitation reaches the soil layer, particularly in semiarid areas (Shakesby and Doerr 2006). The effects that fire has on soils include the temporary loss of soil transpiration, the reduction of aggregate stability, and changes to soil-water repellency. Temperatures of ~175–200° C increase soil-water repellency, producing what are termed 'hydrophobic layers in the soil', whereas temperatures of 200–470° C have the opposite effect, thus increasing water infiltration. Fuel load and fire intensity are thought to be the key factors in soil heating, although results from recent fire experiments at the catchment scale have contradicted the assumed proportional relation between these variables. At the patch scale, little available fuel and low fire intensity may still cause soil heating, and vice versa (Stoof et al. 2013). We note that vegetation type appears to affect fire-induced water repellency (Malkison and Wittenberg 2011), being particularly well researched in chaparral soils (Hubbert and Oriol 2005).

More obvious impacts of fire are losses of vegetation and litter. These losses affect rates of canopy interception, water infiltration, overland flow, and rain-splash erosion. Burning also removes shrubs, trees, or nests of large woody debris that store sediment on hillslopes. Incineration of such vegetation dams may partly explain why dry ravel increases immediately after wildfires, even if the weather remains dry. One way to estimate the postfire yield of sediment impounded by vegetation may thus involve multiplying the density of vegetation with its average sediment trapping efficiency (Lamb et al. 2013). In Arctic permafrost areas wildfires remove insulating vegetation or peat layers, thus increasing the chance of ground thawing and thermokarst, and accelerating the decay of permafrost (Figure 13.2) (Zhang et al. 2015). Reported changes to the thickness of the active layer of

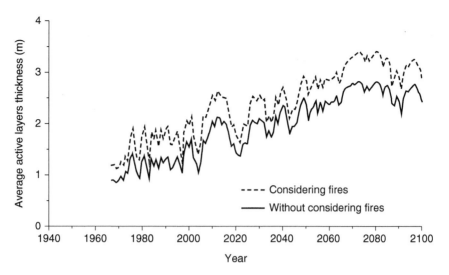

Figure 13.2 Modelled average permafrost active layer thickness with and without fire effects, Northwest Territories, Canada. Active layer thicknesses were calculated as the average for previously burned areas where permafrost exists. From Zhang et al. (2015).

permafrost seem to vary with the thickness of the organic layer removed and the vigour with which vegetation re-establishes in burnt areas.

Yellowstone fires of 1988 Because wildfires benefit ecosystems, many scientists believe that suppressing natural fires in forested areas is undesirable. In 1976 officials in Yellowstone National Park, which covers 9000 km^2 in Wyoming and Montana, instituted a policy of allowing natural fires to burn in wilderness areas of the park as long as human lives, visitor areas, and nearby areas outside the park were safe. Any human-caused fire, however, would be extinguished immediately. Before the summer of 1988, the worst fire in the park's history had burned only about 100 km^2 of forest. This figure stands in stark contrast to over 3200 km^2 of forest consumed during the 1988 fire season, one-third of Yellowstone's total area. The 1988 fires led to a major controversy over the National Park Service natural-burn policy. The problem started when lightning strikes ignited 50 fires in the park in the early summer of 1988. Twenty-eight of the

50 fires were allowed to burn according to the natural-burn policy. These fires quickly expanded under the hot dry summer conditions, fuelled by high winds. In mid-July Yellowstone officials bowed to political pressure and sent fire crews in to fight one of the natural fires. Within four days, many natural fires in and around Yellowstone were being fought, but it was clear that they were beyond the control of the crews. The fires continued until September 10, when rains fell throughout the area. Snows in November finally extinguished the flames. Yellowstone's natural-burn policy was severely criticised during and after the 1988 fires. Although most scientists agree that fire is good for the natural environment, people found it hard to sit and watch the park burn. Critics claimed that the fires would have been smaller if park officials had fought them from the beginning. Others argued, however, that the fires would have been less severe if prior fire-suppression policies had avoided fuel accumulation to hazardous levels in the area (Figure 13.3). Yellowstone park officials still adhere to a natural-burn

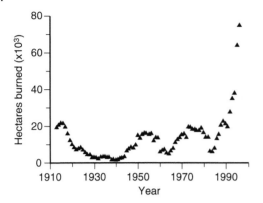

Figure 13.3 Lands burned by wildfire in Arizona and New Mexico. The increase in area burned in the 1990s is due to aggressive wildfire suppression land management that increased fuel loads to the point that major fires were favoured. Modified from Neary et al. (1999).

policy. This policy is preferable because Yellowstone's ecosystems have, through geologic time, adapted to and become dependent on wildfire. The fires of 1988 left much of the park intact, and in burned areas revitalised ecosystems through natural transformations that cycle energy and nutrients through soils, plants, and animals.

13.2.3 Indirect and Postfire Impacts

Images of burned, seemingly lifeless, forests can evoke the view that only intact vegetation reduces erosion in steeplands. Depending on how thoroughly fire destroys or alters vegetation stands, it indirectly reduces the rates at which rainfall infiltrates into soils and increases the rates at which runoff is generated. The result is that peak discharges in river catchments affected by fire become higher and more flashy. In the western United States, for example, wildfires can be responsible for generating an average of 10–20% of the total streamflow in widely burned catchments, thus possibly offsetting projected decreases in streamflow due to atmospheric warming (Figure 13.4) (Wine et al. 2018). The differences in discharge before and immediately after burning can exceed an order of magnitude. Numerical models built on shallow water equations or approximated kinematic waves can handle rainfall time series, soil infiltration rates, and channel geometry to predict entire hydrographs, and thus the timing and characteristics of peak discharge, in burned catchments (Rengers et al. 2016a).

The increased flashiness of postfire surface runoff generates flows that are capable of entraining and transporting more sediment. At the same time, sediment on hillslopes can become more mobile among the burned vegetation, thus material can more readily enter river channels, where the sudden mobilization of stored and saturated sediment can give rise to debris flows (McGuire et al. 2017). Compared to floods, debris flows have a higher runoff coefficient, which is defined as the ratio of peak discharge over the product of catchment area and rainfall intensity (Kean et al. 2016). The reduced capacity for soil infiltration following wildfires allows surface runoff to entrain more sediment and to develop debris flows with coarse bouldery fronts. This mechanism can quickly exhaust material in channels and on hillslopes, and explains in parts why most studies concerned with postfire runoff, erosion, and sediment yield report highly elevated values in the first year following fire (Figure 13.5) (Shakesby and Doerr 2006). Hence, erosion following wildfires may be much higher than the average background erosion under undisturbed conditions, at least in forested areas. Detailed repeat measurements of burnt areas with LiDAR scans showed that most postfire sediment came from hillslopes, whereas channelized areas experienced more incision; while burnt areas generally became rougher as postfire flows carried away fine sediments from the surface (Rengers et al. 2016b). Exceptions to this general observation can arise, however, and streamflow or sediment fluxes may hardly change, especially if rainfall or snowmelt are

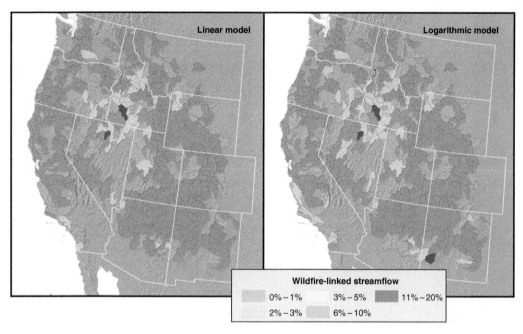

Figure 13.4 Wildfire contribution to streamflow (1986–2015) expressed as a fraction of total streamflow in the western United States. From Wine et al. (2018).

below average following burning (Owens et al. 2013). Regenerating vegetation begins to intercept runoff and sediment, so that the reduced amount of readily erodible material causes sediment yields to decline and return to background levels within five or so years on average. The return of runoff to prefire values may take longer, up to several decades (McClelland et al. 2004). As a rule of thumb, the increases in erosion and sediment yield following wildfires depend largely on how severely the vegetation was burned, how susceptible the terrain is to erosion, and how frequently and intensely rainstorms or snowmelt runoff affect a burnt area (Ryan et al. 2011).

A reduced or missing protective vegetation cover favours increased wind erosion and more shallow landslides and debris flows. Mass movements mostly occur during major rainstorms in the years following the fire. The triggers of debris flows in burnt hilly and mountain landscapes may change from surface runoff in the first few years after a fire to debris slides in later years. Fire-impacted

soils repel water less as the hydrophobic compounds break down, such that surface runoff decreases during storms with time since the fire. At the same time, roots of burnt trees gradually decompose such that they lose their ability to stabilize soils by anchoring or mechanical cohesion within 5–10 years of the fire (Wondzell and King 2003). Repeated high-resolution LiDAR scanning of burnt catchments helps to identify the local sediment sources of postfire debris flows, and Staley et al. (2014) reported that raindrop-impact-induced erosion, ravel, surface wash, and rilling were amongst the most prominent processes delivering sediment from hillslopes to channels. Wildfires in high mountain catchments can also promote more snow avalanches, where damaged vegetation fails to stabilise the snowpack (Sass et al. 2010). Burning of Arctic tundra may give rise to thaw slumps and active layer detachments in some cases. In their study of two major tundra fires that burned 500–1200 km^2 of the North Slope of Alaska between 1880 and 1920 AD, Jones

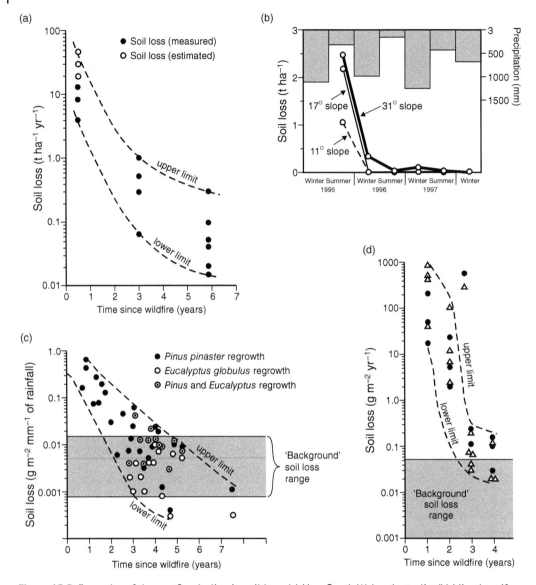

Figure 13.5 Examples of the postfire decline in soil loss. (a) New South Wales, Australia. (b) Mixed conifer, eastern Oregon, USA (note arithmetic scale for soil loss, inclusion of rainfall amounts, and shore timescale compared with other examples. (c) Pine and eucalyptus plantations, north-central Portugal. (d) Pine and oak scrub, Mount Carmel, Israel. Modified from Shakesby and Doerr (2006).

et al. (2013) remarked that the thaw slumps or active layer detachment slides were largely independent of signs of burning-induced permafrost degradation.

Suspended sediment yields of more than 1400 times the prefire background rates have been reported in areas where erosion has increased after burning (Smith et al. 2011); yet most reports quote an order-of-magnitude increase. Burning of mixed conifer forests in Arizona, for example, led to some of the highest specific sediment yields recorded (>41 000 t km^{-2} yr^{-1}) in the first year following the fires (Shakesby and Doerr 2006). Geomorphologists usually associate such rates with landslide-disturbed catchments in tectonically active mountain belts. On the other hand, some reports of postfire sediment yields are orders of magnitude lower than this value, mainly from slowly denuding areas in South Africa and Australia (Lane et al. 2006). Few studies deal with the bedload sediment that rivers transport following fires. In a detailed study of fire impacts on two mountain catchments in southeastern Australia, Lane et al. (2006) measured the material trapped behind weirs, and estimated that bedload made up 30–45% of the total sediment yield in the first two years following burning. In the Transverse Ranges of southern California, detailed records of fire occurrence and rates of erosion show that increasing fire frequencies during the twentieth century may have raised catchment-wide erosion rates by as much as 60% or more on average (Lavé and Burbank 2004). In any case, fire-induced peaks in sediment transport are short-lived and quickly go down in the following years when only less mobile sediment is available, vegetation recovers, and runoff and discharge drop commensurately. However, long-term rates of soil loss are commonly many times higher than the rates required to form new soil, although litter dams and root mats trap eroded soils on hillslopes, causing mismatches between erosion rates and sediment yields, hence compromising net rates of topsoil erosion (Shakesby et al. 2007). Riparian vegetation damaged or killed by fires means that channel banks will also lose some of their shear strength from root networks, which promotes channel-bank erosion and lateral instability (Owens et al. 2013).

The types and rates of geomorphic responses to wildfires can differ widely among catchments. The geomorphic consequences of high fire-induced sediment yields include channel erosion and aggradation, the formation of small alluvial fans, and the development of knickpoints that continue to disturb or destroy aquatic habitats for years to decades. In burnt forested areas, postfire erosion also mobilizes large volumes of woody debris that can build jog jams in streams, trap sediments, and alter flow, but also create new riparian habitats.

Several studies of North American rivers show that such episodic inputs of sediment are integral parts of riverine ecosystems (Benda et al. 2003).

Wildfires are key drivers of the cyclic flushing and replenishing of sedimentary channel fills in headwater streams of the Oregon Coast Range in the western United States. For decades following wildfires some of these catchments still experienced debris-flow activity up to 40% greater than in prefire periods. A dendrochronological study of 13 channels showed that the basin-wide volume of sediment and large woody debris per unit length of headwater channel increased nonlinearly with time since the last wildfire-induced debris flows (May and Gresswell 2003). Mainly triggered by rainfall, debris flows may scour channel fills, exposing the bedrock floor and creating new storage sites for sediment and large woody debris. Large log jams, resulting from fires, landslides, and windstorms, increase local hydraulic roughness, trap incoming sediment, and create new sediment retention space and habitats in steep headwater channels.

Detailed surveys based on terrestrial and airborne laser scanning are useful for tracking changes in postfire erosion and sediment yield on alluvial fans. Orem and Pelletier (2015) used both these methods to show that postfire sediment volumes flushed from two catchments decreased linearly with time following fire, but sediment yields decreased exponentially owing to local storage effects. This model seems applicable to other areas. Nyman et al. (2011) pointed out that, despite the well-acknowledged importance of wildfire in Australian scrub and forest ecosystems, few studies have explored in detail how fire might lead to highly erosive events in landscapes that otherwise denude very slowly. Debris flows in the mountains of southeastern Australia might be more common after burning than previously thought, especially given that previous reports seem to have confused these events with flash floods. Stream-sediment concentrations as

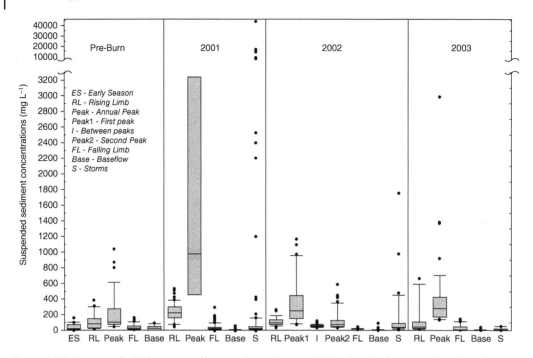

Figure 13.6 Impacts of wildfire on runoff and sediment loads at Little Granite Creek, western Wyoming. Suspended sediment concentrations over a three-year period compared with composite preburn values from 1982–1993. The box represents the 25th to 75th percentiles; the 10th and 90th percentiles are shown by vertical lines; the horizontal line in each box is the median value; and outliers are depicted as dots. From Ryan et al. (2011).

high as 143 g l^{-1} following a fire in southeastern New South Wales show that short-lived flows may carry the bulk of material from burnt catchments. Comparably high sediment concentrations have been measured in heavily denuding catchments on China's loess plateau. Studying fire impacts in the Great Dividing Range in southeastern Australia, Nyman et al. (2015) mapped several hundreds of small debris flows that formed in the first few years following three major fires, and proposed that burn severity, terrain-surface slope, the degree of dryness, and rainfall intensity derived from radar measurements correllated those channel reaches with traces of debris-flow erosion.

The flushing of suspended sediment and burnt vegetation reduces the water quality of streams and water bodies, particularly when mobilizing toxic constituents or trace elements from soils, sediments or spalled-off bedrock (Smith et al. 2011). The elevated amounts of

sediment that streams carry in suspension after a fire are critical, as many of the chemical species attach themselves to sediment grains. Ryan et al. (2011) described highly turbid 'blackwater flows' containing high loads of sediment, ash, and charcoal following wildfires in the Gros Ventre Range of eastern Wyoming (Figure 13.6).

13.3 Geomorphic Tools for Reconstructing Past Wildfires

Wildfires have left their mark in both historic and older sedimentary archives. We can also trace some impacts back into deep geological time. Geomorphologists have been most concerned with the immediate impacts of wildfires on erosion, sediment transport, and deposition, and so most studies have monitored the consequences of burning over several years, or rarely decades. Reconstructing the geomorphic and ecological impacts of wildfire becomes more difficult farther back in time, and the same applies in assessing the role of fire in landscape evolution, for example when tracking the time of emerging vegetation cover in previously glaciated forelands of northern Scandinavia (Carcaillet et al. 2012). Similarly, identifying and dating fire-related deposits in alluvial fans helped Meyer and Pierce (2003) to establish a chronology of late Holocene fires in Yellowstone National Park, United States. Accounting for possible gaps in the deeper layers of this sedimentary archive, they estimated that wildfires caused the episodic deposition of charcoal-rich sediments on one fan every 300–450 years on average over the past 3000 years. These sediments contain characteristic coarse, angular charcoal and dark mottles, and were mostly laid down by debris flows, hyperconcentrated flows, and sheet floods when climate was drier and warmer than today. The fire-related sediments make up some 30% of the volume of this alluvial fan. This fraction can be a reliable measure of the long-term contribution of fire-induced sedimentation locally, but misses out on any amount of sediment that was eroded following fires, but managed to bypass the fan.

In reviewing a handful of similar studies, Shakesby et al. (2007) concluded that fire-related sediments contribute between 5% and >60% of the total long-term sediment yields. The mismatch between immediate postfire sediment yields and those averaged over several millennia can be high, however. In the Payette River of central Idaho, Meyer and Pierce (2003) found that sediment yields within the first year after fire disturbance were nearly 400 times greater than those estimated from cosmogenic ^{10}Be concentrations in river sands. Part of this mismatch may result from the size of the study area, especially if sediment yields are estimated from small hillslope plots that often cover <0.01 km^2 (Shakesby et al. 2007) and may thus fail to capture the overall catchment response to fire.

Soot trapped in ice cores and charcoal trapped in sedimentary layers are among the most widely used proxies of former fires or 'fire regimes'. Ecologists recognize that a fire regime incorporates aspects of fuel type and its distribution, the rate of spread, seasonality, and frequency of fires (including prescribed burning, if any), the size and patchiness, and their impacts on vegetation and soils (Bowman et al. 2009). Charcoal data are acquired during sedimentological, ecological, or archaeological studies and thus complement many other useful palaeoenvironmental proxies, such as pollen, tephra, and geochemical fingerprints. Many well-dated sedimentary archives containing charcoal offer direct links between former climatic and vegetation conditions and fire regimes (Figure 13.7). Establishing fire episodes from charcoal in sediments requires standardized steps to enable comparisons among sites. Counts of larger charcoal fragments are expressed as the number of particles (commonly > 125μm) per sample and are multiplied by the average accretion rate to arrive at a measure of charcoal particle flux [cm^{-2} yr^{-1}], also known as CHAR. Time series of CHAR values must be interpolated to equal intervals and smoothed to obtain long-term trends that serve as the basis for assessing short-term fluctuations. Peaks in such filtered CHAR time series, marking past fire episodes, can be defined using an arbitrarily set threshold. Analyses of sedimentary charcoal records are a prime source of information revealing tight natural bonds between phases of fire activity and climate: less fires tend to occur in colder climates and more fires in warmer climates (Pierce et al. 2004), similar to observations under contemporary conditions. Traces of soot accumulate in ice cores that are routinely used to infer Quaternary climate oscillations and offer insights into more regional patterns of biomass burning. These records offer opportunities to link fire activity to past climates. Other proxies of past fires include pollen records that document the relative abundance of fire-resistant species through time or fire

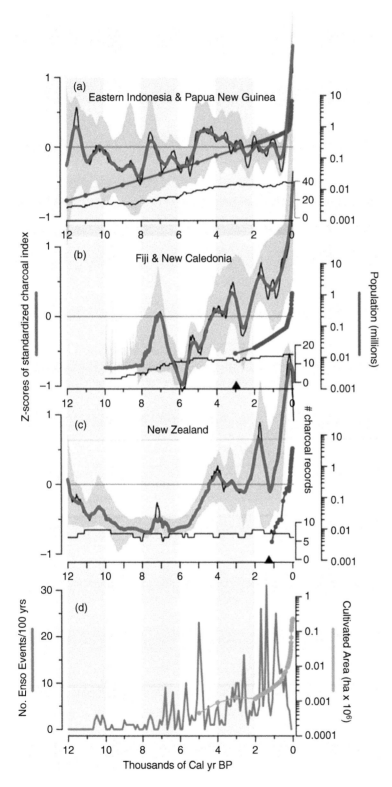

Figure 13.7 Biomass burning trends for A) New Guinea, B) Fiji and New Caledonia, and C) New Zealand, all smoothed with a 500-year (red) and 250-year (black) line and shown with 95% bootstrap confidence intervals. Area-weighted population estimates (blue) are shown for each region. D) A reconstruction of the frequency of El Niño-Southern Oscillation (ENSO) events. From Marlon et al. (2013).

scars on woody plants that are resistant to fire and amenable to annual growth-ring dating (Lynch et al. 2007).

Nevertheless, studies that wish to reconstruct the magnitude and frequency of past wildfires acknowledge almost routinely that humans have long altered background levels of natural fire activity by burning forests to create agricultural lands, promoting grazing, dispersing plants, and actively suppressing fires (Figure 13.8) (Bowman et al. 2011). A synthesis of more than 400 charcoal records from six continents affords detailed insights into climatic and anthropogenic forcing of biomass burning during the past 2000 years (Marlon et al. 2008). Interestingly, the last 150 years of this record show that biomass burning declined, largely because fuel has been reduced in the wake of expanding intensive grazing, agriculture, and fire management. A similar situation holds for a 1500-year record from a sediment core taken from Lake Loon, Oregon Coast Range (Richardson et al. 2018). The scientists identified 23 'event beds', partly derived from hyperpycnal flows, in the sediments. One lesson from this study was that attributing these beds to enhanced erosion pulses following wildfires needs to consider carefully any competing and equally plausible explanations such as large earthquakes, floods, and, more recently, timber harvesting. Independent dates of past earthquakes raised the likelihood that more than half – and also the older to oldest – of the event beds were linked with a seismic origin.

Fire is a key element in shaping Australia's landscapes and ecosystems, and is also the continent's most deadly natural hazard. Much research has focused on when and by how much humans altered natural fire regimes on the continent (Mooney et al. 2011). Interpretations of charcoal records hold that the highest fire frequencies occurred during times of major climatic and environmental instability, including the Holocene climatic optimum between 7000 and 5000 years ago (Figure 13.9). Nonetheless, pronounced peaks in many charcoal records also mark the colonization

of Australia by Europeans (Lynch et al. 2007). Whether the first arrival of humans on the continent at around 60 ka spurred changes in fire regimes remains debated, but it is known that they caused the extinction of many megafauna. Judging from more than 200 sedimentary charcoal records, the fire regimes seemed to have remained unchanged after that time (Mooney et al. 2011). The authors pointed out that the most pronounced increases in inferred fire activity took place in the latter half of the past millennium.

Many studies on fire regimes in the United States have concluded that most methods cannot sufficiently separate human from natural controls. Pierce et al. (2004) determined the ages of fire-related sediment deposits in alluvial fans in central Idaho to reconstruct the Holocene fire history in xeric ponderosa pine forests. In this area, frequent low-severity fires happened during colder periods, likely fuelled by increased understory growth, whereas stand-replacing fires and highly erosive debris flows were common during warmer drier periods. Detailed analyses of soot and char contents in Holocene sediments in Lake Daihai, Inner Mongolia, China, revealed a similar link between lower wildfire activity and cooler periods (Han et al. 2012).

Even cool-temperate rainforests are susceptible to occasional fire, although at much lower frequencies than other vegetation types. The higher fuel loads in southeastern Australia's more moist forests, for example, often lead to more extreme wildfires that recur every 70 years on average (Nyman et al. 2011). Detailed investigations of nearly 500 years of tree fire scars and ring data showed that the fire history of the rainforests of northwestern Patagonia is closely linked to the Southern Annual Mode, which is the most prominent control of extratropical climate variability in the Southern Hemisphere (Holz and Veblen 2011). Only in the second half of the twentieth century did this relationship break down, most likely mirroring an increasing interest in suppressing and managing fires.

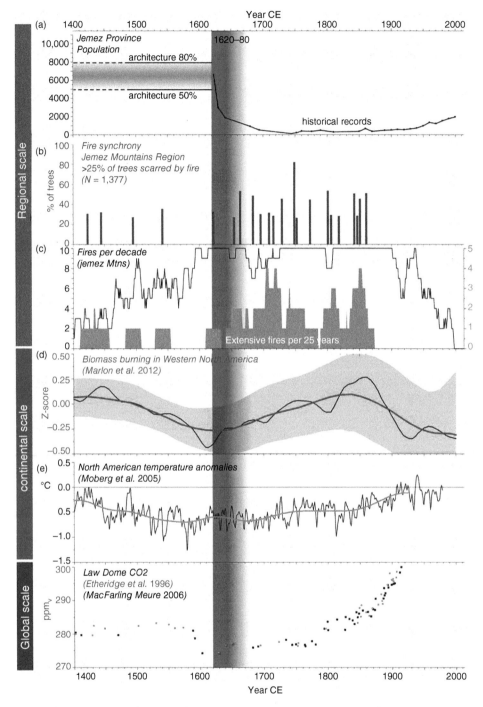

Figure 13.8 Comparison of regional, continental, and global data on climate and fire regimes, CE 1400–2000. Red column indicates period of major demographic decline. (a) Native American population in the Jemez Province, southwest United States. (b) Chronology of regional fire synchrony based on 1377 samples from across the Jemez Mountains region. (c) Fires per decade and extensive fires per 25 year in the Jemez Mountains. (d) Biomass burning in western North America. (e) North American temperature anomalies. (f) Global atmospheric carbon dioxide recorded in Law Dome ice core, Antarctica. From Liebmann et al. (2016).

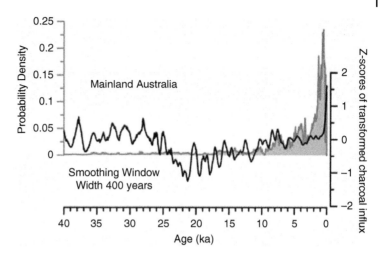

Figure 13.9 Comparison of the composite charcoal curve for sites from the Australian mainland over the past 40 000 years (black curve) compared to probability density estimates of human populations based on radiocarbon-dated archaeological records (green infilled curve). Population data from the AustArch Database. From Mooney et al. (2011).

Numerical modelling of fire ignition and spread has investigated how the underlying controls play out geographically. Many of these models express the statistical relationship between the number of recorded fires or the percentage of burned area and satellite data that record the number of lightning strikes per unit area, the type and thickness of vegetation cover and land use, and the local population density. For example, Amaral-Turkman et al. (2011) modelled the spatial pattern of both ignition and relative area burned in Australia and New Zealand using a Bayesian approach. They used temperature and rainfall data from various sources to capture characteristics of the dry and wet seasons, the human footprint, the state of agricultural land use, and grass and tree cover in 1° tiles of latitude and longitude. They further collapsed other potentially important, although more local, influences such as wind speed and topography into a hidden or latent random effect in the model, arguing that ignoring such additional effects might severely compromise predictions. Lehsten et al. (2010) came to a similar conclusion when modelling contemporary fire patterns and the fraction of areas burned throughout Africa. They noted that data on possibly important predictors such as the local hillslope gradient or aspect were either unavailable or meaningless at the coarse resolution needed for modelling fire at the continental scale, therefore making fire models less representative and reliable.

13.4 Towards More Megafires?

Burning of fossil fuels and wildfires impact air quality at local, regional, and global scales. Besides the climatic effects of carbon dioxide emissions, air pollution is nowadays compromising human health at levels far beyond those of any geomorphic impacts tied to fires. For example, open fires for cooking and providing warmth have long been used in billions of households around the world. Yet household air pollution from fuel stoves and wood fires claimed 4.3 million lives in 2012 alone; more than 3.3 million of these deaths were in Southeast Asia and the western Pacific (Subramanian 2014).

Warmer conditions in general, and droughts in particular, reduce soil and atmospheric moisture and increase the flammability of vegetation. As the atmosphere warms, and assuming that humidity remains unchanged, the severity of droughts increases, thus stressing vegetation for longer intervals. Forests in particular appear to be experiencing increasingly more frequent, larger, and longer wildfires. Here the term 'megafires' refers to extreme wildfires influenced by droughts that depress moisture contents, increase tree mortality, and thus promote flammability. Satellite monitoring and fire databases show that the areas and severity of fires have increased in recent years. Many Mediterranean landscapes with their highly seasonal climates have seen order-of-magnitude increases in annual fire frequencies in the second half of the twentieth century, bearing in mind that satellites have also gained accuracy in detecting more and smaller fires. Pausas et al. (2008) argued that such increases are most harmful where the naturally fire-adjusted or even fire-resilient vegetation stands have been altered to be more vulnerable to fires. They also pointed out that fire-related soil losses in the Mediterranean differed by at least an order of magnitude without any clear trends, though strongly biased by researchers focusing on erosion-prone areas. Drawing on satellite and field studies, Miller et al. (2008) reported that fires of high severity replaced entire forest stands over nearly 120 000 km^2 in California and Nevada between 1984 and 2006. They also noted increases in both the mean and maximum areas burned, and attributed this growth mostly to rising temperatures and precipitation. However, the widespread policy of suppressing wildfires over many decades may also have promoted the accumulation of large amounts of fuel. According to MunichRe, the 2017 wildfire season in California saw the largest fires ever reported, and also the highest insured wildfire losses in the state's history (Figure 13.10). Detailed studies in the American Southwest have shown that the increasing tendency towards drought between about 1980 and 2015 correlated positively with both the area of bark-beetle infestation and the area of wildfire in forests, highlighting the correlations between temperature extremes and biological processes (Millar and Stephenson 2015), and their implications for natural hazards and geomorphic response. From an ecological viewpoint, low-impact fires promote growth of herbaceous flora, increase plant-available nutrients, and thin dense forests. Severe fires change rates of plant succession, alter above- and below-ground species composition, volatilize nutrients, and entrain ash in smoke columns. Wildfires alter mineralization rates and nutrient ratios, for example carbon:nitrogen (C:N) ratios. Nutrient losses can accelerate soil erosion, leaching or denitrification; decrease micro- and macrofauna; and change microbial populations and processes.

Wildfires can be a consequence of climate change, but they can also themselves impact climate. Carbon and aerosol emissions by large wildfires can initiate positive feedbacks that affect climate. About one-third of our planet's total atmospheric soot mass comes from wildfires, and fires in southeast Asia or Russia may contribute twice as much regionally. The amount of radiative forcing of this naturally derived soot and its role in atmospheric warming remain largely unconstrained and thus a major uncertainty in climate models. The way that soot particles aggregate strongly influences its optical properties and its potential contribution to atmospheric warming (Chakrabarty et al. 2014). Wildfires produce aerosols, which in combination with airborne dust, black carbon from industrial emissions, and biofuel burning may interfere with precipitation patterns, particularly in regions with a seasonal monsoonal climate such West Africa (Knippertz et al. 2015) and India. Model studies suggest that, in many regions of the Indian subcontinent and the adjacent Himalayas, the mix of soot and black carbon leads to local or regional heating of the air because particles absorb solar radiation, thus reducing cloud cover and precipitation, particularly during the South Asian summer monsoon season (Ganguly et al. 2012). Thus, reduced precipitation might influence rates of runoff, erosion, and sediment transport that are mostly tied to the monsoonal season.

Black carbon at lower and middle tropospheric levels can form large plumes, known as brown clouds, that contribute to melting of snowpacks and glaciers in the Himalayas. The net effects of black carbon in that area

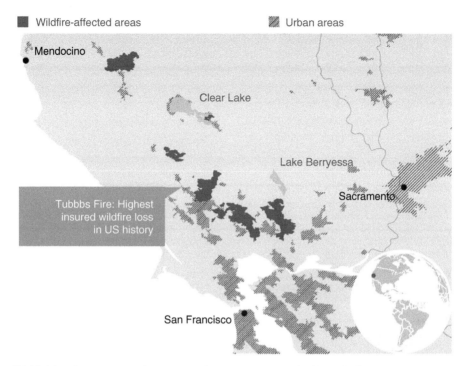

■ Wildfire-affected areas　　　▨ Urban areas

Mendocino

Clear Lake

Lake Berryessa

Tubbbs Fire: Highest insured wildfire loss in US history

Sacramento

San Francisco

Figure 13.10 Wet winters created just enough fuel to exacerbate the 2017 California wildfires, which were the state's largest and costliest in history. Source: MunichRe (2018).

might be as important as atmospheric CO_2 in terms of regional warming (Ramanathan and Carmichael 2008). Widespread surface melt events at high levels on the Greenland ice sheet have been triggered by black carbon release from wildfires during exceptionally warm periods. Dated firn cores contain layers from wildfires during exceptionally warm summers such as 1889 and 2012 (Keegan et al. 2014). The content of black carbon in Holocene lake sediments from China also points to a strong climatic control, at least for the past few thousand years, with lower carbon concentrations linked to cooler periods (Han et al. 2012). These and many more observations emphasize the complexity of the tightly interlinked chain of processes and feedbacks that influence the climate system, which then can affect some natural hazards. The close coupling of wildfires, atmospheric aerosol contents, and snow and glacier melt highlights potential impacts on the regional availability of freshwater resources in mountain belts such as the Andes and the Himalayas (Immerzeel et al. 2010).

References

Amaral-Turkman MA, Turkman KF, Le Page Y, and Pereira JMC 2011 Hierarchical space-time models for fire ignition and percentage of land burned by wildfires. *Environmental and Ecological Statistics* **18**, 601–617.

Benda L, Miller D, Bigelow P, and Andras K 2003 Effects of post-wildfire erosion on channel environments, Boise River, Idaho. *Forest Ecology and Management* **178**(1-2), 105–119.

Boike J, Grau T, Heim B, et al. 2016 Satellite-derived changes in the permafrost landscape of central Yakutia, 2000–2011: Wetting, drying, and fires. *Global and Planetary Change* **139**, 116–127.

Bowman DMJS, Balch JK, Artaxo P, et al. 2009 Fire in the Earth system. *Science* **324**(5926), 481–484.

Bowman DMJS, Balch J, Artaxo P, et al. 2011 The human dimension of fire regimes on Earth. *Journal of Biogeography* **38**(12), 2223–2236.

Carcaillet C, Hörnberg G, and Zackrisson O 2012 Woody vegetation, fuel and fire track the melting of the Scandinavian ice-sheet before 9500 cal yr BP. *Quaternary Research* **78**(3), 540–548.

Chakrabarty RK, Beres ND, Moosmüller H, et al. 2014 Soot superaggregates from flaming wildfires and their direct radiative forcing. *Scientific Reports* **4**, 1–8.

Cochrane MA 2009 Fire in the tropics. In *Tropical Fire Ecology* (ed Cochrane MA), Springer pp. 1–24.

Ganguly D, Rasch PJ, Wang H, and Yoon JH 2012 Climate response of the South Asian monsoon system to anthropogenic aerosols. *Journal of Geophysical Research: Atmospheres* **117**, 1–20.

Han YM, Marlon JR, Cao JJ, et al. 2012 Holocene linkages between char, soot, biomass burning and climate from Lake Daihai, China. *Global Biogeochemical Cycles* **26**(4), 1–9.

Hantson S, Pueyo S, and Chuvieco E 2014 Global fire size distribution is driven by human impact and climate. *Global Ecology and Biogeography* **24**(1), 77–86.

Holz A and Veblen TT 2011 The amplifying effects of humans on fire regimes in temperate rainforests in western Patagonia. *Palaeogeography, Palaeoclimatology, Palaeoecology* **311**(1-2), 82–92.

Hubbert KR and Oriol V 2005 Temporal fluctuations in soil water repellency following wildfire in chaparral steeplands, southern California. *International Journal of Wildland Fire* **14**(4), 439–447.

Immerzeel WW, van Beek LPH, and Bierkens MFP 2010 Climate change will affect the Asian water towers. *Science* **328**(5984), 1382–1385.

Jones BM, Breen AL, Gaglioti BV, et al. 2013 Identification of unrecognized tundra fire events on the north slope of Alaska. *Journal of Geophysical Research: Biogeosciences* **118**, 1334–1344.

Kean JW, McGuire LA, Rengers FK, et al. 2016 Amplification of postwildfire peak flow by debris. *Geophysical Research Letters* **43**, 8545–8553.

Keegan KM, Albert MR, McConnell JR, and Baker I 2014 Climate change and forest fires synergistically drive widespread melt events of the Greenland Ice Sheet. *Proceedings of the National Academy of Sciences* **111**(22), 7964–7967.

Knippertz P, Evans MJ, Field PR, et al. 2015 The possible role of local air pollution in climate change in West Africa. *Nature Climate Change* **5**(9), 815–822.

Lamb MP, Levina M, DiBiase RA, and Fuller BM 2013 Sediment storage by vegetation in steep bedrock landscapes: Theory, experiments, and implications for postfire sediment yield. *Journal of Geophysical Research: Earth Surface* **118**, 1147–1160.

Lane PNJ, Sheridan GJ, and Noske PJ 2006 Changes in sediment loads and discharge from small mountain catchments following wildfire in south eastern Australia. *Journal of Hydrology* **331**(3-4), 495–510.

Lavé J and Burbank D 2004 Denudation processes and rates in the Transverse Ranges, southern California: Erosional response of a transitional landscape to external and anthropogenic forcing. *Journal of Geophysical Research: Earth Surface* **109**, 1–31.

Lehsten V, Harmand P, Palumbo I, and Arneth A 2010 Modelling burned area in Africa. *Biogeosciences* **7**, 3199–3214.

Liebmann MJ, Farella J, Roos CI, Stack, et al. 2016 Native American depopulation, reforestation, and fire regimes in the southwest United States, 1492–1900 CE. *Proceedings of the National Academy of Sciences* **113**, E696–E704.

Lynch AH, Beringer J, Kershaw P, et al. 2007 Using the paleorecord to evaluate climate and fire interactions in Australia. *Annual Review of Earth and Planetary Sciences* **35**(1), 215–239.

Malamud BD, Morein G, and Turcotte DL 1998 Forest fires: an example of self-organized critical behavior. *Science* **281**(5384), 1840–1842.

Malkinson D and Wittenberg L 2011 Post fire induced soil water repellency – Modeling short and long-term processes. *Geomorphology* **125**(1), 186–192

Marlon JR, Bartlein PJ, Carcaillet C, et al. 2008 Climate and human influences on global biomass burning over the past two millennia. *Nature Geoscience* **1**(10), 697–702.

Marlon JR, Bartlein PJ, Daniau A-L, et al. 2013 Global biomass burning; a synthesis and review of Holocene paleofire records and their controls. *Quaternary Science Reviews* **65**, 5–25.

May CL and Gresswell RE 2003 Processes and rates of sediment and wood accumulation in headwater streams of the Oregon Coast Range, USA. *Earth Surface Processes and Landforms* **28**(4), 409–424.

McClelland JW, Holmes RM, Peterson BJ, and Stieglitz M 2004 Increasing river discharge in the Eurasian Arctic: Consideration of dams, permafrost thaw, and fires as potential agents of change. *Journal of Geophysical Research: Atmospheres* **109**, 1–12.

McGuire LA, Rengers FK, Kean JW, and Staley DM 2017 Debris flow initiation by runoff in a recently burned basin: Is grain-by-grain sediment bulking or en masse failure to blame?. *Geophysical Research Letters* **44**(14), 7310–7319.

Meyer GA and Pierce JL 2003 Climatic controls on fire-induced sediment pulses in Yellowstone National Park and central Idaho: a long-term perspective. *Forest Ecology and Management* **178**(1-2), 89–104.

Millar CI and Stephenson NL 2015 Temperate forest health in an era of emerging megadisturbance. *Science* **349**, 823–826.

Miller JD, Safford HD, Crimmins M, and Thode AE 2008 Quantitative evidence for increasing forest fire severity in the Sierra Nevada and Southern Cascade Mountains, California and Nevada, USA. *Ecosystems* **12**(1), 16–32.

Mooney SD, Harrison SP, Bartlein PJ, et al. 2011 Late Quaternary fire regimes of Australasia. *Quaternary Science Reviews* **30**(1-2), 28–46.

Munich RE 2018 Rain fuels wildfire risk. Munich RE annual report. https://www.munichre .com/topics-online/en/climate-change-and-natural-disasters/natural-disasters/wildfires/rain-fuels-wildfire-risks-2018.html

Neary DG, Klopatek CC, DeBano LF, and Ffolliott PF 1999. Fire effects on below ground sustainability: a review and synthesis. *Forest Ecology and Management* **122**, 51–71.

Nyman P, Sheridan GJ, Smith HG, and Lane PNJ 2011 Evidence of debris flow occurrence after wildfire in upland catchments of south-east Australia. *Geomorphology* **125**(3), 383–401.

Nyman P, Smith HG, Sherwin CB, et al. 2015 Predicting sediment delivery from debris flows after wildfire. *Geomorphology* **250**(C), 173–186.

Orem CA and Pelletier JD 2015 Quantifying the time scale of elevated geomorphic response following wildfires using multi-temporal LiDAR data: An example from the Las Conchas fire, Jemez Mountains, New Mexico. *Geomorphology* **232**(C), 224–238.

Owens PN, Giles TR, Petticrew EL, Leggat MS, et al. 2013 Muted responses of streamflow and suspended sediment flux in a wildfire-affected watershed. *Geomorphology* **202**(C), 128–139.

Pausas JG, Llovet J, Rodrigo A, and Vallejo R 2008 Are wildfires a disaster in the Mediterranean basin? - A review. *International Journal of Wildland Fire* **17**(6), 713–723.

Pierce J, Meyer G, and Jull AJT 2004 Fire-induced erosion and millennial-scale climate change in northern ponderosa pine forests. *Nature* **432**(7013), 87–90.

Ramanathan V and Carmichael G 2008 Global and regional climate changes due to black carbon. *Nature Geoscience* **1**, 221–227.

Rengers FK, McGuire LA, Kean JW, et al. 2016a Model simulations of flood and debris flow timing in steep catchments after wildfire. *Water Resources Research* **52**(8), 6041–6061.

Rengers FK, Tucker GE, Moody JA, and Ebel BA 2016b Illuminating wildfire erosion and deposition patterns with repeat terrestrial lidar. *Journal of Geophysical Research: Earth Surface* **121**, 588–608.

Richardson KND, Hatten JA, and Wheatcroft RA 2018 1500 years of lake sedimentation due to fire, earthquakes, floods and land clearance in the Oregon Coast Range: geomorphic sensitivity to floods during timber harvest period. *Earth Surface Processes and Landforms* **43**(7), 1496–1517.

Ryan SE, Dwire KA, and Dixon MK 2011 Impacts of wildfire on runoff and sediment loads at Little Granite Creek, western Wyoming. *Geomorphology* **129**(1-2), 113–130.

Sass O, Heel M, Hoinkis R, and Wetzel KF 2010 A six-year record of debris transport by avalanches on a wildfire slope (Arnspitze, Tyrol). *Zeitschrift für Geomorphologie* **54**(2), 181–193.

Scott AC, Bowman DMJS, Bond WJ, et al. 2014 *Fire on Earth. An Introduction.* Wiley Blackwell.

Shakesby RA and Doerr SH 2006 Wildfire as a hydrological and geomorphological agent. *Earth-Science Reviews* **74**(3), 269–307.

Shakesby RA, Wallbrink PJ, Doerr SH, et al. 2007 Distinctiveness of wildfire effects on soil erosion in south-east Australian eucalypt forests assessed in a global context. *Forest Ecology and Management* **238**(1-3), 347–364.

Smith HG, Sheridan GJ, Lane PNJ, et al. 2011 Wildfire effects on water quality in forest catchments: A review with implications for water supply. *Journal of Hydrology* **396**(1-2), 170–192.

Staley DM, Wasklewicz TA, and Kean JW 2014 Characterizing the primary material sources and dominant erosional processes for post-fire debris-flow initiation in a headwater basin using multi-temporal terrestrial laser scanning data. *Geomorphology* **214**(C), 324–338.

Stocks BJ, Mason JA, Todd JB, et al. 2002 Large forest fires in Canada, 1959–1997. *Journal of Geophysical Research: Atmospheres* **108**, 1–12.

Stoof CR, Moore D, Fernandes PM, et al. 2013 Hot fire, cool soil. *Geophysical Research Letters* **40**(8), 1534–1539.

Subramanian M 2014 Deadly dinners. *Nature* **509**, 548–551.

Wine ML, Makhnin O, and Cadol D 2018 Nonlinear long-term large watershed hydrologic response to wildfire and climatic dynamics locally increases water yields. *Earth's Future* **6**(7), 997–1006.

Wondzell SM and King JG 2003 Postfire erosional processes in the Pacific Northwest and Rocky Mountain regions. *Forest Ecology and Management* **178**(1-2), 75–87.

Zhang Y, Wolfe SA, Morse PD, et al. 2015 Spatiotemporal impacts of wildfire and climate warming on permafrost across a subarctic region, Canada. *Journal of Geophysical Research: Earth Surface* **120**, 2338–2356.

14

Snow and Ice Hazards

Snow and ice are ingredients and some-times catalysts of serious hazards that occur mainly in high-mountain and polar environments, but also along northern rivers with seasonal ice cover. Our percep-tion of snow- and ice-related hazards may be biased by their rare reporting or their less dramatic consequences in comparison to other natural hazards (Whiteman 2011). Nevertheless, melting glaciers, ice sheets, and permafrost continue to make head-lines as the key symptoms of a warming atmosphere, paving the way for major environmental changes and unexpected hazards in polar and mountainous regions in particular. Snow and ice avalanches are among the best publicized and most studied of these hazards, although degrad-ing permafrost in the Arctic and in high mountains is increasingly being studied and is gaining greater public attention. Drifting sea ice and river ice can damage human assets by, respectively, disrupting navigation and causing ice-jam floods. Much progress has been made in under-standing the degradation of permafrost, a slow and gradual process that can compro-mise river-channel and hillslope stability and boost sediment and biogeochemical budgets. Sophisticated engineering mea-sures are required to stabilize permafrost soils disturbed by human activities, thus offering ample applications for applied geomorphology.

14.1 Frequency and Magnitude of Snow and Ice Hazards

Some of the most compelling evidence of atmo-spheric warming has come from documented changes to the cryosphere, particularly melting of ice caps and glaciers, thawing of permafrost, and loss of sea ice. Cryospheric change has dra-matically affected both life and geomorphology on Earth; only 20 000 years ago vast areas of land were covered in glacier ice, and sea level was about 130 m lower than today. Still, the rapidity at which today's cryosphere is chang-ing is disturbing and most likely unparalleled in Earth history. Ice melt in Antarctica alone could add more than a metre to global sea-level rise by the end of the twenty-first century, assuming unchanged greenhouse emissions (DeConto and Pollard 2016). Dwindling moun-tain glaciers have become emblematic of contemporary atmospheric warming, but detailed measurements of glacier mass bal-ance reveal distinct regional differences and it appears that a minority of glaciers is advancing (Fujita and Nuimura 2011).

July 2012 marked the first time in the age of satellite remote sensing that nearly 100% of the surface of the Greenland ice sheet experienced melting (Keegan et al. 2014). A growing num-ber of studies warn that the melting of several outlet glaciers of the West Antarctic ice sheet may have approached an irreversible point,

Geomorphology and Natural Hazards: Understanding Landscape Change for Disaster Mitigation, Advanced Textbook Series,
First Edition. Tim R. Davies, Oliver Korup, and John J. Clague.
© 2021 John Wiley & Sons Ltd. Co-published 2021 by the American Geophysical Union and John Wiley & Sons Ltd.

potentially triggering its eventual catastrophic collapse (Joughin et al. 2014). This possibility arises because most of the West Antarctic ice sheet lies on an inland-sloping bed that is below sea level and is thus prone to marine incursions and instability if melting proceeds further. In contrast, the East Antarctic ice sheet rests on land surfaces above sea level and is, therefore, less prone to instability. In East Antarctica, however, removal of a coastal ice 'plug' at the margin of the Wilkes Basin could change regional ice flow and trigger a self-sustaining discharge of ice from the entire basin, with the result that global sea level would rise 3–4 m (Mengel and Levermann 2014). Numerical models that couple the dynamics of both Antarctic ice sheets and their ice shelves predict largely unstoppable melting and sea-level rise for centuries to millennia to come if the current average surface temperature on Earth rises another 1.5–2 K; for scenarios with high emissions of greenhouse gases, models predict an average global rise in sea level by a total of 0.5–3 m by the late twenty-third century (Golledge et al. 2015).

In spite of these marked changes, the amount of anthropogenic forcing remains controversial, especially because the response of glaciers and ice sheets to changes in climate may be delayed by years or even decades. Model simulations suggest that only 25 ± 35% (*sic!*) of the global glacier mass loss during the period from 1851 to 2010 can be attributed to anthropogenic causes; this percentage is much higher (69 ± 24%) for the ice loss from 1991 to 2010 (Marzeion et al. 2014). Satellite-based remote sensing campaigns have shown that glacial lakes on the Greenland ice sheet are now prolific in areas without lakes in the recent past. Many of these lakes are draining areas of highest mass loss, particularly in southwest Greenland (Selmes et al. 2011). Between 2005 and 2009, an average of 263 supraglacial lakes emptied rapidly annually. These drainage events deliver large amounts of meltwater to the base of the ice sheet and contribute to its increased velocity.

More than 1.4 billion people base their livelihoods on the waters in the Indus, Ganges, Brahmaputra, Yangtze, and Yellow rivers. These rivers have a pronounced seasonal discharge regime dominated by the South Asian and East Asian summer monsoons. Glaciers in the mountains fringing the Tibetan Plateau are an important source of water during low-flow seasons, especially in the Indus and Brahmaputra basins. Changing glacier extent will mainly change the amount of water stored, and likely also the seasonal distribution of river flow. However, ice melt may already have been a major component of runoff over the past centuries. Changes in the summer monsoon and in glacier extent in a warming climate may have important consequences for the soaring energy demands of India and China, both of which are actively expanding their hydropower electricity development. India has plans to build nearly 300 new dams in Himalayan catchments by 2030 to double current hydropower capacity (Grumbine and Pandit 2013). Hydrological modelling that considers scenarios of atmospheric warming predicts that the Indus and Brahmaputra basins are likely to be affected strongly by reduced water flows, putting at risk much more than hydropower projects. Such lower flows could compromise the food security of approximately 60 million people who depend on water from the Himalayas (Immerzeel et al. 2010).

A comparison of historic maps and satellite images reveals the rapid, although geographically uneven, decay of lakes in China both in terms of their number and size. More than 240 lakes completely disappeared and only 60 formed between the 1960s and 2006. Climate change could be responsible for the loss of lakes in the semiarid and arid parts of northern China, whereas land-use activities appear to be more important in the more densely populated southern part of the country (Ma et al. 2010). While only a few of these lakes may set the stage for new ice- or snow-related hazards, prominent and highly publicized attention has focused on the possibility of glacial lake outburst floods (GLOFs) from moraine- or ice-dammed lakes in the Himalayas and adjoining mountains over the past decades (Das et al. 2015). The few published estimates of the economic risks from GLOFs give first glimpses of the some of the possible impacts we can expect from present and future glacier melt. In the valleys of the Nepal Himalayas, for example, single reported GLOFs have caused losses of up to US$500 million (Shrestha et al. 2010).

14.2 Geomorphic Impact of Snow and Ice Hazards

14.2.1 Snow and Ice in the Hazard Cascade

Studying the geomorphic impacts and hazards of glaciers means understanding and estimating the relative contributions of ice, meltwater, and sediment dynamics. Glaciers can release large amounts of water and sediment to landscapes; Bering Glacier in the St Elias Mountains of Alaska alone delivered an estimated 925 km^3 of fine sediment to the Gulf of Alaska over the past 130 kyr, which is more than 7 Mm^3 yr^{-1} on average and among the higher contributions of solid material reported (Montelli et al. 2017). The rapid release of glacier-derived sediment can form unstable deposits and make landforms prone to episodic reworking or collapse. Most glacier-related hazards involve the sudden motion of large masses of water, snow, ice, or debris, or some combination thereof. Many of these abrupt shifts result from the gradual build-up of instabilities as ice slowly deforms at rates that are rarely visible to the naked eye. Surging glaciers are spectacular exceptions, and involve the rapid advance of flowing ice masses over several months to several years (Figure 14.1). In the semiarid Central Andes of Argentina, for example, glaciers advanced at average speeds of up to 35 m d^{-1} that were sustained over several weeks, and pushed the

glacier snout some 3 km downvalley (Bitte et al. 2016).

Given the large area and poor accessibility of the environment in which these hazards operate, remote sensing has become the premier tool for measuring rates, frequencies, and magnitudes of moving snow and ice (Kääb et al. 2005). However, local measurement campaigns that combine modern time-lapse photography, UAV-based photogrammetry, terrestrial laser scanning, and traditional field mapping, can produce short, but detailed, time series of how snow and ice shape landforms. For example, Kochel et al. (2018) recorded the deposition of snow and ice avalanches, rock falls, slush flows, and debris flows on more than a dozen ice-rich debris fans in Alaska and New Zealand (Figure 14.2). These cone-shaped sedimentary landforms form mainly below rock cliffs topped by small ice caps or hanging glaciers. Icy debris fans are common in many mountain regions, but have eluded detailed study. The sediment budget by Kochel et al. (2018) takes in several hundred to more than 2000 deposits that were emplaced on single fans over nine months to two years. These ice-rich deposits covered and reworked the fan surfaces between three and 40 times during that short period, and contributed up to half of a given fan volume in places, also feeding large fractions of ice to valley glaciers below. This study illustrates nicely that we need to learn more about all landforms of the sediment cascade to fully appreciate the rapid turnover of snow, ice, and sediment.

14.2.2 Snow and Ice Avalanches

The spontaneous detachment of snow along inclined weak layers in the snowpack leads to avalanches that range in size from a few cubic metres to several tens of thousands of cubic metres and travel at speeds greater than 1 m s^{-1}. Snow avalanches occur frequently in hilly and mountainous terrain where snow may attain depths of many metres. Deep snow is rarely required for generating destructive

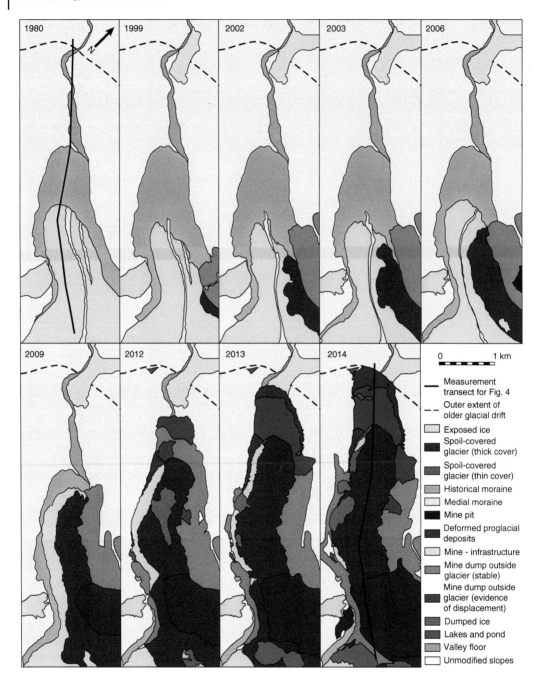

Figure 14.1 Rapid advance of a glacier in response to mine-related debris loading. Time series of land-cover classification for Davidov Glacier, Kyrgyzstan, from 1980 to 2014. After Jamieson et al. (2014).

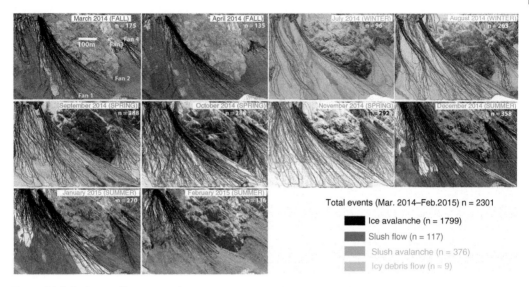

Figure 14.2 Delivery of ice and sediment to a valley glacier decoupled from an ice cap in New Zealand. Monthly summary maps of deposits on icy debris fans at Mueller Glacier, New Zealand, March 2014 to March 2015 using time-lapse photography. From Kochel et al. (2018).

and deadly snow avalanches, although maritime mountains in Europe, North America, Japan, New Zealand, and elsewhere can have seasonal snowpacks more than ten metres thick. Russia's Kuril and Sakhalin islands north of Japan receive abundant snowfall and have been among the most deadly places on Earth in terms of snow avalanches (Qiu 2014). Avalanches can affect transport networks, sometimes closing roads or railway lines for several days, sometimes cutting off entire communities, and causing economic losses. In the worst of incidences, avalanches have buried or severely damaged buildings. Heavy snow loads from snowfall may cause buildings to collapse, and avalanches exert far greater pressures on structures due to impact forces. Snowpacks are multiphase systems that must be thoroughly understood to assess avalanche hazards and to predict the forces that are large enough to uproot and entrain trees, destroy buildings, and transport boulders several metres in diameter.

Field observations show that wet snow avalanches can erode soil and bedrock (Figure 14.3). Specific sediment yields reported for wet snow avalanches range from 10^{-2} to 10^2 t km^{-2} yr^{-1}, and are comparable to moderate rates of soil erosion reported from agricultural landscapes. Soil erosion by avalanches is highly seasonal and in the long term depends on how well we can extrapolate measurements made mostly during a single late winter or early spring (Korup and Rixen 2014). Some researchers thus prefer expressing erosion rates or sediment yields for each deposit (Freppaz et al. 2010). Avalanches can bury sections of river channels and collapsing snow bridges holding up river water may produce rapid slush flows, a special type of water-saturated snow avalanche. Slush has a density between that of water and ice (0.9–1 t m^{-3}) and thus may have sediment concentrations similar to those of debris flows (Decaulne and Sæmundsson 2006).

The repeated physical impacts by snow avalanches that drop onto valley floors from steep hillslopes can create potentially diagnostic landforms such as avalanche plunge pools or avalanche tarns (Smith et al. 1994). Other landforms that result from snow avalanches include protalus ramparts and avalanche cones. Snow avalanches are also an important source of snow and water for many glaciers and rock glaciers (Humlum 2000).

(a)

(b)

(c)

Figure 14.3 Geomorphic impacts of avalanches. Top left: Avalanche tracks slicing through forest in the Kananaskis River valley, Rocky Mountains, Alberta (John Clague). Top right: Avalanche cone, Italian Alps (John Clague). Bottom: Snow-avalanche cone covered by decimetre-thick layer of organic debris and forming an ephemeral bridge over a small mountain stream (Oliver Korup).

Ice avalanches occur from the toes of ice caps and glaciers, especially those terminating on steep rock slopes. Anecdotal evidence holds that, similarly to other types of mass movement, smaller ice avalanches appear to be more frequent than larger ones, though systematic measurements may shed more light on the frequency and magnitude of ice avalanches in relation to the seasonal mass balance and movement rates of the ice sources. Van Der Woerd et al. (2004) reported six large ice avalanches triggered by the 2002 M 7.9 Kekexili earthquake in the Kunlun Range fringing the northern Tibetan Plateau. The volumes of ice that detached during strong seismic ground shaking ranged from 0.1 to 1 Mm^3. Yet evidence of these catastrophic failures was gone about two years after the event.

One of the earliest documented examples is the Allalin Glacier avalanche in the Swiss Alps in 1965 (Evans and Clague 1994). On 30 August of that year, and without warning, about 1 Mm^3 of ice at the terminus of the glacier broke off, cascaded down a steep rock slope, and struck a construction site below; 88 people working and living there were killed. Subsequent investigations showed that the Allalin avalanche occurred following a rapid advance of the glacier that started 2–3 weeks earlier, pushing its toe onto a steep portion of the slope. Mountaineering expeditions frequently report ice avalanches from otherwise remote terrain: in 2014, collapsing seracs (small towers of ice) caused a major ice avalanche that killed 16 climbers in the Khumbu Icefall below Mount Everest, Nepal. However, glaciers on more gentle slopes can also detach catastrophically: a huge ice avalanche from an unnamed glacier in the Aru Range in western Tibet on 17 July 2016 killed nine people and hundreds of animals (Kääb et al. 2018). Satellite imagery acquired shortly after the disaster shows that a large portion of the lower part of the glacier (40–70 Mm^3) slid away from its substrate and moved down a narrow valley before spilling out onto, and largely covering, an alluvial fan (Figure 14.4). The part of the glacier that sheared off had an average surface slope of around 13°, which is very low for a failure of this type. Increased ice flux due to a surge in the months before the avalanche may have trapped water beneath the glacier, elevating basal water pressures and causing the glacier to fail on its bed. Alternatively, conditions at the base of the glacier may have changed from cold-based to warm-based. Many hanging glaciers in the European Alps may be sources of similar, though less common, hazards.

The few detailed accounts of catastrophic ice avalanches or mixed ice–rock avalanches (Schneider et al. 2010) reveal another poorly understood natural hazard in cold regions. This is because ice–rock avalanches are rare, and their deposits are ephemeral and disappear before being detected or investigated in sufficient detail. Large amounts of snow and ice in a ice–rock avalanche alter the basal friction properties and thus eventually runout characteristics (Sosio et al. 2012). The 2002 Kolka–Karmadon ice–rock avalanche in North Ossetia-Alania unexpectedly and spectacularly demonstrated the destructive capability of such mixed mass movements. The avalanche most likely began as a rockslide that shaved off part of the Kolka glacier tongue, entraining large amounts of ice and transforming it into an extremely rapid ice–rock avalanche that travelled 19 km in only a matter of minutes, obliterating the valley floor en route (Evans et al. 2009b). A similar cascade of fall and flow phases involving ice, debris, and water characterizes the few other reported mass movements of similar character. A M 7.9 earthquake in Peru in 1970 caused a rockfall from the summit of Nevados Huascaran, Cordillera Blanca, that impacted a hanging glacier, surrounding snow fields, and a moraine. This mixture of failed materials transformed into a catastrophic rock–snow–ice avalanche, eventually burying the towns of Yungay and Ranrahica about 15 km from the mountain. About 6000 people died; the exact number remains unknown and estimates largely rely on projections of population densities at the time. Investigations revealed that a smaller aseismic event in 1962 had very similar runout dynamics. Although this precursor bypassed Yungay, it devastated the nearby town of Ranrahica. Solid evidence also points to a very much larger rock avalanche from Huascaran around 50 000 years ago. The areas obliterated by these rapid mass movements in 1962 and 1970 have now been resettled, resulting in some unmanageable and possibly unacceptable risks (Evans et al. 2009a).

Confusing moraines and landslides High mountains store many seemingly chaotic deposits that can be hard to tell apart, especially moraines and landslides. Confusing moraines with landslides leads to misinterpretations of past climate and landslide activity. Two examples from the Southern Alps of New Zealand illustrate this.

Firstly, 'The Hillocks' along the Dart River is a hummocky deposit protruding from the floodplain, and was for long taken to be a glacial kame field. If this landform is a remnant of a long-past glaciation, it reveals little, if anything, about hazards. A more recent interpretation is that

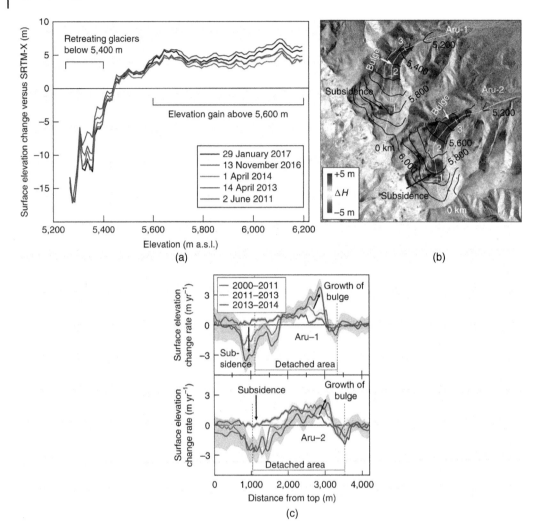

Figure 14.4 Rapid collapse of two Tibetan glaciers. Glacier thickness changes over the Aru region and Aru glaciers since 2000. (a) Glacier surface-elevation changes between SRTM-X (2000) and various TanDEM-X DEM dates over non-surging glaciers plotted against elevation. (b) TanDEM-X-derived surface elevation change (ΔH) map between June 2011 and April 2013. (c) SRTM-X and TanDEM-X surface elevation change rates over Aru glacier profiles (in b) showing bulging and bulge propagation over the period 2000–2014. Simultaneous bulge growth and frontal thinning and retreat steepened the glacier front. The line shading indicates one standard deviation uncertainty. From Kääb et al. (2018).

these hills are part of a rock-avalanche deposit (Figure 14.5) (McColl and Davies 2011), which has a consistently sized source area on the hillslope above. The mounds are linearly aligned, conical, contain highly crushed angular debris, and become consistently smaller away from the assumed landslide source.

Secondly, the Waiho Loop of Franz Josef Glacier has long been assumed to be a terminal moraine emplaced during a glacier advance in the Younger Dryas glacial period ~12 kyr ago. The Waiho Loop is about 100 m high, extends some five kilometres across the Waiho River outwash plain, and has a volume of at

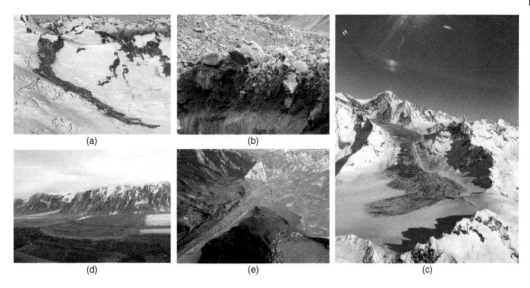

Figure 14.5 Examples of rock avalanches that fell onto or interacted with glaciers. (a) The Jubilee rock avalanche in the southern Coast Mountains, British Columbia, left a tongue-shaped lobe of debris on a glacier (John Clague). (b) Rock avalanche debris resting on surface of Black Rapids Glacier, Alaska (person for scale) (John Clague). (c) 2014 La Perouse rock avalanche, Alaska (John Clague). (d) Rock avalanche deposit on Black Rapids Glacier, Alaska (John Clague). (e) Icy debris of 2002 landslide in Karmadon Depression, view to the north toward the Skalistyl Range (Igor Galushkin). Left foreground: road toward the buried village of Nizhniy Karmadon disappears under the debris mass. Large temporarily dammed lake formed in the village of Gornaya Saniba is visible to the right of centre.

least ~0.1 km³. Tovar et al. (2008) revealed that most of the moraine's constituent rocks were both angular and composed of sandstone, a lithology that outcrops only in the upper Waiho valley. These characteristics suggest that the moraine is made up of rock-avalanche debris. This finding opens up the possibility that the Waiho Loop was caused by a landslide instead of climate cooling. The emplacement of rock-avalanche debris on the ablation zone of the glacier would have reduced total ablation, altering the mass balance and causing the glacier to thicken enough to build the large terminal moraine. The large size of the moraine implies a large rock avalanche and the corresponding availability of a large volume of supraglacial debris to form the moraine.

What does this reinterpretation have to do with hazards? It is likely that the glacier was much farther advanced than today when the loop was emplaced (Shulmeister et al. 2009). Recent work has shown that the terminus was at the Loop position when the rock avalanche happened (Alexander et al. 2014). Modelling suggests that a landslide of some 10 Mm³ covering the glacier's ablation zone with one to two metres of debris could cause the glacier to advance almost to the present-day town of Franz Josef Glacier (Reznichenko et al. 2011). Although such an advance would hardly impact the township and local infrastructure, the associated river aggradation might be unmanageable. The Franz Josef Glacier advanced about 1.5 km between 1982 and 1999, and the forefield aggraded by 15–20 m in response due to sediment being eroded and redeposited by the subglacial water system. Aggradation on this scale would render the township and coastal highway unusable.

Motivated by the historic and sedimentary evidence pointing to at least several such hazard cascades in the Cordillera Blanca, Somos-Valenzuela et al. (2016) specified a similar scenario with coupled numerical models that addressed the sudden detachment of rock and ice avalanches above Lake Palcacocha; the generation and dynamics of resulting impact waves in the lake; the wave overtopping and potential breaching of the terminal moraine that impounds the lake; the routing of thus generated outburst flows downvalley; and the eventual inundation of populated areas. The authors conceded that outputs of this model can be next to impossible to validate with the few reliable field data available, but stress that such outputs offered sufficient detail to define worst-case scenarios at least. For example, the model predicted that an artificial lowering of the lake level by as much as 30 m could reduce the most probable area of inundation by a glacial lake outburst by about a third in settled areas downstream.

The examples highlighted here are part of a larger spectrum of mass movements involving water, snow, ice, and debris. Snow and ice caps on volcanoes can also be melted or mobilized rapidly during even minor eruptions. In 1985 a small eruption of Nevado del Ruiz, Colombia, caused rapid snow and ice melt on the volcano summit, generating a series of highly mobile lahars that devastated the town of Armero, killing more than 20 000 of its 29 000 inhabitants (Voight 1990). In spite of months of warnings of the possibility of such events, the order to evacuate never came because of an impending local election. Such lahars and similar hazards that are native to ice-clad volcanoes underscore the importance of combining research methods of glaciology, volcanology, and geomorphology. Glaciologists generate data on mass balances and melt rates that could feed into models of edifice unloading, whereas volcanologists deliver data on how volcanic slope deformation might tilt local slopes and promote more ice avalanches;

finally, geomorphologists can predict (or reconstruct) the runout and impact of the resulting ice avalanches, lahars, or otherwise mixed mass flows.

14.2.3 Jökulhlaups

Warm-based glaciers discharge meltwater and sediment continuously, but occasionally glaciers do this in rapid bursts that can come without much warning. The Icelandic word 'jökulhlaup' was coined for the sudden release of meltwater from a lake beneath a glacier because of volcanic heating from below (Björnsson 2003). Today the word is applied more generally to a flood resulting from the partial or complete draining of water from any glacier-dammed lake or temporarily-blocked subglacial drainage system. Three mechanisms allow meltwater to drain from glacial lakes. Firstly, drainage can begin at pressures lower than the ice overburden in conduits that grow slowly as the ice walls melt by frictional and sensible heat in the water. Secondly, the level of the lake can rise until the ice dam lifts, allowing water under pressure to escape (Björnsson 2003). Thirdly, a fractured and weak glacier dam can mechanically fail due to the force exerted by the impounded water. The shape of reservoirs and the nature of processes involved during these floods can differ between locations, and Roberts (2005) distinguishes as many as seven different types of jökulhlaups.

Temporary subglacial or englacial water storage and release during joklhlaups are likely responsible for very flashy floods that issue from the mouths of tunnels at the snouts of valley glaciers. These floods can achieve discharges in excess of 1000 $m^3 s^{-1}$ within less than an hour (Russell et al. 2010). Floods caused by subglacial lake bursts have immense power and may have contributed to shaping some Pleistocene glacial river canyons. The spectacular eruption of the Icelandic subglacial volcano Grimsvötn in 1996 provided an opportunity to monitor in detail the processes and changing landforms associated with a large jökulhlaup. In historic times subglacial floods from Grimsvötn have occurred every 1–10 years, producing peak discharges between 600 and 50 000 $m^3 s^{-1}$ at the glacier margin; these floods lasted between two days and four weeks. The total volume of meltwater released during each of the floods ranges from 0.5 to 4.0 km^3. During the 1996 jökulhlaup discharge increased linearly with time and peaked at an estimated 45 000–54 000 $m^3 s^{-1}$ after 17 hours. The flood was over about 27 hours later, having moved 3.2–3.8 km^3 of water. Laser altimetry measurements showed that the jökulhlaup dumped 38 Mm^3 of sediment on the proglacial sandur plain (Magilligan et al. 2002). This volume is roughly half of the average annual lava volume that erupts throughout Iceland, at least judging from data for the past millennium (Thorardson and Larsen 2007). One lesson is that we need to acknowledge more thoroughly the role of rare floods in transporting volcanic sediment and modifying proglacial outwash plains. The Grimsvötn jökulhlaup has spurred some rethinking of the major formative processes and the sedimentology of sandur plains. Deposits of catastrophic floods and bedrock landforms carved by flowing water show that yet higher sediment transfers – of the order of 1 km^3 – may have been laid down during earlier, larger floods in Iceland (Björnsson 2003). The 2010 eruption of Eyjafjalla volcano, some 100 km to the west, released more than 140 jökulhlaups. Rarely did the amount of sediment transported scale with the amount of water released during these flows, mainly because of the highly varying concentration of solids. Detailed terrestrial laser scanning before and after the eruption also highlighted that the many different flows completely filled a former proglacial lake basin with a fan of debris from the floods. Some flows reworked this fan, so that its overall form yielded litle about the relative geomorphic contributions of even the largest jökulhlaups (Dunning et al. 2013).

14.2.4 Degrading Permafrost

Permafrost is frozen ground – more properly a subsurface condition characterized by below-zero temperatures for at least two consecutive years. The definition of permafrost has broadened over the years, but the term still refers to a thermal state rather than a material substrate (Dobinski 2011). Despite a general agreement on the continental-scale distribution of permafrost based on boreholes, excavations, or actively deforming landforms, its regional distribution is much less well known, especially at its equator-ward and lower limits, where frozen ground forms discontinuous or sporadic patches.

Boreholes in permafrost areas offer time series of ground temperatures over periods of many years, but often require a high degree of logistics and maintenance. Hence, scientists mainly infer the distribution of permafrost from spatially extrapolating these data, and also resort to using diagnostic landforms and numerical models. Patterned ground, solifluction lobes, stone stripes, and rock glaciers are among the tell-tale landforms, and their regional and altitudinal distribution has helped to map the distribution of permafrost.

Rock glaciers as thermometers of frozen ground Rock glaciers are among the most striking and impressive landforms in permafrost regions and consist of a mixture of debris and ice that flows downhill under the influence of gravity (Figure 14.6). Elevations of the toes of rock glaciers and potential incoming solar radiation are two of the most widely used predictors for modeling where mountain permafrost occurs, as these two measurements can be determined from most digital elevation models (Brenning and Trombotto 2006). Rock glaciers can move several metres per year, but most advance much slower. Rock glaciers can have various origins, and scientists prefer to see rock glaciers as part

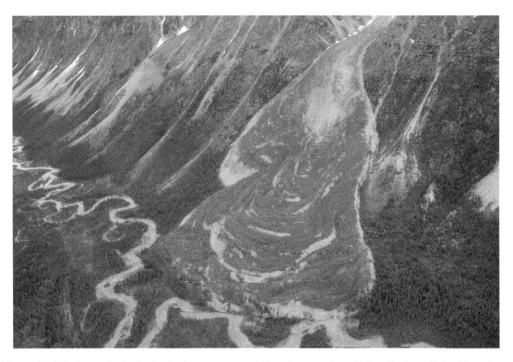

Figure 14.6 Active rock glacier impinging on a meandering river, southern Yukon Territory (John Clague).

of a continuum of flowing ice–debris mixtures that also include clear-ice glaciers and debris-covered glaciers. We know very little about rock glaciers as potential hazards: a handful of studies have reported the sudden collapse of the steep toes or sections of rock glaciers, generating debris flows or flash floods similar to small jökulhlaups (Iribarren Anacona et al. 2015). Collapsing rock-glacier fronts can give rise to landslides and debris flows that form fans dotted with molards, which are 'conical-shaped, often symmetrical debris mounds with a distinctive radial grain size gradation' (Milana 2015). Molards on fans in the semiarid Andes and some central Asian ranges may be a sign of the decay of rock glaciers or debris-covered glaciers.

The toe elevations of actively flowing or deforming rock glaciers found widespread use as a proxy for the lower limit of sporadic mountain permafrost. The accuracy of this proxy hinges on the assumption that only the local freezing level determines the distribution of active rock glaciers, and that ice or debris supply are negligible controls. Another assumption is that rock glaciers are in equilibrium conditions with their surrounding climate. Statistical and numerical models make use of these proxies, but also include direct measurements of solar radiation, topographic hillslope aspect, and temperature fields.

Much of the geomorphic response to degrading permafrost occurs slowly and without any determinable onset (Figure 14.7). Exceptions include sudden slope failures in frozen soil or rock. The large number of rockfalls in many high mountain regions during several exceptionally hot summers in the early twenty-first century has fuelled research on the role of mountain permafrost in slope stability (Gruber and Haeberli 2007). Many of the more recent catastrophic rockfall and rock slide studies in alpine terrain now feature some close scrutiny of temperature and rainfall time series prior to and after the failure(s) happened to check for a connection with possible degrading

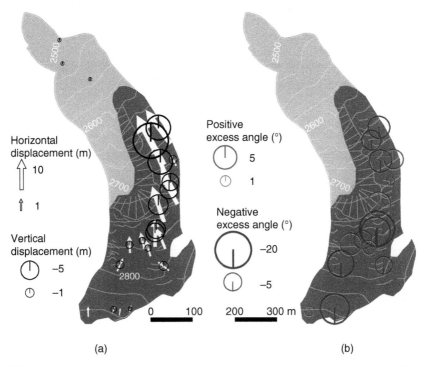

Figure 14.7 Total displacements (in metres) of the Bérard rock glacier, French Alps, and its collapsed mass (light grey) measured by differential GPS between June 2007 and September 2010. (a) Horizontal and vertical components of displacement. (b) 'Excess angle' (difference between the actual slope angle of movement of a block and the local surface slope angle). From Bodin et al. (2016).

permafrost. Adequately testing this hypothesis should also involve studying slope failures that take place in the colder season (Phillips et al. 2016) to avoid slope-stability models which are overly calibrated towards warming trends. Field measurements and numerical models are based on the assumption that permafrost thaw may destabilize rock slopes, although researchers are just beginning to understand in detail the underlying causes and triggers. Laboratory experiments with shear boxes can inform researchers about the stiffness and strength of ice-filled joints and how these strength properties depend on both normal stress and temperature (Davies et al. 2000). Hence the Factor of Safety of a rock slope containing ice-filled joints is likely to decrease with increasing temperature, depending on how deep into the rock thermal changes are able to propagate.

Considering the effect of ice-filled fractures, Krautblatter et al. (2013) proposed a modified version of the classic Mohr–Coulomb failure criterion for rock slopes that involves fracturing of rock bridges, friction along rough fracture surfaces, ductile creep of ice, and detachment mechanisms along the boundary between rock and ice. The model suggests that rock mechanical properties chiefly control larger rock-slope failures, whereas ice mechanical properties are more important for smaller failures in slopes with degrading permafrost. Alpine permafrost may increase shear strength in rock slopes by cryostatic and also potentially hydrostatic pressures and ice segregation. In contrast, creep and fracture of ice in rock joints, failure of rock-ice contacts, friction along rock fractures, and fracture of intact rock bridges may lower the shear strength of alpine rock slopes with permafrost.

Degrading permafrost is one of the motors of geomorphic change in many soil- and sediment-mantled low-relief regions in Arctic Siberia, Alaska, and Canada. The term 'thermokarst' refers to the development of depressions on the land surface due to thawing of ice-rich permafrost or melting of massive ground ice or ice wedges. In parts of the Beaufort Sea coastal plain in Alaska and northwestern Canada, degradation of ground ice has occurred since the early 1980s, leading to the establishment of new water-filled depressions over 4% of the area; in places more than 1300 lakes formed per square kilometre (Jorgenson et al. 2006). Some of these lakes appear to result from explosive gas emissions, creating deep craters (Chuvilin et al. 2020) – a potentially hazardous process to people and industries located on permafrost. Record-breaking increases of 2–5 K in ground temperatures since the 1980s appear to be responsible for this major geomorphic and hydrologic turnover in parts of the Arctic landscape. How these changes might eventually affect natural hazards is unclear and deserves detailed research. More water bodies in Arctic landscapes, for example, could contribute to reducing the areas burned by wildfires.

Thermokarst processes also involve the horizontal temperature-driven erosion of ice, often aided by surface runoff and undercutting by rivers and waves. Wave erosion, in particular, is effective in reshaping many permafrost coasts that make up about one-third of the world's coastlines. Günther et al. (2013) distinguished two types of erosion responsible for the retreat of permafrost-rich Arctic coastal cliffs: on the one hand, 'thermo-denudation' operates at the top of cliffs and is driven by the effects of solar radiation and heat advection; on the other hand, 'thermo-abrasion' attacks cliff toes by wave impacts and the melting of ground ice. These two erosion processes cause frequent slope failures that largely determine the net rates of cliff retreat. In the Laptev Sea of eastern Siberia, for example, these landward losses took place at average rates of 0.5–6.5 m yr^{-1} over the past 50 years, with estimated regional averages closer to the lower value (Günther et al. 2013). Most estimates of current Arctic coastal erosion rates rely on measuring changes using air photos and satellite images. The few detailed field studies concerned with how Arctic coastlines change emphasize that local erosion rates vary considerably in response to ice content, topography, wave regime, and other factors.

Thermokarst also undermines river banks, especially those developed in 'yedoma' soils that contain mostly silts, organics, and ground ice. Once destabilized, these permafrost soils can turn into highly productive point sources of fine sediment. Kanevskiy et al. (2016) calculated from detailed field measurements over a four-year period that a single actively undercut cliff released nearly 0.07 Mt of soil per year into the Itkillik River, a meandering lowland stream in northern Alaska. This is nearly 10% of the bulk sediment mass estimated to reach the Beaufort Sea, if coastal deltas are neglected as major sinks. These estimated erosion rates are some of the highest reported when compared to other Arctic rivers. Most local measurements of rates of river bank retreat span a few years to decades and mostly range from about 1 to 20 m yr^{-1}. The highest rates are associated with substrates containing high amounts of fines and ice; the fraction of ice contained in these volumetric estimates is often approximate. LANDSAT satellite images offer several decades worth of information on such changes, and automated analyses confirm that some coastal cliffs in the Lena Delta retreated by up to 20 m yr^{-1} between 1999 and 2014 (Nitze and Grosse 2016).

A systematic sediment-budget study showed that a small thermokarst gully that formed in 2003 in a 0.9-km^2 catchment in the headwaters of the Toolik River in the foothills of the North Slope of Alaska delivered more sediment to the river than the much larger, adjacent upper Kuparuk River catchment did in 18 years. Measured concentrations of ammonium, nitrate, and phosphate were higher downstream of this gully (Bowden et al. 2008). Similarly, Jolivel and Allard (2013) observed that, between 1957 and 2009, more than 20% of the discontinuous permafrost had vanished in the Sheldrake River catchment in eastern Hudson Bay, Nunavik, Canada, and that during that period the volume of material eroded by gullies and landslides increased by 12–38%.

Another source of sediment involves active layer detachments. These shallow landslides initiate at the interface between frozen and unfrozen ground on slopes as low as 5° (Lewkowicz and Harris 2005), and may feed more sediment and nutrients to Arctic streams, thus changing nutrient loads and affecting aquatic ecosystems. Data on such inputs remain sparse, but will be an important addition in understanding the effects on the sedimentary and biogeochemical impacts of a degrading active layer in the Arctic (Bowden et al. 2008).

14.2.5 Other Ice Hazards

Drifting sea ice is a circumpolar hazard that can disrupt navigation and affect coastal infrastructure. The sudden collapse of shelf ice or the sudden capsizing of icebergs may release sufficient energy to produce tsunamis. A simplified physical model supported by numerical experiments shows that icebergs breaking off Antarctic ice shelf may create tsunami wave heights up to 1% of the initial iceberg height before it capsized. Given the estimated size distribution of icebergs in the region, tsunami waves caused by collapse of shelf ice could be some four metres high (Levermann 2011).

Rivers in the North American and Eurasian Arctic are prone to ice-jam floods. There, the northward advancing melt season during spring and early summer causes the flow of major northward-draining Arctic rivers to be backed up behind still-frozen river ice in their more northern downstream reaches. The resulting ice jams back up the flow, such that it enters the floodplain where it may cause widespread serious flooding both upstream and downstream of the jam. In 2001, an ice-jam flood on the Lena River near Lensk, Russia, for example, affected more than 30 000 people, and destroyed more than 3000 houses. Local measurements made over several years illustrate that, on average, ice-jam floods attained the same inundation level more often than floods without ice (Beltaos and Prowse 2009). Ice forming along river banks can gradually bind soil particles into masses that can subsequently be plucked en masse with the ice. Drifting pieces of river ice can erode and transport sediment at much lower discharges than can rivers without ice cover. This process is known as 'rafting' and may entrain more than 100 kg of sediment per m^3 of ice. Some rivers may transport most of their sediment during the spring season when ice breaks up (Turcotte et al. 2011). In contrast, ice sticking to channel banks and beds may reduce the availability of sediment for parts of the year, thus accentuating the transport of solids over short periods of thawing. Where floodwaters form isolated scour holes or distinct gullies on the floodplain, new channel cutoffs may initiate after several ice-jam floods (Smith and Pearce 2002). If shown to be widespread, this process is at odds with the view that many of the oxbow channels in Arctic floodplains result from slowly meandering channels; instead seasonal ice-jam floods could be the major driver for frequent channel avulsions and chute cutoffs. However, few studies report long-lived effects of river ice on channel geometry. Rapid break-up of ice jams releases flood waves called 'javes' (short for jam-release waves) that carry large amounts of sediment, including boulders several metres in diameter. The few measurements available feature distinct spikes in suspended concentrations approaching 1 g l^{-1}. Despite case studies that highlight the hazard and geomorphic impacts of ice-jam floods, and despite the fact that nearly 60% of rivers in the Northern Hemisphere are prone to seasonal freezing, the amount and dynamics of river ice remains largely unknown (Beltaos and Prowse 2009).

14.3 Geomorphic Tools for Reconstructing Past Snow and Ice Processes

Unraveling the past history of Earth's cryosphere and its relation to palaeoclimate, sea-level change, landscape evolution, and water resources has been a key element of geosciences for decades. Countless papers detail the geomorphic work done by glaciers in shaping high-mountain and high-latitude landscapes. Likewise, many authors have stressed the importance of glaciers in conditioning rates of geomorphic processes long after deglaciation, which is encapsulated in the 'paraglacial' concept in which past glaciations determine adjustments in landscapes and many of the processes operating on them over subsequent millennia (Figure 14.8) (Ballantyne 2002). Exponentially decaying sediment yields from formerly glaciated catchments support this concept, as do assessments of slope stability along glacially oversteepened valleys (McColl 2012). One interpretation is that geomorphic process rates are determined by both the glacially-limited release of material and the work-rate of the denudation processes, and can be described by an exhaustion model. Cosmogenic exposure dating has assisted in attempts to date large rock-slope failures from glacial valleys, which many interpret as a direct geomorphic response to deglaciation (Cossart et al. 2008). However, Ballantyne and Stone (2013) reported 47 cosmogenic nuclide exposure ages ranging from about 17 to 1.5 ka from the deposits from the deposits of 17 catastrophic rock-slope failures in the Scottish Highlands. These ages are random in time without a noticeable exponential decay since the last major glaciation. In general, deglaciation results in the exposure and reworking of unstable sediments that were originally transported by flowing ice. Paraglacial sediment storages may provide large volumes of readily erodible sediment, leading to high rates of sediment transport during periods of glacier or ice-sheet expansion. Few studies concerned with paraglacial geomorphology address how long it takes a formerly glaciated landscape to revert to nonglacial conditions, and exhaustion models rarely pin down when glacial preconditioning of modern erosion, transport, and deposition becomes negligible.

A large body of work has documented the geomorphic impacts of glacial lake outburst floods (GLOFs, see Chapter 11.2.2 for more detail) (Figure 14.9). These non-meteorological floods may completely change downstream channels, widen floodplains, and remove forest. They have caused widespread destruction in many mountain belts mostly because they happen without any alarming precursors let alone sufficient warning time (Figure 14.10) (Clague and Evans 2000). The Himalayas host more than 2000 glacier-dammed lakes, of which dozens have catastrophically emptied since the beginning of the twentieth century. Similar numbers are reported for other less known mountain belts such as the Kyrgyz Tien Shan, where at least 70 lakes have catastrophically burst since the early 1950s (Janský et al. 2010). Assessing the potential for lake outbursts in such vast areas is an effort that requires a combination of remote sensing, numerical modelling, and field study. Despite satellite monitoring of Himalayan glacier lakes, the debate continues about which metrics are the most useful and reliable for predicting future GLOF hazards.

Researchers have stressed also the growing numbers of debris flows associated with the continuing shrinkage of alpine glaciers. Chiarle et al. (2007) documented the important antecedent rainfall conditions for debris flows in the glacierized catchments of the northwestern Italian Alps. Recent events in this region followed intense and prolonged rainfall that saturated glacigenic sediments, culminating in debris flows that mobilised up to 0.8 Mm^3 of sediment. Short, high-intensity rainstorms can destabilize glacial drainage systems, triggering debris flows of the order of 0.1 Mm^3. They can also trigger outbursts from glacial lakes or melt of surface or buried ice.

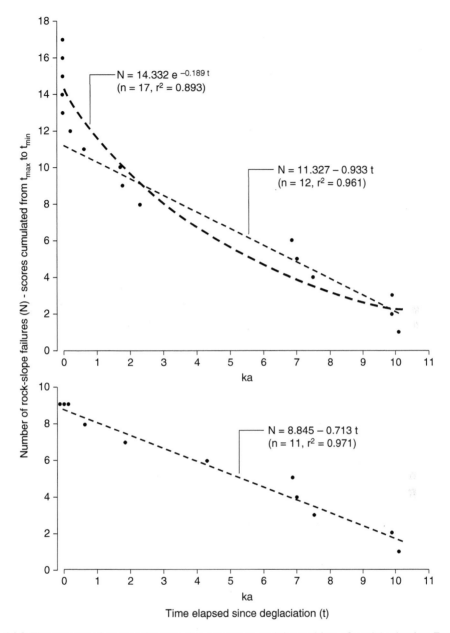

Figure 14.8 The concept of paraglacial rock-slope response and the problem of model selection. Top: cumulated (youngest to oldest) frequency of 17 rock-slope failures in the Scottish highlands plotted against time elapsed since deglaciation. The best-fit model treats the five 'rapid response' ($t \approx 0$) sites as separate from the others, which exhibit roughly constant periodicity. A negative exponential curve has a seemingly poorer fit. Bottom: cumulated (youngest to oldest) frequency of 11 rock-slope failures located within the limits of Younger Dryas glaciation, plotted against time elapsed since deglaciation. From Ballantyne (2013).

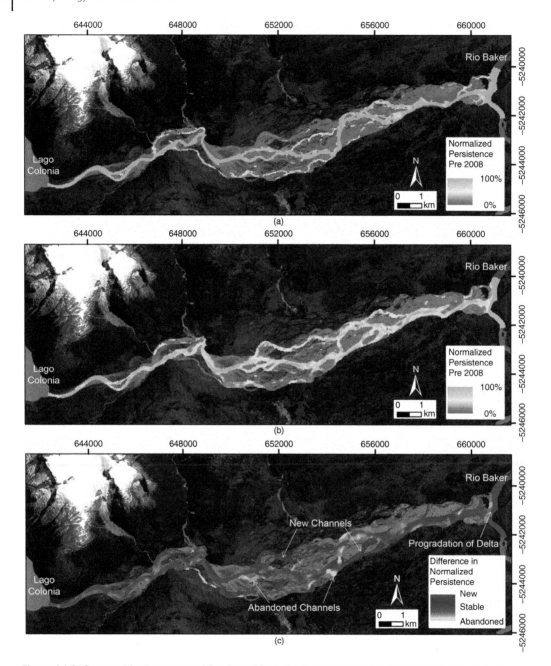

Figure 14.9 Geomorphic changes resulting from 21 glacial lake outburst floods (GLOFs) on Rio Colonia, Chile. The floods began in April 2008. Channels (a) in 2008, prior to the GLOFs and (b) after the GLOF episode. (c) Difference in channel persistence before and after 2008. New channels are indicated by magenta colors. Abandoned channels and other areas that were once water and are now land (e.g. progradation of delta) are indicated by light blue colours. Continuously active channels are indicated by dark blue colours. Background is Landsat 8 panchromatic image acquired on 21 January 2015. From Jacquet et al. (2017).

Figure 14.10 Digital image of Nostetuko Lake and the breached, Little Ice Age moraine of Cumberland Glacier, Canada (view towards southeast). The snout of Cumberland Glacier detached and fell into the lake in July 1983. The displacement wave generated by the collapse overtopped and incised the moraine. The digital image consists of an orthophoto, based on July 1994 aerial photographs, draped on a digital elevation model obtained from the same photos. Changes in surface elevation between July 1981 and July 1994 are shown in colour tones. Dramatic surface lowering is seen at the snout of Cumberland Glacier (the blue and purple area behind Nostetuko Lake). Surface lowering at Nostetuko Lake records the fall in lake level due to failure of the moraine dam. From Clague and Evans (2000).

14.4 Atmospheric Warming and Cryospheric Hazards

The warming cryosphere favours hazards induced by snow and ice melt and the failure of natural dams formed by glaciers or moraines (Figure 14.11). Yet quantitative predictions are rare as researchers are still struggling with collecting sufficiently robust data (Figures 14.12 and 14.13). Although increasing atmospheric temperatures seem to have triggered a notable increase in geomorphic process activity such as rockfalls, glacier lake outburst floods, and coastal erosion in permafrost, the key problems that remain are to check whether these events stand out from a poorly constrained set of background events, and, if so, whether these events can be directly attributed to climate change. One strategy is to use well documented changes over past decades as possible templates for estimating future changes (Figure 14.13).

Thinning and recession of mountain glaciers creates new unvegetated and debris-rich forelands, which are a source of sediment to streams, but also provide abundant space in mountain valleys where sediment can be temporarily stored. Advancing glaciers can have even greater effects in increasing sediment supply to proglacial rivers (Davies et al. 2003). Ice-scoured depressions may fill with sediment and meltwater can be trapped behind moraines or glaciers.

Advancing glaciers in tributary vallies can cross the main valley to abut against the far wall and form a dam across the main river. Most such glacier dams are short-lived given the rates at which glaciers change. Their sudden failure may lead to catastrophic outburst floods and debris flows that cause damage downstream (Korup and Tweed 2007). Satellite images show that Himalayan glacial lakes in Nepal and Bhutan increased in size markedly over the period 1990–2009, whereas lakes in the western Himalayas and Karakoram changed little over the same period (Gardelle et al. 2011). This regional pattern is only partly consistent with changes in the glaciers themselves; satellite-based laser altimetry shows pronounced losses throughout the Himalayas from 2003 to 2008 (Kääb et al. 2013; Bolch et al. 2012). Some lakes are growingly rapidly in size,

Figure 14.11 Schematic diagram of a hazardous moraine-dammed lake. Potential triggers, conditioning factors, and key phases of a glacial lake outburst flood (GLOF) event are highlighted. Possible triggers include: (A) glacier calving, (B) icefall from hanging glaciers, (C) rock/ice/snow avalanches, (D) dam settlement or piping, (E) ice-cored moraine degradation, (F) rapid input of water from a supraglacial, englacial, or subglacial source, and (G) seismicity. Conditioning factors for dam failure include: (a) large lake volume, (b) low width-to-height dam ratio, (c) degradation of buried ice within the moraine, and d) limited dam freeboard. Key stages of a GLOF include: (1) progradation of displacement waves in the lake or piping through the dam, (2) breach initiation and formation, and (3) progradation of the resultant flood wave down-valley. From Westoby et al. (2014).

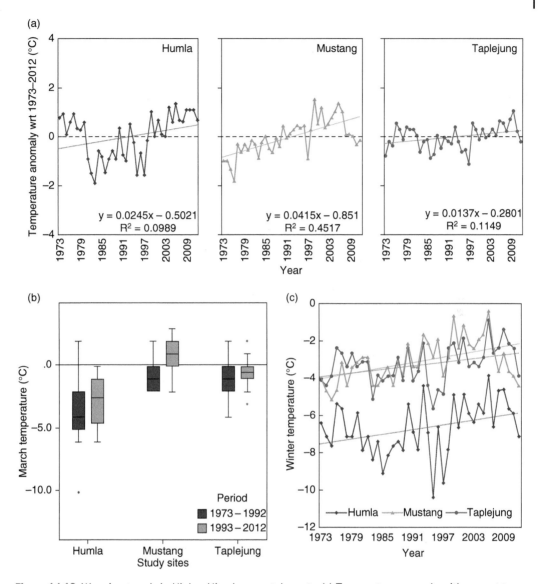

Figure 14.12 Warming trends in Higher Himalayan catchments. (a) Temperature anomaly with respect to 1973–2012 for the Humla, Mustang and Taplejung districts, Nepal. (b) Differences in March temperature (spring onset) between the 1973–1992 and the 1993–2012 period, for each district. (c) Trend in mean winter temperature from 1973 to 2012 for each district. From Uprety et al. (2017).

mainly by glacier melt and calving (Thompson et al. 2012). In the central Himalayas, for example, glacial lake area increased by ~17% between 1990 and 2010 (Nie et al. 2013). Yet the rate of lake growth alone does not reliably predict sudden dam breaks. Watanabe et al. (2009) studied the lake at the front of Imja Glacier, Nepal, which had attracted much

public attention because of its high historic rate of growth. They found that the lake level decreased by some 40 m from 1960 to 2006 and concluded that 'Alarmist prognostications based solely upon rapid areal expansion are counterproductive'; we might add that they are also rarely sufficiently correct (Figures 14.14 and 14.15).

Figure 14.13 Examples of glacier- and moraine-dammed lakes. (a) Moraine-dammed lake near Bishop Glacier, southern Coast Mountains, British Columbia. (b) Rapidly growing lake at the terminus of Klinaklini Glacier, southern Coast Mountains, British Columbia. (c) Breached moraine dam at Nostetuko lake in the southern Coast Mountains, British Columbia. The moraine was breached in 1983 by an overtopping wave triggered by an ice avalanche. (d) Glacier-dammed lake, Yukon (all photos by John Clague).

Recurrent outbursts from some glacier lakes may be linked to repeated opening and closing of subglacial conduits, thus giving the impression of growing outburst flood hazards (Dussaillant et al. 2009). Some researchers have proposed that the volume of water stored behind glacial lakes is one way to identify the most important sites for detailed investigations. Fujita et al. (2013) estimated the potential flood volumes for more than 2000 Himalayan glacier lakes and found that only about 50 of these lakes had volumes >1 Mm^3, which is roughly the threshold volume released during major historic GLOFs.

Several studies attribute apparent increases in outburst floods from glacier lakes to contemporary warming. The average frequency of reported GLOFs in the Yarkant River catchment, Chinese Karakoram, between 1959 and 1986 was 0.4 per year, but this frequency increased to 0.7 per year between 1997 and 2006 (Chen et al. 2010). Similar observations come from the extratropical Andes, where at least 31 lakes have emptied suddenly since 1780, giving rise to more than 100 catastrophic floods (Iribarren Anacona et al. 2015). The trend in these data may seem clear, but remains volatile given the small sample size, possible confusion of GLOFs with other floods triggered by processes independent of climate change, and possible undersampling in remote and sparsely populated mountain areas. Ng et al. (2007), however, used a thermomechanical model to infer that mean air

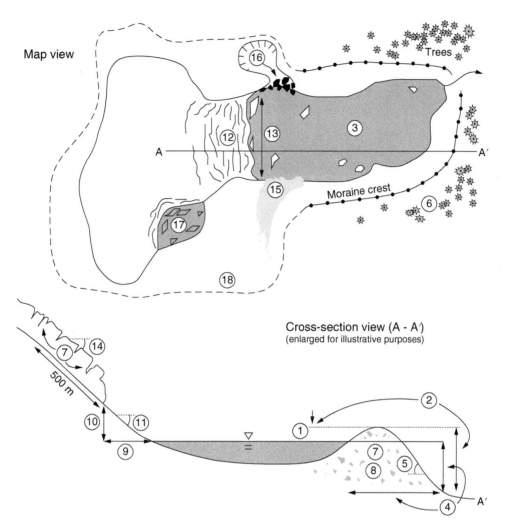

Figure 14.14 Summary of factors relevant to the stability of moraine-dammed lakes, including (1) lake freeboard, (2) freeboard-to-moraine crest height ratio, (3) lake area, (4) moraine height-to-width ratio, (5) moraine downstream slope steepness, (6) moraine vegetation cover, (7) ice-cored moraine, (8) moraine lithology, (9) lake-glacier proximity (horizontal distance), (10) lake-glacier relief (vertical distance), 11) slope between lake and glacier, (12) crevassed glacier snout, (13) glacier calving front width, (14) glacier snout steepness, (15) snow avalanches, (16) landslides, (17) unstable lake upstream, and (18) catchment area. From McKillop and Clague (2007).

temperature affected the peak discharge of recurring outburst flows from Merzbacher Lake, a water body dammed by South Inylchek Glacier in the Kyrgyz Tien Shan. If such a weather-dependent control on the magnitude of jökulhlaups holds for other sites, atmospheric warming may contribute to producing larger outburst events in the future, all other factors being the same.

Buildings on permafrost can be severely damaged if the ground thaws. The State of Alaska spends around 4% of its annual budget repairing damage caused by seasonal and permanent thawing of permafrost (Comiso and Parkinson 2004).

In the Yakutsk region a rise in soil temperature by 2 K has nearly halved the bearing capacity of frozen ground under buildings.

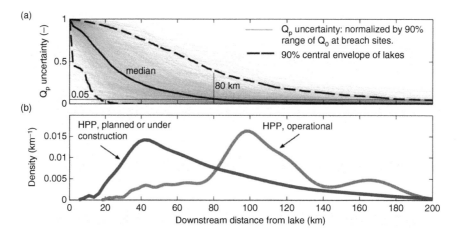

Figure 14.15 Downstream trend of uncertainties in GLOF simulations. (a) Uncertainty about local peak discharge Qp expressed as the 90% range of simulated values normalized to peak discharge Q_0 at the breach site. This uncertainty decreases downstream at different rates. For 90% of all lakes, the range of normalized Qp reduces by 2% to 90% along the first 20 km downstream of the lakes. (b) Distribution of distances of hydro-power project (HPP) sites from glacial lakes. Currently planned or constructed HPP are in river sections with higher uncertainty in peak discharge compared to operational HPP. From Schwanghart et al. (2016).

Differential ground deformation and settlement have affected hundreds of residential buildings, the local airport, and a power generating station to such an extent that Yakutsk city was declared a natural disaster area in 1998. If warming continues, most of the buildings in the city will be affected by permafrost-related ground deformation by the year 2030 (French 2013). On the southern Tibetan Plateau the lowest limit of permafrost shifted upwards by 80 m on average over the past 20 years, and the thickness of the active layer increased by up to 0.5 m in places between 1996 and 2001 (Yang et al. 2010). Permafrost thaw has accelerated also along the Tibet–Lhasa highway and railway, where thaw slumps and ground settlement compromise the stability of foundations. Ground-temperature measurements in the Tien Shan of Central Asia show that permafrost temperatures increased 0.3–0.6 K over the past 30 years, and the thickness of the active layer increased by 23% on average since the 1970s. Numerical models suggest that the lowest permafrost elevations shifted 150–200 m upward in the Tien Shan during the twentieth century (Marchenko et al. 2007).

In mountains throughout the world, concern is growing that the shrinkage of mountain glaciers and degradation of alpine permafrost might destabilize rock slopes and trigger rock falls, rock slides, and rock avalanches. The relation between climate change and landslide occurrence will surely invite much more detailed research in the future. Permafrost in rock is a thermal phenomenon and occurs even without visible amounts of ice, although completely dry slopes are rare in nature. The added cohesion provided by ice lenses in rock masses is debated, but it is likely that episodic freeze–thaw cycles weaken rock masses along discontinuities through expansion and contraction effects. Rising temperatures measured in >100-m deep boreholes drilled into bedrock in the Alps, Scandinavia, and Svalbard shows nonlinear deviations from the linear geothermal gradient, documenting warming of sporadic alpine permafrost (Harris et al. 2002).

August 2003 was a time of record summer temperatures throughout western Europe and frequent rock falls in the European Alps. Similar reports from other mountain belts throughout the world have pointed out the coincidence between enhanced slope instability and warm temperatures (Huggel et al., 2010; 2012). Detailed inventories reveal that historic slope failures commonly have their sources at high elevations and on steep slopes characterized by periglacial weathering and permafrost. In their study of 56 slope failures in the central European Alps between 1900 and 2007, Fischer et al. (2012) concluded 'that not only the changes in cryosphere, but also other factors which remain constant over long periods play an important role in slope failures'. This statement highlights the need for collecting more data to attribute robustly increases in the occurrence of slope failures to climate change. This attribution is highly relevant to

hazard management as catastrophic rock-slope failures with long runouts are a major concern in all mountain belts throughout the world (Guthrie et al. 2012). Such concerns hinge on a solid knowledge of the permafrost landscape: we need to know reliably where frozen ground occurs, and whether regional predictions can be readily scaled down to catchments or hillslopes. Harris et al. (2009) extensively reviewed the history of permafrost monitoring and modelling throughout Europe and concluded that predictions of impacts of permafrost degradation will continue rely on process-based modelling that takes advantage of the growing number of detailed time series, borehole measurements, and high-resolution topographic data.

References

Alexander D, Davies T, and Shulmeister J 2014 Formation of the Waiho Loop terminal moraine, New Zealand. *Journal of Quaternary Science* **29**(4), 361–369.

Ballantyne CK 2002 Paraglacial geomorphology. *Quaternary Science Reviews* **21**, 1935–2017.

Ballantyne CK and Stone JO 2013 Timing and periodicity of paraglacial rock-slope failures in the Scottish Highlands. *Geomorphology* **186**(C), 150–161.

Beltaos S and Prowse T 2009 River-ice hydrology in a shrinking cryosphere. *Hydrological Processes* **23**(1), 122–144.

Bitte P, Berthier E, Masiokas M, et al. 2016 Geometric evolution of the Horcones Inferior Glacier (Mount Aconcagua, Central Andes) during the 2002–2006 surge. *Journal of Geophysical Research: Earth Surface* **121**, 111–127.

Björnsson H 2003 Subglacial lakes and jökulhlaups in Iceland. *Global and Planetary Change* **35**(3), 255–271.

Bodin X, Krysiecki J-M, Schoeneich P, et al. 2017 The 2006 collapse of the Bérard rock glacier (southern French Alps). *Permafrost and Periglacial Processes* **28**, 209–223.

Bolch T, Kulkarni A, Kääb A, et al. 2012 The state and fate of Himalayan glaciers. *Science* **336**(6079), 310–314.

Bowden WB, Gooseff MN, Balser A, et al. 2008 Sediment and nutrient delivery from thermokarst features in the foothills of the North Slope, Alaska: Potential impacts on headwater stream ecosystems. *Journal of Geophysical Research: Biogeosciences* **113**, 1–12.

Brenning A and Trombotto D 2006 Logistic regression modeling of rock glacier and glacier distribution: Topographic and climatic controls in the semi-arid Andes. *Geomorphology* **81**(1-2), 141–154.

Chen Y, Xu C, Chen Y, et al. 2010 Response of glacial-lake outburst floods to climate change in the Yarkant River basin on northern slope of Karakoram Mountains, China. *Quaternary International* **226**(1-2), 75–81.

Chiarle M, Iannotti S, Mortara G, and Deline P 2007 Recent debris flow occurrences associated with glaciers in the Alps. *Global and Planetary Change* **56**(1-2), 123–136.

Chuvili, E, Sokolova N, Davletshina D, Bukhanov B, Stanilovskaya J, Badetz C and

Spasennykh M 2020 Conceptual models of gas accumulation in the shallow permafrost of northern West Siberia and conditions for explosive gas emissions. *Geosciences*, **10**(5), 195.

Clague JJ and Evans SG 2000 A review of catastrophic drainage of moraine-dammed lakes in British Columbia. *Quaternary Science Reviews* **19**(17), 1763–1783.

Comiso JC and Parkinson CL 2004 Satellite-observed changes in the Arctic. *Physics Today* **57**(8), 38–44.

Cossart E, Braucher R, Fort M, et al. 2008 Slope instability in relation to glacial debuttressing in alpine areas (Upper Durance catchment, southeastern France): Evidence from field data and ¹⁰Be cosmic ray exposure ages. *Geomorphology* **95**(1-2), 3–26.

Das S, Kar NS, and Bandyopadhyay S 2015 Glacial lake outburst flood at Kedarnath, Indian Himalaya: a study using digital elevation models and satellite images. *Natural Hazards* **77**(2), 769–786.

Davies MCR, Hamza O, and Harris C 2000 The effect of rise in mean annual temperature on the stability of rock slopes containing ice-filled discontinuities. *Permafrost and Periglacial Processes* **12**(1), 137–144.

Davies TRH, Smart CC, and Turnbull JM 2003 Water and sediment outbursts from advanced Franz Josef Glacier, New Zealand. *Earth Surface Processes and Landforms* **28**(10), 1081–1096.

Decaulne A and Sæmundsson T 2006 Meteorological conditions during slush-flow release and their geomorphological impact in northwestern Iceland: a case study from the Bíldudalur valley. *Geografiska Annaler: Series A, Physical Geography* **88**(3), 187–197.

DeConto RM and Pollard D 2016 Contribution of Antarctica to past and future sea-level rise. *Nature* **531**(7596), 591–597.

Dobinski W 2011 Permafrost. *Earth-Science Reviews* **108**(3-4), 158–169.

Dunning SA, Large ARG, Russell AJ, et al. 2013 The role of multiple glacier outburst floods in proglacial landscape evolution: The 2010 Eyjafjallajokull eruption, Iceland. *Geology* **41**(10), 1123–1126.

Dussaillant A, Benito G, Buytaert W, et al. 2009 Repeated glacial-lake outburst floods in Patagonia: an increasing hazard?. *Natural Hazards* **54**(2), 469–481.

Evans SG and Clague JJ 1994 Recent climatic change and catastrophic geomorphic processes in mountain environments. *Geomorphology and Natural Hazards. Proceedings of the 25th Binghamton Symposium in Geomorphology,* 24–25 September, Binghamton, USA, pp. 107–128.

Evans SG, Bishop NF, Smoll LF, et al. 2009a A re-examination of the mechanism and human impact of catastrophic mass flows originating on Nevado Huascarán, Cordillera Blanca, Peru in 1962 and 1970. *Engineering Geology* **108**(1-2), 96–118.

Evans SG, Tutubalina OV, Drobyshev VN, et al. 2009b Catastrophic detachment and high-velocity long-runout flow of Kolka Glacier, Caucasus Mountains, Russia in 2002. *Geomorphology* **105**(3-4), 314–321.

Fischer L, Purves RS, Huggel C, et al. 2012 On the influence of topographic, geological and cryospheric factors on rock avalanches and rockfalls in high-mountain areas. *Natural Hazards and Earth System Sciences* **12**(1), 241–254.

French HM 2013 *The Periglacial Environment.* Wiley Blackwell.

Freppaz M, Godone D, Filippa G, et al. 2010 Soil erosion caused by snow avalanches: A case study in the Aosta Valley (NW Italy). *Arctic, Antarctic, and Alpine Research* **42**(4), 412–421.

Fujita K and Nuimura T 2011 Spatially heterogeneous wastage of Himalayan glaciers. *Proceedings of the National Academy of Sciences* **108**(34), 14011–14014.

Fujita K, Sakai A, Takenaka S, et al. 2013 Potential flood volume of Himalayan glacial lakes. *Natural Hazards and Earth System Sciences* **13**(7), 1827–1839.

Gardelle J, Arnaud Y, and Berthier E 2011 Contrasted evolution of glacial lakes along the Hindu Kush Himalaya mountain range between 1990 and 2009. *Global and Planetary Change* **75**(1-2), 47–55.

Golledge NR, Kowalewski DE, Naish TR, et al. 2015 The multi-millennial Antarctic commitment to future sea-level rise. *Nature* **526**(7573), 421–425.

Gruber S and Haeberli W 2007 Permafrost in steep bedrock slopes and its temperature-related destabilization following climate change. *Journal of Geophysical Research: Earth Surface* **112**, 1–10.

Grumbine RE and Pandit MK 2013 Threats from India's Himalaya dams. *Science* **339**(6115), 36–37.

Günther F, Overduin PP, Sandakov AV, et al. 2013 Short- and long-term thermo-erosion of ice-rich permafrost coasts in the Laptev Sea region. *Biogeosciences* **10**(6), 4297–4318.

Guthrie RH, Friele P, Allstadt K, et al. 2012 The 6 August 2010 Mount Meager rock slide-debris flow, Coast Mountains, British Columbia: Characteristics, dynamics, and implications for hazard and risk assessment. *Natural Hazards and Earth System Sciences* **12**(5), 1277–1294.

Harris C, Arenson LU, Christiansen HH, et al. 2009 Permafrost and climate in Europe: Monitoring and modelling thermal, geomorphological and geotechnical responses. *Earth-Science Reviews* **92**(3-4), 117–171.

Harris C, Mühll DV, Isaksen K, et al. 2002 Warming permafrost in European mountains. *Global and Planetary Change* **39**(3), 215–225.

Huggel C, Salzmann N, Allen S, et al. 2010 Recent and future warm extreme events and high-mountain slope stability. *Philosophical Transactions of the Royal Society A: Mathematical, Physical and Engineering Sciences*, **368**(1919), 2435–2459.

Huggel C, Clague JJ, and Korup O 2012. Is climate change responsible for changing landslide activity in high mountains? *Earth Surface Processes and Landforms*, **37**(1), 77–91.

Humlum O 2000 The geomorphic significance of rock glaciers: estimates of rock glacier debris volumes and headwall recession rates in West Greenland. *Geomorphology* **35**, 41–67.

Immerzeel WW, van Beek LPH and Bierkens MFP 2010 Climate change will affect the Asian water towers. *Science* **328**(5984), 1382–1385.

Iribarren Anacona P, Mackintosh A, and Norton KP 2015 Hazardous processes and events from glacier and permafrost areas: lessons from the Chilean and Argentinean Andes. *Earth Surface Processes and Landforms* **40**(1), 2–21.

Jacquet J, McCoy SW, McGrath D, et al. 2017 Hydrologic and geomorphic changes resulting from episodic glacial lake outburst floods; Rio Colonia, Patagonia, Chile. *Geophysical Research* **44**, 854–864.

Jamieson SSR, Ewertowski MW, and Evans DJA 2015 Rapid advance of two mountain glaciers in response to mine-related debris loading. *Journal of Geophysical Research: Earth Surface* **120**, 1418–1435.

Janský B, Šobr M, and Engel M 2010 Outburst flood hazard case studies from the Tien-Shan Mountains, Kyrgyzstan. *Limnologica* **40**(4), 358–364.

Jolivel M and Allard M 2013 Thermokarst and export of sediment and organic carbon in the Sheldrake River watershed, Nunavik, Canada. *Journal of Geophysical Research: Earth Surface* **118**(3), 1729–1745.

Jorgenson MT, Shur YL, and Pullman ER 2006 Abrupt increase in permafrost degradation in Arctic Alaska. *Geophysical Research Letters* **33**(2), 1–4.

Joughin I, Smith BE, and Medley B 2014 Marine ice sheet collapse potentially under way for the Thwaites Glacier Basin, West Antarctica. *Science* **344**(6185), 735–738.

Kääb A, Huggel C, Fischer L, et al. 2005 Remote sensing of glacier-and permafrost-related hazards in high mountains: an overview. *Natural Hazards and Earth System Sciences* **5**(4), 527–554.

Kääb A, Berthier E, Nuth C, et al. 2013 Contrasting patterns of early twenty-first-century glacier mass change in the Himalayas. *Nature* **488**(7412), 495–498.

Kääb A, Leinss S, Gilbert A, et al. 2018 Massive collapse of two glaciers in western Tibet in

2016 after surge-like instability. *Nature Geoscience* **11**, 114–120.

Kanevskiy M, Shur Y, Strauss J, et al. 2016 Patterns and rates of riverbank erosion involving ice-rich permafrost (yedoma) in northern Alaska. *Geomorphology* **253**(C), 370–384.

Keegan KM, Albert MR, McConnell JR, and Baker I 2014 Climate change and forest fires synergistically drive widespread melt events of the Greenland Ice Sheet. *Proceedings of the National Academy of Sciences* **111**(22), 7964–7967.

Kochel RC, Trop JM, and Jacob RW 2018 Geomorphology of icy debris fans: Delivery of ice and sediment to valley glaciers decoupled from icecaps. *Geosphere* **14**(4), 1710–1752.

Korup O and Rixen C 2014 Soil erosion and organic carbon export by wet snow avalanches. *The Cryosphere* **8**(2), 651–658.

Korup O and Tweed F 2007 Ice, moraine, and landslide dams in mountainous terrain. *Quaternary Science Reviews* **26**(25-28), 3406–3422.

Krautblatter M, Funk D, and Günzel FK 2013 Why permafrost rocks become unstable: a rock–ice-mechanical model in time and space. *Earth Surface Processes and Landforms* **38**, 876–887.

Levermann A 2011 When glacial giants roll over. *Nature* **472**(7341), 43–44.

Lewkowicz AG and Harris C 2005 Morphology and geotechnique of active-layer detachment failures in discontinuous and continuous permafrost, northern Canada. *Geomorphology* **69**(1-4), 275–297.

Ma R, Duan H, Hu C, et al. 2010 A half-century of changes in China's lakes: Global warming or human influence?. *Geophysical Research Letters* **37**(24), 1–6.

Magilligan FJ, Gomez B, Mertes LAK, et al. 2002 Geomorphic effectiveness, sandur development, and the pattern of landscape response during jökulhlaups: Skeidarársandur, southeastern Iceland. *Geomorphology* **44**, 95–113.

Marchenko SS, Gorbunov AP, and Romanovsky VE 2007 Permafrost warming in the Tien Shan Mountains, Central Asia. *Global and Planetary Change* **56**(3-4), 311–327.

Marzeion B, Cogley JG, Richter K, and Parkes D 2014 Attribution of global glacier mass loss to anthropogenic and natural causes. *Science* **345**(6199), 919–921.

McColl ST 2012 Paraglacial rock-slope stability. *Geomorphology* **153-154**(C), 1–16.

McColl ST and Davies TR 2011 Evidence for a rock-avalanche origin for 'The Hillocks' "moraine", Otago, New Zealand. *Geomorphology* **127**(3-4), 216–224.

McKillop RJ and Clague JJ 2007 Statistical, remote-based approach for estimating the probability of catastrophic drainage from moraine-dammed lakes in southwestern British Columbia. *Global and Planetary Change* **56**, 153–171.

Mengel M and Levermann A 2014 Ice plug prevents irreversible discharge from East Antarctica. *Nature Climate Change* **4**(6), 451–455.

Milana JP 2015 Molards and their relation to landslides involving permafrost failure. *Permafrost and Periglacial Processes* **27**(3), 271–284.

Montelli A, Gulick SPS, Worthington LL, et al. 2017 Late Quaternary glacial dynamics and sedimentation variability in the Bering Trough, Gulf of Alaska. *Geology* **45**(3), 251–254.

Ng F, Liu S, Mavlyudov B, and Wang Y 2007 Climatic control on the peak discharge of glacier outburst floods. *Geophysical Research Letters* **34**(21), 1–5.

Nie Y, Liu Q, and Liu S 2013 Glacial lake expansion in the Central Himalayas by Landsat images, 1990–2010. *PLoS ONE* **8**(12), e83973.

Nitze I and Grosse G 2016 Detection of landscape dynamics in the Arctic Lena Delta with temporally dense Landsat time-series stacks. *Remote Sensing of Environment* **181**(C), 27–41.

Phillips M, Wolter A, Lüthi R, et al. 2016 Rock slope failure in a recently deglaciated permafrost rock wall at Piz Kesch (Eastern Swiss Alps), February 2014. *Earth Surface Processes and Landforms* **42**(3), 426–438.

Qiu J 2014 Avalanche hotspot revealed. *Nature* **509**, 142–143.

Reznichenko NV, Davies TRH, and Alexander DJ 2011 Effects of rock avalanches on glacier behaviour and moraine formation. *Geomorphology* **132**(3-4), 327–338.

Roberts MJ 2005 Jökulhlaups: A reassessment of floodwater flow through glaciers. *Reviews of Geophysics* **43**(1), 1–21.

Russell AJ, Tweed FS, Roberts MJ, et al. 2010 An unusual jökulhlaup resulting from subglacial volcanism, Sólheimajökull, Iceland. *Quaternary Science Reviews* **29**(11-12), 1363–1381.

Schneider D, Bartelt P, Caplan-Auerbach J, et al. 2010 Insights into rock-ice avalanche dynamics by combined analysis of seismic recordings and a numerical avalanche model. *Journal of Geophysical Research: Earth Surface* **115**, 1–20.

Schwanghart W, Worni R, Huggel C, et al. 2016 Uncertainty in the Himalayan energy–water nexus: Estimating regional exposure to glacial lake outburst floods. *Environmental Research Letters* **11**(7), 074005.

Selmes N, Murray T, and James TD 2011 Fast draining lakes on the Greenland Ice Sheet. *Geophysical Research Letters* **38**(15), 1–5.

Shrestha AB, Eriksson M, Mool P, et al. 2010 Glacial lake outburst flood risk assessment of Sun Koshi basin, Nepal. *Geomatics, Natural Hazards and Risk* **1**(2), 157–169.

Shulmeister J, Davies TR, Evans DJA, et al. 2009 Catastrophic landslides, glacier behaviour and moraine formation – A view from an active plate margin. *Quaternary Science Reviews* **28**(11-12), 1085–1096.

Smith DG and Pearce CM 2002 Ice jam-caused fluvial gullies and scour holes on northern river flood plains. *Geomorphology* **42**, 85–95.

Smith DJ, McCarthy DP, and Luckman BH 1994 Snow-avalanche impact pools in the Canadian Rocky Mountains. *Arctic and Alpine Research* **26**(2), 116–127.

Somos-Valenzuela MA, Chisolm RE, et al. 2016 Modeling a glacial lake outburst flood process chain: the case of Lake Palcacocha and Huaraz, Peru. *Hydrology and Earth System Sciences* **20**(6), 2519–2543.

Sosio R, Crosta GB, Chen JH, and Hungr O 2012 Modelling rock avalanche propagation onto glaciers. *Quaternary Science Reviews* **47**(C), 23–40.

Thompson SS, Benn DI, Dennis K, and Luckman A 2012 A rapidly growing moraine-dammed glacial lake on Ngozumpa Glacier, Nepal. *Geomorphology* **145-146**(C), 1–11.

Thorardson T and Larsen G 2007 Volcanism in Iceland in historical time: Volcano types, eruption styles and eruptive history. *Journal of Geodynamics* **43**, 118–152.

Tovar SD, Shulmeister J, and Davies TR 2008 Evidence for a landslide origin of New Zealand's Waiho Loop moraine. *Nature Geoscience* **1**(8), 524–526.

Turcotte B, Morse B, Bergeron NE, and Roy AG 2011 Sediment transport in ice-affected rivers. *Journal of Hydrology* **409**(1-2), 561–577.

Uprety Y, Shrestha UB, Rokaya MB, et al. 2017 Perceptions of climate change by highland communities in the Nepal Himalaya. *Climate and Development* **9**, 649–661.

Van Der Woerd J, Owen LA, Tapponnier P, et al. 2004 Giant, M8 earthquake-triggered ice avalanches in the eastern Kunlun Shan, northern Tibet: Characteristics, nature and dynamics. *Geological Society of America Bulletin* **116**(3/4), 394–406.

Voight B 1990 The 1985 Nevado del Ruiz volcano catastrophe: anatomy and retrospection. *Journal of Volcanology and Geothermal Research* **44**, 349–386.

Watanabe T, Lamsal D, and Ives JD 2009 Evaluating the growth characteristics of a glacial lake and its degree of danger of outburst flooding: Imja Glacier, Khumbu Himal, Nepal. *Norsk Geografisk Tidsskrift - Norwegian Journal of Geography* **63**(4), 255–267.

Westoby MJ, Glasser NF, Brasington J, et al. 2014 Modelling outburst floods from moraine-dammed glacial lakes. *Earth-Science Reviews* **134**, 137–159.

Whiteman CA 2011 *Cold Region Hazards and Risks*. Wiley Blackwell.

Yang M, Nelson FE, Shiklomanov NI, et al. 2010 Permafrost degradation and its environmental effects on the Tibetan Plateau: A review of recent research. *Earth-Science Reviews* **103**(1-2), 31–44.

15

Sea-Level Change and Coastal Hazards

The level of the sea at the shoreline is constantly changing, regardless of whether it is over seconds or millions of years. The largest tidal range on Earth is 17 m and occurs at two places in Canada – the Bay of Fundy, Nova Scotia, and Ungava Bay, Quebec. Tides result from the gravitational attraction of the moon and the sun, causing daily and seasonal fluctuations in sea level as the position of a coast relative to the moon and sun constantly shifts. Changes in wind speed and atmospheric pressure also affect sea level over periods of hours or days. In the open ocean, strong winds pile up water and increase wave height, producing swell. The swell increases both water level and wave heights when it reaches the shore, while waves physically impact on beaches and cliffs, picking up sediment and abrading rock surfaces. Storm surges can temporarily raise local sea level many metres. Other controls such as water temperature change sea level more gradually on top of the cycles of tides, or episodically, as weather does. The position of the sea at the shore, referred to as 'relative sea level', is controlled by the vertical movement of both the land and seawater. These movements can be local, regional, or global in extent (Figure 15.1).

Sea level rises or falls globally when the amount of water in the world's oceans increases or decreases, or when the shape

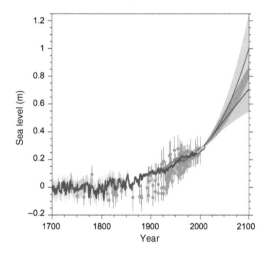

Figure 15.1 Global mean sea level trends combining data from palaeo sea-level studies (purple), tide gauge data (blue, green, red), altimeter data (light blue) and central estimates and likely ranges for projections from the combination of Coupled Model Intercomparison Project Phase 5 (CMIP5) and process-based models RCP2.6 (blue) and RCP8.5 (red) scenarios, all relative to pre-industrial levels. From IPCC (2013).

of ocean basins changes. These global changes in sea level are referred to as 'eustatic'. The shape of oceans changes due to tectonic plate movements over millions of years but the arrangement of continents is fixed in the short term. However, local rates of tectonic uplift and subsidence can strongly influence the rates of relative sea level changes at a given coast. Included here are the effects of large earthquakes and sediment compaction, which

Geomorphology and Natural Hazards: Understanding Landscape Change for Disaster Mitigation, Advanced Textbook Series,
First Edition. Tim R. Davies, Oliver Korup, and John J. Clague.
© 2021 John Wiley & Sons Ltd. Co-published 2021 by the American Geophysical Union and John Wiley & Sons Ltd.

also alter the elevation of the land relative to the sea.

Climate controls sea level in several ways. Some of the more short-lived to seasonal influences include changing atmospheric pressure and wind fields. The average air temperature influences the average temperature of the oceans. Warming seawater expands, while it contracts when cooling. Globally, sea level is currently rising at an average rate of about 3 mm yr^{-1} (Church and White 2011), although this rate varies slightly depending on measurement method, and even more so depending on geographic location. If globally averaged, about half of this rise is thought to be caused by thermal expansion, also known as the 'steric effect'. This volumetric increase in ocean bodies is driven by atmospheric warming. Weather patterns or natural climate fluctuations that dictate seasonal runoff patterns in large rivers can also influence sea level, to a lesser effect though more frequently, than the gradual steric expansion. For example, the 2015–2016 El Niño raised sea level along southern California coasts consistently by 0.15–0.25 m, causing higher tides, storm waves, and net erosion (Figure 15.2) (Young et al. 2018).

Changes in atmospheric temperature affect the amount of snow and glacier ice on land, and hence the available runoff to oceans. The volume of water locked up in alpine glaciers and ice sheets on Greenland and Antarctica responds to the average air temperature over years, decades or centuries. Atmospheric warming over the past century has caused glaciers to melt and thus has increased the amount of water in the oceans. About half of the eustatic sea-level rise in recent years is due to glacier melt, although estimates differ and remain heavily debated. Melting of the Greenland ice sheet, for example, may have contributed as much as 25 mm to global mean sea-level rise in the twentieth century, judging from glacier changes recorded on historic air photos; however, the uncertainty of this estimate is nearly 10 mm (Kjeldsen et al. 2015).

The general equilibrium of the forces tending to elevate or depress the Earth's crust and the underlying mantle is known as 'isostasy'. Isostatic sea-level changes are caused by an increase or decrease in the weight of ice or water on the crust or by a change in the thickness of the lithosphere. The land can rise or fall slowly as the Earth surface slowly responds to changes in glacier cover on the continents. The disappearance of large ice sheets in North America and Eurasia at the end of the Pleistocene removed a huge load on the crust, triggering a phenomenon generally termed glacio-isostatic adjustment (GIA) by the research community working on sea-level changes (Ostanciaux et al. 2012). This unloading was accompanied and followed by hundreds of metres of uplift of the land in some areas and smaller amounts in others. In some areas, such as Hudson Bay and the Arctic Islands in Canada and Scandinavia, glacio-isostatic uplift is still occurring, although at a much slower rate than at the end of the Pleistocene. A similar effect results from unloading of continental shelves due to eustatic sea-level lowering, and depends on where the load of surface water is shifted with respect to a saturated and permeable crust. At the peak of the last Pleistocene glaciation, eustatic sea level was about 130 m lower than at present, and consequently continental shelves around the world had only about one-third of the seawater above them that they have today. Unloading of the shelves due to eustatic sea-level lowering caused 'hydro-isostatic' uplift at the fringes of continents. Conversely, when eustatic sea level rose at the end of the Pleistocene, continental shelves were loaded with more water and subsided at the same time that the continents were being unloaded of ice.

Even for regions that have tide gauge data going back as far as the eighteenth century, scientists who reconstruct sea-level curves must consider different measurement practices, instrumental shifts, and local effects (Figures 15.3 and 15.4) (Zerbini et al. 2017). The human fingerprint has also become

(a) Portuguese Bend Landslide

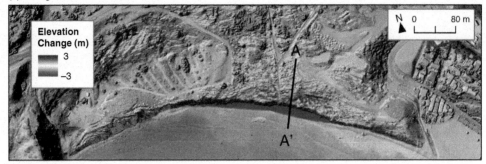

(b) Silver Strand Berm

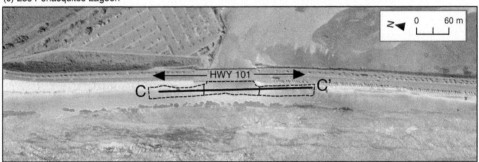

(c) Los Peñasquitos Lagoon

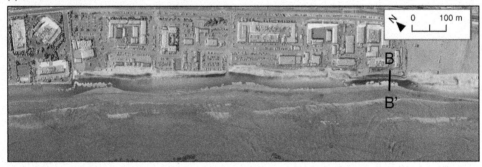

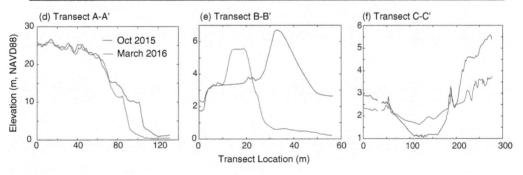

Figure 15.2 Effects of ENSO on southern California coasts. Topographic change (6 October 2015 and 22 March 2016; see color bar) at (a) Portuguese bend landslide in Palos Verdes, (b) Silver Strand berm erosion, (c) Los Peñasquitos lagoon inlet accretion/adjacent beach erosion and the Coast Highway 101, and (d–f) corresponding transect elevation. Dashed black lines in (c) outline the estuary mouth and adjacent beach regions used in estuary change analysis. From Young et al. (2018).

Figure 15.3 Examples of raised beaches. Top: Beaches bordering Kluane Lake, Yukon. These beaches formed as the level of the lake fell from its 17th century high-stand (John Clague). Middle: Raised beaches on the east coast of Lowther Island, Canadian Arctic. These beaches formed as relative sea level fell during the late Holocene (Donald L. Forbes). Bottom: Raised beaches resulting from tectonic uplift at Turakirae Head near Wellington, New Zealand (Lloyd Homer).

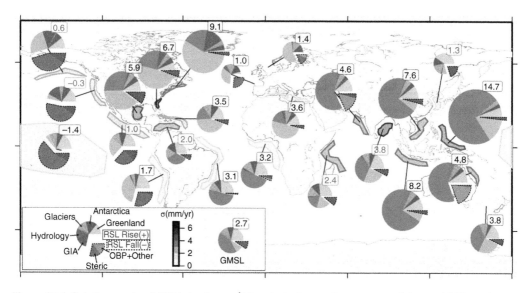

Figure 15.4 Relative sea-level (RSL) rise (mm yr^1) in selected coastal zones around the world. The polygons are coloured relative to the estimated total error. Wedge areas reflect the absolute magnitude of the different contributions, and negative wedges are shaded. Trends with red number are statistically nonsignificant, although the individual components may have significant contributions. From Rietbroek et al. (2016).

important – for example, river sediments that formerly slowly built up the surfaces of deltas on which many of the world's largest cities are located are being intercepted by structures before they reach the coast. Large amounts of groundwater are being extracted from coastal aquifers. Both activities ensure the livelihoods of urban populations but lower the land surface in coastal areas, causing relative sea level to rise (Syvitski et al. 2009).

The more tangible geomorphic impacts of sea-level change happen over at least decades and depend, among other things, on the rate of rise or fall of the sea surface, the geometry and erodibility of the coastline, the mobility of intertidal materials, the temperature field of the ocean, and sediment supply by rivers and longshore currents (Cazenave and Cozannet 2014; Nicholls and Cazenave 2010). The ocean surface sets the base level for much land-surface erosion and thus plays a major role in many geomorphic processes and studies. Base level is the datum at which the effect of flowing water on continental erosion becomes zero, because the water surface slope and flow velocity become zero. The centimetre-scale changes due to continuing sea-level rise influence modern erosion, but are shadowed by the effects of a 130-m sea-level depression during the Last Glacial Maximum, when the courses of continental rivers lengthened to the shelf edge, thus severely 'stretching' their longitudinal profiles in some cases. We note in this context that the effect of base-level change on subaerial river profiles depends on the relative gradients of the subaerial river and the offshore sea bed. If the offshore gradient is steeper, then base-level fall will cause river incision, whereas if the offshore gradient is lower then as sea level falls the river will lengthen, so the reduction in elevation of the mouth is more than compensated for by the increase in length and the river will aggrade (Schumm 1993).

15.1 Frequency and Magnitude of Sea-Level Change

Sea-level change eludes casual observation in large part because it is superimposed on much larger daily tidal cycles. Seasonal or storm-driven changes in sea level may seem ephemeral and minute compared to trends in geological history, but can be destructive nevertheless. Some beach locations in southern California retreated landward by up to 80 m because of elevated sea level and higher storm waves during the 2015–2016 El Niño (Young et al. 2018). Averaged over several decades or centuries, the cumulative effects of sea-level change expose coastal communities to natural hazards such as coastal erosion, flooding, salt intrusions into groundwater, and degradation of ecosystems. Tsunamis, storm surges, and other extreme waves will also have higher runup if sea level rises. Rising sea levels are a direct concern for the more than 600 million people who live less than ten metres above the waterline and share the densely populated spaces of most of Earth's largest megacities. Hanson et al. (2010) estimated that nearly 40 million people in 136 port cities were living beneath the water level of the 100-year coastal flood in 2005, and that this number of people exposed could increase threefold by 2070 (Figure 15.5). Along similar lines, Neumann et al. (2015) concluded that the Asian countries of China, India, Bangladesh, Indonesia, and Vietnam are likely to be most exposed to 100-year storm surges in the coming decades. Extensive flat sedimentary coasts provide much of the easily accessible land and the resources that cities require. Yet these coasts are most susceptible to flooding and erosion caused by even small changes in sea-level. For example, only one-half metre of sea-level rise would affect some 3.8 million people living on the Nile River delta, and potentially destroy

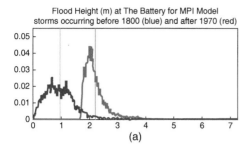

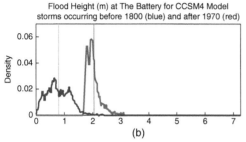

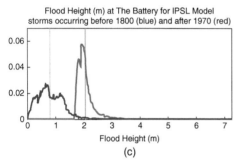

Flood Height (m)
(c)

Figure 15.5 Distributions of flood heights at The Battery, New York City, for the pre-anthropogenic era (blue) and the anthropogenic era (red) for three models. Each distribution is normalized by the number of events it contains. The 99% confidence interval, based on 100 000 bootstrap simulations of the mean of each set, is shown in light blue for pre-anthropogenic era flood events and in light red for anthropogenic era flood events. From Reed et al. (2015).

some 1800 km² of cropland in the region (Figure 15.6) (FitzGerald et al. 2008).

Decades of continuous, though local, tidal gauge readings are now being complemented by high-precision measurements from satellite imagery that cover variations of the sea surface across entire ocean basins. For example,

data from the space-borne Gravity Recovery And Climate Experiment (GRACE) show that the worldwide average sea level rose at a rate of 2.74 ± 0.58 mm yr^{-1} between 2002 and 2014. Close to half of this increase, that is 1.38 ± 0.16 mm yr^{-1}, was because of higher ocean temperatures causing the water to expand (Rietbroek et al. 2016). Thermal expansion of the oceans contributes mostly to sea-level rise around the Philippines and Indonesia, where satellites detected that the ocean surface rose at almost 15 mm yr^{-1} over that period. Melting ice from the polar caps and glaciers contributed 1.37 ± 0.09 mm yr^{-1} to worldwide sea-level rise, whereas reduced runoff to the oceans balanced some of this increase (Rietbroek et al. 2016). Estimated rates of sea-level rise vary strongly across regions regardless of measurement or estimation method, because of large-scale atmospheric patterns, differing ocean temperatures, continuing glacio-isostatic adjustment, and effects of local tectonics. A compilation of hundreds of measurements derived from the difference between satellite-based measurements of the sea surface and local tide-gauge records shows that vertical ground motion, mostly along the world's coasts, is currently between −53 and +35 mm yr^{-1} (Ostanciaux et al. 2012).

Sea-level rise is of direct consequence for coastal engineering; two key metrics used for planning and building infrastructure or coastal restoration are the maximum expected wave height and its return period. These metrics can be obtained from extreme-value statistical analysis of local tide gauge data, and provide a physical measure of wave action. You and Nielsen (2013) reported that a storm:

> on the central coast of New South Wales, Australia on 8 June 2007, produced huge seas with maximum wave heights of up to 14 m, beached a large coal ship, Pasha Bulker of 76.741 tones and 225 m long,

0.5 m sea-level rise

Affected population: 3,800,000
Affected cropland: 1,800 km²

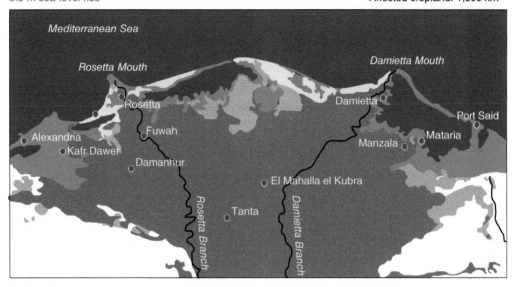

1.0 m sea-level rise

Affected population: 6,100,000
Affected cropland: 4,500 km²

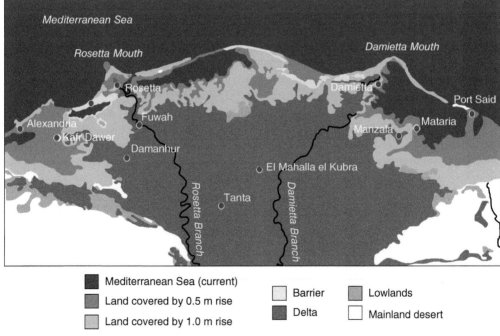

Figure 15.6 Effect of sea-level rise on the populations and croplands of the Nile Delta. From FitzGerald et al. (2008).

on Nobbys Beach in Newcastle [...]. This coastal storm also eroded coastal beaches, damaged coastal properties and dumped flooding rains on the NSW coastal areas, resulting in the total damage of about US\$1.35 billion.

How do vertical increases in local sea level translate to horizontal inundation? One widespread approach is to use the Bruun rule, a geometric model that approximates how far inland sedimentary coasts retreat in response to a given sea-level rise:

$$R = \frac{L_* S}{B + h_*} \qquad (15.1)$$

where R is the horizontal retreat of the coast [m], L_* is the distance of the shoreline from water depth h_* [m], S is the amount of sea-level rise [m], and B is the elevation estimated for the eroded berm or dune [m] at the foreshore (Figure 15.7) (Ruessink and Ranasinghe 2014). Alternatively, we can express R and S in Eq (15.1) as rates of coastal retreat and sea-level rise, respectively. The Bruun rule makes few assumptions about the net sediment flux away from the coast and ignores sediment being replenished by rivers or long-shore drift. Under the latter conditions, the amount of coastal retreat due to sea-level rise will be lower than the model suggests. The Bruun rule, however, appears to hold for a balanced sediment budget, in which the material eroded from the upper part of the shore profile ends up filling the lower near-shore part, thus eventually balancing out the effect of sea-level rise. It follows that the dynamics in the local sediment budget are highly relevant for estimating the

effects of sea-level rise. For example, for barrier islands, an extended version of the Bruun rule is:

$$R = \frac{(L_{*_o} + W + L_{*_b})S}{b_{*_o} - b_{*_b}} \qquad (15.2)$$

where L_{*_o} and L_{*_b} are the active nearshore widths of the ocean and the lagoon [m], respectively, W is the width of the barrier island [m], and b_{*_o} and b_{*_b} are the water depths of the ocean and the lagoon [m], respectively (FitzGerald et al. 2008). Several other extended forms of the Bruun rule account for along-shore changes to the coastal sediment budget or those only indirectly related to sea-level rise. Cooper and Pilkey (2004) emphasized that the Bruun rule has only limited application for predicting coastal response to sea-level rise, but they conceded that many researchers still use the rule because it is simple, easier to use than more sophisticated alternatives, and hard to formally disprove. The Bruun rule can also be adapted for cliff coasts:

$$\frac{dx}{dt} = E_h + \frac{\Delta SLR I}{ch_c + d} \qquad (15.3)$$

where x is a coordinate for the cliff edge [m], t is time [yr], E_h is the past rate of cliff retreat [m yr^{-1}], ΔSLR is the difference between past and future rates of sea-level rise, I is the length of the active profile measured normal to the shore [m], c is the fraction of cliff-derived sediment coarse enough to be stored in the nearshore environment, h_c is the height of the cliff [m], and d is the depth below which sediment transport is negligible measured at points along the shore [m]; the value of d is

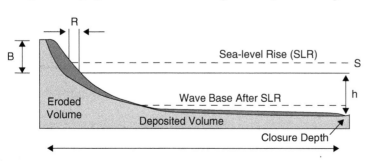

Figure 15.7 The Bruun rule of coastal erosion (see Equation 15.1). From Pilkey and Cooper (2004).

estimated empirically from data on significant wave heights and periods (Limber et al. 2018).

A widespread alternative is to use numerical models that couple the basic dynamics of sedimentary coasts with wave regimes and storm patterns, and simulate how beaches, barrier islands, tidal inlets, and lagoons respond to disturbances. These models bring together several components dealing with erosion rate as a function of shoreface geometry, the mobility of coastal dunes, the water and mass fluxes in tidal prisms, and other key components that determine the coastal sediment budget (Figure 15.8).

Models geared towards coastal protection and risk estimates also include socioeconomic drivers. For example, McNamara and Werner (2008) developed a barrier-island model that describes how sea-level rise and storm-surge heights drive barrier shore-face erosion and gradual longshore diffusion by ocean currents, and also simulates the process of barrier rollover, the growth and migration of dunes, and stabilizing effects of coastal vegetation. They coupled this model with economic and policy components of resort development to see how barrier islands respond to growing tourism infrastructure. They considered aspects such as the location preference, number, size, and investment costs of hotels that agents build on the coastal barrier, and travel costs for tourists, but also replenishment

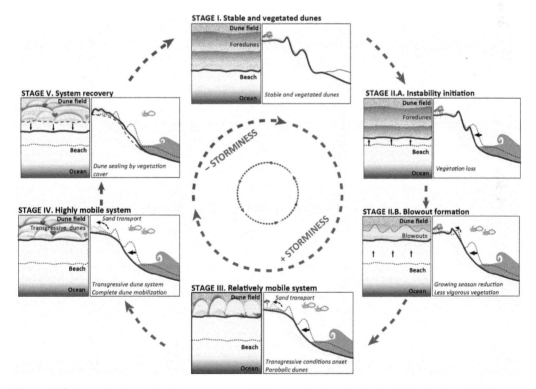

Figure 15.8 Conceptual model of large-scale coastal dune behaviour during the Little Ice Age (LIA). The model includes five stages, from an early stage of stable and vegetated dunes before the onset of the LIA (stage I) towards a series of processes that increased the instability of the dune system, resulting in an unvegetated, highly mobile system (stage IV). The transitions between stages are driven by a combination of storminess and other parameters that could induce pauses and temporary reversals between stages, as indicated by the internal circle. Stages IIA and IIB are early phases of dune instability, with the front edges of dunes supplying the initial pulse of aeolian activity and stage III progressing into stage IV. Stage IV then eventually recedes into stage V of system recovery and stability thereafter. From Jackson et al. (2019).

efforts to maintain the barrier intact. This many-faceted model allows first-order insights into how the impact of sea-level rise is linked to the yearly tourist population density, and predicts that developing resorts along barrier islands could dampen the coastline's natural response to frequent storms because continuing development would prompt commensurate protection measures. In the long-term, though, the model predicts major boom-and-bust cycles, especially because the rarer storm events will likely cause major disasters along heavily developed coasts.

15.2 Geomorphic Impacts of Sea-Level Change

15.2.1 Sea Levels in the Hazard Cascade

Sea-level change is mostly a slow-onset natural hazard, but its geomorphic consequences can pave the way for a range of other hazards. Some of the most frequently reported impacts of sea-level rise are coastal erosion, flooding, river avulsion, and intrusion of saltwater into coastal aquifers. Rising sea level also makes coasts more susceptible to tsunamis, storm surges, and river floods in the long term. For example, Wallace and Anderson (2013) reconstructed the flux of sediment around Galveston Island in the Gulf of Mexico and found that coastal erosion had increased by about 45% in the past two millennia, most likely reflecting the effects of accelerated sea-level rise and the impacts of large storms, particularly in the past few hundred years. A fall in sea level, in turn, can cause deltas to prograde, the downstream reaches of rivers to incise (or aggrade if the offshore gradient is less than the onshore gradient), and former coasts to become stranded, making their harbours unusable. Medieval settlements along the Baltic Sea coast were abandoned because sea level fell due to crustal rebound following the disappearance of the ice sheet that had blanketed Scandinavia and

northern Europe. Many former harbours are now several metres above sea level. On the Greek island of Crete, for example, an earthquake in 365 AD caused up to nine metres of uplift (Stiros 2010). In contrast, submerged historic buildings around the Mediterranean attest to the postglacial rise in sea level outside glaciated areas and its effects on society.

15.2.2 Sedimentary Coasts

The Bruun rule (Eq (15.1)) predicts that changing sea level will affect larger areas on flat coasts than on steep ones. The Volga delta that has been built out into the Caspian Sea, for example, is one of the flattest large deltas in the world, and nourished by the world's largest endorheic (i.e. closed-basin) river. Sea-level fall has enabled the delta apex to prograde for some 700 km downstream between the last glacial and the Holocene, yet the elevation of its apex has changed only about 130 m simply because the basin is so flat (Kroonenberg et al. 1997). The sediments and landforms of the Volga delta offer a view of a coastal response to changing sea level in fast-forward mode.

More generally, the supply of river sediment from the hinterland plays a major role in shaping coastlines and partly determines whether these coasts will be erosive or depositional. For example, major storms may erode parts of a coast, but may also trigger landslides in a mountainous hinterland, liberating large volumes of sediment to rivers that subsequently flush this material to the coast. Longshore drift moves this sediment along the coast, thus potentially balancing or offsetting the losses from wave erosion by creating new beach ridges.

An instructive way of visualizing the stability of a shoreline is to consider its sediment budget. This budget is similar to a bank account. In the case of a bank account, deposits made to the account are input, the account balance is how much money is in the account at a given time, and withdrawals are output. Similarly, we can partition the sediment budget of a beach in terms of input, storage, and output of sand, gravel, and boulders. Beaches both gain and lose material. Sand and gravel are added to a beach by the coastal processes that move sediment along the shore or by local wave erosion. Longshore drift brings sediment from updrift sources, such as rivers. Local wave erosion of sand dunes or sea cliffs also adds sediment to a beach. Wave erosion is greatest, and so is possibly sediment deposition, during storms when waves are largest and reach farthest inland. Sediment in storage on a beach is what you see when you visit a site, but this storage may change seasonally or in the long term. Output of sediment occurs when coastal processes move sand and gravel away from the beach. These coastal processes include littoral drift, a return flow to deep water during storms, and wind erosion. When input exceeds output, the beach will grow – more sediment is stored and the beach widens. When input and output are about the same, the beach will be in equilibrium and remain fairly constant in width. If output of sediment exceeds input, the beach will erode and become narrower.

Storms affect sediment supply, although generally only temporarily. Along the west coast of North America, for example, many beaches lose sediment during winter, when storms are common. This sediment is replenished during the summer when storms are few. On the Atlantic coast, much erosion is caused by hurricanes and by other severe storms known as 'nor'easters'. More subtle effects such as the gradual loss in seasonal sea-ice cover in the Arctic can also change the wave climate. To check whether and how this played out on erosion and sediment transport, Farquharson et al. (2018) studied coastline changes in the Chukchi Sea, northwest Alaska, and found that the annual duration of cover by sea ice decreased by an average of ten days per decade since the late 1980s. They also noted that shoreline changes, expressed as both cumulative erosion and accretion, have become about twice as variable since that time.

Long-term changes of coastlines, and of the sediment budgets of beaches and rocky coasts can be caused by climate change or by human interference with natural shore processes. The continuing erosion of sedimentary coasts has become a serious global problem because of both sea-level rise and the increasing development along shorelines. Analyses of the available time series of satellite imagery produced alarming numbers. Between 1984 and 2015, the world's coasts lost a total of about 28 000 km^2, which was about twice as much as the land that accreted during that time (Mentaschi et al. 2018). Another satellite-based, though independent, study reported that an estimated 20 500 km^2 of sand, mud, or rock flats within the current tidal range had disappeared over almost the same period (Murray et al. 2018). These studies will spur more detailed assessments to check the regional distribution and types of land surfaces lost (or gained), but the overall message in terms of the scope of global sea-level rise is clear.

Large sums of money are spent to control coastal erosion, but many of the benefits are temporary and much of the money is wasted. Most Canadian provinces and territories and the 30 US states adjacent to oceans and the Great Lakes have problems with coastal erosion. Average rates of local landward retreat along some barrier islands in the United States approach 8 m yr^{-1}, but so do local accretion rates (Conery et al. 2018). This is an extreme example, and most reported rates of coastal erosion are much lower, ranging from less than 0.01 m yr^{-1} to about 0.05 m yr^{-1}. Mushkin et al. (2018) proposed that some of the reported historic rates of coastal cliff retreat are biased, because they were derived over short periods that rarely considered the time between episodes of erosion. Collapsed cliff material at the foot of rocky coasts is a natural protection against further erosion, and waves need to remove this debris before the cliff can retreat further (unless its instability is also driven by processes other than wave impact or undercutting). Hence the time needed to evacuate collapsed material should be the minimum period over which to average out rates of cliff erosion. Without this adjustment, short-term rates of cliff erosion appear to be orders of magnitude higher than long-term rates, even at the same location.

Managing rates of land loss requires detailed knowledge about the local sediment budget to make sure that protection measures are effective and sustainable. In computing this sediment budget, we recall that erosion and deposition depend on processes acting at the shoreline, and also beyond. For example, coral or oyster reefs as well as coastal vegetation such as wetlands or mangroves can reduce the physical impacts of waves and thus naturally reduce erosion rates (Chapter 17.2.2).

These and many other examples show that feedbacks between flood hydrology, vegetation cover, sediment supply and storm climatology obscure the net contribution of sea-level change in the hazard cascade. This close coupling requires care when trying to isolate the net geomorphic impacts of sea-level change. In a detailed review of how coastal flooding from tropical cyclones and sea-level rise affects the world's coasts, Woodruff et al. (2014) concluded that:

> Tropical cyclone climatology partly drives the length of recovery time that coastal systems have between storm disruptions. However, extreme-value flood statistics consistently point towards [sea-level rise] as a competing, if not more important, factor in driving the frequency of extreme coastal flooding by tropical cyclones. Thus, although storms provide the dominant mechanism for erosion, it is often an increase in [sea-level rise] and/or a drop in sediment supply that is the true underlying cause of long-term rates of shoreline retreat.

Many models of sea-level rise and its capacity to inundate coastal areas are now integrating the role of wetland vegetation as a natural buffer. The continuing loss of wetlands to coastal development has thus become a major concern in preparing for higher sea levels along sedimentary coasts. At the same time, building dams and reservoirs in rivers has reduced the amount of sediment reaching the sea, thus cutting many sedimentary coasts off from their natural material supply and requiring construction of costly breakwaters and other engineered structures, or beach nourishment, to keep the coast in place. Despite these efforts, many sedimentary coasts are experiencing active erosion. For barrier coasts, this process can take a few decades to centuries, starting with pronounced thinning and breaching of dunes and increased overwash, and culminating in a gradual rollover during which parts of, or even the entire, coastal barrier system breach or migrate landward. The fate of thus destabilized barrier systems depends largely on whether enough sediment is available to replenish the losses; barrier systems that are driven across marsh-filled lagoons are more resilient than those surrounded by open water (FitzGerald et al. 2008). In essence, managing coastal erosion, with or without sea-level rise, requires that we understand and appreciate sediment mass conservation, and use the principle to predict the consequences of a disturbed coastal mass balance.

E-Lines and E-Zones The US National Research Council (NRC), at the request of the US Federal Emergency Management Agency (FEMA), developed several recommendations for managing coastal zones, some of which are:

- Estimates of future erosion should be based on historic changes to the shoreline or on statistical analysis of local wave and wind conditions and sediment supply.
- After average local erosion rates have been determined, maps should be made showing erosion lines and zones, which are referred to, respectively, as E-lines

and E-zones. An E-line is the expected position of the shoreline after a specified number of years; for example, the E-10 line is the expected shoreline in 10 years. E-zones are the areas between the present shoreline and the E-line; the E-10 zone, for example, is the area between the shoreline and the E-10 line.

- The E-10 zone should be off-limits to new habitable structures.
- Movable structures are allowed in the E-10 to E-60 zones, which are deemed to be at intermediate and long-term risk.
- Permanent large structures must have setbacks to beyond the E-60 line.
- All new structures built seaward of the E-60 line, with the exception of those on high bluffs or sea cliffs, should be built on pilings. They should be designed to withstand erosion that could be caused by a storm that recurs, on average, once every 100 years.

These NRC recommendations on setbacks are considered to be minimum standards for state and local coastal erosion management programs.

15.2.3 Rocky Coasts

Rocky coasts are more resistant to erosion than sedimentary coasts, but are steeper and have higher local relief. Rocky coasts may take longer to retreat or otherwise adapt to changing sea levels than sedimentary coasts, and the erosion of cliffs can be more episodic and localized, frequently driven by slope failure and primed by weathering. Erosion rates in cliffs made of unconsolidated sediments or soft rocks can be as a high as those along some sedimentary coasts. A compilation of almost 1700 measurements places coastal cliff-erosion rates between at least 1 mm yr^{-1} and nearly 10 m yr^{-1} (Prémaillon et al. 2018). This spread covers four orders of magnitude and is due to varying influences of marine wave climate, weather conditions and weathering, groundwater flow, and soil- or rock-mass

properties. Material properties and defects of the cliffs seem to be the most important factors in explaining these highly variable cliff-erosion rates (Prémaillon et al. 2018). Most mechanistic models of eroding rocky coasts have a time-dependent 'forcing term', which recognizes, for example, the annual wave energy flux or wave runup, as the driver of the rate of cliff retreat $\frac{dx}{dt}$ [m yr^{-1}] in the long term:

$$\frac{dx}{dt} = K \left(\frac{1}{8} \rho g H_b^2 C_g \right) e^{-\chi w(t)} \qquad (15.4)$$

where K is a negative coefficient [m N^{-1}] describing the effectiveness of the forcing term, ρ is the density of seawater [t m^{-3}], g is acceleration by gravity [m s^{-2}], H_b is breaking wave height [m], $C_g = \sqrt{gd_b}$ is wave-front speed [m s^{-1}] in shallow water as a function of wave-breaking depth d_b [m], χ is a decay constant with values between 0.01 and 0.1 [m^{-1}], and $w(t)$ is the time-dependent width of the surf zone and beach [m]; Limber et al. (2018) discuss this and related models with regard to their forcing terms.

High atmospheric contents of salt-rich aerosols, physical wave impact, and steep pore pressure gradients in the coastal fringe add to the driving forces that destabilize rock cliffs through either reducing apparent shear strength along potential failure planes or transient pressure. In Arctic permafrost regions, sedimentary coasts consolidated by high amounts of ground ice can develop mechanical properties similar to those of rocky coasts, causing episodic slope failure (Günther et al. 2013). Slope failure is a key process in shaping marine cliffs, and can, in turn, constrain sea-level curves and erosion. In the eastern Caspian Sea basin, for example, stranded Pleistocene shorelines are dotted with dozens of slope-failure deposits that are up to several kilometres long, many of them having moved up to some cubic kilometres of bedrock and sediment seaward (Pánek et al. 2016). Several generations of these giant landslides have eroded former coastal cliffs or were instead truncated by marine abrasion and thus might be tied to oscillating sea levels in this shallow inland lake basin.

Detailed and local surveys of coastal cliffs of steep coasts now routinely involve laser scanning (Rosser et al. 2013) and structure-from-motion photogrammetry of drone images. The digital elevation models that both methods offer can resolve decimetre- to centimetre-scale features and provide more extensive coverage compared to traditional surveys of cross-sections, and highlight where cliffs erode and where material is stored (Medjkane et al. 2018). Repeat measurements confirm that the activity in cliff erosion can be highly seasonal, for example mostly happening during winter storms with high waves; the presence and shape of beaches below cliff faces can also be important in creating locally-varying wave energies (Earlie et al. 2018). In extreme cases, waves can run up and inundate cliff tops tens of metres high and dump coarse sediments there (Hansom and Hall 2009).

In California, years during a strong El Niño phase are often times of accelerated coastal cliff erosion, especially at sites of slow moving or reactivating landslides (Hapke and Green 2006). Photogrammetric analysis of air photographs taken in 1942 and 1994 showed that coastal cliff landslides in central California pushed an annual average of 21 000 m^3 per kilometre of coastline into the ocean. The yields differed by an order of magnitude between sites, but were consistently highest in weak and tectonically shattered rocks (Hapke 2005). Wave erosion can gradually create abrasion platforms, notches, and rock cliffs that are prone to failure, so that even contemporary rates of sea-level rise may influence landslide hazard along a coast. Several studies suggested that coastal rockfall or landslide volumes from a given section of cliff approximate to an inverse power-law size distribution that also seems to characterize landslide erosion elsewhere. If used with representative parameters, this statistical approximation can be a useful tool to estimate the rates and volumes of future cliff erosion (Barlow et al. 2012; Gilham et al. 2018).

15.3 Geomorphic Tools for Reconstructing Past Sea Levels

Unravelling the history of relative sea-level change from geological archives has been a long-standing topic of research. Detailed measurement of oxygen isotopes in marine sediments and ice cores has contributed to linking the chronology of global sea-level change with those of terrestrial ice volumes and mean global temperatures during the late Cenozoic Period. An integrated proxy record from the Red Sea now spans the past 5.3 Myr (Rohling et al. 2014). This chronology shows that average rates of sea-level rise during and immediately following periods of rapid deglaciation in the Pleistocene were of the order of two metres per century and hence much higher than contemporary rates. Rates of sea-level rise reached their maximum within a couple of thousand years of the onset of most of these deglaciation phases (Grant et al. 2014).

Many studies of past sea-level change use coastal landforms as local markers of wave erosion. Some of the most widely used proxies of sea-level change include marine terraces, corals, uplifted beaches, erosional notches, fossils or biogenic traces in rocky cliffs, salt-marsh deposits, and archaeological remains (Milne et al. 2009) (Table 15.1). Dating of trees that are recolonizing uplifting shorelines can yield clues about the rate of relative sea-level fall, provided that this rate is high enough to produce measurable differences in slowly expanding tree stands (Motyka 2003). Cosmogenic nuclide exposure dating of wave-cut notches and beach rocks is another option for young rocks that have been only recently uplifted above sea level. Matsushi et al. (2006) measured the concentrations of ^{10}Be and ^{26}Al in soil profiles on hillslope interfluves of the Boso Peninsula, southeast of Tokyo, Japan.

Table 15.1 Selected methods for reconstructing sea level and their precision.

Time period (kyr BP)	Method	Chronology	Max. resolution (yr)	Estimated precision (± m)	Max. rate (mm yr^{-1})
0–470	Oxygen isotopes	AMS^{14}C palaeomagnetism, tuning	200	12	25
0–30	Corals	U/Th	400	5	40
0–20	Sediment facies, microfossils	AMS^{14}C	200	3	40
0–16	Isolation basin stratigraphy	AMS^{14}C	200	0.2-1.0[a]	n/a
0–10	Basal peat	AMS^{14}C	200	0.2-0.5[a]	2
0–7	Microatolls	^{14}C	200	0.1-0.2[a]	2
0–7	Biological indicators on rocky coasts	^{14}C	200	0.05-0.5[a]	1
0–2	Archaeology	Historical documentation	100	0.1-0.5[a]	1
0–0.5	Salt-marsh microfossils	AMS^{14}C,^{210}Pb,^{137}Cs, Pb isotopes, pollen, chemostratigraphy	20	0.05-0.3[a]	2

a) Uncertainties depend on the tidal range. Some methods express the error on inferred past sea level by analysing the distribution of flora and fauna within the tidal zone. In such cases the estimated uncertainty is less in areas with a smaller tidal range. AMS = accelerator mass spectrometry. Adapted from Milne et al. (2009).

Bedrock in the area consists of Pleistocene marine mudstones that have been uplifted up to 350 m above sea level over the past 0.45 Myr.

Most studies emphasize that we need to understand the geomorphic consequences of sea-level change when it comes to assessing natural hazards. In tectonically active regions that are prone to large earthquakes, knowledge of sea-level history is useful for mapping the pattern of long-term coastal deformation, which may include coseismic uplift or subsidence events (Tamura et al. 2010). For example, dating of cryptotephra in uplifted marine and fluvial terraces on northeastern Honshu, Japan, revealed a trend of long-term uplift at odds with contemporary tide gauge records that show subsidence (Matsu'ura et al. 2009). This mismatch characterizes many of Japan's coasts, where long-term Pleistocene uplift is in contrast with contemporary rates of sea-level rise of several mm yr^{-1} obtained from tidal gauges and a dense network of

GPS stations (Kimura et al. 2008). One possible explanation for this difference involves differing rates and directions of interseismic and coseismic deformation along Japan's subduction zones. In these areas vertical crustal motion can include coseismic subsidence juxtaposed with rapid uplift, and thus give a false impression of sea-level changes based on a few years of record only (Ballu et al. 2011).

Deciphering past sea-level change in tectonically active settings requires extra care in areas that have experienced widespread glaciation, for example Iceland and southern Alaska. In these areas glacio-isostatic rebound may have elevated coastlines at high ($\sim10^1$ mm yr^{-1}) rates. Such extreme rates of postglacial rebound due to the loss of Pleistocene ice masses need to be disentangled from earthquake-related deformation. Much of the contemporary extreme uplift in Glacier Bay, southeast Alaska occurs at rates of >30 mm yr^{-1}, and dates to the past 200 years, which was a period of widespread glacier ice loss following the Little Ice Age (Larsen et al. 2005).

Brothers et al. (2013) modelled the impacts of eustatic sea-level rise and added loading stresses to fault systems along continental margins, concluding that this effect could induce bending stresses that might lead to elevated seismicity, which in turn could trigger large submarine landslides. These increased loading stresses may also have been responsible for increases in volcanic activity. Time-series analyses of dated ash deposits in marine sediments along the Pacific Ring of Fire have shown that increases in volcanism lag behind the highest rates of Quaternary sea-level rise by ~4 kyr, which is consistent with numerical predictions of maximum stress changes (Kutterolf et al. 2013).

The geoarchaeology of harbours and coastal infrastructure such as breakwaters or wells informs us about past sea-level change, and how coastal communities have adapted (or failed to adapt). The accuracy of archaeological markers of former sea level depends on how well their functional height can be constrained, and therefore on the vertical position of a given marker relative to mean sea level. Remains of former fish ponds are among the most reliable of such markers, because they were constructed with close attention to the tidal range at the time, such that sluice gates, for example, allowed the replenishment of seawater (Auriemma and Solinas 2009). Similarly, kilns for separating salt from sea water were built close to the winter and spring high-tide level in the Sundarbans area of the outer Ganges-Brahmaputra delta. Hanebuth et al. (2013) used ^{14}C and OSL to date the sudden burial of several such kilns currently lying >1.5 m below sea level, and obtained a relative rate of local sea-level rise of 5.2 ± 1.2 mm yr^{-1} over the past 300 years. Rapid sinking of the delta surface, possibly catastrophically following major earthquakes or due to swift sediment compaction, contributed to more than half of this rate, which is within the range of 3–8 mm yr^{-1} recorded by

several tidal gauges (Pethick and Orford 2013). What destroyed these kilns simultaneously remains a mystery, though Hanebuth et al. (2013) speculated that a combination of large earthquakes and cyclone-driven storm surges was the most likely cause. Large local deviations from such long-term trends are worrying, and water levels at the Ganges–Brahmaputra delta have risen as rapidly as ~17 mm yr^{-1} some 120 km inland, strikingly outpacing rates measured at tidal gauges. Much of this difference may have arisen from the many artificial embankments created along tidal channels to reclaim more land for settlement. These embankments reduce channel cross-sections and the availability of floodplains to store flood sediment and waters, and hence inadvertently amplify the local tidal range (Pethick and Orford 2013). Other effects such as land subsidence or groundwater extraction might also add to the relative sea-level rise. Only by considering robust and long chronologies can we appraise how much recent human-induced changes contribute to this rise.

Other remains of harbours, such as walkways on piers, bollards or clinch rings, are also good indicators of former sea levels that researchers have put to good use. In the Mediterranean, for example, the pace at which Holocene sea level rose slowed down about 6000 years ago, and sediment delivered by rivers initiated a phase of coastline progradation, submerging and burying prehistoric and historic harbours beneath several metres of sediment. Some coastal cities became landlocked or their harbours were gradually relocated (Figure 15.9) (Marriner and Morhange 2007). The Po delta of northeastern Italy advanced at an average rate of 4.5 m yr^{-1} between the sixth century BC and 1600 AD. This rate has accelerated to 70 m yr^{-1} since then, partly due to land-use changes (Walsh 2014). Remains of wave and current control structures and scour marks of dredging in several harbours of the Roman period attest to early engineering efforts to mitigate the build-up of sediment resulting, in part, from

Ancient Harbour Parasequence

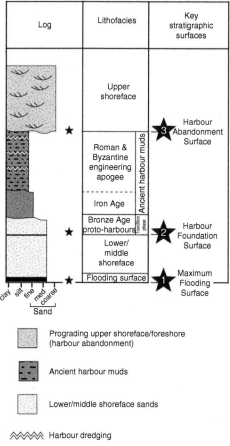

Progant Prograding upper shoreface/foreshore
(harbour abandonment)

Ancient harbour muds

Lower/middle shoreface sands

Harbour dredging

key stratigraphic surfaces

Figure 15.9 Illustration of the impact of human palaeoengineering on coastal deposition in Mediterranean harbours and their vicinity, resulting in distinct lithofacies and key stratigraphic surfaces. From Anthony et al. (2014).

human activity, given that these harbours acted as artificial sediment traps. Eventually, however, sediment overwhelmed these efforts at mitigation and obliterated many Mediterranean harbours during one of the major phases of human-induced delta progradation in the past millennium. As Walsh (2014) pointed out:

> There clearly comes a point in time when people, and engineers in particular, are

aware of the potential environmental problems, such as sedimentation within a harbour [...]. Either it was accepted that this would happen, or the intensity or even the nature of the process was not understood.

The Roman period, when human-induced sediment supply in rivers caused many deltas to prograde, was one of the two major phases of this kind in the past six millennia. The other phase occurred during the Little Ice Age, when pronounced population growth and higher river discharge further enhanced sediment supply to deltas in the Mediterranean (Anthony et al. 2014). This and many other insights from studies that reconstruct sea-level curves are essential for informing us how the geomorphology and exposure of coasts may change under future sea-level rise.

15.4 A Future of Rising Sea Levels

Sea-level rise has become a prominent feature in the media as a consequence of global warming. Projections for the next two millennia show that warming by another 3 K could raise sea levels enough to inundate more than half of the territory of three to as many as 12 nations. About two to three dozen countries would lose some 10% of their current land area, and about 7% of the global population would be displaced (Marzeion and Levermann 2014).

Estimates of the total contribution of continuing ice melt to sea-level rise differ widely. The many attempts to consolidate these estimates rely on a combination of scientific disciplines, including glaciology, remote sensing, and coastal geomorphology. As much as 60% of the contemporary global ice loss could be from mountain glaciers and ice caps, and this meltwater input almost exclusively accounts for the nonsteric component of sea-level rise (Meier et al. 2007). Observed temperature increases in the past several decades have been highest in the polar regions, where precipitation might increase by more than 50% above current levels by the end of the twenty-first century. The Arctic is more sensitive than other areas of the planet to increases in precipitation per unit warming. There, the rate of increase in mean precipitation is estimated at 4.5% K^{-1} (per unit temperature increase) as opposed to the global average of 1.6–1.9% K^{-1} (Bintanja and Selten 2014). An increase in Arctic precipitation is likely to produce higher runoff and river discharge, and thus feed larger amounts of freshwater into the Arctic Ocean. Snowfall over ice sheets could also increase as a consequence, thus affecting their mass balance and ultimately the rate of sea-level rise.

Although most meltwater sources are located in polar and mountainous regions, it is the lower latitude coastal areas, particularly low-lying and low-gradient coasts, that are most exposed to contemporary and future sea-level rise. The simplest method of predicting the extent of inundation from sea-level rise is a simple 'bathtub' model that computes the fraction of land area lost for a given vertical increment of sea-level rise. With Geographic Information System (GIS) software and suitably resolved digital elevation models (DEMs), this method allows planners to determine the effects of sea-level rise on coastal communities globally. The bathtub model can be extended by account dynamic effects of wave action, storms or tides, and it depends on the accuracy of forecasts of future sea-level rise. Forecasts of the amount and rates of sea-level rise are constantly updated as numerical models grow more complex and measurements increase in number and accuracy. Levermann et al. (2013), for example, anticipate a global average rate of sea-level rise of 3.2 m K^{-1} for the next 2000 years(!). This estimate comes from physical modelling and geological proxies, and takes into account the expected future contribution of melting from the Greenland and Antarctic ice sheets, which potentially will outpace ocean warming and glacier melt as the dominant drivers of

sea-level rise (Figure 15.10). With the effects of geoid irregularities, wind fields, ocean currents, freshwater influx, and local isostatic and tectonic effects, the projected sea-level rise will differ between regions, and forecasts will need to consider this variation.

Woodruff et al. (2014) stress the role that geomorphology, together with sustainable land use, can play in facing such problems by noting that these 'impacts can be mitigated partly with adaptive strategies, which include careful stewardship of sediments and reductions in human-induced land subsidence'. A global study of 40 large river deltas that together sustain nearly 300 million people reported that rates of local sea-level rise range from 0.5 to 12.5 mm yr^{-1}. About one-fifth of these deltas have been subject to accelerated subsidence, and eustatic changes dominate the sea-level trends in only five cases (Ericson et al. 2005). Groundwater extraction is among the main contemporary causes of the subsidence of some of the world's major and densely populated deltas. The highest rates of contemporary relative sea-level rise and inferred risk are associated with deltas in southeast Asia and the developing parts of Africa and the Middle East (Tessler et al. 2015). Many deltas with large urban centres have sunk by several metres in the past decades. The city of Manila, Philippines, subsided by more than one metre between 1991 and 2003, which amounts to an average subsidence rate of >80 mm yr^{-1} (Woodruff et al. 2014). The sinking Chao-Phraya River delta, on which Thailand's capital Bangkok is built, has caused more than 1 km of shoreline retreat in places, already submerging telegraph poles (Nicholls and Cazenave 2010). The populations of coastal cities built on those deltas are likely to grow into the twenty-first century, thus raising their exposure to storms and floods.

This high exposure of low-lying population centres to flooding from the sea is far beyond a problem limited to developing countries only. In Japan, for example, some 5.4 million people live in areas below sea level, mostly in

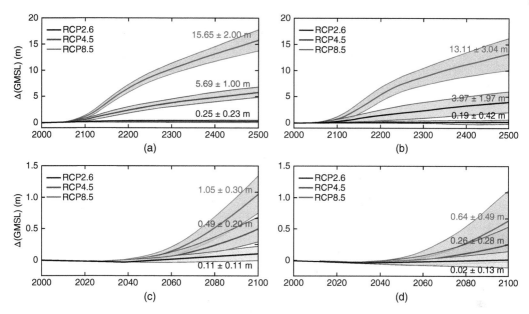

Figure 15.10 Large ensemble models of future Antarctic contributions to global mean sea level (GMSL), tuned to reconstructions of Pliocene sea levels that were higher than today some 3 million years ago. (a) RCP ensembles to 2500. (b) RCP ensembles to 2100 with a Pliocene GMSL target of 10–20 m. Changes in GMSL are shown relative to 2000, although the simulations begin in 1950. (c) and (d) Same as (a) and (b) but with a lower Pliocene GMSL target of 5–15 m. Solid lines are ensemble means, and the shaded areas show the standard deviation (1σ) of the ensemble members. The 1σ ranges represent the model's parametric uncertainty, whereas the alternate Pliocene targets (a and b vs. c and d) illustrate the uncertainty related to poorly constrained Pliocene sea-level targets. Mean values and 1σ uncertainties at 2500 and 2100 are shown. From DeConto and Pollard (2016).

Tokyo, Osaka, and Nagoya. A projected rate of sea-level rise of 0.6 m per decade, forecast in the worst-case scenario by the Intergovernmental Panel on Climate Change (IPCC), could result in a 50% increase in populated areas that are below sea level, although the approximate time frame over which this may occur remains debated. Scenario-based models for the city of Shanghai predict that sea level could rise at rates of ~19 mm yr^{-1} by 2050. The land on which the city is built is subsiding at rates of up to 24 mm yr^{-1}, making this problem even worse. If this trend were to continue without remedial or adaptive measures, almost half of the coastal defences would be overtopped and up to half of the city could be flooded by the year 2100 (Wang et al. 2012).

Sea-level rise along the US Gulf Coast will increase the hazard from storm surges, particularly during hurricanes (Wahl et al. 2014). Both measurements and climate model simulations show that extreme wave heights for a given return period have been increasing in many coastal regions around the world, a trend that will continue according to most studies. In southern California, for example, the projected average land losses by coastal cliff retreat alone are 19–41 m in the twenty-first century simply because of higher sea levels (Limber et al. 2018). This prediction remains to be validated but comes from an ensemble of five different mechanistic models of cliff retreat. Uncertainties of these model outputs were high (between 5 and 15 m), judging from a Monte Carlo simulation that accounted for unknown or poorly constrained parameter inputs. Yet the overall trend shows that future rates of cliff retreat may be twice as high as historic ones. In a similar study concerned with future erosion of a rock cliff in southern England, Gilham et al. (2018) argued that the size distribution of slope failures obtained from laser scanning was (moderately) correlated with significant wave heights. They used this relationship to predict more than 20 m of local cliff retreat by the end of the twenty-first century given the projections of sea-level rise in a climate scenario of medium-level greenhouse emissions.

With this focus on extreme events, we recall that rising sea levels will also have more frequent smaller impacts. William and Park (2014) investigated coastal 'nuisance-level' floods that lie between the mean and the highest tidal levels. They defined nuisance inundations as those surpassing Mean Higher High Water (MHHW) by as much as 0.6 m and found that their annual exceedance probability had increased from the 1920s to 2015, with large contributions from El Niño events along the coast of the western United States.

Finally, attributing all contemporary sea-level rise to atmospheric warming can be misleading. In 2005, the United Nations Environmental Program (UNEP) announced in a press release that inhabitants of the Torres Islands of Vanuatu, southwest Pacific, would become the first 'climate-change refugees', since their islands began experiencing periodic flooding related to a rising sea level. Detailed GPS data confirmed that sea level rose by 150 ± 20 mm between 1997 and 2009. The same GPS data also highlighted that most of this rise is attributable to subsidence of the islands. Large earthquakes before and after the measurement period caused several hundred millimetres of sudden vertical motion, thus adding noise to the climatic signal of sea-level rise around the islands (Ballu et al. 2011). In essence, we must correct local rates of sea-level rise for seismic and tectonic activity in some coastal regions. From a similar, though broader, background Le Cozannet et al. (2014) reviewed studies that were concerned with how rising sea levels could modify coasts worldwide, and recognized two general approaches; one searching for correlations between trends and variability in observed coastal change and sea-level curves, and the other trying to attribute observed coastal changes to sea-level rise via numerical modelling. Their sobering conclusion was that interpretations regarding the role of sea-level rise in driving coastal changes differed widely, mainly because of the large influence of local geomorphic conditions. One lesson to take away from this is that assessments of coastal hazards due to sea-level rise will require mostly site-specific work.

References

Anthony EJ, Marriner N, and Morhange C 2014 Human influence and the changing geomorphology of Mediterranean deltas and coasts over the last 6000 years: From progradation to destruction phase? *Earth-Science Reviews* **139**(C), 336–361.

Auriemma R and Solinas E 2009 Archaeological remains as sea level change markers: A review. *Quaternary International* **206**(1-2), 134–146.

Ballu V, Bouin MN, Simeoni P, et al. 2011 Comparing the role of absolute sea-level rise and vertical tectonic motions in coastal flooding, Torres Islands (Vanuatu). *Proceedings of the National Academy of Sciences* **108**(32), 13019–13022.

Barlow J, Lim M, Rosser N, et al. 2012 Modeling cliff erosion using negative power law scaling of rockfalls. *Geomorphology* **139-140**(C), 416–424.

Bintanja R and Selten FM 2014 Future increases in Arctic precipitation linked to local evaporation and sea-ice retreat. *Nature* **509**(7501), 479–482.

Brothers DS, Luttrell KM, and Chaytor JD 2013 Sea-level-induced seismicity and submarine landslide occurrence. *Geology* **41**, 979–982.

Cazenave A and Cozannet GL 2014 Sea level rise and its coastal impacts. *Earth's Future* **2**(2), 15–34.

Church JA and White NJ 2011 Sea-level rise from the late 19th to the early 21st century. *Surveys in Geophysics* **32**(4-5), 585–602.

Conery I, Walsh JP, and Corbett DR 2018 Hurricane overwash and decadal-scale evolution of a narrowing barrier island, Ocracoke Island, NC. *Estuaries and Coasts* **41**, 1626–1642.

Cooper JAG and Pilkey OH 2004 Sea-level rise and shoreline retreat: time to abandon the Bruun Rule. *Global and Planetary Change* **43**(3-4), 157–171.

DeConto RM and Pollard D 2016 Contribution of Antarctica to past and future sea-level rise. *Nature* **531**(7596), 591–597.

Earlie C, Masselink G, and Russell P 2018 The role of beach morphology on coastal cliff erosion under extreme waves. *Earth Surface Processes and Landforms* **43**(6), 1213–1228.

Ericson JP, Vörösmarty CJ, Dingman SL, et al. 2005 Effective sea-level rise and deltas: Causes of change and human dimension implications. *Global and Planetary Change* **50**(1), 63–82.

Farquharson LM, Mann DH, Swanson DK, et al. 2018 Temporal and spatial variability in coastline response to declining sea-ice in northwest Alaska. *Marine Geology* **404**, 71–83.

FitzGerald DM, Fenster MS, Argow BA, and Buynevich IV 2008 Coastal impacts due to sea-level rise. *Annual Review of Earth and Planetary Sciences* **36**(1), 601–647.

Gilham J, Barlow J, and Moore R 2018 Marine control over negative power law scaling of mass wasting events in chalk sea cliffs with implications for future recession under the UKCP09 medium emission scenario. *Earth Surface Processes and Landforms* **43**(10), 2136–2146.

Grant KM, Rohling EJ, Ramsey CB, et al. 2014 Sea-level variability over five glacial cycles. *Nature Communications* **5**, 1–9.

Günther F, Overduin PP, Sandakov AV, et al. 2013 Short- and long-term thermo-erosion of ice-rich permafrost coasts in the Laptev Sea region. *Biogeosciences* **10**(6), 4297–4318.

Hanebuth TJJ, Kudrass HR, Linstadter J, et al. 2013 Rapid coastal subsidence in the central Ganges-Brahmaputra Delta (Bangladesh) since the 17th century deduced from submerged salt-producing kilns. *Geology* **41**(9), 987–990.

Hansom JD and Hall AM 2009 Magnitude and frequency of extra-tropical North Atlantic cyclones: A chronology from cliff-top storm deposits. *Quaternary International* **195**(1-2), 42–52.

Hanson S, Nicholls R, Ranger N, et al. 2010 A global ranking of port cities with high exposure to climate extremes. *Climatic Change* **104**(1), 89–111.

Hapke CJ 2005 Estimation of regional material yield from coastal landslides based on historical digital terrain modelling. *Earth Surface Processes and Landforms* **30**(6), 679–697.

Hapke CJ and Green KR 2006 Coastal landslide material loss rates associated with severe climatic events. *Geology* **34**(12), 1077.

IPCC (Intergovernmental Panel on Climate Change) 2013 *Climate Change 2013; The Physical Science Basis; Working Group I Contribution to the Fifth Assessment Report of the Intergovernmental Panel on Climate Change.*

Jackson DWT, Costas S, and Guisado-Pintado E 2019 Large-scale transgressive coastal dune behaviour in Europe during the Little Ice Age. *Global and Planetary Change* **175**, 82–91.

Kimura G, Kitamura Y, Yamaguchi A, and Raimbourg H 2008 Links among mountain building, surface erosion, and growth of an accretionary prism in a subduction zone—An example from southwest Japan. *Geological Society of America Special Papers* **436**, 391–403.

Kjeldsen KK, Korsgaard NJ, Bjørk AA, et al. 2015 Spatial and temporal distribution of mass loss from the Greenland Ice Sheet since AD 1900. *Nature* **528**, 396–400.

Kroonenberg SB, Rusakov GV, and Svitoch AA 1997 The wandering of the Volga delta: a response to rapid Caspian sea-level change. *Sedimentary Geology* **107**, 189–209.

Kutterolf S, Jegen M, Mitrovica JX, et al. 2013 A detection of Milankovitch frequencies in global volcanic activity. *Geology* **41**(2), 227–230.

Larsen CF, Motyka RJ, Freymueller JT, et al. 2005 Rapid viscoelastic uplift in southeast Alaska caused by post-Little Ice Age glacial retreat. *Earth and Planetary Science Letters* **237**(3-4), 548–560.

Le Cozannet G, Garcin M, Yates M, et al. 2014 Approaches to evaluate the recent impacts of sea-level rise on shoreline changes. *Earth-Science Reviews* **138**(C), 47–60.

Levermann A, Clark PU, Marzeion B, et al. 2013 The multimillennial sea-level commitment of global warming. *Proceedings of the National Academy of Sciences* **110**(34), 13745–13750.

Limber PW, Barnard PL, Vitousek S, and Erikson LH 2018 A model ensemble for projecting multidecadal coastal cliff retreat during the 21st century. *Journal of Geophysical Research: Earth Surface* **123**(7), 1566–1589.

Marriner N and Morhange C 2007 Geoscience of ancient Mediterranean harbours. *Earth-Science Reviews* **80**(3-4), 137–194.

Marzeion B and Levermann A 2014 Loss of cultural world heritage and currently inhabited places to sea-level rise. *Environmental Research Letters* **9**(3), 034001.

Matsushi Y, Wakasa S, Matsuzaki H, and Matsukura Y 2006 Long-term denudation rates of actively uplifting hillcrests in the Boso Peninsula, Japan, estimated from depth profiling of in situ-produced cosmogenic ^{10}Be and ^{26}Al. *Geomorphology* **82**(3-4), 283–294.

Matsu'ura T, Yamaguchi A, and Saomoto H 2009 Long-term and short-term vertical velocity profiles across the forearc in the NE Japan subduction zone. *Quaternary Research* **71**(2), 227–238.

McNamara DE and Werner BT 2008 Coupled barrier island–resort model: 1. Emergent instabilities induced by strong human-landscape interactions. *Journal of Geophysical Research: Earth Surface* **113**, 1–10.

Medjkane M, Maquaire O, Costa S, et al. 2018 High-resolution monitoring of complex coastal morphology changes: cross-efficiency of SfM and TLS-based survey (Vaches-Noires cliffs, Normandy, France). *Landslides* **15**, 1097–1108.

Meier MF, Dyurgerov MB, Rick UK, et al. 2007 Glaciers dominate eustatic sea-level rise in the 21st century. *Science* **317**(5841), 1064–1067.

Mentaschi L, Vousdoukas M, Pekel JF, et al. 2018 Global long-term observations of coastal erosion and accretion. *Scientific Reports* **8**, 1–11.

Milne GA, Gehrels WR, Hughes CW and Tamisiea ME 2009 identifying the causes of sea-level change. *Nature Geoscience* **2**(7), 471–478.

Motyka RJ 2003 Little Ice Age subsidence and post Little Ice Age uplift at Juneau, Alaska, inferred from dendrochronology and geomorphology. *Quaternary Research* **59**(3), 300–309.

Murray NJ, Phinn SR, DeWitt M, et al. 2018 The global distribution and trajectory of tidal flats. *Nature* **565**, 223–225.

Mushkin A, Katz O, and Porat N 2018 Overestimation of short-term coastal cliff retreat rates in the eastern Mediterranean resolved with a sediment budget approach. *Earth Surface Processes and Landforms* **44**(1), 179–190.

Neumann B, Vafeidis AT, Zimmermann J, and Nicholls RJ 2015 Future coastal population growth and exposure to sea-level rise and coastal flooding – A global assessment. *PLoS ONE* **10**(3), e0118571.

Nicholls RJ and Cazenave A 2010 Sea-level rise and its impact on coastal zones. *Science* **328**(5985), 1517–1520.

Ostanciaux É, Husson L, Choblet G, et al. 2012 Present-day trends of vertical ground motion along the coast lines. *Earth-Science Reviews* **110**(1-4), 74–92.

Pánek T, Korup O, Minár J, and Hradecký J 2016 Giant landslides and highstands of the Caspian Sea. *Geology* **44**(11), 939–942.

Pethick J and Orford JD 2013 Rapid rise in effective sea-level in southwest Bangladesh: Its causes and contemporary rates. *Global and Planetary Change* **111**(C), 237–245.

Pilkey OH and Cooper JAG 2014 *The Last Beach.* Duke University Press.

Prémaillon M, Regard V, Dewez TJB, and Auda Y 2018 GlobR2C2 (Global Recession Rates of Coastal Cliffs): a global relational database to investigate coastal rocky cliff erosion rate variations. *Earth Surface Dynamics* **6**(3), 651–668.

Reed AJ, Mann ME, Emanue, KA, et al. 2015 Increased threat of tropical cyclones and coastal flooding to New York City during the anthropogenic era. *Proceedings of the National Academy of Sciences* **112**, 12610–12615.

Rietbroek R, Brunnabend SE, Kusche J, et al. 2016 Revisiting the contemporary sea-level budget on global and regional scales. *Proceedings of the National Academy of Sciences* **113**(6), 1504–1509.

Rohling EJ, Foster GL, Grant KM, et al. 2014 Sea-level and deep-sea-temperature variability over the past 5.3 million years. *Nature* **508**, 477–482.

Rosser NJ, Brain MJ, Petley DN, et al. 2013 Coastline retreat via progressive failure of rocky coastal cliffs. *Geology* **41**(8), 939–942.

Ruessink G and Ranasinghe R 2014 Beaches. In *Coastal Environments & Global Change* (eds Masselink G and Gehrels R), Wiley Blackwell, pp. 149–177.

Schumm SA 1993 River response to base-level change: implications for sequence stratigraphy. *The Journal of Geology* **101**(2), 279–294.

Stiros SC 2010 The 8.5+ magnitude, AD365 earthquake in Crete: Coastal uplift, topography changes, archaeological and historical signature. *Quaternary International* **216**(1-2), 54–63.

Syvitski JPM, Kettner AJ, Overeem I, et al. 2009 Sinking deltas due to human activities. *Nature Geoscience* **2**(10), 681–686.

Tamura T, Murakami F, and Watanabe K 2010 Holocene beach deposits for assessing coastal uplift of the northeastern Boso Peninsula, Pacific coast of Japan. *Quaternary Research* **74**(2), 227–234.

Tessler ZD, Vörösmarty CJ, Grossberg M, et al. 2015 Profiling risk and sustainability in coastal deltas of the world. *Science* **349**, 638–643.

Wahl T, Calafat FM, and Luther ME 2014 Rapid changes in the seasonal sea level cycle along the US Gulf coast from the late 20th century. *Geophysical Research Letters* **41**, 491–498.

Wallace DJ and Anderson JB 2013 Unprecedented erosion of the upper Texas coast: Response to accelerated sea-level rise and hurricane impacts. *Geological Society of America Bulletin* **125**(5-6), 728–740.

Walsh K 2014 *The Archaeology of Mediterranean Landscapes.* Cambridge University Press.

Wang J, Gao W, Xu S, and Yu L 2012 Evaluation of the combined risk of sea level rise, land subsidence, and storm surges on the coastal areas of Shanghai, China. *Climatic Change* **115**(3-4), 537–558.

William, V S and Park J 2014 From the extreme to the mean: Acceleration and tipping points of coastal inundation from sea level rise. *Earth's Future* **2**, 579–600.

Woodruff JD, Irish JL, and Camargo SJ 2014 Recent shifts in coastline change and shoreline stabilization linked to storm climate change. *Nature* **504**(7478), 569–585.

You ZJ and Nielsen T 2013 *Extreme coastal waves, ocean surges and wave runup.* (ed Finkl CW), Springer Science & Business Media, pp. 677–733.

Young AP, Flick RE, Gallien TW, et al. 2018 Southern California coastal response to the 2015-2016 El Niño. *Journal of Geophysical Research: Earth Surface* **123**(11), 3069–3083.

Zerbini S, Raicich F, Prati CM, et al. 2017 Sea-level change in the Northern Mediterranean Sea from long-period tide gauge time series. *Earth-Science Reviews* **167**, 72–87.

16

How Natural are Natural Hazards?

We have seen that, to assess risk quantitatively, we need to gather information about the values at specific risk, their vulnerability, and the probability that a potentially damaging event will occur in the first place. In many cases, the psychological aspects of risk management, termed aversion, also play a role. We have tacitly assumed that all of these factors remain constant over time. This view may be convenient, but of limited use, to maintain in a world with a rapidly growing and expanding human population that is transforming much of its environment. Our short account on future sea-level changes and the contribution by human activities (Chapter 15.4) was but one example of how we increasingly interfere with natural systems, and hence natural hazards.

Maintaining a static or stationary viewpoint on natural risks may be mathematically convenient, but ignores the many superimposed cycles, oscillations, and feedbacks in Earth's lithosphere, hydrosphere, atmosphere, and biosphere. For example, earth scientists estimate that up to 90% of all natural disasters are linked to meteorological phenomena. Hence, changing weather and climate should result in changing natural hazards and risks. In the light of this change, scientists are increasingly adopting statistical methods that recognize nonstationary processes to counter the reduced capacity for predictions that can arise from time-series data that have varying means and variances. Nevertheless, the more straightforward and traditional predictions based on the simplifying assumptions of stationarity may often give approximate enough results to be understood and used by practitioners (Montanari and Koutsoyiannis 2014). A confounding factor for predicting natural hazards and risks is that humans have interfered with many elements of the Earth system. Most seemingly natural processes bear an anthropogenic fingerprint. One major issue is whether we can still refer to natural hazards as 'natural'. The scientific literature is now teeming with terms and labels such as 'critical thresholds', 'tipping points', 'ecological footprints', 'carrying capacities', and 'planetary boundaries' (Steffen et al. 2015) emphasizing that Earth's natural resources are finite, so there might be irreversible system changes that we need to consider seriously if sustainability is a concern (Figure 16.1). In this chapter we question to what degree natural hazards can still be regarded as 'natural'. Note that this enquiry differs from the assertion that 'natural disasters' – which are potential triggers for harmful outcomes – are partly human-made by definition.

16.1 Enter the Anthropocene

Humans have transformed almost all of Earth's natural environments to some degree; figuratively, our fingerprints can now be traced in Earth's water cycle, landscape, soil, vegetation, atmosphere, and the use of fossil fuels,

Geomorphology and Natural Hazards: Understanding Landscape Change for Disaster Mitigation, Advanced Textbook Series,
First Edition. Tim R. Davies, Oliver Korup, and John J. Clague.

The Human Footprint ver. 2

Global

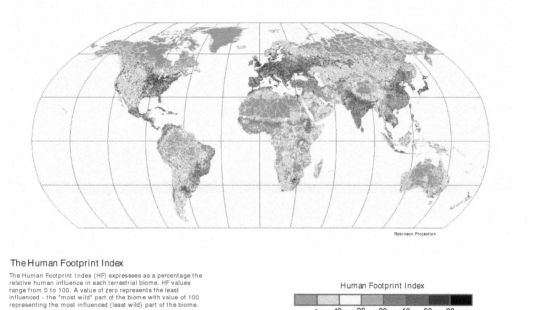

Robinson Projection

The Human Footprint Index

The Human Footprint Index (HF) expresses as a percentage the relative human influence in each terrestrial biome. HF values range from 0 to 100. A value of zero represents the least influenced - the "most wild" part of the biome with value of 100 representing the most influenced (least wild) part of the biome.

Human Footprint Index

1   10   20   30   40   60   80

Figure 16.1 The human footprint on Earth. Human impact is expressed as the percentage of human influence relative to the maximum influence recorded for each biome. Data include human population density, land transformation (including global land cover, roads, and cities), electrical power infrastructure (NOAA night-lights data), and access to the land via roads, navigable rivers, and coastline. From NASA Socioeconomic Data and Applications Center (SEDAC), hosted by CIESIN at Columbia University, https://sedac.ciesin.columbia.edu/maps/gallery/search

rocks, and minerals. Satellite images reveal the dense networks of traffic arteries linking urban centres that stand out as bright islands in Earth's night-time appearance. Many other large built structures can be easily seen from space, including urban sprawl, surface mines, reservoirs, engineered coastlines, and artificial islands. These scenes only hint at the rates at which humans are transforming the natural environment, and thus directly or indirectly altering the boundary conditions for erosion, sediment transport, and deposition.

Whether these human activities are sustainable has engendered a broad body of multifaceted research that attempts to measure resource use under different future scenarios. The global rates of consumption of some 20 renewable (and nonrenewable) resources such as wood, wheat, vegetables, meat, cotton, and peat have increased dramatically between 1960 and 2010 (Seppelt et al. 2014). Societies are approaching the limits of use of these and other resources, if we neglect commensurate changes in technology or other adaptive strategies. Resources that have yet to peak include coal, gas, and oil, but these are nonrenewable over periods that matter to people. A worldwide compilation of natural resources suggests that about one third of currently extractable oil resources, half of the gas reserves, and >80% of the coal reserves must remain untouched from 2010 to 2050 if global warming is to be contained with a maximum temperature increase of two degrees that appears to be required to avert potentially irreversible impacts of climate change (McGlade and Ekins 2015).

From an ecological and biodiversity perspective, the current rate at which humans are directly or indirectly altering, degrading, or destroying habitats has raised the extinction of other species by about 1000 times the estimated background rate, causing losses of both species richness and abundance (Newbold et al. 2015; Pimm et al. 2014). Human-induced global warming is leading to temperatures that are above long-term averages; together with habitat fragmentation, ocean acidification, environmental pollution, overfishing and overhunting, the spread of invasive species and pathogens, and expanding human numbers, this warming exerts more ecological stress than most living species have ever experienced (Barnosky et al. 2011). Current trends are alarming: of the 5–9 million species on our planet, we are losing between 11 000 and 58 000 each year on average. More than 322 terrestrial vertebrate species alone have become extinct through human impacts since 1500 AD (Dirzo et al. 2014), and up to a third of all terrestrial vertebrates are considered to be threatened or endangered. Numbers of individuals of these species have also plummeted, by 28% on average during the past four decades alone. This wave of anthropogenic and global-scale mass extinction has analogues in deep geological time. Some of the consequences of this global defaunation include negative impacts on insect pollination, pest control, water quality, and human health. Native predators in the United States control pests saving Americans US\$ 4.5 billion each year (Dirzo et al. 2014). With further loss of birds, bats, and other insect predators, however, crop losses are likely to grow between two and fourfold in the future. These losses will eventually push up the price of food, while undermining livelihoods based on poorly adapted agricultural practices. Losses of global biodiversity are also likely to increase the number of natural biological hazards and risks.

Our ever-expanding resource use and reworking of our planet's surface may have led Crutzen (2002) to argue that humankind has prompted the beginning of a new geological epoch. He suggested that this epoch be termed the 'Anthropocene'. The dawn of this new epoch was the start of the industrial age with its ever-increasing use of fossil fuels and associated emissions of greenhouse gases. This proposition has generated a lot of attention and research, also well beyond the geosciences. Several new scientific journals have been inaugurated, focusing on this theme or the broader issues of relationships between humans and the environment.

Yet both the proposition and definition of an Anthropocene have triggered much controversy (Lewis and Maslin 2015). Some geologists have argued that, as large as it may be, the human fingerprint on planet Earth is far from a clear and global stratigraphic marker. Waters et al. (2016) rebutted this by presenting an extensive list and chronology of anthropogenic deposits that have the potential to disseminate rapidly and survive in the global geological record (Figure 16.2). These deposits include plastics, concrete, elemental aluminium, polyaromatic hydrocarbons, polychlorinated biphenyls, excess ^{14}C from nuclear bomb tests, and many others. Other critics of the Anthropocene argue that overall the net effects of human interference with natural systems fail to meet the formal requirements needed to define a geological epoch (Autin and Holbrook 2012). On a similar note, Gale and Hoare (2012) concluded that 'much of the work undertaken on the Anthropocene lies beyond stratigraphy, and a stratigraphic definition of this epoch may be unnecessary, constraining and arbitrary'. Other scientists have argued that the Holocene geological epoch already embodies the decisive imprint of humans that renders this epoch different from those of previous interglacials or the Pleistocene altogether.

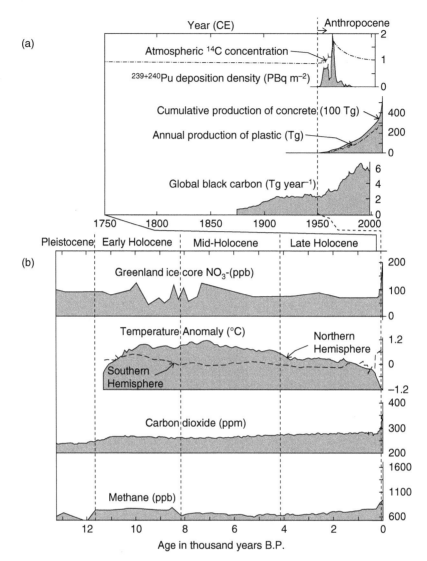

Figure 16.2 Summary of the magnitude of key markers of anthropogenic change that are indicative of the Anthropocene. (a) Novel markers, such as concrete, plastics, global black carbon, and plutonium (Pu) fallout, shown with radiocarbon (^{14}C) concentration. (b) Long-ranging signals such as nitrates (NO_3^-), CO_2, CH_4, and global temperatures, which remain at relatively low values before 1950, rapidly rise during the mid-twentieth century and, by the late twentieth century, exceed Holocene ranges. From Waters et al. (2016).

Aside from the debate over whether human impact is indeed sufficient for future geologists to trace our legacy in sedimentary archives globally, argument arises over when exactly this potentially new geological epoch began. Pinpointing this starting date would need proper and robust geological markers of a synchronous, global, and long-lasting change.

The original definition – based on the viewpoint of an atmospheric scientist – placed the beginning of the Anthropocene at around 1800 AD, coincident with the dawn of industrialization in northwestern Europe. That time of human development was when concentrations of carbon dioxide, among other emissions, began to rise in the atmosphere. However,

Figure 16.3 Comparison of area under land use and methane mixing ratio. Methane mixing ratio records are from Law Dome (red circles) and West Antarctic Ice Sheet (orange circles) ice cores. The shaded area is an uncertainty estimate on global land area under use (blue curve). From Sapart et al. (2012).

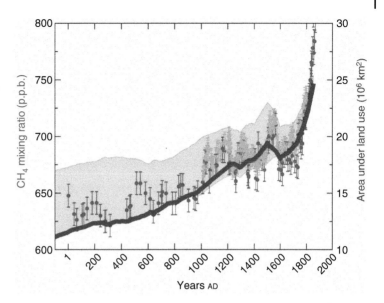

increasing concentrations of atmospheric CO_2 only begin to show up in ice-core records in the nineteenth century (Lewis and Maslin 2015). Other geoscientists have argued that humans began to disturb natural biogeochemical cycles much earlier, and the combined insights from palaeoecological, archaeological, and historical studies have compiled some of the most profound anthropogenic impacts date back some 3000 years at least, if considering that humans may have started extinguishing large land animals, commonly referred to as 'megafauna', on several continents (Ellis et al. 2013; Doughty 2013).

Take, for example, the sharp rise in atmospheric methane (CH_4) concentrations from ~560 ppb to >680 ppb (parts per billion) over the past 5000 years recorded in Antarctic ice cores. This increase in methane appears to coincide with the advent of widespread rice irrigation in eastern Asia in the early Bronze Age. In China the number of archaeological sites with evidence of rice irrigation increased tenfold 6000 to 4000 years ago, offering a possible testimony to several millennia of anthropogenic greenhouse production from one of humankind's most ancient agricultural practices (Ruddiman et al. 2008). According to Ruddiman's hypothesis the trend of increases in atmosphere CO_2 and CH_4 during the Holocene differs strongly from generally declining trends in earlier interglacial periods. Distinct drops in the concentration of CO_2 coincided closely with the downfall of major civilizations, leading to abandoned land use and regenerating forest cover. During the past 2000 years at least, oscillations in the content of atmospheric CH_4 in air bubbles in Greenland ice cores are correlated with some major shifts in human population and land use, such as the decline of the Roman empire or the Chinese Han dynasty, but also with natural climatic oscillations (Figure 16.3) (Sapart et al. 2013). Hence, it appears that the Holocene record of fluctuating and eventually increasing concentrations of methane cannot yet be exclusively attributed to anthropogenic forcing.

Other geoscientists have argued that studying certain prehistoric cultures offers clues about how humans have impacted their environment. Beach et al. (2015) dubbed the term 'Mayacene' to characterize the impressive range of environmental consequences that arose with the Central American Maya culture between 3000 and 1000 years ago. Examples include their advanced techniques of altering river courses, such that the Mayan landscape was criss-crossed and dotted with thousands of artificial water ponds, reservoirs, canals, and wetland fields, together with monumental stone structures including many well-known temples. Widespread forest clearing during the Maya period is documented in thick layers of soil sediment known as the 'Maya Clays'. Concerted deforestation may have been widespread enough to alter local climate, favouring more warm and dry conditions. Mayan agriculture focused on maize and other food crops that altered the carbon isotope signatures of natural ecosystems, which are distinct from the surrounding tropical ecosystem. Although these impacts are regionally contained, Beach et al. (2015) argued that the 'Mayacene' could serve as an instructive small-scale analogue for a possibly global Anthropocene.

In a detailed review of how humans have impacted planet Earth on continental to global scales, Lewis and Maslin (2015) identified only two possible events or phases that would qualify for change sufficiently large to mark the global onset of an Anthropocene. First is the encounter of the Old with the New World, with its associated rise of colonialism, resource exploitation, conversion of natural ecosystems to parcels for agriculture and forestry, transatlantic trade, and exchange of goods, and biota. This event may have been responsible for a large and rapid decrease of CO_2 recorded in ice cores around 1610 AD. Some scientists see this decrease as the result of enhanced carbon sequestration by secondary vegetation that expanded following the rapid decline of Native American populations, farming, and fire use. This chain of events is encapsulated in the 'Orbis hypothesis'. A second candidate for the beginning of an Anthropocene is the nuclear bomb-derived ^{14}C peak in 1964 AD that is recorded worldwide in tree rings; several other widely traceable radionuclides mark the onset of rapid anthropogenic production and spread of artificial and highly toxic substances. In discussing these two candidate events, Lewis and Maslin (2015) concluded that the 'Orbis spike implies that colonialism, global trade, and coal brought about the Anthropocene',

whereas 'the bomb spike tells a story of an elite-driven technological development that threatens planet-wide destruction'. This view may be over-emotive to some, but emphasizes the manifold aspects that the discussion about the Anthropocene involves.

The question of when to place the beginning of the Anthropocene is far more than a purely academic concern, and defining this new geological epoch will likely have impacts beyond the geosciences. Lewis and Maslin (2015) noted that:

> Defining an early start date may, in political terms, 'normalize' global environmental change. Meanwhile, agreeing a later start date related to the Industrial Revolution may, for example, be used to assign historical responsibility for carbon dioxide emissions to particular countries or regions during the industrial era.

A definitive start date for worldwide human impacts on the Earth system would be a benchmark for determining liability, regardless of who is going to judge that liability eventually. Acknowledging the Anthropocene as a distinct geological epoch would be explicitly acknowledging human impacts on the environment as long-lasting enough to persist in the geological record, and consequently raise the question of accountability. In any case, the Anthropocene has stirred a lot of cross-disciplinary discussion and further raised the awareness of humans as geological and geomorphic agents.

In the following sections, we consider other human impacts on the environment, which have had regional to local consequences, and some of which have shaped the surface of our planet for up to several thousands of years.

16.2 Agriculture, Geomorphology, and Natural Hazards

Millennia of agriculture have changed arable soils in many of Earth's fertile landscapes, and we have encountered the example of the 'Mayacene' in the previous section. Yet the origins of agriculture as the most fundamental type of land use have differed between continents and regions such that a single global onset can hardly be used as an informative geological marker of the Anthropocene. In central Europe, for example, the first major phase of deforestation and widespread soil erosion accompanied the establishment of agriculture during the Neolithic, and can be traced widely in pollen records, lake deposits, and fluvial sediments (Kalis et al. 2003). Much of the eroded soil has remained on hillslopes to the present day, where it forms thick colluvial layers rich in organic carbon (Hoffmann et al. 2013). Expanding from ancient agricultural centres in the Middle East 11 000 to 10 000 years ago, the early farmers spread into the loess-rich areas of central Europe. Brown et al. (2013) emphasized that in the United Kingdom floodplain sedimentation rates in most rivers increased following the beginning of widespread agricultural development. However, the transitions from natural to human-influenced river dynamics differed by up to several thousands years from river to river, highlighting some of the uncertainties when trying to resolve some of the geomorphic consequences of agriculture.

The dawn of agriculture in tropical rainforests was about 6000 years ago, and began with shifting use of small parcels of land, likely affecting a much smaller total area than is the case today (Lewis et al. 2015). Even the southern Amazonian rainforests, which may appear to many of us today as emblematic of a pristine vegetation cover, hide geometric earthworks covering some 12 000 km^2 that attest to a period of deforestation long before Europeans arrived in South America. Archaeological and palaeoecological evidence points to an open savanna landscape that the natives actively maintained against a climate-driven expansion of the rainforests about 2000 years ago (Carson et al. 2014). Some of the long-lasting impacts of humans on the Amazonian landscape involved

raising fields for agriculture, building artificial mounds and roads, though these impacts were mostly tied to the river network, which granted the best access, whereas an ongoing debate concerns the degree of human disturbances on forest ecosystems away from the river network (Piperno et al. 2015).

Many of the Pacific islands also lost much of their forest cover, despite their isolation. Reports by the first European settlers mentioned that some islands had degraded forest stands already. Rolett and Diamond (2004) used a multivariate regression to identify some key predictors of this pre-European deforestation on 69 islands, and found that low-lying, drier, more poleward islands with lower nutrient availability and lower rates of plant growth were most prone. Tephra fall from volcanoes in the Pacific Ring of Fire and airborne dust eroded from Central Asian deserts are some of the natural mechanisms to replenish nutrients on Pacific islands, and thus influence both plant regrowth and ecosystem diversity.

Many ecosystems that today appear to be infertile or uninhabitable were modified by humans several millennia ago. Pastoralists could have even entered high Himalayan valleys as early as 4380 cal yr BP, using fire to clear the natural vegetation (Meyer et al. 2009). Nomadic peoples changed the native vegetation in the Nianbaoyeze Mountains of eastern Tibet, causing incipient soil erosion by 3930 cal yr BP, and early human impacts may date to several thousands years earlier than this (Schlütz and Lehmkuhl 2009). Environmental conditions were more conducive to human settlement than today. Early Holocene lake levels in central and southern Tibet were much higher than today, indicating that more precipitation entered the semiarid to arid interiors of the plateau or that evaporation rates were lower.

Other regions that today are largely inhospitable supported human populations in the past. Rock paintings in the Sahara depicting giraffes and crocodiles inform us that the desert was a much more humid place during the early Holocene. Widespread laminated lake sediments containing mollusc shells, diatoms, and bones of aquatic fauna such as crocodile (*Crocodylus niloticus*) or hippo (*Hippopotamus sp.*) attest to former climatic and ecological conditions that are more reminiscent of Africa's modern savanna (Roberts 2014). Deeply incised and winding bedrock river canyons that are now buried beneath large sand sheets are another legacy of periods with higher runoff and discharge. Today the Sahara is devoid of perennial rivers except for the Nile. The collapse of the summer monsoon and winter rainfalls since the Holocene climatic optimum of the African Humid Period resulted in drought and desiccation six millennia ago, which would have gradually forced people out of this previously more humid region (Brooks 2010). Archaeological and palaeoenvironmental studies thus offer glimpses of when humans began to settle and transform landscapes in response to desert margins that shifted by some 800 km within several centuries. These studies also provide important clues as to how these people dealt with past climate changes, and possibly also

natural disasters, over the past 12 000 years (Kuper and Kröpelin 2006).

Cropland currently covers some 12% of the Earth's ice-free land surface; if grazing lands are added, this fraction increases to ~38% (Foley et al. 2012). According to Amundson et al. (2015), this agricultural sprawl has had more consequences for soil resources than Pleistocene glaciations. The near-global use of soils for food production has had many geomorphic and biogeochemical impacts. Twentieth-century agriculture relied increasingly on more and larger machines, so that intensely used land was graded to be accessible for large-scale production. This practice involved, for example, diverting, straightening or infilling natural channels. These changes triggered higher erosion rates and sediment yields, that in some agricultural landscapes rivalled the geomorphic impacts of changing precipitation or crop evapotranspiration patterns (Schottler et al. 2013). The deforestation of tropical lowlands – that usually precedes intensive agricultural use – seems to have very similar impacts (Figure 16.4). Increasing rates of average river-bank erosion allow meanders

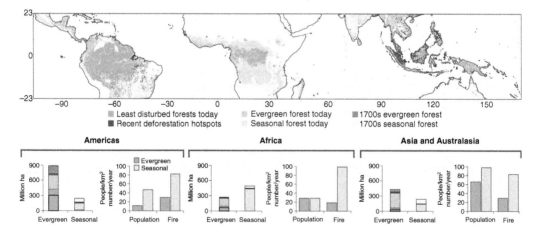

Figure 16.4 Map of the current and historical extent of evergreen and seasonal tropical forest. Grey shading represents the extent of forest before the Industrial Revolution (~ 1700). Green indicates current extent. Darkest green represents currently intact forest landscapes, 95% of which is evergreen forests. Red represents recent intense land-cover change (2000–2012, ≥10% deforestation per 10 km²). Below, for each continent, a pair of bar plots summarize forest data (left) and human population density plus fire numbers within forested areas (right); dark boxes denote a more conservative definition of least-disturbed forests: a 5-km buffer from any high-intensity human influence. From Lewis et al. (2015).

to be more mobile and move faster across their floodplains once the protective vegetation cover and root networks are lost (Horton et al. 2017).

The rates of soil erosion caused by deforestation and agricultural practice have not only substantially raised sediment yields in many rivers (Schmidt et al. 2018), but also have been orders of magnitude higher than the rates of soil replenishment. Landslides, for example, cause long-lasting losses of agricultural productivity by stripping organic-rich topsoils from pastoral steeplands in places like New Zealand. Simulations show how such losses accumulate, but eventually reach a threshold beyond which the rate of increase of losses becomes so low that the economic motivation to build or maintain landslide protection measures is lost (Luckman et al. 1999). Systematically compiling published data, Wilkinson and McElroy (2007) estimated that contemporary rates of erosion under farmland were about 0.6 mm yr^{-1} on average, and occurred mainly in low-lying areas. They further estimated that agricultural erosion from cropland involved up to 75 Gt yr^{-1}, although much of this sediment goes straight into local storage instead of reaching world's oceans directly (Figure 16.5):

> Accumulation of postsettlement alluvium on higher-order tributary channels and floodplains (mean rate ~12,600 m/m.y.) is the most important geomorphic process in terms of the erosion and deposition of sediment that is currently shaping the landscape of Earth. It far exceeds even the impact of Pleistocene continental glaciers or the current impact of alpine erosion by glacial and/or fluvial processes.

Consider the total present-day flux of sediment leaving the terrestrial land surface and entering the oceans through rivers; depending

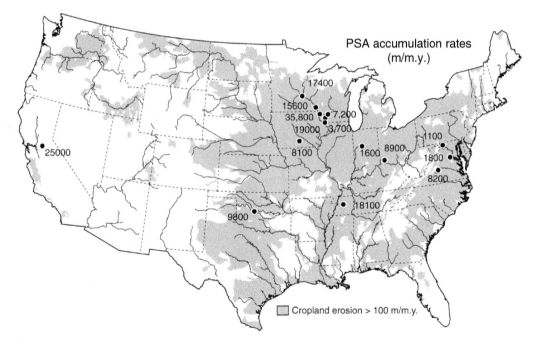

Figure 16.5 The impact of humans on continental erosion and sedimentation. Vertical accumulation rates for 15 reported deposits of post-settlement alluvium (PSA). These occurrences are primarily in valleys of 3rd- to 6th-order catchments directly downslope from areas of accelerated agricultural erosion. Shaded area is the portion of the United States where cropland soil losses exceed 500 m/m.yr. From Wilkinson and McElroy (2007).

on the method of study, this flux is roughly between 10 and 20 gigatons per year. For example, Jennerjahn (2012) estimated this flux to be about 12.6 Gt yr^{-1}, whereas the sediment mass that is eventually reaching the world's oceans from contemporary deforestation and concomitant soil erosion is only 2.3 Gt yr^{-1}. This unsustainable practice has led to widespread and currently irreversible loss of soil resources such as organic carbon. The global reservoir of organic carbon in soils is estimated at ~2300 Gt C, which is much larger than the amount of carbon in the atmosphere and biosphere. Measuring the amount of soil carbon similarly at continental to global scales requires many data sources and extrapolations, and hence uncertainties range as high as ~700 Gt C. For example, Amundson et al. (2015) believe that contemporary soil erosion mobilizes about 40 ± 20 Gt C yr^{-1}, which is about half of the total soil mass loss that Wilkinson and McElroy (2007) suggested independently. Whatever the exact amounts, both cultivation and soil erosion have led to partial oxidation of the terrestrial carbon pool, adding to total CO_2 emissions prior to the second half of the twentieth century. Another direct consequence of soil erosion is that we lose both biological productivity and the capacity for storing soil water. This is also known as 'green water', and an asset, given that soils hold some 65% of the global fresh water resources.

Among the consequences of soil erosion of deforested and agriculturally transformed land we also have to include the commensurate spread and deposition of soil sediment. Many river valleys in the Mediterranean show evidence of widespread aggradation in response to enhanced soil erosion from late Roman to early medieval times. Many of these sedimentary layers contain pottery that testifies to their young ages. This phase of valley filling with reworked soils is only one of three major 'detritic crises' that also contributed to burying river mouths and harbours some 2000 years ago, in late Antiquity, and the Little Ice Age (Marriner and Morhange 2007). Sediments

in an alpine peat mire in Australia document the rapid increase of wind-deposited dust and heavy metals that since the 1880s has exceeded natural background rates by a factor of 10–30 (Marx et al. 2014). The sediments provide a detailed record of the spread of newly arrived European farmers to Australia. On the continent's Southeastern Tablelands, for example, European settlers began to decimate forests, introduce grazing, and eventually transform open woodlands, swampy meadows, and poorly drained chains of ponds to a more gullied landscape with more mobile sediments. Optically stimulated luminescence samples of sediments that infilled former chains of ponds and swamps in this area returned burial dates ranging between 1800 and 1932 CE (Figure 16.6) (Portenga et al. 2016). In tropical Australia, the footprint of European settlement is also documented in a steep rise of sediment volumes flushed offshore, contributing to degrading marine national parks such as the Great Barrier Reef, the world's largest living edifice. With Europeans beginning to clear land for agriculture, and establish cattle grazing and mining at around 1850, suspended sediment yields in rivers have increased more than five-fold. Currently, more than 17 Mt of eroded soils are entering the Great Barrier Reef from a combined onshore catchment area of more than 424 000 km^2 each year (Brodie et al. 2012). This increased sediment load delivers nearly six and nine times the pre-European loads of nitrogen and phosphorus to the reef, respectively, along with high amounts of various pesticides from agricultural runoff. The isotopic fingerprints in some coral species support the historic rise in sediments and nutrients (McCulloch et al. 2003).

Similarly, the arrival of Europeans in New Zealand in the nineteenth century has been associated with unprecedented soil losses in the wake of widespread deforestation that resulted in up to twentyfold increases in sedimentation rates (Glade 2003). Thick layers of 'post-settlement alluvium' – sediments attributed to soil erosion mostly tied to

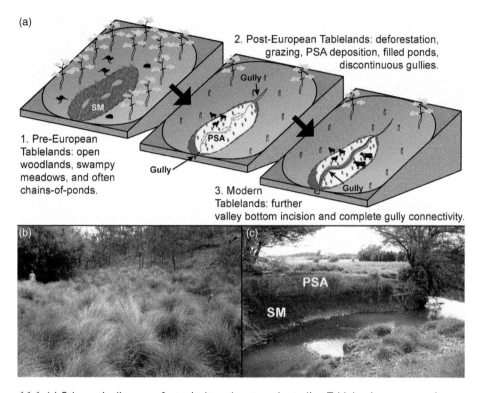

Figure 16.6 (a) Schematic diagram of a typical southeastern Australian Tablelands stream catchment before landscape disturbance (left), after post-settlement alluvium (PSA) deposition during initial gully incision (middle) and after gully connectivity is complete (right). (b) Photograph of a modern swampy meadow (SM) landscape in the headwaters of Wangrah Creek (person for scale). (c) Photograph of a gullied valley bottom at Primrose Valley Creek with PSA and SM sediments exposed in the gully wall. Gully wall is ~ 2 m in height above water level. From Portenga et al. (2016).

European settlement phases – blanket many river floodplains in New Zealand, particularly in Northland, which is tectonically the least active and topographically the most subdued region. These floodplains record increases in sedimentation rates between fivefold and eightfold in the past 1000 years, showing that the pre-European Maori settlers also contributed to increased erosion by decimating natural forests. Some floodplains in the Northland peninsula are still actively aggrading at rates of 3–14 mm yr^{-1}, which is contributing to flooding in many places (Richardson et al. 2014). Similarly, most island ecosystems in the Pacific bear the unmistakable geomorphic imprint of Polynesian and later European settlement. Sediment cores from ponds and lakes in the Pacific islands from Hawaii to Polynesia

and New Zealand document sudden drops in the concentrations of forest pollen with concomitant increases in sedimentation rates that researchers interpret as signs of when humans first started to cut down forests (Rick et al. 2013).

Detailed source-to-sink studies of the Waipaoa River catchment on the North Island of New Zealand have compiled archives of episodic erosion and sedimentation resulting from natural disturbances like heavy rainstorms, volcanic eruptions, and strong earthquakes, and also human land use that involved widespread deforestation. Sedimentary archives reveal that widespread clear-felling caused the largest perturbation of the Waipaoa sedimentary system in the past 2400 years, flushing large amounts of sediment that overwhelmed the storage capacity of the continental shelf, with sediment trapping efficiency dropping from 90% in pre-human times to only ~25% today (Carter et al. 2010). Sedimentation in lakes increased at least ten-fold, while valley floors aggraded by up to several tens of metres, caused by gully erosion in weakly indurated and easily erodible mudrocks in the headwaters of the Waipaoa River catchment. The most recent phase of deforestation, between the 1890s and 1920s, extended into small headwater basins, triggering widespread gully erosion in their weak rocks. About 900 gully complexes with a combined area of ~40 km^2 dotted the Waiapu River, a tributary of the Waipaoa, at the beginning of the twenty-first century, delivering sediment at rates as high as 3×10^4 t km^{-2} yr^{-1} (Kasai et al. 2005). Yet in spite of these dominant sediment sources, the headwaters in this catchment changed little during Cyclone Bola in 1988, the most recent major storm, which triggered at least thousands of shallow soil landslides elsewhere. An explanation for this natural resilience is that earlier storms stripped off enough sediment from hillslopes to exhaust in parts the supply of material.

The 'Himalayan Dilemma' revisited The seemingly intuitive link between deforestation, enhanced erosion, and high sediment yields in many mountain ranges of the world has long been less clear in the Himalayas, mainly because few measurements have been made of rates of natural erosion and sediment yield. In the 1980s some researchers expressed their concern about rapid deforestation in many Himalayan catchments. The loss of protective tree cover appeared to be boosting erosion rates and sediment export, while enhancing flooding in the Himalayan foreland. This 'Himalayan Dilemma' (Ives and Messerli 1989) centred on the question of anthropogenic acceleration of natural erosion, environmental degradation, and loss of sustainable lifestyles in the mountain belt and its immediate foreland. The past three decades, however, have produced a sizeable amount of data on both natural and human-induced erosion rates in the Himalaya that allow an informed re-assessment of this dilemma. Increasingly diverse methods, such as sediment trapping in natural reservoirs, cosmogenic ^{10}Be inventories in river sands, or strontium isotope ratios, have shown that natural denudation rates in the mountain belt has been high over past millennia, mostly of the order of $1-2$ mm yr^{-1}, though locally much higher. These studies call into question the notion that land-use changes are solely responsible for today's high erosion rates (Wasson et al. 2008). At the same time, the natural high background rates may be poor justification for anthropogenic disturbances, as many of the world's most densely forested mountain ranges also feature very high natural erosion rates.

16.3 Engineered Rivers

Humans have now regulated, trained, diverted, or dammed most of the world's rivers (Figure 16.7). The resulting impacts include flooding of valleys, resettlement of large numbers of people, changes to stream flow, sediment trapping and downstream erosion, and destruction of aquatic habitats, including the undesired effect of channel and floodplain aggradation from eroded soils. Some geoscientists have labelled this human imprint the 'third evolutionary stage of biosphere engineering of rivers' (Williams et al. 2014). When we discuss these impacts on rivers, we also need to consider the many impacts on lakes and wetlands that are connected to rivers. Woodward et al. (2014) reported that between 9 and 12% of global wetlands are now located in areas that were formerly forested. This deforestation contributed to raising water yields in catchments by up to 600 mm, especially in tropical and subtropical regions.

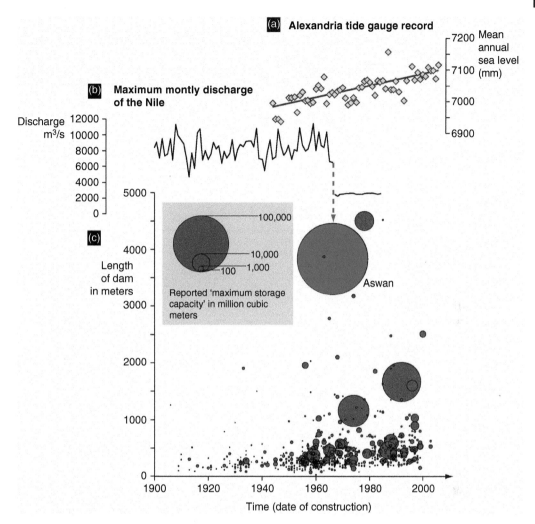

Figure 16.7 Tracking Nile Delta vulnerability to recent environmental change. (a) Alexandria tide gauge record for the period 1944–2006. (b) Maximum monthly discharge of the Nile River during the twentieth century. (c) Bubble plot of dams in southern Europe and the circum-Mediterranean. The Aswan High dam is represented in red. The figure shows the impact of construction of the Aswan dam on Nile River discharge and, indirectly, a large decrease in sediment input to the delta area. In the current context of rising Mediterranean sea level, locally attested by the tide gauge at Alexandria, this post-Aswan decrease in sediment does not allow the delta system to naturally offset sea-level change. From Marriner et al. (2013).

From the perspective of deep geological time, both the oxygenation of the Earth's atmosphere by ~2.4 Ga, and the appearance of vascular plants by ~0.416 Ga were important turning points in how rivers operated. Oxygenation completely altered the weathering regime and mineral composition of rivers, whereas the spread of riparian vegetation transformed dominantly braided rivers to meandering ones.

The extent to which humans have altered river courses, floodplains, and the associated water and sediment fluxes over the past several millennia may be the next such geological turning point, and the notion stresses how important it is to study biotic impacts on river systems. However, it is unclear how long, or even whether, such impacts will be traceable in the future geological record (Williams et al. 2014).

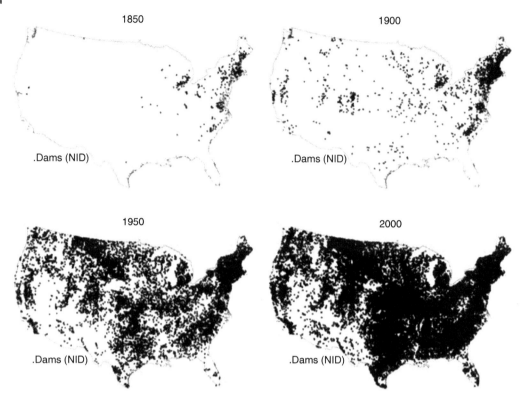

Figure 16.8 Growth of United States dams and reservoirs as recorded in the National Inventory of Dams (NID). There were no dams in 1800. From Syvitski and Kettner (2011).

Large dams in active landscapes, for example, are prone to damage or complete failure if impacted by earthquakes, volcanic eruptions, or landslides. Peng and Zhang (2012) reported that in China alone more than 3460 dams, or 4% of the total, failed between 1954 and 2004. It is unclear, though, how many of these failures occurred because of natural processes. The proliferation of dams also means that humans now have control over large fractions of Earth's natural terrestrial water and sediment (Figure 16.8). A first-order estimate holds that all of Earth's large cities, defined as those having more than 0.75 million inhabitants each, draw water from >40% of the global land surface and require transport of 184 km^3 yr^{-1} of water over a total distance of 27 000 ± 3800 km (McDonald et al. 2014). Yet one-quarter of these cities suffer from water stresses because of environmental or political

constraints. These constraints partly arise from dams that alter the natural transport of water and sediment. While ensuring access to water resources, such dams also cause downstream losses of river sediment.

Estimates of how much river sediment artificial dams trap depend on the choice of statistical models, and large uncertainties remain despite a general consensus that the amount is large. Vörösmarty et al. (2003) estimated that 25–30% of the global river sediment flux is intercepted by less than 700 large dams worldwide. The total number of dams on Earth may be as high as 800 000; these dams could intercept as much as 60% of the total sediment flux in all rivers, that is, as much as 3.7 Gt of sediment per year (Jennerjahn 2012).

China's Yangtze River alone has now some 50 000 dams in its catchment, most built since

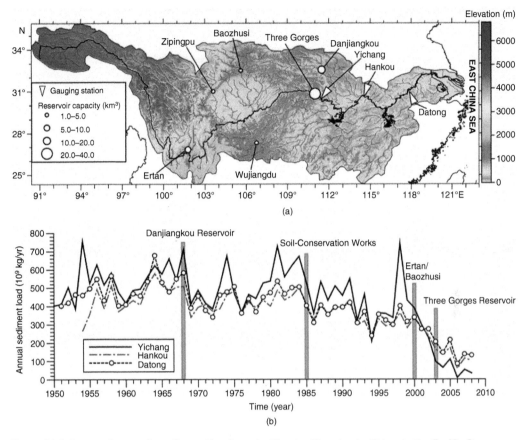

Figure 16.9 Recent changes in sediment flux from the Yangtze River basin, China, to the Pacific Ocean. (a) Major reservoirs and gauging stations. (b) Time-series data (1950s–2008) of annual sediment load recorded at major gauging stations. Since 2002, sediment loads at Hankou and Datong have become larger than at Yichang, indicating notable sediment supply from the middle and lower channel due to bed scour. From Wang et al. (2011).

1950 and including the 185-m high Three Gorges Dam that was completed and commissioned in 2003. The Three Gorges Dam trapped an estimated 1.8 Gt of sediment in the first decade following its construction, retaining as much as 85% of the sediment carried by the river between 2008 and 2012 (Yang et al. 2014). Reaches downstream of the dam are commensurately starved in sediment, and the Yangtze River channel has begun to erode and coarsen its bedload in areas where it had been aggrading previously in its more natural state. The dam has also affected the subaqueous part of the Yangtze delta, which has experienced increased wave erosion because of the reduced supply of fluvial sediment. The switch

from a dominantly aggrading to an incising channel downstream of the Three Gorges Dam is dramatic (Figure 16.9). Detailed time series of suspended sediment discharge show that net channel sedimentation rates averaged ~90 Mt yr^{-1} from the mid-1950s to the mid-1980s, but then gave way to ~65 Mt yr^{-1} of net channel erosion after the dam was completed (Yang et al. 2011). The magnitude of this reversal in the Yangtze's sediment budget is more than twice as large as the estimated total contemporary flux of sediment from the tectonically-active Southern Alps of New Zealand (~63 Mt yr^{-1}), one of the most rapidly, and mostly naturally, denuding mountain belts on Earth.

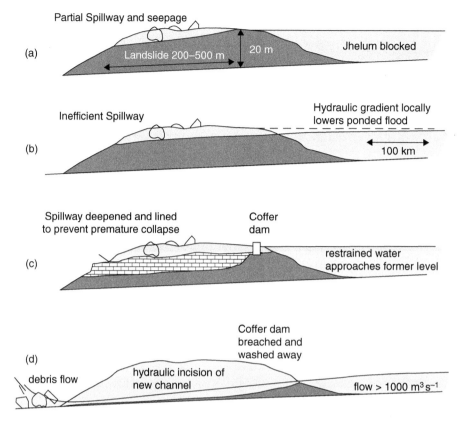

Figure 16.10 Example of historic disaster management. Possible sequence adopted to empty a lake that formed behind the ninth century Jhelum landslide in Kashmir. The initial spillway across the landslide dam, constructed by 'Suyya', King Avantivarman's engineer (a, b), which drained for three days, may have been inefficient and difficult to deepen. As a result, Suyya built a coffer dam (c), and dug and lined a narrow spillway through the bulk of the slide debris, avoiding rocks too heavy to move. By building the spillway (d), he was able to enlist the erosive power of fast-running water to deepen and widen the outflow channel, leading ultimately to the stream channel returning to near its former level. From Bilham and Bikram (2014).

Sediment trapping in large reservoirs Artificial and natural dams interrupt the flux of water and sediment. Although large reservoirs provide the benefits of hydropower generation, irrigation, streamflow regulation, flood attenuation, and drinking water, they also act as large sediment traps and many require ongoing maintenance, such as dredging and periodic sediment flushing, to maintain the water-storage volume (Figure 16.10). Detailed knowledge of a river's short- and long-term sediment budget is essential for determining the extent of such maintenance or for predicting the operational lifespan of a reservoir. Much can be learned in this regard by studying the infilling of natural lake basins of different sizes and environmental settings (Einsele and Hinderer 1997). About half of the contemporary global sediment load in rivers is doomed to be trapped, or at least temporarily stored, in human-made reservoirs (Vörösmarty et al. 2003). The larger reservoirs appear to play a larger role in this effect; the Three Gorges Dam, China, alone had sequestered an estimated 1.8 km³ of sediment delivered by the Yangtze River in the first decade after completion (Yang et al. 2006). The problem downstream of many dams is then a shortage of river sediment loads. Sediment-starved rivers incise their channels, coarsen their load, and rework older valley fills. Engineers have realized the importance of replenishing the natural supply of sediment below dams, and many flushing techniques are now operating to counteract the problem of sediment starvation (Kondolf et al. 2014). In the case of the sediment-starved Yangtze River downstream of the Three Gorges Dam, coastal erosion at the Yangtze delta has increased notably and is likely to increase local subsidence of this heavily populated region in the future, opening the doors for higher rates of sea-level rise, storm surges, and coastal flooding.

A similar fate may await the Mekong River, which is currently experiencing widespread hydropower development, with a projected 10-fold increase in water storage capacity. The resulting sediment trapping efficiency would rise commensurately, such that ~70 Mt, or more than half of the total annual catchment sediment flux, might eventually be stored (Kummu et al. 2010). This estimate is roughly consistent with that by Darby et al. (2016), who inferred that suspended sediment loads to the delta had decreased by 52.6 ± 10.2 Mt between 1981 and 2005. However, they also proposed that more than 60% of this decrease was linked to variations in the intensity of rainfall brought by tropical cyclones. Anthropogenic sediment trapping is also pronounced in Japanese mountain rivers, which are highly regulated and dotted with thousands of small structures. Between 0.02 and 0.35 km^3 yr^{-1} of sediment are intercepted by these. Similar detailed studies of the sedimentation in reservoirs are useful for assessing the flux and composition of sediment integrated within a given catchment area and over several decades, especially where highly episodic erosion by landslides and severe storms may render shorter term measurements unrepresentative (Imaizumi and Sidle 2007).

Large dams, however, are only one means starving rivers of their sediment. Widespread river training works on the Rhine River in the nineteenth and twentieth centuries, for example, have greatly increased the channel's transport capacity, and prompted flushing of large amounts of sandy loads, thus raising rates of bed degradation to up to 3 mm yr^{-1}. This ongoing loss of fine sediment is also causing several problems to river traffic, so that engineers have to replenish sediment artificially at a rate that makes up more than 40% of the overall annual input of gravel and sand (Frings et al. 2014). In other reaches, excess sedimentation requires ongoing channel dredging to keep navigation routes clear.

Countless other engineering structures intended for flood protection or erosion control in the world's rivers have profoundly changed the way that rivers erode and deposit sediment. Most bridges, for example, require foundations in the floodplain, channel banks, and channel bed. Sediment continuity dictates that any such changes to the channel geometry at the bridge site will trigger changes in water and sediment discharge both upstream and downstream. For example, erosion of the local channel bed and banks will take place where river flow is artificially confined (and thus deepened) by access ramps, bridge abutments or piers. This confining effect causes local scouring that can eventually undermine bridge foundations and compromise their structural integrity.

16.4 Engineered Coasts

The lessons from interfering with the natural fluxes of water and sediment in rivers readily extend to coastal environments. Most of humankind is living near water, and many coastal areas bear a clear human footprint that is visible even from space. About 22 000 km of European coastlines are artificially covered with concrete and asphalt, including defences against storm surges and rising sea levels (Figure 16.11) (Kareiva et al. 2007). Harbours are artificial sediment traps that can enhance sedimentation rates 10–20 times above natural rates along prograding coastlines (Marriner and Morhange 2007). Other artificial structures along coasts, such as groynes, jetties or sea walls, can interrupt or redirect the longshore drift, and may starve sections of sediment, leading to coastal erosion. Between 80 and 95% of the contemporary beach erosion in southern Florida may be linked to the stabilization of tidal inlets with jetties (Finkl and Makowski 2013). This leads to loss of recreational space, decreased revenues from tourism, and higher exposure to storm surges. The annual loss rate of sand from the world's beaches by artificially altered sediment budgets or direct mining may now be of the same order of magnitude as the combined input of all rivers to the oceans (Padmalal and Maya 2014).

What strategies should we pursue in dealing with storms and rising sea level? Jungerius (2010) summarized some of the experiences that the Dutch people have had over several centuries of being exposed to these hazards: construction of coastal defence structures; building of dykes; pumping to remove accumulated water; strengthening of dunes; support sand nourishment; and reclaiming new land from the sea. These are essentially risk management strategies aimed at reducing the potential damage from sea-level rise with the added benefit of claiming new land. This practice has transformed large tracts of coastlines worldwide, altering their geomorphology, sediment budgets, and hence vulnerability to natural hazards. For example, in only ten years (1961–1971), over 4000 km of embankments were built along estuarine channels to reclaim 10 000 km^2 of land in the western Ganges–Brahmaputra delta, a densely populated area that is emblematic in its exposure to tropical cyclones and storm surges (Pethick and Orford 2013)

In their global assessment of the risk to large deltas from sea-level rise and flooding, Tessler et al. (2015) proposed that the highest likely changes in risk for deltas are in the eastern Indian Ocean. They argued that both rapidly rising sea level and the high social and economic vulnerability of people living in deltas in that region were the main drivers of increasing risk. They also argued that projected increases in energy prices, labour, and material costs would weigh heavily on maintaining and strengthening coastal defence structures. Seen through this economic lens, sea-level rise will impact most wealthy nations. In contrast, economically less developed nations with large populations on deltas are likely to see future increases in risk due to growing population densities and sprawl of built environments.

This problem is prominent where humans have created or amplified the conditions that raise local sea level: artificial draining of low-lying land has caused widespread subsidence and dykes have added to this effect. Venice, Italy, is emblematic of a coastal community threatened by sea-level rise. The city is built on 17 small islands connected by more than 400 bridges. Its seaside location and canals are part of its attraction, but the city lies only just above sea level and is periodically flooded during storms and exceptional high tides. The problem partly arises from the slow subsidence that the sediments beneath Venice lagoon are experiencing. Although subsidence has been occurring naturally for millions of years, pumping of groundwater from the 1930s to the 1960s increased the rate at which the land and the city are sinking. The pumping and thus the human contribution to this natural hazard ended in the 1970s, but, unfortunately, natural subsidence is still occurring at a rate of 1–2 mm yr^{-1}, enhancing the effect of the rise in sea level driven by global climate warming. The response of local and regional authorities in the past was to raise buildings and streets, but this solution cannot continue indefinitely in the long term.

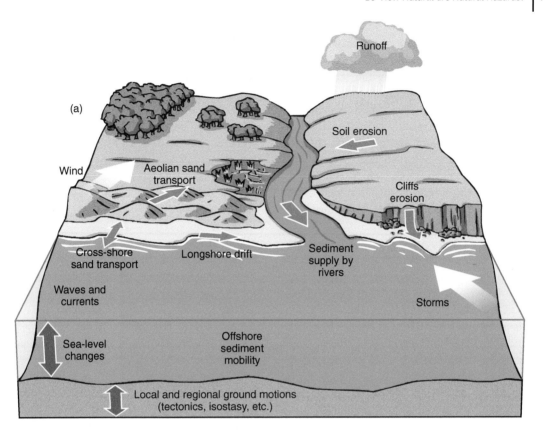

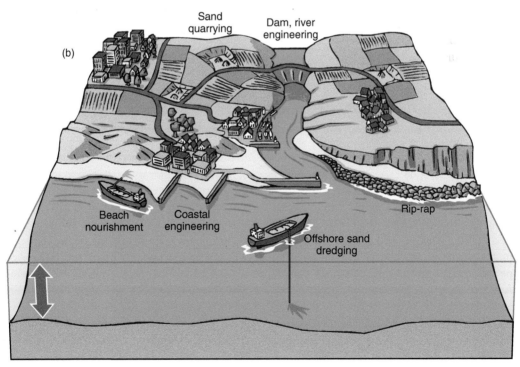

Figure 16.11 Sketches showing processes involved in coastal sediment transport and shoreline changes. (a) Absence of human intervention. (b) Direct and indirect anthropogenic actions affecting sediment mobility. Modified from Cazenave and Le Cozannet (2014).

To combat this flooding, the Italian government in 2003 initiated the 'Mose' (Modulo Sperimentale Elettromeccanico) project aimed at preventing water from the Adriatic Sea entering Venice Lagoon at high tides and during storms. The idea was to construct 78 floodgates across the three tidal inlets that connect Venice Lagoon to the Adriatic Sea. The floodgates swing upward across the tidal inlets when tides reach a threatening level. Originally, the project was estimated to cost US\$2.6 billion, but nearly US\$10 billion has been spent, of which about US\$3 billion has been lost due to corruption. Whether these works will correct the problem in the short term has been questioned. In any case, subsidence and sea-level rise will continue, thus the floodgates are, at best, only a temporary solution. Venice is an example but the problem is global: artificial groundwater extraction in coastal plains and deltas, particularly those that sustain the world's megacities, may be responsible for up to nearly 10% of the relative rise in sea level worldwide, and for greatly accelerating rates locally (Konikow 2011). Fewer studies are concerned with what happens to this extracted water, and where it re-enters the hydrological cycle. Nonetheless, the high net subsidence rates of many densely settled deltas show that aquifers are replenished at negligible rates.

Humans currently withdraw water at an estimated rate of 3800 km^3 yr^{-1} from rivers, lakes, and aquifers globally. Hydrologists call this 'blue water', which currently makes up some 8% of the water carried to oceans by rivers worldwide (Oki and Kanae 2006). Judging from a global survey of more than three million LANDSAT satellite images covering the three decades from 1984 to 2015, Pekel et al. (2016) noted that more than 70% of the water surface areas lost during that period were in the Middle East and Central Asia, most likely because of drought, but also because of river diversion, damming, and water withdrawal. Artificial drainage of swamps, marshes, and other natural wetlands dries out soils such that they compact, and

eventually shrink in volume, leading to land subsidence. The Florida Everglades were originally a marshland covering >16 000 km^2. Systematic drainage and urbanisation since the 1900s made the land subside by up to 4.5 m by the end of the twentieth century; the average rates of subsidence were 25–30 mm yr^{-1} (Finkl and Makowski 2013). Detailed geodetic measurements now afford unprecedented insight into how groundwater extraction can trigger local crustal uplift through unloading. In California's San Joaquin Valley the rates of uplift measured by GPS are 1–3 mm yr^{-1} and exceed any subsidence effects (Amos et al. 2015). Such flexural uplift contributes to reducing the effective normal stress on the nearby San Andreas fault, highlighting a surprising link between human land use, crustal dynamics, and earthquake hazard. However, reducing normal stress may also mean that earthquakes will occur more frequently but will be less severe and future research will be needed to develop the implications for seismic hazard.

16.5 Anthropogenic Sediments

Humans have been directly and indirectly responsible for redistributing large amounts of sediment and water on the Earth's surface. Human and domesticated animal wastes, agricultural fertilisers, mining wastes and other industry products, as well as pathogens and contaminants, have been transported with these sediments. On top of that, nearly half of the global population lived in cities at the turn of the twenty-first century, and produced an estimated 3 Mt of solid waste per day; this waste is bound to double by 2025 (Hoornweg et al. 2013). 'Throwing away' then means adding a moving mass that currently amounts to 1–2 Gt per year, which is roughly 10% of the sediment mass that rivers transport to the world's oceans each year. From a more global perspective of mass turnover or 'mass action', defined as the product of mass, the distance it moves, and its average speed, Haff (2010) argued that humans shift more material across our planet's surface faster than any other natural geomorphic system except for rivers, at least over years to decades. Some of the waste that humans generate may fuse with rocks. Types of plastic, for example, may be incorporated into rocks when heated, and the term *plastiglomerate* describes natural sediment particles fused by plastic (Corcoran et al. 2014).

Sand mining is moving more sediment than rivers You may think that sand is a material that occurs just about everywhere. Yet most of the sand in playground sandboxes that we played in as kids was brought there. Sand mining and shipping has become a voluminous business with mean annual production rates of 0.5–4.5 m^3 per capita in the twentieth century (de Leeuw et al. 2009). The United Nations Environment Programme estimated that 25.9–29.6 km^3 of sand and gravel were mined in 2012 alone, thus doubling the estimated flux of sediments delivered by rivers to the sea. Ocean beaches seem like the ideal location from which to extract sand, as continuous wave action along the shore has conveniently abraded, sorted, and deposited rock particles for many industrial and building purposes. Large sandy rivers and floodplains are also popular targets for extraction. Sand mining in the Mekong River delta involved some 0.2 km^3 between 1998 and 2008, and is now a major cause of widespread and rapid shoreline erosion, adding to the effects of dams that starve the river of sediments, and groundwater extraction which cause the land surface to sink (Anthony et al. 2015). Nonetheless the dynamics of sands in beaches and rivers obeys the physical laws of sediment continuity. Removing material from one location will disturb the surface by creating pits, knickpoints or knicklines; waves or

flowing water subsequently work towards filling those gaps with the sediment that they carry. Heavily mined sand beaches or river reaches expose their underlying bedrock and promote more aggressive wave climates, higher river-flow velocities, dune erosion, or river-bank collapse. Apart from altering these and other coastal and flooding hazards, sand mining also has many negative impacts on surface and groundwater quality as well as coastal and riparian habitats.

Topography has lost part of its role as a barrier to developing the built environment. The mechanical removal of mountain tops has levelled out entire compartments of the landscape. Widely used during strip mining operations in the eastern United States, the cutting of hilltops and concomitant shifting of millions of tonnes of sediment has become a common practice in China. Land creation for the city of Yan'an, Shanxii Province, will involve the artificial flattening of 78.5 km^2 of hilly terrain (Guralnik 2014). Such practices are highly likely to have protracted environmental impacts, such as increased soil erosion and landsliding, a higher frequency of dust storms, and reduced groundwater availability.

Protracted extraction of materials from river channels changes flow widths, depths, roughness and velocities, and can thus alter flood stages, frequencies, and durations. In some cases, this anthropogenic 'erosion' of river sediments has already surpassed the natural rates. In the 1970s, gravel mining and bedrock quarrying in Japanese rivers occurred at rates of about 350–740 t km^2 yr^{-1}, respectively, comparable to the natural fluvial sediment yields (Kadomura 1980). Hydraulic mining in the lower Sacramento Valley in northern California during the nineteenth century gold rush moved at least 3.7 Gt of sediment (James 2004a). This perturbation of the natural sediment budget generated specific sediment yields of ~16 800 t km^{-2} yr^{-1} that today are known from only the most active or naturally disturbed landscapes. Valley floors aggraded and widened rapidly. Thus raised river-channel beds promoted lateral erosion into bedrock spurs, creating new gorges at rates as high as 0.5 m yr^{-1} (James 2004b). More recently, large-scale open-pit mining at Ok Tedi, Papua New Guinea, has created a classic example of how large-scale mining operations change the natural sediment transport in rivers. Mining operations between 1984 and 2013 resulted in the dumping of 2 km^3 of untreated mining wastes into the headwaters of the Fly River. The mining company reported that between 80 and 90 Mt entered the river annually, an amount that surpasses the contemporary estimated sediment flux from the entire rapidly denuding western Southern Alps of New Zealand. The toxic mine wastes raised the channel bed of the Fly River by as much as 10 metres over a length of 1000 km, directly killing $\sim$ 1600 km^2 of rainforest and disrupting the livelihoods of as many as 50 000 people.

Abandoned tailings from Soviet-era uranium mining in the Mailuu-Suu valley of the Kyrgyz Tien Shan are prone to impact by large deep-seated and slow-moving landslides that could bulldoze the radioactive material into the headwaters of Central Asian rivers (Havenith et al. 2006). Similarly, river-blocking landslides have repeatedly impounded ephemeral lakes that may have inundated some of these tailings, eroding contaminated material that is flushed from those lakes. The partial meltdown of the Fukushima-Daichi nuclear power plant following the Tohoku tsunami of 2011 further demonstrates the high vulnerability of modern technologies to natural disasters. Measurements in the Abukuma River, which was contaminated by radioactive fallout, show that nearly 90% of all radiocaesium flux was particulate matter attached to fluvial sediment (Kitamura et al. 2014). Contaminated sediments from the regions of highest fallout in the inland mountains reached the coastal plains only half a year after the disaster, largely due to flushing by a series of tropical cyclones

that struck the area. The radioactive material that was liberated during the meltdown of the reactor now serves as a tracer of the high efficacy of sediment transport in the affected rivers (Chartin et al. 2013). This event underscores once more the importance of measuring the transport of sediment in river systems impacted by natural disasters. Even controlled experiments with radioactive materials can produce distinct geomorphic impacts. According to the Preparatory Commission for the Comprehensive Nuclear-Test-Ban Treaty Organization, the French government conducted 193 nuclear experiments on the Mururoa and Fangataufa atolls in the South Pacific between 1966 and 1996. Nuclear tests in 1979 triggered large landslides at Mururoa atoll, releasing a tsunami that impacted the neighbouring Tureia atoll 105 km to the northeast (Morrison et al. 2013). Numerical modelling results show that a collapse of the structurally weakened coral reef could well be an unaccounted-for hazard to the inhabitants of Tureia.

Oil spills from ships, tankers, and drilling platforms feature regularly in the news, mostly with emphasis on the impacts on marine and coastal ecosystems, the fishery, and tourism industries. Yet understanding and potentially mitigating the spread and depositional fate of organic compounds released by oil spills often depends on a sound understanding and consideration of geomorphic processes. In the case of the Deepwater Horizon oil spill in the Gulf of Mexico in 2010 – the largest oil spill in the history of the United States, covering up to 180 000 km^2 of ocean surface – the high incoming suspended sediment yield from the Mississippi River largely favoured the formation of oil-mineral aggregations that sank to the ocean floor, greatly raising local sedimentation rates (Daly et al. 2016).

Marine debris, defined as manufactured or processed solid waste material entering the oceans, is another example of how artificial sediments interfere with natural geomorphic and ecosystems. Estimating how much plastic is drifting in the world's ocean relies on various observations ranging from trawling with nets to visual observations. Eriksen et al. (2014) analysed data from expeditions across all the major ocean gyres to project a total of some 5.25×10^{12} particles with a total mass of nearly 270 000 t. The Ocean Conservacy's 2015 Global Ocean Cleanup Report (www.oceanconservancy.org) estimates that 5–12 Mt of plastic make their way into the oceans from land-based sources every year. Jambeck et al. (2015) published a very similar estimate of contemporary plastic input to the oceans, though attributed much of the source of this plastic to a 50-km wide strip along the world's coasts. For the Cleanup Report, more than half a million volunteers picked up 7341 t of marine debris from nearly 21 500 km of coastline throughout the world in 2014, averaging to a load of 0.34 t per kilometre of coastline. This number hides a lot of variability, and peak loads were up to 93 and 88 t km^{-1} on the islands of Curaçao and Hongkong, respectively. In an effort of nationwide monitoring, bimonthly beach surveys of some 20 South Korean beaches from 2008–2009 retrieved 105 797 pieces of marine debris with a total mass of 19 t, and a volume of 105 m^3 (Hong et al. 2014). On average, one hundred metres of beach length contained about 480 pieces of debris, amounting to almost 0.5 m^3, or an average of 0.87 t km^{-1}. In this survey, the total mass of plastic and styrofoam debris nearly balanced the mass of woody debris. About a fifth of the debris count consisted of styrofoam buoys and fishing ropes. The flux of plastic materials to the world's oceans is of particular concern, because plastic fragments floating in the sea can break down into nano-sized particles that readily enter marine food webs that support in large part the biological productivity of our planet. The fate of plastic in seawater remains to be fully understood, and measured concentrations often deviate from what ocean circulation models predict. Recent sampling cruises even suggest that the volume of floating plastic debris in ocean gyres is much lower than expected, perhaps

because a lot of the debris had broken down to a nearly undetectable size (Cozar et al. 2014). Surface-trawling plankton nets routinely pick up floating plastic debris of roughly millimetre-size, which may represent about 1% of the mass of plastic debris entering the oceans annually (van Sebille et al. 2015). In any case, major discrepancies in the budgeting of marine plastic debris remain and call for more detailed research. Fewer studies cover plastic sediments in rivers and lakes, let alone how these particles influence geomorphic processes in water bodies. From the few available reports we can glean that river sediment already contains abundant microscopic plastic particles; Hurley et al. (2018) reported peak concentrations greater than half a million particles per square metre in suburban and rural river channels in northeast England (Figure 16.12). The few documented average concentrations are about an order of magnitude lower, but emphasize nevertheless the need to gather more data. Floods readily flush those microplastic loads from surveyed reaches and thus offer an alternative explanation for the high amounts of microplastic in lakes and oceans.

Anthropogenic sediment and waste add to the problems generated by natural disasters. The costs of 'cleaning up' may equal or exceed the documented original losses, and may be incurred for years after the disaster. The 2011 Great Tohoku tsunami entrained an estimated 25 Mt of anthropogenic sediment from the coasts of eastern Japan, carrying the debris into the open Pacific Ocean. Some of the debris already littered the coasts of Hawaii and western North America in the years following the tsunami, while large amounts of the debris are drifting in the North Pacific gyre.

Much of the fluvial contaminant load from human activities eventually reaches the coast and open ocean. Rivers carry large amounts of dissolved load to the sea, and our contribution to acidifying the world's oceans is documented in time series that show an average decrease of 0.1 pH unit since industrialization began. This

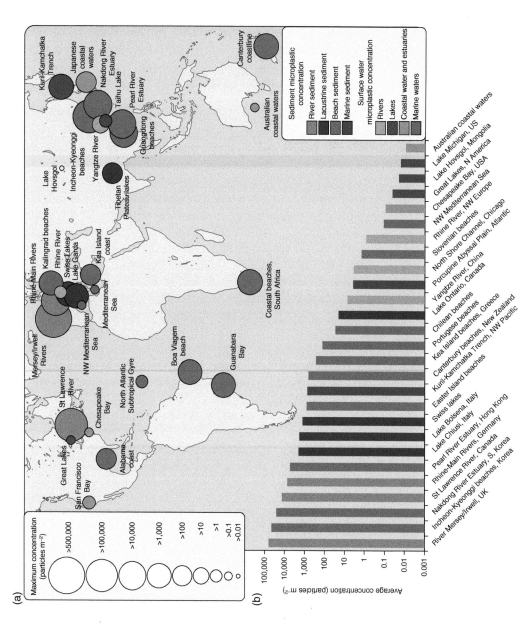

Figure 16.12 Global microplastic (<5 mm) concentrations reported by aquatic and sedimentary environment. (a) Microplastics mapped by maximum concentrations. (b) Microplastics ranked by average concentration. From Hurley et al. (2018).

increase is about 30–100 times more rapid than any known natural oscillation, and a further decline by 0.2–0.3 pH units is likely by the end of the twenty-first century in the absence of any adequate countermeasures (Doney 2010). This ongoing acidification reduces carbon dioxide uptake by the oceans and may thus worsen the consequences of further greenhouse gas emissions, leading to more warming. At the same time, increasing acidification negatively impacts the primary productivity of the world's oceans, thus putting marine, and eventually also terrestrial, ecosystems at risk. Degrading or eroding coral reefs, for example, will lose their hydrodynamic roughness with respect to incoming waves, thus promoting higher run-up and inundation levels for a given wave regime. The overall implication is that coastal hazards may change as oceans acidify further (Quataert et al. 2015).

The relentless movement of human bodies on Earth has also strongly changed the vectors of diseases and the spread of pathogens. Natural disasters can damage processing plants containing oil and gas, structures containing asbestos, or industrial facilities storing highly toxic chemicals and thus release such substances to the environment. Scientists have proposed the term 'hazmat' for such hazardous materials, which can develop into na-tech disasters (Young et al. 2004). Pathogens and diseases may also be on the rise following major river engineering projects. Dams and irrigation projects, for example, are responsible for increases in diseases such as schistosomiasis in parts of Africa and South Asia (Myers et al. 2013). Changing vegetation patterns following high inputs of nitrogen and phosphorus from agricultural fertilisers have led to the spread of malaria. Outbreaks of other mosquito-borne diseases such as dengue, Rift Valley fever, Murray Valley encephalitis, and West Nile virus coincide with pronounced flooding or drought (Anyamba et al. 2014). Rising global temperatures are likely to promote the migration of disease-carrying mosquitoes to new areas (Medley 2010).

16.6 The Urban Turn

Several factors favour an increase in the risk from natural hazards in the twenty-first century. For one, rapid population growth will increase the number of people and assets at risk. Apart from the expansion of land use, the rapid expansion of cities is another key aspect most tightly linked to profound changes in Earth surface processes and landforms. The beginning of the twenty-first century has seen the 'urban turn': By 2007, for the first time in history, more people were living in cities than in rural regions. More than 70% of the world's population was living in rural areas in 1950, but this percentage decreased to ~47% by 2014 (World Risk Report, 2014; https://ehs.unu.edu). In China, the number of people living in cities tripled between 1978 and 2012, reaching an estimated 53% of the total population (Bai et al. 2014). Estimates from a meta-analysis of global population growth hold that between 1970 and 2000, some 58 000 km^2 of land, more than the area of Switzerland, became urbanized. And some 1.5 million km^2 of land are likely to be converted to urban land use by 2030, mostly involving a built environment less compact, though more sprawling (Seto et al. 2011). Current demographic projections assume that two out of three people will be living in cities by 2050. Population densities in cities are on the rise, commensurately increasing the risk from natural disasters. Projections of the exposure of coastal megacities are particularly grim. One scenario foresees that the number of people living in large cities exposed to earthquakes and tropical cyclones will more than double by 2050, as referenced to the period 1975–2005 (Bank 2010). Some 310 million lived in cities at risk from tropical cyclones in 2000, but the projected number for 2050 is 680 million; for earthquakes the increase is from 370 million to 870 million people. This exposure is such that seismologists have dubbed structurally weak buildings 'weapons of mass destruction' (Bilham and Gaur 2013). The highest estimated

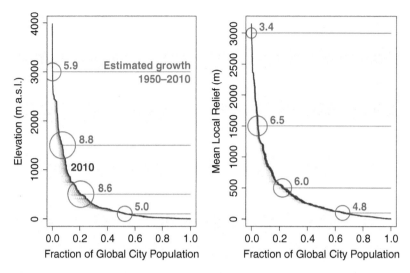

Figure 16.13 Population dynamics of cities with >75 000 residents between 1950 and 2010 as a function of mean elevation (m a.s.l.) and mean local topographic relief. Mean topographic relief is the maximum elevation range within a 10-km radius. About 20% of the global urban population lived above 500 m a.s.l. and on terrain with a mean topographic relief of >500 m; these areas have some of the highest rates of population growth (red horizontal lines and commensurately scaled bubbles).

annual growth rate of exposure to earthquakes is 3.5% in South Asia.

How has this increase in risk from urbanization happened? Dodman and Satterthwaite (2008) offer several explanations. Firstly, economic and political decisions have outweighed considerations of risk, so that attractive, but also potentially hazardous, locations along active coasts or rivers now support urban infrastructure. Secondly, few cities disappear, and once they have developed they persist and grow despite natural events that claim lives and damage infrastructure. Thirdly the attraction of people to cities means that cities must grow, often by expanding into terrain that is more hazardous than the parts of the city that had been settled first. Fourthly, low-income households move into higher-risk areas, which may be unattractive to, or completely avoided by, higher-income households. Wealthier groups and enterprises face different risks than low-income households. Consequently, expanding cities create new risks, but rarely adequate management strategies.

Urban growth also has a decidedly hypsometric footprint. Although only about a fifth of the world's urban population lives in mountainous areas above an elevation of 500 m and with local topographic relief of >500 m, it is those areas that experienced the most rapid rate of population growth between 1950 and 2010 (Figure 16.13). Over that period, cities with populations of more than 750 000 lying between 500 and 1500 m a.s.l. grew more than eightfold in population. Similarly, the highest rates of population increase were in areas with a mean local relief of 1500 m. Assuming for the sake of argument that all other factors in the risk equation (Eq (2.1)) such as hazard, vulnerability, and aversion remained constant during the six decades from 1950 to 2010, the total risk from natural hazards in large cities in mountains has increased threefold to more than eightfold solely due to ongoing urbanization. Such a global projection ignores local detail; some cities, for example, are less vulnerable to natural hazards than others. Yet the implications of the global trend of urban populations pushing farther into mountainous terrain are clear. Bear in mind that these predictions only take into account population growth, irrespective of any effects of climate change.

16.7 Infrastructure's Impacts on Landscapes

Roads connect coastal and inland areas, and proposals for new roads are skyrocketing. By 2050, the global road network will grow by more than 25 million kilometres, which by comparison is more than one-quarter the distance between Earth and the Sun, and nearly 60% greater than the network in 2010 (Laurance et al. 2014). This expanding growth has diverse environmental implications that involve habitat losses and fragmentation, rainfall runoff and streamflow changes, increased wildfire potential, surface compaction, enhanced soil erosion and slope instability, and pollution. Road-related landslide erosion in the tropical Andes of southern Ecuador, for example, is happening at rates nearly twice those on undisturbed hillslopes with comparable geology and vegetation cover (Muenchow et al. 2012).

Sidle et al. (2005) compiled a detailed summary of erosion processes and rates related to forest management in the steeplands of Southeast Asia – a regional hotspot of both logging and logging-related erosion studies – and concluded that road-related rates of soil and sediment loss were often the highest, outweighing the effects of extensive deforestation. They cautioned, however, about comparing erosion rate estimates obtained using different methods and observation times. Thus the erosion rates they collated in this review differ by four orders of magnitude. Based on a review of more than one thousand case studies, Kleinschroth and Healey (2017) reported that half of all logging roads occupied only about 2% of deforested areas in tropical forests, but caused some of the largest disturbances, including:

increased fire incidence, soil erosion, landslides, and sediment accumulation in streams. Once opened, logging roads potentially allow continued access to the forest interior, which can lead to biological invasions, increased hunting pressure, and proliferation of swidden agriculture.

Continuously expanding infrastructure and traffic also have changed the ocean floors. Commercial fishing has had its largest growth in the twentieth century, and large-scale bottom trawling has increased physical and ecological disturbances to the degree that Oberle et al. (2016) claimed that bottom trawling 'is the most widespread anthropogenic activity that impacts the seabed on the continental shelf', drawing on estimates that submerged fishing gear scrapes an area of more than half the world's continental shelf each year (Figure 16.14). The consequences of bottom trawling include stirring, resorting, and resuspension of seafloor sediments, increased water turbidity, triggering of small turbidity currents, and altered sedimentation rates, but detailed data are limited to a few locations. All of these processes add to the negative impacts on benthic habitats and organisms, and increased human mixing of upper seafloor sediment layers adds more oxygen, thus promoting the mobilization of organic carbon that would be otherwise gradually buried and sequestered (Martín et al. 2014). A detailed tracking study along the northwest Iberian Shelf conducted over a single year revealed that trawling vessels reworked more than 11 500 km^2 of seafloor, stirring up between 3.6 and 31 Mt of sediment (Oberle et al. 2016). To put this estimate into perspective, the researchers compared this resuspension flux with the reported influx of sediment from rivers annually (10–20 Gt): globally, bottom trawling may remobilize as much as 21.9 Gt yr^{-1} of sediment. If this initial estimate was to hold, the common practice of neglecting anthropogenic disturbances of the submarine environment

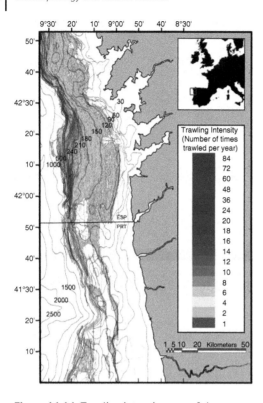

Figure 16.14 Trawling intensity map of the northwest Iberian shelf based on January 2013–January 2014 Automatic Information System (AIS) data (ESP: Spanish territorial waters, PRT: Portuguese territorial waters). The number of times a given location was trawled per year is shown. From Oberle et al. (2016).

could strongly compromise source-to-sink sediment budgets of ocean margins.

16.8 Humans and Atmospheric Warming

At the time of writing this book, 'Global surface temperature in 2016 was the warmest since official records began in 1880. It was the third year in a row to set a new heat record, and the fifth time the record has been broken since the start of the 21st century.' (www.noaa.gov/climate). According to the same source, the year 2016 was the fortieth consecutive year with above-average yearly global temperature over the period of record.

Eleven out of 12 of the warmest years in the 137-year long climate record have occurred in the twenty-first century. The projected probability of extremely hot summers is likely to increase by a factor of five to ten over the next four decades (Figure 16.15) (Barriopedro et al. 2011). The average amount of temperature rise since climate measurements were first made differs regionally; warming has been greatest in the polar regions and in some high mountains (Hinzman et al. 2005).

Although Earth has experienced alternations between glacial and interglacial periods over the past three million years, it is the current rate of warming that seems to be unprecedented in Earth history. This notion assumes, however, that we are able to detect with confidence rapid temperature changes in sufficiently well-resolved prehistoric climate data. However, separating out the effects of anthropogenic forcing of climate through greenhouse emissions and the natural variability of the Earth's climate system is a task that thousands of scientists are tackling. Some research results are surprising, underlining our limited knowledge of all the feedbacks and especially the teleconnections between the components of Earth's climate system. For example, recent models showed that a cooling of sea-surface temperatures in the tropical Pacific may have been responsible for about half of the rapid warming observed in the northeastern Canadian Arctic and Greenland (Bader 2014).

Concentrations of anthropogenic carbon dioxide, methane, and other greenhouse gases are now higher than they have been at any time in the past many millions of years. Humans have now released some 555 Pg C into the atmosphere, mainly by burning of fossil fuels and deforesting land (Lewis and Maslin 2015).

By March 2015 the concentrations of atmospheric CO_2 had reached 400 ppm for the first time in human history. Predictions for future climate in the twenty-first century and beyond are becoming increasingly grim with each successive IPCC report, including the most recent (fifth) assessment, and many of

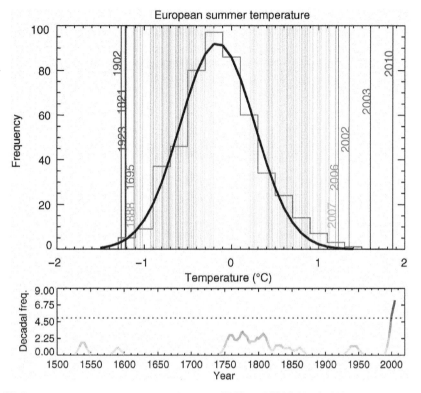

Figure 16.15 European summer temperatures between 1500 and 2010. Vertical lines are reconstructed and instrumental land temperature anomalies (°C relative to the 1970–1999 period), with the five warmest and coldest summers highlighted. Grey histogram shows distribution for the 1500–2002 period, with a Gaussian model fit in black. Lower panel shows smoothed decadal frequency of extreme summers with temperature above the 95th percentile of the 1500–2002 distribution. Dotted line 95th percentile of maximum decadal values that would be expected by random chance. From Barriopedro et al. (2010).

these predictions have consequences for how atmospheric and oceanic processes change, and how they will affect the Earth's surface. Future climate predictions rely on specific scenarios of future CO_2 and other greenhouse gas emissions. Large ensembles of independent numerical models that solve coupled differential equations describing the basic physical and chemical processes in the atmosphere and the oceans offer predictions of temperature, rainfall, and so on for these scenarios. These model predictions will require careful validation with measured data to reliably attribute observed climate changes to natural and anthropogenic forcing.

About 35% of all anthropogenic carbon dioxide emissions since 1850 were due to land-use activities including deforestation

and burning of fossil fuels; commensurate changes in land cover led to further changes in local and regional water and surface energy balances that in turn may have modified regional climate (Foley et al. 2005). Ground- and aircraft-based measurements show that the seasonal amplitude of atmospheric CO_2 concentrations in the Northern Hemisphere has increased by as much as 50% over the past 50 years. Only recently did researchers discover that the increasing productivity of Northern Hemispheric agricultural croplands that sustain extratropical maize, wheat, rice, and soybean could be responsible for nearly a quarter of these seasonal fluctuations (Gray et al. 2015). The rate of increase of atmospheric CO_2 may be set to rise. Oceans absorb a large amount of carbon dioxide from the atmosphere, but the uptake appears to be decreasing due to the increase in acidity of surface waters. Temperature, however, is only one climate variable being affected by anthropogenic greenhouse gas emissions. Detailed extreme-value analyses of heavy precipitation data combined with ensemble climate modelling suggest that anthropogenic forcing has contributed to intensifying extreme precipitation events over nearly two-thirds of the Northern Hemisphere, at least for those areas where data are available (Min et al. 2011). Atmospheric warming may also increase the frequency of extreme El Niño events (Cai et al. 2014), thus possibly raising the impacts on floods, droughts, wildfires, tropic cyclone activity, and many other natural hazards. Numerical modelling suggests that hourly rainfall extremes over much of Europe may increase by some 14% for each degree of atmospheric warming (Lenderink and van Meijgaard 2008). In contrast, southern and southeastern Europe may experience the same level of droughts between two and ten times more frequently on average by 2070 (Lehner et al. 2006). These are just a few studies, and many other projected changes in rainfall mostly share the underlying view that extremes will be more pronounced in coming decades. The increased potential for high and short-lived rainfall leads to higher runoff, more flash floods, soil erosion, and debris lows. In late May 2017, intensive storms delivered high amounts of rainfall anywhere between the 100-year event and the 1000-year event, and triggered flash floods and debris flows in several villages in Southern Germany. Historic photographs and documents show that comparable geomorphic impacts occurred more than a hundred years ago.

When referring to natural hazards and disasters, scientists are nearly unanimous in stating that the frequency and magnitude role of extreme weather events is a key uncertainty in the effects of future climate change. The need to learn more about this uncertainty is one of the reasons why the Intergovernmental Panel on Climate Change (IPCC) issued a nearly 600-page special report on 'Managing the Risks of Extreme Events and Disasters to Advance Climate Change Adaptation' (Change 2012).

16.9 How Natural Are Natural Hazards and Disasters?

This brief overview of the Anthropocene and the many human impacts on landscapes, sediment budgets, contaminant fluxes, and 'na-tech' hazards and risks should explain why some scientists propose that natural disasters have ceased to be natural. The main reason for this proposal is that humans have interfered in too many ways with Earth's natural processes. So can we treat some natural hazards and disasters as 'natural' still? Hazards and disasters involve humans by definition, but if the causes and triggers are also mostly human, then the question of liability will arise. For example, airlines have to cancel flights because of adverse weather conditions or volcanic eruptions every now and then. One practical question is whether these disruptions would warrant refunds for stranded passengers or not. In the case of volcanic eruption, one could argue for a natural, unforeseeable incident, or at least one that was less foreseeable than adverse weather. That nations are paying for their CO_2 emissions already, underlines that these nations are recognizing, and willing to balance out, some the potential impacts of human-induced atmospheric warming. Yet convincingly attributing the well-publicized rise in the number of documented natural disasters to anthropogenic warming still awaits solid evidence. For example, Bouwer (2011) reviewed more than 20 studies that had looked at (mainly weather-related) disaster trends through time and duly corrected these trends for growing populations and wealth. The conclusion was that rapidly increasing population, economic assets, and partly also vulnerability were mostly responsible for the growing losses from natural disasters; any direct and mechanistic links related to measurable increases in natural hazards from anthropogenic warming remained elusive.

Humans are the only species that has deliberately used fire for land management purposes, and this use may date back several tens of thousands of years (Bowman et al. 2011). Accordingly, wildfire is a natural hazard with frequent human fingerprints, such that the qualifier 'natural' has to be questioned. Land-use changes reaching back several tens of thousands of years have modified natural fire frequencies in many parts of the world, thus introducing new fire regimes with characteristic fuel types, new patterns of burning, and altered consequences. Humans have long used fire to clear forest lands for agriculture. Therefore disentangling the impacts of natural from anthropogenic fires promises to be a subject of scientific debate. Marlon et al. (2013) analysed more than 700 charcoal records from five continents to elucidate the role of humans in biomass burning during the Holocene. They had some doubt about whether the most distinct and nearly global increase in fire activity from 3000 to 2000 years ago was due to humans or natural climate oscillations. Although fire frequency increased at the beginning of the Holocene, consistent with a general warming following the end of the Pleistocene, neither the reconstructed global pattern of the human population nor changes in land use seem to match this general trend. In Australia, a long-standing debate centres on the question of whether Aboriginal vegetation burning practices had a measurable effect on the late Quaternary summer monsoon. An experiment using climate modelling suggests that any such impact would be hardly detectable; however, some abnormal rainfall patterns in the austral spring before the monsoon season could be related to altered vegetation cover and reduced evapotranspiration (Notaro et al. 2011).

In the Americas, studies have linked the post-Colombian decline of Native American people to changing fire regimes (Liebmann et al. 2016). According to this 'early Anthropocene burning hypothesis', forests recovered from less frequent burning, thus storing more organic carbon while reducing atmospheric CO_2 emissions and eventually bringing about climate cooling. More recent work in the southwestern United States reconciled detailed population estimates from surveyed ruins with historical, tree-ring and fire-scar data, and showed that the Native American population dwindled almost a century after Europeans arrived in the area. Large fires occurred more frequently thereafter, while Ponderosa pine forests native to the region slowly recovered (Liebmann et al. 2016). Minute reconstructions of this kind prompt us to rethink simple models of fire regimes based on population densities alone, and warrant a more diverse look at how human land use alters burning frequencies and how these affect vegetation, erosion, and the carbon cycle.

Fire management strategies, rural depopulation, and the introduction of flammable tree species such as pines or eucalypts can change natural fire regimes to such an extent that wildfires are frequently caused or at least influenced by humans instead of having formed in the 'wild'. A widely cited illustration of this point is the wildfire in the hills above Berkeley, California, in October 1991. This wildfire destroyed nearly 3800 houses and apartments in the cities of Oakland and Berkeley, claimed 25 lives, and caused more than US$ 1.68 billion damage, making it one of the worst urban disasters in United States history. The fire started on the evening of 19 October, when flames escaped from a cooking fire in a camp of homeless people. Urbanization had reduced open land on the slopes above Berkeley and Oakland from 47% in 1939 to barely 20% in 1988. It also added fuel to the slopes, which previously were grass-covered with scattered oak and redwood trees. The additional fuel included many homes and non-native trees, mainly eucalyptus. In hot and windy weather, the fire quickly became uncontrollable and moved through the urban landscape quickly; during the first hour it consumed a home every five seconds.

One conclusion is indisputable: natural hazards and risks are subject to change, and a static view is unacceptable in the management of natural hazards and reduction of disaster impacts. In several scenarios, contemporary global warming and its anticipated consequences are likely to incur increasing losses of lives and infrastructure, as well as depletion of soil and water resources in a world that has seen its human population increase by more than 50% in the past 40 years. The umbrella term 'global environmental change' encompasses climate and ecosystem changes, the processes and effects of global population growth, expansion and changes in land use, changing lifestyles, socioeconomic changes, and developments in information technology.

So how does the proposition and conceptual framework of the Anthropocene tie in practically with natural hazards and disasters? We recognize at least three avenues of research that have emerged, or are emerging, to address this question:

- Identify those human activities that alter the factors in the risk equation, i.e. hazard, vulnerability, elements at risk, and aversion, the most. All of these factors are prone to change, though some may have more weight than others. To reduce risks it is essential for

us to know which of those factors are most relevant and deserve most attention.

- Estimate how much these human activities contribute to the overall variability in risk so that we can objectively attribute potentially adverse outcomes to human or natural drivers (or a mixture of both). This is a delicate issue that will question responsibilities and raise liabilities. If anything, it will raise or even consolidate awareness about how we interact with the natural environment, a theme that aptly characterises the intensive interdisciplinary debate (and hype) about the Anthropocene (Autin and Holbrook 2012; Lewis and Maslin 2015).

- Reconstruct what the geological record reveals about how societies or cultures interacted with their natural environment, and responded to natural disasters in the distant past. This approach weds geosciences with archaeology and follows the more formal stratigraphic requirements of a geological epoch in that it should have a characteristic marker horizon that future geologists should be able to find globally. It also has the potential to instruct us about the potential shortcomings in how past societies and cultures managed their environment or failed to prepare for both natural and partly man-made disasters.

References

Amos CB, Audet P, Hammond WC, et al. 2015 Uplift and seismicity driven by groundwater depletion in central California. *Nature* **509**(7501), 483–486.

Amundson R, Berhe AA, Hopmans JW, et al. 2015 Soil and human security in the 21st century. *Science* **348**(6235), 1261071–1261071.

Anthony EJ, Brunier G, Besset M, et al. 2015 Linking rapid erosion of the Mekong River delta to humanactivities. *Scientific Reports* **5**, 1–12.

Anyamba A, Small JL, Britch SC, et al. 2014 Recent weather extremes and impacts on

agricultural production and vector-borne disease outbreak patterns. *PLoS ONE* **9**(3), e92538.

Autin WJ and Holbrook JM 2012 Is the Anthropocene an issue of stratigraphy or pop culture?. *GSA Today* **22**(7), 60–61.

Bader J 2014 The origins of regional Arctic warming. *Nature* **509**, 167–168.

Bai X, Shi Pi, and Liu Y 2014 Realizing China's urban dream. *Nature* **509**, 158–160.

Bank TW (ed) 2010 *Natural Hazards, UnNatural Disasters*. The World Bank and The United Nations.

Barnosky AD, Matzke N, Tomiya S, et al. 2011 Has the Earth's sixth mass extinction already arrived? *Nature* **470**(7336), 51–57.

Barriopedro D, Fischer EM, Luterbacher J, et al. 2011 The hot summer of 2010: Redrawing the temperature record map of Europe. *Science* **332**(6026), 220–224.

Beach T, Luzzadder-Beach S, Cook D, et al. 2015 Ancient Maya impacts on the Earth's surface: An Early Anthropocene analog?. *Quaternary Science Reviews* **124**(C), 1–30.

Bilham R and Bikram SB 2014 A ninth century earthquake-induced landslide and flood in the Kashmir Valley, and earthquake damage to Kashmir's Medieval temples. *Bulletin of Earthquake Engineering* **12**, 79–109.

Bilham R and Gaur V 2013 Buildings as weapons of mass destruction. *Science* **341**(6146), 618–619.

Bouwer LM 2011 Have disaster losses increased due to Anthropogenic climate change? *Bulletin of the American Meteorological Society* **92**(1), 39–46.

Bowman DMJS, Balch J, Artaxo P, et al. 2011 The human dimension of fire regimes on Earth. *Journal of Biogeography* **38**(12), 2223–2236.

Brodie JE, Kroon FJ, Schaffelke B, et al. 2012 Terrestrial pollutant runoff to the Great Barrier Reef: An update of issues, priorities and management responses. *Marine Pollution Bulletin* **65**(4-9), 81–100.

Brooks N 2010 Human responses to climatically-driven landscape change and resource scarcity: Learning from the past and planning for the future. In *Landscapes and Societies* (eds Martini IP and Chesworth W), Springer, pp. 43–66.

Brown A, Toms P, Carey C, and Rhodes E 2013 Geomorphology of the Anthropocene: Time-transgressive discontinuities of human-induced alluviation. *Anthropocene* **1**, 3–13.

Cai W, Borlace S, Lengaigne M, et al. 2014 Increasing frequency of extreme El Niño events due to greenhouse warming. *Nature Climate Change* **5**(1), 1–6.

Carson JF, Whitney BS, Mayle FE, et al. 2014 Environmental impact of geometric earthwork construction in pre-Columbian Amazonia. *Proceedings of the National Academy of Sciences* **111**(29), 10497–10502.

Carter L, Orpin AR, and Kuehl SA 2010 From mountain source to ocean sink – the passage of sediment across an active margin, Waipaoa Sedimentary System, New Zealand. *Marine Geology* **270**(1-4), 1–10.

Cazenave A and Cozannet GL 2014 Sea level rise and its coastal impacts. *Earth's Future* **2**, 15–34.

Change IPoC 2012 *Managing the Risks of Extreme Events and Disasters to Advance Climate Change Adaptation* Special Report of the Intergovernmental Panel on Climate Change. Cambridge University Press.

Chartin C, Evrard O, Onda Y, et al. 2013 Tracking the early dispersion of contaminated sediment along rivers draining the Fukushima radioactive pollution plume. *Anthropocene* **1**, 23–34.

Corcoran PL, Moore CJ, and Jazvac K 2014 An anthropogenic marker horizon in the future rock record. *GSA Today* **24**(6), 4–8.

Cozar A, Echevarria F, Gonzalez-Gordillo JI, et al. 2014 Plastic debris in the open ocean. *Proceedings of the National Academy of Sciences* **111**(28), 10239–10244.

Crutzen PJ 2002 Geology of mankind. *Nature* **415**(6867), 23–23.

Daly KL, Passow U, Chanton J, and Hollander D 2016 Anthropocene. *Biochemical Pharmacology* **13**, 18–33.

Darby SE, Hackney CR, Leyland J, et al. 2016 Fluvial sediment supply to a mega-delta reduced by shifting tropical-cyclone activity. *Nature* **539**(7628), 276–279.

de Leeuw J, Shankman D, Wu G, et al. 2009 Strategic assessment of the magnitude and impacts of sand mining in Poyang Lake, China. *Regional Environmental Change* **10**(2), 95–102.

Dirzo R, Young HS, Galetti M, et al. 2014 Defaunation in the Anthropocene. *Science* **345**, 401–406.

Dodman D and Satterthwaite D 2008 Institutional capacity, climate change adaptation and the urban poor. *IDS Bulletin* **39**(4), 67–74.

Doney SC 2010 The growing human footprint on coastal and open-ocean biogeochemistry. *Science* **328**(5985), 1512–1516.

Doughty CE 2013 Preindustrial human impacts on global and regional environment. *Annual Review of Environment and Resources* **38**(1), 503–527.

Einsele G and Hinderer M 1997 Terrestrial sediment yield and the lifetimes of reservoirs, lakes, and larger basins. *Geologische Rundschau* **86**, 288–310.

Ellis EC, Kaplan JO, Fuller DQ, et al. 2013 Used planet: A global history. *Proceedings of the National Academy of Sciences* **110**(20), 7978–7985.

Eriksen M, Lebreton LCM, Carson HS, et al. 2014 Plastic pollution in the world's oceans: More than 5 trillion plastic pieces weighing over 250,000 tons afloat at sea. *PLoS ONE* **9**(12), e111913.

Finkl CW and Makowski C 2013 The Southeast Florida Coastal Zone (SFCZ): A cascade of natural, biological, and human- induced hazards. In *Coastal Hazards* (ed Finkl CW), Springer Science & Business Media, pp. 3–56.

Foley J, DeFries R, Asner G, et al. 2005 Global consequences of land use. *Science* **309**(5734), 570–574.

Foley JA, Ramankutty N, Brauman KA, et al. 2012 Solutions for a cultivated planet. *Nature* **478**(7369), 337–342.

Frings RM, Gehres N, Promny M, et al. 2014 Today's sediment budget of the Rhine River channel, focusing on the Upper Rhine Graben and Rhenish Massif. *Geomorphology* **204**(C), 573–587.

Gale S and Hoare P 2012 The stratigraphic status of the Anthropocene. *The Holocene* **22**(12), 1491–1494.

Glade T 2003 Landslide occurrence as a response to land use change: a review of evidence from New Zealand. *Catena* **51**, 297–314.

Gray JM, Frolking S, Kort EA, et al. 2015 Direct human influence on atmospheric CO_2 seasonality from increased cropland productivity. *Nature* **515**(7527), 398–401.

Guralnik OG 2014 Accelerate research on land creation. *Nature* **510**, 29–31.

Haff PK 2010 Hillslopes, rivers, plows, and trucks: mass transport on Earth's surface by natural and technological processes. *Earth Surface Processes and Landforms* **35**(10), 1157–1166.

Havenith HB, Torgoev I, Meleshko A, et al. 2006 Landslides in the Mailuu-Suu Valley, Kyrgyzstan—Hazards and impacts. *Landslides* **3**(2), 137–147.

Hinzman LD, Bettez ND, Bolton WR, et al. 2005 Evidence and implications of recent climate change in northern Alaska and other Arctic regions. *Climatic Change* **72**(3), 251–298.

Hoffmann T, Schlummer M, Notebaert B, et al. 2013 Carbon burial in soil sediments from Holocene agricultural erosion, Central Europe. *Global Biogeochemical Cycles* **27**(3), 828–835.

Hong S, Lee J, Kang D, et al. 2014 Quantities, composition, and sources of beach debris in Korea from the results of nationwide monitoring. *Marine Pollution Bulletin* **84**(1-2), 27–34.

Hoornweg D, Bhada-Tata P and Kennedy C 2013 Waste production must peak this century. *Nature* **502**, 615–617.

Horton AJ, Constantine JA, Hales TC, et al. 2017 Modification of river meandering by tropical deforestation. *Geology* **45**(6), 511–514.

Hurley R, Woodward J, and Rothwell JJ 2018 Microplastic contamination of river beds significantly reduced by catchment-wide flooding. *Nature Geoscience* **11**, 251–257.

Imaizumi F and Sidle RC 2007 Linkage of sediment supply and transport processes in Miyagawa Dam catchment, Japan. *Journal of Geophysical Research: Earth Surface* **112**, 1–17.

Ives JD and Messerli B 1989 *The Himalayan Dilemma: Reconciling Development and Conservation*. Routledge.

Jambeck JR, Geyer R, Wilcox C, et al. 2015 Plastic waste inputs from land into the ocean. *Science* **347**(6223), 768–771.

James LA 2004a Decreasing sediment yields in northern California: vestiges of hydraulic gold-mining and reservoir trapping. *IAHS Publication* **288**, 235–244.

James LA 2004b Tailings fans and valley-spur cutoffs created by hydraulic mining. *Earth Surface Processes and Landforms* **29**(7), 869–882.

Jennerjahn TC 2012 Biogeochemical response of tropical coastal systems to present and past environmental change. *Earth-Science Reviews* **114**(1-2), 19–41.

Jungerius PD 2010 Sea-level rise and the response of the Dutch people: Adaptive strategies based on geomorphologic principles give sustainable solutions. In *Landscapes and Societies* (eds Martini IP and Chesworth W), Springer, pp. 271–283.

Kadomura H 1980 Erosion by human activities in Japan. *GeoJournal* **4.2**, 133–144.

Kalis AJ, Merkt J, and Wunderlich J 2003 Environmental changes during the Holocene climatic optimum in central Europe-human impact and natural causes. *Quaternary Science Reviews* **22**, 33–79.

Kareiva P, Watts S, McDonald R, and Boucher T 2007 Domesticated nature: Shaping landscapes and ecosystems for human welfare. *Science* **316**(5833), 1866–1869.

Kasai M, Brierley GJ, Page MJ, et al. 2005 Impacts of land use change on patterns of sediment flux in Weraamaia catchment, New Zealand. *Catena* **64**, 27–60.

Kitamura A, Kurikami H, Sakuma K, et al. 2014 Initial flux of sediment-associated radiocesium to the ocean from the largest river impacted by Fukushima Daiichi Nuclear Power Plant. *Scientific Reports* **41**(12), 1708–1726.

Kleinschroth F and Healey JR 2017 Impacts of logging roads on tropical forests. *Biotropica* **49**(5), 620–635.

Kondolf GM, Gao Y, Annandale GW, et al. 2014 Sustainable sediment management in reservoirs and regulated rivers: Experiences from five continents. *Earth's Future* **2**(5), 256–280.

Konikow LF 2011 Contribution of global groundwater depletion since 1900 to sea-level rise. *Geophysical Research Letters* **38**(17), 1–5.

Kummu M, Lu XX, Wang JJ, and Varis O 2010 Basin-wide sediment trapping efficiency of emerging reservoirs along the Mekong. *Geomorphology* **119**(3-4), 181–197.

Kuper R and Kröpelin S 2006 Climate-controlled Holocene occupation in the Sahara: Motor of Africa's evolution. *Science* **313**(5788), 803–807.

Laurance WF, Clements GR, Sloan S, et al. 2014 A global strategy for road building. *Nature* **513**, 229–232.

Lehner B, Döll P, Alcamo J, et al. 2006 Estimating the impact of global change on flood and drought risks in Europe: A continental, integrated analysis. *Climatic Change* **75**(3), 273–299.

Lenderink G and van Meijgaard E 2008 Increase in hourly precipitation extremes beyond expectations from temperature changes. *Nature Geoscience* **1**(8), 511–514.

Lewis SL and Maslin MA 2015 Defining the Anthropocene. *Nature* **519**(7542), 171–180.

Lewis SL, Edwards DP, and Galbraith D 2015 Increasing human dominance of tropical forests. *Science* **349**, 827–832.

Liebmann MJ, Farella J, Roos CI, et al. 2016 Native American depopulation, reforestation, and fire regimes in the Southwest United States, 1492–1900 CE. *Proceedings of the National Academy of Sciences* **113**(6), E696–E704.

Luckman PG, Gibson RD, and DeRose RC 1999 Landslide erosion risk to New Zealand pastoral steeplands productivity. *Land Degradation & Development* **10**, 49–65.

Marlon JR, Bartlein PJ, Daniau AL, et al. 2013 Global biomass burning: a synthesis and review of Holocene paleofire records and their controls. *Quaternary Science Reviews* **65**(C), 5–25.

Marriner N and Morhange C 2007 Geoscience of ancient Mediterranean harbours. *Earth-Science Reviews* **80**(3-4), 137–194.

Martín J, Puig P, Palanques A, and Giamportone A 2014 Commercial bottom trawling as a driver of sediment dynamics and deep seascape evolution in the Anthropocene. *Anthropocene* **7**, 1–15.

Marx SK, McGowan HA, Kamber BS, et al. 2014 Unprecedented wind erosion and perturbation of surface geochemistry marks the Anthropocene in Australia. *Journal of Geophysical Research: Earth Surface* **119**(1), 45–61.

McCulloch M, Fallon S, Wyndham T, et al. 2003 Coral record of increased sediment flux to the inner Great Barrier Reef since European settlement. *Nature* **421**(6924), 727–730.

McDonald RI, Weber K, Padowski J, et al. 2014 Water on an urban planet: Urbanization and the reach of urban water infrastructure. *Global Environmental Change* **27**, 96–105.

McGlade C and Ekins P 2015 The geographical distribution of fossil fuels unused when limiting global warming to 2. *Nature* **517**(7533), 187–190.

Medley KA 2010 Niche shifts during the global invasion of the Asian tiger mosquito, *Aedes albopictus* Skuse (Culicidae), revealed by reciprocal distribution models. *Global Ecology and Biogeography* **19**(1), 122–133.

Meyer MC, Hofmann CC, Gemmell AMD, et al. 2009 Holocene glacier fluctuations and migration of Neolithic yak pastoralists into the high valleys of northwest Bhutan. *Quaternary Science Reviews* **28**(13-14), 1217–1237.

Min SK, Zhang X, Zwiers FW, and Hegerl GC 2011 Human contribution to more-intense precipitation extremes. *Nature* **470**(7334), 378–381.

Montanari A and Koutsoyiannis D 2014 Modeling and mitigating natural hazards: Stationarity is immortal!. *Water Resources Research* **50**(12), 9748–9756.

Morrison RJ, Denton GRW, Tamata UB, and Grignon J 2013 Anthropogenic biogeochemical impacts on coral reefs in the Pacific Islands—An overview. *Deep-Sea Research Part II* **96**(C), 5–12.

Muenchow J, Brenning A, and Richter M 2012 Geomorphic process rates of landslides along a humidity gradient in the tropical Andes. *Geomorphology* **139-140**(C), 271–284.

Myers SS, Gaffikin L, Golden CD, et al. 2013 Human health impacts of ecosystem alteration. *Proceedings of the National Academy of Sciences* **110**(47), 18753–18760.

Newbold T, Hudson LN, Hill SLL, et al. 2015 Global effects of land use on local terrestrial biodiversity. *Nature* **520**(7545), 45–50.

Notaro M, Wyrwoll KH, and Chen G 2011 Did aboriginal vegetation burning impact on the Australian summer monsoon?. *Geophysical Research Letters* **38**(11), 1–5.

Oberle FKJ, Storlazzi CD, and Hanebuth TJJ 2016 What a drag: Quantifying the global impact of chronic bottom trawling on continental shelf sediment. *Journal of Marine Systems* **159**(C), 109–119.

Oki T and Kanae S 2006 Global hydrological cycles and world water resources. *Science* **313**(5790), 1068–1072.

Padmalal D and Maya K (eds) 2014 *Sand Mining. Environmental Impacts and Selected Case Studies.* Springer.

Pekel JF, Cottam A, Gorelick N, and Belward AS 2016 High-resolution mapping of global surface water and its long-term changes. *Nature* **540**(7633), 418–422.

Peng M and Zhang LM 2012 Analysis of human risks due to dam-break floods—part 1: A new model based on Bayesian networks. *Natural Hazards* **64**(1), 903–933.

Pethick J and Orford JD 2013 Rapid rise in effective sea-level in southwest Bangladesh: Its causes and contemporary rates. *Global and Planetary Change* **111**(C), 237–245.

Pimm SL, Jenkins CN, Abell R, et al. 2014 The biodiversity of species and their rates of extinction, distribution, and protection. *Science* **344**(6187), 1246752-1–10.

Piperno DR, McMichael C, and Bush MB 2015 Amazonia and the Anthropocene: What was the spatial extent and intensity of human landscape modification in the Amazon Basin at the end of prehistory? *The Holocene* **25**(10), 1588–1597.

Portenga EW, Bishop P, Gore DB, and Westaway KE 2016 Landscape preservation under post-European settlement alluvium in the south-eastern Australian tablelands, inferred from portable OSL reader data. *Earth Surface Processes and Landforms* **41**, 1697–1707.

Portenga EW, Westaway KE, and Bishop P 2016 Timing of post-European settlement alluvium deposition in SE Australia: A legacy of European land-use in the Goulburn Plains. *The Holocene* **26**(9), 1472–1485.

Quataert E, Storlazzi C, van Rooijen A, et al. 2015 The influence of coral reefs and climate change on wave-driven flooding of tropical coastlines. *Geophysical Research Letters* **42**, 6407–6415.

Richardson JM, Fuller IC, Holt KA, et al. 2014 Rapid post-settlement floodplain accumulation in Northland, New Zealand. *Catena* **113**(C), 292–305.

Rick TC, Kirch PV, Erlandson JM, and Fitzpatrick SM 2013 Archeology, deep history, and the human transformation of island ecosystems. *Anthropocene* **4**, 33–45.

Roberts N 2014 *The Holocene*. Wiley Blackwell.

Rolett B and Diamond J 2004 Environmental predictors of pre-European deforestation on Pacific islands. *Nature* **431**(7007), 443–446.

Ruddiman WF, Guo Z, Zhou X, et al. 2008 Early rice farming and anomalous methane trends. *Quaternary Science Reviews* **27**(13-14), 1291–1295.

Sapart CJ, Monteil G, Prokopiou M, et al. 2013 Natural and anthropogenic variations in methane sources during the past two millennia. *Nature* **490**(7418), 85–88.

Schlütz F and Lehmkuhl F 2009 Holocene climatic change and the nomadic Anthropocene in Eastern Tibet: palynological and geomorphological results from the Nianbaoyeze Mountains. *Quaternary Science Reviews* **28**(15-16), 1449–1471.

Schmidt AH, Gonzalez VS, Bierman PR, et al. 2018 Anthropocene. *Biochemical Pharmacology* **21**, 95–106.

Schottler SP, Ulrich J, Belmont P, et al. 2013 Twentieth century agricultural drainage creates more erosive rivers. *Hydrological Processes* **28**(4), 1951–1961.

Seppelt R, Manceur AM, Liu J, et al. 2014 Synchronized peak-rate years of global resources use. *Ecology and Society* **19**(4), art50.

Seto KC, Fragkias M, Grüneralp B, and Reilly MK 2011 A meta-analysis of global urban land expansion. *PLoS ONE* **6**, 1–9.

Sidle RC, Ziegler AD, Negishi JN, et al. 2005 Erosion processes in steep terrain—Truths, myths, and uncertainties related to forest management in Southeast Asia. *Forest Ecology and Management* **224**(1), 199–225.

Steffen W, Richardson K, Rockstrom J, et al. 2015 Planetary boundaries: Guiding human development on a changing planet. *Science* **347**(6223), 1259855-1–10.

Syvitski JPM and Kettner A 2011 Sediment flux and the Anthropocene. *Philosophical Transactions of the Royal Society, Mathematical, Physical and Engineering Sciences* **369**(1938), 957–975.

Tessler ZD, Vörösmarty CJ, Grossberg M, et al. 2015 Profiling risk and sustainability in coastal deltas of the world. *Science* **349**, 638–643.

van Sebille E, Wilcox C, Lebreton L, et al. 2015 A global inventory of small floating plastic debris. *Environmental Research Letters* **10**(12), 1–11.

Vörösmarty CJ, Meybeck M, Fekete B, et al. 2003 Anthropogenic sediment retention: major global impact from registered river impoundments. *Global and Planetary Change* **39**(1-2), 169–190.

Wang H, Saito Y, Zhang Y, et al. 2011 Recent changes of sediment flux to the western Pacific Ocean from major rivers in east and Southeast Asia. *Earth-Science Reviews* **108**, 80–100.

Wasson RJ, Juyal N, Jaiswal M, McCulloch M, et al. 2008 The mountain-lowland debate: Deforestation and sediment transport in the upper Ganga catchment. *Journal of Environmental Management* **88**(1), 53–61.

Waters CN, Zalasiewicz J, Summerhayes C, et al. 2016 The Anthropocene is functionally and stratigraphically distinct from the Holocene. *Science* **351**(6269), aad2622-1–10.

Wilkinson BH and McElroy BJ 2007 The impact of humans on continental erosion and sedimentation. *Geological Society of America Bulletin* **119**(1-2), 140–156.

Williams M, Zalasiewicz J, Davies N, et al. 2014 Humans as the third evolutionary stage of biosphere engineering of rivers. *Anthropocene* **7**, 57–63.

Woodward C, Shulmeister J, Larsen J, and Jacobsen GE 2014 The hydrological legacy of deforestation on global wetlands. *Science* **346**, 844–847.

Yang SL, Milliman JD, Li P, and Xu K 2011 50,000 dams later: Erosion of the Yangtze River and its delta. *Global and Planetary Change* **75**(1-2), 14–20.

Yang SL, Milliman JD, Xu KH, et al. 2014 Downstream sedimentary and geomorphic impacts of the Three Gorges Dam on the Yangtze River. *Earth-Science Reviews.* **138**(C), 469–486.

Yang ZS, Wang HJ, Saito Y, et al. 2006 Dam impacts on the Changjiang (Yangtze) River sediment discharge to the sea: The past 55 years and after the Three Gorges Dam. *Water Resources Research* **42**(4), 1–10.

Young S, Balluz L and Malilay J 2004 Natural and technologic hazardous material releases during and after natural disasters: a review. *Science of The Total Environment* **322**(1-3), 3–20.

17

Feedbacks with the Biosphere

17.1 The Carbon Footprint of Natural Disasters

The negative outcomes of natural disasters, climate change, and extreme weather involve more than fatalities, structural damage, and geomorphic impacts. We must also consider any damage to the biosphere, biotic recycling of human and societal wastes, because these consequences add to indirect and intangible losses. Natural disasters can affect habitats, food webs, agriculture, forestry, fishery, or more generally access to natural resources, and thus compromise the livelihoods of humans. Such disturbed ecosystems may accelerate some rates of geomorphic processes, thereby inducing a new cascade of disruption. Landforms and sediments may preserve information about such landscape disturbances, possibly informing us about their geomorphic and biogeochemical impacts (Morris et al. 2015). Assessing these effects on the biosphere requires a solid understanding of the natural variability of ecosystems and how humans treat them.

The field of disturbance ecology has long focused on the effects of rare and destructive events on ecosystems. Weather-related extremes such as storms or droughts can severely damage vegetation, and today the main currency that we use for evaluating this damage is either the amount of organic carbon mobilized or the changes to biodiversity. In this chapter we focus on the carbon cycle, because

it shares many aspects and processes with geomorphic systems. Scientists from several disciplines have long investigated in detail the implications of natural disasters on the carbon cycle (Figure 17.1) (Reichstein et al. 2013). The bulk of this research has addressed climate extremes, however. Nevertheless, many of these extremes have geomorphic consequences that may damage ecosystems further and for longer.

How do natural hazards and disasters link up with the carbon cycle? Many natural disasters involve the sudden erosion of soils and biomass, and thus liberate organic carbon to rivers, lakes, oceans, and the atmosphere. If this mobilized carbon is kept from entering any nearby sinks, it decomposes and oxidizes, and eventually adds (in a natural way) to greenhouse concentrations in the atmosphere. Gradual erosion processes can add to this impact, making soil erosion or the thawing of carbon-rich permafrost potent, though often elusive, natural hazards that affect the carbon cycles over years to decades, and in a way that we cannot fully measure yet. Carbon in soils and sediments is a good marker for radiocarbon dating and hence for establishing for how long organic deposits, or the landforms they are encased in, have been in place. Measuring carbon in river sediment loads offers geochemical

Geomorphology and Natural Hazards: Understanding Landscape Change for Disaster Mitigation, Advanced Textbook Series,
First Edition. Tim R. Davies, Oliver Korup, and John J. Clague.
© 2021 John Wiley & Sons Ltd. Co-published 2021 by the American Geophysical Union and John Wiley & Sons Ltd.

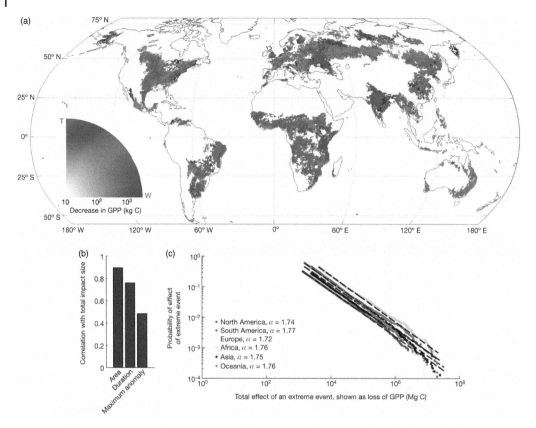

Figure 17.1 Global impact of extreme events on the carbon cycle. (a) Global distribution of extreme events impacting the terrestrial carbon cycle, defined as contiguous regions of extreme anomalies of the fraction of absorbed photosynthetically active radiation (fAPAR) from 1982 to 2011. The hundred largest events on each continent are shown, along with whether they can be associated with water scarcity (blue, W), extreme high temperatures (red, T), both (pink) or neither (grey). The colour reflects the intensity of the extreme event in terms of integrated loss of gross primary productivity (GPP), as indicated in the inset. (b) Correlation of event impact size with maximum spatial extent, duration and maximum intensity. (c) Size distribution of event impacts, following a power law with similar scaling exponents across all continents. α is the scaling exponent. From Reichstein et al. (2013).

indicators of how much carbon comes from the biosphere and how much comes from bedrock sources. Knowing carbon concentrations in soils and sediments further allows us to assess and balance up the major sources, fluxes, and sinks in the carbon cycle, and eventually to inform guidelines for soil management and remediation. Finally, putting monetary values to net carbon emissions by natural disasters offers a way to compare some of their indirect costs.

Understanding the changes in the global turnover of carbon due to such disturbances is important because carbon makes up nearly half of the dry weight of all living things on Earth. The motion of carbon approximates the flow of energy around the planet and is a proxy for the metabolism of natural, human, and industrial systems (Houghton 2007). The global carbon cycle has attracted a lot of attention because many greenhouse gases contain carbon, and because their atmospheric effects and contribution to warming can be estimated numerically on the basis of chemical and physical principles. The availability of carbon dioxide and methane as greenhouse gases depends largely on where in the Earth system carbon is being sequestered and released (Houghton 2007). Balancing these sources and sinks has revealed that components of the global carbon budget match to the first order, but has also highlighted many inconsistent or missing terms that likely reflect unappreciated sinks, such as inland waters, estuaries, continental shelves (Bauer et al. 2013), and vegetation stands (Battin et al. 2009). Sediment erosion, transport, and deposition operate in all these sinks. Hence, filling in the numbers or refining those estimates of carbon stocks and fluxes can inform our knowledge of how geomorphic processes operate in cycling carbon, and better gauge the contributions of natural and human disturbances.

The continuing loss of forests in many parts of the worlds is perhaps the most visible and widely known example of how land use leaves its mark on the carbon cycle. Although forests cover less than a third of the Earth's land surface, they and their soils store about 75% of the terrestrial carbon. Contemporary rates of deforestation in the tropics, for example, involve about 1–2 Gt C yr^{-1} (1 Gt or 10^9 t is equal to one billion tonnes, also written as 1 Pg). Between 1980 and 2012, about one million km^2 of tropical rainforest, an area about half the size of Greenland, was converted to farmland, in particular to oil palm and soy plantations. Between 2000 and 2005 alone, about one-fifth of the area of all tropical rainforest was selectively logged for the most valuable trees (Lewis et al. 2015). The relevance of these numbers becomes clearer when we consider that one-third of the total metabolic activity on Earth's land surface is taking place in tropical forests (Malhi 2011).

Taken together, contemporary natural and humanmade disturbances to forests affect an estimated area of 0.4–0.7 million km^2 per year, mainly in the form of fire, windstorms, logging, cultivation, flooding, landslides, and avalanches (Frolking et al. 2009). These and many other estimates tell only part of the story, however, as soil organic matter contains more organic carbon than our planet's vegetation and atmosphere together (Lehmann and Kleber 2015). Soils and their erosion are also key issues that many geomorphologists work on, and most natural disasters also involve the erosion and deposition of soil materials. This link further motivates study of the geomorphic and carbon footprints of natural disasters together. Contemporary atmospheric warming adds to the balance, and projections hold that 1 K of warming could reduce the global stock of organic carbon in upper soil layers of 203 ± 161 Gt C by more than 10%, depending on the rate of warming and the resulting effects on plants, microbes, and animals that govern the exchange of carbon with the atmosphere. The role of physical erosion in these projections awaits further attention. High-latitude areas of the northern hemisphere are particularly prone to such climate-driven losses in soil organic carbon (Crowther et al. 2016). These and other losses might need some refinement after including nonbiotic processes of erosion and sedimentation.

Peatlands are another important storage of organic carbon and are highly susceptible to drying, fires, and land instability. Estimates hold that peatlands contain about a quarter of the world's soil organic carbon, although they cover only a few percent of the land surface (Turetsky et al. 2015). Artificial draining adds to peatland losses, and promotes rapid land subsidence at rates than can exceed 0.5 m per year as another negative side effect, mainly caused by consolidating and compacting ground and oxidizing peat (Hooijer et al. 2012). Mass movements in peat, including 'bog bursts' or 'peat slides', also reduce terrestrial carbon storage and release it to the environment. Documented slope failures in the UK mobilized up to several million cubic metres of peat (Warburton 2015), locally choking river channels and contaminating stream habitats and reservoirs. Measuring the amount of organic carbon that peat slides mobilize is yet in its infancy.

These and many other types of biomass losses in the wake of natural disturbances (or disasters) negatively affect regional to global carbon budgets, if vegetation regrowth is limited, and carbon dioxide drawn from the atmosphere. Recovering (secondary) tropical forests, for example, can effectively take up carbon at much faster rates than comparable old-growth stands, and accumulate up to 90% of the original above-ground biomass in several decades to less than a century (Bongers et al. 2016). Data on regrowing forests can inform geomorphic studies concerned with the decadal recovery or restoration of river channels, floodplains, and hillslopes that were disturbed by natural disasters such as tropical cyclones or strong earthquakes.

17.1.1 Erosion and Intermittent Burial

The awareness that natural Earth surface processes, in part linked to climatic extremes (Reichstein et al. 2013) or land-use practices (Wang et al. 2017), control the release and sequestration of carbon has grown in recent years (Hilton et al. 2015; Smith et al. 2015), and so has the recognition that natural disasters can perturb regional carbon cycles. A large amount of available data concern soil erosion because soils may store 2–3 times as much carbon as the living vegetation they support, which alone is estimated at roughly 500 ± 100 Gt C. Soil erosion is ubiquitous on the Earth's surface and may mobilize an estimated 4–6 Gt C every year, if assuming an average soil organic content of 2–3% and that about 10% of the eroded soils enters the drainage network (Lal 2003). At the same time, patches of soil cleared of vegetation provide surfaces for regrowth and new carbon fixation in plants. Li et al. (2015) hypothesized that areas subject to high rates of erosion would have an equilibrium of high organic carbon losses driven by soil erosion and high carbon gains owing to high net primary production.

Local field measurements on the Chinese Loess Plateau seem to support this hypothesis, although further testing in other rapidly eroding landscapes is required.

Actively eroding landscapes – whether driven by natural processes or unsustainable land-use practice – produce fresh sedimentary deposits that are the parent for new soils. These deposits can also bury soils, thus potentially locking their organic carbon stocks for several centuries (Wang et al. 2015). Most quantitative estimates of global soil carbon stocks focus on the upper few centimetres of the soil column, but rarely include deeper layers of organic carbon. Rapid sedimentation may tuck away such layers and contribute to sequestering soil organic carbon for several millennia, provided that those carbon stocks are safe from decomposition, mineralization, or re-incision.

Volcanic settings are particularly suited to sequester organic carbon in buried soils (Figure 17.2), while alluvial and loess deposits outside permafrost regions are also large organic carbon repositories (Chaopricha and Marín-Spiotta 2014). Tephra falls may affect vegetation growth and biological soil production rates (Figure 17.3). Field data from a peatland in northern Japan showed that moderate to large tephra loads might suppress carbon accumulation, favouring instead peat humification. Yet, the establishment of *Sphagnum* mosses on reworked tephra beds may trigger very high carbon accumulation rates, of the order of >100 g C m^{-2}yr^{-1} (Hughes et al. 2013). Clearly, we also have to assess carbon stocks in areas beyond those used for agriculture. Degrading and eroding Arctic permafrost, for example, liberates large amounts of biogenic carbon into rivers, lakes, oceans, and the atmosphere. Scientists estimated that Arctic permafrost soils contain at least 400 Gt C (Maslin et al. 2010), although uncertainties are large, especially for methane concentrations (O'Connor et al. 2010). A more recent estimate projected 1035 ± 150 Gt C for the upper three meters in northern hemispheric permafrost soils alone (Schuur et al. 2015), and gives an

(a) (b)

Figure 17.2 Examples of buried soils and their carbon storage. (a) Coastal bluff, Bell Block, North Island, New Zealand. Peat units are interbedded with lahar deposits derived from Taranaki volcano. (b) Two soils, one reddened and the other with an subfossil rooted tree in growth position, west fork of the Nostetuko Riverv valley, southern Coast Mountains, British Columbia (all photos by John Clague).

idea how much we still need to learn when attempting regional to global carbon budgets. Many permafrost soils in Siberia and Alaska are derived from metre-thick deposits of Pleistocene loess and hold large amounts of grass roots and animal bones. The organic carbon content of these deposits is 2–5% on average, or 10–30 times higher than the content of deeper layers in mineral soils outside Arctic permafrost areas (Zimov et al. 2006). The carbon stocks in these frozen aeolian deposits alone might be 500 Gt C. Releasing only 10% of methane from the global reservoir of methane hydrates stored in permafrost and in frozen sea-floor sediments to the atmosphere could trigger changes similar to those produced by a tenfold increase in carbon dioxide (Archer 2007). Accounting for the organic carbon stored at greater depths in permafrost areas could easily more than double the current estimates of carbon stocks in high-latitude regions (Schuur et al. 2008), but exact numbers are elusive.

A survey among nearly 100 experts on permafrost landscapes and their ecology revealed that the amount of organic carbon released through erosion by Arctic rivers and coastlines could increase by 75% by the end of the twenty-first century, while wildfires could release four times as much (Figure 17.4)

(Abbott et al. 2016). Although predictions differ widely, the consensus is that the Arctic permafrost region will change from a carbon sink to a carbon source by the end of the twenty-first century, possibly releasing a total of the order of 220 Gt C from permafrost soils by then. Most experts also agree that the positive effects of a warmer Arctic, including a longer growing season, CO_2 fertilization, and more available nutrients from decomposing soil organic matter, might offset these carbon inputs for some time before becoming less efficient and giving way to net fluxes to the atmosphere.

One of the largest sources of ice-bound organic carbon is the East Siberian Arctic Shelf, where thermal collapse, enhanced wave erosion, and increasing ocean temperatures continue to liberate an estimated 0.044 ± 0.010 Gt C yr^{-1} from coastal and shallow marine sediments; about two-thirds of this mass may be lost directly to the atmosphere (Vonk et al. 2012). This study also mentioned the high uncertainties that remain in measuring the flux of organic carbon from Arctic coasts (Figure 17.5). Many estimates hinge on the extrapolation of measurements of hundreds of soil and sediment samples that are rarely randomly distributed in the landscape and hence poorly representative. Detailed

Figure 17.3 Volcanic disturbance of hillslope and floodplain forests, Chaitén Volcano, southern Chile. Tephra fall and lahars killed large tracts of temperate rainforest and mobilized clusters of tree trunks and woody debris. Much of the area shown in the photos was dense rainforest before the 2008 eruption of the volcano (all photos by Oliver Korup).

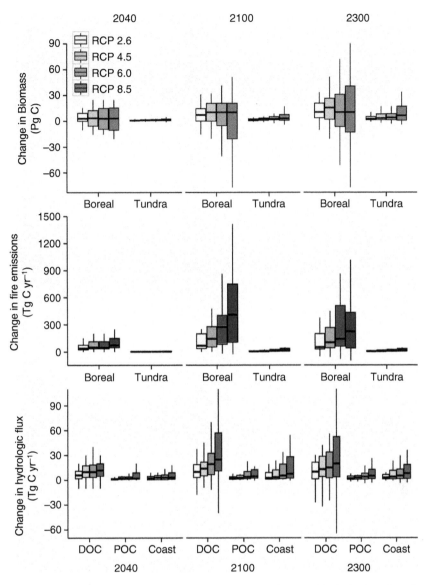

Figure 17.4 Estimates of change in nonsoil biomass, wildfire emissions, and hydrologic carbon flux from Earth's permafrost region for four warming scenarios at three time points. All values represent change from current pools or fluxes. Biomass includes above- and below-ground living biomass, standing deadwood, and litter. Dissolved and particulate organic carbon (DOC and POC, respectively) fluxes represent transfer of carbon from terrestrial to aquatic ecosystems. 'Coast' represents POC released by coastal erosion. Representative concentration pathway (RCP) scenarios range from aggressive emissions reductions (RCP2.6) to sustained human emissions (RCP8.5). Box plots represent median, quartiles, and minimum and maximum within 1.5 times the interquartile range. From Abbott et al. (2016).

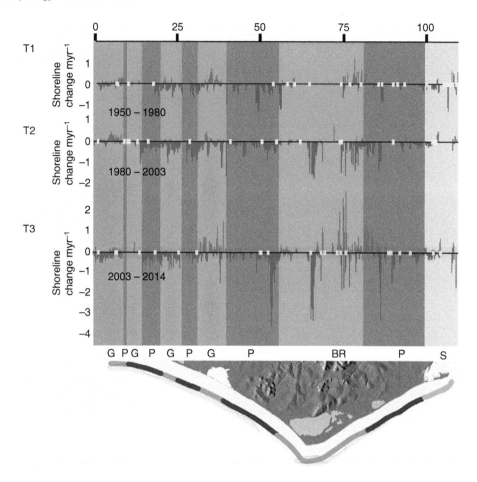

Figure 17.5 Shoreline change (m yr⁻¹) at Cape Crusenstern, Alaska, during each of the three study periods: T1, 1950–1980, T2, 1980–2003, and T3 2003–2014. Each histogram column depicts a transect point along shore. Yellow dots show where no change was measured. Transects are arranged northwest to southeast. Colours in the background correspond to the main types of coastal geomorphology: pink = gravel bars; grey = permafrost bluffs; blue = welded gravel bars and beach ridges; green = spit. From Farquharson et al. (2018).

measurements along the coastline of the Laptev Sea revealed that coastal cliffs retreated at rates of up to more than a metre per year. Günther et al. (2013) obtained these detailed measurements from satellite images acquired between 1965 and 2011, and found that the average organic carbon losses were 88–800 t C for each kilometre of coastline, mainly driven by wave undercutting and cliff-top melting.

Thermokarst lakes in degrading permafrost release large amounts of organic carbon, mostly by exhaling methane, to the atmosphere, although the importance of this process remains disputed. Estimated methane emissions from such thaw lakes in northern Siberia rose by up to nearly 60% between 1974 and 2000, based on the expansion of sampled lakes fed by degrading permafrost that formed mostly during the Pleistocene (Walter et al. 2006). Extrapolating detailed field measurements yields regional carbon emissions of the order of 0.0038 Gt C yr⁻¹ for northern Siberia, which is up to five times higher than previous estimates. Using robust methods for extrapolation is key to making these and many other field-based estimates of carbon emissions

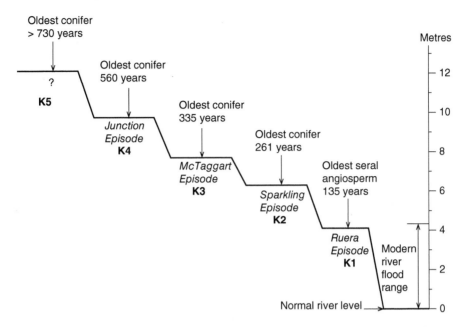

Figure 17.6 The terrace sequence at Karangarua River, south Westland, New Zealand. Successive terraces were abandoned at times of four episodes of catchment-wide disturbance. Oldest tree ages include a colonization delay adjustment of 28 years. From Wells et al. (2001).

reliable. Plant regrowth on thawing permafrost may offset some of the releases of old carbon in the short term, but field measurements give the impression that the decadal pattern of permafrost degradation is bound to cause a net transfer of old carbon into the atmosphere in the long run (Schuur et al. 2009). Other studies suggest that thermokarst lakes, which are common throughout Arctic Siberia, Alaska, and Canada, may have acted as both net sources and sinks of atmospheric methane and carbon dioxide. Detailed field studies of how organic carbon is stored in and released from relict thermokarst lakes shows that they have shifted from net carbon sources to net carbon sinks during the Holocene (Anthony et al. 2015). The contribution and fate of these thermokarst lakes depends also on how quickly they infill with sediment from degrading permafrost soils in the surrounding landscape.

Forest biomass has organic carbon stocks comparable to those in soils, at least to an order of magnitude. Hence episodic disturbances mobilize both plant debris and soil.

Earthquakes are a major nonclimatic disturbance to the biosphere, and strong local ground shaking may fell trees, but more importantly exposes trees to physical impact and erosion by coseismic slope failures (Figure 17.6). Detailed plot studies following a M_w 6 earthquake in the Southern Alps of New Zealand revealed that between 0.6 and 24% of forest trees were killed in a study area of nearly 300 km^2. In three of the four cases studied, landslides triggered by the earthquake were responsible for the total stem biomass mortality (Allen et al. 1999). Plot-to-plot variations can be large when assessing the biomass and carbon content of forest vegetation, and a solid error propagation is needed when budgeting changes to biomass and organic carbon stocks (Holdaway et al. 2014). Plots may also fail to capture larger and rarer disturbances such as single catastrophic landslides that can destroy large cohorts of trees on hillslopes and on floodplains. For example, the 2010 Mount Meager landslide in British Columbia had an estimated volume of 51 Mm^3, and destroyed ~0.11 Mm^3 of mixed western hemlock, silver fir, subalpine fir, and western red cedar that had an average log market value of CAD\$8.7 million at the time (Guthrie et al. 2012). Against this background it is necessary to look at the rates at which biomass and soils regrow and reform, and thus sequester organic carbon, following such disturbances. For example, detailed measurements of soil carbon stocks in pasture lands of North Island, New Zealand, show that total contents in landslide scars increased logarithmically with time after slope failure. These data were collected over a period of 70 years. During that time most of the landslide sites had lower soil carbon contents than uneroded sites. De Rose (2012) proposed that carbon stocks C_{tot} [kg m^{-2}] recover on average as a logarithmic function of landslide-scar age t [yr]:

$$C_{tot} = a + b(1 - e^{-ct}), \tag{17.1}$$

where a and b are constants [kg m^{-2}], and c is a rate exponent [yr^{-1}]. This model implies that pre-erosion contents of soil carbon will be lost. The field data from a range of North Island soils indicate suggest that total carbon stocks in soils may only recover to about 75% of their pre-erosion levels given at least a century of undisturbed conditions.

17.1.2 Organic Carbon in River Catchments

Current debates about natural and human contributions to terrestrial carbon sources and sinks have focused on rivers and wetlands, because these environments favour the production and transport of biogeochemical species over land surfaces or directly into the atmosphere. For example, artificial impoundments such as farm dams are an important local sink for organic carbon, especially where runoff from intensely used agricultural lands is rich in fertilizers. In such settings the rate of organic carbon burial scales crudely with incoming sediment yield, although the rates may differ by up to two orders of magnitude (Downing et al. 2008). Currently, sediment

storage in rivers, lakes, estuaries, and coastal waters may sequester about half of the organic carbon mass – up to about 0.5 Gt C – that land-use generates and transfers to rivers each year on average (Regnier et al. 2013). Other estimates hold that rivers flush an estimated 0.3–0.5 Gt C yr^{-1} into the world's oceans. This flux is comparable to that arising from deforestation in the Amazon, which is approximately 0.34 Gt C yr^{-1} (Pearson et al. 2014), but an order of magnitude lower than the estimated rates of organic carbon mobilized by soil erosion (Lal 2003).

If accounting for equal parts of organic and inorganic carbon, rivers likely carry as much as 1 Gt C in total annually. Only about 20% or less of the total river-transported organic carbon reaches the deep seas, mainly as dissolved carbon. The bulk of the particulate organic matter is deposited in continental margin settings. Although organic carbon is the focus of these and many other studies, rivers also transport divalent elements such as calcium produced by terrestrial silicate weathering; these elements fix carbon dioxide upon entering the ocean (Gislason et al. 2006).

Understanding how rivers and wetlands mobilize, transport, and deposit biogeochemical constituents adds to our understanding of the carbon cycle beyond its atmospheric or oceanic components (Aufdenkampe et al. 2011). Temperate and tropical river floodplains are major carbon sinks, at least temporarily. There, the average residence times of carbon depend strongly on how rapidly rivers rework their valley-fill sediments. Nearly 1500 sedimentary archives document that floodplains in central Europe have accumulated an estimated average of 0.7 ± 0.2 g C m^{-2}yr^{-1} of soil sediment since the onset of anthropogenic soil erosion about 7500 years ago, when Neolithic settlers began to cultivate land (Hoffmann et al. 2013). Colluvial sediment wedges lodged on hillslopes contain more than half of this volume of soil material per unit area, and should be considered when budgeting carbon stocks (Figure 17.7). In tectonically active settings,

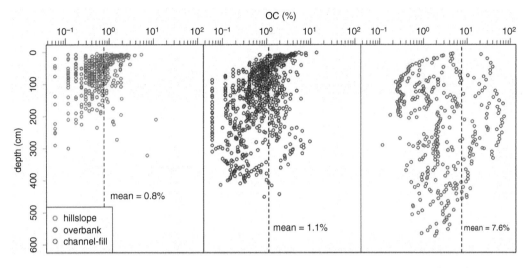

Figure 17.7 Carbon burial in soil sediments resulting from Holocene agricultural erosion. Organic carbon (OC) concentration versus depth of hillslope, overbank, and channel-fill deposits in Central Europe. From Hoffmann et al. (2013).

trapping rates of organic carbon on floodplains can be higher, but still small compared to the total export from a given catchment. For example, Gomez et al. (2003) estimated that ~2.7 g C m^{-2} yr^{-1}, or only about 4% of the mean annual yield of particulate organic carbon of New Zealand's Waipaoa River, enters floodplain storage. Global projections based on measurements from soils and sediments and numerical modelling suggest that soil erosion and burial following agriculture has drawn some 78 ± 22 Gt C from the atmosphere since the beginning of the Neolithic. This storage may offset more than one-third of the estimated CO_2 emissions attributed to human land-use changes (Wang et al. 2017). Much of this drawdown occurs because the bulk of eroded upland soils is swiftly redeposited in hillslope and alluvial sediments, lakes, reservoirs, and rivers. This rate of carbon burial may have grown more than fourfold in the past two centuries, partly reflecting the increased trapping of sediments by human activities (see Chapter 16.3). Inland waters are another sink for terrestrial organic carbon, and current estimates project that about 5 Gt C enter lakes and reservoirs per year. This estimate has nominally increased fivefold within a decade

of research. Part of this increase reflects uncertainties in extrapolation, although a yearly net increase of about 0.3 Gt C seems plausible (Drake et al. 2017).

A key assumption linking biogeochemical budgeting and geomorphologic research is that suspended sediment yield is a reliable proxy for particulate organic carbon loads in rivers (Komada et al. 2004). Compiled data on particulate organic carbon yields [tC km^{-2} yr^{-1}] from both the biosphere and rocks scale nonlinearly with suspended sediment yields. Galy et al. (2015) projected from this relationship that physical erosion dominated the export of particulate organic carbon from living vegetation stands with average fluxes of 0.157 Gt C yr^{-1} in a sample of rivers worldwide. In contrast, rivers carried only an average of 0.043 Gt C yr^{-1} from petrogenic sources according to this regression.

Mountain streams draining cool temperate and glaciated catchments feed large amounts of sediment to lowland rivers and oceans, and often also carry large amounts of organic carbon. In Oceania, rivers that drain small and densely forested mountain catchments with little sediment storage capacity may account for as much as 21–38% of the total global organic global flux (Carey et al. 2005).

New Zealand's rivers alone deliver an estimated 4 ± 1 tC km^{-2} yr^{-1} of dissolved, and 10 ± 3 tC km^{-2} yr^{-1} of particulate, organic carbon to the ocean. These values are 2–6 times the global average and 40% of the national fossil fuel emissions (Scott et al. 2006). Catchments in tectonically more quiescent and less humid settings may produce comparably high yields, though only during extreme floods (Rathburn et al. 2017).

A growing number of studies have highlighted the role of landslides, mostly triggered in large numbers by tropical cyclones or earthquakes, as the major mechanisms of delivering particulate organic carbon to streams (Hilton et al. 2008a; West et al. 2010). Wet snow avalanches also entrain and push soils into streams during winter and spring at rates that rival some of the highest yields known in mountain rivers (Korup and Rixen 2014), and surpassing, for example, the estimated release of dissolved organic carbon from melting alpine glaciers (Singer et al. 2012). Local measurements confirm that the amount of dissolved organic carbon in rivers may depend on how much of a catchment is covered by glacier ice (Hood and Scott 2008). Thus we expect that high fluxes of organic carbon enter oceans directly along mountainous coasts or via fjords and other inlets. Smith et al. (2015) estimated that 0.018 Gt C yr^{-1} accumulate in fjords throughout the world, mostly owing to the high fluxes of fine glacigenic sediment. This projected burial rate of carbon in fjords is about twice that of the global marine average, and thus fjord sediments are strongly enriched in organic carbon.

Erosion can be highly selective and, for example, preferentially entrain the lighter fraction of organic carbon, that is organic carbon with a bulk density of <1.8 t m^{-3} (Lal 2003). Carbon fluxes contain both modern plant-derived and old bedrock-derived organic carbon that are mobilized, respectively, by shallow landslides and deep-seated gully erosion (Gomez et al. 2010). A study of mountain catchments in California and New Zealand reported that catchments with higher fluvial sediment yields mobilized larger amounts of old organic carbon derived from bedrock, whereas catchments with lower sediment yields had much higher fractions of young organic carbon (Leithold et al. 2006). One explanation for this observation is that the mechanically weak rocks – soft sandstones, siltstones, melange, and marine sedimentary rocks – in the catchments studied are prone to deep-seated gully erosion and earthflows, which deliver largely unweathered sediment and organic carbon to the drainage network. While helping to refine carbon budgets, this biogeochemical fingerprint identifies sources and potential mechanisms of erosion in river catchments. Refractory old soil organic carbon can be a tracer for estimating the relative contributions of tributary catchments. Galy and Eglinton (2011) estimated that one-fifth of the bulk biospheric organic carbon transported by the Ganges–Brahmaputra River is, on average, 15 000 years old. Most of this old carbon comes from the semiarid and sparsely vegetated headwaters draining the Tibetan Plateau. The abundance of organic carbon from these sources suggests local erosion rates that are higher than previously thought.

Organic carbon loads in rivers draining cold regions are less well known and still await representative measurements. Arctic rivers may currently transport as much as 0.12 Gt C yr^{-1}, but uncertainties remain high (Abbott et al. 2016). The Mackenzie River, Canada, for example, delivers some 0.0022 Gt C to the Arctic Ocean, while much of the eroded biospheric carbon has resided in the catchment for several millennia, judging from radiocarbon dating (Hilton et al. 2015). More data are available for tropical streams. For example, the Fly–Strickland River draining the densely forested mountains of Papua New Guinea delivers an estimated 0.0012–0.0024 Gt C to the western Pacific each year. These fluxes rival the organic carbon yields reported from the Amazon River that has a catchment area roughly ten times larger (Alin et al. 2008).

The high carbon flux of the Fly–Strickland River is of the same order of magnitude as the annual carbon emissions resulting from tropical deforestation in countries like Guyana or Suriname (Pearson et al. 2014). Rivers draining Indonesian peatlands have, like many other rivers in the lowland tropics, a characteristic dark orange-black colour owing to the high concentrations of humic acids. If we extrapolate point samples from these 'blackwater rivers' to their entire catchments, we obtain fluxes of 0.018–0.021 Gt C yr^{-1} to the ocean, mostly in a dissolved form (Moore et al. 2011). These fluxes make up about one-tenth of the world's total fluvial carbon export to the oceans, depending of course on the method of estimation. The reasons for these high yields are high rates of terrestrial primary production, high amounts of soil organic carbon, and the high discharge of steep rivers. Episodic flooding during tropical cyclones flushes large amounts of organic carbon from these tropical rivers, but longer observations, ideally over decades, are needed to confirm their sediment and biogeochemical fluxes.

Small tropical volcanic islands with steep dissected slopes that receive several metres of rainfall per year deliver much dissolved organic carbon to the oceans. Lloret et al. (2011) estimated that flash floods are responsible for up to 60% of the total dissolved organic carbon flux, and 25–45% of the dissolved inorganic carbon flux from the island of Guadeloupe in the French West Indies.

Agricultural land is very prone to losses of organic carbon. In a nested study of small headwater catchments of the Mekong River, Chaplot et al. (2005) found that reworked and redeposited soils (or soil sediment) had more than twice as much organic carbons as undisturbed soils. They also found that estimates of yields of sediment and soil organic carbon from plots of 1 m^2 differed by two orders of magnitude from those at the outlet of the headwater catchment (0.62 km^2). Part of this mismatch arises from the intermittent storage of eroded sediment and soil organic carbon on hillslopes and valley floors. Data from plot studies may elucidate in great detail how organic carbon forms and resides in the soil column, but may poorly represent how eroded organic carbon travels through the drainage network.

17.1.3 Climatic Disturbances

Storms can have major carbon footprints where they hit forested terrain. For example, high wind speeds were responsible for about half of the annual average of 35 Mm^3 of wood that was damaged in European forests between 1950 and 2000 (Schelhaas et al. 2003). Moreover, the hazard of injury or death due to tree fall remains for some time after storms have passed. The high wind speeds associated with tropical cyclones also damage or kill trees. An analysis of satellite images of tree cover in the Gulf of Mexico shows that Hurricane Katrina in 2005 may have reduced the total biomass by 0.092–0.112 Gt C by killing more than 320 million trees. This sudden loss of vegetation cover is roughly 50–140% of the net annual carbon added to forests in the United States (Chambers et al. 2007). The high uncertainty associated with this estimate, however, underlines the need for refined and accurate methods that are best validated by field data. All trees die eventually, so what matters in terms of carbon release is how rapidly and for how long tree mass is changing globally. This question also needs to include the balance between deforestation and planting of forests.

The importance of these and other catastrophic impacts on tree stands has been well-documented in many studies in disturbance ecology. Storms, earthquakes, avalanches, and landslides are important processes for generating large woody debris, especially in dense mountain forests (Figure 17.8). This general statement masks a lot of variation, depending on geographic regions, biomes, species, or disturbing processes involved. Crausbay and Martin (2016) concluded from a review of ecological studies that tropical cyclones, for example, had caused only limited

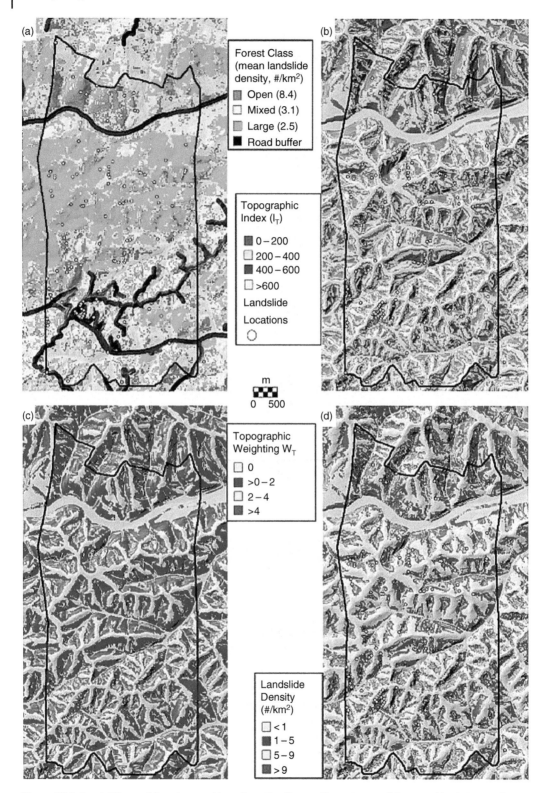

Figure 17.8 Landslides and forest cover. Maps from the Oregon Department of Forestry Mapleton study area showing the (a) topographic and land cover context for field-mapped landslides, (b) topographic index (I_T), (c) topographic weighting term (W_T), and (d) density of landslide initiation points calculated as the product of the topographic weighting term and the mean landslide density. Study site boundary is indicated by the heavy black line. From Miller and Burnett (2007).

mortality rates, largely below 7%, in tropical mountain forests worldwide. Nonetheless, the authors cite one example of how these disturbances can initiate a hazard cascade:

> The high fuel loads left by Hurricane Gilbert (1988) in the lowland forests of the Yucatán [peninsula] and a subsequent drought in 1988–1989 led to exceptionally severe and large fires in 1989 [...].

In terms of natural hazard appraisals, it is important to learn how forests respond to and are resilient to disturbances. For example, landslides and snow avalanches widely impact *Nothofagus* (southern beech) forests that cover about a quarter of the land area of New Zealand (Davis et al. 2002). These disturbances occur without any evidence that their present rate is strongly affected by humans. The large woody debris thus produced is an important carbon pool for regenerating forest stands in many temperate rainforests, and provides habitat for pioneering species. Field data show that the valley-floor confinement of rivers may control to first order the carbon stocks in vegetation. In the Colorado Front Range, United States, unconfined valleys have greater potential for storing wood from old-growth forests than confined valleys (Wohl 2013). Large woody debris derived from landslides is a major element, and sometimes also hazard, in many mountain catchments. Piles or nests of large wood can obstruct channels, block culverts, force sedimentation, create scour and undercutting when jammed up at bridge piers, and compromise navigation and the operation of reservoirs (Figure 17.9). A budgeting study of 42 large reservoirs in Japanese mountain catchments with densely forested headwaters showed that large woody debris had to be removed at rates of 0.09–5 t km^{-2} yr^{-1}, averaged over survey periods of 5–15 years (Seo et al. 2012).

Hydrologic and geomorphic feedbacks between forest stands and slope stability can determine patterns of sediment and carbon

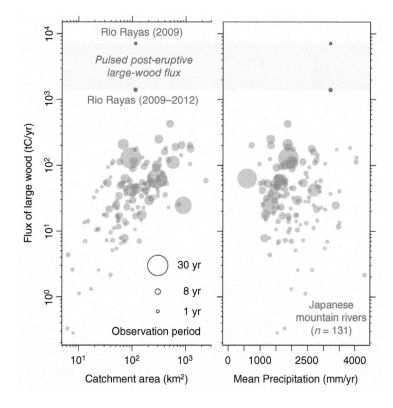

Figure 17.9 Pyroclastic eruption boosts organic carbon fluxes into Patagonian fjords. Comparison of rates of large wood flux in the Rayas River following the 2008 eruption of Chaitén volcano, southern Chile, and 131 densely forested mountain rivers in Japan that are likewise prone to frequent volcanic eruptions, earthquakes, and landslides. From Mohr et al. (2017).

export. Landslides can deliver large loads of organic carbon from densely forested mountain slopes to steep and powerful rivers that swiftly export their loads to the sea (Hilton et al. 2008b, 2010), especially in the wake of major earthquakes or tropical cyclones. Ren et al. (2009) suggested that the 2008 Wenchuan earthquake may have mobilized up to 0.235 Gt C through coseismic landslides that scraped biomass from densely forested mountain hillslopes. This figure hinges on a reliable extrapolation of detailed field measurements, but only occasionally are the associated uncertainties well documented. Ramos Scharrón et al. (2012) estimated that shallow landslides triggered by Hurricane Mitch in 1998 mobilized 0.043 Mt C from mountain forests in eastern Guatemala. Depending on the average return period of storms of this size, the long-term carbon flux rate in that region could be 8–33 t C $km^{-2} yr^{-1}$, assuming that soils and biomass would regenerate sufficiently rapidly to be available in future storms, and that these storms would be similarly efficient in triggering landslides.

Very high carbon fluxes during storms have been reported from Taiwan. Measurements of particulate organic carbon in the Choshui River during the passage of Typhoon Mindulle in 2004 revealed that, on average, each square kilometre of the catchment lost ~1.7 t C hr^{-1} at peak flow (Goldsmith et al. 2008). The measured suspended sediment yields were more than two orders of magnitude higher than 'background' values and produced hyperpycnal flows where the river enters Taiwan Strait. Such high-density flows are a mechanism for transporting sediment that is rich in particulate organic carbon far beyond the near-shore environment into deep basins, thus enabling rapid burial of carbon. Allison et al. (2010) estimated that as much as 85% of the global carbon buried at continental margins is on river-dominated continental shelves. As yet some of the linkages and feedbacks related to carbon burial in these environments remain to be fully understood. Tropical cyclones may also cause sufficient wave action to resuspend large amounts of sediment from continental shelves, thus exposing particulate and dissolved organic carbon to remineralization and re-introduction into coastal food webs.

Infrequent disturbances are now known to be vital in ecosystem dynamics and to affect carbon fluxes. A global assessment of four different data sets on gross primary production over the past 30 years emphasized that rare, weather-related extremes, mainly drought and fire, curtailed the photosynthetic uptake of carbon by as much as 3.5 Gt C yr^{-1}. These extremes may explain nearly 80% of the annual variability of gross primary production, although they happened in less than 10% of the observation period (Zscheischler et al. 2014). Satellite images documenting changes in the normalized difference vegetation index (NDVI) – a measure of greenness and proxy for the photosynthetic potential of vegetation – and anomalies in land-surface temperature show that weather extremes such as floods or droughts may be responsible for 10–80% of the difference in agricultural production per given year and region (Anyamba et al. 2014).

Natural fires produce CO_2 emissions of the order of 2–4 Gt C annually, which is about half the emissions from the burning of fossil fuels, or ~7.2 Gt C yr^{-1} (Bowman et al. 2009). The widespread Indonesian rainforest and peat fires during the 1997–1998 El Niño caused the nation's largest CO_2 emissions in the 42 years of systematic measurements. A ten-year satellite-based study of fires on Borneo between 1997 and 2006 confirmed that during the El Niño phase, fire-affected areas were on average three times larger than during normal weather conditions. About one-fifth of the island area was burnt during these ten years (Langner and Siegert 2009). The economic costs of Indonesia's 1997–1998 fires were about $US 8.8–9.3 billion, with at least $US 1 billion arising from the adverse health effects of smoke haze (Bowman et al. 2009). Assuming that emissions incur costs of

US\$ 20 per ton of carbon, the economic impact of these fires may be even higher, in the range of \$US 60–190 billion, unless partly offset by forest regrowth (Cochrane 2009). Estimating how wildfires might change biogeochemical cycles in the future depends largely on how accurately we can predict changes in vegetation and on wildfire management strategies, which eventually determine how much, and how, flammable fuel becomes available.

Wildfires in the Arctic are increasingly tapping organic carbon stored in boreal forests, the tundra vegetation, and the degrading permafrost soils below. On average nearly 90 000 km^2 of boreal forests and tundra burn in the Arctic each year, releasing about 0.26 Gt C (Abbott et al. 2016). Fires in boreal forests and the tundra have become larger over the past several decades, thus liberating more organic carbon than smaller burns. The Anaktuvuk River fire burned about 1000 km^2 of tundra vegetation on Alaska's Arctic Slope in 2007, releasing an estimated 0.0021 Gt C into the atmosphere (Mack et al. 2012). Averaged over a 25-year period, this mass is roughly equal to the estimated net annual carbon sink in the entire Arctic biome. However, these fire-driven contributions to atmospheric CO_2 are ephemeral if the vegetation regrows and recaptures carbon dioxide from the atmosphere. Yet fires also return some of the mobilized carbon directly back to soils. Ash deposition from wildfires, for example, may reach several tens of tonnes per hectare. Such deposits are a largely overlooked source of pyrogenic carbon that could contribute to long-term carbon sequestration if the ash is incorporated into the soil (Santín et al. 2012).

17.2 Protective Functions

Undisturbed ecosystems provide benefits that reduce losses from some natural disasters. The concept of 'ecosystem services' is a way of measuring – often in monetary terms – the benefits gained by maintaining intact natural environments to counter losses from disasters (Nel et al. 2014). Studying such protective functions and using them in managing natural hazards and risks goes beyond a single discipline, but calls for contributions by geomorphologists also concerned with biological and ecological processes. One such interface, for example, concerns the activity of 'ecosystem engineers', i.e. species that exert a disproportionately larger impact on sediment budgets than other species (Stallins 2006). Viewing the situation through a geomorphic lens can reveal some of the limits that ecosystem-based protection measures might have (Gedan et al. 2010). Most research on the protective role of ecosystems has been on how floodplains, coasts, and, to a lesser degree, hillslopes, can mediate the hydrologic and geomorphic impacts of disturbances. The philosophy of this approach fits well with a more natural (in the true sense of the word) and sustainable approach to hazard mitigation. Yet many of these studies were rooted in a field now variably known as 'eco-geomorphology', 'ecohydrogeomorphology', or 'biomorphodynamics' (Murray et al. 2008). These labels mean to emphasize research that asks how biological processes (and disturbances) influence geomorphic processes and landscapes, and vice versa (Figure 17.10). Indeed, the role of disturbances on ecosystems is among a recently identified set of the 100 most prominent questions in modern ecology (Sutherland et al. 2012).

17.2.1 Forest Ecosystems

We emphasize studies of ecosystem processes because land-cover and land-use changes are among the key manipulators of geomorphic processes. Forests, for example, have well known and well publicized protective functions against natural hazards, and mostly against floods, soil erosion, and shallow landslides. Unfortunately, this protective function becomes clear mostly after forests have been cleared. In a study of floods and forest cover in 56 developing countries, for

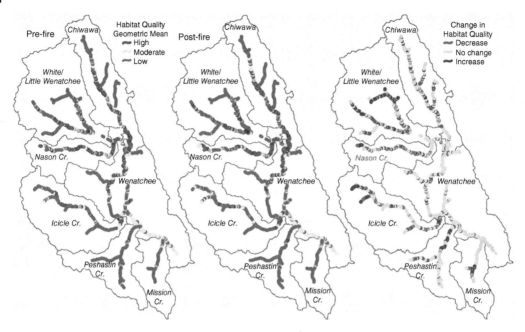

Figure 17.10 Prefire, postfire, and changes due to fire in the geometric mean of habitat quality (combining the egg/fry, juvenile, and adult life stages) for spring Chinook salmon throughout the modeled spatial extent of the Wenatchee River subbasin, Washington. From Flitcroft et al. (2016).

example, Bradshaw et al. (2007) reported that flood frequency between 1990 and 2000 was consistently lower in areas with remaining natural forest. Flood frequency also correlated with the average rates at which forests were lost, irrespective of additional controls such as climate or topography. Such global to regional comparisons may confirm that it is important to maintain healthy forest cover for protection against floods and other natural hazards, but offer few clues as to how to put this into practice for a given catchment or river reach.

Some river scientists have emphasized the importance of capturing the full spread of channel and floodplain forms rather than relying on mean properties such as is done for hydraulic geometry, for example. One approach that takes into account more complexity involves 'riverscapes', which are ecological perspectives that portray 'rivers as a combination of broad scale trends in energy, matter, and habitat structure as well as local discontinuous zones and patches'

(Carbonneau et al. 2012). Some aspects of riverscapes emphasize correlations between variables of channel geometry and habitats, and how these correlations change downstream. In this sense, riverscapes are closely connected to spatial statistical methods.

Other work has looked both empirically and conceptually at the feedbacks between forests and catchment processes. Floods or debris flows, for example, are ecological disturbances that can destroy riparian soil and vegetation along steep, narrow, low-order channels, or completely bury floodplain forests beneath coarse sediment (Nakamura et al. 2000). In turn, both riparian vegetation and large wood change local flow hydraulics and thus the erosion and sedimentation rates of river sediment on valley floors (Gomi et al. 2004). The succession of vegetation, and eventual establishment of forest stands, on active channel beds or floodplains is thus in a dynamic feedback with flood-driven channel processes. Vegetation in river beds can establish and perish rapidly enough to be of concern for river management.

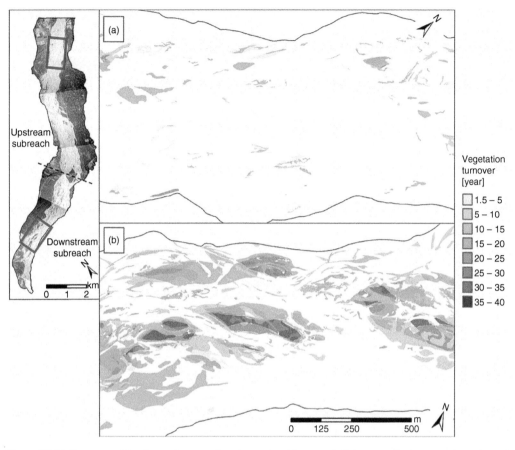

Figure 17.11 Turnover of in-channel vegetation determined from aerial photographs in two subreaches of the Tagliamento River, Italy. (a) Upstream subreach. (b) Downstream subreach. From Surian et al. (2014).

The Tagliamento River, Italy, for example, is one of the few remaining nearly natural gravel-bed rivers in Europe. A study on the turnover of in-channel vegetation in this river showed that half of the plants remained in place for only 5–6 years on average, and that only 10% had survived in the active river bed after nearly two decades (Figure 17.11) (Surian et al. 2014).

It is useful to define an interval required for vegetation to grow on recently disturbed surfaces such as freshly aggraded channel beds (or floodplains), T_{veg}. We can then compare T_{veg} to the time required for a channel-bed to be completely reworked by erosion and deposition, T_{ch} (Gran et al. 2015). This simple concept predicts that channels should remain largely

mobile with only subdued stabilizing effects of vegetation, as long as $\frac{T_{veg}}{T_{ch}}$ <1, and many field studies seem to support this. Flume-channel experiments that simulated vegetation cover using alfalfa sprouts (*Medicago sativa*) showed how plants provide the necessary ingredients for forming meandering channels from previously nonvegetated braided channels (Tal and Paola 2010). The experiments included distinct flood pulses and illustrated how plant cover stabilizes the banks such that inner-bank accretion rates keep pace with outer-bank erosion rates, while providing potential for channel avulsions once distinct vegetated levees had formed. The simulated variable flood regime also cleared new patches for vegetation to re-establish, thus distinctly partitioning the floodplain into stable and abruptly eroded portions.

Introducing exotic vegetation into landscapes can alter some of the original protective functions against natural hazards. Scenario-based simulations using climate, hydrological, and land-cover models for the Cape coast of South Africa suggest that the uncontrolled spread of invasive alien tree species, for example, may alter river flow regimes. Projected changes in the management of forest plantations could cause flood levels currently associated with a 100-year return period to occur once every 80 years on average in the future. In the worst case, monthly river flows could only be half as high during very dry spells. These lower flows during droughts would increase water shortages, and also roughly double fire intensities (Nel et al. 2014). A five-year study of 27 small headwater catchments (<1 km^2) in Indonesia, Laos, Philippines, Thailand, and Vietnam showed that soil erosion and sediment yields depended on the topography of the catchment and on the choice of crop under cultivation. Fallow catchments and those with fruit-tree plantations and planted fodder had less erosion than catchments with extensive or uniform maize or cassava plantations (Valentin et al. 2008).

Forest logging can compromise slope stability, and many studies have documented increased landslide activity after forestry operations. Reported rates of increase range widely, but often involve an order of magnitude. For example, forest clearing on Vancouver Island, Canada, increased the frequency of landslides by factors of between two and 24, and the number of landslides that reached channels increased two- to twelvefold (Guthrie 2002). Besides logging itself, the roads constructed to access cutting blocks cause further slope instability. In a study of several decades of erosional responses to rainforest logging of the Segama River, Malaysia, Walsh et al. (2011) identified road-related landslides and steeplands as the main sources of sediment despite affecting

only small fractions of the catchment area. In areas with distinct seasonal climates such as south-central Chile, the season during which logging is done may affect the amount of water and sediment delivered from clear-cut catchments; even without much landslide activity, suspended sediment yields from logged areas can be twice as high as from undisturbed forest stands (Mohr et al. 2014). Studies of sediment storage and yields following forest harvesting operations in steep mountain catchments in Japan also showed that the hydrogeomorphic response to tree cutting scales partly with the time since forest harvesting or the age of forests, but also depends on rainfall characteristics, previous episodes of soil erosion, and local sediment storage (Imaizumi and Sidle 2012).

Modelling future deforestation based on rates recorded over the past decades may shed light on how slopes are likely to behave in the future. In using a simple deterministic slope-stability model, however, we assume that many of its input parameter values, such as substrate or the overall hillslope geometry, remain unchanged and without being affected by previous slope failures (Vanacker et al. 2003). Some of these models incorporate different forestry practices and highlight the higher impacts of clear-cutting as compared to partial removal of trees from a given catchment (Dhakal and Sidle 2003). Different types of forest vegetation also influence rates of sediment supply and recharge to debris-flow channels. Knowing these rates is important because they partly control the recurrence of debris flows. Jakob et al. (2005) found that clear-cut gullies in coastal British Columbia recharged at much slower rates than those sustaining old-growth forest and thus questioned the practice of assessing debris-flow hazard only on the basis of their frequency and magnitude, especially in settings where sediment recharge limits their occurrence. Istanbulluoglu et al. (2004) parameterized the geomorphic conditions of a model catchment to simulate the effects of wildfire from hundreds to thousands of years. Their simulations predicted that forest vegetation lowers the more frequent sediment yields, while enhancing sediment yields during extreme events. Wildfires also appear to surpass forest harvesting as the dominant cause of enhanced sediment delivery in the model runs.

People living in the European Alps have long realized and appreciated the value of forests as natural barriers against rockfall, soil slides, and snow avalanches. Many protection forests have averted or reduced damage from these mass movements, buffered runoff during intense rainstorms, and stabilized steep hillslopes (Dorren et al. 2004). Some of these protection forests have had legal protection status for several centuries. Mapping landslides from air photos or high-resolution satellite images of forested terrain is a widespread technique to assess how tree cover may suppress slope instability. Using this method to establish links between forest types and landslide frequency is prone to sampling bias, however, because the tree canopy can hide landslide scars, and because the smallest slope failures will remain elusive. In a survey of forests in the Oregon Coast Range, Miller and Burnett (2007) found that landslide density varied with size of study area and type of forest cover. Smaller study areas of several tens of square kilometres with old forests had some of the highest landslide densities, whereas larger study areas (>500 km^2) covered by the same forest type had much lower landslide densities. These low densities were still about one-third lower than in recently harvested forests, though. Similar work on densely forested mountains in British Columbia showed that up to 85% of all landslides were hidden by foliage, so that the total landslide volumes were underestimated by as much as 30% (Brardinoni et al. 2003). These insights call for carefully planned strategies for measuring instabilities in forested terrain, as well as for sustainable replanting or reforestation in landslide-prone terrain.

The emerging field of landslide ecology addresses some of these issues and contributes practical findings about vegetation types that are suitable for remediating and stabilising landslide scars (Fusun et al. 2013). Ecological models now recognize that landslides may influence the total amount, and loss of, biomass and biodiversity in tropical montane forests (Dislich and Huth 2012; Restrepo et al. 2003), and that these effects may differ substantially between landslide sites or entire landscapes. Similar findings may be applicable to other forest ecosystems in steep terrain. Landslides can provide new habitats for both flora and fauna by generating a characteristic microtopography dotted with small water-filled depressions. For example, Geertsema and Pojar (2007) noted that deposits of low-gradient, deep-seated landslides in British Columbia had been colonized by beaver and other water- and shrub-loving fauna. Landslides also unroot trees and push them into nearby channels, so increasing the loads of large woody debris in the drainage network, thus altering or creating new riparian habitats. Field measurements showed that hillslopes eroded by landslides triggered by the 2008 Wenchuan earthquake in China had lower saturated water content, capillary moisture capacity, field water capacity, total porosity, and capillary porosity than hillslopes that were unaffected by landslides. Bulk densities were also higher on the impacted slopes, mainly because of compaction and moisture loss. These soil conditions can partly determine the succession of vegetation. Local reforestation experiments using two species showed that Japanese cedar (*Cryptomeria fortunei*) was better able to cope with these soil conditions than *Cupressus funebris* (Chinese weeping cypress) (Cheng et al. 2012).

The type of forest on a hillslope may or may not make a difference about whether it remains stable. Dense canopies with abundant foliage can reduce the intensity of rain falling on the soil for a time, thus reducing water pressures that might destabilize slopes in all but long-duration storms (Keim and Skaugset 2003). Tropical mountains covered by dense, rain-saturated forests can accumulate masses of up to nearly 1000 t ha^{-1} in their organic layer to the point that this additional load compromises slope stability (Vorpahl et al. 2012). In southeast Alaska, the widespread decline of yellow cedar (*Chamaecyparis nootkatensis*) has caused a nearly fourfold increase in landslide frequency, mainly because of an increase in soil saturation and a loss of soil strength as tree roots decayed (Johnson and Wilcock 2002). Root strength and depth, in particular, are important factors in the stability of forested hillslopes (Tang et al. 2015), and several models simulate the added cohesive strength from single roots and entire root bundles (Schwarz et al. 2010). Direct mechanical field measurements of root strength in fresh landslide scars suggest that root strength can be predicted by mapping the distribution and characteristics of trees on potentially unstable slopes (Roering et al. 2003). The broad range of reported root-strength measurements also highlights the importance of lateral root strength and root-thickness distributions for slope stability (Schwarz et al. 2010). Neglecting these factors in appraisals of landslide susceptibility, by assuming that root cohesion is spatially uniform, can give overoptimistic results, and underestimate the occurrence of landslides. To overcome this issue, Hwang et al. (2015) used high-resolution LiDAR measurements of forest-canopy heights that correlated with below-ground biomass as a proxy for local root strength. Adding this spatial variance to landslide susceptibility assessments may produce more realistic outcomes than those assuming uniform shear resistance throughout a study area. Treeless hillslopes can also gain shear strength from turf mat membranes that add surface cohesive strength of the order of 0.2–7.6 kPa (Preston and Crozier 1999). From these studies it appears that objectively capturing the diversity in forest structure may add physically and ecologically meaningful variance to slope-stability models and thus offer more rich results for interpretation.

Effect of vegetation change on landslide magnitude and frequency A popular concept is that losses of forest cover in upland terrain increase erosion, whereas gains in forest cover reduce erosion. For example, severe storms cause many more landslides on grassed than on forested hillslopes. However, a longer-term view of this concept reveals some subtle detail. Soil or regolith forms in contact to bedrock that actively weathers by biological, chemical, and physical processes. Over many millennia, regolith will be eroded at the rate it is created. If erosion rate exceeds regolith formation rate, bedrock will be exposed, and if the converse applies, regolith depth will become too deep for maintaining a sloping surface. Hence the maximum soil depth able to remain stable beneath grass and tree cover is important: deeper and stronger tree roots can hold more soil on a slope. Hence, for two comparable slopes with grass and tree cover that both create new regolith at the same rate, erosion will create differing results. In the long term the erosion of the two slopes is equal, controlled by the weathering rate. The forested slope fails much more rarely, through in larger events, whereas the grassed slope fails more frequently in smaller events. This concept also explains a common observation following deforestation: when forest is cleared, the residual soil is too deep to be stable, so rainstorms create high erosion after several years, when the remaining tree roots have decayed (Figure 17.12) (Sidle and Ochiai 2006), and the soil depth reduces towards its equilibrium. A grassy slope that is reforested, however, will see little erosion until its soil depth has built up to the stable limit under forest.

17.2.2 Coastal Ecosystems

Many of the protective functions of forests also apply to coastal vegetation. A growing number of scientists believe that ecosystem restoration offers a more sustainable, cost effective, and ecologically sound alternative to conventional

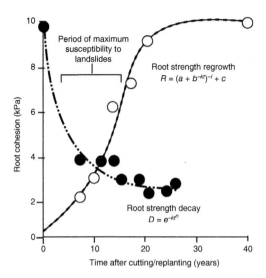

Figure 17.12 Root strength decay and recovery curves for Sugi (Japanese cedar, *Cryptomeria japonica*) based on uprooting tests of different ages of stumps and live trees, respectively. The simultaneous decay and regrowth of root strength offers a brief period of enhanced susceptibility to landslides. From Sidle et al. (2005).

engineering for protecting coastal environments (Figure 17.13) (Temmerman et al. 2013). The wealth of case studies reporting the negative consequences of perturbed coastal ecosystems endorses this view. Removal of protective dunes and artificial consolidation of the shoreline surface may cause higher storm waves, and is likely make rare wave heights more frequent (Nel et al. 2014). Coral reefs offer protection against incoming storm waves, although the performance of these natural barriers is unclear if the current rate of sea-level rise exceeds the rate of coral growth (Villanoy et al. 2012). Coral, oyster, and other living reefs may be able to dissipate >95% of the incoming wave energy. Thus the livelihoods and structural foundations of many island nations depend heavily on the health of such natural barriers (McCauley et al. 2015). However, anthropogenic acidification of oceans worsens the living conditions of coral reefs and degrades them. Atmospheric warming and pronounced El Niño events cause widespread coral bleaching, which slows growth and may kill the coral. The resulting loss of bed roughness on the sea floor promotes higher wave run-up and flooding along shorelines (Quataert et al. 2015). These increases will add to projected increases in wave heights as storminess rises in a warmer climate. Harris et al. (2018) emphasized that the structure and complexity of coral reefs determine their protective function against wave erosion and coastal flooding, and proposed a reef health index based on vertical erosion or accretion rates and the diversity of coral assemblages (Figure 17.14). The authors used a one-dimensional model of wave dissipation to test how rising sea level, increasing wave heights, reef accretion, and reef complexity played out against each other. These simulations showed that wave heights in the back-reef zone were consistently lowest for intact and healthy reefs. While projected increases in coastal erosion depend on sea-level rise, they also depend on the integrity and diversity of coral reefs. Other natural marine barriers such as oyster reefs can also increase drag resistance, alter wave currents, and reduce smaller wave heights, especially if the tops of these reefs are close – usually less than a metre – to sea level (Wiberg et al. 2019). These desirable properties have encouraged experiments that artificially build such reefs to reduce coastal erosion and to restore tidal marshes. Most of these experiments seem to work, albeit only on coasts with low wave energies.

Going beyond reefs, wetland vegetation stabilizes many sedimentary coasts and makes them more resistant to wave erosion and rising sea level. Most researchers would agree that natural coastal vegetation reduces incoming storm-wave and surge run-up heights, inland inundation, and wave-impact energy. Surface roughness and continuity of coastal wetlands further contribute to moderating storm surge heights (Figure 17.15) (Barbier et al. 2013). Even small coastal wetlands may provide some protection (Gedan et al. 2010). Salt marshes attenuate wave heights as a function of marsh width while stabilizing shorelines through accretion and reduced lateral erosion (Shepard et al. 2011). Plant communities

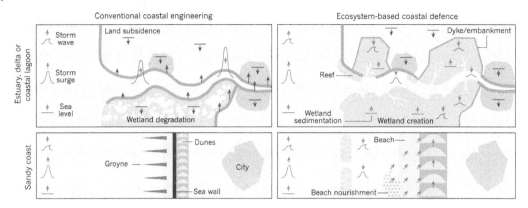

Figure 17.13 Conventional coastal engineering compared with new ecosystem-based defence. The schematic maps illustrate global and regional changes that increase the risk of coastal flood disasters (blue arrows indicate an increase or decrease in intensity of storm waves, storm surge and sea level), and the basic principles of flood protection by conventional coastal engineering (left) and new ecosystem-based defences (right) for an estuary, delta or coastal lagoon (top) and a sandy coast (bottom). In the case of conventional defences, red arrows indicate the need for maintenance and heightening of dykes, embankments and sea walls with sea-level rise. In an engineered estuary, delta or coastal lagoon (top left), embankment of wetlands stimulates the landward heightening of storm surges and exacerbates land subsidence (brown arrows) due to inhibited sediment supply and soil drainage. In the case of ecosystem-based defence in an estuary, delta or coastal lagoon (top right), wetland and reef creation attenuate landward storm surge propagation and storm waves, and stimulate wetland sedimentation (green arrows) with sea-level rise. For an engineered sandy coast (bottom left), groynes and sea walls may provoke dune degradation due to hindered sand supply, whereas for ecosystem-based defence along a sandy coast (bottom right), reefs help to attenuate storm waves and surge, and offshore sand nourishment stimulates beach and dune sedimentation with sea-level rise (orange arrows). From Temmerman et al. (2013).

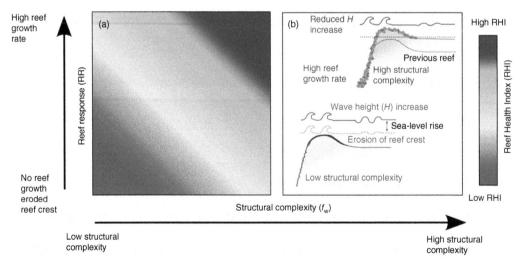

Figure 17.14 Conceptual diagram showing the future scenarios of coral reef structural complexity and vertical reef accretion. The reef health index (RHI) measures the capacity of a coral reef to accrete vertically and maintain structurally complex coral communities, with red indicating a low RHI and blue indicating a high RHI. From Harris et al. (2018).

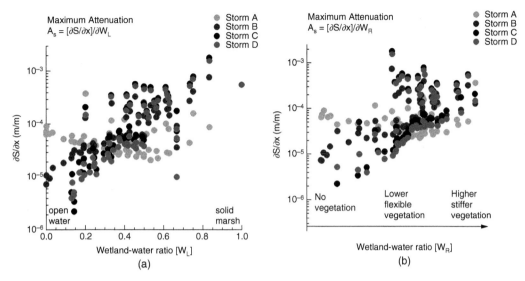

Figure 17.15 Attenuation (A$_S$) of storm surge (S) as a function of wetland continuity (W$_L$) and roughness (W$_R$) along a storm track segment of distance (x) in m for four hurricanes in the Caernarvon Basin of southeast Louisiana. (a) maximum attenuation as influenced by wetland continuity, A$_S$, where W$_L$ is represented by the wetland/water ratio ranging from open water (W$_L$ = 0) to solid marsh (W$_L$ = 1). (b) maximum attenuation as influenced by wetland roughness, A$_S$, where W$_R$ is represented by Manning's n for bottom friction caused by degrees of wetland vegetation ranging from no vegetation (W$_R$ = 0.02) to high dense vegetation (W$_R$ = 0.045). Storm A = Central pressure of 96 kPa, radius to maximum winds of 67 km, forward speed of 20.5 km/hr. Storm B = Central pressure of 93 kPa, radius to maximum winds of 47 km, forward speed of 20.5 km/hr. Storm C = Central pressure of 96 kPa, radius to maximum winds of 46 km, forward speed of 20.5 km/hr. Storm D = Central pressure of 93 kPa, radius to maximum winds of 33 km, forward speed of 11.1 km/hr. From Barbier et al. (2013).

such as mangroves and sedges that grow in salt marshes and other coastal wetlands trap and accumulate suspended sediment when they are inundated by tides or during storms. The elevation gain resulting from sediment accretion may allow the plants to keep pace with sea-level rise: the wetland forms a natural platform that rises in concert with sea level. Captured sediment takes up decaying plant matter and turns it into soil. But without continued accretion and inundation, the wetland surface may gradually compact and subside, causing relative sea level to rise locally. Detailed measurements of plant growth and sediment accretion in coastal wetlands have identified a critical range of rates of sea-level rise at which vegetation may still keep pace without being killed by inundation. These threshold rates differ with wetland topography and vegetation type. Some IPCC scenarios hold

that a rise in sea level of 5 mm yr^{-1} could be critical, but some intact salt marshes accrete sediment at more than 7 mm yr^{-1} (FitzGerald et al. 2008). Mangroves may be similarly capable of keeping up with sea-level rise, although a regional study of these tropical coastal forests in the Indian and Pacific Oceans revealed that sea level was rising faster than accreting sediment could compensate for at nearly 60% of all sites surveyed (Lovelock et al. 2015). All of these measurements were made over periods shorter than 16 years, though, so that the contribution of infrequent large storms or pulses of sediment from the hinterland remains unknown in many budget studies. The local topography and land use of the coastline also largely control whether mangroves (or other sediment-trapping wetlands) are able to migrate landward in response to rising sea levels. Wetlands and mangroves growing on coasts with higher tidal ranges may need longer times to respond to changes in sea level than those with a smaller tidal range.

How coastal vegetation and soils interact may decide the fate of large tracts of shoreline. Howes et al. (2010) reported that the brackish wetlands of the Louisiana coastal plain have weak and only slightly cohesive layers along the base of shallow roots, whereas the more saline marshes have deeper penetrating roots and are structurally less weak. These different soil properties became apparent when the storms surges driven by Hurricane Katrina eroded mostly the less salty wetlands in 2005. Differences in coastal vegetation structure do matter: after Hurricane Mitch killed large numbers of mangrove trees on the Bay Islands of Honduras in 1998, large areas of peat collapsed at rates of up to 40 mm yr^{-1} as the roots decomposed and soils compacted, slowly losing their strength (Cahoon et al. 2003). In contrast those mangroves that sustained little or moderate damage gained elevation at rates

of up to 5 mm yr^{-1} because their roots were growing faster than sediment could accumulate. The effects of disturbed coastal vegetation may also propagate throughout the coastal environment. For example, if rapid sea-level rise were to drown marshes in tidal inlets, the volume of water flowing through the inlets during half of a tidal cycle, known as the tidal prism, would likely increase and with it the volume of sand stored in ebb- and flood-tidal deltas (FitzGerald et al. 2008).

Coastal wetlands can be remarkably resilient to environmental change even over periods that go beyond the purposes of coastal engineering and protection (Figure 17.16). For example, marshes on the Yangtze River delta, China, have prograded seaward since the seventh century in spite of subsidence rates of the order of 50 mm yr^{-1} (Kirwan and Megonigal 2013). A similar case can be made for intact mangroves along tropical coasts. The extensive root systems of mangroves trap sufficient fine sediment in the intertidal zone to keep pace

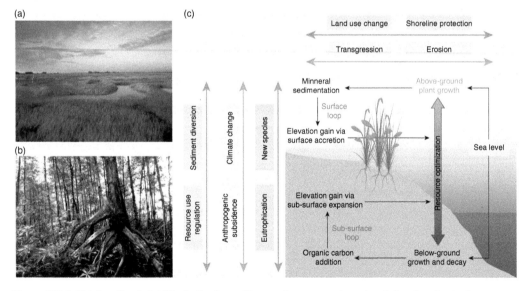

Figure 17.16 Tidal wetland stability in the face of human impacts and sea-level rise. Feedbacks between marshes (top left) and mangroves (bottom left) operate both horizontally and vertically at different scales and with distinct sets of processes to influence wetland stability. In the case of vertical elevation change, feedbacks operate through natural processes above and below ground. These natural processes can be perturbed by local factors (green), such as eutrophication and new species; large-scale climatic and geomorphic processes (blue); and political, social, and economic factors (orange), which affect the other processes. From Kirwan et al. (2013).

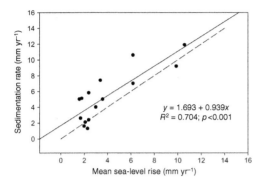

Figure 17.17 Relationship between sediment accretion rates in various mangrove forests and mean sea-level rise. Solid line is the linear regression; the dashed line is a 1:1 relationship. From Alongi (2008).

with sea-level rise (Figure 17.17) (Alongi 2008). Sedimentation rates in mangroves can be as high as 10 mm yr^{-1}, and in the Gulf of Papua they trap an estimated 5.4–56 Mt of sand and mud, or 2–14% of the average annual fluvial sediment load (Walsh and Nittrouer 2004). Mangroves also offer some protection from tsunamis (Danielsen et al. 2005). Numerical models that parameterize tropical coastal forest characteristics show that mangrove belts that are at least 100 m wide can reduce tsunami forces (Alongi 2008). Many of these models deal with fixed conditions, and it remains unclear how this protective function scales with the width of the mangrove belt. According to numerical simulations (clear-water) flow velocities can be reduced up to 50%, but local topography and flow roughness largely determine tsunami inundation depths and flow velocities (Kaiser et al. 2011). How much wave energy mangroves are able to dissipate depends on their stem and root density, stem diameter, the gradient of the shore, foreshore bathymetry, the spectral characteristics of incident waves, and the tidal state during the tsunami. Numerical model simulations revealed some weaknesses in empirical approaches that rely on only roughly constrained input variables such as the drag coefficient or resistance offered to waves by vegetation, as some of these inputs vary with flow conditions (Maza et al.

2015). Numerical simulations can inform us how the distribution of mangrove trees modulates wave energies, thus usefully informing reforestation projects. Simulations calibrated with field data from tsunami-damaged mangroves can show biophysical limits to the protection that mangroves offer. For example, Yanagisawa et al. (2010) estimated that the tsunami inundation height needed to nearly completely destroy a ten-year old, 500-m wide strip of mangrove forest is about four metres (Figure 17.18). In contrast, a 30-year old mangrove could tolerate inundation heights of up to five metres with minimal losses (~20%), while absorbing half of the tsunami's hydrodynamic force. In essence, all numerical estimates of mangrove protection values (or their vulnerability) rely on how well the local bathymetry and coastal topography are captured, and on how local flow roughness changes with vegetation structure. Field survey data remain essential to determine mangrove damage as a function of vegetation structure, topography, and mechanic properties such as the maximum bending stresses that the trees can absorb.

Another important ecosystem service of mangroves is that they are carbon sinks, mainly owing to their high primary productivity of at least 0.2 Pg C yr^{-1} ha^{-1} (Bouillon et al. 2008). Nutrients, however, are in constant demand but often undersupplied in mangroves. The consequence is that nutrient recycling can be an important factor that leads to varying estimates about how efficiently mangroves sequester organic carbon (Jennerjahn 2012). Nevertheless, scientists have attempted to monetize these ecosystem services. A global estimate holds that mangrove forests currently cover between 147 000 and 160 000 km^2 of coastlines worldwide and have an average economic value of around US\$ 0.2 million ha^{-1} per year (Lovelock et al. 2015), with annual values as high as US\$ 0.9 million ha^{-1} in some areas (Alongi 2008). In total, mangroves provide at least US\$ 1.6 billion in ecosystem services annually (Polidoro et al. 2010). The protective function

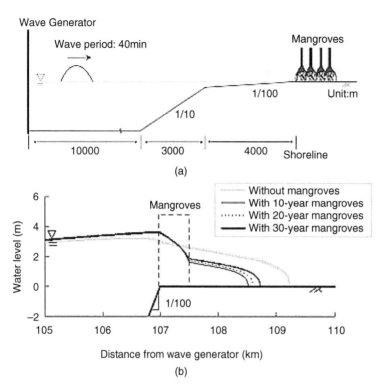

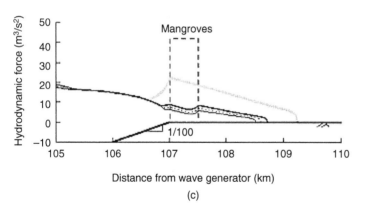

Figure 17.18 Tsunami damage reduction performance of a mangrove forest in Banda Aceh, Indonesia. (a) Numerical conditions; the incident wave's period is assumed to be 40 minutes, and the coastal landform is simplified from the nearshore of Banda Aceh. (b) Maximum tsunami inundation depth. (c) Hydrodynamic force with and without mangrove forest. From Yanagisawa et al. (2010).

against storm and tsunami waves makes up the majority of the monetary value of mangroves (Marois and Mitsch 2015). A worldwide survey of experts in ecosystem service valuation showed that the coastal protective function of mangroves in currency of 2007 International Dollars (a hypothetical financial currency with the same purchasing power as the US dollar had in the United States for a specific year) is about Int\$ 8500 ha^{-1} yr^{-1} (Mukherjee et al. 2014). Other research suggests that the coastal protection value of mangroves exceeds direct-use values such as wood harvesting or aquaculture by as much as 97% (Sanford 2009).

Still, mangrove forests are being lost at annual rates of 1–2% to coastal development, shrimp farming, and fisheries. Pollution and sediment starvation prompting higher rates of coastal erosion are also contributing to mangrove decline (Giri et al. 2015). Currently, eleven of the 70 known mangrove species are at risk of extinction (Polidoro et al. 2010). Deforestation and degradation of mangroves releases carbon dioxide into the atmosphere. This release comes from loss of biomass and the remobilization of carbon-rich sediment previously trapped in vegetated coastal wetlands. This 'blue carbon' is captured within oceans and coastal sediments and has only recently attracted attention (Pendleton et al. 2012). Analysis of LANDSAT satellite images has shown that coastlines in Indonesia, Malaysia, Thailand, Burma (Myanmar), Bangladesh, India, and Sri Lanka that were impacted by the 2004 Indian Ocean tsunami lost 12% of their mangrove forests between 1975 and 2005, mainly due to expanding agriculture, aquaculture, and urban development (Giri et al. 2008). The Food and Agriculture Organization (FAO) had previously estimated twice this amount of mangrove loss. Local differences complicate this picture; in some areas, such as in the Irrawaddy delta, the area of mangroves increased by nearly 3% between 1989 and 2009 (Shearman et al. 2013).

Development along the coasts of the Gulf of Mexico is also responsible for losses of coastal wetlands. Among the few encouraging signs in this region is the growing awareness of the savings provided by intact coastal vegetation following hurricanes. Along the Louisiana coastline, for example, increasing the continuity of wetlands by 10% could reduce storm-surge losses by US$ 99–133 per metre length of coastline and restoring vegetation roughness by 0.1% could lower damages by US$ 24–43 per metre (Barbier et al. 2013). Such projections hinge on how well one can estimate the roughness of wetland vegetation (in this case approximated by Manning's n, a measure well-known to geomorphologists trying to capture the hydraulic roughness of channel beds). The key point, however, is that vegetation in inland areas also has protective values that can be monetized. Dymond et al. (2012) estimated that each ton of soil loss avoided in New Zealand would save NZ$ 1, compared to NZ$ 73 for each ton of sequestered carbon. The protective functions of vegetation can be especially high in flood-prone and rapidly eroding landscapes. Barbier et al. (2008) proposed that ecosystem-based management of coasts goes beyond strict conservation measures. They advocated instead a mix of both environmental protection and targeted development. Their review of field data showed that mangroves, salt marshes, sea-grass beds, coral reefs, and dunes attenuate wave energy in a highly non-linear manner that had rarely been considered when estimating their protective values. Taking into account more integrated scenarios that favour coastal vegetation both as an ecosystem service and space for economic development may provide practical options for land use.

We also recognize some limits to an ecosystem-based approach in mitigating losses from natural disasters. A large proportion of seismic risk, for example, is due to the vulnerability of buildings and other infrastructure and will change little regardless of whether ecosystems are intact or not. However, the potential damage from other elements in the seismic hazard cascade such as landslides, excess sediment transport, unstable river channels, and increased flooding may be reduced notably with intact vegetation.

References

Abbott BW, Jones JB, Schuur EAG, et al. 2016 Biomass offsets little or none of permafrost carbon release from soils, streams, and wildfire: an expert assessment. *Environmental Research Letters* **11** (3), 1–13.

Alin SR, Aalto R, Goni MA, et al. 2008 Biogeochemical characterization of carbon sources in the Strickland and Fly rivers, Papua New Guinea. *Journal of Geophysical Research: Earth Surface* **113**, 1–21.

Allen RB, Bellingham PJ, and Wiser SK 1999 Immediate damage by an earthquake to a temperate montane forest. *Ecology* **80** (2), 708–714.

Allison MA, Dellapenna TM, Gordon ES, et al. 2010 Impact of Hurricane Katrina (2005) on shelf organic carbon burial and deltaic evolution. *Geophysical Research Letters* **37** (21), 1–5.

Alongi DM 2008 Mangrove forests: resilience, protection from tsunamis, and responses to global climate change. *Estuarine, Coastal and Shelf Science* **76** (1), 1–13.

Anthony KMW, Zimov SA, Grosse G, et al. 2015 A shift of thermokarst lakes from carbon sources to sinks during the Holocene epoch. *Nature* **511** (7510), 452–456.

Anyamba A, Small JL, Britch SC, et al. 2014 Recent weather extremes and impacts on agricultural production and vector-borne disease outbreak patterns. *PLoS ONE* **9** (3), e92538.

Archer D 2007 Methane hydrate stability and anthropogenic climate change. *Biogeosciences* **5**, 521–544.

Aufdenkampe AK, Mayorga E, Raymond PA, et al. 2011 Riverine coupling of biogeochemical cycles between land, oceans, and atmosphere. *Frontiers in Ecology and the Environment* **9** (1), 53–60.

Barbier EB, Koch EW, Silliman BR, et al. 2008 Coastal ecosystem-based management with nonlinear ecological functions and values. *Science* **319**, 321–323.

Barbier EB, Georgiou IY, Enchelmeyer B, and Reed DJ 2013 The value of wetlands in protecting southeast Louisiana from hurricane storm surges. *PLoS ONE* **8** (3), e58715.

Battin TJ, Luyssaert S, Kaplan LA, et al. 2009 The boundless carbon cycle. *Nature Geoscience* **2** (9), 598–600.

Bauer JE, Cai WJ, Raymond PA, et al. 2013 The changing carbon cycle of the coastal ocean. *Nature* **504** (7478), 61–70.

Bongers F, Aide TM, Zambrano AMA, et al. 2016 Biomass resilience of Neotropical secondary forests. *Nature* **530** (7589), 211–214.

Bouillon S, Borges AV, Castañeda-Moya E, et al. 2008 Mangrove production and carbon sinks: A revision of global budget estimates. *Global Biogeochemical Cycles* **22** (2), 1–12.

Bowman DMJS, Balch JK, Artaxo P, et al. 2009 Fire in the Earth system. *Science* **324** (5926), 481–484.

Bradshaw CJA, Sodhi NS, Peh KSH, and Brook BW 2007 Global evidence that deforestation amplifies flood risk and severity in the developing world. *Global Change Biology* **13** (11), 2379–2395.

Brardinoni F, Slaymaker O, and Hassan MA 2003 Landslide inventory in a rugged forested watershed: a comparison between air-photo and field survey data. *Geomorphology* **54** (3-4), 179–196.

Cahoon DR, Hensel P, Rybczyk J, et al. 2003 Mass tree mortality leads to mangrove peat collapse at Bay Islands, Honduras after Hurricane Mitch. *Journal of Ecology* **91**, 1093–1105.

Carbonneau P, Fonstad MA, Marcus WA, and Dugdale SJ 2012 Making riverscapes real. *Geomorphology* **137** (1), 74–86.

Carey AE, Gardner CB, Goldsmith ST, et al. 2005 Organic carbon yields from small, mountainous rivers, New Zealand. *Geophysical Research Letters* **32** (15), 1–5.

Chambers JQ, Fisher JI, Zeng H, et al. 2007 Hurricane Katrina's carbon footprint on U.S. Gulf Coast forests. *Science* **318** (5853), 1107–1107.

Chaopricha NT and Marín-Spiotta E 2014 Soil burial contributes to deep soil organic carbon storage. *Soil Biology and Biochemistry* **69** (C), 251–264.

Chaplot VAM, Rumpel C, and Valentin C 2005 Water erosion impact on soil and carbon redistributions within uplands of Mekong River. *Global Biogeochemical Cycles* **19** (4), 1–13.

Cheng S, Yang G, Yu H, et al. 2012 Impacts of Wenchuan Earthquake-induced landslides on soil physical properties and tree growth. *Ecological Indicators* **15** (1), 263–270.

Cochrane MA 2009 Fire in the tropics. In *Tropical Fire Ecology* (ed Cochrane MA), Springer, pp. 1–24.

Crausbay SD and Martin PH 2016 Natural disturbance, vegetation patterns and ecological dynamics in tropical montane forests. *Journal of Tropical Ecology* **32**, 384–403.

Crowther TW, Todd-Brown KEO, Rowe CW, et al. 2016 Quantifying global soil carbon losses in response to warming. *Nature* **540** (7631), 104–108.

Danielsen F, Sørensen MK, Olwig MF, et al. 2005 The Asian tsunami: A protective role for coastal vegetation. *Science* **310**, 643.

Davis MR, Allen RB, and Clinton PW 2002 Carbon storage along a stand development sequence in a New Zealand Nothofagus forest. *Forest Ecology and Management* **177** (1), 313–321.

De Rose RC 2012 Slope control on the frequency distribution of shallow landslides and associated soil properties, North Island, New Zealand. *Earth Surface Processes and Landforms* **38** (4), 356–371.

Dhakal AS and Sidle RC 2003 Long-term modelling of landslides for different forest management practices. *Earth Surface Processes and Landforms* **28** (8), 853–868.

Dislich C and Huth A 2012 Modelling the impact of shallow landslides on forest structure in tropical montane forests. *Ecological Modelling* **239**, 40–53.

Dorren LKA, Berger F, Imeson AC, and Maier B 2004 Integrity, stability and management of protection forests in the European Alps. *Forest Ecology and Management* **195**, 165–176.

Downing JA, Cole JJ, Middelburg JJ, et al. 2008 Sediment organic carbon burial in agriculturally eutrophic impoundments over the last century. *Global Biogeochemical Cycles* **22** (1), 1–10.

Drake TW, Raymond PA, and Spencer RGM 2017 Terrestrial carbon inputs to inland waters: A current synthesis of estimates and uncertainty. *Limnology and Oceanography* **3** (3), 132–142.

Dymond JR, Ausseil AGE, Ekanayake JC, and Kirschbaum MUF 2012 Tradeoffs between soil, water, and carbon - A national scale analysis from New Zealand. *Journal of Environmental Management* **95** (1), 124–131.

Farquharson LM, Mann DH, Swanson DK, et al. 2018 Temporal and spatial variability in coastline response to declining sea-ice in northwest Alaska. *Marine Geology* **404**, 71–83.

FitzGerald DM, Fenster MS, Argow BA, and Buynevich IV 2008 Coastal impacts due to sea-level rise. *Annual Review of Earth and Planetary Sciences* **36** (1), 601–647.

Flitcroft RL, Falke JA, Reeves GH, et al. 2016 Wildfire may increase habitat quality for spring Chinook salmon in the Wenatchee River subbasin, WA, USA. *Forest Ecology and Management* **359**, 126–140.

Frolking S, Palace MW, Clark DB, et al. 2009 Forest disturbance and recovery: A general review in the context of spaceborne remote sensing of impacts on aboveground biomass and canopy structure. *Journal of Geophysical Research: Biogeosciences* **114**, 1–27.

Fusun S, Jinniu W, Tao L, et al. 2013 Effects of different types of vegetation recovery on runoff and soil erosion on a Wenchuan earthquake-triggered landslide, China. *Journal of Soil and Water Conservation* **68** (2), 138–145.

Galy V and Eglinton T 2011 Protracted storage of biospheric carbon in the Ganges–Brahmaputra basin. *Nature Geoscience* **4** (10), 1–5.

Galy V, Peucker-Ehrenbrink B, and Eglinton T 2015 Global carbon export from the terrestrial biosphere controlled by erosion. *Nature* **521** (7551), 204–207.

Gedan KB, Kirwan ML, Wolanski E, et al. 2010 The present and future role of coastal wetland vegetation in protecting shorelines: Answering recent challenges to the paradigm. *Climatic Change* **106** (1), 7–29.

Geertsema M and Pojar JJ 2007 Influence of landslides on biophysical diversity — A perspective from British Columbia. *Geomorphology* **89** (1-2), 55–69.

Giri C, Long J, Abbas S, et al. 2015 Distribution and dynamics of mangrove forests of South Asia. *Journal of Environmental Management* **148** (c), 101–111.

Giri C, Zhu Z, Tieszen LL, et al. 2008 Mangrove forest distributions and dynamics (1975–2005) of the tsunami-affected region of Asia. *Journal of Biogeography* **35** (3), 519–528.

Gislason SR, Oelkers EH, and Snorrason Á 2006 Role of river-suspended material in the global carbon cycle. *Geology* **34** (1), 49–52.

Goldsmith ST, Carey AE, Lyons WB, et al. 2008 Extreme storm events, landscape denudation, and carbon sequestration: Typhoon Mindulle, Choshui River, Taiwan. *Geology* **36** (6), 483–486.

Gomez B, Baisden WT, and Rogers KM 2010 Variable composition of particle-bound organic carbon in steepland river systems. *Journal of Geophysical Research: Earth Surface* **115**, 1–9.

Gomez B, Trustrum NA, Hicks DM, et al. 2003 Production, storage, and output of particulate organic carbon: Waipaoa River basin, New Zealand. *Water Resources Research* **39** (6), 1–8.

Gomi T, Sidle RC, and Swanston DN 2004 Hydrogeomorphic linkages of sediment transport in headwater streams, Maybeso Experimental Forest, southeast Alaska. *Hydrological Processes* **18** (4), 667–683.

Gran KB, Tal M, and Wartman ED 2015 Co-evolution of riparian vegetation and channel dynamics in an aggrading braided river system, Mount Pinatubo, Philippines. *Earth Surface Processes and Landforms* **40** (8), 1101–1115.

Günther F, Overduin PP, Sandakov AV, et al. 2013 Short- and long-term thermo-erosion of ice-rich permafrost coasts in the Laptev Sea region. *Biogeosciences* **10** (6), 4297–4318.

Guthrie R 2002 The effects of logging on frequency and distribution of landslides in three watersheds on Vancouver Island, British Columbia. *Geomorphology* **43**, 273–292.

Guthrie RH, Friele P, Allstadt K, et al. 2012 The 6 August 2010 Mount Meager rock slide-debris flow, Coast Mountains, British Columbia:

characteristics, dynamics, and implications for hazard and risk assessment. *Natural Hazards and Earth System Sciences* **12** (5), 1277–1294.

Harris DH, Rovere A, Casella E, et al. 2018 Coral reef structural complexity provides important coastal protection from waves under rising sea levels. *Science Advances* **4**, 1–7.

Hilton RG, Galy A, and Hovius N 2008a Riverine particulate organic carbon from an active mountain belt: Importance of landslides. *Global Biogeochemical Cycles*.

Hilton RG, Galy A, Hovius N, et al. 2008b Tropical-cyclone-driven erosion of the terrestrial biosphere from mountains. *Nature Geoscience* **1** (11), 759–762.

Hilton RG, Galy A, Hovius N, et al. 2010 Efficient transport of fossil organic carbon to the ocean by steep mountain rivers: An orogenic carbon sequestration mechanism. *Geology* **39** (1), 71–74.

Hilton RG, Galy V, Gaillardet J, et al. 2015 Erosion of organic carbon in the Arctic as a geological carbon dioxide sink. *Nature* **524** (7563), 84–87.

Hoffmann T, Schlummer M, Notebaert B, et al. 2013 Carbon burial in soil sediments from Holocene agricultural erosion, Central Europe. *Global Biogeochemical Cycles* **27** (3), 828–835.

Holdaway RJ, McNeill SJ, Mason NWH, and Carswell FE 2014 Propagating uncertainty in plot-based estimates of forest carbon stock and carbon stock change. *Ecosystems* **17** (4), 627–640.

Hood E and Scott D 2008 Riverine organic matter and nutrients in southeast Alaska affected by glacial coverage. *Nature Geoscience* **1** (9), 583–587.

Hooijer A, Page S, Jauhiainen J, et al. 2012 Subsidence and carbon loss in drained tropical peatlands. *Biogeosciences* **9** (3), 1053–1071.

Houghton RA 2007 Balancing the global carbon budget. *Annual Review of Earth and Planetary Sciences* **35** (1), 313–347.

Howes NC, FitzGerald DM, Hughes ZJ, et al. 2010 Hurricane-induced failure of low salinity wetlands *Proceedings of the National Academy of Sciences*, pp. 14014–14019.

Hughes PDM, Mallon G, Brown A, et al. 2013 The impact of high tephra loading on late-Holocene carbon accumulation and vegetation succession in peatland communities. *Quaternary Science Reviews* **67** (C), 160–175.

Hwang T, Band LE, Hales TC, et al. 2015 Simulating vegetation controls on hurricane-induced shallow landslides with a distributed ecohydrological model. *Journal of Geophysical Research: Biogeosciences* **120** (2), 361–378.

Imaizumi F and Sidle RC 2012 Effect of forest harvesting on hydrogeomorphic processes in steep terrain of central Japan. *Geomorphology* **169-170** (C), 109–122.

Istanbulluoglu E, Tarboton DG, Pack RT, and Luce CH 2004 Modeling of the interactions between forest vegetation, disturbances, and sediment yields. *Journal of Geophysical Research: Earth Surface* **109**, 1–22.

Jakob M, Bovis M and Oden M, 2005 The significance of channel recharge rates for estimating debris-flow magnitude and frequency. *Earth Surface Processes and Landforms* **30** (6), 755–766.

Jennerjahn TC 2012 Biogeochemical response of tropical coastal systems to present and past environmental change. *Earth-Science Reviews* **114** (1-2), 19–41.

Johnson AC and Wilcock P 2002 Association between cedar decline and hillslope stability in mountainous regions of southeast Alaska. *Geomorphology* **46** (1), 129–142.

Kaiser G, Scheele L, Kortenhaus A, et al. 2011 The influence of land cover roughness on the results of high resolution tsunami inundation modeling. *Natural Hazards and Earth System Sciences* **11** (9), 2521–2540.

Keim RF and Skaugset AE 2003 Modelling effects of forest canopies on slope stability. *Hydrological Processes* **17** (7), 1457–1467.

Kirwan ML and Megonigal JP 2013 Tidal wetland stability in the face of human impacts and sea-level rise. *Nature* **504** (7478), 53–60.

Komada T, Druffel ERM, and Trumbore SE 2004 Oceanic export of relict carbon by small mountainous rivers. *Geophysical Research Letters* **31** (7), 1–4.

Korup O and Rixen C 2014 Soil erosion and organic carbon export by wet snow avalanches. *The Cryosphere* **8** (2), 651–658.

Lal R 2003 Soil erosion and the global carbon budget. *Environment International* **29** (4), 437–450.

Langner A and Siegert F 2009 Spatiotemporal fire occurrence in Borneo over a period of 10 years. *Global Change Biology* **15** (1), 48–62.

Lehmann J and Kleber M 2015 The contentious nature of soil organic matter. *Nature* **528**, 60–68.

Leithold EL, Blair NE, and Perkey DW 2006 Geomorphologic controls on the age of particulate organic carbon from small mountainous and upland rivers. *Global Biogeochemical Cycles* **20** (3), 1–11.

Lewis SL, Edwards DP, and Galbraith D 2015 Increasing human dominance of tropical forests. *Science* **349**, 827–832.

Li Y, Quine TA, Yu HQ, et al. 2015 Sustained high magnitude erosional forcing generates an organic carbon sink: Test and implications in the Loess Plateau, China. *Earth and Planetary Science Letters* **411** (C), 281–289.

Lloret E, Dessert C, Gaillardet J, et al. 2011 Comparison of dissolved inorganic and organic carbon yields and fluxes in the watersheds of tropical volcanic islands, examples from Guadeloupe (French West Indies). *Chemical Geology* **280** (1-2), 65–78.

Lovelock CE, Cahoon DR, Friess DA, et al. 2015 The vulnerability of Indo-Pacific mangrove forests to sea-level rise. *Nature* **526** (7574), 559–563.

Mack MC, Bret-Harte MS, Hollingsworth TN, et al. 2012 Carbon loss from an unprecedented Arctic tundra wildfire. *Nature* **475** (7357), 489–492.

Malhi Y 2011 The productivity, metabolism and carbon cycle of tropical forest vegetation. *Journal of Ecology* **100** (1), 65–75.

Marois DE and Mitsch WJ 2015 Coastal protection from tsunamis and cyclones provided by mangrove wetlands – a review. *International Journal of Biodiversity Science, Ecosystem Services & Management* **11** (1), 71–83.

Maslin M, Owen M, Betts R, et al. 2010 Gas hydrates: Past and future geohazard?. *Philosophical Transactions of the Royal Society A: Mathematical, Physical and Engineering Sciences* **368** (1919), 2369–2393.

Maza M, Lara JL, and Losada IJ 2015 Tsunami wave interaction with mangrove forests: A 3-D numerical approach. *Coastal Engineering* **98**, 33–54.

McCauley DJ, Pinsky ML, Palumbi SR, et al. 2015 Marine defaunation: Animal loss in the global ocean. *Science* **347** (6219), 1255641-1–1255641-7.

Miller DJ and Burnett KM 2007 Effects of forest cover, topography, and sampling extent on the measured density of shallow, translational landslides. *Water Resources Research* **43** (3), 1–23.

Mohr CH, Korup O, Ulloa H, and Iroume A 2017 Pyroclastic eruption boosts organic carbon fluxes into Patagonian fjords. *Global Biogeochemical Cycles* **31**, 1626–1638.

Mohr CH, Zimmermann A, Korup O, et al. 2014 Seasonal logging, process response, and geomorphic work. *Earth Surface Dynamics* **2** (1), 117–125.

Moore S, Gauci V, Evans CD, and Page SE 2011 Fluvial organic carbon losses from a Bornean blackwater river. *Biogeosciences* **8** (4), 901–909.

Morris JL, McLauchlan KK, and Higuera PE 2015 Sensitivity and complacency of sedimentary biogeochemical records to climate-mediated forest disturbances. *Earth-Science Reviews* **148** (C), 121–133.

Mukherjee N, Sutherland WJ, Dicks L, et al. 2014 Ecosystem service valuations of mangrove ecosystems to inform decision making and future valuation exercises. *PLoS ONE* **9** (9), e107706.

Murray AB, Knaapen MAF, Tal M, and Kirwan ML 2008 Biomorphodynamics: Physical-biological feedbacks that shape landscapes. *Water Resources Research* **44** (11), 1–18.

Nakamura F, Swanson FJ, and Wondzell SM 2000 Disturbance regimes of stream and riparian systems - a disturbance cascade perspective. *Hydrological Processes* **14**, 2849–2860.

Nel JL, Le Maitre DC, Nel DC, et al. 2014 Natural hazards in a changing world: A case for ecosystem-based management. *PLoS ONE* **9** (5), e95942.

O'Connor FM, Boucher O, Gedney N, et al. 2010 Possible role of wetlands, permafrost, and methane hydrates in the methane cycle under future climate change: A review. *Reviews of Geophysics* **48** (4), 1–33.

Pearson TRH, Brown S, and Casarim FM 2014 Carbon emissions from tropical forest degradation caused by logging. *Environmental Research Letters* **9** (3), 034017.

Pendleton L, Donato DC, Murray BC, et al. 2012 Estimating global "Blue Carbon" emissions from conversion and degradation of vegetated coastal ecosystems. *PLoS ONE* **7** (9), e43542.

Polidoro BA, Carpenter KE, Collins L, et al. 2010 The loss of species: Mangrove extinction risk and geographic areas of global concern. *PLoS ONE* **5** (4), e10095.

Preston NJ and Crozier MJ 1999 Resistance to shallow landslide failure through root-derived cohesion in east coast hill country soils, North Island, New Zealand. *Earth Surface Processes and Landforms* **24**, 665–675.

Quataert E, Storlazzi C, van Rooijen A, et al. 2015 The influence of coral reefs and climate change on wave-driven flooding of tropical coastlines. *Geophysical Research Letters* **42**, 6407–6415.

Ramos Scharrón CE, Castellanos EJ, and Restrepo C 2012 The transfer of modern organic carbon by landslide activity in tropical montane ecosystems. *Journal of Geophysical Research: Biogeosciences* **117**, 1–18.

Rathburn SL, Bennett GL, Wohl EE, et al. 2017 The fate of sediment, wood, and organic carbon eroded during an extreme flood,

Colorado Front Range, USA. *Geology* **45** (6), 499–502.

Regnier P, Friedlingstein P, Ciais P, et al. 2013 Anthropogenic perturbation of the carbon fluxes from land to ocean. *Nature Geoscience* **6** (8), 597–607.

Reichstein M, Bahn M, Ciais P, et al. 2013 Climate extremes and the carbon cycle. *Nature* **500** (7462), 287–295.

Ren D, Wang J, Fu R, et al. 2009 Mudslide-caused ecosystem degradation following Wenchuan earthquake 2008. *Geophysical Research Letters* **36** (5), 1–5.

Restrepo C, Vitousek P, and Neville P 2003 Landslides significantly alter land cover and the distribution of biomass: an example from the Ninole ridges of Hawai'i. *Plant Ecology* **166**, 131–143.

Roering JJ, Schmidt KM, Stock JD, et al. 2003 Shallow landsliding, root reinforcement, and the spatial distribution of trees in the Oregon Coast Range. *Canadian Geotechnical Journal* **40** (2), 237–253.

Sanford MP 2009 Valuating mangrove ecosystems as coastal protection in post-tsunami South Asia. *Natural Areas Journal* **29** (1), 91–95.

Santín C, Doerr SH, Shakesby RA, et al. 2012 Carbon loads, forms and sequestration potential within ash deposits produced by wildfire: new insights from the 2009 'Black Saturday' fires, Australia. *European Journal of Forest Research* **131** (4), 1245–1253.

Schelhaas MJ, Nabuurs GJ, and Schuck A 2003 Natural disturbances in the European forests inthe 19th and 20th centuries. *Global Change Biology* **9**, 1620–1633.

Schuur EA, Bockheim J, Canadell JG, et al. 2008 Vulnerability of permafrost carbon to climate change: Implications for the global carbon cycle. *BioScience* **58** (8), 701–714.

Schuur EAG, McGuire AD, Schädel C, et al. 2015 Climate change and the permafrost carbon feedback. *Nature* **520** (7546), 171–179.

Schuur EAG, Vogel JG, Crummer KG, et al. 2009 The effect of permafrost thaw on old carbon release and net carbon exchange from tundra. *Nature* **459** (7246), 556–559.

Schwarz M, Lehmann P, and Or D 2010 Quantifying lateral root reinforcement in steep slopes - from a bundle of roots to tree stands. *Earth Surface Processes and Landforms* **35** (3), 354–367.

Scott DT, Baisden WT, Davies-Colley R, et al. 2006 Localized erosion affects national carbon budget. *Geophysical Research Letters* **33** (1), 1–4.

Seo JI, Nakamura F, Akasaka T, et al. 2012 Large wood export regulated by the pattern and intensity of precipitation along a latitudinal gradient in the Japanese archipelago. *Water Resources Research* **48** (3), n/a–n/a.

Shearman P, Bryan J, and Walsh JP 2013 Trends in deltaic change over three decades in the Asia-Pacific region. *Journal of Coastal Research* **290**, 1169–1183.

Shepard CC, Crain CM, and Beck MW 2011 The protective role of coastal marshes: A systematic review and meta-analysis. *PLoS ONE* **6** (11), e27374.

Sidle RC 2005 Influence of forest harvesting activities on debris avalanches and flows. In *Debris-Flows and Related Hazards* (eds Jakob M and Hungr O), Springer-Praxis Books, pp. 387–409.

Sidle RC and Ochiai H 2006 Landslides. Processes, Prediction, and Land Use, Water Resources Monograph 18. American Geophysical Union.

Singer GA, Fasching C, Wilhelm L, et al. 2012 Biogeochemically diverse organic matter in Alpine glaciers and its downstream fate. *Nature Geoscience* **5** (10), 710–714.

Smith RW, Bianchi TS, Allison M, et al. 2015 High rates of organic carbon burial in fjord sediments globally. *Nature Geoscience* **8**, 450–453.

Stallins JA 2006 Geomorphology and ecology: Unifying themes for complex systems in biogeomorphology. *Geomorphology* **77** (3-4), 207–216.

Surian N, Barban M, Ziliani L, et al. 2014 Vegetation turnover in a braided river:

frequency and effectiveness of floods of different magnitude. *Earth Surface Processes and Landforms* **40** (4), 542–558.

Sutherland WJ, Freckleton RP, Godfray HCJ, et al. 2012 Identification of 100 fundamental ecological questions. *Journal of Ecology* **101** (1), 58–67.

Tal M and Paola C 2010 Effects of vegetation on channel morphodynamics: results and insights from laboratory experiments. *Earth Surface Processes and Landforms* **35** (9), 1014–1028.

Tang Y, Bossard C, and Reidhead J 2015 Effects of percent cover of Japanese cedar in forests on slope slides in Sichuan, China. *Ecological Engineering* **74**, 42–47.

Temmerman S, Meire P, Bouma TJ, et al. 2013 Ecosystem-based coastal defence in the face of global change. *Nature* **504** (7478), 79–83.

Turetsky MR, Benscoter B, Page S, et al. 2015 Global vulnerability of peatlands to re andcarbon loss. *Nature Geoscience* **8** (1), 11–14.

Valentin C, Agus F, Alamban R, et al. 2008 Runoff and sediment losses from 27 upland catchments in Southeast Asia: Impact of rapid land use changes and conservation practices. *Agriculture, Ecosystems & Environment* **128** (4), 225–238.

Vanacker V, Vanderschaeghe M, Govers G, et al. 2003 Linking hydrological, infinite slope stability and land-use change models through GIS for assessing the impact of deforestation on slope stability in high Andean watersheds. *Geomorphology* **52** (3-4), 299–315.

Villanoy C, David L, Cabrera O, et al. 2012 Coral reef ecosystems protect shore from high-energy waves under climate change scenarios. *Climatic Change* **112** (2), 493–505.

Vonk JE, Sánchez-García L, van Dongen BE, et al. 2012 Activation of old carbon by erosion of coastal and subsea permafrost in Arctic Siberia. *Nature* **488** (7414), 137–140.

Vorpahl P, Dislich C, Elsenbeer H, et al. 2012 Biotic controls on shallow translational landslides. *Earth Surface Processes and Landforms* **38** (2), 198–212.

Walsh JP and Nittrouer CA 2004 Mangrove-bank sedimentation in a mesotidal environment with large sediment supply, Gulf of Papua. *Marine Geology* **208** (2-4), 225–248.

Walsh RPD, Bidin K, Blake WH, et al. 2011 Long-term responses of rainforest erosional systems at different spatial scales to selective logging and climatic change. *Philosophical Transactions of the Royal Society B: Biological Sciences* **366** (1582), 3340–3353.

Walter KM, Zimov SA, Chanton JP, et al. 2006 Methane bubbling from Siberian thaw lakes as a positive feedback to climate warming. *Nature* **443** (7107), 71–75.

Wang Z, Hoffmann T, Six J, et al. 2017 Human-induced erosion has offset one-third of carbon emissions from land cover change. *Nature Climate Change* **7** (5), 345–349.

Wang Z, Van Oost K, and Govers G 2015 Predicting the long-term fate of buried organic carbon in colluvial soils. *Global Biogeochemical Cycles* **29**, 1–15.

Warburton J 2015 Peat landslides. In *Landslide Hazards, Risks, and Disasters* (ed Davies TRH), Elsevier, pp. 159–190.

Wells A, Duncan RP, and Stewart GH 2001 Forest dynamics in Westland, New Zealand: The importance of large, infrequent earthquake-induced disturbance. *Journal of Ecology* **89**, 1006–1018.

West AJ, Lin CW, Lin TC, et al. 2010 Mobilization and transport of coarse woody debris to the oceans triggered by an extreme tropical storm. *Limnology and Oceanography* **56** (1), 77–85.

Wiberg PL, Taube SR, Ferguson AE, et al. 2019 Wave attenuation by oyster reefs in shallow coastal bays. *Estuaries and Coasts* **42**, 331–347.

Wohl E 2013 The complexity of the real world in the context of the field tradition in geomorphology. *Geomorphology* **200** (C), 50–58.

Yanagisawa H, Koshimura S, Miyagi T, and Imamura F 2010 Tsunami damage reduction performance of a mangrove forest in Banda Aceh, Indonesia inferred from field data and a numerical model. *Journal of Geophysical Research: Oceans* **115** (C6), 1–11.

Zimov SA, Schuur EA, and Chapin, III FS 2006 Permafrost and the global carbon budget. *Science* **312** (5780), 1612–1613.

Zscheischler J, Mahecha MD, von Buttlar J, et al. 2014 A few extreme events dominate global interannual variability in gross primary production. *Environmental Research Letters* **9** (3), 035001.

18

The Scope of Geomorphology in Dealing with Natural Risks and Disasters

We began this book by emphasizing that quantifying hazard is an essential building block for putting numbers to risk. The process of measuring risk is a multidisciplinary framework that brings together expertise from different scientific fields. This framework offers many opportunities for considering the geomorphic consequences of natural disasters that are beyond the interest of academic experts. On the one hand, are we concerned with estimating the probabilities of hazardous events occurring. On the other hand, we are also concerned with anticipating the impacts of hazard events on people, so we must also consider the vulnerability of people, built structures, and the geomorphic impacts on the environment. In a separate step, we must assess the nature and values of the elements at risk. Finally, one essential step in evaluating risks is to acknowledge the different risk perceptions of individuals and groups. Our discussion about ecosystem services, for example, has highlighted that parts of our environment that we may take for granted, such as clean water or intact forests, are increasingly evaluated in monetary terms and thus included in cost–benefit analyses in natural risk management (Chapter 17.2).

In this chapter we summarize the scope of geomorphology in dealing with natural hazards and risks by going beyond quantitative hazard assessments. We take into account how geomorphology can possibly contribute to constraining better the remaining factors the risk equation. Much of the philosophy of modern risk management is encapsulated in the current paradigm of disaster risk management as set out, for example, in the 2015 Sendai Framework for Action (https://www.unisdr .org). We emphasize, however, that this is a rather general framework that has been forged at an international level. At the local level this approach alone is unable to reduce the impacts of the next disaster on a given community. The overall consequences of future disasters will be the sum of the impacts on many local communities, and this is a fundamental limitation to risk-based planning. This limitation requires much more work to become better understood. At the same time this recognition is, however, useful in raising awareness; we have seen that increased awareness can be both an efficient and effective way to reduce vulnerability.

Is cost–benefit analysis imprecise? Economic considerations often motivate us in how we feel or decide about how to invest in measures of natural disaster reduction. This approach can also involve human-made hazards such as soil erosion (Posthumus et al. 2013). Probabilistic risk estimates involve inaccuracy by definition, and these uncertainties must enter decision-support tools such as cost–benefit analyses. Probability distributions estimated from data on past disturbances often rely on sufficient sample numbers to

Geomorphology and Natural Hazards: Understanding Landscape Change for Disaster Mitigation, Advanced Textbook Series, First Edition. Tim R. Davies, Oliver Korup, and John J. Clague.
© 2021 John Wiley & Sons Ltd. Co-published 2021 by the American Geophysical Union and John Wiley & Sons Ltd.

reduce prediction errors. The number of damaging events that occur in a planning period can be minute; if they happened more frequently they would be avoided or mitigated, and thus be less damaging. The upshot is that the costs and benefits associated with any proposed mitigation strategy are also liable to errors. Davies (2015) suggested that, when predicting into a sample of 100 events (as in analysing one-year floods over a century) the second error on its own is about ±6%, while when predicting ten events (ten-year floods over a century) it is about ±16%. To these must be added errors of the first kind, and also errors in estimating financial factors such as interest rates up to a century into the future. The key factor in a cost–benefit analysis is the net benefit, which is the difference between the costs and the benefits of the scheme. This net benefit is the difference between two inaccurate numbers, and thus necessarily carries higher inaccuracy. Assume that a town suffers $1 million of damage per year (average annual cost), and that the annualized cost of a flood protection scheme is $850 000 per year; then the net annual benefit is $150 000. This is apparently a worthwhile investment. However if it is recognized that each figure has an imprecision of, say, ±10% associated with it, then the average annual damage is between $900 000 and $1 100 000, while the annualized costs are between $935 000 and $765 500. Hence the net benefit is between $335 000 and −$35 000. It is doubtful whether the investment will be beneficial.

18.1 Motivation

Today, managing natural hazards is largely in the hands of local or regional government authorities, and frequently runs under the title of 'risk management' or even 'disaster risk management'. This overall top-down approach to dealing with natural disasters has received criticism. Many scientists are calling for a more mutual interaction between acting bottom-up and top-down, more attention to combining local knowledge together with scientific outcomes, and a better inclusion of stakeholders (Gaillard and Mercer 2012). This participatory aspect has become highly important and, together with the top-down approach of intervening in cases of natural disasters, forms a framework that has emerged under the umbrella of 'social capacity building' and 'risk governance'. Kuhlicke et al. (2011) distinguish five major types of social capacities, dealing with knowledge, motivation, networks, economy, and procedures. At the national and international scales, non-structural measures such as agendas or policies can provide some overall guidance by expressing the diverse efforts of risk management in a few key messages. At the local scale, authorities often use conventional concepts such as statistical return periods (e.g. the 100-year flood) when approving and designing defence works. Furthermore authorities may largely consider only the use of physical structures to modify the dynamics of natural systems, and authorities do so in response to a societal wish to continue business as usual. Consultation with affected communities might take place only after technical investigations offer only one option without any alternatives. While this situation is changing in some jurisdictions, the efforts to reduce disaster risk remain largely uncoordinated. The authorities responsible for disaster risk management may have few, if any, sources of guidance on how to achieve a certain degree of optimality, sustainability, or indeed community acceptance in exercising that responsibility. The highly diverse geomorphic impacts of natural disasters outlined in previous chapters are incentive enough to develop and offer such guidance.

We consider effective management of disaster risk as reducing risk to a level that is acceptable to those affected (Figure 18.1). Effective management of risk requires adequate knowledge of the natural and societal

Figure 18.1 Risk–cost diagram illustrating the optimization of mitigation measures by using the margin-cost criterion. The intersection of the risk–cost curve by a line with a tangent of 1 marks the economical optimal combination of measures; CHF is swiss francs. From Bründl et al. (2009).

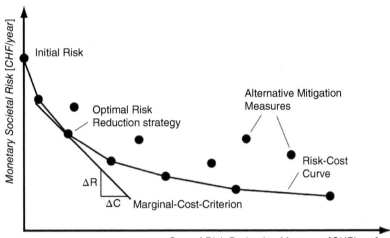

settings in which disasters occur, and also agreement on what is meant by 'acceptable'. The previous chapters have outlined many physical aspects of natural disasters, and we refer to publications that put more and appropriate weight on the viewpoint of the people affected or those dealing with solutions to the problem (Norris et al. 2007). Considering the societal aspects of natural disasters must involve:

- involvement of all stakeholders throughout the process that is risk management.
- awareness of the dynamics and feedbacks between the natural and human systems;
- a comprehensive and robust risk management process that highlights all options for mitigation and adaptation, and explicitly acknowledges and accepts or addresses any residual risk;
- effective establishment and maintenance of the agreed on risk-management strategy;
- ongoing monitoring, evaluation, and refinement to avoid obsolete risk estimates;

In previous chapters we have emphasized that understanding human and natural systems are equally important for reducing disaster risk. Yet documenting how groups, communities, or nations respond to natural disasters involves standards and protocols that differ from those used to document the

physical aspects of natural disasters. From the viewpoint of geoscientists, this difference compromises our ability to put what we know about natural system dynamics to good use. Hence, when developing strategies for disaster risk reduction, thoroughly and genuinely involving all people who might be impacted by a disaster is the best way to ensure that these strategies address the needs of communities. Specifically, the scope of geomorphology (or any other scientific discipline, for that matter) in dealing with natural hazards, risks, and disasters can only improve with major input from stakeholders, meaning those people who must accept and live with these strategies during their planning and operation. This participatory approach requires that affected communities be adequately aware of both natural system dynamics and risk reduction tools. This will likely be the case only if these stakeholders are fully engaged in the process and issues at play. Modern disaster risk management thus requires procedures to guarantee that all stakeholder groups are involved throughout the investigation, analysis, management, and monitoring phases. Below we outline how geomorphology can, and should, contribute to disaster risk management and offer a systematic procedure for incorporating the relevant data into the decision-making processes.

To develop or not? Cheekye River drains a small (60 km^2) but steep catchment on the west flank of Mount Garibaldi, a dissected Plio-Pleistocene stratovolcano in coastal southwest British Columbia. The river has built a very large debris flow fan in the Squamish River valley ~3 km north of the town of Squamish, which is located between Vancouver and Whistler, the centre for the 2010 Winter Olympics. Squamish is one of the fastest growing communities in British Columbia, and pressure to allow development to proceed from the fringes of Cheekye fan towards its apex has mounted in recent years. However, the District of Squamish, which has responsibility for approving new development in the valley, has been reluctant to approve further development on the fan because of the hazard posed by debris flows. But what is the risk? Is the risk within what Canadians would consider acceptable? Or is the risk unacceptable to society?

To address this question, geomorphologists and geologists have, over the past several decades, extensively studied the size and frequency of debris flows on the fan. Based on this work, a frequency–magnitude curve features events with return periods up to about 10 000 years. A problem, however, is the large uncertainties for events with average return periods of thousands of years; yet it is these events that commonly dictate what society considers 'acceptable'. The decision as to what level of risk to accept, and accordingly what mitigation of the hazard might be required, is informed by geomorphologists and geologists. Yet the public, through its representatives, has to decide on further action based on this information.

Possible large, albeit rare, debris flows create residual risk, which is the risk remaining after mitigation. Society, at its discretion, may choose to tolerate some level of residual risk. If it opts to allow further development on Cheekye Fan, the District of Squamish, in partnership with other stakeholders, must select the upper threshold of the return period that it deems acceptable. The choice of this threshold will affect the resulting risk assessment and remedial measures taken to reduce risk to below the acceptable value. Because any choice will involve the acceptance of some residual risk, the trade-off between increased residual risk and uncertainty in the estimates of hazard probability must be considered.

This example illustrates the decision-making process involved in modern quantitative risk assessments in environments subject to natural hazards, whether they be debris flows, floods, earthquakes, or other processes. The process is grounded in robust frequency–magnitude analyses, which are the purview of scientists, but the critical decisions are made by others, based on political, economical, and other societal considerations.

18.2 The Geomorphologist's Role

Geomorphologists, like all other scientists, are obliged to communicate their results to the expert community and the wider public.

Consider a site with either present or planned development that might be at risk from geomorphic processes and for which a natural hazard assessment is required. In risk management, the geomorphologist needs to define the range of processes expected to occur at and around the site during the future period of interest, together with their magnitudes and frequencies.

The first step in this process is to determine the causes that make a site hazard-prone or suspected of being hazard-prone. This step requires collecting and analysing data about topography, uplift, erosion, soils, past disturbances, and likely many other controls.

The second step involves establishing the local and regional history of geomorphic processes based on historical, cultural, geological, and landform evidence. Local communities come into play here by adding valuable information that may have escaped systematic recording or measurement.

The third step is to develop several possible future scenarios of geomorphic activity and its impacts. These scenarios simulate the dynamics of natural systems and societal behaviour set within the range of data analysed, but can also offer projections beyond this data range. In most cases, these scenarios include 'business-as-usual' and 'worst-case' variants. Appreciating the nature and scale of the largest possible event is central to the latter scenario, though possibly at the limits of experience and methods. Scenarios also describe the residual impacts that the community will have to choose whether or not to accept.

At this point in the process, an understanding of the causes of natural hazards may be sufficient to establish whether engineered risk mitigation is likely to be achievable, effective, and sustainable. The range of mitigation measures that a community considers realistic and affordable is limited to those measures that offer acceptable societal development and return on investment. In economic terms, risk management is only successful if the investment costs for countermeasures remain below, or at least equal to, the expected costs from natural disasters over a period of interest. In short, investments must measurably reduce risk. From a geomorphic perspective, this purely economic thinking must also consider how the parts of thus engineered landscapes will respond to the proposed risk-management strategy in the future. For example, putting check-dams into a mountain river may turn out to be a cost-effective option for reducing debris-flow risk over several years to decades, but may alter the flux of sediment such that larger and rarer debris flows occur in the long term. Similarly, solutions without engineering structures must also be considered, particularly in disturbed mountain rivers that are prone to delayed or protracted channel adjustments in the future. A wide range of different strategies and designs must be tested against the hazard and impact scenarios. If experts choose impact reduction measures shown to be ineffective, unsustainable, or unacceptable to stakeholders, these measures must be modified or abandoned. This step is iterative and calls for repeatedly and systematically analysing plans for how we develop and modify landscapes. We cannot do this without anticipating the economic, social, and environmental impacts, nor without involving all stakeholders.

The advantage of bringing nonscientific stakeholders to the table from the start is that we have already decided as to when to start involving society at large. Consultation and decisions are social processes driven by community wishes, constrained by local and national rules, laws, and customs, and informed by science. In short, the risk management combines scientific expert and stakeholder inputs to strategy development. Many nations have begun to formalize this approach; see, for example NZS 9401:2008 'Managing Flood Risk – A Process Standard' (SNZ 2008). The process of community involvement in disaster reduction and its underlying decision making is the subject of current research in New Zealand. Experience to date is that it is a time- and energy-consuming process, which nevertheless empowers communities in their dialogue with officials and scientists and shows promise of leading to more effective adaptations to risk scenarios.

18.3 The Disaster Risk Management Process

Modern risk management can best be viewed as a process incorporating two major stages – risk assessment and risk evaluation (Bründl et al. 2009).

What is disaster risk management? The overall purpose of disaster risk management is to (i) improve understanding of risk by persons, communities, and governments, (ii) reduce the risk to society from disasters to a level as low as reasonably practicable in the circumstances, and (iii) enable society to function in the future with confidence that disaster risks are acceptably well managed. Managing land use and landscapes provides a framework within which to deal with natural disaster risk. Such risk management should be sustainable and be able to reduce conflicts between natural and social systems over the longer term. Changes in natural processes, hazards, exposed values, and their vulnerabilities should be identified through an adaptive management process and addressed in a timely manner. Appropriate risk management entails a broad assessment of strategies and options, anticipation of change, and awareness of residual risks. Comprehensive strategies for reducing and adapting to risk include mitigation, readiness, response, recovery, and renewal.

Disaster risk reduction is probabilistically-based and, therefore, can only apply reliably to reducing the risk ensembles of many disasters. In contrast, disaster impact reduction aims at devising means to reduce the impacts of a small number of disasters (expected to occur in a given location within a reasonable planning period, though in practice often the next disaster) on a defined local community. In many cases, the next disaster to affect a community is the one to whose impacts resilience needs to be achieved.

18.3.1 Identify Stakeholders

At least three groups of people must be involved in the development and application of a disaster mitigation strategy – geoscience professionals, government employees charged with managing risk, and the public. This would involve the person on the street, farmers, shopkeepers, tourism operators, captains of industry, planners, engineers, councillors, hazards managers, organization CEOs, scientists, and government ministers. Local residents in particular may be the most familiar with some of the natural processes and hazards at play in their immediate environment, and strong traditions may help to preserve knowledge of past disasters. Hence it may be desirable to integrate better such local knowledge – if reliable and robust – with scientific insights (Mercer et al. 2010).

Expectations of each stakeholder will also differ beyond how to deal with a given set of hazards, but also regarding the role of the scientists and their input (Vogel et al. 2007). For example, a person might focus on her or his risk first and that of the larger community second. Scientists might be mainly concerned with getting a paper published in a top-ranked journal; and officials might be mainly focused on fitting in with local planning and political agendas. All three groups are equally important for managing disaster risk, although in practice the underlying responsibilities differ. Hence there is a need to develop well-defined and agreed-upon roles and responsibilities for each stakeholder. With more and more people having access to information technology, active cooperation with stakeholders during the modelling process in risk and disaster studies opens new pathways (Figure 18.2) (Voinov et al. 2016). For example, visualizing possible planning and mitigation outcomes or structural measures on a computer screen can be an instructive experience for laypersons, and offer new ways of interactive information sharing and scenario modelling. In contrast, an overload of technical detail might confuse laypersons (and some experts) unnecessarily. Regardless of technological insight, however, all three groups must interact and be fully engaged at the outset, and remain so throughout the investigation, analysis, operation, and monitoring processes. This requires that all three groups agree on a

Figure 18.2 A human model for science and decision evaluation. The cycle starts with setting goals and ends with evaluation of results. Either end it here or loop back into the cycle. From Voinov et al. (2016).

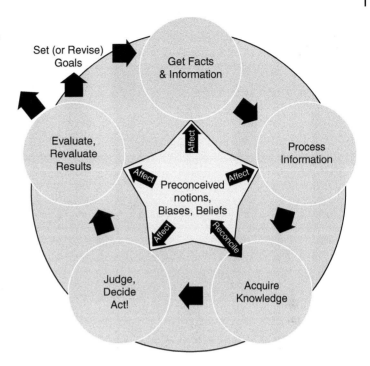

set of priorities. This procedure is referred to as the community-based participatory approach. The assumption behind this procedure is that society and its members know best what they want, so their involvement will ensure that their wishes and attitudes are compatible with the ultimate solutions. Having been involved in the process of developing a solution, it becomes hard to legitimately object to it.

How do geomorphologists come into play here? In New Zealand, standards such as AS/NZ 4360 and SAA/SNZ HB 436 are good examples of how to embed geomorphology in risk management and provide a robust and proven basis for assessing risks associated with management options. The objective of standards such as these is to establish a common approach that the public, private and community organizations, and other stakeholders can use to manage risks and reduce impacts. The framework further requires that all stakeholders communicate and collaborate closely at all stages. Each stakeholder, whether researcher, professional, local or central government, community or citizen, has a different point of

view and a different role that depends on many factors, notably knowledge of, vulnerability to, interest in, and ability to control or mitigate the overall risk. Those in professional positions who are responsible for disaster management especially need to understand how each stakeholder can contribute. To achieve this level of mutual understanding, trust and acceptance, some form of expert facilitation will probably be required.

18.3.2 Know and Share Responsibilities

Along with risk comes responsibility, and it is the sharing of this responsibility among all involved that is challenging, but also essential. Some stakeholders, particularly professionals and governments, have legal obligations, but all have an ethical obligation to be fully involved and to encourage and welcome the involvement of others.

Sharing responsibility in the risk management process should involve maintaining disaster control works; following planning rules and local authority guidance; seeking information about local disaster risk and vulnerable areas; obtaining insurance if possible; being prepared for plausible natural disaster scenarios by acquiring information, assembling and renewing emergency kits, and developing, practising, and revising evacuation plans; seeking advice and making considered choices about living in areas that might be subject to hazard; and learning about disaster emergency management measures.

Professionals are required by law to use high, but reasonable, standards of skill, care, and diligence in executing their responsibilities. They must be aware of current developments in disaster management practice. Professionals also should be aware of land-use developments that are likely to be proposed in the future, and they should be able to anticipate that natural disasters may occur without any known precedents. Professionals play a key role in the collaborative process, which entails understanding points of view of all stakeholders and communicating technical and complex and also lay information in a way that everyone can understand.

Officials, professionals, lay people, engineers, and developers can best reduce risks from hazardous natural processes by ensuring that all risks are properly identified and considered during the design and construction of infrastructure, in particular also critical lifelines (such as communication and traffic networks, water and electricity mains, hospitals) and countermeasures. Natural disasters must always be included in the risks considered, and where a potential hazard cannot be fully understood, let alone controlled, contingency planning needs to be undertaken to ensure that the basic necessary societal functions can be maintained during and after a natural disaster.

Communities must take responsibility for managing risks associated with, and affected by, their collective decisions. They should recognize the need for local authorities to balance benefits, risk, and costs, and they should

support organized processes around disaster anticipation, reduction, and preparation, and emergency management.

Governments at all levels have mandated responsibilities for risk management. These responsibilities need to be made clear to, and understood by, all stakeholders. Leadership, communication, and coordination from governments are crucial in effectively managing risk. For example, effective approaches to disaster preparedness, response, evacuation, health, welfare, and community recovery all require leadership and coordination by governments. In particular, local authorities need to be proactive in setting up procedures for involving all stakeholders from the outset, and in all aspects of the risk management decision-making process.

Principles of disaster risk management
Several principles are the basis for successfully dealing with risk from natural disasters.

- Continued engagement of communities and stakeholders is essential to ensure that decision making includes cultural factors, social values, infrastructure requirements, and community aspirations and needs. Ancillary benefits are that the community will fully understand the risks, and the benefits, consequences, and costs of different options for reducing risk; and that all parties commit to the chosen measures.
- Understanding natural systems, particularly where, how, and how quickly Earth's surface changes, is fundamental to managing risk in a sustainable way. Landscapes set the stage for all human development, but are very sensitive to intervention, and the potentially serious consequences of intervention need to be understood.
- Decision making at the local level is the most effective way to manage risk and

ensure that solutions balance private and public, as well as local and national, interests.

- All possible forms and levels of management should be considered for present and proposed development within hazardous areas. Preference should be given to hazard avoidance where new development is proposed.
- Residual risks before and after mitigation must be explicitly recognized and managed through readiness, response, and recovery activities, and long-term monitoring and planning.

Managing risk is never a one-time operation, but rather a continuing process. Through this process, understanding is gained and shared, options considered, and improvement sought, while all parties continuously communicate and collaborate with one another. These interactions must be sought, managed, and moderated to prevent confusion, and to encourage learning. Such collaboration needs appropriate communication, and requires that people exchange their views and insights continuously. A collaborative environment is essential to ensure that all risks are identified and understood, and that solutions are accepted by all parties. All these groups should strive to express themselves in ways that others can understand. Many collaborations fail because one or more parties may misunderstand what the others are saying, leading to mistrust. To ensure that communication is successful, it may be necessary to involve professional facilitators from the outset.

Ongoing monitoring, review, and adaptation are key elements of the risk management process. At each stage the team must reflect on the situation and, if necessary, reconsider or adjust the strategy. Each step must to be documented and understandable to all involved, so that the process is efficient in terms of time and resources.

18.3.3 Understand that Risk Changes

Recall that disaster risk is a function of four elements: the probability, intensity, and areal extent of extreme events; exposed assets and their vulnerabilities; the number of persons, and cultural, environmental, social, and economic values exposed; and the perception of disasters. All four elements are critical and need to be cast in numbers. Changes in any of these elements will affect the level of risk. More specifically, understanding disaster risk involves knowing the available resources, concerns, expectations, interests, attitudes, and vulnerabilities of those affected by an extreme event.

- Which infrastructure and community resources might be affected?
- What are the cascade effects of the loss of critical infrastructure on communities and stakeholders?
- Which strategies should be used to reduce risk?
- Should we rely on engineered solutions, better preparedness, or simply improved response and recovery after a natural disaster? Or on a combination of these?
- Which countermeasures can realistically be achieved in balancing all the interests involved?
- Cost efficiency aside, what are the geomorphic and ecological consequences of the implemented risk reduction measures?
- How confident is the community that its risk is being managed satisfactorily?

Risk reduction can have regulatory and engineering elements, but also involves emergency management planning. Such planning, if well thought out and designed, can reduce the vulnerability of communities. Planning measures might include raising stakeholders' awareness, developing community strategic and emergency response plans, refining community land-use plans, designing and building community response and recovery plans, and adopting appropriate land management practices.

The requirements for risk management evolve in response to changes in the disaster–risk profile that is, the spectrum of risks of varying severity that the community is exposed to, vulnerabilities, and community expectations, as well as changes in climate, vegetation, and surface processes. The human population is increasing almost everywhere, commercial activity is accelerating, and society is becoming ever more complex and interconnected, thus the overall vulnerability of society to disasters is increasing.

Risk thus evolves, and communities may over time become less resilient or become more dependent on community-focused and national risk-reduction measures. Different risks also evolve differently, and we can learn from mutual cooperation between those involved in studying and mitigating disaster risks and those concerned with risks from climate change. Schipper and Pelling (2006) pointed out that, despite the many intimate linkages between natural disasters and climate change, two somewhat detached communities and their agendas seem to tackle these problems from different ends. For example, climate change policy focuses mostly on climate-related hazards and their impacts, and largely ignores others. This contrast is partly due to the different periods that people consider or even worry about, as natural disaster impacts are mostly abrupt and tied to communities, regions, or nations. In contrast, the impacts of atmospheric warming often take longer to manifest, so that we need to consider how social and economic systems will change over the same time period. That disaster risk reduction has been largely a local and national issue, grew mainly out of pragmatism and necessity, whereas climate change risks warrant by definition a more global perspective. As risk is being transferred among organizations, governments, and people, communities will increasingly expect that the risk will be managed to their satisfaction by government bodies and organizations, and that the community will have a large contribution to risk management decision-making.

18.3.4 Analyse Risk

At the start of risk analysis geomorphologists need to identify what type and size of disaster could happen in a given place, what might cause it, how frequently it can happen, and what geomorphic consequences it might have. The occurrence of such events can be approximated with statistical tools, but some (mostly aleatoric) uncertainty will remain. Likewise, the many feedbacks among complex natural and human systems are poorly understood, so that mostly epistemic uncertainty limits predictions of possible societal impacts of natural disasters. Both types of uncertainties are unavoidable and must be accepted. Risk analysis will always have to live with, and will always entail, some unknowns and subjective evaluation. Some scientists may be familiar with these limitations in their everyday work, but the broader public and laypersons are generally not. Because uncertainty and subjectivity are inevitable, it is necessary to include all parties in discussions about risk identification so that the full societal spectrum of perspectives is tabled.

The probabilities and consequences of potential disasters must be rigorously analysed (including uncertainties) and compared with the likely consequences of risk management strategies and activities. A precautionary approach is needed to avoid underestimating risk. In general, the consequence of understating a risk is an unexpected disaster; the consequence of overestimating a risk is that the expected disaster may fail to happen. The latter may seem preferable to the former in most cases.

Many natural processes and societal structures change slowly, thus it is important to communicate that geomorphic effects may can lag behind their causes. For example, several tributaries of a large river can flood at the same time, thus contributing to a super-design event or structural failure; or a severe earthquake can trigger a large landslide that dams a river, leading to a downstream outburst flood. In any case, the sensitivity of the risk analysis to underlying assumptions should be tested. This is important, as it may be necessary to gather additional information to ensure that the analytical results are robust and reliable.

Getting the message across Many of the problems that risk managers face are the result of poor communication among stakeholders. For example, Citizen K on being told that his house has been identified as being in a rockfall hazard zone, says: *'Rubbish! I have lived here for ten years and never seen any rocks fall. This is academic nonsense. These big rocks by my house have always been there? Professor X told me they were transported by a glacier'.*

Citizen K, on being told he must pay for a hazard assessment for his house, responds: *'No way! This is completely unnecessary. The Government is ripping me off again. It's not my responsibility!'*

Citizen K, on being told he must pay for rockfall catch fences, says: *'What!? This is the limit! I shall write to my MP! I shall start a citizens' protest group! I shall refuse to pay! I am willing to take the unseen risk of your imaginary rockfalls!!'*

Citizen K, on reporting that his house has been damaged by rockfall: *'Why didn't someone tell me? I might have been killed! The Government is to blame and must pay for the damage. I have a building consent for the house: they said it's safe, and it's their responsibility! I demand that a protection fence be erected immediately!'*

This reaction, of course, contradicts the previous ones. It is easy to label Citizen K as ignorant, inconsistent, or arrogant because of this contradiction. However, he sees his behaviour as completely logical and consistent. What happened was that Citizen K's life experience changed. He experienced the hazard in real life as a disaster. Before it happened, it was outside of his belief system; afterwards it was part of it. Because his experience changed, his beliefs changed, and so did his behaviour. Our personalities and behaviour change with our experiences.

18.3.5 Communicate and Deal with Risk Aversion

Recent history contains many examples of natural disasters that were foreseen and even predicted, including the Vajont landslide disaster in 1963, the Huascaran landslide disaster in 1970, the lahar that devastated Armero, Colombia, in 1987, the Indian Ocean tsunami in 2004, and Hurricane Katrina in 2005. What was done to inform those at risk? In each case there was scientific evidence of a pending disaster, and this evidence was available to the civil authorities responsible for public safety. In each case the opportunity to take action was lost, with the result that large numbers of people died. Are we just biased in hindsight or could we have done better? Would more or better science have allowed the authorities to take preventive action such as evacuating people? Would better communication of science have prevented the disaster? If so, how can we communicate science better? What are the stumbling blocks that prevent the effective use of relevant scientific information in disaster management, and can these stumbling blocks be removed?

If the unwillingness of authorities to act upon scientific advice is indeed an issue, we should expect the situation to improve over time as the amount and quality of science increases. Foreseeable landslide disasters are still taken less seriously than they should be more than 50 years after Vajont. Perhaps scientific information alone is not the key issue. Today we know much more about large landslides and their causes and triggers than we did in 1963. Furthermore, we are able to mechanistically model and predict the future behaviour of large slopes, which was less possible 50 years ago. These advances and improvements rarely guarantee that all natural hazards and disasters will be researched to the degree that modern science allows. Science depends on resources (funding mostly) and continues to advance and improve the quantity and quality of knowledge that can be applied to a specific situation, and such improvements are well funded by governments on the assumption that improvements in disaster prevention will justify the cost of the research. It seems contradictory when a government organization (e.g. a Ministry for Science funding research) will fund general improvements in science whereas other government organizations (e.g. a Ministry for Infrastructure) may refrain from funding their practical application. It appears, therefore, that the lack of availability of scientific information hardly explains the reluctance of authorities to take action on potential disasters.

Communicating science to such diverse audiences is a skill and an art in its own right (Gluckman 2014). Government officials are rarely familiar with the processes and language of science, and scientists are rarely well-acquainted with government officials and how they operate. Politicians at all levels may make decisions that seem counter to scientific evidence. Yet they may see scientists as making bewildering demands for funding that they barely understand and that, if granted, might result in unwelcome information becoming public. This lack of common ground and understanding between scientists and government officials limits mutual trust. It is clear that improving the communication of science is a complex, though necessary, task. It is occasionally recognized by governments through the appointment of science advisors to high-level politicians and through the presence of scientists in local government organisations. Better communication of science to decision makers lies beyond rephrasing descriptions

of what is in the minds of scientists so that decision makers can understand it. Comprehension goes far beyond agreement on the meanings of words and phrases.

Many geoscientists are trained to tackle research questions that involve thinking about processes operating, and landforms forming, from tens of thousands to millions of years. By contrast, lay people are accustomed to thinking about days, months, years, and rarely a few decades at most. Societal decision making is commonly delegated to elected bodies whose time frame of interest is seldom longer than a small number of years, mostly constrained by the length of legislative periods. Someone standing for a policy of restricting land use because of a landslide that might take another hundred to several thousands of years to happen, but could kill someone tomorrow, is unlikely to be re-elected. This is particularly the case when local economic issues dominate the political agenda and when immediate fiscal pressures might be relieved by approving development of sites vulnerable to rare, but disastrous events.

Thus balancing short-term needs against the need for long-term safety from loss of life and damage is a key problem. Ironically, in many such cases cost–benefit analyses show a clear long-term net economic benefit in avoiding vulnerable areas because the costs are so high when the disaster occurs. Unfortunately, such analyses have little impact because the short-term benefits are accrued assuming that the event is unlikely to happen in the short term. This, we believe, is a major reason that hazard science finds little fertile soil in the minds of decision makers. A further reason lies in the aversion that most people have to the risk from natural disasters. Decision makers have to consider whether to take short-term risks for long-term gain. Yet many decision makers will have little experience with the types of disasters scientists are talking about and so simply cannot be deeply affected by the prospect of one happening.

18.3.6 Evaluate Risks

Once quantitative risk estimates are available, the next task is to evaluate these risks. The purpose of risk evaluation is to gauge the estimated risk against the desired level of safety or protection. This approach allows for some remaining risk that cannot be eliminated. Hence, risks need to be kept at an acceptable level; the choice of this level follows mathematical models and recommendations, but eventually remains largely a political decision under uncertainty (Vrijling et al. 1998). Published information on acceptable risks from natural disasters is sparse, and confusing definitions shroud the concept.

Tsunamis from landslides and acceptable risk Milford Sound in southwest New Zealand is one of the world's popular tourist destinations. It combines fjords, rain forest, and sea life, drawing many thousands of tourists every day during the austral summer. It is geologically similar to the Norwegian fjords, which are well known for their enticing landscapes and touristic amenities, but less for rare and sometimes disastrous landslide-generated tsunamis. In 1934, for example, 2 Mm^3 of rock fell into Tafjorden in Norway and triggered a tsunami that devastated a village five kilometres away, killing 48 residents. Today the hazard is managed by identifying, investigating, and monitoring potential landslide sites. One assumption that underpins the warning system is that in Norway's low-seismicity setting, any large rockfall or landslide is likely to be preceded by an acceleration in the creeping motion of the unstable slope and possibly by precursory rockfalls.

Milford Sound also has the potential for large rockfalls and rock slides into the sea, but, unlike Norway, it is seismically active. Strong (*M* ~8) earthquakes occur on the nearby Alpine Fault several times per millennium, likely having triggered many of the large postglacial rockfall or rock-slide deposits that litter the floor of Milford Sound. Some of these presumably caused tsunamis. Given that tourists at Milford Sound are always within ten metres of sea level unless aboard an aeroplane or helicopter, reliable risk estimates are needed.

Precursory slope movements or minor rockfalls may give little, if any, warning of a pending coseismic landslide at Milford Sound. The only warning of a landslide tsunami might be the earthquake shaking itself. The estimated likelihood of a strong earthquake on the Alpine Fault in the next 50 years is about 29% (Cochran et al. 2017). The probability of a landslide sufficiently large to generate a tsunami occurring in any specific earthquake is unknown, and perhaps about 20–30%, judging from the number and size of postglacial landslide deposits.

Most definitions consider the annual risk of death by disaster is acceptable to an individual, if in the range of 10^{-6} to 10^{-3}. The risk to an individual is equal to the average number of deaths per year expected from the event divided by the number of people exposed to the risk. The annual expected number of deaths due to a landslide-induced tsunami at Milford Sound, New Zealand, is 0.38. At least half a million tourists visit the area per year, thus the annual risk to an individual is about 10^{-6}, and thus apparently acceptable for events such as landslides and dam failures. Along similar lines, Finlay and Fell (1997) outline the acceptability of geotechnical disasters involving different numbers of fatalities. From their perspective, 45% of the 500 people exposed to a landslide-induced tsunami at Milford Sound, or 225 people, would be expected to die. The upper tolerable risk limit for this number of deaths in a single event according to Finlay and Fell (1997) is about 5×10^{-4} per year. The annual risk of 225 deaths occurring at Milford Sound is $0.38/225 = 1.7 \times 10^{-3}$, which is unacceptable by any standard. The risk to society is unacceptable in this case, whereas the risk to a single tourist is not. One way to reduce this risk is to abandon Milford Sound and nearby Doubtful Sound as tourist destinations. Yet Milford Sound is likely to remain attractive to a growing number of people, given its world fame and its UNESCO Natural World Heritage status.

One could ask whether, if the people concerned were content to accept the risks, should society also be? Here risk aversion comes into play. The answer lies in the nonlinear relationship between the number of deaths and their impact on society, including businesses and government. A single death by natural process at Milford Sound would hardly alter the numbers of visitors, nor would a single boat overturning with several people dying. Two hundred deaths in a tsunami, however, would make headlines around the world and would likely result in reduced visitor numbers. It could also result in court proceedings on the grounds that the government, being aware of the situation, should have done something about it. Once the number of deaths is so large that the anticipated loss increases sharply, the risk becomes less acceptable.

Note the distinction between a risk that is acceptable for a single person but unacceptable to society. In considering the risks a person is facing, he or she may ignore the risks others are taking. A person may be indifferent if other people die with him or her at the same time, but it greatly matters if he or she dies. In contrast, society may be neutral about who dies, and care only about how many people die. The two acceptabilities are therefore very different. If a foreseen natural disaster could cause a large number of deaths with a known probability per unit time, the relevant criterion is the societal (or political – since governments are inevitably involved in major disasters) acceptability of that number of fatalities annually.

It is perhaps useful to consider these two aspects of facing risk – personal and societal acceptability – thus emphasizing the two very different aspects of our lives. This distinction highlights two very different groups of responsibilities, one to ourselves, our families, and perhaps our close friends, and the other to society in general. When exercising these responsibilities, it is usual to ensure that both are met – being socially irresponsible to one's family is a sign that something is wrong. Acceptable risk needs to be determined so that both private and public criteria are met. However, it is unreasonable to require a person to consider the public acceptability of the risks that he or she is facing when making a personal decision. Rather it is society's responsibility to oversee the public acceptability of risks, which it generally does by way of legislation. Herein, however, often lies a source of confusion: public risk acceptability is determined on the basis of what is acceptable to a person, leading to unacceptable public risks (Dai et al. 2002).

One approach is to treat as acceptable risks that are 'as low as reasonably practicable' (ALARP). The ALARP principle addresses the level of acceptability or tolerance of risk and is a pragmatic, albeit somewhat subjective, tool for decision makers. It is fairly straightforward to assess technological disasters using the ALARP principle, but it is much more demanding for society to assess risk from natural disasters in the same way. In any case, determining ALARP is largely equivalent to determining acceptable risk. The ALARP criterion offers the public assurance that all reasonably practical steps have been taken to avert disaster and to ensure their safety, and requires that the question 'Is the risk as low as reasonably practicable?' is asked at the evaluation stage. In answering the question, it is important to consider the likely cost effectiveness of further risk reduction measures and to seek the views of all stakeholders. Almost always some stakeholders will feel that more can and should be done, while others may believe the costs are higher than reasonable.

Some local risk reduction measures and costs may have to be deferred as part of a long-term management plan. Options need to be fully costed, and impacts assessed and communicated to stakeholders before a consensual decision on an overall management solution is reached. This process helps to make explicit any risk redistribution within a community and associated risk acceptability criteria. The level of residual risk must also be evaluated and clearly communicated to the community to ensure that the level finally selected is acceptable to all who bear that risk. The evaluation process leads to the selection and further development of a range of reduction, readiness, response, and recovery options that are accepted by all stakeholders.

Risks and risk perceptions will change over time, and require adaptive management to ensure that protection and safety measures remain appropriate. Adaptive management means that the selected risk reduction measures are monitored and reviewed. Such monitoring and reviewing should acknowledge the different times it takes for landscapes or landforms to respond to and recover from human and natural disturbances, including the legacy of natural disasters. It is also important to monitor the plethora of changes that can affect risk reduction, including, for example, climate and land-use changes; changes in risk perception, community aspirations, or policies and plans; and changes in standards and best practices. Locally, possible side effects of risk reduction measures must be monitored, and possible new risks introduced by the management scheme itself must be dealt with. Monitoring, in part, requires evaluating whether current and future risks are properly understood. The information obtained through monitoring will form the basis for adapting the risk management system so that it continues to meet community requirements.

Another key goal of risk evaluation is to inform decision makers about mitigation options with an optimum balance among costs, benefits, and stakeholder acceptance.

Risk cannot be completely eliminated and most risk reduction measures have associated costs. In many cases, however, cost–benefit analyses are themselves based on estimates and commonly ignore intangible human values. Different risks need to be linked with stakeholder aspirations and then ranked to identify priorities for their management.

18.3.7 Share Decision Making

Risk management decisions require collaboration at all levels. Although the principle that local risk is best managed locally is sound, disasters with widespread regional effects do require regional and national responses. Central governments normally carry the responsibility for managing national risks that, because of their nature, scale, or complexity require management at senior levels of government. Examples include national defence, biosecurity, terrorism, and pandemics, where a government is responsible for providing risk management frameworks to ensure its internal functions and to support lower level government in addressing hazards. The government portfolio also includes risks associated with events that cannot be handled by local authorities. Therefore the process of risk management needs to work across jurisdictions and within and between different administrative organisations, let alone a plethora of private, corporate, community, and differing interests. Everyone's responsibilities must be understood and coordinated. Having a systems perspective is essential, as is accepting responsibility for defined roles. Knowing one's role within an organization ensures consistency and priority. Good communication between organizations provides clarity about roles and responsibilities, identifies gaps that need to be managed, and establishes a framework that can provide the best assurance of sustainable outcomes in disaster management.

Throughout the world, the business community is increasingly characterized by a few gigantic organizations, many medium-sized ones, and countless small ones. Smaller independent organizations are burgeoning due to the devolution of governance, the shifting of public and private responsibilities, and the limited embrace of social concerns by the traditional private sector. Small organizations may have limited capacity to deal with major natural disasters, yet social and commercial systems are becoming increasingly interconnected and complex, driving a need to manage interactions among organizations. Yet this need is hardly met. The Global Assessment Report on Disaster Risk Reduction 2013 (www .unisdr.org/we/inform/publications/33013) has recognized this shortcoming and emphasizes the many opportunities that natural disasters provide, especially for small businesses. The report advocates the idea of both shared risk and shared value and makes a 'business case' for disaster risk reduction.

Dealing with natural risks in a UNESCO World Heritage site The rapidly growing community of Franz Josef Glacier, a major international tourist destination on the west coast of South Island, New Zealand, is rich in natural hazards and has an increasing number of successful, growing businesses. It is a good example of modern commercial society developing and growing in a hazardous place. The range of natural hazards in Franz Josef derives from its location on the Alpine Fault, the active boundary between the Pacific and Australian tectonic plates. This fault unleashes $M \sim 8$ earthquakes several times per millennium. Scientists believe the last big earthquake happened in 1717 AD. The town is located where the Waiho River exits the Southern Alps. Each year the Waiho catchment receives between five and ten metres of precipitation and an average tectonic uplift of 10 mm. A permanent population of about 500 welcomes thousands of visitors each day in the eight-month tourist season, many of whom stay overnight and visit Franz Josef Glacier.

Expansions of the township have been driven by demand for tourist accommodation, and have ignored risks from debris flows, river flooding and aggradation, an Alpine Fault earthquake, a rock avalanche from the hillslopes above the town, a rapid advance of Franz Josef Glacier triggered by a landslide over its ablation zone, or a rock-ice avalanche triggered by a large rock-slope failure onto the glacier (Davies et al. 2003). Most of these scientific concerns still await acceptance by the decision makers.

A $M > 8$ Alpine fault earthquake has an estimated probability of occurrence of about 29% in the next 50 years (Cochran et al. 2017), but its risk to the community awaits detailed elaboration. The local government is very poorly funded, with a rating base of about 30 000 inhabitants over an area of some 10 000 km^2, and limited in its ability to constrain development or fund research. Nevertheless, a disaster to a flagship location of the tourism industry of New Zealand, with the possibility of many deaths and long-term interruption of visitors on the West Coast, should be of great concern to the nation.

Is the present growth of the township, if continued until the next disaster, sufficiently lucrative to offset the expected losses? Conversely, can the future of the township be modified so that the risk is reduced, and does this benefit justify the loss of income from modifying development? Answers to these questions, founded on a belief that disaster is inevitable, would provide the information required for progress. The answers also depend on when the next destructive event occurs, what damage the township will experience in its future, more developed state, and what benefits would accrue by developing resilience.

This situation illuminates the dilemma of anticipating a natural disaster. Because such a disaster seems unlikely to many, politicians and the community of Franz Josef are unwilling to invest in costly measures that will quite possibly yield little, if any, return in their lifetimes. Many residents remain in the area for a few years only, and incentives are low to invest in the future safety of the town for the benefit of their descendants. The solution to this state of affairs remains to be found. A first step in remedying these unknown risks is to invest time and effort in communicating with the residents of Franz Josef. What is required is a well thought out and systematic plan to provide locals with the geoscientific background in a manner and a form understandable to them. If they trust the purveyors of this unwelcome hazard information they will be prepared to make the effort to deal with the hazards they face.

Postscript: As this book goes to press, the Franz Josef story has moved on a little. A minor Waiho River flood in 2016 destroyed a hotel, fortunately without casualties – an event that triggered government funding for an investigation into whether to avoid the hazards by large-scale relocation, to reduce them by minor land-use zonation or just to live with them. To date, however, this process has yielded no decision.

18.4 The Future – Beyond Risk?

We have argued that geomorphology is more than the science of landforms and the processes that shape them. We have argued that geomorphology is an entire toolkit for predicting and reducing the impacts of natural disasters. One widespread way to do this is via predicting expected losses via the risk concept.

18.4.1 Limitations of the Risk Approach

For using the mathematical definition of risk we require data to learn the probability that a

potentially damaging event happens in a specific area and period. The data that we obtain most readily, for example earthquake intensities, flood levels, wind speeds, or wildfire areas, are likely to be mostly of frequent events of moderate size. Still stronger, destructive events might remain absent in our measurement period, however. Hence, our limited sample makes characterizing the probability distribution of rare events more uncertain. Extreme-value theory is one statistical method that uses data on smaller events to extrapolate to larger ones. This extrapolation is purely mathematical and may inaccurately capture and predict the rarer events. This seemingly obvious point is often ignored and the corresponding uncertainties have only gradually entered risk analyses. The more serious events that trigger disasters may occur only a small number of times, if at all, in the period of interest to a community, region, or nation. This observation means that, regardless of how accurate the statistics describing these disastrous events, their occurrence in the planning period is very likely to depart from the probabilistically predicted occurrence.

Many other uncertainties compromise planning for disaster risk reduction. One of the most important is grounded in complexity, which means that the nature and intensity of the most important disaster-triggering event at any location cannot be known with certainty. The role of geomorphology that we outlined in Chapter 3 appears to apply mainly to frequent events, and less so to rare and disastrous ones. How then can geomorphology contribute to disaster impact reduction?

18.4.2 Local and Regional Disaster Impact Reduction

We have outlined that the contribution of geomorphology to disaster reduction lies in its ability to clarify the types and potential magnitudes of surface processes that can occur at a given locality (see Chapter 3). However, from the perspective of people who will suffer the effects of the next disaster, the nature and intensity of the triggering event are of less interest than its effects, that is what the event will do to their community. It is less important whether the trigger is a tsunami, a typhoon, a landslide, or an earthquake, than how it affects the buildings in the community, the roads that connect the community to the outside world, the community's power supply, communications with the outside world, and, most importantly, the people in the community and its economic continuity.

Geomorphology has a crucial role to play in assessing many of these effects. Every community is unique with respect to its local and regional 'hazardscape', because of its unique geography and the events that could impact it. Its commercial connection to the wider region and nation depend on regional and national infrastructure. Thus, apart from knowing the types and possible magnitudes of events that could trigger local disasters, it is necessary to know the cascading regional and national impacts of these events. Likewise, the possible impacts of external events on the local community must be understood. It follows that disaster impact reduction cannot be solely limited to a consideration of local effects, but also must include important regional effects. It might be that risk assessment can play a useful role in understanding and mitigating these regional effects because there will be many more disasters in a region or nation in a century than in any given community. The effects include:

- damage to buildings and infrastructure;
- loss of communication, power, and accessibility due to damaged lifelines;
- loss of social and economic services;
- deaths and injuries;
- psychological effects.

Sediment flux and engineering The summer of 1987 was a very wet one in Switzerland, and several floods occurred during heavy rainstorms. One particular disaster in that year resulted from sediment delivery during a flood that was an order of magnitude greater than any such event in the past 300 years. The disaster happened at Val Varuna in Ticino, southern Switzerland. Rickenmann and Zimmermann (1993) analysed the records of floods in the area dating back to 1750, and concluded that mitigation measures that had been taken since the mid-1800s to prevent flooding had caused the disaster.

Val Varuna is an Alpine catchment that drains via a steep gully onto an alluvial fan, which seemed an ideal place to locate a town – the fan is above the level of the flooding on the main valley floor, and has well-drained soils and good views. However, from time to time alluvial fans are subject to floods and sedimentation. Development on these fans renders them unacceptable places for sediment delivery and storage, and measures are taken to either intercept sediment before it reaches the fan or to deliver sediment in engineered channels to the valley floor. These measures, in turn, lead to more extensive development, as the hazards are perceived to be a thing of the past. In Europe, preventative measures commonly involve placing a series of check-dams along the stream to trap sediment and to prevent the stream from incising into its bed and destabilizing the adjacent hillslopes.

The cumulative sediment delivery over time from Val Varuna from 1750 to 1987 derived from written accounts attests to a reduction in sediment delivery from ~1000 m^{3}yr^{-1} to negligible amounts in about 1930, corresponding to the time when check-dams were built. In 1987, a rainstorm-fed flood flushed ~0.18 Mm^3 of sediment onto the fan. The huge flush of sediment overwhelmed the check-dams and escaped the channel, damaging buildings in the town. Extrapolated past sediment yields are very close to the total amount of sediment delivered in the 1987 event. Thus, the 1987 event released all of the sediment that the check-dams had intercepted during their lifetime. We conclude that, even in a tectonically stable region such as the Swiss Alps, erosion is governed by weathering and precipitation, and may be slowed or halted for a while by engineered works, but eventually will restore the long-term geological rate of sediment delivery. A lesson learned from this event is that future disasters can be avoided by allowing sediment to pass through the village by not rebuilding the check-dams, but by restricting development on the fan. This recommendation, however, has gone unheeded, as the check-dams have been rebuilt.

This example illustrates the dilemma of disaster reduction. We know what caused the disaster, but by aiming to continue our way of life without disruption we recreate the conditions that caused the disaster. In the case of Val Varuna, the check-dams were rebuilt so that the densely developed village could continue with business as usual. This decision guarantees that the disaster will be repeated at some time in the future. In the interim, businesses will reap disaster-free profits, and improved engineering may extend the disaster-free period, which of course means that the next disaster will be worse still. This gloomy scenario is predicated on the assumption that the new check-dams will fail at some time in the future.

One could argue that the increased income from economic development in the disaster-free years offsets or exceeds the costs of the inevitable disaster. The problem, however, is that a small number of people reap the income while a larger group pays the costs. Quite apart from questions of morality, real-life pragmatism suggests that this situation will be unacceptable to the latter group.

To reduce disaster impacts, communities, regions, and countries need to anticipate these effects, develop means to reduce their severity before they occur, and develop and operate remedial measures so that the effects will be as small and brief as possible. Again, the effects of the disaster are strongly conditioned by the nature of the local landscape, which owes its morphology to the spectrum of processes and intensities it has experienced far back in time. Measures that can be taken to reduce the effects of disasters include:

- Robustness, relocation, and redundancy – increasing the resistance of buildings and services to damage, relocating them to less hazardous places, or having back-up assets.
- Planned provision of emergency transport.
- Availability of emergency communications, power, social and medical services and supplies.
- Anticipation of social and economic disruption and minimisation by communication.

18.4.3 Relocation of Assets

Hazard avoidance might be impossible, especially in a fully developed urban area because land to which a vulnerable asset can be relocated is unavailable. Nevertheless, such obstacles may be more a perception than reality (Figure 18.3); prolonged negotiation and sufficient time to change entrenched attitudes can sometimes solve these issues. In choosing a relocation site, a complete picture of the geomorphic processes that are, or might be, active in the area is required, and the least vulnerable site selected. The relative frequencies and intensities of damaging processes can be assessed, and informed judgment then used to choose the location that will be least likely to be affected, and ideally informed by a conventional risk assessment. For example, in an area with evidence of major earthquakes occurring every few hundred years and the last one having happened several hundred years ago, it would be wise, though rarely fully feasible or realistic, to avoid locations vulnerable to severe shaking and to any of its geomorphic

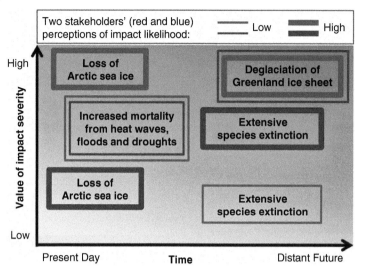

Figure 18.3 Climate risk space. Conceptual map of climate risk perceptions held by multiple stakeholders. Risks vary across a spectrum of severity and time scale beyond the limits of adaptation over which impacts may be realized with increasing global temperature. Two hypothetical stakeholders (red and blue) may value impacts differently (vertical position) and perceive impact likelihood differently (box-line thickness). Four possible combinations are illustrated. From Luers and Sklar (2013).

consequences, such as coseismic landslides, ocean or lake tsunamis, river sedimentation, and liquefaction (Robinson and Davies 2013).

It is also important to consider that the next disaster at a given locality might be decades away, so immediate relocation might be less desirable than gradual redevelopment as properties become available. This strategy is contingent on a social concord that relocation is desirable – then it can take place in a planned fashion with a minimum of surprises. However, relocation will always be a contentious issue and is often seen as a last resort, especially in highly developed areas. Nevertheless, in the case of very hazardous areas it is worthwhile to bring it to the table. In a location where damage to lifeline assets would have serious consequences, consideration might be given to installing redundant or backup assets in less vulnerable places. Many old assets were sited without considering hazards and are correspondingly vulnerable. It might be possible, rather than prioritizing relocation of the asset, to develop a complementary asset that, following a disaster, can stand-in for the older asset and subsequently be developed to become its replacement. The complementary asset can be sited with due consideration of known hazards, preventing

complete lifeline outage and serving as a basis for redevelopment.

18.4.4 A Way Forward?

Insurance is one method that allows costs to be distributed over a large number of years and people. The insurance company spreads the risk by receiving regular payments from its clients. However, natural disasters also have uninsurable impacts, so that another strategy is needed that will cope with hazards ranging from the smallest potentially damaging up to the maximum possible event. Currently these two extremes are dealt with by different, but related strategies.

The first strategy is *hazard management*. For events with return periods up to about 10 years over a 100-year planning period, damage prevention should be possible at reasonable cost through construction of protective works following a reliable cost–benefit analysis. For example, a flood-prone property in the community with river control banks capable of containing a 10-year return period flow might then be flooded a small number of times, if at all, during its occupancy by one owner. However, it is important to note here the mounting evidence that efforts to control natural systems have intrinsic and subtle problems in that the natural system reacts to the control and alters its behaviour. An example is the use of flood banks to constrain rivers and contain floods, which often causes long-term riverbed aggradation that increases the flood risk gradually over time (Davies et al. 2003; Beagley et al. 2020). In essence the engineered structures introduce an element of nonstationarity that makes the data on which the structures were designed invalid for planning purposes. The increase in development on land perceived to be flood-free following structural flood mitigation measures causes the risk from the inevitable super-design events to increase over time. The issue with this approach lies in the choice of the highest discharge that the containment measures are designed to contain. This choice is often based on the scale of works that maximize average annual net benefit, thus maximizing the difference between annualized cost of works and annualized damage reduction. However, if the maximum flood that the works are designed to contain is exceeded only a few times during the scheme life (for example, if the works are intended to contain a 50-year flood over a period of 100 years), then the number of exceedances is likely to be different from the number predicted probabilistically. For example, if the probability analysis says that two exceedances are most likely, but three occur, then the total damage over the scheme life is 1.5 times the expected damage and the net benefit is altered drastically, perhaps even becoming a net loss.

The second strategy involves *warning and evacuation*. This strategy requires detailed planning and coordination, and is prone to the effects of inevitable false alarms that can make people more oblivious to warning signals in general. When super-design events do occur, deaths and some damage could be prevented, or at least reduced, by establishing a flood a routine embracing readiness, warning, and evacuation informed by rainfall and river flow monitoring.

The third strategy is based on *scenario-based resilience*. Event scenarios can be of value for reducing the impacts of future disasters and of much greater value in the case of the less frequent, larger events to which risk management cannot be applied. Such scenarios must be carefully chosen and scientifically credible to depict events that, although rare, can occur, along with their plausible, local and regional impacts on communities and infrastructure. Importantly, these impacts can last for many years or decades. Such scenarios, although unlikely to be realised in detail, serve two important functions: firstly as a source of information to the populace about what sorts of effects can be expected from a major event at any time in the future; and secondly as specific events to which the populace and its governance can attempt to increase future resilience.

Becoming more resilient to a rare large event will automatically increase a community's resilience to all events that have impacts of the same type. A further advantage of this strategy is that a society with increased intrinsic resilience is one that is better protected against the impacts of unforeseen hazard events of all types in the future. In most cases the chosen scenario event will lie in magnitude between a frequent (perhaps ten-year) event and the maximum credible event (worst-case scenario), although in some cases the latter may be a useful scenario. If a community alters its way of life to be less vulnerable to a specific impacts scenario that is relevant to them, the community will be to some degree less vulnerable to all impacts of that type. Further, because it is the effects of the disaster that are important to the community, and effects vary much less than the types of disasters that might happen, reduced vulnerability to the scenario disaster implies increased resilience to all disasters.

We reiterate that risk management decisions have economic outcomes but also affect other aspects of the lives of the inhabitants. Methods for formally including social, cultural, and environmental issues in decision making are crude relative to the apparent sophistication attainable through traditional cost–benefit methods. As we have commented, however, much of this latter sophistication is questionable. We recognize a strong case for giving at least equal weighting to the intangible impacts, because we cannot offer the same level of accuracy for less frequent events in conventional decision-making processes. It is clear that current risk management methods have serious limitations for the purpose for which they are intended, except for frequent small events that occur several times during a planning time period. Statistical information on less frequent hazard events with smaller probabilities can contribute to instead of overriding community decision making.

In summary, this chapter has dealt with the requirements for communication among all involved in decision making for reduced disaster impacts and risks. The 'ideal-world' situation is far beyond the horizon, and future disaster reduction processes will continue to be suboptimal, in part due to the limitations of universal communication. The situation is in any case at the mercy of aspects of societal and political processes that lie far beyond disaster reduction. Nevertheless, we maintain that cognisance of these issues is beneficial to disaster reduction practitioners, and is likely eventually to lead to better understanding among all concerned: public, officials, scientists and politicians.

References

Beagley R, Davies T and Eaton B 2020 Past, present and future behaviour of the Waiho River, Westland, New Zealand: a new perspective. *Journal of Hydrology* **59**(1), 41–61.

Bründl M, Romang H, Bischof N, and Rheinberger CM 2009 The risk concept and its application in natural hazard risk management in Switzerland. *Natural Hazards and Earth System Sciences* **9**, 801–813.

Cochran UA, Clark KJ, Howarth JD, et al. 2017 A plate boundary earthquake record from a wetland adjacent to the Alpine fault in New Zealand refines hazard estimates. *Earth and Planetary Science Letters* **464**, 175–188.

Dai F, Lee C, and Ngai YY 2002 Landslide risk assessment and management: an overview. *Engineering Geology* **64**, 65–87.

Davies T 2015 Developing resilience to naturally triggered disasters. *Environment Systems and Decisions* **35**, 237–251.

Davies TRH, McSaveney MJ, and Clarkson PJ 2003 Anthropic aggradation of the Waiho River, Westland, New Zealand: microscale modelling. *Earth Surface Processes and Landforms* **28**(2), 209–218.

Finlay PJ and Fell R 1997 Landslides: Risk perception and acceptance. *Canadian Geotechnical Journal* **34**, 169–188.

Gaillard JC and Mercer J 2012 From knowledge to action. *Progress in Human Geography* **37**(1), 93–114.

Gluckman P 2014 The art of science advice to government. *Nature* **507**, 163–165.

Kuhlicke C, Steinführer A, Begg C, et al. 2011 Perspectives on social capacity building for natural hazards: outlining an emerging field of research and practice in Europe. *Environmental Science and Policy* **14**(7), 804–814.

Luers AL and Sklar LS 2013 The difficult, the dangerous, and the catastrophic: Managing the spectrum of climate risks. *Earth's Future* **2**, 114–118.

Mercer J, Kelman I, Taranis L, and Suchet-Pearson S 2010 Framework for integrating indigenous and scientific knowledge for disaster risk reduction. *Disasters* **34**(1), 214–239.

Norris FH, Stevens SP, Pfefferbaum B, et al. 2007 Community resilience as a metaphor, theory, set of capacities, and strategy for disaster readiness. *American Journal of Community Psychology* **41**(1-2), 127–150.

Posthumus H, Deeks LK, Rickson RJ, and Quinton JN 2013 Costs and benefits of erosion control measures in the UK. *Soil Use and Management* **31**, 16–33.

Rickenmann D and Zimmermann M 1993 The 1987 debris flows in Switzerland: documentation and analysis. *Geomorphology* **8**, 175–189.

Robinson TR and Davies TRH, 2013 Potential geomorphic consequences of a future great (M_w = 8.0+) Alpine Fault earthquake, South Island, New Zealand. *Natural Hazards and Earth System Sciences* **13**(9), 2279–2299.

Schipper L and Pelling M 2006 Disaster risk, climate change and international development: scope for, and challenges to, integration. *Disasters* **30**(1), 19–38.

SNZ (Standards New Zealand) 2008 Managing Flood Risk – A Process Standard. NZS 9401:2008.

Vogel C, Moser SC, Kasperson RE, and Dabelko GD 2007 Linking vulnerability, adaptation, and resilience science to practice: Pathways, players, and partnerships. *Global Environmental Change* **17**(3-4), 349–364.

Voinov A, Kolagani N, McCall MK, et al. 2016 Modelling with stakeholders – Next generation. *Environmental Modelling & Software* **77**(C), 196–220.

Vrijling JK, van Hengel W, and Houben RJ 1998 Acceptable risk as a basis for design. *Reliability Engineering & System Safety* **59**, 141–150.

19

Geomorphology as a Tool for Predicting and Reducing Impacts from Natural Disasters

In previous chapters we have outlined how predicting Earth surface processes and landforms is indispensable in natural hazard assessments and also in risk management. We summarize in closing how geomorphology as an applied and increasingly integrative discipline of Earth science can contribute to predicting and reducing *impacts* from natural disasters.

19.1 Natural Disasters Have Immediate and Protracted Geomorphic Consequences

Regardless of type or mechanism, most natural disasters alter the Earth's surface. Natural disasters cause extensive damage, fatalities, and physical and psychological injury. We have emphasized that natural disasters also destroy, reshape, or create landforms. This may hardly surprise a geomorphologist, but is rarely acknowledged explicitly in the literature on natural hazards, risks, and disasters. Most immediate geomorphic consequences of natural disasters are tied to how landscapes respond to transient stresses. These stresses can come from seismic ground acceleration, sudden vertical or horizontal crustal motions, or the physical impacts by moving water, wind, and ice. Animals and plants can mediate or amplify many of these responses. The chapters in this book highlight many examples of such geomorphic consequences of natural disasters,

including river-channel changes during a flood, coseismic subsidence or uplift, volcanic flank collapse during an eruption, or enhanced soil erosion following wildfires. History has shown that these and many other geomorphic consequences can cause much greater damage than the initiating process alone (for example, earthquakes, storms, droughts), regardless of whether these geomorphic consequences have been rapid or gradual. Nevertheless, many damage statistics of natural disasters commonly attribute losses to what seems the initiating process, even if this process is responsible for only a fraction of the reported loss. While it is useful to recognize the cause of a natural disaster, its effects are usually what we need to deal with.

Many of these effects arise from the unexpected motion of rock, sediment, soil, nutrients, and contaminants. Immediate geomorphic impacts include the physical impacts by sediment in floods or debris flows, the collapse of river banks, or the storm-wave erosion of barrier beaches. The underlying processes of erosion, transport, and deposition that reshape the disaster landscape respond to this disturbance according to the principles of conservation of mass, momentum, and energy. Thus, disturbed landscapes adjust until a new equilibrium is attained. Such adjustment may remain incomplete before the next disturbance occurs. This adjustment also means that the geomorphic response to natural disasters can be drawn out over years, centuries, or

Geomorphology and Natural Hazards: Understanding Landscape Change for Disaster Mitigation, Advanced Textbook Series, First Edition. Tim R. Davies, Oliver Korup, and John J. Clague.

even millennia. This protracted response may involve rates of erosion, transport, and sedimentation far from their averages and with negative effects on humans, their infrastructure, and land use. Examples of protracted geomorphic responses to natural disasters are the inundation and salinization of land following major storm surges, the increased likelihood of lahars from hillslopes mantled by fresh sheets of unconsolidated tephra, and river-channel aggradation following widespread typhoon- or earthquake-triggered landsliding.

19.2 Natural Disasters Motivate Predictive Geomorphology

Geomorphology draws on many tools for predicting both the immediate and protracted impacts of natural disasters on landscapes. By casting the principle of mass conservation into the form of geomorphic transport laws, geomorphologists can inform practitioners about the rates at which disaster impacts propagate through the landscape and how long it will take them to dissipate. Most geomorphic disturbances also impact ecosystems and habitats. In this regard, the geomorphology of natural disasters offers direct links to disturbance ecology. Both fields can assist in anticipating and measuring indirect and intangible losses that arise long after a natural disaster has happened. For example, we can estimate the transport and residence time of large woody debris in rivers after a wildfire by using data on channel geometry or water and sediment discharge. In well-constrained cases, sediment continuity allows a geomorphologist to predict how many years it will take a river channel of specified geometry and ambient load to regain most of its initial form after a million tonnes of excess sediment has been introduced to it. Geomorphologists can solve diffusion and advection equations to estimate the elevation of the channel bed at a given location and

time after a disturbance. The same equations support forecasts of the transport and concentration of contaminants attached to the river sediments.

Recall that by 'prediction' we mean to 'make informed statements about unobserved phenomena'. The geomorphic legacy of natural disasters offers opportunities for estimating the frequency and magnitude of past disturbances far back in time. We can also learn how these past disturbances changed the rates of erosion, transport, and deposition. Where geoarchaeological evidence allows, we might even glean how people have responded to sudden erosion or siltation in the wake of natural disasters hundreds to thousands of years ago. Landforms and sediments offer means to test independently chronologies of past natural disasters derived by other methods. Palaeoseismology, for example, relies heavily on stratigraphic and geochronologic data obtained by trenching across fault traces, but geomorphic archives now complement studies of prehistoric earthquakes. Landforms can be proxies for local and regional disturbances. A single lake dammed by a lava flow can be the result of one or several volcanic eruptions, whereas dozens of lakes of the same age dammed by landslides point to a regional trigger such as a strong earthquake or an extreme rainstorm, or a set of these close in time.

Geomorphic archives can inform us first-hand about the damage or intensity of past natural disasters. Some of these archives can point at disturbances that are unprecedented in human memory. For example, thick valley fills burying former soils may be the sole result of episodes of catastrophic input of sediment from regional landsliding. The thickness of these fills is a measure of how buildings and infrastructure might be impacted by similar episodes in the future. Sequences of parallel beach ridges can contain information about the timing and run-up of past tropical cyclones, or earthquakes, and marking the limits of inundation. Driftwood, vegetation trimlines, or slackwater deposits similarly demarcate the limits of large floods and thus possible damage to bridges or other critical infrastructure during future flood disasters. In interpreting the legacy of these past events, geomorphology enables predictions of how natural and engineered landscapes might respond to future earthquakes, volcanic eruptions, tsunamis, floods, or landslides. At the very least, we can forecast the general direction of the geomorphic response to disasters.

19.3 Natural Disasters Disturb Sediment Fluxes

Humans have been reshaping parts of the Earth surface for millennia. The first phase of clearing forests began in the early Holocene when agriculture and animal domestication began to spread. Soil erosion resulting from land clearing in that and subsequent cultural periods transformed many landscapes and altered the fluxes of sediment and biogeochemical species. Many hillslope and valley deposits in Central Europe record this human-made increase in sediment mobility. Today many rivers and coasts are engineered and have sediment budgets that have been pushed far from their natural states. Concrete now reinforces and seals many natural surfaces so that the physical conditions for geomorphic processes and natural disasters are now different. These non-natural sediment budgets are responding to human disturbances by causing undesirable erosion and sedimentation that undermine and bury land used for housing and development.

For example, large sea walls are intended to dampen storm or tsunami waves, but dampen the regime of lesser energetic waves. Dams and reservoirs provide hydropower and irrigation water and attenuate floods, but retain much of the sediment load of rivers, thus starving downstream reaches to the point that erosion increases and rivers incise their channels. Entire cascades of check-dams reduce flooding and local erosion, but accumulate material to

be released during rare and destructive debris flows. Dykes and levees reduce the natural capacity of river cross-sections, but promote channel-bed aggradation even if the supply rate from upstream sediment source remains unchanged.

Our ongoing and accelerating transformation of the Earth's surface has also paved the way for na-tech disasters, and many of these are intimately linked to sediment flux. Rivers, winds, and ocean currents can transport contaminants and pathogens released during oil spills or chemical leaks. The continued flushing and dispersal of radioactive particles in rivers years after the tsunami-induced meltdown of the Fukushima power plant underlines how important it is to understand how particles move across landscapes, oceans, and the sea floor. Geomorphology offers the toolkit to determine the degree to which landscapes have been altered from their natural state, and to project the degree to which they might change in the future. Geomorphology also provides the methods to predict future impacts of erosion, sediment transport, and sediment deposition on humans, their built environment, land use, and natural resources. The concept of sediment budgets is a natural benchmark against which we can gauge human disturbances of mass fluxes across the Earth's surface. The buffering capacity and resilience of natural ecosystems against unwanted mass fluxes (Alexander 2013; Park et al. 2013; Zhao et al. 2009) is instructive and can offer guidelines for a more natural and cost-effective mitigation of impacts of natural disasters.

19.4 Geomorphology of Anthropocenic Disasters

The debate about whether humankind has left a distinct, global footprint in the geological record continues as we write these lines. Much of this debate revolves around whether the rapidly growing human pressure on our planet merits a new geological epoch termed the Anthropocene. The alternative would be to stick to the Holocene as the epoch already defined by pronounced human impacts on the environment. Regardless of its formal approval, however, the term 'Anthropocene' triggered a discussion that is important and useful in several ways.

Firstly, the discussion urges us to reflect on the human presence and its consequences for our planet. These consequences arise from our consumption of natural resources, including those that some of us take for granted, like clean air and fresh water. The idea of 'sustainable development' has become both buzzword and strategy in political agendas, but has also highlighted the many dependences between humans and their environments. To this end, the discussion about the Anthropocene demands that scientists communicate beyond their fields of expertise: the term 'Anthropocene' at least has spread rapidly far beyond the earth sciences to social, economic, historic, and many other disciplines.

Secondly, the search for a globally detectable human signal in geological records requires a thorough review and comparison of scientific work that records human impacts on the environment. Proponents of the Anthropocene and their counterparts continue to synthesize highly detailed records of past human activities from regional to global case studies. This detailed summary of where we stand offers perspective on our resource use and its effect on biogeochemical cycles, climate change, and natural hazards.

Thirdly, this systematic recollection of data shows how human influence on the Earth surface differs regionally and with respect to natural rates of change. The resulting patterns also show how well we can attribute environmental changes to human activity. One problem is to discriminate changes that are reversible from those that are not: what seems irreversible in a lifetime may vanish in geological time. Another problem is to make informed predictions about future environmental changes. Studying past changes to landscapes and landforms together with simulating future changes using basic geomorphic transport laws can assist us in making these predictions.

Fourthly, natural disasters have rarely been mentioned in the discussion about the Anthropocene. Geoscientists are mostly concerned with clear sedimentary and geomorphic fingerprints of humans, but whether this also means that some 'natural disasters' are still 'natural' remains unanswered. Yet disasters with a human fingerprint might change legal accountabilities, affect risk financing, or even trigger political conflicts.

We have changed in many ways how surface processes operate and how landscapes look today. By far the most important and persistent impact of humans is our growing cities that concentrate many societal and economic assets. According to the Wealth Health Organization (WHO) more than half of the human population was living in urban areas in 2014, and the growth of cities is accelerating. An unavoidable consequence of this amassing of people and wealth in cities is that disaster impacts will worsen. Natural erosion, transport, and deposition of materials also operate in cities and affect people and resources, so that the field of urban geomorphology has become timely and highly relevant. Geomorphology so far has contributed to reconstructing and predicting how Earth surface processes in natural landscapes adversely affect people. In the future, however, this focus will be increasingly insufficient, because how people influence natural geomorphic processes in engineered landscapes will need much more attention. As a result of this background, geomorphology has become an interdisciplinary science, with strong links to engineering, tectonics, climatology, ecology, biogeochemistry, and disaster research. Within this network, geomorphology has a prime role when it comes to understanding, measuring, reconstructing, and predicting the natural and anthropogenic fluxes of mass across Earth's surface; and in particular, in anticipating their impacts on human society.

References

Alexander DE 2013 Resilience and disaster risk reduction: an etymological journey. *Natural Hazards and Earth System Sciences* **13**(11), 2707–2716.

Park J, Seager TP, Rao PSC, et al. 2013 Integrating risk and resilience approaches to catastrophe management in engineering systems. *Risk Analysis* **33**(3), 356–367.

Zhou H, Wang J, Wan J, and Jia H 2009 Resilience to natural hazards: a geographic perspective. *Natural Hazards* **53** (1), 21–41.

Glossary

accumulated cyclone energy (ACE) a measure of the activity of tropical storms, cyclones, or cyclone seasons, and computed as the sum of each tropical storm's maximum sustained wind speed squared, evaluated at standard six-hour intervals; the minimum wind speed to qualify for a tropical storm, and thus be relevant for ACE, is 35 knots or $\sim$65 km h^{-1}.

active layer detachment shallow landslide initiating at the boundary between upper nonfrozen and lower frozen soil in Arctic permafrost terrain.

ALARP short for 'as low as reasonable practicable', and often used in terms of residual risk.

Anthropocene refers to the proposal of a new contemporary geological epoch characterized by global-scale human footprints in the Earth system that are long-lasting enough to be detected in the geological record; initially proposed to embrace the time span since humans have begun to alter global climates noticeably.

aleatoric uncertainty something unknown to us that eludes learning about it, because of some intrinsic randomness in a given process.

aridity index the ratio between mean annual precipitation and mean annual potential evapotranspiration.

asteroid rocky minor body that circles the Sun between the orbits of Earth and Jupiter.

atmospheric river a stream of concentrated water vapour in the atmosphere, and a possible predictor of flooding events.

availability bias to make judgements about the frequency of probability of an event based on how readily one can think of examples of the event.

black carbon forms from the anthropogenic and natural burning of fossil fuels and biomass, consisting mainly of very fine-grained soot.

black swan an event that is in the extreme of a specified magnitude–frequency distribution, with a very low probability of occurrence.

bolide a meteorite that is heated to incandescence by atmospheric friction, and that explodes in the Earth's atmosphere. A bolide may also impact the Earth.

channel avulsion the lateral shifting of a river channel (or a distributary channel) across a floodplain, fan, or delta by abandonment of an existing channel and formation of a new one/reoccupation of an older one.

confirmation bias the preference of confirming rather than falsifying evidence.

debris avalanche a large and extremely rapid landslide normally on a volcanic edifice; note that in Canada 'debris avalanche' is sometimes used to characterize a rapid debris flow.

debris flow a slurry-like flow of sediment and water, in which sediment has a

Geomorphology and Natural Hazards: Understanding Landscape Change for Disaster Mitigation, Advanced Textbook Series,
First Edition. Tim R. Davies, Oliver Korup, and John J. Clague.
© 2021 John Wiley & Sons Ltd. Co-published 2021 by the American Geophysical Union and John Wiley & Sons Ltd.

concentration of about 70–90% by weight or 47–77% by volume.

dense rock equivalent (DRE) a measure of the volume [km^3] of magma just before it erupts, and thus corrected for low-density contributions of liquids and gases.

desertification the United Nations Convention to Combat Desertification defines this as 'land degradation in arid, semiarid and dry subhumid areas resulting from various factors, including climatic variation and human activities'.

dragon king an event that lies outside the known probability distribution of such types of event, and thus cannot be predicted probabilistically.

ecosystem services a generic terms for the benefits of nature; in terms of protective functions against natural hazards, ecosystem services refer to the estimated damage value that an intact ecosystem may help to avoid.

elusive hazard a potentially damaging process that has no abrupt or well-defined onset or termination.

epistemic uncertainty refers to something presently unknown to us that we can learn eventually.

equifinality the concept that different geomorphic processes may give rise to very similar looking landforms.

exposure in some definitions this refers to the product of vulnerability and the value of elements at risk.

false negative an event is predicted to occur, but does not occur.

false positive an event is predicted to occur, but does not occur.

fault trench method for directly investigating vertical and horizontal displacements perpendicular to the trace of a seismic surface rupture.

fire regime an ecological term that characterizes the fire type (ground, surface, and crown), occurrence (rate of spread, seasonality, and frequency), pattern (size and patchiness), and consequences (impacts on vegetation and soils) associated with wildfires.

flash-flood magnitude index a metric for the abruptness of water discharge in a river; often expressed as the standard deviation of the log-transformed annual maximum series of a given river.

flood the stage of a river discharge at which water overtops the channel banks; note that this definition can have problems when applied to anabranching (i.e. multichannel) rivers or bedrock gorges.

Fujita scale a six-grade scale ranging from F0 ('light damage') to F5 ('incredible damage') designed for measuring the intensity of tornadoes, based mainly on the observed damage to infrastructure and vegetation.

glacial lake outburst flood (GLOF) the sudden emptying of a water body dammed by a glacier or its deposits; many of the resulting 'floods' have high sediment concentrations, so that 'hyperconcentrated flood' or 'debris flow' is a more appropriate term.

groyne a rigid hydraulic structure designed to modify the flow of water and thus regulate the amount of sediment it transports; often used along river banks and sedimentary coasts to trap sediments while modulating local flow velocities.

Gt gigaton: 1 Gt = 10^9 t; a unit mass that is equivalent to petagram (1 Pg = 10^{15} g).

Gutenberg–Richter relationship an empirical size distribution that links the (log-transformed-frequency) number of earthquakes above a given magnitude with that magnitude in the form of a simple linear model; the relationship predicts that smaller earthquake magnitudes are systematically more frequent than larger ones.

hazard a natural event with the potential to adversely affect society: or the probability of a potentially damaging (natural) event specified for a given area and region; hazard is often specified as an annual probability.

heat wave a positive deviation (e.g. measured in standard deviations) from the climatological mean summer temperature.

hindsight bias the case when people estimate a much higher probability of a given event after they have learnt about its outcome, compared to people without any advance knowledge about the outcome.

homogenite homogeneous, structureless, and fine-grained reworked deep-sea sedimentary beds that differ strongly from nearshore tsunamiites. Type A: a locally derived pelagic turbidite from liquefied unconsolidated late Quaternary sea-floor sediments; Type B: megaturbidites of distant, shallow-water provenance; homogenites were initially described in the Mediterranean.

hurricane synonymous with tropical cyclone; commonly used in the North Atlantic.

hyperconcentrated flow a type of water-sediment flow intermediate between a 'clearwater' flood and a debris flow; sediment concentration is 40–70% by weight or 20–47% by volume.

hysteresis a measure of the time-lag between peaks of water and sediment discharge arising from the same trigger, e.g. a rainstorm.

ice-jam flood a type of flooding that occurs mostly in Arctic poleward-draining rivers in which the meltwaters generated in headwaters during the spring seasons are partly dammed, diverted, and suddenly released downstream in the colder and still frozen ice-covered reaches.

ignimbrite deposit from pyroclastic density currents; may occur in welded and nonwelded variants and may mantle entire landscapes.

indirect losses damage that arises from shortages of services or goods in the aftermath of a natural disaster.

intangible losses damage that cannot be readily expressed in monetary terms.

intensity for many natural hazards such as earthquakes, tsunamis, or storms the term intensity refers to a measure of the observed damage to buildings or infrastructure; intensity scales provide a means to estimate the energy released during impact.

jökulhlaup *[Icelandic]* subglacial outburst flood triggered by a subglacial volcanic eruption; nowadays the term is used more broadly to characterize subglacial outburst floods in various settings.

lahar *[Indonesian]* volcanic mudflow, or more precisely, a sediment-laden water flow containing pyroclastic material.

levee a natural dyke along a river or debris-flow channel formed by preferential deposition of coarse sediment after the fluid had overtopped the channel; in the USA it can also be an artificial flood-bank.

lifelines a generic term for critical infrastructure such as traffic, water, power, and communication networks or public services.

machine learning a generic term for a collection of computational methods for detecting and analysing relationships in data sets; the term 'statistical learning' is sometimes used synonymously.

meteorite an extraterrestrial piece of rock that has fallen from the sky and impacted on the Earth surface.

Modified Mercalli Scale an twelve-category system to rank people's response and damages observed during and after earthquakes.

Monte Carlo simulation a statistical resampling technique that uses a high number of iterations during which random numbers drawn from statistical distributions replace fixed parameter values; Monte Carlo simulation is a common method for propagating and expressing uncertainties about natural hazards data.

na-tech disaster a combination of a natural and a technological disaster, the former usually triggering the latter.

natural catastrophe a term used to describe a particularly destructive natural disaster.

natural disaster the harmful and damaging societal consequence of a natural process or a combination of natural processes on human assets; definitions of natural disasters often entail inability of local resources to cope, or a minimum of financial damage or mortality.

overfitting the undesired effect of fitting a(n often too complex) model to data noise instead of data.

palaeoflood hydrology the scientific discipline of inferring the pre-instrumental occurrence and magnitude of floods from their deposits and erosional features.

palaeoseismology the scientific discipline of reconstructing the frequency and magnitude of pre-instrumental earthquakes.

palaeotempestology the scientific discipline of inferring the pre-instrumental occurrence of storms from their deposits.

Pg petagram: $1 \, Pg = 10^{15}$ g; also known as gigaton ($1 \, Gt = 10^9$ t).

phreatomagmatic refers to explosive volcanic eruptions during which ascending magma comes into contact with groundwater.

reef blocks boulder-sized fragments of coral reefs transported to fore- and backshore areas by wave action; reef blocks are often used as diagnostic of tsunami or storm waves.

regional flood frequency a method for determining flood magnitudes in ungauged river catchments, where flood discharge for a given return period in specific regions is inferred via statistical regression of catchment characteristics.

resilience the capability of a social group to anticipate and adapt to damage from natural disasters, thus reducing future disaster impacts.

Richter scale a physical measure expressing the energy released by an earthquake; the Richter scale is log-based and without an upper bound.

risk a measure of the expected damage from a given natural hazard, often expressed as the product of annual event probability and event consequence.

risk aversion a psychological measure of the extent to which a person or a group is inclined to overestimate an objective expected loss.

risk cycle concept of recurring disasters that mark successive phases of immediate response, rehabilitation, and preparation.

risk management any measure or policy that aims at reducing risk by either reducing hazard, vulnerability, the elements at risk, risk aversion, or any combination of these factors of the risk equation resulting in a net decrease of risk.

river metamorphosis refers to changes of channel pattern.

rock avalanche an extremely rapid landslide involving the flow-like motion of rock particles; rock avalanches frequently mobilize more than $1 \, Mm^3$ of rock and are characterized by much longer (= excess) runout than conventional Coulomb friction would predict.

run-up landslide run-up is defined as the vertical elevation difference that the moving mass travels upwards against the force of gravity; tsunami run-up is defined as the maximum elevation of the wave above mean sea level.

sabo *[Japanese]* sediment retention dam used for reducing the physical impact of debris flows.

sackung type of slow-moving, deep-seated landslide that often involves the full hillslope relief, often characterized by antiscarps.

sediment cascade a complex of local sources, pathways, and sinks for sediments that travel from the highest mountains to the ocean (or inland basins) eventually.

sediment wave an oscillation of an alluvial river-bed elevation resulting from the downstream passage of excess material.

slackwater deposits fine-grained, often silt-sized, sediments dropping from suspension during floods in backwater or low-velocity reaches.

slump a rotational landslide, i.e. a slope failure with an upwardly concave failure plane.

slush flow a rapidly moving mixture of snow, meltwater, and sediment.

Southern Oscillation Index (SOI) the difference of standardized air pressure at sea level measured at Papeete, Tahiti and Darwin, Australia.

stopbank an artifical dyke built along a river channel and designed to contain flood levels up to a specified design flow.

sturzstrom (*German*) a rock avalanche.

tempestite a sediment or sedimentary rock deposited by a storm.

thermokarst the landforming process involving the thawing of ice-rich permafrost or the melting of massive ice; thermokarst comprises thermokarst subsidence and horizontal thermoerosion.

toreva block giant intact piece of volcanic rock dislodged by a debris avalanche and remaining in the proximal part of the deposit.

tree avalanche a type of rapid mass movement largely involving organic soil and tree stands with negligible amounts of debris or rock.

true negative refers to the successful prediction of an absent event, or 'non-event'.

true positive refers to the successful prediction of an event.

tsunami *[Japanese: 'harbour wave']* a transient impulse wave in oceans or enclosed inland water basins; tsunami have long wave lengths, and can be triggered by earthquakes, landslides, volcanic eruptions, or asteroid impacts.

tsunamiite a tsunami-worked or -reworked sediment that may have been deposited by either the tsunami itself or tsunami-induced currents.

turbidite the deposit from a turbidity current.

turbidity current an underwater sediment gravity flow.

twister synonymous for tornado.

typhoon synonymous for tropical cyclone; commonly used in the western Pacific.

urban turn is a an estimated date after which more people were living in cities than in rural areas.

vog short for 'volcanic smog'.

Volcanic Explosivity Index (VEI) a relative measure of the explosiveness of volcanic eruptions

vulnerability the fraction of the total value lost during a natural disaster.

wall cloud nearly vertical mass of clouds surrounding the calm and low-pressure eye of a tropical cyclone.

water spout a tornado travelling over water.

Index

Geomorphology and Natural Hazards: Understanding Landscape Change for Disaster Mitigation, Advanced Textbook Series,
First Edition. Tim R. Davies, Oliver Korup, and John J. Clague.
© 2021 John Wiley & Sons Ltd. Co-published 2021 by the American Geophysical Union and John Wiley & Sons Ltd.